Lecture Notes in Artificial Intelligence 4095

Edited by J. G. Carbonell and J. Siekmann

Subseries of Lecture Notes in Computer Science

Stefano Nolfi   Gianluca Baldassarre
Raffaele Calabretta   John C. T. Hallam
Davide Marocco   Jean-Arcady Meyer
Orazio Miglino   Domenico Parisi (Eds.)

From Animals to Animats 9

9th International Conference
on Simulation of Adaptive Behavior, SAB 2006
Rome, Italy, September 25-29, 2006
Proceedings

 Springer

Series Editors

Jaime G. Carbonell, Carnegie Mellon University, Pittsburgh, PA, USA
Jörg Siekmann, University of Saarland, Saarbrücken, Germany

Volume Editors

Stefano Nolfi
Institute of Cognitive Science and Technology ISTC-CNR
Via S. Martino della Battaglia 44, 00185 Rome, Italy
E-mail: s.nolfi@istc.cnr.it

Gianluca Baldassarre
E-mail: gianluca.baldassarre@istc.cnr.it

Raffaele Calabretta
E-mail: raffaele.calabretta@istc.cnr.it

John C. T. Hallam
E-mail: john@mip.sdu.dk

Davide Marocco
E-mail: davide.marocco@istc.cnr.it

Jean-Arcady Meyer
E-mail: jean-arcady.meyer@lip6.fr

Orazio Miglino
E-mail: orazio.miglino@unina.it

Domenico Parisi
E-mail: parisi@istc.cnr.it

Library of Congress Control Number: 2006931575

CR Subject Classification (1998): I.2.11, I.2, I.6, F.1.1-2, K.4, H.5, J.4

LNCS Sublibrary: SL 7 – Artificial Intelligence

ISSN   0302-9743
ISBN-10   3-540-38608-4 Springer Berlin Heidelberg New York
ISBN-13   978-3-540-38608-7 Springer Berlin Heidelberg New York

Springer is a part of Springer Science+Business Media

springer.com

© Springer-Verlag Berlin Heidelberg 2006
Printed in Germany

Typesetting: Camera-ready by author, data conversion by Scientific Publishing Services, Chennai, India
Printed on acid-free paper   SPIN: 11840541   06/3142   5 4 3 2 1 0

Preface

This book regroups the articles that were presented at the ninth international Conference on the Simulation of Adaptive Behavior (SAB 2006) held at the Italian National Research Council, Rome, on September 25-29, 2006. The objective of the biennial SAB conference is to bring together researchers in computer science, artificial intelligence, alife, complex systems, robotics, neurosciences, ethology, evolutionary biology, and related fields so as to further our understanding of the behaviors and underlying mechanisms that allow natural and artificial animals to adapt and survive in uncertain environments.

Adaptive behavior research is distinguished by its focus on the modelling and creation of complete animal-like systems, which—however simple at the moment—may be one of the best routes to understanding intelligence in natural and artificial systems. The conference is part of a long series that started with the first SAB conference, which was held in Paris in September 1990, and was followed by conferences in Honolulu 1992, Brighton 1994, Cape Cod 1996, Zurich 1998, Paris 2000, Edinburgh 2002, and Los Angeles 2004. In 1992, the MIT Press introduced the quarterly journal *Adaptive Behavior* now published by SAGE Publications. The establishment of the International Society for Adaptive Behavior (ISAB) in 1995 further underlined the emergence of adaptive behavior as a fully fledged scientific discipline. The present proceedings are a comprehensive and up-to-date resource of the latest progress in this exciting field.

The 35 papers and 35 poster summaries published here were selected from 140 submissions after a two-pass review process designed to ensure high and consistent overall quality. The articles cover all main areas in animats research, including perception and motor control, action selection, motivation and emotion, internal models and representation, collective behavior, language evolution, evolution and learning. The authors focus on well-defined models, computer simulations or robotic models, that help to characterize and compare various organizational principles, architectures, and adaptation processes capable of inducing adaptive behavior in real animals or synthetic agents, the animats. We hope that these articles will provide stimulating reading material, with a good overview of the latest developments in this exciting field.

The conference and its proceedings would not exist without the substantial help of a wide range of people. Foremost, we would like to thank the members of the Program Committee, who thoughtfully reviewed all the submissions and provided detailed suggestions on how to improve the articles. We are also indebted to our sponsors.

The enthusiasm and hard work of numerous individuals were essential to the conference's success. Above all, we would like to acknowledge the significant contributions of Diana Giorgini, Gisella Pellegrini, and all the members of the Laboratory of Autonomous Robots and Artificial Life, ISTC-CNR, Rome, for their help with the local arrangements. Finally, once again, we would like to warmly thank Jean Solé for the artistic conception of the SAB 2006 poster and the proceedings cover.

We invite readers to enjoy and profit from the papers in this book, and look forward for the next SAB conference in 2008.

September 2006   Stefano Nolfi
Program Chair
SAB 2006

Organization

From Animals to Animats 9, The Ninth International Conference on the Simulation of Adaptive Behavior (SAB 2006) was organized by the Institute of Cognitive Sciences and Technologies of the Italian National Research Council (CNR), and ISAB (International Society for Adaptive Behavior).

Executive Committee

Conference Chair:	Stefano Nolfi, Institute of Cognitive Sciences and Technologies, CNR, Italy
Conference Co-chairs:	Gianluca Baldassarre, Institute of Cognitive Sciences and Technologies, CNR, Italy
	Raffaele Calabretta, Institute of Cognitive Sciences and Technologies, CNR, Italy
	Davide Marocco, Institute of Cognitive Sciences and Technologies, CNR, Italy
	Orazio Miglino, University of Naples II and Institute of Cognitive Sciences and Technologies, CNR, Italy
	Domenico Parisi, Institute of Cognitive Sciences and Technologies, CNR, Italy
General Chairs:	John Hallam, University of Southern Denmark, Denmark
	Jean-Arcady Meyer, University of Paris 6 - CNRS, France
Local Organization Chairs:	Diana Giorgini, Institute of Cognitive Sciences and Technologies, CNR, Italy
	Gisella Pellegrini, Institute of Cognitive Sciences and Technologies, CNR, Italy

Program Committee

H. Abbass	L. Berthouze	R. Calabretta
R. Arkin	A. Billard	A. Cangelosi
A. Arleo	E. Bilotta	J.M. Carmena
M. Asada	A. Bonarini	A. Clark
G. Baldassarre	A. Borghi	T. Collett
C. Balkenius	J.J. Bryson	H. Cruse
R. Beer	S. Bullock	K. Dautenhahn

Sponsoring Institutions

Applied AI Systems, Inc. - AAI
The European Network for the Advancement of Artificial Cognitive Systems -
    euCognition
Institute of Cognitive Sciences and Technologies - ISTC-CNR
Italian Society for Cognitive Science - AISC
The International Society for Adaptive Behavior - ISAB
Kteam
Laboratoire Informatique de Paris 6 - LIP6
The University of Edinburgh
Université Pierre et Marie Curie - UPMC
The University of Southern Denmark

Table of Contents

The Animat Approach to Adaptive Behaviour

Perception and Motor Control

Action Selection and Behavioral Sequences

Navigation and Internal World Models

Learning and Adaptation

Evolution

Collective and Social Behaviours

Adaptive Behavior in Language and Communication

Applied Adaptive Behavior

The Animat Approach to Adaptive Behaviour

Emotions as a Bridge to the Environment: On the Role of Body in Organisms and Robots

Carlos Herrera Pérez[1], David C. Moffat[2], and Tom Ziemke[1]

[1] University of Skövde
School of Humanities and Informatics
SE-541 28 Skövde, Sweden
{carlos.herrera, tom.ziemke}@his.se
[2] Glasgow Caledonian University
School of Computing and Mathematical Sciences
G4 0BA. Glasgow, Scotland
d.c.moffat@gcal.ac.uk

Abstract. Adaptive agents exhibit tightly coupled interactions between nervous system, body and environment. Parisi recently suggested that the current focus on sensorimotor interaction between agent and environment needs to be complemented by an "internal robotics", i.e. modeling of the interaction between internal physiology and nervous system in, for example, emotional mechanisms. The dynamical systems notion of "collective variables" can help understanding such interactions. In emotions physiological states are key parameters that trace the global dynamic concern relevance of the situation. Such variables may be key, in adaptive systems, to monitoring and controlling the agent's interaction with the external environment. We show in a simple robotic simulation that the neural controller can self-organize to exploit the dynamical regularities traced by these variables. We conclude this can prove to be a useful technique in robots and animats, towards evolving emotion-based adaptive behaviors.

1 Introduction

The dynamical systems (DS) approach to cognition and adaptive behavior (e.g. [1, 2, 3]) is based on the hypothesis that an autonomous agent's adaptation to the environment, through ontogeny and phylogeny, results in a tightly coupled dynamical system. A natural agent's body and its biological niche form an inextricable whole. Nervous systems control and integrate bodies in their niche producing a rich range of adaptive behaviors. For the advocates of this approach, the mind, once considered a private and independent control realm, emerges from this system as a whole, and is extended in time and space to reach as far as cognition can go (e.g. [1]). Hence, we are not dealing with a closed system whose activity can be reduced to the mapping of sensory inputs to motor outputs. In natural agents, nervous system, body and environment are not three components that interact in a synchronic way. In order to understand adaptive behavior, it may be necessary to consider brain, body and environment as tightly coupled parts of a whole [4], even if the exact mechanisms of integration in the dynamic process of interaction are at least partly unknown. The

S. Nolfi et al. (Eds.): SAB 2006, LNAI 4095, pp. 3–16, 2006.
© Springer-Verlag Berlin Heidelberg 2006

complexity of a dynamical system, such as one that includes a nervous system, a complex physiology and an unpredictable environment, makes the problem of mathematically formalising the system more or less intractable. Nevertheless, dynamical system ideas, presented conceptually, offer a range of tools available to the theorist that aims to understand adaptive processes as they unfold in time [5].

Natural adaptation is the result of a self-organising process. It is therefore commonly assumed that artificial adaptation should also be based on principles of self-organisation, such as artificial evolution and neural computing. However, can DS enlighten how self-organisation may be induced in animats?

Parisi [6] recently suggested an "internal robotics", i.e. an integral approach that takes into consideration the essential role of the organism's body (physiology) and its dynamic, largely holistic, relationship with the nervous system. This approach can be complemented with attention to the nervous system / environment relationship and the body / environment interaction. The study of the relationship between nervous system and rest of the body may enlighten, as Parisi argues [6], the emergence of the private aspects of the mind, in particular emotion.

Emotion, in turn, is among the essential provisions of natural agents for their flexible adaptation to uncertain environments (e.g. [7 , 8, 9, 10]) and may be usefully integrated into the mechanisms underlying adaptive behaviour in animats (e.g. [11, 12, 13, 14]). In the following sections we will try to integrate Parisi's proposal for an "internal robotics", with concepts from DS and emotion theory to provide a simple model of the mechanisms underlying the *appraisal* process; and discuss how these mechanisms can be understood in DS terms and modelled in animats.

2   Internal Robotics and Emotion

In recent years there has been a decisive turn towards aspects of situatedness and embodiment in cognitive science and AI (e.g. [1, 15, 16, 17]). Much recent robotics research has focused on the sensorimotor interaction between control systems, their robotic bodies and their external environment. For example, it has been demonstrated that control may be highly distributed and, for instance, morphology plays an active role in control [16]. Furthermore, Parisi ([6] p. 325) has recently argued that "to understand the behavior of organisms more adequately we also need to reproduce in robots the inside of the body of organisms and to study the interactions of the robot's control system with what is inside the body". Parisi coins the term *internal robotics* to denote the interactions between the control system (neural) and the rest of the body (Figure 1).

Parisi sees in these forms of interaction the distinction between *cognition* and *emotion*: "The cognitive components of behavior emerge from the interactions of the nervous system with the external environment, whereas its emotional or affective components emerge from the interactions of the nervous system with the rest of the body" ([6] p. 332). The role of physiological mechanisms in the generation of emotional processes is supported by much research on emotion (e.g. James, Damasio, Frijda), and, with the exception cognitive theories of emotion (e.g. [18]), most approaches to emotion in psychology, philosophy, cognitive science or psychiatry generally take into account the physiological substrate (e.g. [7, 8, 9, 10]).

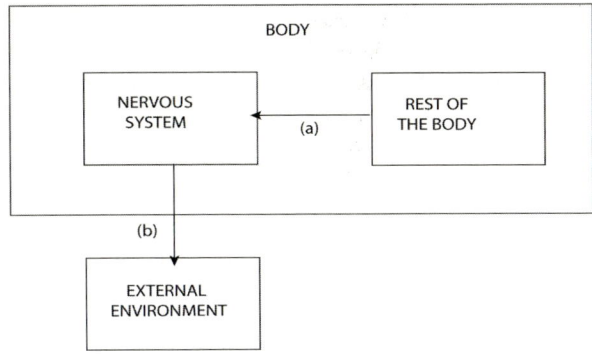

Fig. 1. Internal robotics [6]: The nervous system is in constant interaction with the rest of the body. The body, through mostly chemical channels, produces diffuse effects on the neural system.

The association of emotion with the interaction between nervous system and the rest of the body is consistent with Damasio's theory of feeling. In short feeling is associated with neural patterns that emerge from changes in the body as mapped in the brain [10]. Damasio argues that some brain structures are dedicated to process such changes, either as a perceptual or simulative mechanism, and such structures are integral parts of the emotion process. Damasio distinguishes between feeling and emotion [10]. Whilst feeling springs from the relationship between the nervous system and the rest of the body, other aspects of emotion emerge from the interaction between nervous system, body and environment, as, for instance, appraisal of certain aspects of the environment and the modification of relational behavior. Emotions may be experienced literally 'in the flesh', but they are relational in the sense that they 'connect' agent and environment. Emotional experience can be awareness of autonomic arousal, but also awareness of "action tendency" and/or "situational meaning structure" [7]. Emotion and affect are therefore modes of interaction of the nervous system with the environment, an interaction that is mediated by the body.

Conceptually, it could be argued that interaction with the body is de facto interaction with the environment, as body and environment are tightly coupled. Feedback from the body into the neural system is not only an internal signal, but it is to a certain extent significant of the relationship of the agent to the environment. Such feedback is the origins of embodied appraisals - perceptions of the body, but, through the body, they also allow us to literally perceive danger, loss, and other matters of concern (e.g. [19]). In the rest of the paper we try to unveil the mechanism for such phenomenon and, following Lewis [5], how it may be modelled using a DS approach.

3 A Dynamical Systems Approach to Emotion

There are several reasons why DS are the most appropriate methodology to study emotions [5]. Intuitively, emotions are dynamic and strongly rely on holistic and relational properties. As discussed in the previous section and consistent with much of emotion theory (e.g. [7, 8]) the thesis of situatedness states that emotions are relational phenomena, changes in the ongoing relationship with the environment. For Frijda ([7] p. 466), emotions can be defined as changes in disposition towards some aspects of the world. Readiness change comes in different forms: (a) readiness for action as such (activation), (b) cognitive readiness (attention arousal), (c) readiness for modifying relationships with the environment (action tendencies), (d) readiness for specific concern-satisfying activities (desires and enjoyments).

Change in readiness is a dynamic process that involves the articulation of many physiological and neural systems. This change in readiness is motivated by the *relational* status towards the environment in reference to some of the agent's "concerns". *Concerns* are those conditions essential for the well-being of the agent. It is important to note that a concern need not be represented by the agent; and neither is explicit reasoning necessary for appraisal of concern-relevance. The basic process of response to concern-relevance and readiness generation is called "primary appraisal". Where appraisal involves cognitive articulation, such as "situational meaning structures" [7] or "coping potential" [8], we call the process "secondary appraisal". Together, these phenomena offer an example of grounded cognition, and DS models may help bridging the gap between emotion and cognition.

Much of emotion theory has focused in clarifying the relationship between emotion and cognition. In the cognitivist paradigm, they are considered independent processes, and therefore much of the discussion is about what comes first. Appraisal theories argue that emotions involve an implicit evaluation of the relation with the environment. *Post-cognitive theories* take this evaluation to be explicit and best described in information processing terms. *Perceptual theories* stress how emotions (and the physiological changes associated) shape our perceptions and subsequent cognition of the world. Models of the emotion process, mainly focused on cognitive aspects, are normally linear. Emotional phenomena, nevertheless, are not a linear succession of events. Cognition, behavior, physiological and neural patterns are all participants of emotion, and their interactions are continuous and dynamic.

There has been great discontent in emotion theory with the lack of tools to describe emotional phenomena in non-linear terms [20]. Scherer [21] has suggested that emotion can be defined as an episode of temporary synchronization of all major systems of organismic functioning represented by five components (cognition, physiological regulation, motivation, motor expression, and monitoring/ feeling). Lazarus argues that an approach to synthesizing the causal roles between different aspects of emotion requires a "multivariate system, which consists of a number of causal antecedents, mediating processes, immediate emotional effects, and long-term effects, all acting interdependently" ([8] p. 208).

Therefore emotion theory, although largely focused on cognitive phenomena, seems to suggest that DS are the proper framework for the study of emotional phenomena. Lewis ([5], p. 169) argues in detail that DS offer a way to bridge the cognitive level of analysis with phenomena at other levels. "DS principles stipulate higher-order wholes emerging from lower-order constituents through bidirectional causal processes – offering a common language for psychological and neurobiological models." The DS approach offers a range of notions, such as *trigger, self-amplification, self-stabilization phases*, which Lewis uses to study "neural structures and functions involved in appraisal and emotion, as well as DS mechanisms of integration by which they interact. These mechanisms include nested feedback interactions, global effects of neuro-modulation, vertical integration, action-monitoring, and synaptic plasticity, and they are modeled in terms of both functional integration and temporal synchronization." ([5], p. 169).

Lewis' model of the appraisal process attempts to link higher-order psychological phenomena with neurological elements. The neural dimension forms the system's state space, and certain neural structures form wholes that are associated to psychological phenomena. The goal is to have a simplified system that allows us to trace the dynamic patterns present in the system. The need for reduced dimensionality in the system has two reasons: (1) high dimensionality becomes intractable, (2) not all variables are significant for certain overall dynamics. Main tools for the DS reduction are the consideration of "collective variables" and "control parameters". Collective variables or 'order parameters', are numerically measured quantities that supervene on the behavior of a system's lower-level constituents [2]. In other words, a collective variable is a quantitative measurement that allows us to distinguish qualitative changes in the overall dynamics.

Mathematical analysis can identify control parameters, which 'lead the system through different patterns, but that (unlike order parameters) are not typically dependent on the patterns themselves' ([2], p. 16). The property of being independent allows DS to mathematically study bifurcations, but the concept of control derives from the fact that control parameters can be changed independently, to a certain extent, of the overall patterns. The physiological system, therefore, is a control parameter that is dependent on the mechanisms of the physiological system. The relevance of these concepts for emotion modeling can nevertheless go beyond tools for reducing dimensionality in analysis. In the next section we present the idea that the nervous system may establish a body-mediated adaptive relationship with the environment through attunement to some order parameters within the agent's physiology, that reflect concern-relevance of situations. The fact that the body dynamics may provide sufficient information to track overall relational properties would allow a nervous system to: (1) appraise concern relevance (embodied appraisal), (3) generate action readiness through change in control parameters, (3) generate action tendency through sensory-motor coordination given a state of action readiness.

4 Embodied Appraisal and Response

Our thesis is that emotions are embodied appraisals which actively use proprioceptive feedback to generate perceptions of subject-related features of the environment [22]. The thesis is twofold. On the one hand body states are indicators of salience in a situation, that is, collective variables. On the other hand, they affect behavior dynamics is a distributed way, possibly provoking change in the stability of the system; that is, they are control parameters.

According to Prinz [19], emotions are perceptions of the body, but, through the body, they also allow us to literally perceive matters of concern. "Perceptions of changes in our somatic condition … are also appraisals … as any representation of an organism-environment relation that bears on well-being" [19]. Our interest is to show that the body may be considered an indicator of salient environmental states, or, more precisely, agent/environment relational dynamics. For Prinz this is possible for body parameters "by figuring into the right causal relations" [19].

What do these causal relations consist of? They are complex patterns of interaction between the nervous system, the body and the world, which ultimately may be described as a dynamical system. We know from emotion theory that some body states guard some special relation to certain overall patterns, such that they are indicators of concern-relevance - this relationship is defined in DS theory as a collective variable. The "right causal relation" that allows to carry an evaluation based on physiological feedback is thus that certain physiological patterns are collective variables of concern-related global states. Or, given that physiology and nervous system are tightly coupled, it would be more precise to say that physiology provides the nervous system with patterns that allow appraisal and response to concern relevant situations.

Certainly physiological changes, such as hormonal, visceral or muscular, allow us to trace the relational aspects of the interaction with the environment. The role of neuro-biology of emotion is to unveil the relationship between physiological states and possible emotions. The fact that there is not a one-to-one or linear correlation – an argument classically used to deny the relevance of physiological states – makes DS models a form of analysis that may reveal the intrinsic dynamics of emotion [5].

The first part of our thesis is that some form of processing is based on the dynamical correlation between body states and concern-relevance. Through some stable interaction, the neural system is capable of processing physiological together with sensory information and memory, so to be able to trace some global feature of the relationship. For example, in the case of fear in humans, we can argue that the amygdala circuitry is essential for such processing. In this case, physiology acts as a collective variable. In James-Lange theory, we would say that we feel something is wrong when some physiological indicator goes beyond some point.

The story is nevertheless not so simple, as the nervous system is capable of effecting control over physiological parameters, and participates in the process from the beginning. We cannot understand the relevance of physiological parameters if they are not related to the neural mechanisms that produce them. In other words, physiological parameters are active control parameters at the disposition of the nervous system. Therefore the "right causal relations" are complex dynamic interactions (nervous, physiological and environmental), in which physiology acts as a control parameter at the service of the nervous system to (1) appraise concern relevant global features of the situation, and (2) guide the behavior of the system.

As discussed above, DS description always depends on the level of analysis we adopt, in the order and control parameters we find or choose [2]. But the relevance of collective variables for the study of emotion goes beyond analytical tools. If some physiological states or patterns allow the nervous system to process concern-relevance, the use of such processing for cognitive and behavior generation might be an essential adaptive component for the agent in question. As Schachter and Singer [24] demonstrated, a change in physiological activity may be sufficient for some form of appraisal. But physiological variables may not only allow the agent to appraise, but are parameters under the control of the nervous system, and changes in physiological parameters subsequently change the cognitive-behavioral relationship with the environment.

The active role of feeling as a form of monitoring physiological processes (for example as body maps, [10]), a process that also results in the generation of action readiness, can be therefore understood with the background of DS theory. The basic relationships and dynamical connections between nervous system, body and world in emotion are as follows: (1) Physiological parameters are collective variables of agent/environment relationship (concern-relevant situations).   (2) Physiological variables are also control parameters under the control of the nervous system (change in action readiness). (3) Quantitative change in physiological variables (activation) is followed by qualitative change in relationship and activity of the nervous system (change in action tendency).

Emotions, being perceptual, are not only an appraisal system, but a coordinated system of response. Physiology does not only reflect global dynamics, but is an active participant of them. Physiological activation is therefore the product of not only interaction with the environment, but the interaction with the nervous system. The evolution of physiological variables is "created by the coordination between the parts, but in turn influences the behavior of the parts" ([2], p. 16).

5   Implementing Primary Appraisal

In this section we introduce the experimental approach taken. We should note that we are here not trying to model some complex mechanisms that resemble human or animal emotion systems. What we are trying is to design for the emergence of phenomena that are called emotional, that means, our aim is to produce a coupled agent-environment interaction in which we can observe the essential features of emotion at the level of primary appraisal: concern, concern-relevance and appraisal, action readiness and action tendency. Whether specific mechanisms allow the emergence of such phenomena depends ultimately on environmental coupling.

Our hypothesis is that emotional phenomena can be achieved by a system with the means to process information coming from the rest of the body, used to trace and exploit overall relational dynamics. This approach also may prove a powerful tool for the generation of adaptive flexible behavior in robots, although this issue is out of the scope of this paper. In [25] we define a measure of behavior flexibility based on DS concepts, and evaluate the following experiments.

A robotic implementation, also, provides a concrete set of phenomena for the demonstration and investigation of the principles exposed. So, on one hand theory of emotional phenomena motivate and justify the model, and on the other the model and resulting behavior validate the theory. In summary, the following sections explore how the principle that through the processing of collective physiological the agent has the means to (a) appraise the concern-relevance of the situation, and (b) control action readiness.

5.1   The Model

Following the internal robotics approach, the model illustrated in Figure 2 is intended to represent the relationships between nervous system, body and world. Some aspects of the relationship to the environment are concern-relevant. A number of collective

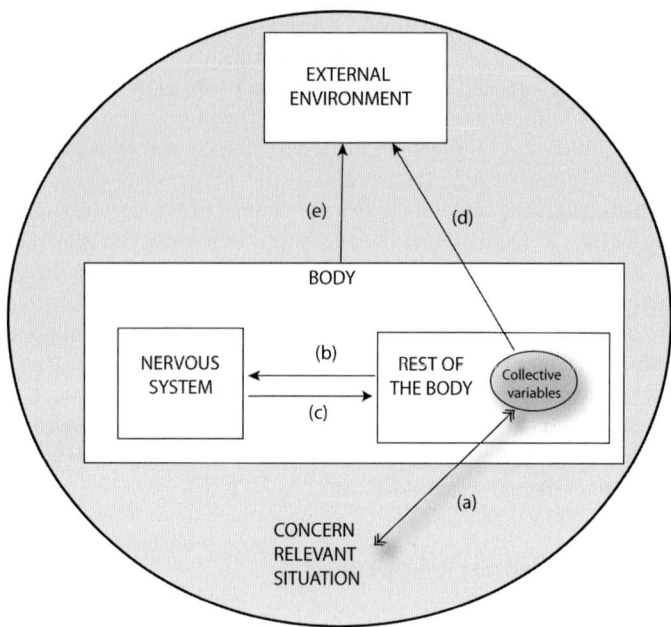

Fig. 2. Model for a dynamical appraisal system of embodied interaction

variables in the body allow us to trace the dynamics of concern-relevant situations. Such collective variables are computed by a complex non-linear physiological system, which serves as an input to the nervous system, in the form of neurally processed body maps (feeling), and as modulator of the activity of the nervous system. This activation is integrated with current cognitive, perceptual and sensory-motor processes. The nervous system also participates in the homeostatic balance in the body, and therefore the collective variables are to a certain extent control variables. A change in these control variables produces a change in action readiness, reflected in the dynamic relationship towards concern-relevant aspects of the situation. Sensory-motor activity (relationship of the nervous system to the environment), in conjunction with further nervous processing (secondary appraisal), produce a change in action tendency. A behavior, embodying this tendency, is the result of this process.

Secondary appraisal, the process by which the system is able to appraise some relevant dimensions (a situational meaning structure, in Frijda's terms), results from a deeper integration of perceptual, behavioral and cognitive abilities, and it is beyond the scope of this paper. When we speak of emotion, we are therefore constrained to the phenomenon of primary appraisal, the appraisal of relevance in the situation and the generation of action readiness.

5.2 Experimental Setup

We apply a evolutionary robotics approach to the implementation of this model, evolving connection weights in the neural controllers of simulated Khepera robots, thus exploiting the principles of self-organization. A prey/predator scenario partly replicates Nolfi and Floreano's experiments reported in [26]. Two Khepera robots, equipped with infrared sensors, are placed in random positions. The predator, which also has a camera, is rewarded for catching the prey, which in turn is rewarded for avoiding the predator. Both robots are controlled by a feedforward network that takes as inputs sensory values and produces as output activation of the motors.

In the Khepera robotic platform, we do not find a complex physiology containing suitable collective variables. To circumvent this, we manually define a recursive function that simulates, or stands in for, the activity of a simple physiological system. The output of this function is intended to be a collective variable of the interaction, i.e. produce a function that allows us to trace concern-relevant situations, and may have an effect on the generation of behavior. In order to achieve this, we will feed it to the neural controller as an extra input. We will refer to this function as CVS (collective variable simulator).

It is worth noting that this initial experimental setup makes (at least) two oversimplifications that further experiments should avoid. First, we have approached the question of the robot's "attunement" to the dynamics by using a manual shortcut, the definition of a CVS. It would be interesting, in later experiments, to let the CVS and CVS-based behavior-generating system to co-evolve or co-develop, therefore investigating the role evolution and learning may play in emotional attunement. This would involve research into the capacities of neural networks to learn patterns of concern-relevance, and the problems of time integration. Second, we have considered the physiological system only in its relationship to the nervous system, and not in its relation to body dynamics. This allowed us to functionally replace the physiological system by a CVS. More complex robotic physiologies, exploring relationships between body states (such as energy but also other related to the system functioning) and their relationship to sensors, motors, and nervous system should be investigated.

5.3 Definition of the Collective Variable

We now need to find a way to define the CVS. We must first pay attention to what situations we want to trace. What emotions are concern-relevant in the prey/predator scenario? If we focus on the prey, we can argue that the situations that are concern-relevant for the prey are those in which there is danger of being caught by the predator, and if some emotion springs from such situations, it should be fear. So, what we need to ask is what situations are especially dangerous for the prey. In order to solve this problem we attend to Nolfi and Floreano's experiments and analyze the scenario to see if we can extrapolate, to the new scenario, some parameter that allows us to trace such situations. Despite several possible strategies of approach and avoidance, given the fastest abilities of the prey, the problem is mostly near walls. Analysis of sample evolved individuals shows that the prey, if no walls interfere, can produce optimal escape behaviors. Danger is present when the prey is caught between the predator and walls.

We therefore suppose this fact may be extrapolated to other prey/predator instances, and we define a CVS that allows us to trace such situations. If we consider that, if a robot is near a wall and a predator, the sum of the activation of all sensors will be larger than when only predator or walls are present. We therefore define a recursive function to which we add, each cycle, the sum of the activation of the sensors, and is multiplied by a decay factor (1).

$$E_t = (E_{t-1} + \sum_{i=1..8} S_i)/2 \tag{1}$$

E_t is the activation of the additional neuron, E_{t-1} represents at any time the activation in the previous time cycle, and ΣS represents the sum of all 8 infrared sensors values. This function can be considered a non-realistic simulation of a system that secretes a hormone in direct relationship to the level of sensory activation, producing an activation level. The hormone which is absorbed at a linear rate, and its level influences the system by being fed back in to the neural controller as an extra input. Essentially, this architecture is equivalent to a simple feedforward network to which we add an internal neuron with a recurrent connection (see Fig. 3). The only difference is that weights between sensors and the internal neuron, and the neuron's recurrent connection, have been fixed.

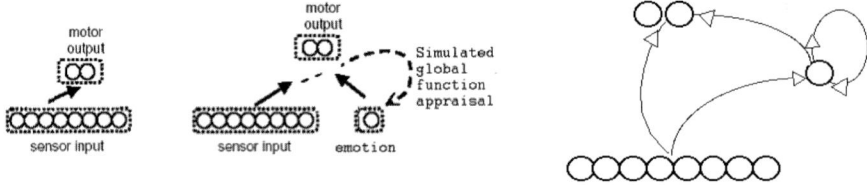

Fig. 3. Neural network controllers. A simple feedforward network (left), a feedforward network with an extra neuron computing the collective variable (center) and the equivalent recurrent network with an internal node (right).

We then start an evolution process in which the weights are evolved. If, as we assumed, the collective variable will be significant of a class of situations (danger), then the robot controller can be expected to use it a resource for the evolution of adaptive emotional behavior. In the next section we evaluate the results.

5.4 Results

As with other experiments in co-evolution, the evolutionary process does not converge to an optimal performance level. Performance of prey and predator are co-dependent, are performance cycles can be normally observed in these cases. It is possible to analyze cross generational strategies and fitness. Due to limited space, we will only analyze the behavior of a single generation (generation 100). In this analysis we will verify whether: (a) the simulated function is a collective variable, that is, it allows us to track situations in which the prey is between wall and predator, (b) the collective function influences the neural controller so as to generate an action tendency that changes the relationship to the environment so that concern is safeguard (i.e. does the prey escape?)

Fig. 4 shows the value of the CVS throughout an interaction in which two dangerous situations are present (points 1 and 3). We can trace the concern-relevance

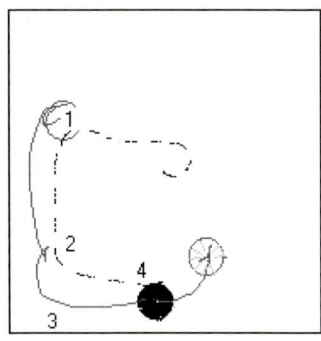 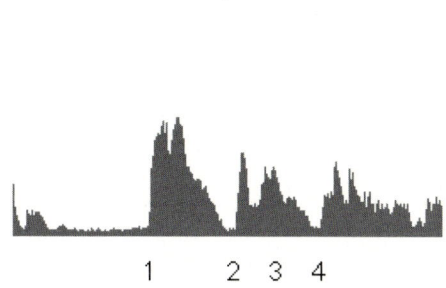

Fig. 4. Interaction between prey and predator. Left: screenshot from simulator. Right: value of collective variable allows to trace the dynamics of interaction.

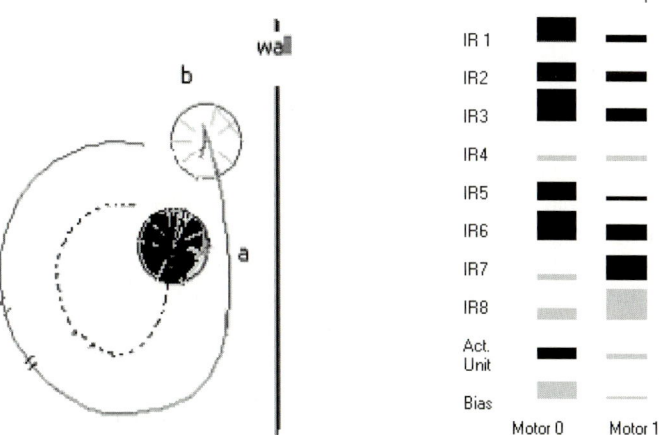

Fig. 5. Left: Escape behavior. Right: values for the connections between sensors (including extra neuron, marked as activation unit) and motors. A square placed in line with IR-i and above motor-x represents the weight of the connection between the unit associated to such IR sensor and the one connected to that motor. The size of the square represents the amount of the weight, while the color represents either positive (grey) or negative (black) value.

of the situations (danger) by looking at the graph representing the activation of the collective function over time. In other words, the value of the CVS allows us to track the dynamics of the interaction. The predator's (black robot, discontinuous trace) approach is reflected as a change in the CVS (1). An escape tendency is generated (continuous trace). A straight, forward movement, at a high speed, allows the prey to avoid crashing against predator or wall, stopping when sufficient distance to the predator is gained (point 2, which can be observed as a cross in the trace of the prey). To analyze the generation of this tendency, we can look more closely at the behavior (Fig. 5, left) and the evolved weights of the neural connections in the control system (Fig. 5, right). The robot relies mostly on one side to produce escape behaviors. Such movements are backwards and therefore are ruled by negative activation. When the emotion unit is highly activated, the right motor activation is compensated to give a movement in a straight trajectory. Therefore the activation unit provokes a negative activation of the left motor and a positive activation of the right motor, balancing out the activation of other sensors to produce straightforward fast motion.

In terms of emotion theory, we can conclude that the prey is capable of appraising concern-relevant situations, by means of attunement through an appropriate collective variable represented by a simulated physiological function. Such appraisal involves the generation of an action tendency whose function is to resolve the concern-relevant situation, producing a fast escape behavior.

6   Discussion

Animal and human agents demonstrate a tight coupling to the environment by means of emotion, an amalgam of processes subject to a variety of dynamic interactions. The DS approach to cognition offers a number of relevant notions that overcome the limitations of linear causal models, inappropriate to describe the dynamics of emotional phenomena (cf. [5]). The concept of collective or global variables, fundamental for dimensional reduction for the analysis of complex systems, may also help enlighten the organizational principles underlying emotional phenomena.

In this paper we have argued that the appraisal process and the generation of action tendencies are based on the interaction of the nervous system with the agent's body. The hypothesis is that concern-relevance is a relational property of the agent-environment interaction, which can be traced by collective variables found in the body. The nervous system is capable of processing such collective variables; therefore also able to appraise the dynamics these variables allow tracing. In turn, these collective variables are also control parameters of the interaction: a change in the activation of the physiological functions picked up by these variables produces a change in the interaction with the environment.

It is therefore necessary to pursue an understanding of the processes that fall under Parisi's label of "internal robotics", the relationship between the nervous system and the rest of the body, as significant of the larger relationship between body and environment. The simple fact that an agent's control system may have access to these internal variables, which are collective variables of some relational aspects of the situation, may provide the means for monitoring and controlling the agent's interaction with the external environment, producing adaptive behavior.

The simple robotic implementation reported here illustrates the basic notions of this DS approach to emotion. The evolutionary process allows the robot to exploit a collective variable input to the network to generate suitable action tendencies. This model has some obvious limitations that derive from the fact that the physiological functions present in a natural agent are here highly abstracted and only the suggested input to the neural system (the collective variable itself) is simulated. Further development should also model the underlying physiology that 'computes' such variables, and, as their natural counterparts, configures action readiness by affecting the nervous, sensory and motor systems. It should also be important to achieve the attunement of the agent to concern-relevant situations through self-organization, possibly through evolution and/or learning.

A further development of the model may show different possibilities in the utilization of feedback from the body in the generation of adaptive behavior, thus enhancing our understanding of situated activity.

Acknowledgements

This work was supported by the project *"Integrating Cognition, Emotion and Autonomy"* (IST-027819), funded by the European Commission as part of the EC *Cognitive Systems* initiative. The experiments were carried out while the first author was at Glasgow Caledonian University [25].

References

1. Clark, A.: Being There - Putting Brain, Body and World Together Again. Cambridge, MA: MIT Press (1997)
2. Kelso, S.: Dynamic Patterns. MIT Press (1995)
3. van Gelder, T. J. Dynamic approaches to cognition. In R. Wilson & F. Keil (eds). The MIT Encyclopedia of Cognitive Sciences. MIT Press (1999) 244-6.
4. Chiel, H.J. & Beer, R.D.: The brain has a body: adaptive behavior emerges from interactions of nervous system, body and environment. Trends in Neurosciences 20(12) (1997) 553- 557.
5. Lewis, M.L.: Bridging emotion theory and neurobiology through dynamic systems modeling. Behavioral and Brain Sciences 28, (2005) 169–245
6. Parisi, D.: Internal robotics. Connection Science 16(4) (2004) 325-338
7. Frijda, N.H.: The emotions. Cambridge University Press (1986)
8. Lazarus, R.S.: Emotion and adaptation. Oxford University Press (1991)
9. Damasio, A.: Descartes' error: Emotion, Reason and the Human Brain. Picador, Cambridge, MA, USA. (1994)
10. Damasio A.: The Feeling of What Happens: Body and Emotion in the Making of Consciousness. Harcourt Brace, New York (1999)
11. Rolls, E.T.: Emotion Explained. Oxford University Press, Oxford (2005)
12. Simon, H.A.: Motivational and emotional controls of cognition. Psychological Review, 74, (1967) 29-39.
13. Sloman, A. and Croucher, M.: Why robots will have emotions. In Proceedings 7th International Joint Conference on AI. Morgan-Kaufman. (1981)
14. Picard, R.: Affective Computing. MIT press (1997).
15. Pfeifer, R. And Scheier, C.: Understanding Intelligence. MIT press (1999)
16. Pfeifer, R.: Dynamics, morphology, and materials in the emergence of cognition. Proc. KI-99, Lecture Notes in Computer Science. Berlin: Springer (1999) 27-44.
17. Ziemke, T. (Ed.): "Special issue on Situated and Embodied Cognition". Cognitive Systems Research Volume 3, Issue 3, (2002) 271-554.
18. Ortony, A., Clore, G.L. & Collins, A.: The cognitive structure of emotions. Cambridge University Press. Cambridge, UK (1988).

19. Prinz, J.: "Embodied Emotions" In Solomon R. C. and Harlan L. C. (Eds.): Thinking about Feeling: Contemporary Philosophers on Emotion. Oxford University Press (2004) 44-58
20. Frijda, N.H.: The Place of Appraisal in Emotion. Cognition & Emotion (7)(3&4) (1993) 357-388
21. Scherer, K. R.: Neuroscience projections to current debates in emotion psychology. Cognition and Emotion, 7 (1993) 1-41.
22. Herrera C.: Emotions And Perception: On The Role Of Proprioceptive Feedback. IASTED 2002. Special session on Perception and Emotions (2002)
23. Weiskopf Daniel A.: "The Place of Time in Cognition". The British Journal for the Philosophy of Science, 55 (2004) 87–105.
24. Schachter, S & Singer, J.: Cognitive, social and physiological determinants of emotional state. Psychological Review, 65 (1962) 379-399.
25. Herrera C.: The synthesis of emotion in artificial agents. Doctoral dissertation,. Glasgow Caledonian University, School of Computing and Mathematical Sciences, 2006
26. Nolfi S. & Floreano D.: Co-evolving predator and prey robots: Do 'arm races' arise in artificial evolution? Artificial Life, 4 (4) (1998) 311-335

Some Adaptive Advantages of the Ability to Make Predictions

Daniele Caligiore, Massimo Tria, and Domenico Parisi

Institute of Cognitive Sciences and Technologies
National Research Council
44 Via San Martino della Battaglia, 00185 Rome, Italy
dan.cal@libero.it
tria@unisi.it
domenico.parisi@istc.cnr.it

Abstract. We describe some simple simulations showing two possible adaptive advantages of the ability to predict the consequences of one's actions: predicted inputs can replace missing inputs and predicted success vs. failure can help deciding whether to actually executing a planned action or not. The neural networks controlling the organisms' behaviour include distinct modules whose connection weights are all genetically inherited and evolved using a genetic algorithm except those of the predictive module which are learned during life.

1 Introduction

Organisms respond to current sensory input from the environment with movements that change the environment or their body's physical relation to the environment. These changes at least partially determine the successive inputs that the environment sends to the organism's sensory organs but this causal relation is ignored by purely reactive organisms which only respond to current input. In contrast, more complex organisms can predict what the next sensory input from the environment is going to be, given the current sensory input and the movement with which they plan to respond to this sensory input. (For possible neural structures underlying the ability to predict in primates, see [1], [2], [3].) What are the adaptive advantage(s) of this predictive ability? What can organisms with a predictive ability do that organisms without this ability cannot do?

The possible adaptive advantages of being able to predict the sensory consequences of one's movements have already been discussed in the literature. For example, Clark et Grush [4] propose that responding to the predicted proprioceptive input resulting from one's movements may allow organisms to move faster because they don't have to wait for the actual proprioceptive input. In this paper we describe some simple simulations that address this question by demonstrating two possible roles of the ability to predict: predicted inputs can replace missing inputs from the environment and predictions of success or failure can help the individual to take decisions. If any thing prevents some critical input from reaching the organism's sensors, the organism can still behave appropriately by responding to a predicted input that replaces the missing input. If an organism can predict whether or not a planned response will

S. Nolfi et al. (Eds.): SAB 2006, LNAI 4095, pp. 17–28, 2006.
© Springer-Verlag Berlin Heidelberg 2006

produce some desired result, the organism can decide to actually execute the response in case of predicted success and avoid executing the response in case of predicted failure. In Sections 2 and 3 we describe some simulations using the first scenario and the second scenario, respectively. In Section 4 we draw some brief conclusions.

2 Predicted Inputs Replace Missing Inputs

To survive and reproduce an organism must reach (and eat) the food elements that are randomly distributed in the environment. At any given time the organism's sensory organs encode the position of the single nearest food element and the organism must respond by turning towards and approaching the food element. The organism's behaviour is controlled by a sensory-motor neural network with one input unit encoding the location of the food element which is currently nearest to the organism, one output unit encoding the movement with which the organism responds to the sensory input, and two internal units (Figure 1a). An initial population of organisms is generated by assigning random connection weights to the neural network that controls each organism's behaviour and a genetic algorithm is used to evolve in a succession of generations networks which have the appropriate connection weights that allow them to perform the task.

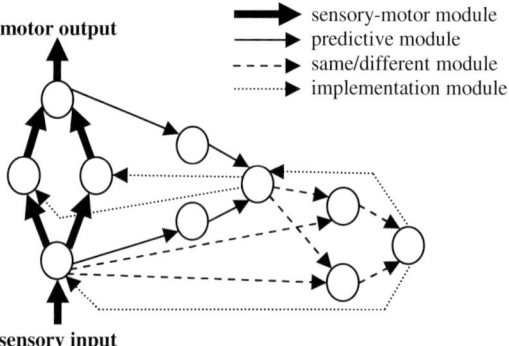

Fig. 1. (a) Sensory-motor network (or module) (thick arrows). (b) Predictive module (thin arrows). (c) Same/different module (broken arrows). (d) Implementation module (dotted arrows).

Now imagine that for a variety of reasons (failures of attention on the part of the organism, something going across between the food and the organism, etc.) in some cycles the input from the nearest food element is replaced by some other, irrelevant, input. We simulate all these different circumstances by assigning a randomly generated activation level to the neural network's input unit in a certain percentage of input/output cycles. If the organism's neural network is a simple network mapping sensory input into motor output, in these 'blind' cycles the organism is lost. The input which replaces the input from food is randomly generated but the organism has no way of knowing this and it responds to the randomly generated input as it were input from food. We expect that in these circumstances the organism's overall behaviour will be significantly less effective. But consider a somewhat more complex organism with a neural network composed of two sub-networks or modules: the module that we have already described which maps sensory input into motor output and a new module that predicts what the next sensory input will be, given the current sensory input and the planned movement with which the organism will respond to the current input (Figure 1b). This predictive module has one input unit encoding the current sensory input from the environment and another input unit encoding the planned motor response of the organism to the current sensory input, two internal units, and one output unit encoding the predicted sensory input from the environment that will appear in the next cycle, i.e., after the planned movement is physically executed. (For other simulations using this neural model of the ability to predict, see [5], [6]; for other models of learning to predict, see [7], [8]; Ackley and Littman's [9] work on evolved reinforcement-producing neural networks that guide learning is also relevant here.)

While the network's entire architecture is fixed and the connection weights of the sensory-motor module evolve and are genetically inherited, the connection weights of the predictive module are learned during life. (The weights of both modules could evolve and be genetically inherited but learning to make predictions during life tends to increase the flexibility of one's predictive abilities.) The predictive module's weights are randomly generated at birth and, early in its life, each individual organism learns to predict the next sensory input using the backpropagation procedure. In each input/output cycle the predicted input is compared with the actual input (which functions as teaching input) and the discrepancy between the two (error) is used to gradually change the predictive module's connection weights in such a way that after a certain number of learning cycles the predictive module is able to make correct predictions.

How is this predictive ability used? When a 'blind' cycle occurs, the organism replaces the missing input from food with the predicted input and responds to the predicted input rather than to the randomly generated input. We assume that early in life the organism has learned to generate correct predictions, which implies that the missing input and the predicted input are more or less the same. Therefore, the organism can respond to the predicted input as it would have responded to the actual input from food, with similar results. We expect that an organism endowed with this predictive ability will behave more or less as effectively in the world with 'blind' cycles as in the world without 'blind' cycles.

How can the organism know when the current input originating in the environment is from food and therefore is the input to which it should respond, and when the input is not from food but from some other source and therefore it should respond to the predicted input rather than to the input originating in the environment? We imagine that the organism's neural network includes two additional modules: a same/different module and an implementation module. The same/different module judges whether the current input from the environment is the same or different with respect to the predicted input. If the two are the same, this means that the current input is from food and the sensory-motor module should respond to the actual input from the environment. If the current input and the predicted input are different, this means that the current input is from some other sources and the sensory-motor module should respond to the predicted input rather than to the current input from the environment. The implementation module implements this judgment by telling the sensory-motor module which input to use. We will now describe these two modules.

The same/different module (Figure 1c) has one input unit encoding the current input from the environment and one input unit encoding the predicted input which was the output of the predictive module in the preceding cycle. In response to these two inputs the same/different module generates an output that encodes a judgment as to whether the two inputs are the same or different. (This same/different task can be interpreted as a continuous XOR task.)

The implementation module (Figure 1d) relays this same/different judgment to the sensory-motor module. To make it possible for the predicted input, rather than the actual input from the environment, to control the organism's behaviour, the output unit of the predictive module, which encodes the predicted input, has connections linking it to the two internal units of the sensory-motor module. Through these connections the predicted input can determine the organism's behaviour by replacing the actual input from the environment. The implementation module has an input unit encoding the judgment "same or different" of the same/different module and this unit sends connections to both the input unit of the sensory-motor module and the output unit of the predictive module (Figure 1d). In this way the implementation module can evolve weights for these two connections that tend to inhibit the output unit of the predictive module (encoding the predicted input) when the judgment is "same" (the current input is from food) and to inhibit the input unit of the sensory-motor module (encoding the actual input from the environment) when the judgment is "different" (the current input is randomly generated).

In the simulations that we will describe the connection weights of the sensory-motor module, those of the judgment module, and those of the implementation module, are all genetically inherited and they are developed using a genetic algorithm. Only the connection weights of the predictive module are learned during life using the backpropagation procedure.

The simulation scenario is the following. We start with a population of 100 individuals whose behaviour is controlled by a neural network with random connection weights. The total duration of an individual's life consists of 3500 input/output cycles of the individual's neural network. These 3500 cycles are divided up into 70 episodes of 50 cycles each and, at the beginning of each episode, the individual is placed all alone in a bidimensional continuous environment of 100x100 spatial units, in a randomly chosen position and with a randomly chosen orientation. (The division of life into separate episodes was introduced to increase variability.) The environment contains 20 randomly distributed food elements. When the individual happens to be within 2 spatial units from a food element, the individual eats the food element. The food element disappears, the individual's fitness is increased by one unit, and a new food element is introduced in a randomly selected location in the environment, so that the total number of food elements is always 20.

An individual has a facing direction and a visual field of 180 degrees. The neural network controlling the individual's behaviour has one input unit, two internal units, and one output unit. The input unit encodes the location of the nearest food element in the individual's visual field as a continuous value ranging from 0.2 to 0.8, with a value of 0.5 when the food is right in front of the organism, a value of 0.2 when the food is 90 degrees to the right, and a value of 0.8 when the food is 90 degrees to the left. The distance of the food is not encoded and the organism can see a food element whatever the distance. The input unit sends one connection to each of two internal units and the two internal units send their connections to the single output unit (Figure 1a). The output unit encodes the individual's movements, and more specifically the individual's turning to either left or right. The output unit's activation value is continuously mapped into the interval between 0.2 and 0.8, with 0.2 encoding a maximal right turn of 90 degrees, 0.8 a maximal left turn of 90 degrees, and 0.5 the preservation of the current facing direction. In all cycles, after the turning movement has been executed, the individual moves forward 0.5 spatial units in the new facing direction.

At the end of life each individual is assigned a fitness which corresponds to the number of food element eaten by the individual and the 10 individuals with the highest fitness generate 10 offspring each. An individual has a genotype which encodes the connection weights of the individual's neural network as real numbers and each offspring inherits a copy of its single parent's genotype. The value of each connection weight is mutated with a probability of 20% and the mutation consists in adding to or subtracting from the weight's current value a number randomly selected between 0 and 1. The 10x10=100 offspring constitute the second generation. All simulations last for 1000 generations and all simulations are replicated 10 times.

2.1 Simulation 1

Simulation 1 is a baseline simulation in which a population of organisms possessing only a sensory-motor module evolves in two different types of environments: an environment without periodic random inputs and an environment with periodic random inputs. We expect that the population that evolves in the second environment will have a significantly worse performance than the population that evolves in the first environment.

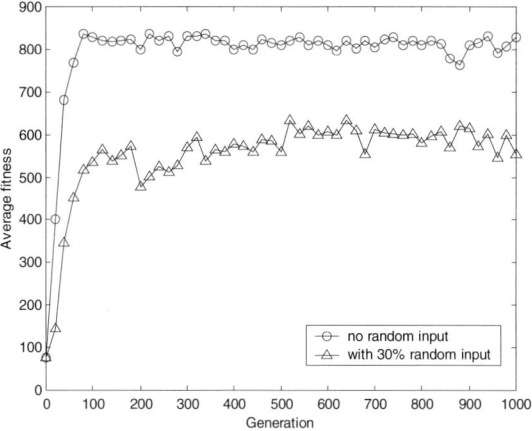

Fig. 2. Average fitness of a population living in an environment in which all inputs are from food and a population living in an environment in which inputs from food are replaced by random inputs 30% of the times

In the first environment an individual receives input from the nearest food element in all cycles. In the second environment in each cycle there is a 30% probability that the input from food will be replaced by a random input. Therefore, in the cycles in which the input from food is missing and is replaced by a random input, the organism will respond in a way which will tend to reduce its fitness.

The results show that, in fact, the average fitness of the population living in an environment where all inputs are from food is higher than that of the population living in the environment where some inputs can be random (Figure 2).

2.2 Simulation 2

In this and the following simulations the population lives in an environment where some inputs can be random. However, the organisms' neural network is more complex than that of Simulation 1. In Simulation 2 the organism's neural network includes a predictive module in addition to the sensory-motor module and each individual learns early in its life how to predict correctly the next sensory input given the current input and the planned response to the current input. In Simulation 2 it is the researcher who, in the cycles with random input, substitutes the current input with the predicted input.

The results of the simulation show that the organisms are very fast at learning to predict correctly the next sensory input from food given the current input from food and the turning movement with which the organism plans to respond to the current input. The prediction error goes to almost zero after only four episodes of an individual's life, which means that during most of its life an organism is able to generate correct predictions of the next input from food. Since in the cycles in which the input from food is missing the researcher replaces the random input with the predicted input, this has the consequence that random inputs cannot disrupt the organism's performance. In fact, the results of the simulation indicate that the performance of these organisms in an environment where some inputs are random tends to be as good as the performance of the organisms living in an environment without random inputs (Figure 3).

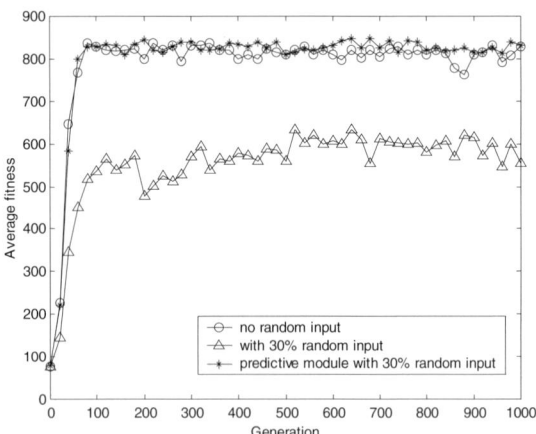

Fig. 3. Average fitness of a population living in an environment with 30% random inputs when the organisms learn early in life to predict the correct input from food and the random input is replaced by the predicted input from food. The two curves of Figure 1 are also shown for comparison.

2.3 Simulation 3

In Simulation 2 the organisms learn to predict the next input from food but it is the researcher who substitutes random inputs with predicted inputs. In Simulation 3 we add a same/different module to the organisms' neural network which gives the organisms more autonomy. The same/different module judges whether the input from the environment is "same or different" with respect to the predicted input, allowing the organism to know if the current input from the environment is from food or random. The connection weights of the same/different module are also encoded in the inherited genotype and they evolve together with the connection weights of the sensory-motor module. However in Simulation 3 it is still the researcher who, if the same/different module's output is "same", causes the sensory-motor network to respond to the input from the environment, whereas if the judgment module's output is "different", he or she substitutes in the sensory-motor module the actual input from the environment with the predicted input generated as output by the predictive module.

The results of the simulation show that the genetic algorithm is able to develop appropriate connection weights for the same/different module, allowing the organism to decide most of the time correctly whether the predicted input and the actual input are the same or different. The researcher replaces the input from the environment with the predicted input if the judgment is "different" and it allows the sensory-motor module to respond to the input from the environment if the judgment is "same". Since the evolved weights of the same/different module are not perfect, the organisms' performance tend to be less good than that of the organisms living in the environment without random inputs but significantly better than the performance of the purely sensory-motor organisms living in the environment with random inputs (Figure 4).

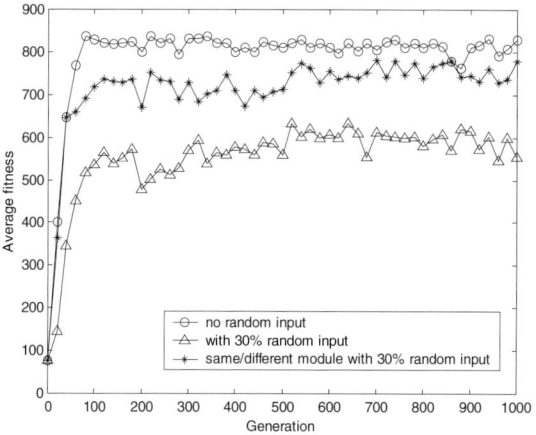

Fig. 4. Average fitness of a population living in an environment with 30% random inputs when the organisms learn early in life to predict the correct input from food and are able to judge if the input from the environment is "same" or "different" with respect to the predicted input. If the judgment is "same", the researcher will cause the organisms to respond to the input from the environment, whereas if the judgement is "different", the researcher causes the organisms to respond to the predicted input rather than to the input from the environment. The two curves of Figure 1 are also shown for comparison.

2.4 Simulation 4

This is the final simulation in which, unlike the preceding simulations, the researcher has no role in determining the organism's behaviour, the organisms are completely autonomous, and every aspect of their behaviour emerges spontaneously through evolution and learning. The implementation module is added to the organisms' neural network and the genetic algorithm is responsible for all the connection weights of their network, except those of the predictive network which are learned during the individual's life and therefore are not genetically inherited.

The results of the simulation show that it is possible to develop completely autonomous organisms that know when it is appropriate to respond to the input from the environment and when it is appropriate to ignore the input from the environment and respond to the predicted input. After a certain number of generations the implementation module develops the appropriate connection weights that allow the implementation module to inhibit the actual input from the environment and to cause the predicted input to determine the organism's behaviour in the cycles in which the input from the environment is random and therefore is different from the predicted input. On the other hand, when the input from the environment is from food and therefore is the same as the predicted input, the implementation module's connection weights allow the module to inhibit the predicted input and to leave to the actual input from the environment control on the organism's behaviour. These entirely autonomous organisms also perform significantly better than the purely sensory-motor organisms living in the environment with random inputs (Figure 5).

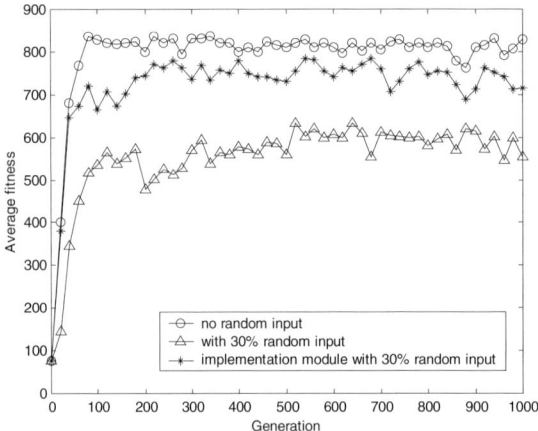

Fig. 5. Average fitness of a population living in an environment with 30% random inputs when the organisms learn early in life to predict the correct input from food and they are able both to judge if the input from the environment is "same" or "different" with respect to the predicted input and to use this judgment to decide whether to respond to the actual input from the environment or to the predicted input. The two curves of Figure 1 are also shown for comparison.

We have also done a control simulation aimed at clarifying a question which inevitably arises with organisms that are able to predict the next input from the environment and to respond to this input rather than to the actual input from the environment. If the organisms' predictions are generally correct, why should the organisms ever want to respond to inputs from the environment instead of simply responding to predicted inputs? An organism which can predict correctly the next sensory input from the environment which will result from its actions, might pay attention and respond only to the very first input from the environment and then ignore all subsequent inputs, always responding to the predicted inputs rather than to the actual inputs. Such an organism would live in a mental world rather than in the real world but its performance in the real world would be as successful as that of an organism responding to the real world.

This is not very plausible, however. Real organisms cannot live entirely in their mental (predicted) world, completely ignoring the inputs from the external environment. The reason is not only that the real world is much more variable and unpredictable than their mental (simulated) world but also that their prediction abilities are not perfect. In fact, even in our very simple and predictable world it is not possible for our simple organisms to always live in their mental world, ignoring the real world. Even if their predictions are generally correct, they are not completely correct - as indicated by the fact that the error in the backpropagation learning procedure never goes exactly to zero - and the errors of successive predictions tend to be cumulative. To demonstrate this point we have run another simulation in which the organisms are allowed to receive an input from the environment (from food) only in the single first cycle of each episode and they respond to the predicted inputs in all subsequent cycles of the episode. The results show that the average fitness at the end of the simulation is less than 200 points compared to almost 600 points of the population in which the organisms have access to the actual input from the environment in more than two/thirds of the input/output cycles (Figure 2).

3 Predicted Success or Failure Help to Take Decisions

In our second scenario, to survive and reproduce an organism has to throw a stone towards a prey animal in such a way that the stone reaches and hits the prey. Stones can be of 10 different weights and the prey can be at 10 different distances. Therefore, in any given occasion to hit the prey the organism has to throw the stone with the force appropriate to the weight of the stone currently in its hand and to the current distance of the prey. The organism's behaviour is controlled by a sensory-motor neural network (Figure 6a) with one input unit discretely encoding the weight of the stone (10 numbers equally spaced between 0.1 and 1.0), another input unit discretely encoding the distance of the prey (10 numbers between 0.1 and 1.0), and one output unit continuously encoding the force of the throwing behaviour (between 0.1 and 1). An output value which is less than 0.1 is interpreted as a refusal to throw the stone in that trial. A table defines the "physics" of the situation by specifying, for each pair of stone weights and throwing forces, the distance covered by the stone. The prey is considered as hit by the stone if the stone falls within a threshold distance from the prey.

The network's connection weights are genetically inherited and are evolved using a genetic algorithm with the same parameter values of our preceding simulations. We compare this simulation with another simulation in which the organism's neural network includes a predictive module and an implementation module (Figure 6b). The predictive module generates a yes/no prediction as to whether or not the planned force with which the stone will be thrown will allow the stone to actually hit the prey. The implementation module relays this prediction/judgment to the sensory-motor module, inhibiting the throwing behaviour if the prediction is "failure" and allowing its physical execution if the prediction is "success". This more complex neural network represents an advantage for the organism if executing physical movements implies an expenditure of both time and energy for the organism. By not executing throwing behaviours that would result in failures, the organism will spare both time and energy (and perhaps avoid the flight of the prey) and therefore would increase its fitness. To implement this idea an individual's fitness is decreased by a fixed quantity for each physically executed throw.

The results of the simulation show that this is actually the case. Compared with organisms with a simple sensory-motor network, organisms with added predictive and implementation modules reach a higher fitness at the end of the simulation (5000 generations) (Figure 7).

As in the preceding simulations, both the weights of the sensory-motor module and the single weight of the implementation module are genetically inherited and they evolve in a succession of generations, whereas the weights of the predictive module are learned during life.

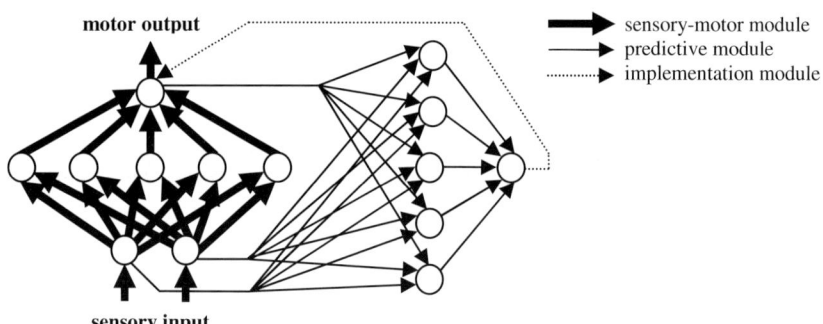

Fig. 6. (a) Sensory-motor module (thick arrows). One input unit encodes the weight of the stone currently in the organism's hand, another input unit encodes the distance of the prey, and the output unit encodes the force with which the stone will be thrown. (b) Predictive module (thin arrows) and implementation module (dotted arrows). The predictive module has three input units, respectively encoding the stone's weight, the distance of the prey, and the force of the planned throwing behaviour. The module's output unit encodes a yes/no prediction on the success or failure of the throwing behaviour. The implementation module is made up of a single connection linking the output unit of the predictive module to the output unit of the sensory-motor module. The implementation module inhibits the throwing behaviour if the prediction is "failure" and it releases the execution of the behaviour if the prediction is "success".

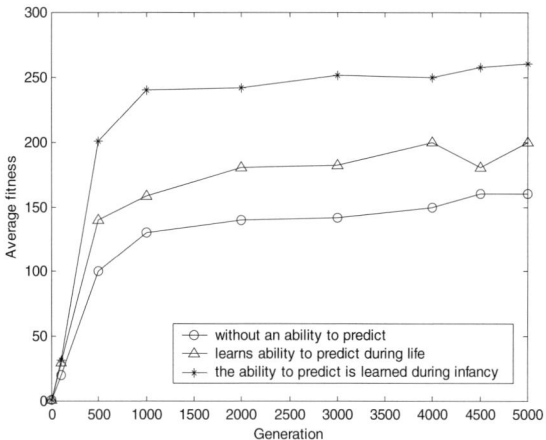

Fig. 7. Average fitness (number of successful throws) across 5000 generations for a population without an ability to predict if a planned throw will be a success or a failure, a population which learns this ability during life, and a population in which the ability to predict is learned during a period of life in which the individual's fitness is not being measured (infancy)

The model that we have described may also suggest a possible evolutionary explanation for the emergence of "infancy". The interpretation of infancy as a "safe" period of learning has been proposed and discussed in variety of context and by many authors, e.g., in evolutionary psychology [11], attachment theory [12], and in Hurford's [13] model of early language learning periods. In the present context infancy can be defined as the initial period of an individual's life in which the fitness of the individual is not being evaluated by the selection mechanism because the individual is provided with the needed resources by other individuals (parents) so that the individual is free to learn some abilities (e.g., the ability to make predictions) that will be useful when the individual becomes an adult and its behaviour will be crucial for the individual's survival and reproduction. To test this model we have compared two simulations. In one simulation an individual's fitness is measured since the individual's birth, and therefore it includes the period of the individual's life in which the individual has not yet learned to make correct predictions and therefore cannot exploit the fitness advantages of being able to predict (not executing throws that would result in failures). In other words, there is no infancy. Individuals are born as adults in the sense that no one takes care of them and their fitness is evaluated from birth. In the other simulation we add infancy. The individual learns to predict during a number of additional input/output cycles that precede its regular life as an adult. The individual's fitness is measured only when the individual becomes an adult and it already knows how to correctly predict the consequences of its actions. The results of the new simulation, also shown in Figure 7, demonstrate that learning useful abilities during a period in which other individuals provide the individual with the needed resources, i.e., during infancy, leads to a higher fitness. This may be an important selective pressure for the emergence of infancy.

4 Discussion

Complex organisms may be able to predict what sensory input will result from their planned but still non executed motor responses to the current sensory input. Why should organisms develop this capacity? What might be its adaptive value? In this paper we have described some simple simulations aimed at providing some answers to these questions. The ability to predict the next sensory input might allow an organism to replace a missing input with the predicted input. If for some reason the appropriate input from the environment is missing (due to obstacles, distractions, or other reasons), the organism can respond to the predicted input which corresponds to the missing input. Another adaptive advantage of the ability to predict is to be able to judge whether or not a planned action will produce the expected result enabling the individual to avoid physically executing expensive actions whose predicted result is not the desired one. Given these, and other (see, e.g., [9]), advantages of being able to predict the results of one's actions we can expect that organisms possessing the appropriate prerequisites, such as the ancestors of humans, will evolve neural architectures (such as the very simplified architecture of Figure 1) that make it possible for them to predict the results of their actions and to use their predictions to generate more effective behavior. The pressures for evolving an ability to predict may have also been pressures for the emergence of infancy as the initial period of an individual's life especially dedicated to learning to predict.

References

1. Blakemore, S-J., Frith, C.D., Wolpert, D.M.: The cerebellum is involved in predicting the sensory consequences of action. NeuroReport **12** (2001) 1879-1884
2. Nixon, P.D., Passingham, R.E.: Predicting sensory events. The role of the cerebellum in motor learning. Experimental Brain Research **138** (2001) 251-257
3. Schubotz, R.I., von Cramon, D.Y.: Predicting perceptual events activates corresponding motor schemes in lateral premotor cortex: An fMRI study. NeuroImage **15** (2002) 787-796
4. Clark, A., Grush, R.: Towards a Cognitive Robotics. Adaptive Behavior **7** (1999) 5-16
5. Parisi, D., Cecconi, F., and Nolfi, S.: Econets: Neural networks that learn in an environment. Network **1** (1990) 149-168
6. Nolfi, S., Elman, J.L., and Parisi, D.: Learning and evolution in neural networks. Adaptive Behavior **3** (1994) 5-28
7. Jordan, M.I., Rumelhart, D.E.: Forward models: supervised learning with a distal teacher. Cognitive Science **16** (1992) 307-354
8. Sutton, R.S.: Learning to predict by the method of temporal differences. Machine Learning **3** (1988) 3-44
9. Ackley, D.H., Littman, M.L.: Interactions between learning and evolution. In C.G. Langton, C. Taylor, C.D. Farmer, and S. Rasmussen (eds.), Artificial Life II. Reading, MA, Addison-Wesley (1992), pp. 487-509
10. Nolfi S., Tani J.: Extracting regularities in space and time through a cascade of prediction networks: The case of a mobile robot navigating in a structured environment. Connection Science **11** (1999) 129-152
11. Buss, D.M. (ed.): The Handbook of Evolutionary Psychology. New York, Wiley, 2005
12. Bowlby, J.: Attachment and Loss. Volume 1: Attachment. London, Hogarth Press, 1969
13. Hurford, J.: The evolution of the critical period for language acquisition. Cognition **40** (1991) 159-201

Perception and Motor Control

Early Perceptual and Cognitive Development
in Robot Vision

Xing Zhang and Mark H. Lee

Department of Computer Science, University of Wales, Aberystwyth, SY23 1LN, UK
{xiz02, mhl}@aber.ac.uk

Abstract. This paper describes research on the design of a robot vision
system that is able to develop its capabilities to perceive objects and
cognise some fundamental relations between objects. Through a developmental approach, our robot vision system is adaptive to environmental
changes and follows autonomous self development. The system can develop its vision gradually from motion detection, to analysis of static
features of objects, to object individuation, to identifying object unity,
to tracking objects and finally to understanding some fundamental object
relations.

1 Introduction

In recent years, more and more attention has been directed to research into developmental robotics, an intersection of robotics and developmental psychology.
On one hand, researchers in robotics attempt to build cognitive systems and
adaptive behaviours with inspirations from infant development [4], [11], [15],
[20]. On the other hand, psychologists tend to embody ideas about infant cognitive development into artificial agents or robots, in attempts to acquire insights
into models and mechanisms of infant development [16], [18].

Studies in infant development reveal a rich source of information about how
infants start cognitive development from some basic reflexes; how they gradually
achieve different stages of sensorimotor competencies; and how representational
abilities emerge by the end of sensorimotor development [6], [17]. Developmental
approaches are inspired, by the way that infants develop cognition gradually, to
build robots that are beyond task-specific goals [21] and able to handle general
complexity [14]. Infant achievement in sensorimotor development also encourages
researchers to build robots that are capable of infant-like competences such as
saccading, visually guided reaching, visual manipulation [1], [13], [5].

Perceptual development is tightly linked with cognitive development in infants. It is regarded that perception is the necessary starting point for higher
level, more clearly cognitive operations, such as reasoning, inferring, and problem solving [6]. However, robotic researchers have not given much attention to
perceptual development in infants. Mostly, perception is investigated in terms
of its interaction with action and how interaction helps cognitive development
[12]. But perception, in its own right, has an important developmental role in
infancy and can shed light on building advanced perceptive robots.

S. Nolfi et al. (Eds.): SAB 2006, LNAI 4095, pp. 31–39, 2006.
© Springer-Verlag Berlin Heidelberg 2006

Visual perception is often considered the most important sensory channel because of its acuteness and amount of information handled. From birth, infants develop their visual perception from the stage of being unable to separate the object from the background, to the stage of being able to perceive objects as units, and then to the stage of being able to reason about objects [6], [10]. It would be interesting to build a robotic perception system that is able to develop its perceptual ability gradually in its environment. This perceptual development could give the robot a high degree of adaptability and autonomy.

In this paper, we present a robot vision system that develops its vision to individuate objects, identify object unity, track objects and understand some fundamental object relations. The next section explains the research problem, followed by the architecture of the vision system, and the experimental results. In the discussion we address some essential issues that we observed about the development process.

2   Research Problem

Our current research is investigating the design of a robot vision system that is able to develop its capabilities to perceive objects and cognise some fundamental relations between objects. A major purpose is to understand more about how to build robot vision systems that can develop their perception and cognition. Within such development in a vision system, we wish to learn: where the development should be initiated; where and how capabilities extend from lower to higher degrees; how successful are the different degrees; and in what order or sequence should the development pass through.

3   Architecture of the Vision System

Our vision system consists of several modules, each of which can work independently either on desired input information or on results generated by other relevant modules. The modules include: motion, shape, colour, experiential pattern, correlation, tracking, and consequence understanding.

The motion module uses an optical flow technique [7] to detect the motion of moving objects in the environment. From a sequence of picture captured from a camera, this module can also detect optical flow fields and locate a region occupied by an object.

The shape module and colour module, respectively, can extract boundary features [8] and a colour distribution pattern [19] of a region when they are provided with rough information about the position and edges of the region.

The experiential pattern module can use knowledge of shape [8] and colour [19] to identify unknown objects. Essentially, this module can dilate, contract and rotate the known shapes of previously observed objects to match newly extracted shapes and see if there is any match. It is also capable of matching colour patterns.

The correlation module can correlate different features together when those colour or/and shape features are found from the same region or from regions where there are consistent optical flow.

The tracking module can observe and record trajectories of object movement in chain code format. The consequence-understanding module can correlate object movements, in terms of spatiotemporal aspects. When the occurrence of an object movement is spatially and temporally close enough to the end of another object movement, the consequence-understanding module will try to establish relations between objects and their movements. In the experiments, we can see an example of this correlation.

4   Hypothetical Development Process

Motion is found as the most salient cue to infant attention. In about 2 months from newborn, infants are able to segregate moving objects from the background, while they can not segregate static objects from the background [2]. A further development after this stage, is that infants start to segregate static objects from background. But those objects must have similar features, such as colour, shape, to the moving objects that have been observed before [2]. Then, after this stage, the infant visual perception will advance to segregate static objects well from the background, and perceive object unity while objects are occluded [10], [9]. The perception and cognition on relations between objects starts to develop after objects can be perceived as units by infants.

So, firstly, the vision system learns to individuate and segregate objects. When an object is moving in the environment, the motion module detects the optical flow in the images. Therefore, the boundary region of an object in the image can be found. In the sense of perception, the object is individuated, but only when the object is moving.

From the regional information provided from motion, the shape and colour modules can extract the boundary and colour distribution features of the object. After this, even when the object is not moving or is in a different orientation from before, the experiential modules can use the previously extracted boundary and colour features to locate the object. The experiential module can dilate, contract or rotate previously known boundaries to match the regions segmented from images and tell if there are any other objects similar to those that have been known before. By this stage, the vision system is able to individuate objects even when the object is static.

Secondly, the vision system is presented with the object unity problem. An object is partly occluded by a cover in the environment and the task is to tell whether unoccluded parts, which are visible to the vision system, belong to a single object or several different objects. For instance, a known object is occluded by a cover, and the vision system can only see the head and tail of the object. Initially, the object is moving left and right behind the cover. Two regions with optical flow are detected and the shape and colour features are correlated together because the two regions are in consistent motion. Then, the experiential module compares features of the perceived tail and head to features of the object, which have been extracted previously. When the experiential module finds a match of the tail and head, the robot vision system can tell that the tail and head are parts of a unity, the object. Here, it is the consistent motion that provided the initial cue to unity and it is the experienced knowledge of objects that confirmed the unity.

Finally, the vision system develops a cause-and-effect understanding of objects. An object A is moving towards another object B and when A bumps B, A stops and B starts to move in the same direction as A moved. The tracking module can track and record the moving trajectories. Because of the temporal continuity and spatial contact between the movements of A and B, the consequence module can correlate these two movements together and record them as a cause-and-effect movement pair.

5   Experiments

Following the implementation of the vision system, some experiments have been performed to show the results of development. Firstly, the vision system observed a moving glue stick and generated a rough contour of the glue stick from optical flow. The optical flow is salient enough to give a good approximation of the shape of the stick. To compare, direct segmentation has also been applied to the same images in attempt to obtain the region of the glue stick. Fig. 1 shows the result derived from optical flow and the region obtained from direct segmentation. As seen, the region gained from direct segmentation misses some area of the glue stick. Fig. 2 gives a quantified comparison of the missing areas lost during motion and direct segmentation.

When given a good approximation of the outline of an object in the image, the vision system could perceive shape and colour features and save these features as models of objects, which were used later to individuate static objects. At this stage, the experiential module could individuate static objects from each other as well

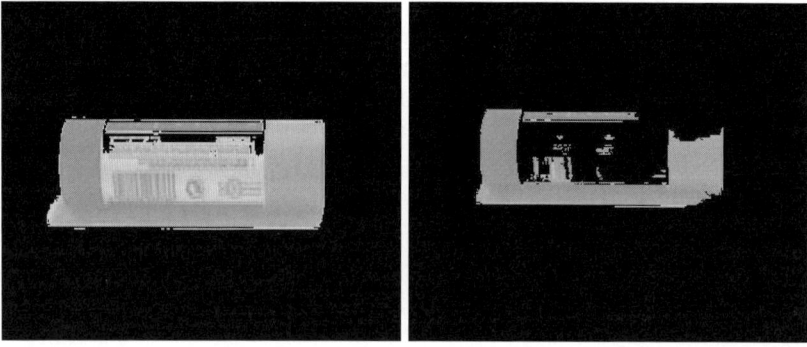

Fig. 1. Regions of a moving object. The left is derived from optical flow and the right is obtained from direct segmentation.

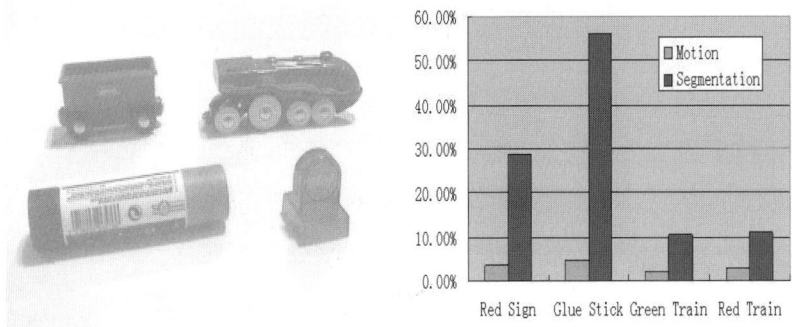

Fig. 2. A comparison of missing area errors by motion and direct segmentation. The rates are calculated by dividing whole object areas by the missing area. As shown in the figure, regions derived from motion tend to have little and stable rates of missing areas.

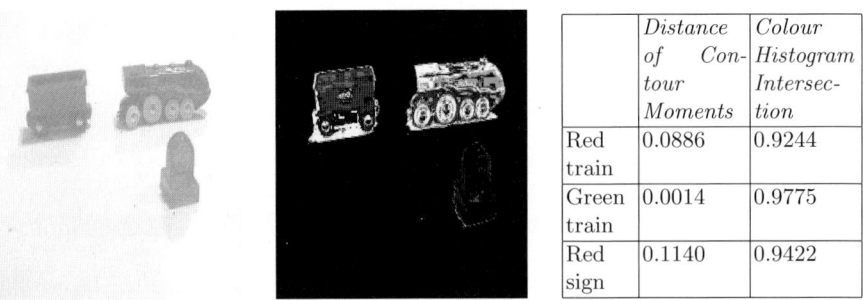

	Distance of Contour Moments	Colour Histogram Intersection
Red train	0.0886	0.9244
Green train	0.0014	0.9775
Red sign	0.1140	0.9422

Fig. 3. On the left is an image with 3 objects. The middle is the resulting image in which the region corresponding to the green train is the most prominent. On the right is a table of comparison results between the green train, respectively, and models of red train, green train and red sign. The smaller the Distance of Contour Moments, the more similar the two objects are. The greater the value of Colour Histogram Intersection, the more similar the two objects are. As shown in the table, the green train, with the smallest contour moments and the greatest intersection value, is picked up from the other two objects.

from the background. But at early stages of its development, only objects with same or similar features of shape or colour can be well individuated. Fig. 3 shows an image with 3 objects, a red train, a green train and a red sign, and the results of individuating the green train out from the background and the other 2 objects.

Next the vision system was tested on some occlusion images. A glue stick was occluded by a lid with only its head and tail visible. When the glue stick was sliding left and right, the optical flow of its head and tail was consistent. Then the experiential module was triggered to perform matching on the visible head and tail, and models built from previous perception development. Fig. 4 gives the results.

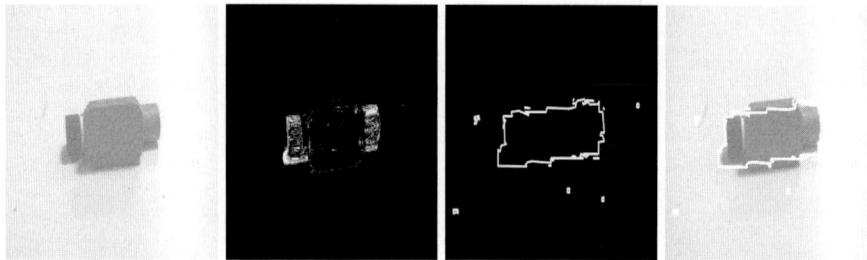

Fig. 4. From left, the first image is a glue stick occluded by a lid; the second image has the optical flow fields of the head and tail; the third image is the best found matching contour; the fourth is the matching result

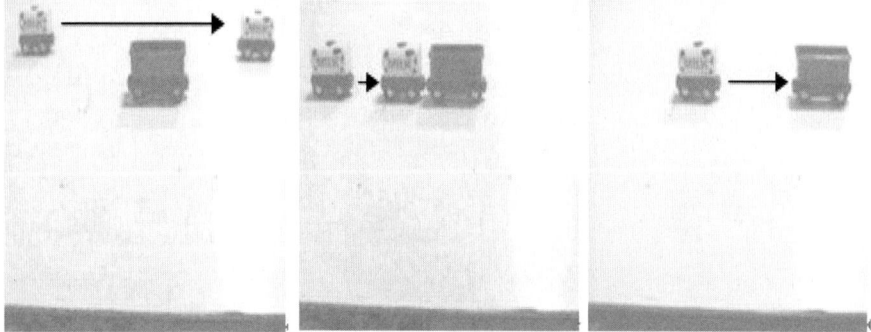

Fig. 5. The left image shows the trajectory of the milk train, without collision. The middle image gives the collide-and-launch event. The right image shows the end positions of both trains.

When the perception of the vision system has reached a fine level of individuation, the development advances toward cognition of object relations, in particular developing a cause-and-effect understanding of objects. In experiments, the vision system first observed a milk train moving from left to right and associated the movement trajectory with the milk train. Then in a later event, a red train was placed on the trajectory of the milk train. During the next movement, the milk train collided with the red train, the milk train stopped at the colliding point and the red train was launched to move in the same trajectory and stops on the right side.

The vision system first learned the trajectory of the milk train and then learned the collide-and-launch event. In the events, the two movements of the two trains were observed and recorded separately. But as there exists a tight spatiotemporal continuity between the two moments, the consequence module linked them together. As a result, the vision system is able to predict the end position of the trajectories of trains, shown in Fig. 5, judging from the relevant positions of the milk train and the red train.

6   Discussions

6.1   From Motion to Static Features

The first point arising from our work is that motion is a better starting point, than static features of objects, for the development of visual perception and cognition. A major reason is that motion is a more salient feature than static features such as boundary or colour or other features [2], [3]. Once the object is moving, its motion can make the object stand out of the environment and separate it from other objects. In contrast, the static features of an object tend to mingle with features of the environment and other objects. Using static features it is not easy to segregate one object from others, unless the vision system has been given prior knowledge about the objects and environment.

Prior knowledge for development of perception and cognition is another crucial issue. From the perspective of development, the starting point is biased towards little or no prior knowledge. To start development with motion, the vision system only needs to detect motion of objects, irrespective of static features of objects. However, to start development with static features of objects, the vision system either needs much prior knowledge, which is less adaptive, or needs to try every possible object shape in every possible region, which requires enormous computation.

Adaptability is another advantage brought by little or no prior knowledge. The world of robots is complex and adaptation is important for a robot to survive in the changing world. With motion, the attention of the vision system can be led to new objects and start to develop knowledge about new objects, including features and relations with other object. In this way the vision system can always further develop its perception and cognition of the world as the world changes. However, if the development starts from static features, it is not so easy to adapt to new changes. In this case, specific and relevant knowledge has to be built into the vision system prior to the development process. When the world changes, e.g. the appearance of completely novel objects, it may entail more knowledge from external sources, usually the system programmer has to re-build the system again to add on more knowledge. Re-building is not appropriate for adaptability.

In sum, the dynamical aspects of vision, in particular object motion detection, seems to enable vision systems to better develop perceptual and cognitive capabilities from little or no prior knowledge.

6.2   From Partial to Holistic

Perceptual and cognitive development in infants is a matter of gradual and cumulative gains in competence [9], [22]. This should also apply to any process of development in artificial vision systems. Initially, our vision system develops its perceptions using one or more aspects of objects, e.g. motion, boundary shape and colour, and then moves on to correlate these aspects together as a synthesis of the whole object. Also, it is only after the stage of object individuation when the vision system can start to perceive and cognise relations between objects. From the interactions between objects, and objects and the environment, the vision system starts to cognise more and more aspects of its world as a whole.

The perception of objects progresses from individual aspects to multiple aspects of objects, and to the whole of objects. This way of progressing is a natural choice in the case when the perceptual development begins from motion. Then the vision system transits its perception from motion to other aspects such as shape and colour features of objects. These individual aspects are correlated together, according to certain mechanisms, to represent the perceived object as a whole. This correlation enables the vision system to search for and locate previously perceived objects and lays the foundation for further cognitive development.

It is after the individuation of objects that the vision system is able to perceive and cognise relations of objects. In the cognitive development on trajectories, starting and ending positions of object movements, objects need to be regarded as individuals rather than regions in an image. Object individuation is the second big turning point, after motion, for the progress of development. After this, development can advance to a higher level of cognition: object relations.

6.3   Passive but Not Active

Essentially, our vision system develops passively. The focus of the described research is development in the robot vision system. The origins of the research comes from studies on perceptual development of infants who are normally younger than 4 months. They are so young that their motor systems do not have many interactions that directly contribute to their visual development. However, infants can still develop their vision to a surprisingly high degree of perceptive ability [6]. Hence, for our study, active sensing is not necessary for exploring the parameters of development of our vision system.

7   Conclusion

The research described has investigated how to build a robot vision system that is able to develop its perception and cognition of objects. Motion is advocated to be the starting point for development in an autonomous and adaptive robot vision system. The development gradually and cumulatively advances from single and simple aspects to multiple and complex aspects of objects and to relations of objects.

Acknowledgements

We are grateful for the support of EPSRC through grant EP/C516303/1.

References

1. Berthouze, L., Bakker, P., Kuniyoshi, Y.: Learning of oculo-motor control: a prelude to robotic imitation. In IEEE/RSJ Int. Conf. on Robotics and IntelligentSystems (IROS'97), Osaka, Japan (1996) 376-381
2. Carey, S., Williams, T.: The role of object recognition in young infants' object segregation. Journal of Experi-mental Child Psychology, Vol. 78, No. 1 (2001) 55-60

3. Cohen, L.B., Cashon, C.H.: Infant object segregation implies information integration. Journal of Experimental Child Psychology, Vol. 78, No. 1 (2001) 75-83
4. Drescher, G. L.: Made up minds: a constructivist approach to artificial intelligence. MIT Press (1991)
5. Fitzpatrick, P., Metta, G., Natale, L., Rao, S., Sandini, G.: Learning About Objects Through Action: Initial Steps Towards Artificial Cognition. In 2003 IEEE International Conference on Robotics and Automation (ICRA), Taipei, Taiwan (May 12-17, 2003)
6. Harris, M., Butterworth, G.: Developmental Psychology. Psychology Press Ltd (2002)
7. Horn,B.K.P., Schunck,B. G.: Determining Optical Flow. Artificial Intelligence. Vol 17 (1981) 185-203
8. Hu, M.: Visual Pattern Recognition by Moment Invariants. IRE Transactions on Information Theory, Vol. 8, No. 2 (1962) 179-187
9. Johnson, S. P.: Theories of development of the object concept. Theories of infant development. Cambridge, MA: Blackwell (2003) 174-203.
10. Johnson, S. P., Cohen, L. B., Marks, K. H., Johnson, K. L.: Young infants' perception of object unity in rotation displays. Infancy, Vol 4 (2003) 285-295
11. Kaplan, F.: Bootstrapping awareness. Proceedings of Second International Symposium on Imitation in Animals and Artifacts, SSAISB (2003)
12. Lungarella, M., Metta, G., Pfeifer R., Sandini, G.: Developmental Robotics: A Survey. Connection Science, Vol. 15, No. 4 (2003) 151-190
13. Metta, G., Sandini, G.,Konczak, J.: A developmental approach to visually-guided reaching in artificial systems. Neural Networks, Vol. 12 (1999) 1413-1427
14. Metta, G., Sandini, G., Natale, L., Panerai, F.: Development and Robotics. In IEEE-RAS International Conference on Humanoid Robots, Tokyo, Japan (November 22-24, 2001) 33-42
15. Natale, L., Metta, G., Sandini, G.: A developmental approach to grasping. In Developmental Robotics AAAI Spring Symposium, Stanford, CA (March 21-23 2005)
16. Pfeifer, R.: Robots as cognitive tools. Int. J. of Cognition and Technology, Vol. 1, No. 1 (2002) 125-143
17. Piaget, J.: The child's conception of the world. Translated by J. and A, Tomlinson, Paladin, London (1973)
18. Schlesinger, M.: A lesson from robotics: Modeling infants as autonomous agents. Adaptive Behavior, Vol. 11 (2003) 97-107
19. Swain, M. and Ballard, D.: Color indexing. Int. Journal Computer Vision, Vol. 7, No. 11 (1991) 11–32
20. Weng, J., McClelland, J., Pentland, A., Sporns, O., Stockman, I., Sur, M., Thelen, E.: Autonomous Mental Development by Robots and Animals. Science, Vol. 291, No. 5504 (Jan. 2000) 599-600
21. Weng, J.: Developmental Robotics: Theory and Experiments. International Journal of Humanoid Robotics, Vol. 1, No. 2 (2004)
22. Westermann, G., Mareschal, D.: From Parts to Wholes: Mechanisms of Development in Infant Visual Object Processing Infancy, Vol. 5 (2004) 131-151

Visual Control of Flight Speed and Height in the Honeybee

Emily Baird, Mandyam V. Srinivasan, Shaowu Zhang, Richard Lamont,
and Ann Cowling

ARC Centre for Excellence in Vision Science, Research School of Biological Sciences,
Australian National University, P.O. Box 475, Canberra, ACT 2601
Emily.Baird@anu.edu.au

Abstract. The properties of visually guided flight speed and height control were investigated by training honeybees (*Apis mellifera* L.) to fly through a tunnel in which the visual cues in the lateral and ventral visual fields could be varied by changing the patterns on the walls and floor of the tunnel. The results show that honeybees regulate their flight speed by keeping the velocity of the image of the environment in their eye constant. The results also show that honeybees use visual information from the ground to control their height above the ground. The findings of this study reveal that the mechanisms of flight speed and height control in the honeybee are perfectly adapted for extracting information from a complex visual environment using simple sensors and computations. Consequently, the techniques of visual guidance that are reported here suggest insect-inspired strategies for the control of aircraft flight.

1 Introduction

Although the brain of a honeybee comprises less than one million neurons, it is able to process with extraordinary accuracy the complex sensory information necessary for a variety of orientation and navigation tasks. Honeybees employ a range of computationally simple techniques to aid flight control and navigation in order to overcome the limitations of their small brain.

Honeybees rely heavily on information from the visual system to navigate. However, information about the 3-D structure of the world is generally required for collision-free flight. Despite the perceptual limitations of immobile eyes, fixed focus optics, low spatial resolution and a lack of stereo vision, honeybees are able to acquire extract range information from cues based on image motion. During flight, the image of the environment moves across the retina, creating a pattern of apparent image motion called optic flow [1]. Properties of optic flow, such as the direction and velocity of certain objects in the visual scene, are useful cues for detecting course deviations or the proximity of objects in the environment. Honeybees are known to use information that has been extracted from optic flow to stabilize flight, estimate the range of objects, negotiate narrow gaps and estimate the distance flown to a food source [2], [3].

Previous studies have indicated that flying insects use visual cues to regulate flight speed (tethered bees: [4]; freely flying fruit flies, David [5]). A study by Srinivasan

S. Nolfi et al. (Eds.): SAB 2006, LNAI 4095, pp. 40–51, 2006.
© Springer-Verlag Berlin Heidelberg 2006

et al. [2] found that bees flying though a tapered tunnel slowed down as the distance between the walls narrowed, and sped up as it widened. This result suggested that the bees were adjusting their flight speed so as to hold constant the velocity of the image generated by the patterns on the walls of the tunnel on their eyes.

In the current study, we present, in part, findings from earlier experiments that tested directly and rigorously the hypothesis that honeybees control their flight speed by maintaining a constant rate of image motion (Baird et al. [6]). The purpose of presenting this data here (as Experiments 1 and 2) is to provide a context for novel data (Experiments 3 and 4) that investigate whether flying honeybees use visual cues to control both their flight speed and their height above the ground.

2 General Experimental Procedures

The experiments were carried out in an All Weather Bee Flight Facility at the Australian National University's Research School of Biological Sciences. The temperature inside the facility was maintained at 24 ± 5 °C during the day and 17 ± 3 °C at night. A beehive mounted on the wall of the facility supplied the bees (*Apis mellifera* L.) used in the experiments.

2.1 Experimental Setup

All of the experiments were conducted in a rectangular tunnel that had clear Perspex walls, which allowed bees flying through the tunnel to view a variety of stationary or moving visual patterns (see below). The tunnel was 320 cm long, 20 cm high and 22 cm wide. A clear Perspex ceiling permitted observation and filming of the bees as they flew in the tunnel (Fig. 1). For each experiment, up to 20 bees were individually marked and trained to fly through the tunnel to a feeder containing sugar solution placed at the far end of the tunnel. In Experiments 1 and 2, flights to the feeder were filmed at 25 frames per second in the central segment of the tunnel by a digital video camera positioned 2.5 m above the tunnel floor (Fig. 1). Due to the limitations of this camera set up, it was necessary to leave the floor of the tunnel blank in Experiments 1 and 2 so that the bees could be easily distinguished from the background.

In Experiments 3 and 4, two cameras were mounted 125 cm above the tunnel floor (Fig. 1.) and positioned such that they had parallel views of the tunnel. Flights of bees were captured directly into a computer from each camera simultaneously via a capture card at 30 frames per second. By tracking the position of the bee in both camera views it was possible, through triangulation, to calculate the three-dimensional position of a bee flying in the tunnel. All of the patterns that were used in these two experiments consisted of dark red-and-white elements because it was necessary to have patterns with a high contrast but it was not possible to track the positions of the bees against the black areas of a black-and-white pattern. The red color in the patterns was considered to be a suitable substitute for black as bees do not posses red color receptors and therefore, they would perceive the red parts of the patterns as a dark shade of grey.

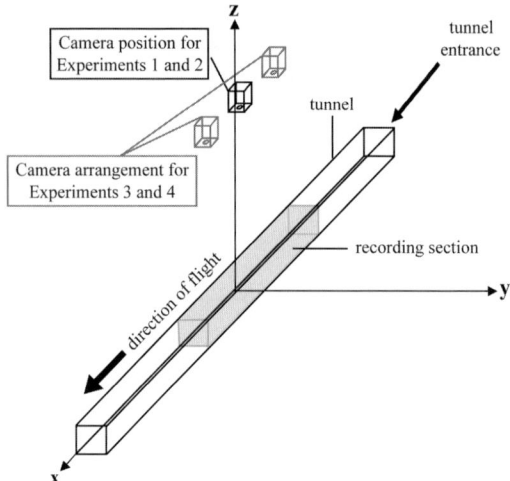

Fig. 1. Illustration of tunnel coordinates and camera position. Flight speed was calculated as the projection of the flight vector along the x axis. Height was measured along the z axis.

2.2 Analysis of Flight Trajectories

An automated tracking program was developed, using Matlab, to track individual bees and analyze the recordings of flights obtained in each experiment. For each flight, the program identified the position of the bee in consecutive frames. The position of the bee was defined in relation to the tunnel co-ordinates x, y and z, where x denotes axial direction, y the transverse direction and z the vertical direction (Fig. 1). In Experiments 1 and 2, only x and y could be measured, as the system used a single camera (rather than a stereo pair). In these experiments, z was assumed to be constant.

The camera configuration that was used in Experiments 3 and 4 allowed the measurement of the position of the bee in x, y and z coordinates. To do this, the position of the bees in x and y pixel coordinates from each camera view were entered into a database in which the three-dimensional coordinates of the bee's position were calculated using a triangulation algorithm.

To generate values of flight speed, the data was analyzed to calculate the component of the flight velocity in the axial (x) direction (V_x). Preliminary analysis revealed that the lateral component of flight velocity (V_y) was much smaller in magnitude compared to that of the axial component (V_x). Given this, it follows that V_x provides a good approximation of the actual magnitude of the flight speed. To generate values of height in Experiments 3 and 4, the average z position of each flight was calculated using the z coordinates of the bee's three-dimensional position.

2.3 Statistical Analysis

Statistical models accounting for multiple levels of variation were developed to assess whether covariates such as treatment (e.g. different patterns), time, temperature, light intensity or humidity affected bee flight speed, and to eliminate their effects. To account for the two principal levels of variation in the study -- variation between bees and variation within bees -- linear mixed models [7] were used, with bee identity as a random effect. For further details of the statistical analysis used see Baird et al. [6].

3   Optic Flow Cues in the Lateral Visual Field Affect Flight Speed

Experiment 1 was designed to examine the contribution of optic flow cues (i.e. cues that generate image motion in the eye of a honeybee) in the lateral visual field to the control of flight speed. The influence of optic flow cues was examined by recording flight speeds when the tunnel walls were lined with two different types of stationary pattern: chequerboard and axial stripes. The chequerboard pattern consisted of alternating black and white checks of 3 x 3 cm. The axial pattern consisted of alternating black and white, horizontally oriented stripes, each with a width of 4 cm. The chequerboard pattern was used in this experiment because the alternating black and white checks would provide strong image motion cues to a bee flying along the tunnel. The axial pattern, on the other hand, was used to create a condition in which the optic flow cues were very weak. This is because flight in the direction of the stripes would produce very little apparent motion of the images of the walls on the retina. In this experiment, the floor of the tunnel was blank white with no discernable optic flow cues.

The results of Experiment 1 are shown in Fig. 2. Interestingly, when optic flow cues are weak (when the tunnel is lined with axial stripes), bees fly considerably faster than when optic flow cues are strong (when the tunnel is lined with a chequerboard pattern). When the tunnel walls were lined with a chequerboard pattern, the mean flight speed was 54 cm s^{-1} but when the tunnel walls were lined with axial stripes, the bees flew significantly faster at a mean flight speed of 97 cm s^{-1} (two sided t-test, $t_{109} = 8.67$, $p < 0.0001$).

Experiment 2 was designed to examine the effect on flight speed of image motion in the lateral visual field. In this experiment, a motorized conveyor belt was placed along the length of the tunnel on each side. Each belt was white in color and carried a pattern of randomly positioned black dots, 2 cm in diameter, on its surface. The conveyor belt system allowed the pattern to be moved toward or away from the closed end of the tunnel, at a range of speeds. The influence of image motion was examined by recording flight speeds when the patterns on the walls of the tunnel were moving at six pattern velocities in each direction, and for one condition in which the pattern was static. When the pattern was moved in the direction of flight to the feeder, the highest pattern speed was limited by the maximum speed of the motor. The speeds used for pattern motion in this direction were 15, 22, 30, 37, 45 and 52 cm s^{-1} (these velocities were regarded as positive).When the pattern was moved against the direction of flight to the feeder, at high pattern speeds, the bees were unable to enter the tunnel. The maximum speed used in this condition was therefore limited to the highest speed at which the bees could enter the tunnel and fly to the feeder. The speeds tested in this condition were 6, 12, 18, 24, 30 and 36 cm s^{-1} (these velocities were regarded as negative).

The results of Experiment 2 showing the dependence of flight speed on pattern velocity are shown in Fig. 2. The results indicate that when the pattern is moved in the direction of flight (decreasing the velocity of the perceived image motion) flight speed increases (as indicated by the data points on the right hand side of the graph). When the pattern is moved against the direction of flight (increasing the velocity of the perceived image motion) flight speed decreases (as indicated by the data points on the left hand side of the graph.

If bees regulate their flight speed by keeping the rate of optic flow (i.e. the velocity of the image on the eye constant), flight speed should vary linearly with pattern velocity and the change of flight speed should be equal to the change of pattern velocity. Thus, the equation for the hypothesized flight speed adjustment takes the form:

$$y = mx + c \qquad (1)$$

where c is the flight speed when the pattern is static, x is the pattern velocity and, if the bees maintain a constant rate of optic flow in the eye, m = 1. This calculation assumes that at zero pattern velocity, flight speed is set to achieve the desired optic flow.

An analysis of the data indicates that a model which includes three lines of different slopes provides a good approximation of the effect of large positive pattern velocities, large negative pattern velocities and small positive and negative pattern

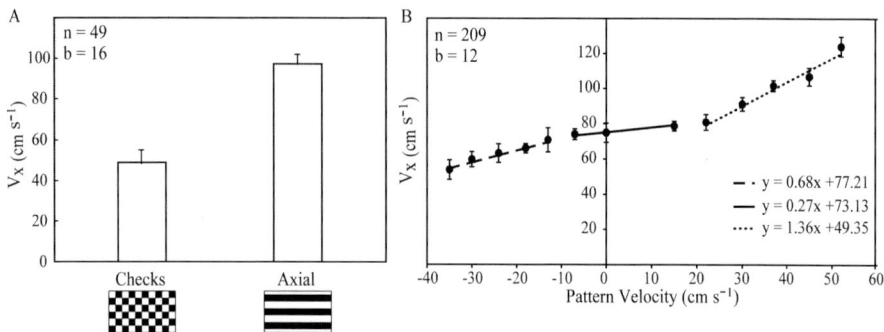

Fig. 2. (A) Experiment 1 – effect of optic flow cues in the lateral visual field on flight speed. Comparison of mean flight speeds when the walls of the tunnel are lined with a chequerboard pattern (producing strong optic flow cues) or axial stripes (producing weak optic flow cues). (B) Experiment 2 – effect of pattern motion on flight speed. The graph shows mean axial flight speed (V_x) when the pattern on the walls was static (0 pattern velocity), moved in the direction of flight (positive pattern velocity values) or against the direction of flight (negative pattern velocity values). The black circles represent V_x values for various pattern speeds. The dashed line represents a model of the flight speed data for large negative pattern velocities; the slope of this line is slightly smaller than 1. The solid line represents a model of the flight speed data for the positive and negative pattern velocities near zero. The slope of this line is not significantly different from zero. The dotted line represents a model of the flight speed data for large positive pattern velocities. For positive pattern velocities, the slope of the regression line was slightly greater than 1. The equations for each regression are shown. The error bars represent the standard error of the mean, n denotes the number of flights and b denotes the number of bees.

velocities (including zero pattern velocity) on flight speed. To fit this model, the pattern velocities were classified into three categories: high positive, near zero and high negative. A separate line was fitted within each class. For details of the development of the model, please see Baird et al. [6].

For large positive pattern velocities, the model revealed a slope of m = 1.36 (dotted line, Fig. 2). There is some evidence that this slope is significantly greater than 1 (two-sided t-test, t_{189} = 1.86, p = 0.06). This result suggests that when the pattern is moved in the direction of flight the bees respond by increasing their flight speed by slightly larger amount. Thus, when the pattern moved in the direction of flight, the bees were, to a small extent, over compensating for the changes in pattern speed and, as a result, experiencing a slightly decreased rate of optic flow.

For large negative pattern velocities, the slope of the model was m = 0.68 (dashed line, Fig. 2). There is some evidence that this slope is significantly different from 1 (two-sided t-test, t_{189} = 1.78, p = 0.08). Thus, when the pattern moved against the direction of flight, the bees were not making a complete adjustment of flight speed to counter the changes in pattern speed: they were experiencing a slightly increased rate of optic flow.

For small pattern velocities about zero, the slope of the model was m = 0.27 (solid line, Fig. 2). This slope is not significantly different from zero (two-sided t-test, t_{189} = 0.63, p = 0.53). Thus, at low pattern speeds, the bees were not adjusting their flight speed to compensate for the small changes in the rate of optic flow.

4   Optic Flow Cues in the Ventral Visual Field Affect Flight Speed and Height

Experiment 3 was designed to investigate the contribution of optic flow cues in the ventral visual field to the control of flight speed and height. The influence of optic flow cues was examined by recording the flight speed and height of bees when the tunnel floor was lined with two different types of pattern: chequerboard and axial stripes. The chequerboard pattern consisted of alternating red-and-white checks of 3 x 3 cm^2. The axial pattern consisted of alternating red-and-white stripes 4 cm in width, oriented along the longitudinal axis of the tunnel. In both Experiment 3 and 4, the walls of the tunnel were lined with a chequerboard pattern of 3 x 3 cm so that there would be strong optic flow cues in the lateral regions of the bee's visual filed The aim of this arrangement was to ensure that any changes in flight speed or height were a result of the changes in the patterns placed on the floor of the tunnel.

The results are shown in Fig. 3. When the tunnel floor was lined with an axial pattern, the bees flew at a mean flight speed of 60 cm s^{-1} and a mean height of 14 cm. When the tunnel floor was lined with a chequerboard pattern, the mean flight speed was 44 cm s^{-1} and the mean height was 19 cm. The data indicate that bees fly significantly faster (t_{81} = 4.06, p < 0.001) and lower (t_{81} = 3.85, p < 0.001) when the optic flow cues in the ventral visual field are weak (axial stripe patterns) than when the optic flow cues in the ventral visual field are strong (chequerboard pattern). This suggests that the mechanisms that mediate flight speed and height control in the honeybee are influenced by optic flow cues in the ventral region of the visual field

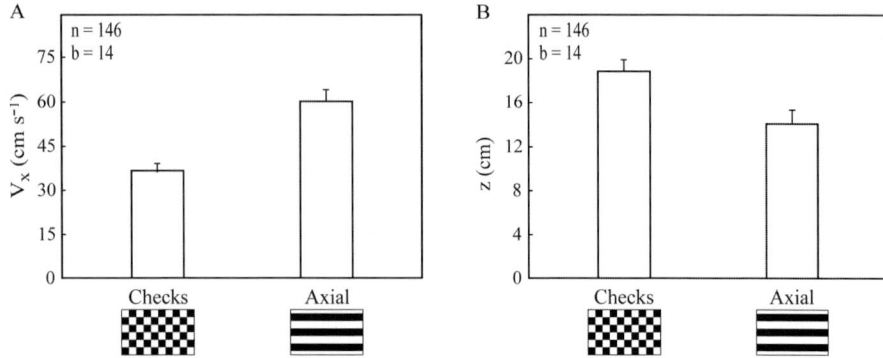

Fig. 3. Experiment 3 – effect of optic flow in the ventral visual field on flight speed and height. (A) Comparison of mean flight speeds when the floor of the tunnel is lined with either a chequerboard pattern (strong optic flow cues) or axial stripes (weak optic flow cues). (B) Comparison of height when the floor of the tunnel is lined with either a chequerboard pattern or axial stripes. Other details are as in Fig. 2.

even when the optic flow cues in the lateral region of the visual field are strong. Interestingly, flight speed is slightly lower when the floor is blank, than when it is lined with axial stripes. This could be an effect of the visual phenomenon known as "contrast adaptation", as discussed in [6].

Experiment 4 was designed to investigate whether the flight speed and height of bees is affected by changes in the spatial frequency (the number of changes in contrast over a given distance) of the pattern in the ventral visual field. Flight speed and height were measured when the tunnel floor was lined with chequerboard patterns of various check sizes: 1.5 x 1.5 cm, 3 x 3 cm and 6 x 6 cm. For a bee flying along the

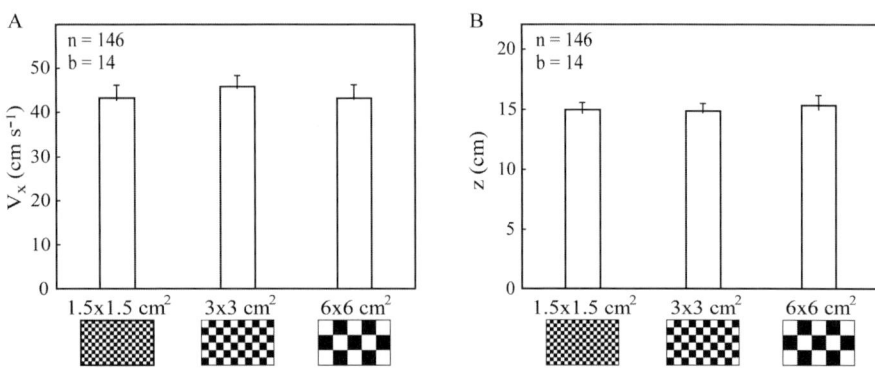

Fig. 4. Experiment 4 – effect of pattern texture in the ventral visual field on flight speed and height. Comparison of mean flight speed (A) and mean distance from the floor of the tunnel (B) when the floor of the tunnel is lined with chequerboard patterns with check sizes of 1.5 x 1.5 cm, 3 x 3 cm and 6 x 6 cm. Other details are as in Fig. 2.

midline of the tunnel, the dominant spatial frequency of the checks on the floor of the tunnel as seen by the ventral field of the eye would be 0.06 cycles deg^{-1}, 0.03 cycles deg^{-1} and 0.01 cycles deg^{-1}, respectively.

The results are shown in Fig. 4. The data indicate that neither the speed nor the height of flight is significantly influenced by changes in the spatial frequency of patterns in the ventral visual field. The average flight speed of bees was 45.5 cm s^{-1} with 3 x 3 cm checks, 43.5 cm s^{-1} with 1.5 x 1.5 cm checks ($t_{130} = 0.89$, $p = 0.37$ when compared with the 3 x 3 cm checks) and 43.8 cm s^{-1} with 6 x 6 cm checks ($t_{130} = 0.75$, $p = 0.46$ when compared with the 3 x 3 cm checks). The average flight height was 14.4 cm with 3 x 3 cm checks, 14.8 cm with 1.5 x 1.5 cm checks ($t_{130} = 0.58$, $p = 0.56$ when compared with the 3 x 3 cm checks) and 15.2 cm with 6 x 6 cm checks ($t_{130} = 1.26$, $p = 0.21$ when compared with the 3 x 3 cm checks). Therefore, the mechanisms that mediate control of flight speed control and height appear to be relatively robust to variations in the spatial frequency of patterns in the ventral visual field.

5 Discussion

The results shown here clearly demonstrate that flight speed and height control in the honeybee are regulated using optic flow. In different visual environments, flight speed is regulated so as to hold constant the speed of the image on the retina. This finding supports the hypothesis first proposed by Srinivasan et al. [2] that honeybees use the rate of optic flow to regulate their flight speed. In addition, this study has shown for the first time that height control in the honeybee is mediated by optic flow cues in the ventral region of the visual field. Earlier work in fruit flies [5], moths [8] and beetles [9] has shown that the flight speed of insects following odor plumes at different heights increases with their distance from the ground. From this work however, it is not possible to determine whether visual cues in the ventral region of the visual field influence the height at which an insect flies, as all of the insects in these experiments were following odor plumes at a set height. By using freely flying honeybees it was possible, in the present study, to test directly whether the properties of the optic flow on the ground influence height and flight speed. Until now, no study has demonstrated that flight height in insects can be influenced by the properties of visual features on the ground.

5.1 Flight Speed Control

Experiment 1 demonstrates that optic flow cues play an important role in the regulation of flight speed. When optic flow cues are weak, (when the walls are lined with an axial stripe pattern), bees fly much faster than when optic flow cues are strong (when the walls are lined with a chequerboard pattern). The reason for the difference in flight speed between these two conditions is likely to be related to the fact that, as the axial pattern carries no strong horizontal optic flow cues, it elicits weak image motion signals, thus causing the bees to fly faster. Interestingly, flight speed also increases when the optic flow cues on the floor of the tunnel are weak, even though the walls provide strong optic flow cues, as shown in Experiment 3. This indicates that the visual information in the ventral region of the bee's eye is important for the regulation of flight speed. These results are consistent with those of Barron and

Srinivasan [10] who found that when the walls and floor of a tunnel are lined with axial stripes, bees fly three times faster than when the tunnel is lined with a chequerboard pattern.

Interestingly, flight speeds are lower (60 cm s^{-1}) when the floor of the tunnel carries an axial pattern and the walls a chequerboard pattern, than when the walls carry axial patterns and the floor is blank (97 cm s^{-1}). Similarly, bees fly slower (44 cm s^{-1}) when the chequerboard pattern covers all three surfaces of the tunnel, as compared to when it lines only the walls (54 cm s^{-1}). These observations suggest that the mechanism of flight speed control averages the perceived velocity of image motion from the lateral as well as the ventral regions of the visual field.

The hypothesis that flight speed in honeybees is regulated by optic flow was tested directly and rigorously in Experiment 2. Here we found that the bees adjusted their flight speed so as to hold the speed of the image on the retina constant. When the patterns on the walls of the tunnel were moved in the direction of flight, the bees increased their flight speed by an amount that was slightly greater than the speed of the pattern. When the patterns on the walls of the tunnel were moved against the direction of flight, the bees decreased their flight speed by an amount that is slightly lower than the speed of the pattern. When the patterns on the walls of the tunnel are moved at slow speeds either with, or against the direction of flight, there is no associated change in flight speed. This result indicates that the system that mediates flight speed control only responds to changes in the velocity of the image of the visual environment that exceed a certain threshold. This threshold is estimated to lie between 10 and 15 deg s^{-1}. Once the deviation in perceived image velocity exceeds this threshold, flight speed is adjusted so as to return the deviation to a level that is below threshold.

In Experiment 4 we showed that the flight speed of honeybees is not affected by changes in the spatial frequency of the image in the ventral visual field. This result is consistent with the findings of our earlier work [6] which showed that the flight speed of bees was not affected by changes in the spatial frequency pf patterns in the lateral visual field.

We have shown here that honeybees control their flight speed by holding the rate of image motion across their eyes constant. What are the consequences of maintaining a constant image velocity during flight? One outcome would be that, because perceived image velocity is related to the distance of the viewer from the substrate, flight speed is adjusted according to the proximity of objects and surfaces in the environment. For example, flight speed would tend to be high when flying in an open field and low during a flight through dense vegetation. Thus, maintaining a constant image velocity in the eye would ensure that the speed of flight is automatically adjusted to a level that is safe and appropriate to the environment. Our findings also suggest that the visual pathways that control flight speed are capable of measuring and regulating the velocity of the images of the walls, largely independently of the spatial structure of the environment.

5.2 Height Control

Maintaining a constant ground speed may affect the height at which bees fly, but it is not possible to extract absolute height information solely from the rate of optic flow. This is because the perceived velocity of motion of the image of the ground will depend on the speed, as well as the height, of flight. A given ground image velocity can be achieved by slow flight at a low altitude or faster flight at a higher altitude. So, what cues do honeybees use to control the height at which they fly above the ground? In this study we attempt to address this question by investigating whether honeybees rely on optic flow in the ventral region of the visual field to regulate their height, and whether flight height is influenced by the texture of the visual environment.

The data from Experiment 3 suggest that flight height is influenced by optic flow cues in the ventral region of the visual field. Bees fly at a lower height when the pattern on the floor of the tunnel carries weak optic flow cues (axial stripes), than when it provides strong optic flow cues (chequerboard pattern). This finding is interesting because it suggests that the system that mediates height control relies on optic flow cues in the ventral region of the visual field, even when there are strong vertical optic flow cues in the lateral region of the visual field (as provided by the chequerboard pattern on the walls of the tunnel). This result makes sense, because it is only optic flow cues in the ventral field of view that can provide useful information on flight height. Optic flow cues in the lateral visual fields will depend primarily on the distances to objects in the lateral field, which is irrelevant to the estimation of height above the ground.

The results of Experiment 4 indicate that the mechanisms that mediate height control are not sensitive to the spatial texture of the environment in the ventral field. This is consistent with the properties of the system that mediates flight control. This finding makes sense from a real world perspective where the texture of the ground can vary substantially, thus making it desirable to have a system for controlling flight height that is robust to variations in the visual texture of the ground.

The findings from this study, as well as those from our previous investigation [6] suggest that the mechanisms of flight speed and height control in the visual pathway of the honeybee measure and regulate the velocity of the images in both the lateral and ventral regions of the visual field and that these mechanisms are insensitive to the contrast or the spatial texture of the visual environment. The advantages of a system that relies on the measurement of image velocity to control flight speed and height are that these behaviors will be regulated according to the proximity of objects in the environment rather than their visual features. Similar properties have been observed in the visual pathways that mediate other flight behaviors such as the centering response [11] and the visual odometer [12]. The movement-detecting mechanisms that mediate the behaviors discussed above seem to have properties that are rather different from those of the well-studied optomotor response in insects. The optomotor response is a behavior in which a flying insect generates motions to compensate for unwanted body rotations by measuring the associated rotations of the image in the eye [13]. The movement-detecting mechanism that mediates the optomotor response appears to be sensitive to changes in the contrast, spatial frequency and temporal frequency of the moving image. As a result, this system does not seem to encode image velocity in a manner that is robust to variations in these parameters. The visual pathways that control flight speed and height, mediate the centering response and generate the odometric signal have properties that are different from the pathway that drives the optomotor response. There is extensive literature on the anatomy and physiology of movement-detecting neurons whose response properties reflect the characteristics of the optomotor response [reviewed by 14]. However, there is relatively little evidence for the existence of motion-sensitive neurons whose response properties reflect the properties of the system that mediates flight speed and height control, the centering response and the visual odometer. The movement-detecting mechanisms that underlie these behaviors must have the capacity to measure image velocity independently of contrast and spatial texture. There is some evidence that velocity-tuned neurons exist in the visual systems of insects but it is not clear whether these neurons participate in the behaviors discussed above [15].

5.3   Implications and Applications of the Present Findings

The groundspeed of a flying agent is determined by many parameters. Pilots of modern aircraft rely on the measurement of multiple parameters including thrust, local airspeed and global position, to calculate and regulate their groundspeed. The honeybee, on the other hand, appears to use only a single measurement from the external environment (namely the measurement of image velocity) to regulate groundspeed. Although this strategy will not achieve a constant groundspeed -- the groundspeed will depend upon the distances to objects and surfaces in the lateral fields of view -- it will ensure that the groundspeed is automatically adjusted to suit the environment through which the flight occurs. Thus, groundspeed will be high in an open environment and slow in a densely cluttered environment.

The present study was conducted in a controlled environment with no interference from air currents. How would this system of flight control respond in the natural environment, where winds are commonly prevalent? In a strong head wind, it would be difficult for bees to maintain a constant groundspeed and therefore, their preferred image velocity. To compensate, bees would have to fly closer to the ground, thus restoring the image speed to its original value. Interestingly, a reduction of altitude would be likely to reduce the velocity of the headwind that the bee experiences, thus enhancing the bee's ability to compensate for the headwind, decreasing the required thrust and reducing the energy expended for the flight [16], [17]. On the other hand, a strong tailwind would cause the bee to fly higher, in attempting to regulate the image velocity. This in turn would enable the bee to catch a stronger tailwind, thus increasing groundspeed, decreasing the required thrust and again reducing energy consumption.

Our findings in relation to the control of flight speed and height in honeybees suggest insect-inspired strategies for the control of aircraft flight. In the design of guidance systems for autonomous aerial vehicles, there is a growing need to avoid sensors that are heavy or expensive, and which use active devices such as radar, sonar or lasers [18], [19]. The techniques of visual guidance that are employed by flying insects, such as those reported here, suggest   relatively light, inexpensive and computationally simple ways of achieving some of the desired functions like control of flight speed, and terrain following.

Acknowledgements. This work was partly supported by U.S. AFOSR Contract F62562, Australian Research Council Grants FF0241328, CE0561903 and DP 020863, and the Army Research Office MURI ARMY-W911NF041076, Technical Monitor Dr Tom Doligalski.

References

1. J. J. Gibson, The Perception of the Visual World, Houghton Mifflin, Boston, 1950.
2. M. Srinivasan, S. Zhang, M. Lehrer and T. Collett, Honeybee navigation en route to the goal: visual flight control and odometry, J Exp Biol, 199 (1996), 237-44.
3. M.V. Srinivasan, S.W. Zhang, J.S. Chahl, E. Barth and S. Venkatesh, How honeybees make grazing landings on flat surfaces, Biol Cybern, 83 (2000), 171-83.
4. H. Heran, Versuche über die Windkompensation der Bienen, Naturwissenscaften, 42 (1955), 132-133.
5. C.T. David, Compensation for height in the control of groundspeed by *Drosophila* in a new 'Barber's Pole' wind tunnel, J Comp Physiol A, 147 (1982), 485-493.
6. E. Baird, M.V. Srinivasan, S. Zhang and A. Cowling, Visual control of flight speed in honeybees, J Exp Biol, 208 (2005), 3895-905.
7. C. E. McCulloch and S. Searle, Generalized, linear, and mixed models, John Wiley & Sons, New York, 2001.
8. L.P.S. Kuenen and T.C. Baker, Optomotor regulation of ground velocity in moths during flight to sex pheromone at different heights, Physiol Entomol, 7 (1982), 193-202.
9. H.Y. Fadamiro, T.D. Wyatt and M.C. Birch, Flying beetles respond as moths predict: Optomotor anemotaxis to pheromone plumes at different heights, J Insect Behav, 11 (1998), 549-557.
10. A. Barron and M.V. Srinivasan, Visual regulation of ground speed and headwind compensation in freely flying honey bees (*Apis mellifera* L.), J Exp Biol, 209 (2006), 978-84.
11. M. V. Srinivasan, M. Lehrer, W. H. Kirchner and S. W. Zhang, Range perception through apparent image speed in freely flying honeybees, Vis Neurosci, 6 (1991), 519-35.
12. A. Si, M. V. Srinivasan and S. Zhang, Honeybee navigation: properties of the visually driven 'odometer', J Exp Biol, 206 (2003), 1265-73.
13. W. Reichardt, Movement perception in insects, in W. Reichardt, ed., Processing of Optical Data by Organisms and Machines, Academic Press, New York, 1969, 465-493.
14. K. Hausen, Decoding of retinal image flow in insects, Rev Oculomot Res, 5 (1993), 203-35.
15. M.R. Ibbotson, Evidence for velocity-tuned motion-sensitive descending neurons in the honeybee, Proc R Soc Lond, 268 (2001), 2195-2201.
16. J. R. Riley and J. L. Osborne, Flight trajectories of foraging insects: observations using harmonic radar, in D. R. Reynolds, C. Thomas and I. H. Wolwod, eds., Insect movement: mechanisms and consequences, Proceedings of the Royal Entomological Society's 20th Symposium, CABI Publishing, 2001, 129-157.
17. F. Ruffier and N. Franceschini, Optic flow regulation: the key to aircraft automatic guidance, Robotics and Autonomous Systems, 50 (2005), 177.
18. A. I. Peterson, Launched to return, Unmanned Vehicles, 2003, 13.
19. O. Shakerina, Y. Ma, T. J. Koo, J. Hespanha and S. S. Shastry, Vision guided landing of an unmanned air vehicle, Proceedings of the 38th Conference on Decision and Control, Phoenix, Arizona, 1998, 4143-4148.

Visual Learning of Affordance Based Cues

Gerald Fritz[1], Lucas Paletta[1], Manish Kumar[1], Georg Dorffner[2], Ralph Breithaupt[3],
and Erich Rome[3]

[1] JOANNEUM RESEARCH Forschungsgesellschaft mbH,
Institute of Digital Image Processing, Computational Perception Group,
Wastiangasse 6, Graz, Austria
[2] Österreichische Studiengesellschaft für Kybernetik,
Neural Computation and Robotics, Freyung 6, Vienna, Austria
[3] Fraunhofer Institute for Autonomous Intelligent Systems,
Robot Control Architectures, Schloss Birlinghoven, Sankt Augustin, Germany

Abstract. This work is about the relevance of Gibson's concept of affordances [1] for visual perception in interactive and autonomous robotic systems. In extension to existing functional views on visual feature representations, we identify the importance of *learning* in perceptual cueing for the anticipation of opportunities for interaction of robotic agents. We investigate how the originally defined representational concept for the perception of affordances - in terms of using either optical flow or heuristically determined 3D features of perceptual entities - should be generalized to using *arbitrary* visual feature representations. In this context we demonstrate the learning of causal relationships between visual cues and predictable interactions, using both 3D and 2D information. In addition, we emphasize a new framework for cueing *and* recognition of affordance-like visual entities that could play an important role in future robot control architectures. We argue that affordance-like perception should enable systems to react to environment stimuli both more efficient and autonomous, and provide a potential to plan on the basis of responses to more complex perceptual configurations. We verify the concept with a concrete implementation applying state-of-the-art visual descriptors and regions of interest that were extracted from a simulated robot scenario and prove that these features were successfully selected for their relevance in predicting opportunities of robot interaction.

1 Introduction

The concept of affordances has been coined by J.J. Gibson [1] in his seminal work on the ecological approach to visual perception: "*The affordances of the environment are what it offers the animal, what it provides or furnishes, either for good or ill ... something that refers both to the environment and the animal in a way that no existing term does. It implies the complementarity of the animal and the environment.*" In the context of ecological perception, visual perception would enable agents to experience in a direct way the opportunities for action. However, Gibson remained unclear about how this concept could be used in a technical system. Neisser [2] replied to Gibson's concept of direct perception with the notion of a perception-action cycle that shows the reciprocal relationship of the knowledge (i.e., a schema) about the environment

S. Nolfi et al. (Eds.): SAB 2006, LNAI 4095, pp. 52–64, 2006.

directing exploration of the environment (i.e., action), which samples the information available for pick up in the environment, which then modifies the knowledge, and so on. This cycle describes how knowledge, perception, action, and the environment all interact in order to achieve goals.

Our work on affordance-like perception is in the context of technical, i.e., robotic systems, based on a notion of affordances that *'fulfill the purpose of efficient prediction of interaction opportunities'*. We extend Gibson's ecological approach under acknowledgment of Neisser's understanding that visual feature representation on various hierarchies of abstraction are mandatory to appropriately respond to environmental stimuli. We provide a refined concept of affordance perception by proposing (i) an interaction component (*affordance recognition*: recognizing relevant events in interaction via perceptual entities) and (ii) a predictive aspect (*affordance cueing*: predicting interaction via perceptual entities). This innovative conceptual step enables firstly to investigate the functional components of perception that make up affordance-based prediction, and secondly to lay a basis to identify the interrelation between predictive features and predicted event via machine learning technology.

The outline of this paper is as follows. Section 2 describes the relevance of affordance-like representations in robot perception and argues for the importance to learn the features of perceptual entities. Section 3 focuses on the issues of affordance recognition, in contrast to the predictive aspect of affordance-like representations in affordance cueing presented in Section 4. Section 5 illustrates the experimental results that strongly support the proposed hypothesis on the relevance of generalized features that must be learned for successful affordance-like perception in robot control systems. Section 6 concludes with an outlook on future work.

2   Affordance Perception and Learning

Affordance-like perception aims at supporting control schemata for perception-action processing in the context of rapid and simplified access to agent-environment interactions. In this Section we argue that previous research has not yet tackled the relevance of learning in cue selection, and present a framework on functional components that enables to identify relevant visual features.

2.1   Related Work

Previous research on affordance-like perception focused on heuristic definition of simple feature-function relations to facilitate sensor-motor associations in robotic agents. Human cognition embodies visual stimuli and motor interactions in common neural circuitry (Faillenot et al.[3]). Accordingly, the affordance-based context in spatio-temporal observations and sensor-motor behaviours has been outlined in a model of cortical involvement in grasping by Fagg and Arbib [4], highlighting the relevance of vision for motor interaction. Reaching and grasping involves visuomotor coordination that benefits from an affordance-like mapping from visual to haptic perceptual categories (Wheeler et al.[5]). Within this context, the MIT humanoid robot Cog was involved in object poking and proding experiments that investigate the emergence of affordance categories to choose actions with the aim to make objects roll in a specific way (Fitzpatrick et al.[6]). The research of Stoytchev [7] analysed affordances on an object level, investigating new concepts of object-hood in a sense of how perceptions of objects are connected with visual events that arise from action consequences related to the object itself. Although this work innovatively demonstrated the relation between affordance triggers and meaningful robot behaviours, these experiments involve computer vision still on a low level, and do not consider complex sensor-motor representation of an agent interaction in less constrained, even natural environments. In addition, they are restricted to using vision rather than exploiting the multi-modal sensing that robots may perform. In the biologically motivated cognitive framework of Cos-Aguilera et al. [15], object based affordances are set in the context of motivation driven behaviour selection. In contrast to our work, they do not learn visual feature extraction in a purposive manner (Section 2.2) but rather match sensory input with stored object features in a classical sense [16] and then associate object identities with appropriate interaction patterns.

Affordance based visual object representations are per se function based representations. In contrast to classical object representations, functional object representations (Stark and Bowyer [8], Rivlin et al. [9]) use a set of primitives (relative orientation, stability, proximity, etc.) that define specific functional properties, essentially containing face and vertex information. These primitives are subsumed to define surfaces and from the functional properties, such as 'is sit-able' or 'provides stable support'. Bogoni and Bajcsy [10] have extended this representation from an active perception perspective, relating observability to interaction with the object, understanding functionality as the applicability of an object for the fulfillment of some purpose. However, so far function based representations were basically defined by the engineer, while it is particularly important for affordance based representations to *learn* the structure and the features themselves *from experience* (Section 4).

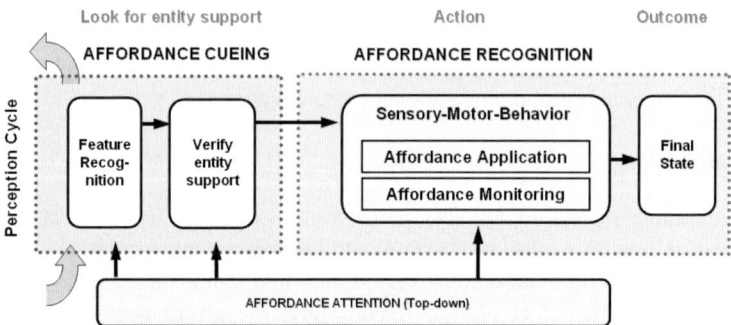

Fig. 1. Concept of affordance perception, depicting the key components of affordance cueing and recognition embedded within an agent's perception-action cycle (most left). While affordance cueing (left) provides a prediction on future opportunities of interaction on the basis of feature interpretation, affordance recognition (right) identifies the convergence of a perceptual patterns in a sensory-motor behavior towards the outcome of the overall process.

2.2 Predictive Features in Affordance Based Perception

Fig. 1 depicts the innovative concept of feature based affordance perception presented in this paper. We identify first the functional component of affordance recognition, i.e., the recognition of the affordance related visual event that characterizes a relevant interaction, e.g., the capability of lifting (*lift-ability*) an object using an appropriate robotic actuator. The recognition of this event should be performed in identifying a process of evaluating spatio-temporal information that leads to a final state. This final state should be unique in perceptual feature/state space, i.e., it should be characterized by the observation of specific feature attributes that are abstracted from the stream of sensory-motor information.

The second functional component of affordance cueing encompasses the key idea on affordance based perception, i.e., the prediction aspect on estimating the opportunity for interaction from the incoming sensory processing stream. In particular, this component is embedded in the perception-action cycle of the robotic agent. The agent is receiving sensory information in order to build upon arbitrary levels of feature abstractions, for the purpose of recognition of perceptual entities. In contrast to classical feature and object recognition, this kind of recognition is *purposive* in the sense of selecting exactly those features that efficiently support the evaluation of identifying an affordance, i.e., the perceptual entities that possess the capability to predict an event of affordance recognition in the feature time series that is immediately following the cueing stage of affordance based perception. The outcome of affordance cueing is in general a probability distribution P_A on all possible affordances (Section 4.1), providing evidence for a most confident affordance cue by delivering a hypothesis that favors the future occurrence of a particular affordance recognition event. This cue is functional in the sense of *associating* to the related feature representation a specific *utility* with respect to the capabilities of the agent and the opportunities provided by the environment, thus representing *predictive features* in the affordance based perception system.

The relevance of attention in affordance based perception has first been mentioned by developmental psychologist E.J. Gibson [11] who recognized that attention strategies are learned by the early infant to purposively select relevant stimuli and processes in interaction with the environment. In this context we propose to understand affordance cues and affordance hypotheses as fundamental part in human attentive perception, claiming that – in analogy – purposive, affordance based attention could play a similar role in machine perception as well.

There are affordances that are explicitly innate to the agent through evolutionary development and there are affordances that have to be learned [1]. Learning chains of affordance driven actions can lead to learning new, more complex affordances. This can be done, e.g., by imitation, whereby it is reasonable to imitate goals and sub goals instead of actions [12]. In the context of the proposed framework on affordance based perception (Fig. 1), learning should play a crucial role in determining predictive features. In contrast to previous work on functional feature and object representations [8, 9], we stress the fact that functional representations must necessarily contain *purposive features*, i.e., represent perceptual entities that refer to interaction patterns and thus must be selected from an existing pool of generic feature representations.

Feature selection (and, in a more general sense, feature extraction) must be performed in a machine learning process and therefore avoid heuristic engineering which is always rooted in a human kind understanding of the underlying process, a methodology which is necessarily both, firstly, error prone due to failing insight into statistical dependencies and, secondly, highly impractical for autonomous mobile systems. Our work highlights the process of learning visual predictive cues which to our understanding represents one of the key innovative issues in autonomous learning for affordance based perception.

3 Affordance Recognition

By affordance recognition we particularly refer to the process of identifying the relevant interaction events from perception that actually 'motivate' an agent to develop/learn perceptual cues for early prediction. In early infant development, the monitoring of affordances such as '*grasp-ability*' of objects or '*pass-ability*' of terrain [1] must be crucial to obtain an early as possible classification of the environment so that interaction behaviors can be initiated as fast and as robust as possible. In analogy, autonomous robotic systems should possess a high degree of flexibility and therefore be capable of perceiving affordances and therefore select appropriate functional modules as early as possible with respect to the goals of the robotic system. In this sense, goals and affordances are intimately related and make up a fully purposive perception system.

Fig. 2 illustrates the various stages within the affordance based perception process, in particular affordance recognition, for the example of the affordance '*fill-able*' in the context of the opportunities for interaction with a coffee cup.

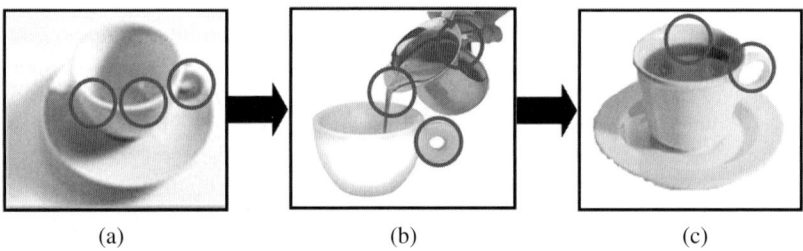

(a)   (b)   (c)

Fig. 2. Affordance recognition (Section 3) in affordance based perception for the example of the affordance *fill-able* with respect to the impact of selecting appropriate features. The seemingly simple interaction of *filling up a coffee cup* can be partitioned into various stages in affordance based perception, such as, (a) affordance cueing by predictive features that refer to a *fill-able* object, (b) identifying perceptual entities that represent the process of the affordance related interaction (e.g., flow of coffee), and (c) recognizing the final state by detecting perceptual entities that represent the outcome of interaction (e.g., level of coffee in cup).

Fig. 2(a) schematically illustrates the detection of perceptual entities that would provide affordance cues in terms of verifying the occurrence of a cup that is related to the prediction of being *fill-able* in general. Fig. 2(b) shows in analogy entities that would underlie the process of interaction of an agent with the cup by actually filling it up. Finally, Fig. 2(c) represents the entities corresponding to the final state of the interaction with the outcome of a successfully filled coffee cup. These figures illustrate that affordance cueing and affordance recognition must be conceptually separated and would involve different perceptual entities in general. While affordance *recognition* actually involves the recognition of the interaction process and its associated final state, affordance *cueing* will be solely determined by the capability to reliably predict this future event in a statistical sense.

4 Visual Cueing of Affordances

4.1 Feature Based Cueing for the Prediction of Affordances

Early awareness of opportunities for interaction is highly relevant for autonomous robotic systems. Visual features are among the ones among multiple modalities from sensory processing that operate perception via optical rays and therefore support early awareness from rather remote locations. Although the necessity of affordance perception from 3D information recovery, such as optical flow, has been stressed in previous work [1], we do not restrict ourselves to any specific cue modality and intend to generalize towards the use of arbitrary features that can be derived from visual information, restricting only on the constraint that they enable reliable prediction of the opportunity for interaction processes from an early point in time.

The outcome of the affordance cueing system is in general expected to be – given a perceptual entity in the form of a multimodal feature vector - a probability distribution over affordance hypotheses,

$$P_A = P(A \mid F_t), \tag{1}$$

with affordance hypothesis set A, and feature vector F_t at time t. It is then appropriate to select an affordance hypothesis $A_{max}(P_A(\cdot)=P_{Amax}(\cdot))$, with Maximum A Posteriori (MAP) confidence support for further processing.

From the viewpoint of a technical system using computer vision for digital image interpretation, we particularly think that complex features, e.g., local descriptors, such as the Scale Invariant Feature Transform (SIFT [13]), could support well the construction of higher levels of abstraction in visual feature representations. SIFT features are derived from local gradient patterns, and provide rotation, translation and – to some degree – viewpoint and illumination tolerant recognition of local visual information, and are therefore well suited for application in real world scenarios for autonomous robotic systems. Among other cues, such as color, shape, and 3D information, we are therefore interested to investigate the *benefit of using visual 2D patterns* for their use in affordance cueing.

Fig. 3 shows the application of local (SIFT) descriptors for the characterization of regions of interest in the field of view. For this purpose, we first segment the color based visual information within the image, and then associate integrated descriptor responses sampled within the regions to the region feature vector. The integration is performed via a histogram on SIFT descriptors that are labeled with 'rectangular'

[Figure 3 images - two panels labeled (a) and (b)]

(a)   **(b)**

Fig. 3. Categories of local descriptor classes supporting affordance cueing. Classes of SIFT descriptors [13] occurring on (a) *rectangular* (favored by descriptors represented by squares) and (b) *circular* (favored by descriptors represented by circles) region boundaries, respectively. It should be noted that the descriptor classes support the classification of segmented regions. These classes are mandatory to discern affordance cues from 2D features.

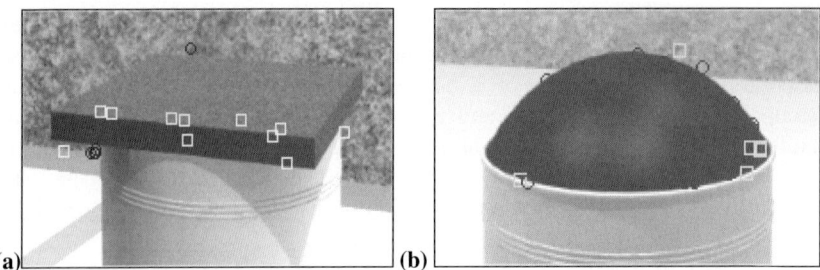

colour	G	R	M	R	Y	B	Bl	Gr
SIFT category	R	R	C	C	R	R	R	N
shape L/W	L	L	L	L	P	P	P	L
T/B	T	T	T	T	B	B	B	N
LIFTABLE	Y	Y	N	N	Y	Y	N	N
NOT LIFTABLE	N	N	Y	Y	Y	Y	Y	N

Fig. 4. Cue-feature value matrix depicting attribute values of 2D features (color G/green, R/red, M/magenta, etc., or SIFT category R/rectangular, C/circular, etc.)) and interaction results (left column) in dependence on various types of visual regions (top row). From this we conclude a suitable feature value configuration (i.e., SIFT categories to discriminate *lift-able/non lift-able* predictions) to support the hypothesis on *lift-able* object information.

(a) and 'circular' (b) attributes, respectively. The labeling is derived from a k-means based unsupervised clustering over all descriptors sampled in the experiments, then by selecting cluster prototypes (centers) that are relevant for the characterization of corresponding rectangular/circular shaped regions, and finally by determining histograms of relevant cluster prototypes that are typical in a supervised learning step (using a C4.5 decision tree [14]).

Fig. 4 shows a sample cue-feature value matrix (in the context of the experiments, see Section V) that visualizes dependencies between feature attributes of the region information and a potential association to results of the affordance recognition process. We can easily see that the SIFT category information (*rectangular=R* and *circular*=*C* region characterization) together with a geometric feature (*top*=*T* region, i.e., representing a region that is located on top of another region) provides the discriminative feature that would allow to predict the future outcome (e.g., *lift-able/non lift-able*) of the affordance recognizer. The latter therefore represents the identification of the affordance and thereby the nature of the interaction process (and its final state) itself.

4.2 Learning of Relevant Feature Cues from Decision Trees

The importance of machine learning methodologies for the selection of affordance relevant features has already been argued in Section 2.2. The key idea about our idea of applying learning for feature selection is based on the characterization of extracted perceptual entities, i.e., *segmented regions* in the image, via a feature vector representation. Each region that would be part of the final state within the affordance recognition process can be labeled with the corresponding affordance classifications. The regions can be back-tracked using standard visual tracking functionality to earlier stages in the affordance perception process. The classification label together with the feature attributed vectors of the region characterization build up a training set that can be input to a supervised machine learning methodology (using a C4.5 DT [14]).

5 Experimental Results

The experiments were performed in a simulator environment with the purpose of providing a proof of concept of successful learning of predictive 2D affordance cues, and characterizing affordance recognition processes.

The scenario for the experiments (Fig. 5) encompassed a mobile robotic system (Kurt2, Fraunhofer AIS, Germany), equipped with a camera stereo pair and a magnetizing effector, and some can-like objects with various top surfaces, colors and shapes. The purpose of the magnetizing effector was to prove the nature of the individual objects by lowering its rope-end effector down to the top surface of the object, trying to magnetize the object (only the body, *not* the top surface of the can are magnetizable) and then to lift the object. Test objects with well magnetizable geometry (with slab like top surfaces, in contrast to those with spherical top surface) are subject to a lifting interaction, while the others were not able to be lifted from the ground. This interaction process was visualized for several test objects and sampled in a sequence of 250 image frames. These image frames were referenced with multimodal sensor information (e.g., size of magnetizing and motor current of the robot, respectively).

5.1 Simulation

The scenario is split up into two phases (a) a *cueing phase,* i.e., the robot is moving to the object, and (b) a *recognition phase,* i.e., the robot tries to lift an object like shown in Fig. 5. In both phases parts of the objects are described by their regions and any region has different features like color, center of mass, top/bottom location and the

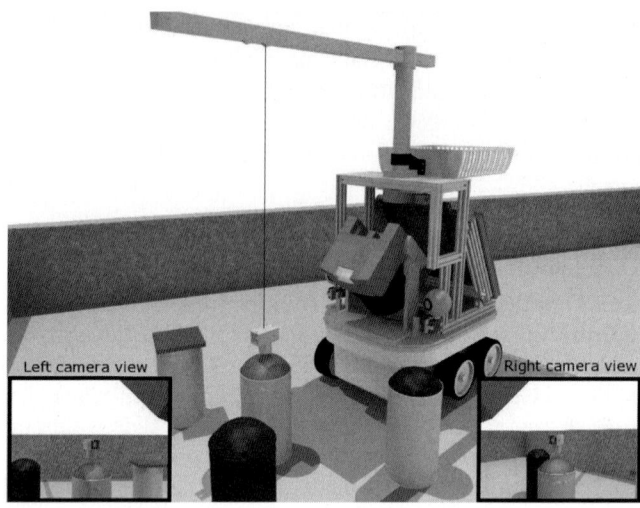

Fig. 5. Scenario of affordance based robot simulation experiments (Section 3). Birds view illustrating robot Kurt2 within a scene of objects of colored cans, using a magnetic effector at the end of a rope for interaction with the scene, described in more detail in Section V. The lower left/right corner shows the field of view of the left and right camera, respectively.

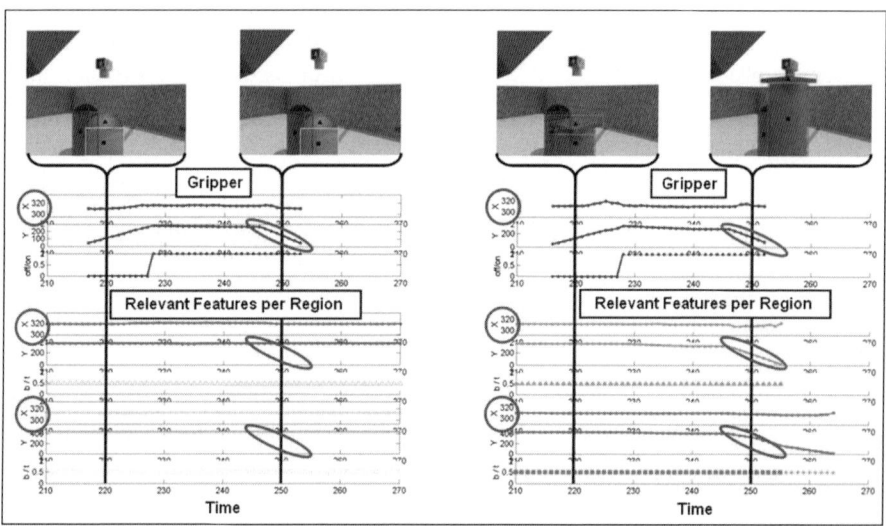

Fig. 6. Example of an affordance recognition process (here referring to '*lift-able*'): The upper image shows the right camera views of the robot while trying to lift a test object by means of a magnetizing effector at the lower end of a rope. The diagrams visualize the observation of robot relevant sensor information (e.g., status of gripper, magnet [on/off] and various features of test objects within the focus of attention. Using this sensor/feature information, the relevant channels to discriminate regions of interest that are associated to *lift-able* and *non-lift-able* objects are identified (highlighted by ellipsoids).

shape description (rectangular, circular) already described in Section IV. Those features are extracted from the robot camera imagery. Additional information, such as, effector position are provided by the robot. Regions are the entities used in the experiment, no explicit object model is generated for the can-like objects.

5.2   Affordance Recognition

The *recognition* of an affordance is crucial for verifying a hypothesis about an affordance A associated with a entity E. These entities are extracted out of the images as follows. Firstly, a watershed algorithm is used to segment regions of similar color together. After merging of smaller parts, every entity is represented by the average color value, the position in the image and the relation to adjacent regions (top/bottom). This information is also used for tracking entities over time. To verify whether or not an entity becomes '*lift-able*', the magnetizable effector of the robot is lowered until the top region of the object under investigation is reached, the magnet is switched on and the effector is lifted up. Fig. 6 shows the features of the effector (position and magnet status) over time (diagram of gripper features). If the entity is *lift-able* (Fig. 6, right column), a common motion between effector and region can be recognized. Additionally the magnet has to be switched on and the effector has to be placed in the center of the top region. These rules build up the affordance recognizer looking for *lift-able* entities in the recognition phase of the experiment.

5.3   Affordance Cueing

Cueing and recognition can require extraction of different kinds of features. Section IV already emphasized the need for some structural description of the top region, to separate the unequal shape of the top regions. In order to get structural information about an entity a histogram over prototypical SIFT descrptors is used to discriminate between circular and rectangular regions.

Classification of Relevant Descriptors. All local SIFT descriptors extracted in the region of the entities are clustered using the k-means (k = 100) method. For each specific entity, we generate a histogram over cluster prototypes, using a NN-approach to get the cluster label for each SIFT descriptor in that region. In a supervised learning step, every histogram is labeled whether it is or isn't associated with a rectangular or circular entity. A C4.5 decision tree of size 27 is then able to distinguish between these two classes. The error rate on a test set with 353 samples is ~ 1.4%. Table 1 shows the resulting confusion matrix for the test set.

Table 1. Confusion matrix for C4.5 based structure classification

Classified as			
Rect.	*Circ.*		
256	1	*Rect.*	Class
4	92	*Circ.*	

Decision Tree Used for Affordance Cueing. The objects tested for the affordance *'lift-able'* in the recognition phase are members of the training set. The outcome of the recognition provides the class label (*'lift-able'* or *'non lift-able'*). The bottom region of the object is marked 'unknown' because this entity is not tested directly. As mentioned earlier, there exists no object model yet, therefore only *entities* exist in the system. Backtracking the object's entities over time allows additional training samples to be used with little more memory effort to remember the data. In our experiment 30 frames are used from the beginning of the affordance recognition back, that means a recall of ~2.5 seconds from the past (12 fps are captured by the robot during simulation). The entity representation for the cueing phase contains the following features: (a) average color value of the region in the image, (b) top/bottom information, (c) the result of the structure classification, (d) the size of the segmented region. Fig. 7 depicts the structure of the decision tree. It is important to note that as a result from learning, the *relevant attributes* in the cueing process are *on top of the tree*, these are *'top'/'bottom'* and *'circ'/'rect'* here. The size attribute is located on the lowest level and only useful to separate 6 non *lift-able* samples from 474 *lift-able* ones. The error rate on the test set, containing the remaining entities which where not used for training, is 1.6%. Table 2 shows the confusion matrix for these data.

```
tb = bottom: unknown (1086.0)
tb = top:
|   structure = circ: non lift-able (552.0)
|   structure = rect:
|   |   size > 1426 : lift-able (402.0)
|   |   size <= 1426 :
|   |   |   size <= 1410 : lift-able (72.0)
|   |   |   size > 1410 : non lift-able (6.0)
```

Fig. 7. Structure of the C4.5 decision tree that maps attributes of the affordance feature vector f(A,t) to affordance capabilities (*lift-able*, *non lift-able* unknown). The number of samples that support the corresponding hypothesis are denoted in brackets.

Table 2. Confusion matrix for C4.5 based descriptor classification

	Classified as			
lift-able	*non lift-able*	*unknown*		
95	11	0	*lift-able*	
3	319	0	*non lift-able*	class
0	0	471	*unknown*	

6   Conclusions

This work presented the perceptual cueing to opportunities for interaction of robotic agents in a general sense, in extension to the classical functional view on feature representations. The new framework for cueing and recognition of affordance-like visual entities is verified with a concrete implementation using state-of-the-art visual descriptors on a simulated robot scenario and proved that features are successfully selected that are relevant for prediction towards affordance-like control in interaction. The simulation was chosen in a realistic way so that major elements of a real world scenario, such as shadow events, noise in the segmentation, etc., characterized the results and thus enable a fundamental verification of the theoretical assumptions.

Future work will focus on extending the feature based representations towards object based prediction of affordance-based interaction, routing in the work on the visual descriptor information presented here, and demonstrating the generality of the concept. Furthermore, we think that the presented machine learning component implemented by a decision tree can be enhanced by using reinforcement learning methodology to learn relevant events in state space for cueing to the opportunities for interaction.

Acknowledgments

This work is funded by the European Commission's projects MACS (FP6-004381) and MOBVIS (FP6-511051) and by the FWF Austrian joint research project Cognitive Vision under sub-projects S9103-N04 and S9104-N04.

References

[1] J.J. Gibson, *The Ecological Approach to Visual Perception*, Boston, Houghton Mifflin, 1979.

[2] U. Neisser, *Cognition and Reality. Principles and Implications of Cognitive Psychology*, San Francisco, Freeman & Co., 1976.

[3] E.J. Gibson, Exploratory behavior in the development of perceiving, acting and the acquiring of knowledge. *Annual Review of Psychology*. 39, 1-41. 1988.

[4] Faillenot, I., Toni, I., Decety, J., Grégoire, M.-C., & Jeannerod, M., Visual pathways for object-oriented action and object recognition: functional anatomy with PET. *Cerebral Cortex*, 7, 77-85. 1997.

[5] Fagg, A. H. and Arbib, M. A., Modeling parietal-premotor interaction inprimate control of grasping. *Neural Networks*, 11(7-8):1277-1303. 1998

[6] Wheeler S.D. and Fagg H.A. and Grupen R.A., Learning Prospective Pick and Place Behavior, *Proc. 2nd International Conference on Development and Learning*, Pages 197-202, IEEE Computer Society, Cambridge, MA, June, 2002.

[7] Fitzpatrick, Paul, Giorgio Metta, Lorenzo Natale, Sajit Rao and Giulio Sandini. "Learning About Objects Through Action - Initial Steps Towards Artificial Cognition*", In Proc. IEEE International Conference on Robotics and Automation (ICRA)*, Taipei, Taiwan, May 12 - 17, 2003

[8] Stoytchev, A., "Behavior-Grounded Representation of Tool Affordances", *In Proc. IEEE International Conference on Robotics and Automation (ICRA)*, Barcelona, Spain, April 18-22, 2005

[9] Stark L. and Bowyer, K. W., ``Function-based recognition for multiple object categories", *Image Understanding*, 59(10), 1--21.

[10] Rivlin, E., Dickinson, S.J., and Rosenfeld, A., "Recognition by functional parts," *Computer Vision and Image Understanding*, 62, pp. 64–176, 1995.

[11] Bogoni L. and Bajcsy R., "Interactive Recognition and Representation of Functionality", Computer Vision and Image Understanding: CVIU, 62(2), 194-214, 1995.

[12] M.G. Edwards, G.W., Humphreys, and U. Castiello, Motor facilitation following action observation: a behavioural study in prehensile action In *Brain Cognition*, volume 53, pp. 495-502, 2003.

[13] D. Lowe, Distinctive image features from scale-invariant keypoints. *International Journal of Computer Vision*, vol. 60(2), pp. 91-110, 2004.

[14] J.R. Quinlan, *C4.5 Programs for Machine Learning*. Morgan Kaufmann, San Mateo, CA, 1993.

[15] I. Cos-Aguilera, L. Cañamero, G. M. Hayes, and A. Gillies, Ecological integration of affordances and drives for behaviour selection. In Bryson J. et al. (eds.), *Proc. Workshop on Modeling Natural Action Selection*, pp. 225-228, AISB Press, 2005.

[16] I. Cos-Aguilera, L. Cañamero and G. M. Hayes, Using a SOFM to learn Object Affordances", *Proc. Workshop of Physical Agents*, WAF'04, Girona, Catalonia, Spain. March, 2004.

Modelling the Peripheral Auditory System of Lizards

Lei Zhang[1], John Hallam[1], and Jakob Christensen-Dalsgaard[2]

[1] Mærsk Institute, University of Southern Denmark
Campusvej 55, DK–5230 Odense M, Denmark
{lzhang, john}@mip.sdu.dk
[2] Institute for Biology, University of Southern Denmark
Campusvej 55, DK–5230 Odense M, Denmark
jcd@biology.sdu.dk

Abstract. Lizards, such as *Mabuya macularia* or *Gecko gecko*, have a relatively simple peripheral auditory system structured as a pressure difference receiver with a strong broadband directional sensitivity. In this paper we take a lumped-parameter model of the lizard auditory system, convert the model into a set of digital filters implemented on a TDT StingRay digital signal processing module carried by a small mobile robot, and evaluate the performance of the robotic model in a phono-taxis task. The complete system shows a strong directional sensitivity for sound frequencies between 1350–1850 Hz and is successful at phonotaxis within this range. The performance of and assumptions underlying the model are also discussed.

1 Introduction and Related Work

Lizards, such as *Mabuya macularia* or *Gecko gecko*, have a relatively simple periph-eral auditory system [1,2] — see the schematic diagram in figure 1. A tympanum on each side of the head connects via wide internal tubes to the central cavity, which also vents to the nasal passages. Sound impinging on the left ear, for instance, is thereby able to travel internally to the right side and affect the vibration of the right tympanum as well as cause vibration of the left tympanum. Because the internal and external sound waves arrive on opposite sides of the tympani, their contribu-tions subtract and the resulting motion of each tympanum is generated by the dif-ference of the instantaneous sound pressures. Recent experiments have shown that the acoustical interaction converts the ear to a pressure difference receiver with the highest directionality reported for any vertebrate [3].

Pressure difference receiver ears have been quite widely studied both theoret-ically and experimentally. They occur not only in lizards [3], but also in crick-ets [4,5], frogs and birds [1,6]. The principal effect of the system is to convert small, hard-to-sense differences of time-of-arrival of a sound — which encode the direction from which the sound appears to originate — into substantial, easier-to-sense, differences in the perceived amplitude of the sound at the two ears. In simplistic terms, the sound appears louder on the side facing the source, and

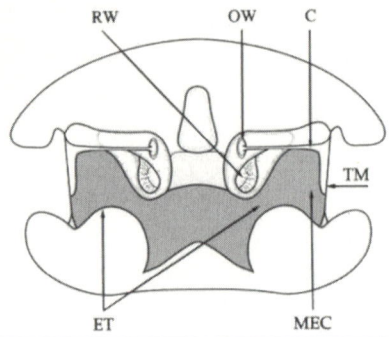

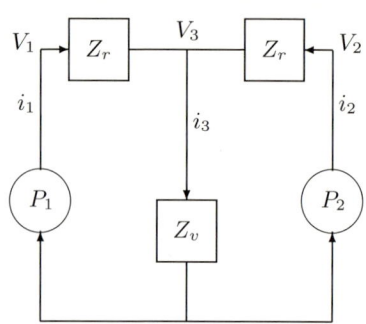

Fig. 1. Schematic Diagram of Lizard Ear Structure from [2], redrawn and altered: TM is the tympanal membrane, ET the Eustachian tubes, MEC the middle ear cavity, C the cochlea, RW the round window and OW the oval window

Fig. 2. Lumped-Parameter Circuit Model of Lizard Ears: sound pressure $P_{\{1,2\}}$ is represented by voltage inputs V_n while tympanal motion maps to current $i_{\{1,2\}}$. Further details in section 2.

quieter on the far side. (This is not actually as obvious as it appears. Since the auditory systems in question are smaller in size than the sound wavelength, the physical amplitude of the sound at the two ears is therefore essentially the same: the sound easily diffracts around the animal's head and body.)

The cricket system has been extensively studied and also modelled using robotic modelling techniques. Webb and her collaborators have investigated the basic mechanisms underlying cricket phonotaxis using a model of the cricket peripheral auditory system constructed with electronic hardware [7]. In this device, the acoustic ports on the cricket body (of which there are four significant ones) are modelled using small capacitative microphones carefully placed so that their physical separation matches that of the cricket's ports. Programmable broadband amplifiers and delays model the internal connecting tubes which join the acoustic ports. Transduction at the tympani is represented using a broadband conversion to amplitude of a weighted sum of contributions from the delayed acoustic signals. The resulting amplitude envelope can be digitised, if desired, and passed to a model of the cricket's internal neural structures and processes (see, for instance, [8,9]). The device can operate in two- or four-port mode and has generally been used for modelling the narrowband phenomenon of cricket phonotaxis, where female crickets travel toward a male using the male's calling song as a guide.

The lizard auditory system has a number of differences from that of the cricket, which make it an interesting system to model in its own right. First, the lizard has only two significant acoustic apertures (the two tympani) [3] whereas the cricket has four. Second, the lizard receiver demonstrates a strong directionality over a relatively wide range of frequencies whereas cricket directional hearing is restricted to a narrow frequency band, as shown by biophysical and behavioural experiments [4,5].

The neural processing of directional information in lizards is not well-studied, but binaural interaction (contralateral inhibition and excitation) has recently been found at the level of the first auditory nuclei in the brain stem. Also, sound localization behavior (interception of calling crickets [10]) has only been reported in one lizard species. However, sound localization is fundamentally useful for any auditory system.

The rest of the paper describes work modelling the lizard peripheral auditory system. In section 2, a simple theoretical model [11] of the lizard auditory system is presented and analysed. This results in a description of the auditory system as a set of coupled filters, which can be implemented using a digital signal processing unit carried by a small mobile robot: the equipment and setup is described in section 3.

Using the robotic model, a number of experiments have been carried out to determine the performance of the auditory model. These are described in section 4 and their implications discussed in section 5.

2   Theoretical Model of the Lizard Peripheral Auditory System

In this paper, we take a lumped-parameter approach to modelling the lizard auditory apparatus, quite closely following [3,11]: we model the components of the auditory system — tympani, tubes and cavities — in terms of their total effect rather than their precise physical structure. (An alternative approach in which the detailed spatial properties of a bat's external ear are modelled by numerical simulation can be seen, for example, in [12].)

Given this lumped-parameter strategy, the standard model of the pressure-difference receiver auditory system reduces to the equivalent electrical circuit shown in figure 2. The sound pressure sources P_1 and P_2 at the left and right ears are modelled by voltage signals V_1 and V_2. Tympanal motion is modelled by the currents i_1 and i_2. In general, the relationship between an electrical voltage and a current is determined by an *impedance*: the model comprises three impedances — Z_r represents the total effect of tympanal mass and stiffness and the tubes connecting the spaces behind the tympani to the central cavity, which is represented by the impedance Z_v. (Note that all these parameters, at any given frequency, are complex numbers so that the amplitude and phase of the signal can be represented simultaneously by the single value.)

The Z_r impedance appears twice, since we assume for modelling that the auditory system is symmetrical (we discuss this assumption later). The sound pressure at the central point is represented by the voltage signal V_3 which drives an associated current i_3 through the Z_v impedance (modelling the movement of air as the pressure in the central cavity changes). The values of the complex impedances Z_r and Z_v are in general dependent on the frequency of the sound presented to the model.

Using Kirchoff's rules and Ohm's law, the behaviour of the circuit model can be described by four equations.

$$V_3 = i_3 Z_v \qquad i_3 = i_1 + i_2$$
$$V_1 - V_3 = i_1 Z_r \qquad V_2 - V_3 = i_2 Z_r \tag{1}$$

With a little gentle algebra, we eliminate V_3 and i_3 and express the two currents (the model outputs) in terms of the voltages (the inputs), discovering that the auditory system can be represented using two equations thus,

$$i_1 = G_I \cdot V_1 + G_C \cdot V_2$$
$$i_2 = G_C \cdot V_1 + G_I \cdot V_2 \tag{2}$$

where

$$G_I = \frac{Z_r + Z_v}{Z_r (Z_r + 2Z_v)} \quad \text{and} \quad G_C = -\frac{Z_v}{Z_r (Z_r + 2Z_v)} \tag{3}$$

are a pair of frequency-dependent gain factors (recall that the impedances Z_r and Z_v depend on frequency; this dependence is suppressed in the above equations, for clarity). In other words, the auditory model comprises two *filters*: G_I models the effect of ipsilateral sound pressure on motion of the tympanum; and G_C represents the effect of contralateral sound pressure on tympanal motion.

How can we now determine the direction to a sound source? Consider figure 3, in which a sound from a source at A propagates to ears at B and C. The distance travelled by the sound to the two ears differs depending on the angle θ between straight ahead and the source direction, the extra distance being illustrated by the line BD in the figure. Recall that the physical amplitude of the sound at the two ears is essentially the same because of diffraction effects: thus the sound pressure signals V_1 and V_2 for the two ears have the same amplitude, but differ in phase.

As mentioned above, the key idea is that the perceived loudness of the sound (i.e. the magnitude of the tympanal motion) on the side closest to the source should be greater than on the opposite side. Consider therefore the magnitude ratio of the two currents i_1 and i_2 which represent the tympanal motion at the two ears.

$$\left|\frac{i_1}{i_2}\right| = \left|\frac{G_I \cdot V_1 + G_C \cdot V_2}{G_C \cdot V_1 + G_I \cdot V_2}\right| = \left|\frac{G_I + G_C \cdot \frac{V_2}{V_1}}{G_C + G_I \cdot \frac{V_2}{V_1}}\right| \tag{4}$$

Since V_1 and V_2 have the same amplitude, their ratio depends only on the relative phase of the two signals. Thus the current magnitude ratio depends on the two frequency-dependent filter gains and the relative phase of V_1 and V_2, this last encoding the arrival direction of the sound.

Given appropriate measurements from an animal [3], we can determine G_I and G_C in equation 3 as a function of frequency, and compute the behaviour of the model system. Figure 4 shows the magnitude ratio of the currents at the two ears for frequencies (y-axis) between 0 and 5 kHz when the sound source is at bearings (x-axis) between $[-\pi, \pi]$ radians with respect to the forward direction.

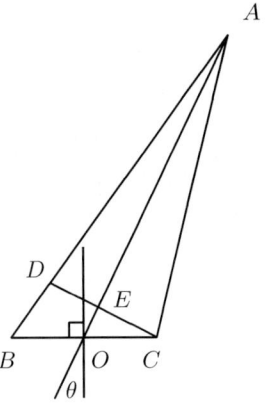

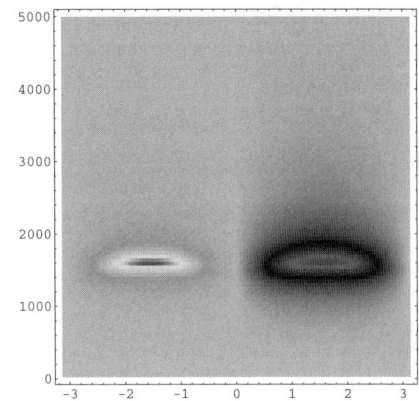

Fig. 3. Sound Arriving at the Ears from a Distant Source (see text for details)

Fig. 4. Model Response Versus Source Direction and Frequency (see text for details)

Brighter colours (greys) indicate a larger ratio. Clearly, the system is strongly directional around 1600 Hz; but it is also quite directional in the band from approximately 1200–2100 Hz.

To summarise, we have now seen how a lumped-parameter model of the peripheral auditory system, in the form of an electrical circuit analogue, leads to a representation of the system as a pair of frequency-dependent gains, or filters, that relate ipsi- and contra-lateral sound pressure to tympanal motion. Using numerical measurements from the lizard, the filters can be defined and their behaviour computed. The system demonstrates a significant broadband directional response. The next question is whether a similar broadband directionality can be demonstrated in a robotic model of the system.

3   A Robotic Model of the Lizard Auditory System

In this paper, a robotic model system is described. The system comprises two microphones which were used to simulate the lizard's ears. The signals from the microphones were preamplified and input to a small, portable, digital signal processor (StingRay, Tucker-Davis Technologies, Florida, USA). The StingRay implemented the ear model and generated 2-bit control signals which selected between three behaviours (forward, left-turn and right-turn) programmed in an RCX processor brick (LEGO).

3.1   Robotic Model System

Figure 5 is the block diagram of the robotic model system. The microphones transduce sound and send analog signals to the preamplifiers. The preamplifiers match the signal to the input range of the StingRay, which processes the amplifier

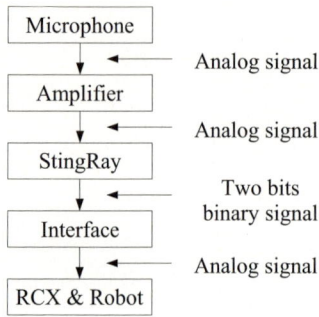

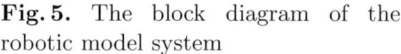

Fig. 5. The block diagram of the robotic model system

Fig. 6. A photograph of the robotic model system

output signals according to the ear model and outputs a two-bit binary signal. This signal was passed through an interface-decode circuit connected to one of the analog sensor ports on an RCX processor brick that controlled the motors of the robot. The ear model could then select between up to 4 pre-programmed motor behaviours ('go-forward', 'turn-left' and 'turn-right' are used here). Figure 6 is a photograph of the robotic model system.

3.2   StingRay DSP System

In the system, the StingRay was used to implement the ear model and generated the robot control signal. The StingRay DSP was programmed using a graphical programming system (RPvds) developed by Tucker-Davis Technologies.

Figure 7 shows the program implementing the ear model on the StingRay. In the figure, there are two A/D converters in column 1. They are used to capture the analog signals output by the microphone preamplifiers and change them into digital ones. In column 2 there is a FIR filter which is used to compensate the two microphones (see discussion). A short delay is used to compensate the delay caused by the FIR filter. Four IIR filters are shown in column 3. These filter two microphone signals in two channels, corresponding to the left and right sides of the auditory system. The outputs of the two 'Sum' nodes are the signal of the two channels, representing the left and right tympanal movements.

In column 4 there are two copies of 'Absval' and 'Smooth' which calculate the signal power in each channel and two copies of 'Log10' and 'ScaleAdd' which change the power into dB and calculate the power ratio. The output of the 'Sum' node is that ratio. The circuit in column 5 uses the ratio value to determine the selected behaviour, which is encoded into a two-bit binary signal. This means up to four distinct control signals could be sent to the robot. In the experiment, three of them were used to control the robot to turn right, turn left and go forward.

The circuit at the top of the figure, part 6, is used to cut noise. When the signal from one microphone is strong enough, the circuit in column 5 is enabled. If not, the output of that circuit is the signal that instructs the robot to go forward.

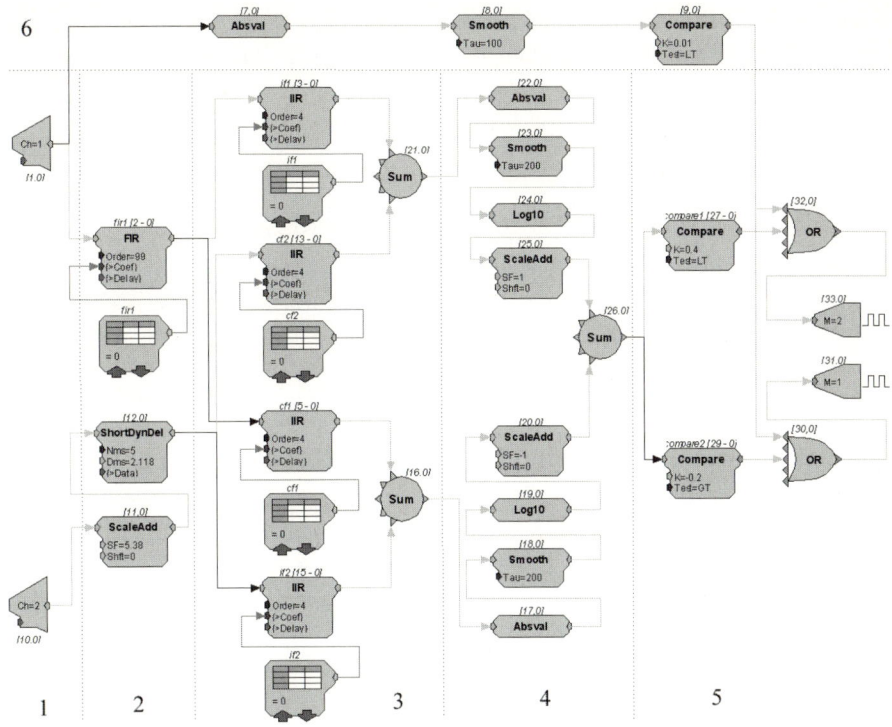

Fig. 7. The circuit loaded into StingRay

4   Experiments and Results

The robotic model has been evaluated in a number of experiments to assess its performance. We were interested in 3 factors: first, does the model work at all, that is, can the robot approach a loudspeaker emitting a suitable sound; second, over what range of frequencies does it work; and third, does the performance depend on the frequency within the operating range.

4.1   General Methods

The same basic design was used to investigate the three questions just listed. A loudspeaker was set on the floor of a room whose walls were lined with 2 cm deep anechoic tiles. The layout was as follows: the robot was 2 metres in front of the loudspeaker and started 1 metre to the left or right of the centre line, travelling directly toward the centre line — thus it must react if it is to reach the loudspeaker. The speaker emitted a continuous sound, whose amplitude was set to be 85.5 dB SPL at 2.2 metres directly in front of the speaker using a calibrated microphone. The robot moved at a speed of 11.8 cm/s.

Ten trials were performed from each starting position. The outcome of each trial was categorised as a *hit* if the robot hit the loudspeaker, a *near* if the robot passed within 20 cm of the speaker, and a *miss* if neither these conditions held. The robot was stopped when it hit or passed the speaker or reached the boundary of the experimental arena. For recording the actual tracks taken, the robot drew with a pen on a sheet of white-coated hardboard, and the track data was measured manually from the drawn trace.

4.2   Does the Robotic Model Approach Sounds?

Table 1 gives the experimental results for 10 trials over the frequency range 1000–2200 Hz. From the table, it is clear that the robot works very consistently over the range 1350–1750 Hz. All 10 trials from both sides hit the speaker. This means in this broad range of frequencies the robot could find the speaker consistently and well.

Table 1. Experiment results for 10 approach trials from left and right at frequencies from 1000–2200 Hz (see text for details)

frequency(Hz)	right			left			frequency(Hz)	right			left		
	hit	near	miss	hit	near	miss		hit	near	miss	hit	near	miss
1000	0	0	10	6	2	2	1650	10	0	0	10	0	0
1050	0	0	10	3	3	4	1700	10	0	0	10	0	0
1100	0	0	10	4	6	0	1750	10	0	0	10	0	0
1150	0	0	10	4	3	3	1800	7	3	0	6	2	2
1200	10	0	0	4	1	5	1850	9	1	0	5	3	2
1250	9	1	0	3	3	4	1900	3	4	3	1	7	2
1300	6	4	0	0	3	7	1950	9	1	0	0	0	10
1350	10	0	0	10	0	0	2000	9	1	0	0	0	10
1400	10	0	0	10	0	0	2050	6	4	0	0	0	10
1450	10	0	0	10	10	0	2100	8	1	1	0	0	10
1500	10	0	0	10	0	0	2150	2	5	3	0	0	10
1550	10	0	0	10	0	0	2200	0	2	8	0	0	10
1600	10	0	0	10	0	0							

4.3   What Is the Working Range of Frequencies?

Having demonstrated that the robot is able to approach the loudspeaker successfully for a broad range of frequencies, but appears to work better at some frequencies than others, it is natural to ask what is the useful range of frequencies for the model. This depends on the experimenter's definition of 'success'; for this paper we define successful operation to mean that the robot *hits* the loudspeaker in at least 5 of the 10 trials. Thus the *near* and *miss* tracks count as failures.

Determining Operating Limits. To estimate the operating limits of the model, the robot was run for 10 trials from the left and from the right with frequencies in the ranges 1000–1400 Hz and 1700–2200 Hz, spaced at 50 Hz, for left and right sides. The results in table 1 indicate that the model was successful in the range 1350–1850 Hz. The robot was more sensitive when stimulated from the left side at low frequencies but more sensitive to stimulation from the right side at high frequencies. As seen in the table, the robot had very stable performance in the 1350–1750 Hz range, and still worked at 1800 and 1850 Hz with degraded performance (*near* and *miss* trials).

In the experiment, the model was very sensitive to the power of the sound in the ranges 1200–1300 Hz and 1800–2000 Hz. If the power were changed a little, the experiment result was very different especially at high frequencies.

Track Directness Comparisons. An alternative method of comparing the performance of the model at different frequencies is to use a 'directness' statistic [9]. The track followed by the robot is measured from the hardboard sheet and segmented at turns into a set of n vectors of lengths l_i. For each segment its heading θ_i relative to the loudspeaker is determined. An average vector is then calculated by averaging the segments (each in its individual loudspeaker-relative frame of reference)

$$\underline{v}_{\text{avg}} = \frac{1}{\sum_1^n l_i} \left(\sum_1^n l_i cos\theta_i , -\sum_1^n l_i sin\theta_i \right) \tag{5}$$

and this average is plotted on a polar plot. If the robot moves directly from starting point to loudspeaker, the average vector will have length 1 and direction 0. (Note that this differs from the calculation described in [9] only in omitting the rescaling by the ratio of minimal to actual path time.)

The resulting polar plots for the model for left and right sides at frequencies of 1650 Hz (theoretical best frequency), 1400 Hz and 1900 Hz, are given in figure 8.

Table 2. Mann-Whitney U value for track directness comparisons (see text)

U(p)	1650Hz-R	1400Hz-R	1900Hz-R	1400Hz-L	1650Hz-L
1650Hz-R					
1400Hz-R	14 (0.0026)				
1900Hz-R	13 (0.0019)	34 (0.1237)			
1400Hz-L	0 (0.0000)	2 (0.0000)	15 (0.0034)		
1650Hz-L	0 (0.0000)	0 (0.0000)	4 (0.0001)	10 (0.0008)	
1900Hz-L	0 (0.0000)	0 (0.0000)	0 (0.0000)	0 (0.0000)	0 (0.0000)

To test how well the model works at different frequencies, the 'Directness' is examined statistically. The distances from the points shown in figure 8 with '+' to the ideal one (0,1) are used as the populations and the Mann-Whitney U-test is used to test these data. $u_1 = u_2 = 10$. The statistic result is shown in table 2 ordered from left to right. For example, when the frequency of the sound was 1650 Hz and the robot started from right front of the loudspeaker (denoted '1650 Hz-R'), the statistic result is the best.

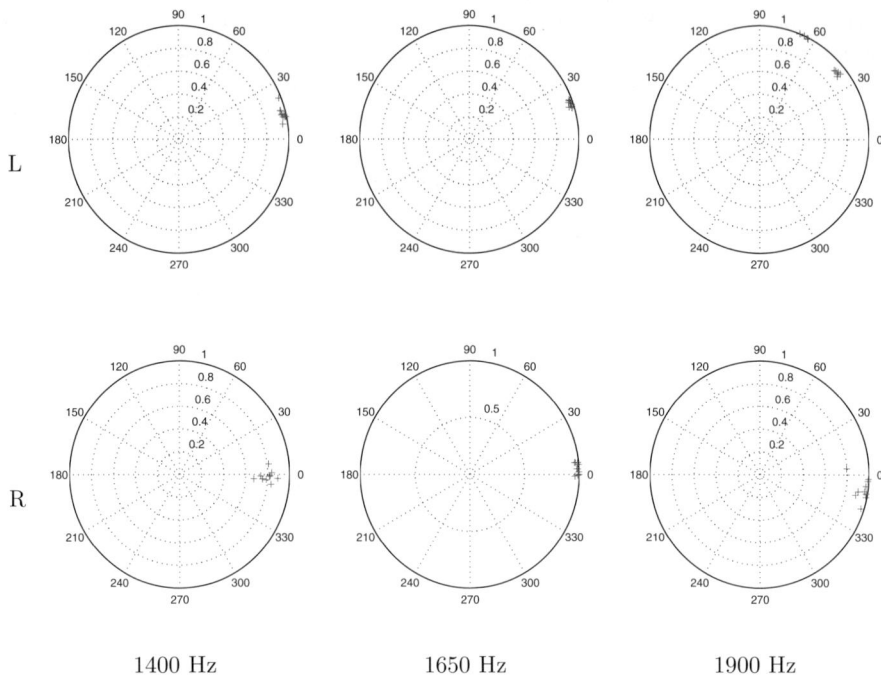

L

R

1400 Hz   1650 Hz   1900 Hz

Fig. 8. 'Directness' (see text) plots for the robot tracks

For fixed frequency, the right side is always better than the left side. At 1650 Hz, compare the right and the left sides — $(U = 0, p \approx 0)$. This means these two groups of data are apparently different and the right side is better than the left. The same result may be found at 1400 Hz, $(U = 2, p \approx 0)$ and 1900 Hz $(U = 0, p \approx 0)$. So the model is more directional from the right side than the left side.

For the right side, the model works best at 1650 Hz. Compared to 1400 Hz $(U = 14, p < 0.0026)$ and 1900 Hz $(U = 13, p < 0.0019)$, the model is more directional at 1650 Hz. But comparing 1400 Hz and 1900 Hz $(U = 34, p < 0.1237)$ shows no significant difference, so the directness is similar in these two cases.

For the left side, the model works best at 1400 Hz, better than both 1650 Hz $(U = 10, p < 0.0008)$ and 1900 Hz $(U = 0, p \approx 0)$. Tracks for 1650 Hz are more directional than those for 1900 Hz $(U = 0, p \approx 0)$.

5   Discussion and Conclusions

The results presented demonstrate that the robotic model of the lizard auditory system exhibits the behaviour predicted from the theoretical analysis on which it is based. The system as described exhibits successful and reliable phonotaxis behaviour over a frequency range of approximately 1350–1850 Hz.

However, this successful behaviour is contingent on the robotic model meeting the expectation of symmetry built into the theory. In the first tests of the model, a strong bias to one side was observed which was traced to a difference in the frequency-response characteristics of the two miniature microphones used. It was necessary to correct for this difference using a digital filter computed from the ratio of the microphone spectra, since the difference in average group delay of the uncorrected microphones was about $0.2rad$ which corresponded to a source azimuth error of 56° at 1000 Hz and 15° at 2000 Hz.

This strong dependence of the model on the symmetry of the physical system is not surprising, given the modelling assumptions. However, symmetry will not hold in general for the robotic model nor will it for the lizard. How then does the lizard compensate for this? We offer two suggestions, to be studied in further work: perhaps differences between the two ears are minimised during the development of the lizard ear, or, more likely, the lizard learns to compensate for the asymmetry in the subsequent neural processing of the tympanal motion signals.

Implementation of the lumped-parameter model as a set of digital filters is an interesting alternative to the representation technique of broadband delay lines used for the cricket auditory system [7]. The filters implement appropriate delays through their group delay properties, but unlike the broadband arrangement, they allow the delays to vary depending on frequency. The delay-based model accounts for the length of the interconnecting tubes but not the frequency-dependent properties of the tympani and cavity. Furthermore, implementing the model as software means that these frequency-dependent effects are programmable in the model: more possibilities for varying the signal-processing properties are available for manipulation. For a narrowband signal, the two implementations are exactly equivalent; for a broadband signal, the software implementation is more flexible.

The model developed makes it possible to investigate a number of questions. Apart from the key question of symmetry, one could study the relationship between model parameters and observed behaviour. How much does error in parameter values affect the performance? Are the parameters measured in the animal 'optimal' in any sense, or are better parameter choices available? (This gives an indication of how much the evolution of the physical structure may be constrained by other factors than directional hearing performance.) What kinds of neural processing models are possible based on the pre-processing done by the peripheral auditory model? For instance, multiple frequency-band neural models, or models that attempt to learn the mapping between tympanal response and sound source direction in the absence of perfect system symmetry are both interesting. These, and other questions, will be topics for future work with the robotic model.

Acknowledgements

We thank Widex A/S, Denmark for providing the Knowles microphones. The study was supported by the Danish National Research Foundation and the Danish National Science Foundation.

References

1. J. Christensen-Dalsgaard, 2005. Directional hearing in non-mammalian tetrapods. In: A. N. Popper & R. R. Fay (eds.) *Sound Source Localization*. Springer Handbook in Auditory Research, New York: Springer-Verlag, pp. 67–123.
2. E. G. Wever, 1978. *The Reptile Ear: Its Structure and Function*. Princeton University Press.
3. J. Christensen-Dalsgaard & G. A. Manley, 2005. Directionality of the Lizard Ear. Journal of Experimental Biology **208** pp. 1209–1277.
4. A. Michelsen & A. V. Popov & B. Lewis, 1994. Physics of Directional Hearing in the Cricket *Gryllus bimaculatus*. Journal of Comparative Physiology A **175** pp. 153–164.
5. A. Michelsen, 1998. Biophysics of sound localization in insects. In: R. R. Hoy & A. N. Popper & R. R. Fay (eds.) *Comparative Hearing: Insects*. New York: Springer-Verlag, pp. 18–62.
6. G. M. Klump, 2000. Sound localization in birds. In: R. J. Dooling & R. R. Fay & A. N. Popper (eds.) *Comparative Hearing: Birds and Reptiles*. New York: Springer-Verlag, pp. 249–307.
7. H. H. Lund & B. H. Webb & J. Hallam, 1997. A Robot Attracted to the Cricket Species *Gryllus bimaculatus*. Proceedings of the Fourth European Conference on Artificial Life, pp. 246–255.
8. B. Webb & T. Scutt, 2000. A Simple Latency Dependent Spiking Neuron Model of Cricket Phonotaxis. Biological Cybernetics, **82**(3) pp. 247–269.
9. R. Reeve & B. Webb & A. Horchler & G. Indiveri & R. Quinn, 2005. New Technologies for Testing a Model of Cricket Phonotaxis on an Outdoor Robot. Robotics and Autonomous Systems **51** pp.41–54.
10. S. K. Sakaluk and J. J. Belwood, 1984. Gecko phonotaxis to cricket calling song: A case of sattelite predation. Animal Behaviour **32** pp. 659–662.
11. N. H. Fletcher, 1992. *Acoustic Systems in Biology*. Oxford University Press.
12. R. Müller & J. C. T. Hallam, 2004. From bat pinnae to sonar antennae: Augmented obliquely truncated horns as a novel parametric shape model. In S. Schaal, A.-J. Ijspeert, A. Billard, S. Vijayakumar, J. Hallam & J-A. Meyer (eds.) *From Animals to Animats 8*, pp. 87–95. MIT Press.

A Model of Sensorimotor Coordination in the Rat Whisker System

Ben Mitchinson[1], Martin Pearson[2], Chris Melhuish[2], and Tony J. Prescott[1]

[1] The University of Sheffield
t.j.prescott@shef.ac.uk
[2] The University of the West of England

Abstract. The rat has a sophisticated tactile sensory system centred around the facial whiskers. During normal behaviour, rats sweep their longer whiskers (macrovibrissae) through the environment to obtain large-scale information, whilst gathering small-scale information with the sensory apparatus around their snout. The macrovibrissae are actively and differentially controlled. Using high-speed video recording, we have observed that temporal and spatial parameters of whisking pattern generation are modulated to match environmental features such as the position and orientation of nearby surfaces. Whisking is also closely co-ordinated with head and body movements, allowing the animal to locate and orient to interesting stimuli detected through whisker contact. In this paper, we present a hybrid (spiking-neuron/arithmetic) model of the neural systems underlying these observed adaptive sensorimotor behaviours, and demonstrate its performance in a simulated robot with rat-like morphology. We also report progress towards embedding these control systems in a physical robot with biomimetic whiskers.

1 Introduction

The rat possesses an impressively acute tactile sensory system, the sensors of which include large mobile whiskers on either side of the snout [1]. Tactile information is gathered by sweeping these whiskers forwards and backwards at 5-25 Hz, and there is now strong evidence that this behavior is generated by output from a 'whisking pattern generator' (WPG) located in the rat hindbrain [2]. Interestingly, the parameters of whisking appear to be controlled independently on each side of the snout in response to changes in the environment and/or the motivation of the animal [3,4,5,6], presumably to optimise perception. Furthermore, studies have long shown that rodents orient their snout towards novel stimuli, apparently to bring more rostral sensory apparatus (small immobile whiskers, teeth, tongue, lips and nose) to bear on items of possible interest. Computational modelling of these aspects of sensorimotor co-ordination in the rat's tactile perception system is the focus of this study.

Using high-speed video recording [7] we have observed patterns of asymmetrical contact with walls and objects suggesting that rats regulate their whisker movements so as to control the nature of these contacts. Specifically, the whiskers

S. Nolfi et al. (Eds.): SAB 2006, LNAI 4095, pp. 77–88, 2006.

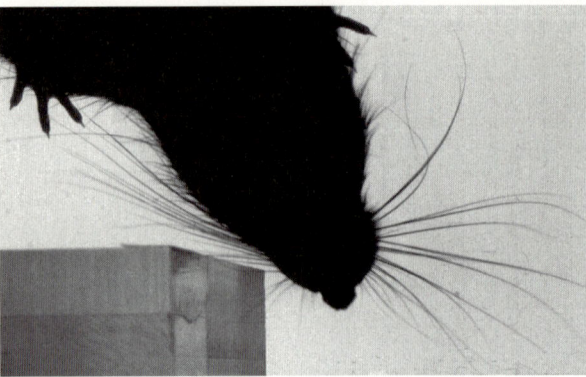

Fig. 1. Still from high-speed video of genetically-blind rat encountering obstacle uni-laterally. Ipsi-/contra-lateral whiskers are retracted/protracted in response.

appear to be actively moved forwards to meet more rostral objects ('maximal contact'), and, at the same time, actively restrained from pushing unduly against more caudal objects ('minimal impingement'). This control strategy is intuitively satisfying as it would tend to maximize the number of contact/detach events between the whiskers and the environment that have been found to lead to robust sensory responses (known as 'ON' and 'OFF' responses) in the primary afferent nerves [8]. Whilst maximizing the rate of information collection, this scheme would also minimize the distortion that could arise through overdriving the sensory apparatus (since the whisker deflections generated by such events will tend to be small). This interpretation is consistent with previous observations of rat whisking behaviour (though see [4,5] for further discussion).

The minimal impingement element of this hypothesized control strategy can be implemented through negative feedback that inhibits protraction (forward motion) of the whiskers when contact occurs. Direct projections from trigeminal sensory nuclei to the facial motor nucleus, both located in the hindbrain, have been identified [9,10] that could provide a substrate for negative feedback in the form of a simple, closed sensorimotor loop (see [11] for further functional anatomical information). Maximal contact, on the other hand, requires knowledge of something located outside the range of the normal whisker sweep. In the genetically blind animals that we study it is thus only observed in response to contact events from earlier whisks, or to contact events on the opposite side of the snout (see [12] for evidence of increased protraction given prior knowledge of rostrally-located items). The pose depicted in Figure 1 is typical, with the whiskers ipsilateral to an obstacle swept back, and those contralateral swept forward. Memory for past sensory events most likely requires cortical pathways, therefore, whilst hindbrain, contralateral positive feedback could provide the substrate for a reactive maximal contact control pathway, some cortical modulation of the WPG is presumably required when memory is involved. In the current study we therefore investigate direct, contralateral positive feedback only.

Accurate orienting of the head/snout to a point of whisker contact requires more advanced circuitry than that proposed for hindbrain feedback control. The midbrain superior colliculus (SC) is known to be essential for the expression of orienting responses to somatosensory stimuli [13], and projections from trigeminal sensory neurons to SC have been identified repeatedly, e.g. [14]. We chose, therefore, to model a sensorimotor loop through SC to mediate orienting. SC is known to use a retino-centric coordinate system, which we approximate as head-centric – what remains, then, is to specify the nature of the transform from the whisker-centric encoding in trigeminal to a head-centric reference frame. Two strategies for instantiating this transformation have been proposed: temporal decoding using neuronal phase-locked loops, and spatial decoding, through the integration of information from contact receptors with that from whisking angle/phase receptors [15]. We chose the latter, for its simplicity.

Below, we present a simulated mechanical environment ('WhiskerWorld') which we use for testing our control models. We also outline our earlier model of whisker sensory transduction [16] which is used to generate biologically accurate spiking input signals for the new models studied here (Section 2). We then detail a hybrid (spiking/non-spiking) computational simulation model of the above aspects of active perception in rat (Section 3), and illustrate its performance in WhiskerWorld (Section 4). We conclude by outlining the proposed embodiment of these computational features in a mobile robot (Section 5).

2   Simulated Environment

Our model environment consists of a two-dimensional simulation of mechanical interactions between six inflexible whiskers and assorted circular obstacles (Figure 7). The whiskers are carried three on each side of the snout of a simulated robot platform with complete freedom of mobility in the plane. The positions of the whisker bases are computed from the location of the platform and its neck angle. For each whisker we also specify its length, the more rostral whiskers being shorter, and the positive acute angle it makes with the symmetry axis of the head, $\theta_{1...6}$. One simulated muscle drives each row of whiskers to protract (there is evidence of row-based motor circuits in the rat [17]), whilst intrinisic elastic forces drive them to retract [3]. Interactions with the immobile elastic obstacles also drive the whiskers as appropriate. Whiskers that temporarily intersect obstacles are considered 'deflected', to a degree and direction, x, concomitant with the intersection and its location along the whisker. The control loops to be discussed (a) drive the simulated protraction muscles such that maximal contact and minimal impingement are elicited, and (b) drive the wheels and neck of the mobile platform such that the robot's 'foveal zone' (indicated in Figure 7) is brought to bear on obstacles in a biologically convincing movement.

When a model whisker makes contact with an object the resulting deflection x is input to a simulation of the mechanical properties of the rat whisker follicle [16]. The output of this mechanical model is used to generate spike trains in model sensory neurons whose response properties were derived from an ex-

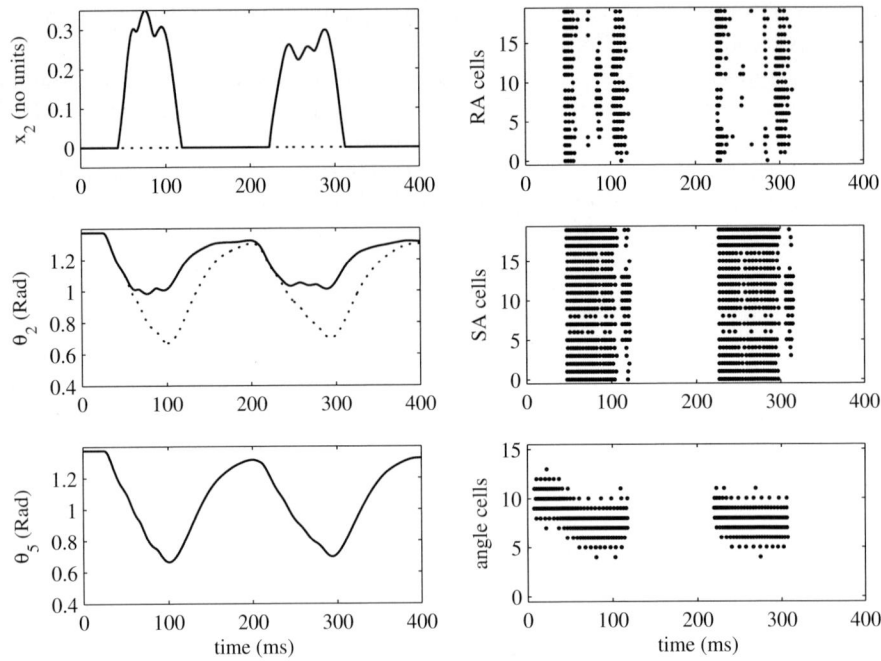

Fig. 2. Two whisks with a unilateral whisker-field obstruction recorded in Whisker-World with no sensorimotor feedback. θ_2 is the angle of the obstructed whisker, θ_5 the angle of the corresponding unobstructed whisker on the other side of the snout; with an obstacle present (solid) and without (dotted). x_2 is the deflection of the obstructed whisker. Right-hand panels show responses of SA, RA, and angle neurons to the obstructed whisks, where each dot represents a spike.

tensive review of relevant electrophysiological studies. Each simulated whisker drives 20 'slowly-adapting' (SA) and 20 'rapidly-adapting' (RA) neurons that together encode the deflection of the whisker. Additionally, 16 model neurons encode the whisker angle θ in analogy to the 'angle/phase' afferents recently discovered in the rat [15]. During a typical obstructed (and unmodulated) whisk, θ decreases (with protraction) until contact occurs, θ is then arrested whilst x displays a pulse resulting from the deflection of the whisker against the obstacle, finally the whisker detaches from the obstacle and θ increases as it falls back to its rest position. This sequence is illustrated in the left panels of Figure 2. In the right panels of that figure are shown typical responses of model sensory neurons to the signals on the left. The RA cells are all similar (save for some structural noise) except each responds most strongly to a different direction of whisker deflection. The neuron with index 0 is tuned to deflection in the positive x direction, and the remainder are uniformly spaced around the circle in x/y (although we do not simulate deflections in the y direction, that is, into the simulated plane, here). The cells that respond most strongly to a positive x deflection (around index 0) spike one to five times in response to contact (the

ON response); those that respond most strongly to negative x deflection (around index 10) respond similarly to detach (the OFF response). However, all RA cells in this example express ON and OFF responses to the substantial deflection signals. The response of the SA cells is similar, except that they show continued responses (with firing rates as high as 400Hz) throughout the deflection period. The angle cells encode θ in a distributed way: the cells have preferred values of θ linearly related to their index, and respond more strongly as θ approaches this value. The cells used here do not respond at all during retraction (as was found for a majority of angle/phase cells in [15]).

3   Simulated Control Loop Models

3.1   Whisking Pattern Generator and Pattern Modulation

The core of the whisking pattern generator (WPG) is a self-resetting integrator (Figure 3) built from two spiking neuron populations. Activity in an 'integrator' population builds up spontaneously over time, then, at some threshold, excitatory drive from these cells to a 'reset' population causes the latter to become active; the integrator neurons are quickly silenced by inhibition from the reset population; activity in the reset population then dies away, and the cycle begins again. This core generator, which runs at around 5Hz in all simulations, provides excitation to two 'output-buffer' populations of spiking cells that are subject to diffuse modulation from all SA cells and are thus the site of modulation by sensory signals. Specifically, the SAs provide ipsilateral inhibition and contralateral excitation to output-buffer neurons, implementing, respectively, the required negative and positive feedback. As implemented, this modulation is linear, so excitation tends to raise the set-point of the output activity whilst inhibition tends to reduce and delay output activity (see insets in figure). Activity in each output population is converted to a scalar rate by driving a leaky integrator (time constant 10ms) with the sum of all cell spikes. The resulting two signals are then used as muscle drive forces in the physical simulation. Note that whilst the pattern generators for the two sides are coupled in phase throughout, the whisk patterns they generate are able to differ in set-point and amplitude.

3.2   Coincidence Detector and Orienting Behaviour

The coincidence detector (CD) consists of 6 banks of 16 spiking cells each, with the mth cell in the nth bank receiving excitation from all RA cells, and from the mth angle cell, associated with the nth whisker. Since we do not deal with deflections out of the simulation plane in the current model, there is no need to distinguish which RA cells from the whisker were stimulated. This connectivity is sufficient to perform coincidence detection, but is not robust against noisy inputs, generating both false positives and false negatives. The addition of strong surround inhibitory connections within each cell bank greatly improves noise resistance. Specifically, we used inhibition with relative strength given by the inverted Gaussian $w = 1 - \exp(-(\theta_i - \theta_j)^2/\Delta\theta^2)$, with $\theta_{i,j}$ the preferred

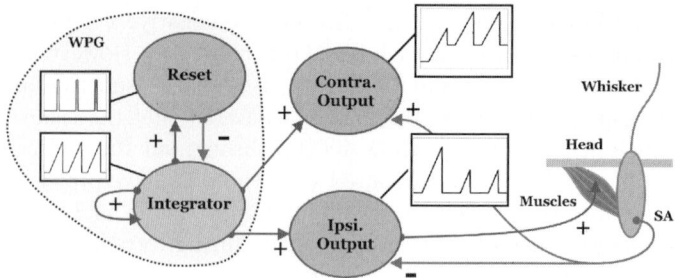

Fig. 3. (Left) Core generator consists of two reciprocally connected cell populations ('integrator' and 'reset'). (Right) Output populations driven by core generator also accept modulation from sensory afferents (SAs). Each population consists of 40 cells.

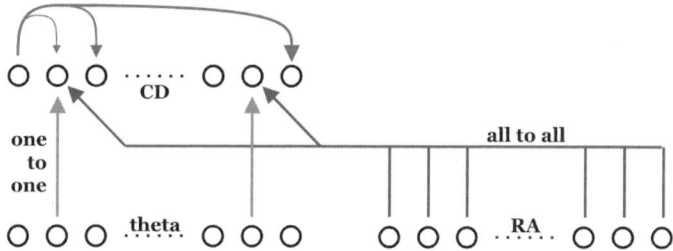

Fig. 4. Summary of connectivity between one whisker follicle and one bank of the coincidence detector. Red represents excitatory afferents, whilst blue represents recurrent surround inhibition, with strength related to separation in θ.

angles of connected cells, and $\Delta\theta$ a parameter. The connectivity is illustrated in Figure 4.

The CD operates as follows. Activity in the RA cells serving the nth whisker coincident with activity in the angle cells serving the nth whisker and tuned to around $\theta = \theta_0$, results in activity in the nth bank of the CD, in the cells corresponding to those angle cells. This activity rapidly and effectively silences sub-threshold activity in other cells in the bank, leaving a well-defined locus of activity in one or a few cells. The identity of these active cells encodes the location at which a whisker contacted the environment in a head-centric coordinate frame. The direction of whisker-sweep (forwards/backwards) is encoded by the identity of the cells within a bank, and in the transverse direction (left/right) by the identity of the bank, though the latter is encoded more indirectly since there is no guarantee that contact occurred at the whisker tip, or that multiple whiskers will not encounter the same or different obstacles. Having implemented a transform from whisker-centric to head-centric contact data, it is straightforward, then, to assign each of the 96 cells in the CD to a region in the simulation plane, relative to the head, wherein contact will typically lead to activity in that cell. These locations are illustrated in the head-centric representation of Figure 5.

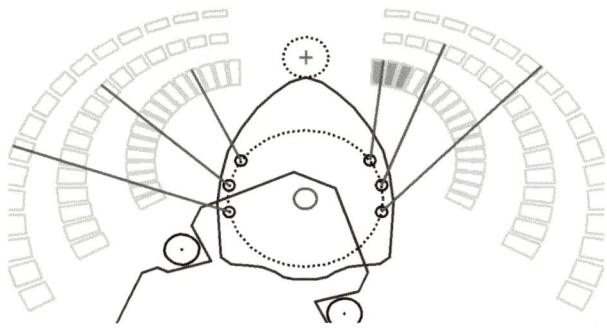

Fig. 5. Typical contact regions for each cell in the CD mapped on to a head-centric coordinate system. Active (filled) regions represent activity recorded concurrently with the last panel of Figure 7.

Orienting itself is implemented arithmetically. Around the point of maximum protraction, the reset population of the WPG pulses briefly. If any single CD cell has over-threshold activity at that time, an average is taken across the assigned locations of each over-threshold cell, and that location is deemed 'interesting'. A path-following algorithm then moves the neck and wheels of the mobile platform such that the foveal zone of the robot follows a direct path to the interesting location, using the neck as much as possible, and moving the wheels only as much as necessary. This algorithm is satisfying, firstly because the calculations are simple (neck follows nose, tail follows neck) and secondly because animals display this 'recruitment' of joints as a movement progresses [18]. Forcing the foveal point to follow a path that loops backwards somewhat (to form a slight 'U') is an alternative strategy that might reduce the incidence of collisions.

4   Performance of the Simulated Model

We repeated the simulation reported in Figure 2, incorporating feedback modulation of the WPG (see Figure 6). The effect of the modulation on the whisking range of whisker 2 (ipsilateral) can be seen as it works more towards the back of the head (θ_2 increases). Furthermore, the protraction force applied to the ipsilateral whiskers ceases soon after contact occurs, so that both contacts are briefer and cleaner than those returned without feedback. The RA cells for whisker 2 now clearly show ON and OFF responses, and nothing else – the spurious additional responses seen in Figure 2 are absent. The SA cells show a much briefer response than before, as a result of the briefer contact – their overall response profile is thus quite similar to that of the RA cells. The response of the angle cells is largely unchanged. The contralateral whiskers (represented here by whisker 5) are swept substantially further forward than in the unobstructed reference plot (θ_5 decreases), illustrating the effect of the contralateral positive feedback. The contralateral whisk amplitude is also reduced, since the mechan-

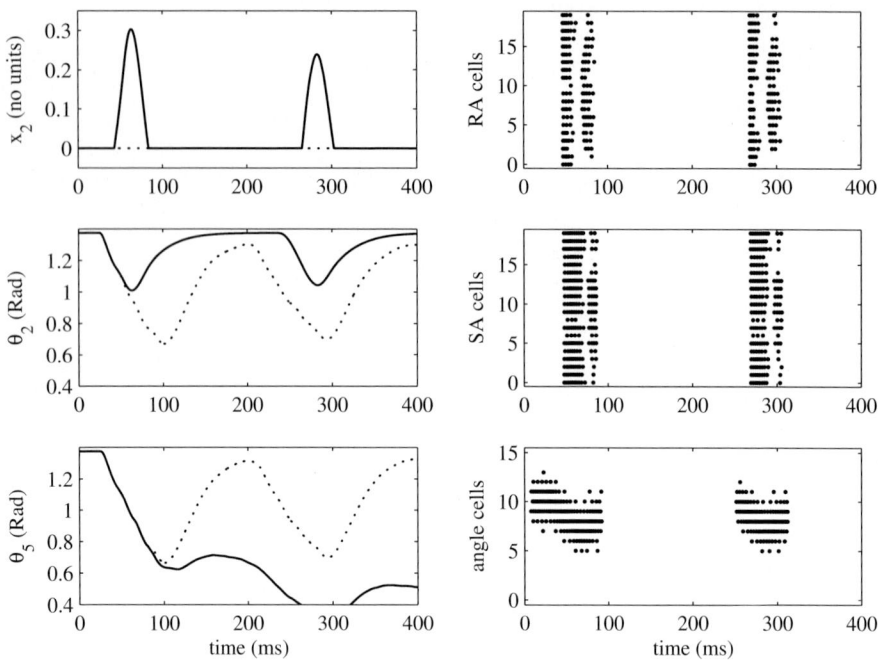

Fig. 6. Two whisks with unilateral whisker-field obstruction recorded in WhiskerWorld *with* sensorimotor feedback. Panels are exactly as for Figure 2.

ical advantage of the muscles is less at this whisking angle – see [3] for a fuller description of the whisking mechanics on which our simulation is based, along with the analogous biological result of amplitude reduction during forward-swept whisking.

Next, we repeated the same simulation, incorporating the CD and orienting response. Soon after contact with the obstacle, the simulated robot 'foveates' to the point of contact, as shown in Figure 7. At $t = 0$, the robot is in its initial state. At $t = 65ms$, the ipsilateral output population of the WPG has been silenced by negative feedback and the ipsilateral whiskers have reached peak protraction. At $t = 110ms$, orienting has just begun with a neck movement; the movement is still almost entirely of the neck until $t = 165ms$ when the body begins to be recruited. At $t = 300ms$ the orienting movement is complete, the second whisk is at around full protraction, and the forward-most contralateral whisker contacts the obstacle, thus illustrating the functional justification for their forward sweep. Note that the orient is specifically to the point of contact, rather than to the nearest face of the obstacle (say). This illustrates both the high resolution of the system despite the paucity of cells involved, and that this whisker sensory system detects surfaces rather than objects. The performance of the WPG and afferents during this orienting movement is similar to that shown in Figure 6, except that the second contact occurs on a different whisker.

5   Towards a Robot Implementation

The control models detailed above will soon be implemented on a mobile robot platform so that we can investigate their performance in a noisy real-world environment. The physical platform will have a mechanical architecture as shown in Figure 7 (the simulation is modelled on the robot design), driven by two of three omni-directional wheels and with a driven neck joint, and is intended to coarsely reflect the morphology of the rat. The six artificial whiskers are of moulded glass fibre, 100-200mm long, and weigh around half a gram. Additional whisker rows can be stacked vertically in the future to more closely emulate the array of vibrissae of the rat mystacial pad. All whiskers share the same taper profile, so the base diameter (1-2mm) is proportional to the length. At the base of each whisker, four strain gauges are mounted longitudinally at ninety-degree intervals, and wired into 2 half-bridge configurations. Each opposing pair measure the strain in opposite faces of the whisker, so that the bridge outputs are monotonic (almost linear) with the deflection of the whisker in each dimension (x, studied in simulation above, and y, up and down). This transduction configuration has proven to have very low noise (better than 60dB SNR) and is extremely sensitive to mechanical deflections in the whisker shaft at any point along its length. The whiskers are driven using a biomimetic system of 'shape metal alloy' wire (BioMetal©) to provide a protraction force analogous to that of muscles, with retraction driven by a spring analogous to the elasticity of the facial tissues. θ is measured directly using an optical quadrature shaft encoder mounted onto the whisker spindle. Full details of the whiskers and whisker drive mechanism (Figure 8) are given in [19]. The x/y deflection signals are passed to the follicle model [16]: a reduced form of the mechanical part of this model is embedded in a DSP processor, whilst nearly 400 primary afferent models are embedded us-

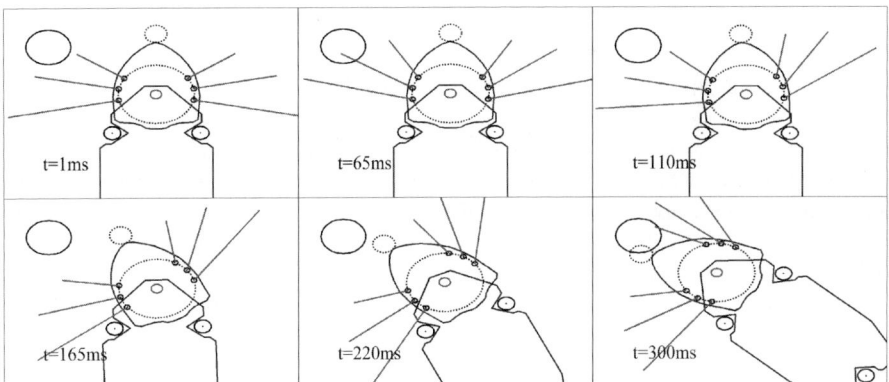

Fig. 7. Simulated evironment: non-interacting robot platform carries six whiskers which interact mechanically with environmental obstacles; dotted circle in front of snout indicates 'foveal zone'. Each panel represents the situation at the time shown during an orient-to-stimulus behaviour.

Fig. 8. Developing hardware: carrier (mechanically analogous to follicle) bears whisker with bonded strain gauges (contact receptors), biowire (marked with black square, muscles), return spring (tissue elasticity), and shaft encoder (angle receptors)

ing a custom built FPGA module, employing a pipelined parallel computation architecture to achieve real time perfomance. The output from this, a wide bus of individual bit streams carrying the spike/no-spike state of each afferent, is distributed to an array of further FPGA modules housing the models discussed above. This neural processing system will share space onboard the robot with an x86-based PC, which will house the remainder of the software system. The PC will also log software and hardware states during short experiments – in this way we hope to perform 'virtual electrophysiology' on the robot, realising previously reported biological experiments *in silico* for direct comparison.

6   Discussion

We have demonstrated a simple spiking-neuron model of whisking motor pattern generation that incorporates sensorimotor feedback. We have shown that the model adaptively modulates whisking to suit the environment, and we have illustrated how two types of feedback have the potential to improve the performance of the sensory system in gathering information: ipsilateral negative feedback by ensuring clean stereotypical contacts, contralateral positive feedback by generating contacts that would not otherwise have occurred. The pattern generator presented here is constrained to generate synchronous whisking – whiskers on each side of the snout move in phase – and our model of sensorimotor modulation does not directly affect pattern generation, only its efferent copy. Asynchronous whisking has been previously observed, though only rarely. In our own behavioural work, however, we have formed the impression that asynchrony is frequently triggered by obstacle contact, with synchrony recovering once the rat moves away from the contacted surface. In ongoing work, we are exploring the possibility that loosely-coupled bilateral whisking pattern generators, modulated by sensory input in a similar way to that described above, could reproduce this coupling between asymmetry and asynchrony in an emergent fashion. In the current paper, we have also described a model of orient-to-stimulus behaviour implemented using a mixture of spiking-neuron computation and arithmetic techniques, and have illustrated how this can be used to bring 'foveal' sensors to bear on a whisker contact point. This model required that a whisker-centric to head-centric coordinate transform be applied to the raw sensory data. We have demonstrated how a previously discussed algorithm (detection of activity coincidence between contact and angle afferents) can be implemented for this purpose using a network of spiking neurons. Taken together, these simulations show that sensorimotor coordination alone can explain several major aspects of the observed investigative whisking behaviour of rats. In the future, we will extend our model to cover signal propagation into thalamus [20] and beyond.

We have outlined our continuing progress towards the incorporation of these simulated models in the control system of a mobile robot platform. In the future, we will report on the performance of this embedded implementation. There have been several previous artificial whisker implementations [21,22,23,24] that have concentrated on what can be achieved functionally with 'bio-inspired' whiskers and whisker arrays, and have shown interesting results in relation to texture and shape discrimination and learning. In most of this existing work the whiskers have been actuated using unmodulated, symmetrical whisking patterns. In contrast, the focus of our research has been to analyse, using high-speed video recording, the 'active whisking' strategies of freely moving rats and to design an artificial system capable of replicating them. In this endeavour, our system aims to remain faithful to the biology wherever possible. In particular, our whiskers are tapered, length-scaled through the row, and measure deflections in two dimensions, and the morphology of the robot/simulation well reflects that of the animal. We have carefully examined signal transduction in the rat whisker follicle, and have designed our artificial transduction system to mimic this. Processing elsewhere in the model architecture is also largely neural, and modeled where possible on identified neural substrates. With this approach, we hope to maintain a direct correspondence between the animal and the engineered system, facilitating knowledge transfer between biology and engineering and vice versa.

References

1. Waite, P.: Trigeminal sensory system. In Paxinos, G., ed.: The Rat Nervous System. Elsevier (2004) 817–851
2. Gao, P., Bermejo, R., Zeigler, H.: Whisker deafferentation and rodent whisking patterns: Behavioral evidence for a central pattern generator. J Neurosci **21** (2001) 5374–5380
3. Berg, R., Kleinfeld, D.: Rhythmic whisking by rat: Retraction as well as protraction of the vibrissae is under active muscular control. J Neurophysiol **89** (2003) 104–117
4. Sachdev, R., Berg, R., Champney, G., Kleinfeld, D., Ebner, F.: Unilateral vibrissa contact: Changes in amplitude but not timing of rhythmic whisking. Somatosens Mot Res **20** (2003) 163–169
5. Hartmann, M., Towal, R.: Bilateral asymmetries in whisking patterns of freely behaving rats. Society for Neuroscience, Washington, Board 625.2 (2005)

6. Mitchinson, B., Prescott, T., Gurney, K., Pearson, M., Gilhespy, I., Pipe, A.: A computational model of a brainstem loop for whisker pattern generation. Society for Neuroscience, Washington, Board 625.4 (2005)
7. Prescott, T., Mitchinson, B., Redgrave, P., Melhuish, C., Dean, P.: Three-dimensional reconstruction of whisking patterns in freely moving rats. Society for Neuroscience, Washington, Board 625.3 (2005)
8. Shoykhet, M., Doherty, D., Simons, D.: Coding of deflection velocity and amplitude by whisker primary afferent neurons: Implications for higher level processing. Somatosens Mot Res **17** (2000) 171–180
9. Pinganaud, G., Bernat, I., Buisseret, P., Buisseret-Delmas, C.: Trigeminal projections to hypoglossal and facial motor nuclei in the rat. J Comp Neurol **415** (1999) 91–104
10. Nguyen, Q., Kleinfeld, D.: Positive feedback in a brainstem tactile sensorimotor loop. Neuron **45** (2005) 447–457
11. Kleinfeld, D., Berg, R., O'Connor, S.: Anatomical loops and their electrical dynamics in relation to whisking by rat. Somatosens Mot Res **16** (1999) 69–88
12. Carvell, G., Simons, D.: Biometric analyses of vibrissal tactile discrimination in the rat. J Neurosci **10** (1990) 2638–2648
13. Di Scala, G., Schmitt, P., Karli, P.: Unilateral injection of GABA agonists in the superior colliculus: Asymmetry to tactile stimulation. Pharmacol Biochem Behav **19** (1983) 281–285
14. Veinante, P., Deschênes, M.: Single- and multi-whisker channels in the ascending projections from the principal trigeminal nucleus in the rat. J Neurosci **19** (1999) 5085–5095
15. Szwed, M., Bagdasarian, K., Ahissar, E.: Encoding of vibrissal active touch. Neuron **40** (2003) 621–630
16. Mitchinson, B., Gurney, K., Redgrave, P., Melhuish, C., Pipe, A., Pearson, M., Gilhespy, I., Prescott, T.: Empirically inspired simulated electro-mechanical model of the rat mystacial follicle-sinus complex. Proc R Soc Lond B Biol Sci **271** (2004) 2509–2516
17. Klein, B., Rhoades, R.: Representation of whisker follicle intrinsic musculature in the facial motor nucleus of the rat. J Comp Neurol **232** (1985) 55–69
18. Dean, P., Redgrave, P., Sahibzada, N., Tsuji, K.: Head and body movements produced by electrical stimulation of superior colliculus in rats: Effects of interruption of crossed tectoreticulospinal pathway. Neuroscience **19** (1986) 367–380
19. Pearson, M., Gilhespy, I., Melhuish, C., Mitchinson, B., Nibouche, M., Pipe, A., Prescott, T.: A biomimetic haptic sensor. Int J Adv Robotic Sy **2** (2005) 335–343
20. Yu, C., Derdikman, D., Haidarliu, S., Ahissar, E.: Parallel thalamic pathways for whisking and touch signals in the rat. PLoS Biol **4** (2006) e124 DOI: 10.1371/journal.pbio.0040124
21. Fend, M., Bovet, S., Yokoi, H., Pfeifer, R.: An active artificial whisker array for texture discrimination. In: Proc of the IEEE/RSJ Int Conf on Intelligent Robots and Systems (IROS). (2003) Las Vegas.
22. Kim, D., Möller, R.: A biomimetic whisker for texture discrimination and distance estimation. In: Proc of the Int Conf on the Sim of Adap Behav (SAB). (2004)
23. Seth, A., McKinstry, J., Edelman, G., Krichmar, J.: Texture discrimination by an autonomous mobile brain-based device with whiskers. In: Proc of the IEEE Int Conf on Robotics and Automation (ICRA). (2004)
24. Schultz, A., Solomon, J., Peshkin, M., Hartmann, M.: Multifunctional whisker arrays for distance detection, terrain mapping, and object feature extraction. In: Proc of the IEEE Int Conf on Robotics and Automation (ICRA). (2005)

Biological Actuators Are Not Just Springs
Investigating Muscle Dynamics and Control Signals

Thomas Buehrmann and Ezequiel Di Paolo

Centre for Computational Neuroscience and Robotics (CCNR),
Department of Informatics, University of Sussex, Brighton, BN1 9QG, UK
{T.Buehrmann, Ezequiel}@sussex.ac.uk

Abstract. While there is a trend in current robotics towards more biologically inspired actuators, most work emphasizes the elastic property of muscles and tendons. Although elasticity plays a major role in many forms of movements, particularly walking and running, other features of animal muscles might also affect or even dominate movement dynamics. In this paper we use the Hill-type muscle model, common in biomechanics, to investigate the relationship between muscle dynamics and control signals in simple goal-directed movements. We find that the various non-linearities of the model lead to desirable properties with regard to controllability, such as increased stability and robustness to noise, independence of position and stiffness, or near linearity in search space. We conclude that in our attempt to create robots exhibiting the same flexibility and robustness as animals we have to seek a balance between the complexity of actuators and the extent to which their natural dynamics can be exploited in a given task.

1 Introduction

It is not unreasonable to assume that the qualitative properties found in the majority of animal muscles have been selected for by evolution for their adaptive advantage with respect to the generation of movement. The benefits of elasticity for instance have been recognized for a while and robotic systems have been built that incorporate springs in order to gain higher force fidelity, low impedance, low friction, good force control bandwidth or energy conservation [9]. More generally, the concept of 'preflexes' [5] summarizes the idea that the intrinsic dynamics of a musculoskeletal system alone can be sufficient for self-stabilization [11] and that they could be tuned such that higher level motor control becomes easier because joint level dynamics don't have to be accounted for. One of the tenets of the embodied approach to behavior, Pfeifer's notion of 'morphological computation' [7], also expresses the idea that materials can sometimes take over some of the processes normally attributed to control mechanisms. In view of this embodied perspective, in this paper we propose to investigate the different qualitative properties of muscles and their effect on motor control. We are interested in the question to what extent it would be useful to deal with the complexities of biological muscles and whether it makes the task of designing robotic control systems

S. Nolfi et al. (Eds.): SAB 2006, LNAI 4095, pp. 89–100, 2006.

easier or harder. Specifically, we use a simulation of an antagonistic muscle pair acting on a hinge-joint to ask how the non-linear visco-elastic properties of muscle affect stability, how they allow for fast but appropriately damped movements and how they relate to control signals.

2   Muscle Model

Muscles are different from any current robotic actuator in many ways. Not considering effects such as hysteresis, the instantaneous force a muscle produces is a complex non-linear function not only of its activation but also of its length and velocity. In biomechanics, the most commonly used model of skeletal muscle is the so-called Hill-type model [12]. Its basic configuration (fig.1) uses dimensionless constitutive relationships to describe the muscle's visco-elastic properties. Specifically, it consists of a contractile element (CE) producing force as a function of length and velocity; a parallel elastic element (PE) exponentially resisting stretch beyond resting length; and a series elastic element (SE), the tendon. Figure 2 depicts the non-linear nature of these elements. As is shown, active force

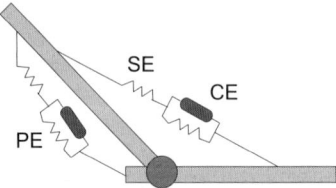

Fig. 1. Schematic of antagonistic muscle pair acting on hinge-joint

production (F_a) peaks at an optimal length of L_0, and decreases to either side. Passive tension (F_p) is the result of an exponential spring with a slack length also around L_0. The dependence of muscle force on velocity (F_v) is described by Hill's relationship. If a muscle is shortening, its force generation ability drops as velocity increases and reaches zero at velocity v_{max} (heavier loads can be lifted less rapidly). If it is lengthening in contrast, force increases when compared to the maximum in statics F_0^M.

The overall force output of the muscle model is specified by

$$F^M = a F_0^M F_v(v^M) F_a(L^M) + F_p(L^M) \tag{1}$$

where activation a scales both F_a and F_v, which themselves combine in a multiplicative way. The term activation refers to the low-pass filtered (neural) input to the muscle and describes its slow activation dynamics. F_0^M is the muscle's maximal force at zero velocity.

The result of eqn.(1) is a normalized force-length-velocity surface scaled by activation as depicted in figure 3. Any movement in this picture corresponds to a 'walk' on that surface. Already complex, in any realistic scenario muscle length,

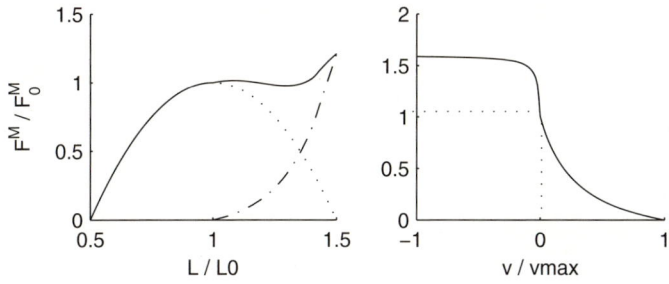

Fig. 2. Dimensionless muscle force as a function of length (left) and velocity (right). The former is the sum of a passive exponential-to-linear elasticity resisting lengthening of the muscle (dash-dotted) and a quadratic function with a maximum at resting length describing the active generation of force (dotted).

velocity and force would probably feed back into the control signal, thereby creating a continuously changing surface. The complete system simulated in the following experiments consists of two antagonistic muscles around a single hinge-joint. Its dynamics are completely determined by the torque produced at the joint through the muscles, as well as the physical properties of the limbs that the joint connects (such as inertia). For simplicity and because we're mainly interested in the effects of the various non-linearities of the muscle model on its dynamics we have made several abstractions. Moment arms of the muscles relative to the joint they actuate are assumed to be constant; gravity in some experiments is zero, which allows us to remove static parts from the control signals and focus on the dynamics instead; no tendon is present. The latter can be justified in elbow movements for example, where the ratio of tendon to muscle lengths is such that tendons have only minimal effect [12]. [1]

2.1   Muscle Dynamics

With regards to the behavior of the muscle model several observations can be made from fig.2 and 3. Firstly, in Hill's force-velocity relationship the slope, that is the rate of change of force, is highest around zero velocity. Hence muscles have the desirable property of being damped the most at rest, while being less damped when moving fast. The slope, and thus damping, will also increase with the level of activation. The same holds for the active force-length relation, hence stiffness will increase with activation as well. It should be noted that the parallel elasticity (the 'spring') of two antagonistic muscles could in principle provide for stability if those muscles were arranged such that over most of the joint's workspace their lengths would be greater than the spring's slack length. Obviously though, this

---

[1] Because we're only interested in qualitative results here, and not particularly simulations of human movements, details about the implementation of the non-linear functions as well as values for the muscle specific parameters F_0, L_0 and v_{max} and limb dimensions are ommitted.

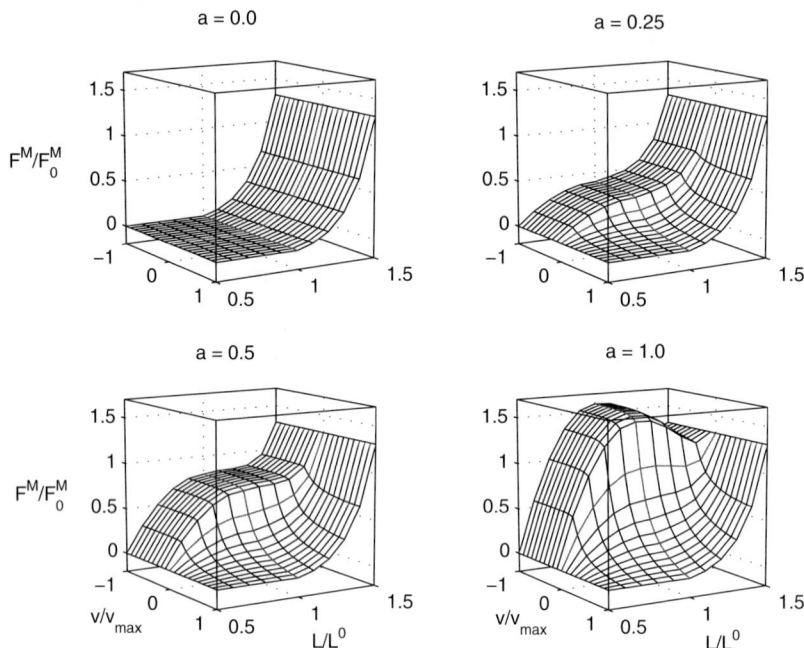

Fig. 3. Force-length-velocity surfaces resulting from eqn.1 for activations of 0.0, 0.25, 0.5 and 1.0. Horizontal axes correspond to normalized length (length divided by resting length) and normalized velocity (velocity divided by maximum velocity).

would make movements rather inefficient because the system would always pull against the opposing muscle's resistance. Indeed, in nature many muscles seem to operate mostly on the ascending limb of the force length curve [3,6], indicating lengths being in a range where the passive elasticity won't dominate the muscle's dynamics. Stability therefore must have a different origin.

Control of position and stiffness. Figure 4 shows how the force-length relationship of two antagonistic muscles interact to create an equilibrium position (EP). The muscles are arranged symmetrically and such that their length, measured over the range of joint positions θ, varies between 0.6 and 1.1. Because the resting positions of both muscles are shifted towards the joint extremes, no elastic resisting forces are created in the midrange. Hence, without activation the system is truly passive and and does not behave like a spring. Activation however creates a stable equilibrium position. While increased co-contraction stiffens the joint at the EP, differential activation will shift its position. The model thus allows for independent control of joint position and stiffness. Note that if the two muscles were springs whose force production depended linearly on displacement and whose resting length was shifted by activation, no increase in stiffness would result from co-activation. The net force produced would only depend on the individual spring stiffnesses and the amount of perturbation from equilibrium.

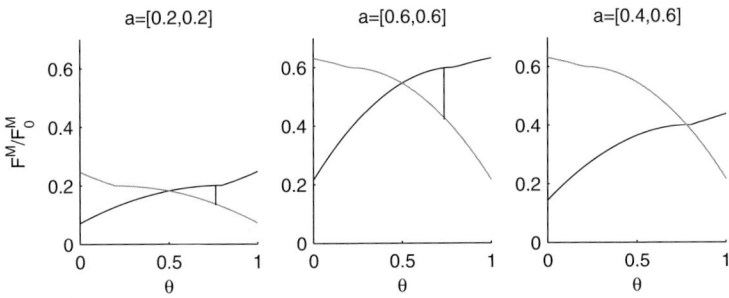

Fig. 4. Force-length relationships of an antagonistic muscle pair. Intersections correspond to the systems equilibrium point (EP). Shifts from the EP will be resisted by a net force equal to the difference between the two curves (vertical line). While co-activation doesn't change the EP, the slope of both force-length curves increases, corresponding to increased stiffness of the joint. A change in the difference between activations will move the EP position.

Rejection of perturbations. From the model it is easily predicted that the viscous element will enhance the system's stability with respect to perturbations. This effect is illustrated in fig.5, where transient loads are applied to the joint at increasing levels of co-activation. Not surprisingly, the model including viscosity shows damping without oscillations and resistance to loads that increases with co-contraction. This rejection of perturbations results from an interplay of both the F_v as well as F_a relationships. While co-contraction creates the EP and stiffens the joint as described above, the steeper slopes in F_v increase its viscosity. Hence the net rejecting force will generally be comprised of a dynamic F_v contribution as well as a static F_a element.

Fast movements. A desirable feature of muscle models including Hill's viscosity term is that they allow for the generation of movements that are fast but appropriately damped so as not to produce significant overshoot. This can be demonstrated for example within the context of the equilibrium point hypothesis (EP-hypothesis) [2]. The general idea is that the neuro-musculoskeletal system creates an equilibrium point that can be shifted using central commands in a simple, e.g. linear, fashion without needing to take into account the actual dynamics. Specifically, in the λ-formulation of the EP-hypothesis, central commands are believed to set the reflex threshold for motoneuron recruitement, such that muscles are activated proportionally to the amount of muscle stretch beyond the setpoint. In this formulation, muscle activation a is described by:

$$a = [L - \lambda + \mu v]^+ \tag{2}$$

where λ is the commanded reflex threshold, v the muscle's velocity and μ the gain of reflex damping. Figure 6 presents a trajectory for the antagonistic muscle pair controlled using this proportional-derivative mechanism (PD). Note that for both, step and ramp shaped commands, the muscle's non-linear relations and

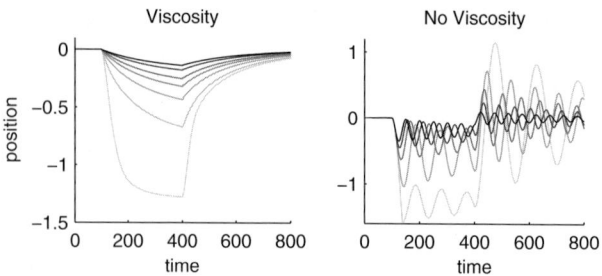

Fig. 5. Response of antagonistic muscle pair to transient external load (between t=200 and t=400) at different levels of co-activation. Left: joint position of model including viscosity, Right: without viscosity. Increased co-activation (darker plots) leads to increased level of stiffness and damping. Only when viscosity is present however, is the resisting force appropriate for self-stabilization. Otherwise oscillations arise.

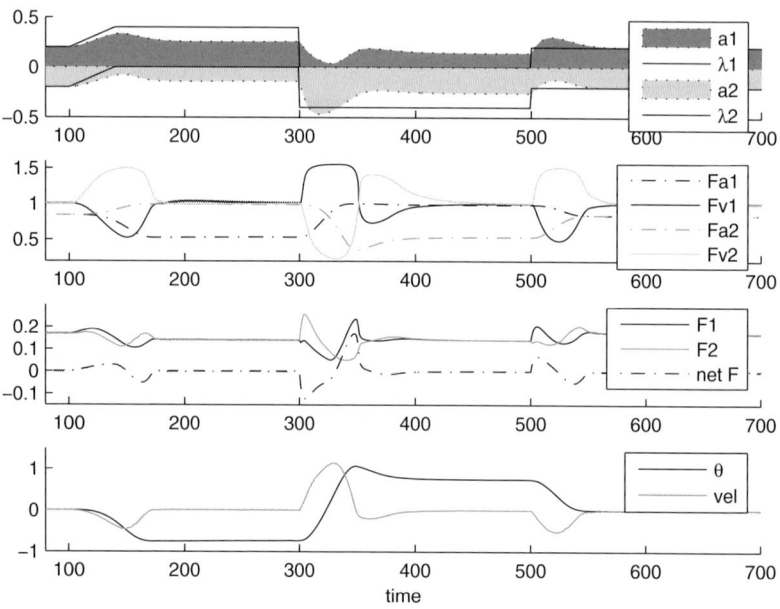

Fig. 6. Trajectories over time of muscles and joint controlled by λ-model. Indices in legend indicate agonist (1) and antagonist (2). Top: input signal λ and resulting activations. Second row: values of F_a and F_v. Third plot: force output of muscles as well as net force. Bottom: joint position and velocity. Commanded reflex thresholds λ change in form of a ramp at time t=100 and as step-functions of different amplitudes at t=300 and t=500. The system starts with initial co-activation of 0.2.

low-pass characteristic combine to sculpt a bi-phasic force trajectory with one peak for acceleration and another for deceleration of the joint. While a pure PD controller would only produce net breaking forces in the case of overshoot, here the breaking starts significantly before overshoot occurs. It is thus fair to say that the muscle model can generate complex trajectories from simple control signals that produce fast but smooth shifts in joint position (similar results with a different model are presented in [4]).

3   Control Signal Optimization

The gain in stability described above also has interesting consequences for the optimization of control signals. To illustrate this point we implemented a simple control strategy that activates each of the two muscles using a square pulse of a given amplitude (a_1, a_2) and duration (d_1, d_2). A fifth parameter (t_2) specifies the time between the onset of agonist activation (fixed) and that of the antagonist. This allows us to look at the space created by evaluating all possible strategies against an optimality criterion, hereafter called the fitness landscape.

Flexibility. Figure 7 shows a fitness landscape in which the criterion consisted of reaching and stopping at a target position of 45° flexion (starting from 0°) at any point during a 2 second trial. In order to show the whole search space we somewhat arbitrarily fixed the amplitudes a_1 and a_2 to values of 0.2. Several interesting observations can be made from this case of unconstrained goal-directed movement. Firstly, the region of good performance spans a considerable range in each of the three remaining dimensions. One can pick almost any value for one of the parameters and will find a combination for the other two that produces a good strategy. In other words, there is a continuum of valid strategies all of which will move the joint towards the desired position, but each having different kinetic or kinematic properties. Movements will differ in terms of velocity, stiffness or energy required. In fact, the point marked B corresponds to the fastest movement in this space, while point C marks the one using least energy (measured as the integral over muscle activation). Thus, compared to the stereotypical behavior of e.g. a PD controller, by using this model one gains immense flexibility with respect to details of a movement, while introducing only few parameters to be chosen (by either a controller or a more constrained optimization procedure). Secondly, although the model is highly non-linear in all its properties, good performance within the fitness landscape is found along rather linear regions. This simple relationship between parameters should make it easy to create a controller that finds and moves along the range of all optimal strategies.

Robustness. In terms of control signal optimization the viscous property of the Hill-type muscle model also shows as increased robustness to noise or increased 'searchability' of the fitness landscape, a property of interest for evolutionary robotics for instance. Fig.8 compares the fitness landscapes of the muscle model with and without the viscosity term for an opimization that maximizes velocity while minimizing overshoot. The slices shown were produced by finding for each model the global peak in the 5-dimensional parameter space $(a_1, a_2, d_1, d_2, t_2)$ and subsequently fixing two of them (amplitudes a_1, a_2) to the found values. The resulting slices show the fitness landscape around the optimum in the remaining

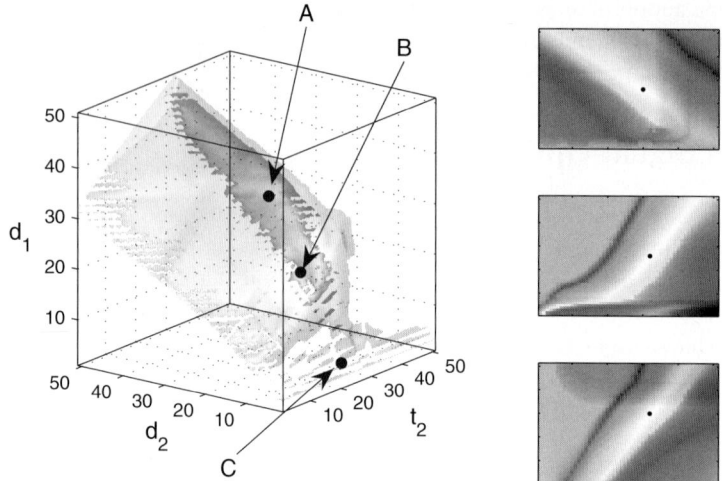

Fig. 7. Left: isosurface of fitness landscape at fitness levels of 95% (dark) and 80% (bright). Point A shows the overall peak of the surface. B corresponds to a movement that maximizes velocity, while C minimizes energy. Right: slices through the peak of the same fitness landscape.

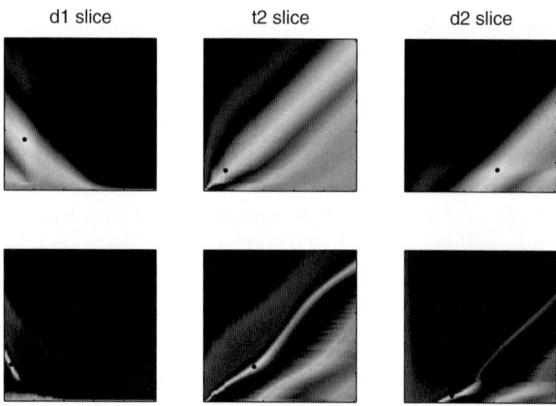

Fig. 8. Slices through peaks of fitness landscapes for maximizing velocity while minimizing overshoot. Top row includes viscosity, bottom row does not.

three dimensions. As can easily be seen, without viscosity the regions of good fitness become much more narrow. For the optimization procedure this means increased difficulty of finding the global optimum. It can also be interpreted though as robustness to noise in the control signal. In the viscous model a slight perturbation away from the optimum will still produce relatively good results, while in the non-viscuous case performance is easily lost completely. Intuitively this is easy to understand. In the non-viscous case, the antagonist activation has

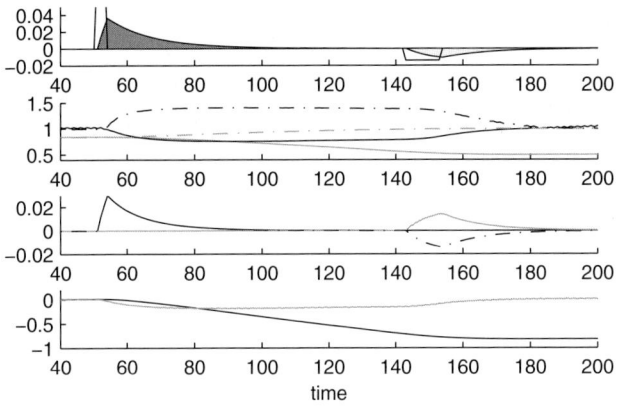

Fig. 9. Trajectory resulting from typical control signals evolved to minimize energy. Plots correspond to fig.6, i.e. top: activations, 2nd plot: F_a, F_v, 3rd plot: muscle forces, bottom: position and velocity.

to be precisely timed and scaled such that at the target position forces cancel out exactly and the joint comes to a stop. Any remaining forces not counteracted completely by the antagonist will move the joint away from the target. In the viscous model however, because of its damping effect, small remaining forces will fade quickly and the joint will come to a stop near the target position.

Efficiency. Motorized actuators have to be powered throughout a movement. Even compliant actuators will have to make motors move to simulate a zero force trajectory, i.e. a purely passive swing. Muscles, however, allow for more efficient movement through bi/tri-phasic pulse patterns. Minimal muscle activations are sufficient to accelerate and decelerate the joint towards a desired position. This is possible, however, only because antagonistic muscles don't work like springs. That is, in their passive state they don't have to work against each other's resistance. Figure 9 is an example of control signals optimized for minimal energy use. Clearly, throughout a large part of the movement neither muscle produces any force and the joint is passively swinging towards its desired position.

Multijoint movement. The movements and control signals presented so far are clearly oversimplified when compared to natural movements involving many interacting joints. It is striking though that simple square pulses, appropriately scaled and timed, allow for well-behaved movement trajectories when combined with non-linear muscle properties. In order to investigate if the increased robustness and flexibility also translates to more complex scenarios we used the same approach of control signal optimization to generate motions of two joints (elbow and shoulder). We also enabled gravity and included a static part in the control signals that could compensate for its effect. Figure 10 presents optimized trajectories in two different conditions. The elbow joint is always required to flex to a position of 45°. However, in scenario 1 the shoulder moves in the opposite

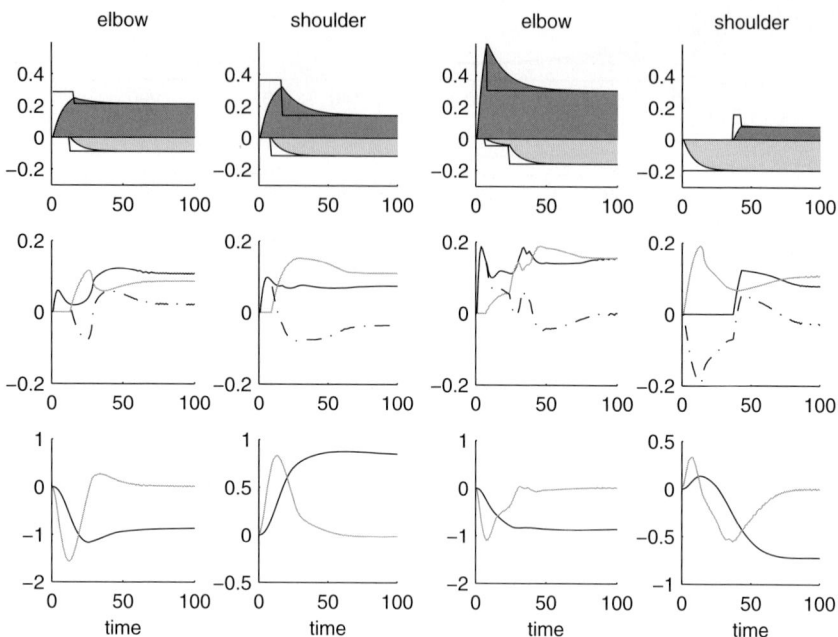

Fig. 10. Shoulder and elbow trajectories optimized for goal-directed 2-joint movements maximizing velocity while minimizing energy. First two columns correspond to scenario 1 (synergistic), the columns on the right to scenario 2. Top: activations, middle: muscle forces, bottom: position and velocity.

direction while in scenario 2 it moves in the same direction. Both cases were easily evolved and produced trajectories whose final positions corresponded to the desired targets. The figure shows that in the first case the velocity profiles resemble smooth bell-shapes, while they are more jerky in the second case. The reason for this effect are the interaction torques arising from the mechanical coupling of the two joints. In the first scenario movement of the shoulder creates interaction torques in the elbow that are 'synergistic', i.e. support the intended movement, while in the second case the torques counteract movement in the desired direction. It is thus obvious that the simple open-loop control used here is insufficient in some circumstances. In fact, it is one of the big open questions in motor control whether the (human) central nervous system uses an internal model of the body to calculate control signals that account for its dynamics or if a well-designed neuro-musculoskeletal system itself could perform the neccessary 'morphological computations'. Independent of the case of human movements, both a feedback controller such as the λ-model as well as appropriate open-loop signals created using internal models could generate the desired movements. While the former stands out for its simplicity, the latter more easily could exploit interaction torques rather than resist them. In both cases though a muscle's intrinsic stability should be beneficial.

4   Conclusion

As shown above, the non-linear behavior of an antagonistic muscle pair produces many desirable properties with respect to embodied motor control; properties which at first could seem surprising, given the complexity of the model. It allows for independent control of position and stiffness; its flexibility allows for a continuum of control strategies differing in various kinematic and kinetic features; it creates 'nice' search spaces for the optimization of control signals; non-linear damping allows for fast movements with precise breaking; it can produce efficient movements by using only phasic activation; and simple control signals can create complex but well behaved trajectories. In addition, the global dynamics of the system can be tuned with only few parameters that describe the shape of the muscle's length and velocity dependence as well as their geometry. We thus argue that versions of the Hill-type or similar muscle models strike an ideal 'ecological balance' [8]. Robots equipped with similar actuators should be able to trade an increase in the complexity of morphology for a reduction in the complexity of the control system. Several attempts exist at building such muscle-like actuators for robots and prostheses. Most prominent are series elastic actuators [9], McKibben-style pneumatic actuators [10] and electroactive polymer actuators [1]. More research seems necessary though to achieve the right combination of viscoelastic properties in an efficient package applicable to multiarticulate robots.

The fact that hill-type muscle models are generic in the sense of needing only few parameters to implement particular types of muscles make them particularly interesting for evolutionary robotics. In fact, if one is looking at more complex motor behaviors such as multijoint reaching or walking, it is not only possible but necessary to co-evolve or co-adapt muscle morphology, skeletal geometry and the control system. Only then is it possible to tune the system's dynamics to be appropriate for a given task.

We'd like to emphasize that the experiments presented herein are not meant to be models of how animals actually control their movements. In particular, we do not make any claim about the relative importance of the brain or the musculoskeletal system in motor control, but emphasize that the latter exhibits intrinsic dynamics which, unless entirely suppressed by the brain, do play a role in natural movements. The forms of control have been chosen only for their simplicity to show the dynamics of the muscle model in the most general context. However, in ongoing work we are concerned with neural control in more realistic tasks. To this end we are evolving dynamical neural 'reflex' networks that try to exploit the natural dynamics of the musculoskeletal system in order to simplify control as seen from higher levels. It can be shown that the right muscle model coupled to a controller in a closed loop can function as a source of force, as a spring, a servo of position or velocity or combinations of these. We believe that well-designed neuro-musculoskeletal systems enable higher levels to set up the right global dynamics for the task at hand, effectively choosing between different modes of control, such that subsequent control signals can make use of the particularities of task, environment and body.

References

1. Bar-Cohen, Y. (2001). Electroactive Polymer (EAP) Actuators as Artificial Muscles - Reality, Potential and Challenges. SPIE Press, Vol. PM98.
2. Feldman, A., Ostry, D., et al. (1998). Recent tests of the equilibrium-point hypothesis (lambda model). Motor Control 2: 189205.
3. Garner, B. A. and Pandy, M. G. (2003). Estimation of Musculotendon Properties in the Human Upper Limb. Annals of Biomedical Engineering 31(2): 207-220.
4. Gribble, P. L., Ostry, D. J., Sanguineti, V. and Laboissiere, R. (1998). Are Complex Control Signals Required for Human Arm Movement? J. Neurophysiology 79: 1409-1424.
5. Brown, I.E. and G.E. Loeb. A reductionist approach to creating and using neuromusculoskeletal models. In: Biomechanics and Neural Control of Posture and Movement. J.M. Winters and P.E. Crago (ed.) Springer Verlag, New York, 2000.
6. Murray, W. M., Buchanan, T. S., et al. (2000). The isometric functional capacity of muscles that cross the elbow. Journal of Biomechanics 33(8): 943-952.
7. Pfeifer, R. and Iida, F. (2005). Morphological computation: Connecting body, brain and environment. Japanese Scientific Monthly 58(2): 48-54.
8. Pfeifer, R. (1996). Building 'fungus eaters': Design principles of autonomous agents. In: From Animals to Animats 4: Proceedings of the Fourth International Conference on Simulation of Adaptive Behavior. Maes, P., Mataric, M.J., Meyer, J.A., Pollack, J. and Wilson, S.W. (ed.) MIT Press.
9. Pratt, J.E. and Krupp, B.T. (2004). Series Elastic Actuators for Legged Robots. SPIE 2004.
10. Verrelst, B. et al. (2005). The pneumatic biped 'lucy' actuated with pleated pneumatic artificial muscles. Autonomous Robots, vol. 18, pp. 201213, 2005.
11. Wagner, H. and Blickhan, R. (1999). Stabilizing function of skeletal muscles: an analytical investigation. J Theor Biol 199: 163-179.
12. Zajac, F. E. (1989). Muscle and tendon: properties, models, scaling, and application to biomechanics and motor control. CRC Crit. Revs biomed. Engng 17: 359-411.

Investigation of Reality Constraints: Morphology and Controller of Two-Link Legged Locomotors for Dynamically Stable Locomotion

Kojiro Matsushita, Hiroshi Yokoi, and Tamio Arai

The University of Tokyo, Dept. of Precision Engineering,
7-3-1 Hongo, Bunkyo-ku, Tokyo 113-8656 Japan
{matsushita, hyokoi, arai-tamio}@prince.pe.u-tokyo.ac.jp

Abstract. In this paper, we purpose to reveal effective design components for morphological functionality and reality constraints by analyzing simple loco-motors in both virtual and real worlds. Firstly, we assumed that human experiences and techniques contained important design components so that we conducted edutainment course to acquire locomotors, which were heuristically designed. Then, we analyzed two remarkable locomotors in both virtual and real worlds. As a result, we have known that symmetrical design played an important role on dynamically stable locomotion because its design enabled to exploit its own dynamics as passive dynamics and also widened its controllability. Addition to it, the locomotors in both virtual and real worlds demonstrated similar characteristics.

1 Introduction

In the field of conventional legged robotics, implementation of control is main focus for achieving dynamically stable locomotion so that many types of gait control and balance control are theorized and applied to legged robots [6]. Common characteristics of these legged robots are made of rectangular-solid materials in mechanism, actuated with many high-drive electrical motors, and controlled with high-speed information processing. Then, with these configurations, these legged robots are able to keep its balance (balance control) and their gaits are basically generated in the same methodology as robot arm manipulators: stance legs are regarded fixed links and swing legs are regarded as effecters during one step motion. By iteration of the manipulation, the robots achieve legged locomotion.

Meanwhile, there have been designed dynamically stable legged robots with biologically inspired knowledge such as anatomy, physiology, and neuroscience [1,9]. For example, passive dynamic walker [3] is a robot, which does not have any motors and any sensors but well-designed for bipedal locomotion - straight legs, curved feet with passive hip joints and latch knee joints as human body components. Then, the robot achieves dynamically stable bipedal locomotion on a specific incline by driving force from not motor actuation but gravity through its own morphology. Thus, it is known that morphological characteristics help to exploit its own physics and contribute achieving dynamically stable locomotion and it can be regarded as morphological functionality in embodied artificial intelligence [5].

S. Nolfi et al. (Eds.): SAB 2006, LNAI 4095, pp. 101–112, 2006.
© Springer-Verlag Berlin Heidelberg 2006

Especially in evolutionary robotics, there have been some works to reveal morphological functionality. The researchers assumed that morphological functionality is emerged by interdependence between morphology and controller and, then, use simulations of evolutionary processes to perform coupled optimization of both the morphologies and controllers of simulated robots (evolutionary designs). One of the most successful of these applications was the work of Sims [7], in which artificial creatures were automatically designed in a three-dimensional physics simulation. The simulation generated a variety of locomotive creatures with unique morphologies and gaits, some of which have no analogy in the biological world. This suggested that the interdependence between morphology and control plays an important role in dynamically stable locomotion.

However, evolutionary design suffers from unknown design components: coupled evolution is conducted only in simulation and how to best represent morphology, controller, environment, and fitness function is not clear yet; differences between virtual and real worlds have not been shortened so that results in virtual world are not always transferable to the real world, especially in the case of dynamic systems, although work is focusing on this problem. Therefore, the work of Lipson [4], who demonstrated automated manufacture of evolved simulated robots, has not focused on dynamical stability and, then, the evolutionary design system including reality constrain generated only static locomotion. Thus, designing reality constrain should be important factor to design dynamical stable locomotion in real world.

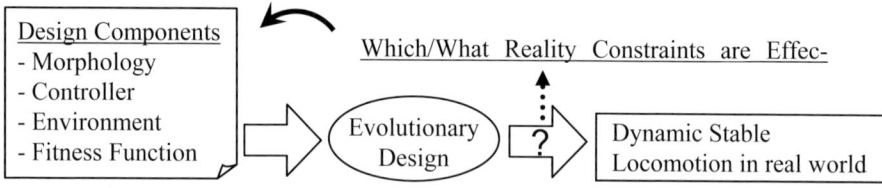

Fig. 1. Conceptual figure of evolutionary process. We have not known design components to achieve dynamic locomotion in real world.

In our research, we purpose to implement evolutionary design system, which generates legged locomotors (locomotive robots), in three dimensional simulation and those locomotors should have morphological functionality to achieve dynamically stable locomotion and, moreover, be effective design to real world. Therefore, in this paper, we mainly focus to reveal effective design components to reality constraints as shown in fig.1. We propose to extract design components from human experiences and techniques (heuristic design) and evaluate those design components in order to reduce reality gaps in both virtual and real world. Concretely, we conducted edutainment (educational entertainment) course to acquire variety of unique locomotors (diverse heuristic design) and analyzed two remarkable locomotors in both virtual and real world, in terms of morphological functionality and reality constraints.

Firstly, we explain our edutainment. Then, we investigate morphology and controller of two-link models on locomotion, which are designed in the edutainment course, in three-dimensional simulation. Finally, we conclude this paper.

2   Robot Edutainment as Diverse Heuristic Design

We conducted robot edutainment course to acquire variety of unique locomotors (diverse heuristic design) for the purpose of investigation of design components on those locomotors. This robot edutainment course is conducted for 20 students in master course at engineering department and aimed at teaching importance of morphological functionality. Main characteristic of this course is that students are able to design both mechanism and controller of robots. So, students designed their own locomotors in rapid proto-typing method and implement its control algorithm based on sample programs, which we provided.

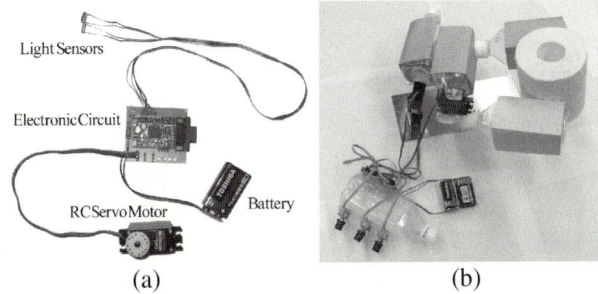

(a)   (b)

Fig. 2. Edutainment tools. (a) controller (b) plastic-bottle robot arm.

2.1   Electric Circuit: Simple Oscillation Controller

Electric circuit Controllers are built on microchip H8/3664 (16MHz) and function to control three RC-servo motors, to read two light sensors and two touch sensors, to communicate with a computer through serial port (fig.3). The following oscillation algorithm for RC servo motors is downloaded as sample program: motor axis moves to two different angle-position alternately at an certain cycle. We instructed students to modify the angle-positions and the cycle to their own locomotors. The students could also use light detect sensor, however, we do not describe the sensor in this paper.

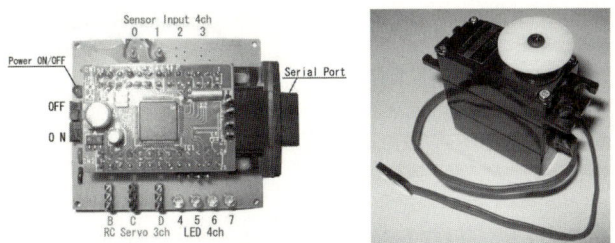

Fig. 3. Robot Controller (left) and RC servo motor (right)

Table 1. Specification of Standard RC Servo Motor GWS S03T/2B

Spec.		Torque	Speed
	6.0V	8.0kg-cm	0.27sec/60°
Weight	46 g		
Size	39.5 × 20.0 × 39.6mm		

2.2 Rapid Proto-typing Method: Plastic Bottles Based Robots

In the robot edutainment course, we applied rapid proto-typing method: plastic bottles are used as body parts, RC-servo motors as actuated joints, and those are connected with hot glue. The advantage of this method is easy-assembling, easy-modifying, and economizing machining time so that students are able to build robots without any technical difficulties. As disadvantage, those robots do not move precisely comparing to metal materials. However, it is enough to realize desired behaviors so that we apply this method to build proto-type robots.

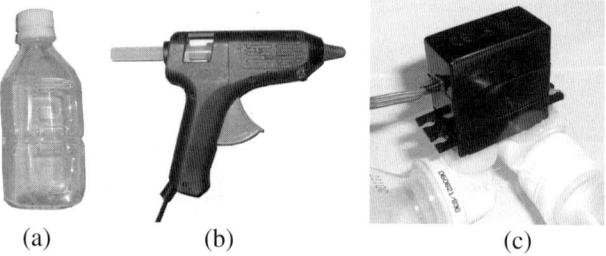

(a)   (b)   (c)

Fig. 4. Material for robot structure. (a) plastic bottle (b) hot glue gun (c) example of connection between plastic bottles and RC servo motor.

2.3 1 Meter Race for Evaluation of Morphological Functionality

This robot edutainment course purposes to design locomotors to exploit its own dynamics through morphology as passive dynamic walkers. Therefore, students are allowed to use only one or two RC servo-motors as design condition and implement autonomous control program freely to design robots. Moreover, all the locomotors are measured time for 1 meter forward locomotion (fig.5) to evaluate their locomotion ability because we assumed that faster locomotors exploit more morphological functionalities.

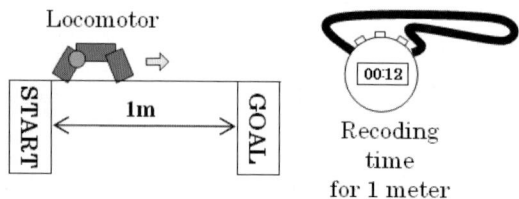

Fig. 5. Competition regulation. Locomotors, which are designed by students, travel 1 meter and traveling time is recoded.

2.4   Results: Locomotors Designed by Students

Locomotors designed by students as shown in fig.6, were qualitatively categorized into three types at the criteria of locomotion pattern: shuffling, kicking-ground, and falling-over. Shuffling type moved forward by a high-frequently motor oscillation - approximately 4Hz. We assumed that the locomotion is generated by friction forces at contact points. Kicking Ground (KG) type moved forward by directly kicking the ground with its rear leg. Falling-Over (FO) type moved forward by falling over after its fore leg was lifted up. In short, the FO type changes its body shape and, then, exploits gravity to fall over to forward.

   Table 2 indicates the results of 1 meter race. The shuffling type moved much slower than other types and was difficult to move straight forward so that they normally could not reach the goal. Meanwhile, the best KG type and FO type reached 1 meter at approximately 10 seconds (fig.7). As for morphological functionality, we assumed that the shuffling type contacted the ground all the time and its crawling locomotion was interfered by its floor condition. Meanwhile, the KG and FO types have legged morphology that utilize actuation force to driving force more efficiently than the shuffling time (we do describe on the KG and FO type in next section).

Fig. 6. Locomotors designed by students

Table 2. Ranking of locomotors designed by students

Ranking	Traveling Time for 1 meter	Locomotion Type
1	7.4 sec	Kicking Ground
2	8.1 sec	Falling Over (Legged)
3	11.0 sec	Kicking Ground (Legged)
4	90 sec	Falling Over
5 - 20	More than 5 min / No goal	Others

3   Investigation of Two Remarkable Locomotors in Real World

In this section, we analyze on locomotion characteristic of two remarkable locomotors as shown in fig.7, which acquired in the robot edutainment course as heuristic design.

In order to investigate effective design components for morphological functionality and reality constraints, we analyze the locomotors in both virtual and world. Firstly, we observed their ground contact information during locomotion in real world, which represents their gaits. Secondly, those robots are modeled and simulated in three-dimensional world and, then, their ground reaction force and center of gravity on z axis, which represents characteristic of rhythmic movement, are recorded and analyzed.

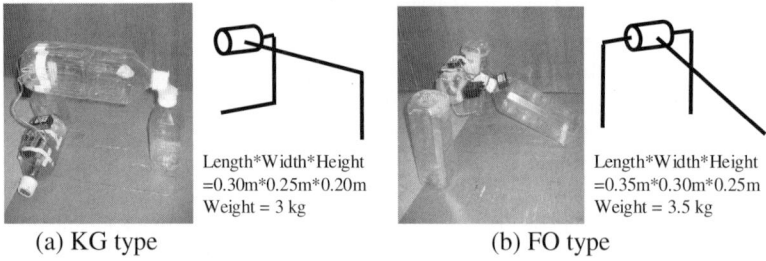

Length*Width*Height
=0.30m*0.25m*0.20m
Weight = 3 kg

Length*Width*Height
=0.35m*0.30m*0.25m
Weight = 3.5 kg

(a) KG type　　　　　　　　　　　　　(b) FO type

Fig. 7. Two remarkable locomotors. (a) kicking ground type (b) falling over type.

3.1 Analysis of Locomotors in Real World

In the previous section, we categorized that the KG type moves forward by kicking the ground (fig.8a) and the FO type moves forward by falling-over (fig.8b).

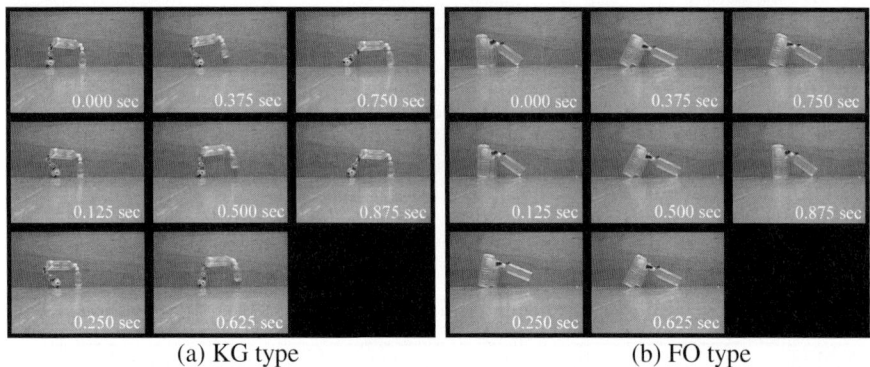

0.000 sec	0.375 sec	0.750 sec
0.125 sec	0.500 sec	0.875 sec
0.250 sec	0.625 sec	

(a) KG type　　　　　　　　　　　　　(b) FO type

Fig. 8. Locomotion scene of two locomotors. (a) kicking ground type (b) falling over type.

So, we quantitatively analyzed the locomotion by recording ground contact information. As results, we have macroscopically seen similar gaits between the KG type and the FO type: their rear legs were all the time on the ground and their fore legs are taken off the ground for approximately 0.5 second when the types opened their legs (fig.9a). However, they have microscopically indicated difference characteristics. In the case of the KG type, the fore leg took the ground off approx. 0.2 sec before and approx. 0.2 sec after the KG type started opening its legs (fig.9b). We assumed that the center of gravity (CoG) should be on the rear leg at the time for

0.2 sec and, then, the rear leg kick the ground for 0.2 sec. On the contrary, in the case of the FO type, the fore leg took the ground off 0.1 sec after the legs started opening. It seems that the fore leg was lifted up for 0.1 sec and, then, the rear leg fell over forward for 0.2 sec. Thus, the ground contact information implies their locomotive patterns. However, with the analysis of the real robots, it is difficult to reveal internal states such as ground reaction force, which possibly indicates exploitation of its own physics. Therefore, we modeled the KG and FO types for analysis of their locomotive patterns.

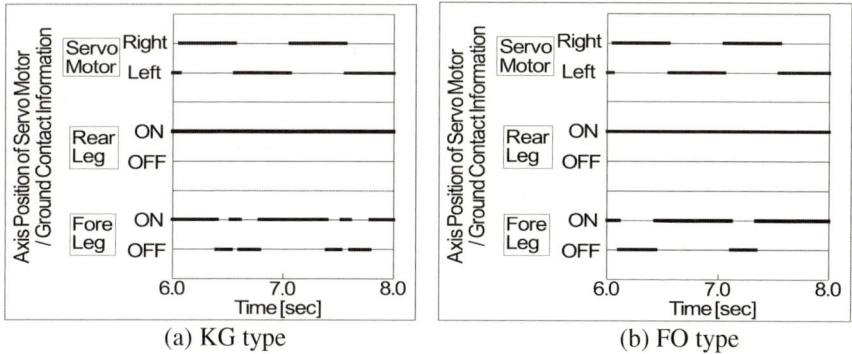

(a) KG type   (b) FO type

Fig. 9. Locomotion gaits of two locomotors. (a) kicking ground type (b) falling over type.

4   Investigation of Two Remarkable Locomotors in Virtual World

In this section, we modeled the locomotors and simulated in three-dimensional world (implemented with three-dimensional library – Open Dynamic Engine [8]) in order to observe their internal states.

4.1   Analysis of Locomotors on Gaits

Fig. 10 shows the modeled KG type and FO type in simulation. These simulated models are slightly different design from the real locomotors because of simplification of analysis and, then, have the same morphological setup: the locomotors are made of two links as mechanical parts and one ball as weight of controller unit. In short, the two locomotors completely consists of the same properties, however, parts assembly is different as shown in fig.10. Moreover, we implement that motors in the simulation are actuated at rotational speed $\pi/2$[rad/s], which is similar to real RC servo motors as reality constraints.

The morphological and control parameters are searched heuristically because these are simple models so that it is not necessary to apply evolutionary design to search optimal locomotion. Fig.11 shows locomotion scenes of the simulated models, which are similar to locomotion patterns of the real locomotors. We observed their ground contact information including ground reaction force (GRF).

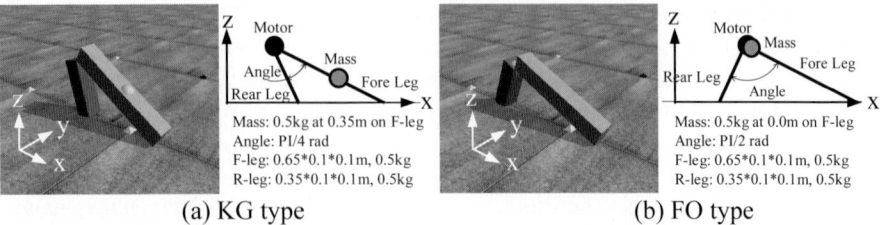

Mass: 0.5kg at 0.35m on F-leg
Angle: PI/4 rad
F-leg: 0.65*0.1*0.1m, 0.5kg
R-leg: 0.35*0.1*0.1m, 0.5kg

(a) KG type

Mass: 0.5kg at 0.0m on F-leg
Angle: PI/2 rad
F-leg: 0.65*0.1*0.1m, 0.5kg
R-leg: 0.35*0.1*0.1m, 0.5kg

(b) FO type

Fig. 10. Simplified locomotor. The locomotors are built with the same morphological parts. (a) kicking ground type (b) falling over type.

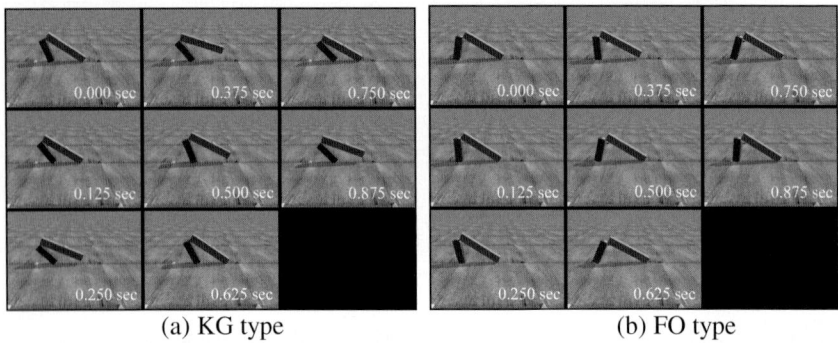

(a) KG type   (b) FO type

Fig. 11. Locomotion scene of two locomotors. (a) kicking ground type (b) falling over type.

The fig.12 and fig.13 indicates the following states of the KG type: the CoG was on the rear leg (GRF at the foreleg gradually decreased and GRF at the rear leg gradually increased), the rear leg started kicking the ground (GRF at the rear leg shows its peak and decreased and, then, GRF at the rear leg is zero for 0.1 sec), stayed in the air (GRF at both legs are zero for 0.05 sec), the fore leg landed (GRF at the fore leg is peak) for one locomotion cycle. Meanwhile, the FO type is similar to crawling because the rear leg keeps relatively constant GRF and even the fore leg keeps constant GRF except its landing peak.

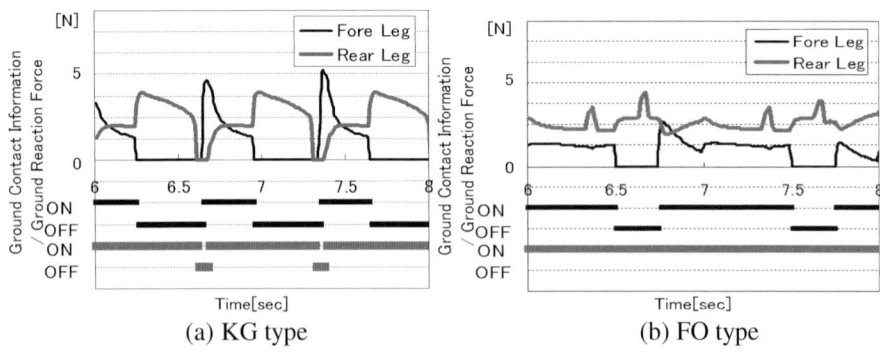

(a) KG type   (b) FO type

Fig. 12. Gait and ground reaction force of two locomotors. (a) kicking ground type (b) falling over type.

4.2 Analysis of Locomotors on Plot Phase of CoG-Z

Fig.13 and fig.14 shows plot phase of center of gravity on z axis (CoG-Z) of two locomotors. The difference is that fig.13 is actuated at $\pi/2$[rad/s] rotational speed, fig.14 is π [rad/s] rotational speed. The remarkable points are that the faster actuation forms more smooth limit cycle, which means less poles and become hemi-sphere shape. In other words, the faster actuation realizes more dynamical stability. Thus, the characteristics indicate the rotational speed relates to dynamical stability.

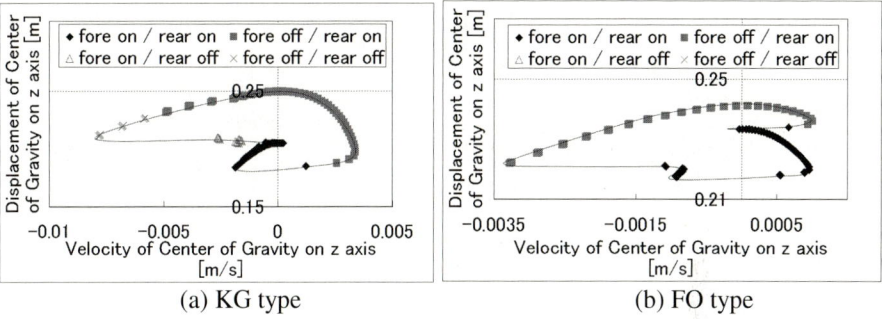

(a) KG type   (b) FO type

Fig. 13. Phase plot of center of gravity on z axis (CoG-Z) of two locomotors. (a) kicking ground type (b) falling over type.

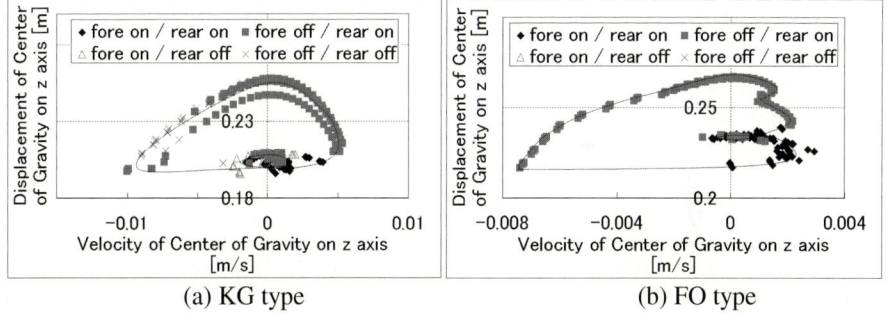

(a) KG type   (b) FO type

Fig. 14. Phase plot of center of gravity on z axis (CoG-Z) of two locomotors in the case of applying double rotational speed. (a) kicking ground type (b) falling over type.

4.3 Analysis of Locomotors on Control Parameters and Distance Traveled

So far, we have shown internal states during locomotion. In this section, we purpose to reveal their morphological characteristics so that we have measured distance traveled forward at some ranges in amplitude and frequency as control parameters. Fig.15 shows results: both types showed similar wave-type solution space. This is because the maximum speed of motors is $\pi/2$ [rad/sec] and, for that reason, the rotational direction switches before the motor reaches the desired angle-position. Therefore, the area at less than 1.5 [sec] on cycle axis is independent from amplitude. As for the best control parameters, both types traveled forward the most at approx. 0.5 [sec] on cycle axis, traveled forward most. Meanwhile, the area where is more than 2 [sec] on cycle axis and more than $3\pi/8$ [rad] on amplitude axis, shows different characteristic. This indicates that the KO type achieves moving forward by direct kicking the ground, therefore, it can kick the ground at both fast and slow cycle. Oppositely, the FO type moves forward by lifting the fore leg up so that the lift-up motion requires fast cycle because the fore leg is not lifted up and it does not move forward if the cycle is slow. Thus, the KG type morphologically achieves static locomotion and has more controllability than the FO type. Meanwhile, the FO type requires morphologically requires appropriate speed to achieve locomotion.

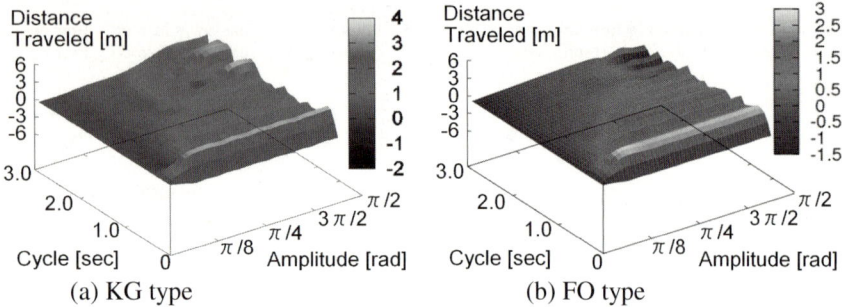

Fig. 15. Control parameters space to two locomotors. (a) kicking ground type (b) falling over type.

5   Investigation of Same Length Type Locomotor

In the previous section, we analyzed two types of two-link locomotors. However as for two-link model, there is another type, which is the same length (SL) type. In order to cover all the two-link morphology, we also investigated the SL type.

5.1   Analysis of SL Type Locomotor on Plot Phase of CoG-Z and Gait

Firstly, we have measured distance traveled forward of the SL type at some ranges in amplitude and frequency. As a result, it indicated hill type, which is completely different solution space from the KG and FO type. Moreover, the SL traveled forward 5 meters at one peak on the solution space where is 1 [sec] on cycle axis and $\pi/4$ [rad] on amplitude axis. The remarkable point is that the locomotion speed is slower than KG and FO type but it achieved longer distance traveled. We qualitatively assume that the SL type exploits pitch oscillation as passive dynamics. Addition to it, it seems that the SL type has better controllability than other types because the hill-type indicates widely stable control area.

Secondly, we have observed its GRF at remarkable control parameters. Then, the SL type has shown symmetrical characteristics in its gaits at a specific control parameters as shown in fig17b: the locomotor starts iteration of switching swing leg and support leg at one spot in simulation and its GRF is similar to M letter shape, which represents human walking gait [2]. We assume that this locomotion is close to bipedal locomotion and utilizing pitching oscillation indicates exploiting its own physics through morphology. It seems that three-link morphology can control its pitching oscillation better so that the three link locomotors should be more appropriate design for passive dynamics.

Therefore, it clarifies that the interdependence between symmetrical design as morphology and appropriate control parameters achieves smooth transition as dynamical locomotion and better locomotion ability. We also design plastic-bottle robot in real world as shown in fig.18. As a result, the locomotor showed the pitching oscillation movement similar to the result of simulation.

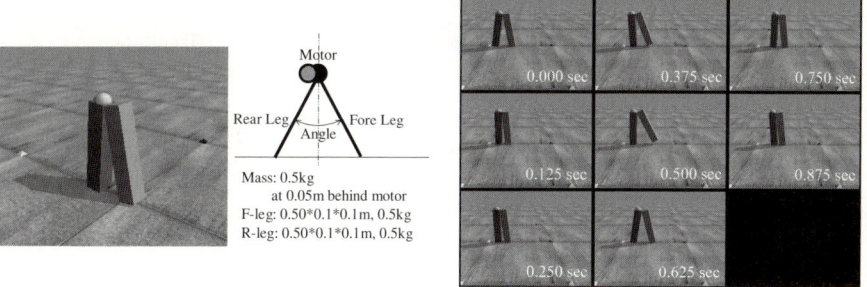

Fig. 16. Qualitative analysis on locomotion pattern of two locomotors. (a) kicking ground type (b) falling over type.

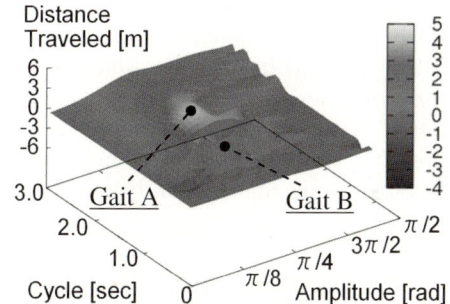

Fig. 17. Control parameters space to SL type locomotor

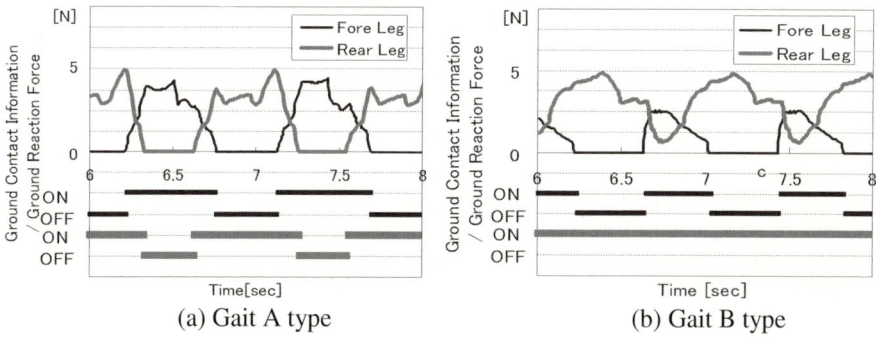

(a) Gait A type   (b) Gait B type

Fig. 18. Gait and ground reaction force of two locomotors. (a) Gait A type (b) Gait B tye.

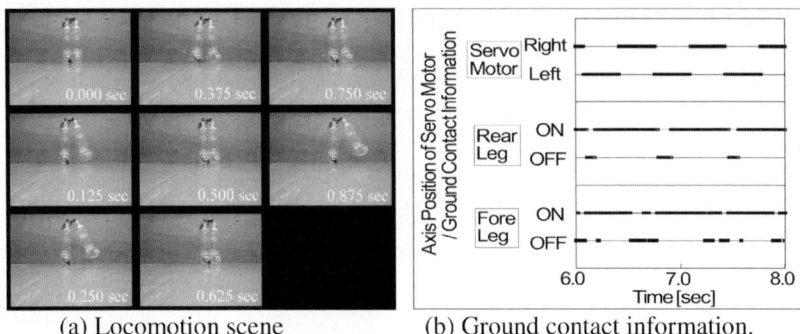

(a) Locomotion scene   (b) Ground contact information.

Fig. 19. Locomotion scene and ground contact information of the SL type in real world. (a) Locomotion scene (b) Ground contact information.

6 Conclusion

In this paper, we investigated reality constraints and morphological functionality for dynamically stable locomotion. At first, variety of locomotors was heuristically designed in edutainment course and two remarkable locomotors were analyzed in both virtual and real worlds. As results, the modeled locomotors have shown two different gaits: hopping and crawling. Furthermore, we have designed the third locomotor: the locomotor exploits its own dynamics - pitching oscillation (passive dynamics); its ground reaction force is similar to M letter shape, which is characterized human dynamic walking; the symmetrical design realizes better smooth controllability because of hill-type control solution space; the locomotors in virtual and real worlds demonstrates similar locomotion characteristic. Thus, we found design components for reality constraints and morphological functionality for dynamic stable locomotion.

References

1. Alexander, R.: Principles of Animal Locomotion. Princeton University Press, New Jersey (2002)
2. Komi,P.V., Gollhofer,A., Schmidtbleicher,D., Frick,U.: Interaction between man and shoe in running: considerations for a more comprehensive measurement approach. Int. J. of Sports Medicine (1987) 8:196-202
3. McGeer,T.: Passive dynamic walking. Int. J. Robotics Research (1990) 9(2):62–82
4. Lipson, H., Pollack, J. B.: Automatic design and Manufacture of Robotic Lifeforms. Nature (2000) 406: 974-978.
5. Pfeifer,R., Scheier,C.: Understanding Intelligence. Cambridge, MA: MIT Press (1999)
6. Raibert, M. H.: Legged Robots That Balance. MIT Press, Cambridge, MA (1986)
7. Sims, K.: Evolving 3D morphology and behavior by competition. In R. Brooks and P. Maes, editors, Artificial Life IV Proceedings. MIT Press (1994) 28-39
8. Smith, R.: Open Dynamics Engine. URL: http://ode.org/ (2000)
9. Vukobratovic,M., Juricie,D.: Contribution to the synthesis of biped gait. IEEE Tran. On Bio-Medical Engineering (1969) 16(1):1-6

Synchronization and Gait Adaptation in Evolving Hexapod Robots

Mariagiovanna Mazzapioda and Stefano Nolfi

Institute of Cognitive Sciences and Technologies, National Research Council (CNR)
Via San Martino della Battaglia, 44 , Rome, Italy
mariagiovanna.mazzapioda@istc.cnr.it,
stefano.nolfi@istc.cnr.it

Abstract. In this paper we present a distributed control architecture for a simulated hexapod robot with twelve degrees of freedom consisting of six homogeneous neural modules controlling the six corresponding legs that only have access to local sensory information and that coordinate by exchanging signals that
diffuse in space like gaseous neuro-trasmitters. The free parameters of the neural modules are evolved and are selected on the basis of the distance travelled
by the robot. Obtained results indicate how the six neural controllers are able to
coordinate so to produce an effective walking behaviour and to adapt on the fly
by selecting the gait that is most appropriate to the current robot/environmental
circumstances. The analysis of the evolved neural controllers indicates that the
six neural controllers synchronize and converge on an appropriate gait on the
basis of extremely simple control mechanisms and that the effects of the physical interaction with the environment are exploited to coordinate and to converge
on a tripod or tetrapod gait on the basis of the current circumstances.

1 Introduction

In this paper we describe a method for developing the control system for a simulated
hexapod robot with twelve degrees of freedom that has to exhibit a walking behaviour. The architecture proposed is fully distributed and consists of six identical neural
modules in which each module is located in the corresponding leg and in which neural modules coordinate by producing and detecting signals that diffuse in space like
gaseous neuro-trasmitters.

This implies that as in other related models [1],[2] leg coordination does not arise
from a centralized gait generator, but rather from the interactions between the neural
modules controlling the corresponding legs. More precisely, leg coordination is mediated by the physical robot/environment interaction and by the signals produced by
neural modules located nearby. However, contrary to the other models referenced
above, each neural module influences and is influenced by the neural modules located
nearby in the same way (i.e. the control system is constituted by a set of identical neural modules).

In more general terms, we assume that the control system of our robot is composed
by a number of homogeneous neural modules that, each separately, exhibit a limit cycle (i.e. a periodic behaviour). Our problem, therefore, is that to define the rules that
determine the conditions in which signals are produced and the way in which detected
signals affect nearby neural modules so that the resulting closed loop system exhibits

S. Nolfi et al. (Eds.): SAB 2006, LNAI 4095, pp. 113–125, 2006.

a coordinated limit cycle behaviour that allow the robot to walk effectively (for a re-lated approach see [3]). This problem have been attacked by using a self-organized technique based on artificial evolution [4] in which the free parameters of the neural modules are encoded in a population of evolving genotypes, and variations introduced through genetic operators are retained or discarded on the basis of the overall behaviour exhibited by the neural modules embodied in the robot and tested in the environment.

The goal of this paper is not to understand the biological basis of locomotion control in natural organisms but rather to build real-time walking machines. In par-ticular, we are interested in investigating whether robots that have a modular struc-ture (i.e. that are constituted by repeated homologous body elements) can exhibit coherent and effective behaviour on the basis of modular control systems (i.e. on the basis of distributed control system in which repeated parts of the robots body are controlled by corresponding repeated control units). Progresses toward this ob-jective, in fact, might have a significant impact on robotics with particular refer-ence to self-reconfigurable robotics [5], [6] and evolutionary robotics techniques that allow to co-evolve and co-adapt the robots' control system and body structure [7], [8], [9], [10].

2 The Experimental Setup

In this section we describe the simulated hexapod robot used in the experiments, its control system, and the evolutionary algorithm used to set the free parameters of the robot's control system. The characteristics of the simulated robot are identical to that described in a previous work [11]. In this paper, however, we present an extended version of the control system and new experimental results that show, in particular, how evolved robots are able to adapt their gait on the fly on the basis of the current circumstances.

2.1 The Hexapod Robot

The simulated robot (Fig. 1) consists of a main body (with a length of 20 cm, a width of 4 cm, and a height of 1.5 cm) and 6 legs.

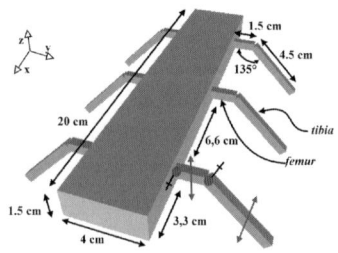

Fig. 1. The simulated hexapod robot. The grey circles shown on the bottom-right side of the picture indicate the position of the joints, while the grey arrows indicate their rotational axis.

Each leg consists of two segments (a "femur" and a "tibia" with a length of 1.5 and 4.5 cm respectively) and has two motors controlling two corresponding joints (the body-femur and the femur-tibia joints). The femur and the body-femur joint allow the robot to raise its central body from the ground and to move the tibia up and down. The body-femur joint is a motorized hinge joint with rotational axis parallel to the x-axis that can rotate from - $\pi/16$ to + $\pi/16$ rad. The femur-tibia joint allows it to move the tibia forward or backward. It is a motorized hinge joint that rotates from - $\pi/8$ to + $\pi/8$ rad with respect to its own axis (i.e. an axis rotated of $\pi/4$ rad with respect to yz-plane). The motors controlling the joints can apply a maximum torque of $0.03Nm$ at maximum speed of 3100 rpm in both directions. For each leg, two simulated position sensors detect the current angular position of the corresponding joint. The total weight of the simulated robot is 387g. Gravity force is –9.8 m/sec^2. The environment consists of a flat surface. The robot and the robot/environment interaction were simulated by using the VortexTM toolkit (Critical Mass Labs, Canada), that allows to realistically simulate the dynamics and collisions of rigid bodies in 3D.

2.2 The Control System

The robot is controlled by a distributed control system consisting of six homogeneous neural modules, located at the junction between the main body and the legs, that control the six corresponding legs (Fig. 2).

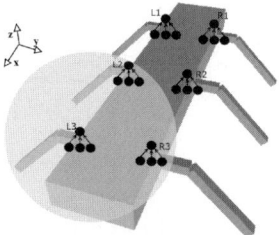

Fig. 2. The robot and its control system consisting of 6 neural modules. L1, L2, and L3 indicate the front, middle and rear leg located on the left side of the robot. R1, R2, and R3 indicate the front, middle and rear leg located on the right side of the robot. The grey circle represent a possible range of diffusion of the signal produced by one neural module (i.e. the neural module controlling the L3 leg).

The six neural modules are identical (i.e. have the same architecture and the same free parameters) and have access to local sensory information only. More specifically, each neural module has access to the current angular position and controls the frequency of oscillation of the two joints of the corresponding leg. Neural modules communicate between themselves by producing signals and by detecting the signals produced by other neural controllers located within a given Euclidean distance. Signals thus are similar to gaseous neuro-transmitters such as nitric oxide that are released by neurons and affect other neurons located nearby in a diffuse manner (see [12], [13], [14]). Signal transmission is instantaneous.

Each of the twelve motors neurons produces a sinusoidal oscillatory movement with a variable frequency of the corresponding joint, within the joint's limits. More specifically, the current desired position of a corresponding joint is computed according to the following equation:

$$pos(t) = \sin (V(t) \cdot t + \varphi) \tag{1}$$

where $pos(t)$ indicates the desired angular position of the joint at time t, $V(t)$ (that ranges between 7 and 14 Hz) indicates the current frequency of the oscillator, and φ indicates the starting position of the joint. The output of the neurons is normalized within the range of movement of the corresponding joint and is used to encode the desired position of the corresponding joint. More precisely, motors are activated so as to reach a speed proportional to the difference between the current and the desired position of the joint (maximum motor speed is 3100 rpm, maximum torque is 0.03Nm). We decided to use an high frequency range ([7, 14] Hz) and to update the state of the sensors and motors at an high rate (every 1.5 ms) to avoid instabilities arising from the calculation of the dynamic of the robot/environmental interaction and to reduce the time required to test individuals' behaviour in simulation.

Each neural module has six input neurons directly connected to six output neurons (Fig. 3).

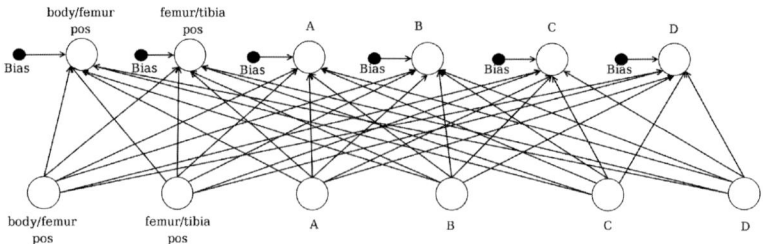

Fig. 3. The topology of each neural module. The six input neurons indicated in the bottom part of the picture encode the current angular position of the two joints of a leg and signal A, B, C and D (see text). The six output neurons are indicated in the top part of the picture. The first two modulate the frequency of oscillation of the two corresponding motorized joints and the others four determine whether or not the signal A, B, C and D are produced.

The input neurons encode the current angular positions of the two joints of the corresponding leg (normalized in the range [0.0, 1.0]) and whether signals, produced by other neural modules, are detected. Each neural module can produce four different signals (A, B, C and D) that diffuse and can be detected up to a certain distance (D_a, D_b, D_c and D_d, in the case of signal A, B, C and D, respectively). The intensity of the detected signal is linearly dependent from the distance of emitting neural module, and vary within 0 (when distance from emitting neuron is D) and 1.0 (when distance from emitting neuron is 0). Furthermore, detected signal is linearly proportional to the number of neural modules that are currently producing the corresponding signal located within the corresponding maximum diffusion distance.

The activation of output neurons is computed by using a standard logistic function. The first two output neurons determine how the frequency of oscillation of the two corresponding joints varies. More specifically, each time step (i.e. each 1.5ms), the frequency of oscillation of a joint can vary by an amount whose range is [-1.4Hz, +1.4Hz] according to the following equation:

$$V(0) = Val \qquad\qquad 7 \leq Val \leq 14$$

$$V(t) = V(t-1) + \begin{cases} (Out-0.75) \cdot 1.4 & Out \geq 0.75 \\ 0 & 0.25 < Out < 0.75 \\ (Out-0.25) \cdot 1.4 & Out \leq 0.25 \end{cases} \qquad (2)$$

Where Val indicates the initial value of frequency of a joint that is randomly set within the range, Out indicates the output of the corresponding motor neuron, and $V(t)$ indicates the current frequency, $V(t-1)$ indicates the frequency at the previous time step. Frequency is bounded in the range [7Hz, 14Hz], i.e. variations that exceed the limits are discarded. This means that each leg oscillates at a given frequency (within a range) and that each neural module can accelerate or decelerate the frequency of oscillation of the corresponding leg by a fixed amount each time step. The other four output neurons determine whether or not signal A, B, C and D are produced. More specifically, signal A, B, C and D are produced when the output of the corresponding output neuron exceeds the corresponding threshold (T_a, T_b, T_c and T_d in the case of signal A, B, C and D, respectively).

2.3 The Evolutionary Algorithm

The free parameters of the neural modules are evolved through an evolutionary algorithm. Robots were selected for the ability to walk along a straight direction as far as possible. Each robot was allowed to "live" for 5 trials, each lasting 3000 ms (i.e. 2000 time steps of 1.5 ms). The state of the sensor and motor neurons, the torque applied to the motors, and the dynamics of robot/environment interaction are updated each time step (i.e. each 1.5 ms). At the beginning of each trial: the main body of the robot is placed at a height of 3.68 cm with respect to the ground plane (i.e the whole robot floats in the air at 0.5 cm from the ground). The initial position of the twelve joints and the initial desired velocity of each corresponding motor is set randomly within the corresponding range. The fitness of each robot is computed by measuring the Euclidean distance between the initial and final position of the centre of mass of the robot during each trial. The total fitness is computed by averaging the distance travelled during each trial.

The initial population consisted of 100 randomly generated genotypes that encoded the connection weights and the biases of a neural module, the maximum distance of diffusion of the four signals (D_a, D_b, D_c and D_d), and the thresholds that determine when signals are produced (T_a, T_b, T_c and T_d). Each parameter is encoded as real number. Connection weights and biases, diffusion distances of signals, and thresholds that determine signal emission are normalized within the following ranges: [-15.0, +15.0], [0.0, 10.0], [0, 1.0], respectively. Each genotype is translated into 6 identical neural modules that are embodied in the robot and evaluated as described above. The 20 best genotypes of each generation were allowed to reproduce by generating five copies each, with 3% of their genotype value replaced with a new randomly selected value (within the corresponding range). The evolutionary process lasted 300 generations (i.e. the process of testing, selecting and reproducing robots is iterated 300 times). The experiment was replicated 10 times starting from different, randomly generated, genotypes.

3   Obtained Results

By analysing the results of the evolutionary experiments we observed that evolved robots display an ability to walk effectively, in all replications of the experiment. In particular, evolved robots display an ability to quickly coordinate the phases and the frequency of oscillation of their twelve motorized joints by converging toward a tripod gait independently from the initial position of the joints (see Fig. 4).

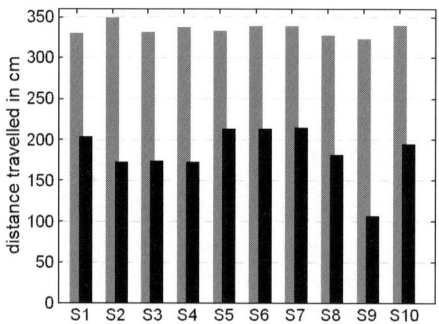

Fig. 4. Average distance travelled by the best robot of each replication in a normal and in a test condition (grey and black histograms respectively) in which the robot is loaded with an additional weight corresponding to 1.5 times the robot's body weight. Average results for 100 trials each lasting 3sec. The robots of all replications display a tripod gait when tested in a normal condition. In the test condition, the robots of replication S1, S5, S6, S7, and S10 display a tetrapod gait. The robots of the other replications, instead, by not being able to select an appropriate gait when loaded with additional weight, display lower performance in this condition.

Surprisingly, we observed that evolved robots generalize their ability to walk in situation in which they have to carry a weight equal to 1.5 of robots' body weight (see Fig. 4). Interestingly, in some of the replications, evolved robots converge on a tripod gait (when they are not loaded with additional weight) and on tetrapod gait (when they are loaded with the additional weight, see Fig. 5).

This implies that evolved robots, as real insects, select a tripod gait in normal conditions and a tetrapod gait when they are loaded with additional weight.

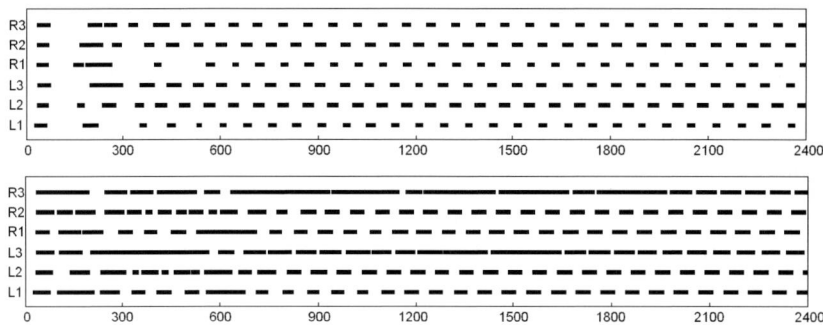

Fig. 5. A typical behaviour exhibited by an evolved robot of one of the best replications during two trials in which the robot is tested in a normal condition or in a test condition in which it is loaded with an additional weight (top and bottom figures, respectively). At the beginning of the trial the position of the joints and frequency of oscillation are randomly initialised within limits. The black lines indicate the phases in which the tibia of the corresponding leg touch the ground. Legs are labelled with L for left and R for right and numbered from 1 to 3 starting from the front of the insect. The horizontal axis indicates time in milliseconds. Gaits remain stable after 2400 ms (results not shown for space reasons). Please notice that these pictures do not indicate the trajectories of the robot in space but only the phase in time during which the legs touch the ground.

The ability to converge on a tripod or a tetrapod gait in the two circumstances play a functional role since the tripod gait is faster when the weight of the robot is not too high but is ineffective when the robot is loaded with additional weight. Indeed, as shown in Fig.4, the robots that are not able to switch to a tetrapod gait when they are loaded with additional weight display significantly worse performance. This can be explained by considering that in the tripod and in the tetrapod gait robots are supported by at least three or four legs, respectively. In the tetrapod gait therefore, the robot can exploit the power produced by four legs rather than three legs at the same time. In the tripod gait, in fact, the front and rear leg of one side and the middle leg of the other side perform they swing movement at the same time and the three other legs are in anti-phase. In the tetrapod gait, instead, a "wave" of swing movements passes along the body from rear to front.

The dynamical behaviour produced by the walking robots does not only result from the interaction between the six neural modules that control the six corresponding legs but also from the dynamics originating from the interaction between the robot body and the environment. Indeed, the way in which the actual position of the joints varies in time (Fig. 7) is influenced not only from the variation of the desired joint position (Fig. 6) but also from the forces arising from the collision between the legs and the ground. These forces are influenced by several factors such us the actual orientation of the robot with respect to the ground, the total weight of the robot, the current velocity of the robot, the characteristics of the ground, etc.

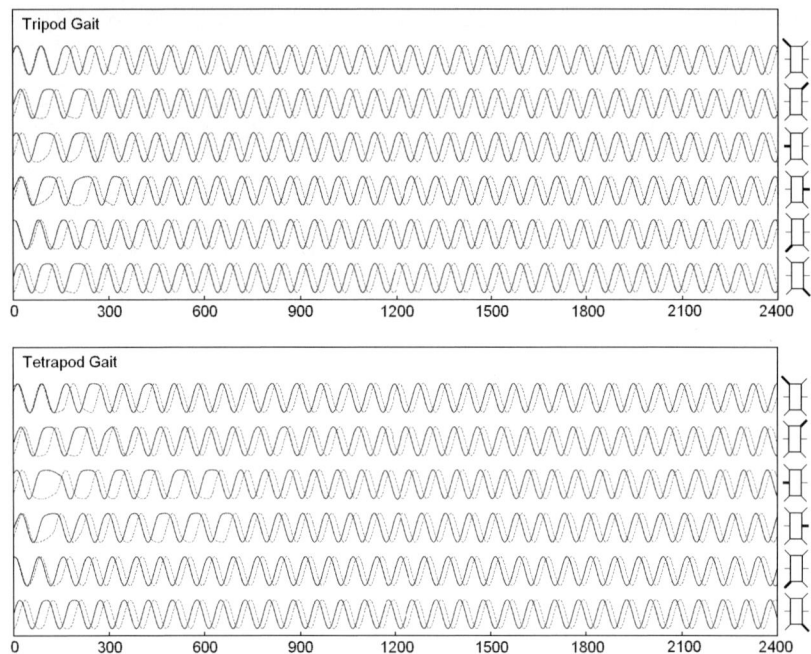

Fig. 6. Desired angular position of the twelve joints during the same trials shown in Figure 5 (top: data for the test in a normal condition, bottom: data for the test with additional weight). Each line indicates the desired angular position of the joints of the leg indicated with a dark line in the right part of the Figure. Full lines and dotted lines indicate the position of the body-femur and femur-tibia joints, respectively. High values indicate positions in which the femur is elevated with respect to the main body and positions in which the tibia is oriented toward the front of the robot.

The results described above refer to robots that have been evolved in a normal condition (in which they were never loaded with additional weight) and have been tested in a normal and over-weight condition (in which in half of the trials they were loaded with additional weight and in half of the trials they were not). By evolving and testing the robots in a normal and over-weighted condition we observed that evolved robots displayed lower performance on the average (with a loss of about 20% with respect to the results shown above). Moreover, in this new experiment evolved robots always displayed tetrapod gaits and were not able to adapt their gait on the fly so to select a tripod gait when tested in normal conditions. This results can be explained by considering that in the latter experiment evolving robot converge on a local minima, i.e. a simple solution that allow them to reach good but sub-optimal performance on the basis of simple control mechanisms. This hypothesis is also supported by the analysis reported in the next sections that suggest how tetrapod gait tend to easily emerge as a result of the effects of the collisions with the ground without necessarily requiring control mechanisms that allow the legs to effectively coordinate and synchronize through signals.

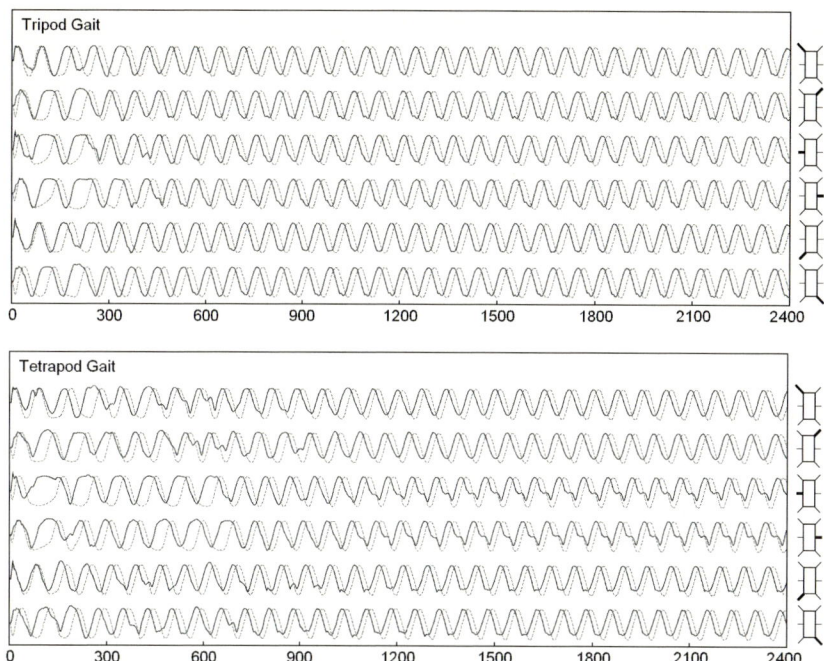

Fig. 7. Actual angular position of the twelve joints during the same trials shown in Fig. 5 and 6 (top: data for the test in a normal condition, bottom: data for the test with additional weight). Each line indicates the actual position of the joints of the leg indicated with a dark line in the right part of the Figure. Full lines and dotted lines indicate the position of the body-femur and femur-tibia joints, respectively. High values indicate positions in which the femur is elevated with respect to the main body and positions in which the tibia is oriented toward the front of the robot.

In another replication of the experiment we verified that the role of space in modulating the effect of signals was really necessary to achieve effective results. More precisely, by running a replication of the experiment in which the signals produced by each module affected all other modules in the same way (independently from the distance between modules) we observed that evolved robots displayed significantly lower performance and were never able to converge on stable gaits.

4 Analysis of the Mechanisms That Lead to Leg Coordination and Gait Selection

To understand the mechanisms that lead to the synchronization of the twelve joints, we analysed the behaviour of each neural module and the interaction between different neural modules mediated by signals (i.e. the conditions in which signals are produced and the effects of signals detected). Here we report the analysis conducted in the case of the evolved individuals already described in Figure 5-7. The analysis conducted on individuals of other replications showed qualitatively similar, although in some case slightly more complex, strategies. We will first described how the six legs converge toward a tripod gait in the normal condition and then how they converge on a tetrapod gait when the robot is loaded with additional weight.

As could be expected, the synchronization between the two joints of each leg is achieved within each single neural controller. More specifically: (a) the body-femur joint decelerates when it is elevated and the tibia is oriented toward the rear, and (b) the femur-tibia joint decelerates when the body-femur joint is lowered and the tibia is oriented toward the rear of the robot. Deceleration results both as the effect of output of the corresponding neural module and as a result of the effect of collisions. The combination of these two mechanisms leads to a stable state, that correspond to stable phase observed after coordination, in which the protraction movement of the tibia is performed when the body-femur joint is elevated the retraction movement is performed when the body-femur joint is lowered.

4.1 Analysis of the Mechanisms That Lead to a Tripod Gait

Although neural modules can produce and detect up to four different signals, the evolved individual shown in Fig. 5-7 only produces one of the four signals: signal B. In the other replications of the experiment, evolved robots use 1 or 2 signals. Interesting however, significantly lower performance have been observed in other experiments in which neural modules were allowed to produce and detect only two signals (result not shown). This suggest that the possibility of using many signals plays a crucial role during the first evolutionary phases despite only 1-2 signals are exploited by evolved individuals.

Since the maximum distance of diffusion of signals is 7.81 cm, in the case of the robot shown in Fig. 5-7, the signal produced by each leg affects the contra-lateral leg of the same segment, the previous and succeeding legs of the same segment, and the previous and succeeding legs of the contra-lateral segment (when present). Since the amount of the signal detected is proportional to the distance (4.0cm, 6.6cm, and 7.71cm respectively) the impact of produced signal is larger on the contra-lateral leg of the same segment, smaller but still significant on the previous and succeeding legs of the same segment, and almost negligible on the previous and succeeding legs of the contra-lateral segment. If we ignore the negligible effect on previous and succeeding legs of the contra-lateral segment, this means that the signal produced by a leg of one group ([L1,L3,R2] or [R1,R3,L2]) affects only the legs of the other group that should be in anti-phase in a tripod gait. The legs that are affected by a signal are 2 out of 3 legs in the case of legs [L1,L3,R1,R3] and 3 out of 3 legs in the case of legs [L2,R2].

To explain how the six legs coordinate we should explain why uncoordinated states are unstable and lead to coordinated phases (through relative acceleration/deceleration of the joints) and why coordinated states are stable.

The latter aspect can be explained by considering that, during coordinated phases, legs belonging to the two groups (A and B) are in phase within the group and in anti-phase between groups. Legs of group A produces a signal when their tibia are oriented toward the rear and their femurs are elevated (i.e. at the starting of the protraction phase) and they reduce their velocity only when their tibia are oriented toward the rear and they detect a signal produced by legs of group B. Since the legs of the two groups are in anti-phase and the signals are produced in an alternate way by the legs of the two groups, in coordinated phases signals do not produce acceleration/deceleration effects.

To explain the former aspect (i.e. why uncoordinated phases are instable) let us consider the case in which the legs of groups A and B, during a retraction phase, have both their tibia oriented toward the rear but the legs of group A are slightly advanced with respect to the legs of group B. Since the interval in which legs emit the signal is larger than the interval in which the legs decelerate when they detect a signal, the deceleration effect of the legs of the group A on the legs of the group B is longer than viceversa. This implies that phase distance between the legs of the two groups tend to increase when they are in phase or almost in phase until the legs of the two groups reach the stable state described in the previous paragraph.

4.2 Analysis of the Mechanisms That Lead to a Tetrapod Gait

To understand the mechanisms that allow robots to switch from a tripod to a tetrapod gait when loaded with additional weight we should consider that the additional weight increases the intensity of friction in particular on the femur joints that are no longer able to reach their extreme posterior position (Fig. 7, bottom). Since legs emit the signal when the femur is elevated, this implies that the overall speed of the legs tend to decrease due to the fact that signals are produced for a longer time period.

Another factor to be considered is that when the body of the robot is not perfectly aligned with respect to the ground plane, different legs are subjected to stronger of weaker friction forces. For example, when the body of the robot is inclined toward the front, with respect to the rostro-caudal axis, the frontal legs are subjected to a stronger friction and, as a consequence, these legs produce signals for a longer time period.

The legs that are particularly stressed (due to current inclination of the body and to the fact that are performing a retraction movement) tend to slow down nearby legs that are also performing a retraction movement. This implies that legs that are under stress tend to recruit to their phase nearby legs. The final result is that the legs of the same side and of adjacent segments do not move in perfect anti-phase, as in the tripod gait, but with a partially overlapping phase.

5 Conclusions

In this paper we present a distributed control architecture for a simulated hexapod robot with twelve degrees of freedom consisting of six homogeneous neural modules controlling the six corresponding legs that only have access to local sensory information and that interact by producing and detecting signals that diffuse in space.

The free parameters of the homogeneous neural modules, that regulate the frequency of oscillation of the corresponding leg and the signals that are emitted on the basis of the current position of the leg and of the signals detected have been set through an evolutionary method in which variation of the free parameters are retained or discarded on the basis of the global behavior exhibited by the robot in the environment. This method allows evolving robots to select solutions that exploit properties emerging from the interaction between the neural modules and between the robot and the environment.

The analysis of the evolved neural controllers indicates that the six homogeneous neural controllers converge on an appropriate gait on the basis of extremely simple control mechanisms. In some of the replications, in particular, coordination and gait selection is achieved on the basis of a single signal. This implies that a single rule (that accelerates or decelerates the frequency of oscillation of nearby legs depending on the state of the leg that detect the signal) is sufficient to converge on an a stable and effective gait.

Finally, we observed that evolved robots generalize their ability to produce an effective walking behaviour also when they are loaded with additional weight by displaying an ability to select a tripod or a tetrapod gait in the normal condition and in test conditions in which they are loaded with an additional weight, respectively.

Overall, the obtained results suggest that an hexapod robot can be controlled on the basis of fully homogeneous distributed control system in which the interaction between neural modules is only regulated by the distribution in space.

Acknowledgments

The research has been supported by the ECAGENTS project funded by the Future and Emerging Technologies programme (IST-FET) of the European Community under EU R&D contract IST-1940.

References

1. Gallagher, J., Beer, D.R., Espenschied, K., Quinn, R.D.: Application of evolved locomotion controllers to a hexapod robot. Robotics and Autonomous Systems. (1996) 19: 95-103.
2. Ijspeert, A.J., Crespi, A., and Cabelguen, J.M.: Simulation and robotics studies of salamander locomotion. Applying neurobiological principles to the control of locomotion in robots. Neuroinformatics. (2005) 3(3):171-196.
3. Hulse, M., Wischmann, S., Pasemann F.: Structure and function of evolved neurocontrollers for autonomous robots. Connection Science. (2004) 16 (4): 249-266.
4. Nolfi, S., Floroeano, D.: Evolutionary Robotics: The Biology, Intelligence, and Technology of Self-Organizing Machines. Cambridge, MA: MIT Press/Bradford Books (2000)
5. Kamimura, A., Murata, S., Yoshida, E., Kurokawa H., Tomita K., Kokaji S.: Self-reconfigurable modular robot - experiments on reconfiguration and locomotion. In T. J. Tarn et al. (Eds), Proceedings of the IEEE/RSJ International Conference on Intelligent Robots and Systems. (2001) New York: IEEE Press.
6. Yim, M., Zhang, Y., Duff, D.: Modular robots. IEEE Spectrum. (2002) 30-34.
7. Sims, K.: Evolving 3D morphology and behavior by competition. In R. Brooks & P. Maes (Eds), Proceedings of Fourth Conference on Artificial Life. (1994) Cambridge, MA:MIT Press.
8. Bongard, J., Pfeifer, R.: Evolving complete agents using artificial ontogeny. In Hara F. & R. Pfeifer (Eds), iMorpho-functional Machines: The New Species: Designing Embodied Intelligence. Berlin: Springer Verlag (2003)
9. Hornby, G. S., Pollack, J. B.: Creating high-level components with a generative representation for body-brain evolution. Artificial Life. (2002) 8(3):223-246

10. Bianco R., Nolfi S. (2004). Toward open-ended evolutionary robotics: evolving elementary robotic units able to self-assemble and self-reproduce. *Connection Science*, 4: 227-248.
11. Mazzapioda, M., Nolfi, S.: Synchronization whitin Homogeneous Neural Modules Controlling a Simulated Hexapod Robot. Proceedings in AlifeX. (2006) In press.
12. Elphick, M.R., Kemenes, G., Staras, K., O'Shea, M.: Behavioural role for nitric oxide in chemosensory activation of feeding in a mollusc. Journal of Neuroscience (1995) 15(11):7653-7664
13. Elphick, M.R., Williams L., O'Shea M.: New Features of the locust optic lobe:evidence of a role for nitric oxide in insect vision. Journal of Experimental Biology (1996) 199: 2395-2407
14. Husbands, P., Philippides A., Smith T.M.C., O'Shea M.: Volume Signalling in Real and Robot Nervous Systems. Theory in Biosciences (2001), 120: 253-269

Computer Simulation of a Climbing Insectomorphic Robot*

Yury F. Golubev and Victor V. Korianov

Keldysh Institute for Applied Mathematics,
Russian Academy of Sciences,
Miusskaya pl. 4, Moscow, 125047 Russia
golubev@keldysh.ru, korianov@keldysh.ru

Abstract. Algorithms for control of a six legged insectomorphic robot able to overcome a sequence of high obstacles are developed. The sequence of obstacles includes vertical upright cylindrical column, shelf with a horizontal supporting plate standing right up to the column, narrow horizontal beam, connecting the shelf with the same another one on the supporting plate level, vertical right corner (corner of a building). The robot doesn't have the special contact equipment in its feet (vacuum suckers). Developed algorithms were worked-out by means of computer simulation of robot's 3D-dynamics.

1 Introduction

The ability of a walker to overcome a terrain with a conglomeration of obstacles can be formed by teaching the robot to overcome isolated obstacles as well as reasonable combinations of obstacles step by step. This approach to training is widely used in a sport like mountain climbing, competitions of firemen and so on. Some examples of overcoming a terrain with small roughness are given in [1]. Overcoming isolated obstacles by means of walker's jump is considered in [2,3]. Also machines with vacuum suckers are developed intensively as they allow moving along vertical walls [4,5]. The requirement of static stability is unimportant for this machines. But they need complicated devices for creating vacuum. It seems that in some particular cases walking machines could move along vertical constructions of significant height simply using Coulomb friction forces as animals does it. Methods for overcoming of vertical column by means of friction forces are presented in [6]. The design of insectomorphic walker motion for surmount combination of two obstacles was presented in [9,10]. This combination consisted of a vertical column, a lofty horizontal shelf edged by a vertical wall (a precipice), two lofty shelves connected by a narrow horizontal beam. In this paper the next step of designing robot's motion is presented for amount its ability to overcome the combination of the previous obstacles with the vertical right angle (angle of a building).

---

* This work was supported by the Russian Foundation for Basic Research, projects nos. 04-01-00065 and 04-01-00105.

2   Robot's Kinematics

The robot consist of parallelepiped-shaped rigid body with dimensions a — side of the body (length), b — front or rear edge (width), c — height. Six identical insectomorphic legs are symmetrically attached to the sides of the body. Points in which the legs are attached (legs attachment points) are located uniformly along the sides. Each leg consists of two links: hip, length l_1 and shank, length l_2 (Fig. 1).

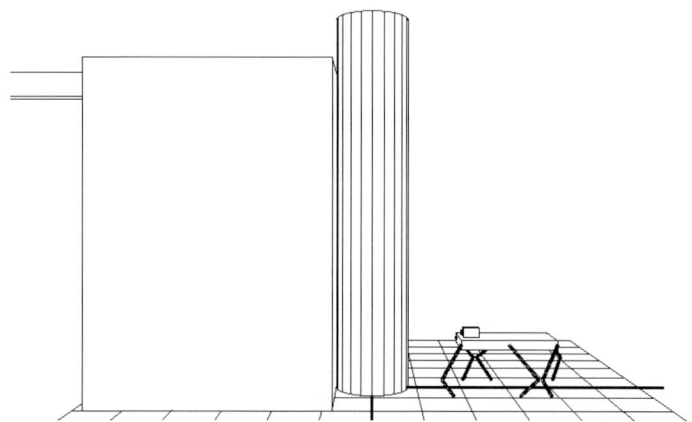

Fig. 1. Robot and the sequence of obstacles

The body and the links of the legs have some volume and mass by gravity. The position of a leg is determined by three joint angles, two of which (α_i, β_i) defines the position of a hip relatively to the body, and thrid (γ_i) — of a shank relatively to the hip. Thus, the total degrees of freedom of the robot is 24. The joint angles of the leg numbered i can be found unambiguously of vector $\mathbf{r}_i$, directed from the attachment point to the foot on the assumption of an orientation of the knee. By default the knee is oriented so that if the foot moves forward the knee moves forward. Joint angles are determined by the inversion of the following correlations:

$$r_{xi} = x_i - p_{xi} = [(-1)^i l_1 \sin\beta_i - l_2 \sin(\beta_i + \gamma_i)] \sin\alpha_i,$$
$$r_{yi} = y_i - p_{yi} = [(-1)^i l_1 \sin\beta_i - l_2 \sin(\beta_i + \gamma_i)] \cos\alpha_i,$$
$$r_{zi} = z_i - p_{zi} = [(-1)^{i+1} l_1 \cos\beta_i + l_2 \cos(\beta_i + \gamma_i)],$$

where (x_i, y_i, z_i) — coordinates of the i-th foot, $(p_{xi}, , p_{yi}, p_{zi})$ — coordinates of the i-th attachment point in the body reference frame.

Dimensions of the robot meet the following condition:

$$a : b : c : l_1 : l_2 : r = 1 : 0.5 : 0.1 : 0.5 : 0.33 : 0.4,$$

where r — radius of the column.

3   Servo-Control System

We assume that robot is equipped with the electromechanical drives in the joints and has full access to the following information: the geometry of the obstacles, own position relatively to the obstacles, joint angles and velocities. The programmed values of the joint angles are generated by the algorithm of control. The algorithm is not strictly fixed, the information about the actual robot configuration during the motion essentially used. For realization of programmed values of joint angles the servo-control method is used [7]. Required motion generates as a servo-constraint, which robot will be aimed to realize by setting the control voltage on the drives. Transient process for elimination the difference between the programmed and the actual values of an angles is realized by the linear regulator:

$$M = U - c_e\dot{\varphi}, \quad U_0 = -\chi_1(\varphi - \varphi_0) - \chi_2\dot{\varphi}, \quad U = \begin{cases} \text{sign}(U_0)U_{\text{max}}, & |U_0| > U_{\text{max}}, \\ U_0, & |U_0| \leqslant U_{\text{max}}, \end{cases}$$

where M — control joint torque, U — the moment due to voltage on the drive, $\chi_1 > 0$, $\chi_2 > 0$ — coefficients of amplification, providing the speed and the quality of the transient process, c_e — coefficient due to the self-induction, $\varphi_0 = \varphi_0(t)$ — programmed value of some angle, φ, $\dot{\varphi}$ — its actual value and velocity.

4   Interaction with a Supporting Surface

The robot can contact with a supporting surface only by feet. The model for reaction of the surface has the following form:

$$\mathbf{F} = \mathbf{F}_\tau + \mathbf{N}, \quad \mathbf{F}' = c_n(\mathbf{r}_T - \mathbf{r}_C) - d_n\varepsilon\mathbf{v}_C, \quad \mathbf{N} = \mathbf{n}(\mathbf{F}' \cdot \mathbf{n}),$$
$$\mathbf{F}'_\tau = \mathbf{F}' - \mathbf{N}, \quad \mathbf{F}_\tau = \begin{cases} \mathbf{F}'_\tau, & F'_\tau/N < k_f, \\ \mathbf{F}'_\tau Nk_f/F'_\tau, & F'_\tau/N \geqslant k_f, \end{cases}$$

where $\mathbf{F}_\tau$ — tangential component of the reaction, $\mathbf{N}$ — its normal component, $\mathbf{F}'$ — elastic-viscous force of interaction, c_n — coefficient of the elastic part, $\mathbf{r}_T$ — radius-vector of the initial point of interaction with a surface, $\mathbf{r}_C$ — current position of a foot during the contact, d_n — constant coefficient, $\varepsilon > 0$ — the quantity of the deepening of a foot into a surface.

5   Motion Design

Legs transfers are realized on the base of the plane step cycles (Fig. 2) [8].
During the transfer the feet are in the planes $C'_i x'_i y'_i$, and

$$\boldsymbol{\rho}_i = \mathbf{r}_i(t_1) - \mathbf{r}_i(t_0), \ C'_i x'_i \| \boldsymbol{\rho}_i, \ C'_i y'_i \| \boldsymbol{\rho}_i \times (\mathbf{e}_i \times \boldsymbol{\rho}_i), \ \mathbf{r}_{Ci} = (\mathbf{r}_i(t_0) + \mathbf{r}_i(t_1))/2,$$
$$x'_i = -d_x \cos\varphi, \ y'_i = (d_y + y_s)\sin\varphi - y_s, \ y_s = |AD|, \ \varphi = \varphi(t), \ x^* = |CB|,$$

where $\mathbf{r}_i$, $\mathbf{r}_i(t_1)$ — radii-vectors of initial and final foot positions respectively, $\mathbf{e}_i$ — unit vector of the orientation of the step cycle plane.

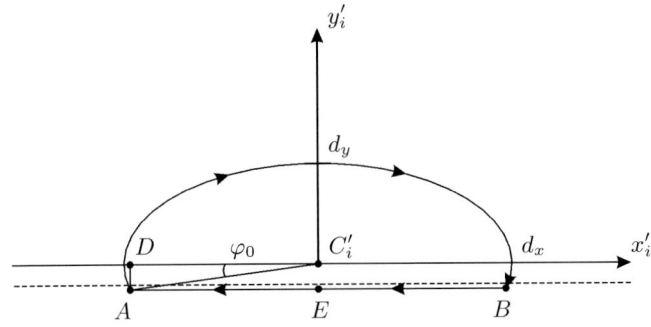

Fig. 2. Step cycle

Described above method for trajectory design is used on the regular gaits. In irregular cases there are additional smoothing elements near the beginning and the end of the transfer:

$$\mathbf{r}_i(t) = (1 - \lambda)\mathbf{r}_i(t_0) + \lambda\mathbf{r}_i(t_1) + \widehat{k}(y' + y_s)\frac{\boldsymbol{\rho}_i \times ((\mathbf{e}_i/|\mathbf{e}_i|) \times \boldsymbol{\rho}_i)}{|\boldsymbol{\rho}_i \times ((\mathbf{e}_i/|\mathbf{e}_i|) \times \boldsymbol{\rho}_i)|},$$

where $0 \leqslant \lambda \leqslant 1$, $\widehat{k}$ — coefficient of stretching. Function $\lambda(t)$ has the special form [9].

5.1 Climbing a Vertical Corner

Consider the problem of climbing from a horizontal plane to a vertical two-sided corner that has the opening angle equal to $\pi/2$. Initially the robot is located on a horizontal plane. The robot has to move to the corner, gripe it symmetrically relative to the bisector plane of the corner by legs from two sides, and continue the motion in the upward direction. Then the robot has to climb on the horizontal top of the corner. The total motion pattern in upward direction is determined by the gallop gait. Theoretical analysis shows that realization of such kind of motion is impossible with friction coefficient less than or equal to 1 [10]. Together with this, the results of computer simulation have shown that the friction coefficient 1.1 turns out to be sufficient for execution of the posed motion problem.

Let us take an absolute right Cartesian coordinate system $O\xi\eta\zeta$. The initial O placed on a horizontal supporting plane along the bisector of the corner, and has the distance to the vertex equal to r. The $O\zeta$ axis is directed upward. The bisector plane of the vertical two-sided angle coincides with the plane $O\eta\zeta$.

For a robot, the algorithm for climbing the corner is performed in several stages. For the stage with the number n, we take t_{n-1}, and t_n are the moments of its beginning and end, respectively, $t_{n+1} = t_n + t^*$. Denote by $\mathbf{r}_i(t_k) = (r_{\xi i}(t_k), r_{\eta i}(t_k), r_{\zeta i}(t_k))$ and $\mathbf{p}_i(t_k) = (p_{\xi i}(t_k), p_{\eta i}(t_k), p_{\zeta i}(t_k))$ the absolute radii-vectors of the feet and the attachment points with the number i at the time instance t_k. The computation of the programmed joint angles based on these data is performed as in [9,10].

At the initial instant t_0, the body is oriented along the axis $O\eta$:

$$\mathbf{p}_i(t_0) = \left((-1)^i b/2, \ \eta_f + a\lfloor(6-i)/2\rfloor/2, \ \zeta_f\right).$$

Here and in what follows, $\lfloor \dots \rfloor$ is the integer part of the number in the brackets, $\xi_f = 0$, η_f, ζ_f are the absolute coordinates of the front point of the body. The legs are at the initial position of the gallop gait. Let x_f be half the width of the robot track, and x^* be the length of the supporting part of the step cycle.

The body motion is constructed so that the point of the front side of the body moves in the upward direction, and the point of the back side is forced to move horizontally. Let $\tau = t - t_0$. Then, the motion law of the front point of the body for the first, second, and third stages is expressed by the formula $\zeta_f = \zeta_f(t_0) + \zeta_s$, where

$$\zeta_s = \begin{cases} \zeta_s', \ \zeta_s' < a, \\ a, \ \zeta_s' \geqslant a, \end{cases} \quad \zeta_s' = v_s \frac{x^*}{t^*} \begin{cases} \tau, & 0 < \tau \leqslant 1.5t^*, \\ 1.5t^*, & 1.5t^* < \tau < 2t^*, \\ \tau - 0.5t^*, \ \tau \geqslant 2t^*, \end{cases} \quad \eta_s = \sqrt{a^2 - \zeta_s^2}.$$

$$(1)$$

Stage 1: $(0 \leqslant \tau < t^*)$. The dimensionless coefficient v_s specifies the velocity of body motion. In the computations, we took $v_s = 4.4$. Stage 2 corresponds to the time $t^* \leqslant \tau < 2t^*$. It is clear that, in the second part of this stage, the body is motionless, and, at the third stage, the motion of the body continues.

The motion law of the points at which the legs are fixed at stages 1–3 has the form

$$\mathbf{p}_i(t) = \left(p_{\xi i}(t_{n-1}), \ p_{\eta 6}(t_{n-1}) + \eta_s\lfloor(6-i)/2\rfloor/2, \ p_{\zeta 1} + \zeta_s\lfloor(i-1)/2\rfloor/2\right).$$

The front legs at the first stage are moved on the supporting two-sided corner, and the radii–vectors of their supporting points are

$$\mathbf{r}_i(t_1) = \left((-1)^i x_f, \ 0, \ \zeta_2\right), \quad \zeta_2 = \zeta_f(t_0) + a - 5x^*, \quad i = 5, 6,$$

Stage 2: $(t^* \leqslant \tau < 2t^*)$. The front side of the body continues to move in the upward direction as was described above up to the time instant $t_0 + 1.5t^*$. The back legs are moved in the forward direction on the horizontal plane. Note that

$$\mathbf{r}_i(t_2) = \left(r_{\xi i}(t_1), \ r_{\eta i}(t_1) - 4.75x^*, \ r_{\zeta i}(t_1)\right), \quad i = 1, 2.$$

Stage 3: The body continues to move according to formula (1). The middle legs are moved on the corner at the supporting points, and the radii-vectors are

$$\mathbf{r}_i(t_3) = \left((-1)^i x_f, \ 0, \ \zeta_1\right), \quad i = 3, 4,$$

where $\zeta_1 = \zeta_2 - 0.7a + x^*$. The source motion parameters are chosen so that $\zeta_1 > 0$ and the corresponding supporting points are located as low on the angle as possible, but well out from the middle legs.

Stage 4: By the moment of the beginning of this stage, the body has already been moved to a vertical position. The front legs are moved in the upward direction to the supporting points

$$\mathbf{r}_i(t_4) = \left(r_{\xi i}(t_3), \ r_{\eta i}(t_3) + 9x^*, \ r_{\zeta i}(t_3)\right), \quad i = 5, 6.$$

This increases the arm between the middle and front supporting points to provide the motion of the back legs.

Stage 5: This stage is the most difficult from the standpoint of maintaining equilibrium. To prepare it, the velocity of body motion is chosen so that, by the moment of the beginning of the fifth stage, the body is in a vertical position, the back legs remain in the initial position on the horizontal plane locating at a position that is sufficiently close to the corner, and the front and middle feet are established on the edges of the corner at points that are substantially distant each from other in height. This provides a sufficiently large arm to compensate the moment of the gravitational force by the support reactions. The back legs are moved onto the corner at the supporting point

$$\mathbf{r}_i(t_5) = ((-1)^i x_f, \, 0, \, \zeta_0), \quad \zeta_0 = \zeta_f(t_0) - 3x^*, \quad i = 1, 2.$$

The legs are carried in all stages according to plane step cycles. At stages 1–4 and 6–8, the ordinate axes of step cycles are parallel to the axis $O\xi$, and, at stage 5, they are specified by unit direction vectors

$$\mathbf{e}_i = ((-1)^i, \, 0, \, 0.9)/|\mathbf{e}_i|.$$

At stages 6–8, the robot is steered to the initial position for moving upward on the angle. In the course of motion, it is provided that the position of the body relative to the corner is leveled by the formula

$$p_{\eta i}(t) = [\eta_f - p_{\eta i}(t_{k-1})](t - t_{k-1})/t^* + p_{\eta i}(t_{k-1})$$

simultaneously with the motion of the body upward

$$p_{\zeta i}(t) = x^*(t - t_{k-1})/t^* + p_{\zeta i}(t_{k-1}).$$

Stage 6: The middle legs are carried to the supporting points with ζ-coordinates $r_{\zeta i}(t_6) = r_{\zeta 5}(t_5) - a/2 - 5x^*$, $i = 3, 4$.

Stages 7 and 8: At stage 7, the front legs are carried, and, at stage 8, the back legs are moved

$$r_{\zeta i}(t_n) = r_{\zeta i}(t_{n-1}) + 2x^*, \quad i = \begin{cases} 5, 6, \, n = 7, \\ 1, 2, \, n = 8. \end{cases}$$

After the end of stage 8, the robot moves upward with a modified gallop gait. The schedule of movements of the front and back legs is the same as in the standard gallop gait [7]. The middle legs change their position with a rate two times greater than the front and back ones so that their supporting points will be close to the supporting points of the next carried pair of legs. This reduces the load on the supporting legs.

5.2   Climbing on the Corner's Horizontal Top

The maneuver start when the ζ-coordinate of the planned position of the front feet is greater than $h - \varepsilon$, where h is the height of the corner, ε prevent the legs from touching the strict edge, on which the outer normal is ambiguous.

Stages 1-4: On the first four stages the position of the robot transforms into the initial position of the modified gallop gait. Positions of feet are moved equally down relatively to the edge. Thus the independence of the subsequent maneuver from the initial position of stage 1 is achieved. At the initial moment of time the difference $\delta = r_{\zeta 5}(t_0) - (h - \delta_k x^*)$, $\delta_k < 2$ is determined. The planned points of feet have the form $r_{\zeta i}(t_n) = r_{\zeta i}(t_{n-1}) - \delta$, with the corresponding modification for the additional middle legs transfer. The body moves so that at the end of stage 4 its center goes down by δ.

Stages 5, 6: First, the front legs are moved above the top of the corner:

$$\mathbf{r}_i(t_5) = (r_{\xi i}(t_4) + 0.05a\mu_{\xi i},\ r_{\eta i}(t_4) + 0.05a\mu_{\eta i},\ h + 0.1a), \quad i = 5, 6,$$

second, they are moved on the top's surface:

$$\mathbf{r}_i(t_6) = (r_{\xi i}(t_5) - 0.01a\mu_{\xi i},\ r_{\eta i}(t_5) - 0.01a\mu_{\eta i},\ h), \quad i = 5, 6.$$

Here $\boldsymbol{\mu}_i$ — outer normal to the side of the corner.

Stages 7, 8: On those stages the body moves up by $3x^*$ (on each stage) and the feet are moved up: middle by $2x^*$, stage 7 (Fig. 3), rear by $7x^*$, stage 8.

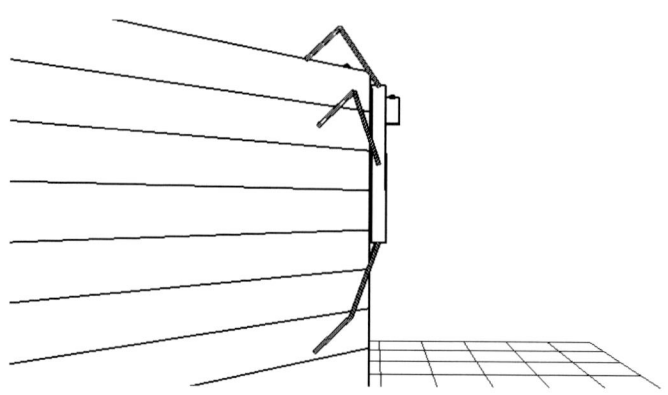

Fig. 3. Final position of stage 7

Stage 9: The body moves up by $2.5x^*$, the middle legs are moved to

$$\mathbf{r}_i(t_9) = \mathbf{r}_i(t_8) + (0.05a\mu_{\xi,7-i},\ 0.05a\mu_{\eta,7-i},\ 2.5x^*), \quad i = 3, 4,$$

located nearer to the body to gain the possibility of supporting polygon on the subsequent stages.

Stage 10: The body moves up by $2.5x^*$, front legs are moved on the top farther from the sides of the corner (Fig. 4). On that stage $\mathbf{e}_i = (0, 0, 1)$.

$$\mathbf{r}_i(t_{10}) = (r_{\xi i}(t_9) - 0.1a\mu_{\xi i},\ r_{\eta i}(t_9) - 0.1a\mu_{\eta i},\ h), \quad i = 5, 6.$$

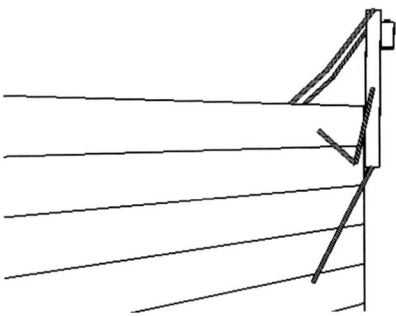

Fig. 4. Final position of stage 10

Stages 11, 12: One those stages the body moves from the vertical position to the horizontal one. Trajectory of motion body's center C is the arc of an ellipse:

$$\xi_C = 0, \, \eta_C = p_{\eta 3}(t_{10}) \cos \tau, \, \zeta_C = p_{\zeta 3}(t_{10}) + (h + \zeta_f/1.5 - p_{\zeta 3}(t_{10})) \sin \tau, \, \tau = (\pi/2)\lambda,$$

where ζ_f is ζ-component of the front point of the body on the initial position on the horizontal plane.

Absolute coordinates of the attachment points has the form

$$\mathbf{p}_i(t) = (p_{\xi i}(t_{10}), \, \eta_C - p_{xi} \sin \tau, \, \zeta_C + p_{xi} \cos \tau).$$

The rear legs are moved to the points (Fig. 5)

$$\mathbf{r}_i(t_{11}) = \mathbf{r}_i(t_{10}) + (0.05 a \mu_{\xi, 3-i}, \, 0.05 a \mu_{\eta, 3-i}, \, 8x^*), \quad i = 1, 2.$$

On the stage 12 the programmed motions are not assigned, but the initial position of the next stage is fixed.

Stages 13, 14: As on stages 5, 6 middle legs are moved to the top of the corner.

Stage 15: The body moves up to the level $h + \zeta_f + 0.1a$, to avoid the mutual intersection of the legs, rear legs are moved to the points

$$\mathbf{r}_i(t_{15}) = (p_{\xi i})(t_{14}) + (-1)^i 0.01 a, \, p_{\eta i}(t_{14}), \, r_{\zeta i}(t_{14})), \quad i = 1, 2.$$

As a result, the rear legs become straight and vertical. It's required to avoid the degeneration of the joint angles. On this stage $\mathbf{e}_i = (0, 1, 0)$.

Stage 16: Rear legs are moved to the narrow track:

$$\mathbf{r}_i(t_{16}) = ((-1)^i b/4, \, r_{\eta i}(t_{15}), \, h), \quad i = 1, 2.$$

At this stage the orientation of the knees on this stage is inverse relatively to the default orientation, i.e. if a foot moves forward the knee moves backward, as on motion along the horizontal beam.

Stage 17: The rear legs are moved to the top (Fig. 6), $\mathbf{e}_i = (0, 0, 1)$:

$$\mathbf{r}_i(t_{17}) = (r_{\xi i}(t_{16}), \, r_{\eta i}(t_{16}) - 0.3a, \, h), \quad i = 1, 2.$$

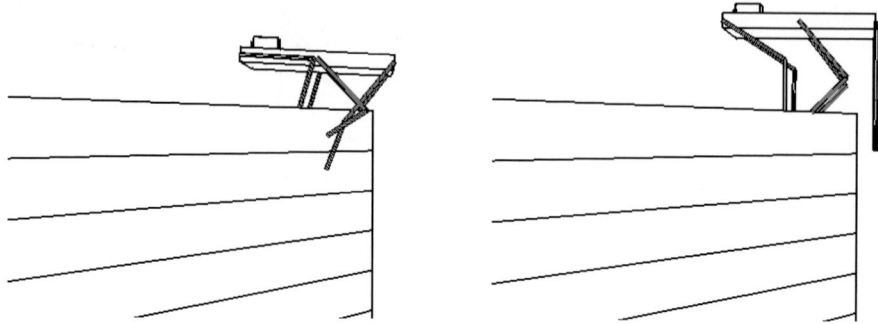

Fig. 5. Final positions of stages 11 and 15

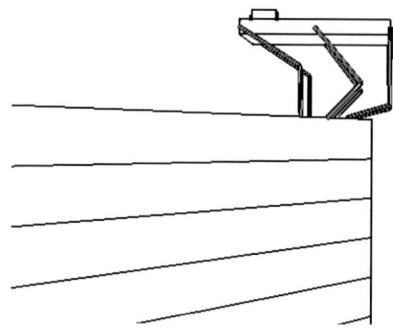

Fig. 6. Robot on the top of the corner

Thus, all six legs are placed on the top of the corner, and several gallop steps of front and middle legs (with the corresponding motion of the body to gain the static stability) may be performed. After that, the rear legs can be moved back to the wide track, and the robot will come to the comfortable position to perform future tasks.

6   Motion on a Horizontal Beam

6.1   Motion Design

Consider an obstacle composed by two parallel shelves of the same height with a horizontal top area. These top areas are connected at the same level by a narrow beam that is perpendicular to the vertical walls of the shelves. Suppose that, at the initial moment, the robot has a symmetric pose on one of the shelves before its edge, and the planes of the legs are perpendicular to the longitudinal axis of the body. The robot has to go from one shelf to another along the beam. The center O of the right absolute coordinate system $O\xi\eta\zeta$ is placed on the projection of the middle line of the beam on the supporting horizontal plane. The axis $O\eta$ is parallel to the middle line, and the axis $O\zeta$ is directed upward. The assumption that the beam is narrow means that, in the motion on the beam, the transversal size is approximately equal to the margin of static stability that is necessary for maintaining the robot balance using only the support reactions. Therefore, to provide robot balance on the beam, additional facilities for controlling its configuration are applied. The front and back legs are still applied to provide the support of the robot on the surface, and the middle legs work as a flywheel in order to provide the robot stability when the conditions of static stability are violated. We first consider the construction of the motion of the front and back pairs of legs.

Since the beam width is supposed to be considerably smaller than the body width, the tracks are to be under the robot body in the course of motion. For a small distance between the tracks, because of the danger of mutual intersection of symmetric legs, it is advisable to admit that the knees move in the direction opposite to the direction of movement of the feet.

The motion is executed according to the following stages: (1) the robot configuration is changed for the motion with a narrow track; (2) the robot goes on the beam with a four-legged diagonal gait to the another shelf; (3) the robot configuration is changed in the reverse way in order to go with a wide track.

Stage 1: The narrow track is determined by the beam width. The legs in triples are sequentially carried on a narrow track using step cycles whose plane is perpendicular to the longitudinal axis of the body. After that, the middle legs are straightened and steered to the horizontal position in order to have the opportunity to provide balance by them. After that, up to the termination of the motion on the beam, we set $\alpha_3 = \alpha_4 = 0$.

Stage 2: In the motion on the beam, the body moves in the forward direction only on the intervals when the front and back legs stand still. The legs are carried by a diagonal gait in pairs (1, 6) and (2, 5). At the initial time instant of the motion on the beam, the projection of the center of mass on the beam surface is at the point of intersection of the segments that connect the pairs of feet (1, 6) and (2, 5). When the pair (1, 6) is carried in the forward direction, only two legs provide the support, and there is no static stability. When this pair of legs has been carried, the body moves so that its center reaches the point of intersection of the segment that joins the supporting points of the pair (1, 6) and the longitudinal axis of the beam. Then, the pair (2, 5) is carried to the supporting points symmetrical to the supporting points of the pair (1, 6). Next, the whole process is repeated, but it is started from the pair (2, 5).

6.2   Motion Stabilization Under a Violation of Static Stability

When the diagonal pair of legs is supporting, the body is a physical pendulum fixed on the axis that passes through the supporting points and located at the upper unstable equilibrium position. To stabilize this position, we can apply the theorem on variation of the angular momentum, and, as a flywheel, we use the middle legs that execute a coordinated rotation in the plane perpendicular to the longitudinal body axis. As the measure of deflection of the body from the

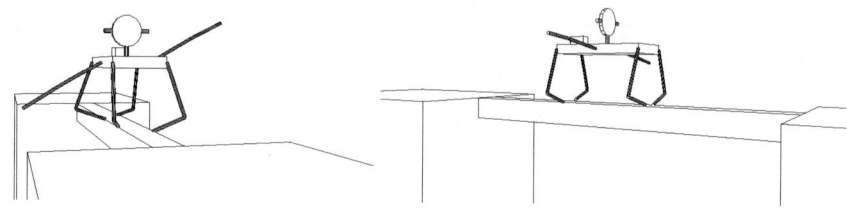

Fig. 7. Balance on the beam

vertical axis, we can take the projection ξ_c of the robot center mass on the axis $O\xi$. The control torque M_3^β relative to the angle β_3 is performed by the formula

$$M_3^\beta = -(c_1\xi_c + c_2\dot{\xi}_c)/2 - M_g, \quad M_g = (m_1 + m_2)gl_c \sin\beta_3,$$

where M_g is the moment of the gravitational force of the third leg relative to the attachment point; l_c is the distance from this point to the center of mass of the straightened leg; and c_i, $i = 1,\ldots,6$ are the feedback gains.

The motion of the fourth leg relative to the body is skew-symmetric to the motion of the third leg, and the control moment is

$$M_4^\beta = c_3(\beta_4 - \beta_3) - c_4\dot{\beta}_4.$$

The values of the angles $\gamma_3 = 0$, $\gamma_4 = \pi$ correspond to the straightened legs.

The angular velocity of the middle legs accumulating in the loss of static stability is eliminated in the course of joint standing of the front and back legs. It is worth noting that the middle legs are controlled to return them into the horizontal position with zero angular velocities in accordance with the formula

$$M_3^\beta = c_5(\beta_3 - \pi/2) - c_6\dot{\beta}_3 - M_g.$$

In the same way, the fourth leg tracks the motion of the third leg in the skew-symmetric manner.

As in [9,10], computer simulation of 3D-dynamics of the robot was fulfilled by means of the program complex Universal mechanism [11].

7   Conclusions

The results of computer simulation were obtained with using a specially designed virtual environment that allows to experiment with a model of robot dynamics in the same way as it can be done in physical experiments.

In the problem of a robot climbing a corner with the help of dry-friction forces with the opening angle $\pi/2$ it was shown that there exists a motion that solves the problem under the friction coefficient 1.1.

In the problem of a six-legged robot walking on a narrow horizontal beam, a control for the flapping motion of the middle legs was found so that it provides the dynamic stability without violation of kinematics and design constraints.

In this paper the solution of the existence problem for required motions with realizable friction coefficients was received. Results of the computer simulation show that robot is not sensitive to the reasonable random initial conditions and minor errors of program trajectories. Influence of measurement noise and variation of environments will be investigated at the next stage of the work.

References

1. Pugh D.R., Ribble E.A., Vohnout V.J., Bihari T.E., Walliser T.M., Patterson M.R., Waldron K.J. Technical Description of the Adaptive Suspension Vehicle. International Journal of Robotics Research, vol.9, No. 2, 1990, pp.24-42.
2. Wong H.C., Orin D.E. Control of a Quadruped Standing Jump and Running Jump Over Irregular Terrain Obstacles. Autonomous Robots, vol. 1, 1995, pp. 111-129.
3. De Man H., Lefeber D., Vermeulen J. Design and Control of a One-Legged Robot Hopping on Irregular Terrain. In Proc. Euromech 375: Biology and Technology of Walking, 1998, pp. 173-180.
4. Nishi A., Wakasugi Y., Watanabe K. Design of a robot capable of moving on a vertical Wall, Advanced Robotics, 1, 1986, pp. 33-45.
5. Gradetsky V., Kalinichenko S., Kravchuk L., Pushkin M. Stability Motion Problem for Wall Climbing Robot with Transition Possibilities. Proc. of the Third International Conference CLAWAR-2000, pp. 305-314.
6. Golubev Yu.F., Korianov V.V. Motion design for six-legged robot overcoming the vertical column by means of friction forces. Proc. of the 6-th International Conference CLAWAR-2003, pp. 609-616.
7. Yu. F. Golubev. Mechanical Systems with Servoconstraints. Elsevier Science Ltd. J. Appl. Math Mechs, Vol. 65, No. 2, pp. 205-217, 2001.
8. D. E. Okhotsimsky, Yu. F. Golubev. Mechanics and Motion Control for an Automatic Walking Robot. Nauka, Moscow, 1984, in Russian.
9. Yu. F. Golubev, V. V. Koryanov. Construction of Motions of an Insectomorphic Robot that Overcomes a Combination of Obstacles with the Help of Coulomb Friction Forces. Pleiades Publishing, Inc. J. of Computer and Systems Sciences Intl., Vol. 44, No. 3, pp. 460–472, 2005.
10. Yu. F. Golubev, V. V. Koryanov. A Control for an Insectomorphic Robot in Motion along a Vertical Corner and a Horizontal Beam. Pleiades Publishing, Inc. J. of Computer and Systems Sciences Intl., Vol. 45, No. 1, pp. 144–152, 2006.
11. http://www.umlab.ru

Adaptive Four Legged Locomotion Control Based on Nonlinear Dynamical Systems

Giorgio Brambilla, Jonas Buchli, and Auke Jan Ijspeert

Biologically Inspired Robotic Group (BIRG),
Ecole Polytechnique Fédérale de Lausanne (EPFL),
Station 14, CH-1015 Lausanne, Switzerland
giorgio@alwaysnet.net, {jonas.buchli, auke.ijspeert}@epfl.ch
http://birg.epfl.ch/

Abstract. Dynamical systems have been increasingly studied in the last decade for designing locomotion controllers. They offer several advantages over previous solutions like synchronization, smooth transitions under parameter variation, and robustness. In this paper, we present an adaptive locomotion controller for four-legged robots. The controller is composed of a set of coupled nonlinear dynamical systems. Using our controller the robot is capable of adapting its locomotion to the physical properties of the robot, in particular its resonant frequency. Our approach aims at developing an on-line learning system that attempts to minimize the energy necessary for the gait. We have implemented the model both in a simulated physical environment (Webots) and on a Sony Aibo robot. We present a series of experiments which demonstrate how the controller can tune its frequency to the resonant frequency of the robot, and modify it when the weight of the robot is changed.

1 Introduction

Nonlinear dynamics is ubiquitous in the physical and in the biological world. Nonlinear dynamics theory has provided us with new tools to understand complex phenomena that were difficult to explain before. It can be used to model competition in predator-prey systems, emergent behavior in collective systems, growth of biological organisms, the production of rhythmic patterns in the heart [10] and in the spinal cord for locomotion [9,7], to name a few examples.

In this article, we explore how a nonlinear dynamical system implemented as a system of coupled oscillators can be designed (1) to control walking gaits of compliant four-legged robots, and (2) to continuously tune important parameters such as the frequency of oscillations to (possibly time-varying) properties of the body. In particular, we aim at designing adaptive controllers in which the adaptive process is embedded in the dynamical system (i.e. expressed as differential equations) rather than in a separate learning or optimization algorithm.

To produce locomotion, a controller must be capable of generating a rhythmic and coordinated behavior. Nonlinear dynamical systems such as systems of coupled oscillators present several advantages over alternative approaches (e.g. gait

S. Nolfi et al. (Eds.): SAB 2006, LNAI 4095, pp. 138–149, 2006.

tables or sine-based trajectory generators) to generate gaits for a robot. They allow to harmoniously interact with the environment, they can create limit cycle behavior, they allow smooth modulations of the trajectories, and they make the emergence of new behaviors possible. Moreover, some dynamical systems, like Adaptive Frequency Oscillators (AFOs) [5,15] are interesting because of their adaptive properties.

The controller we propose is suitable for robots with compliant (i.e. elastic) components. It adapts the walking frequency to resonant frequencies of the robot in order to minimize the amount of energy required to move forward. Hopf oscillators and adaptive frequency oscillators are used as building blocks in the controller. Adaptation is embedded in the dynamical systems, and no external optimization is required. Moreover, adaptation is not a batch process, and the controller adapts its parameters online using proprioceptive signals.

The robot locomotion is based on two different kinds of joints: knees - passively controlled by springs, and hips - actively controlled by servos. The robot swings on the knees behaving like an inverted pendulum. A Hopf oscillator [11] controls each hip. Each of these oscillators is coupled to the other hips for inter-limbs gait coordination. Moreover, each hip is coupled in phase to the relative knee. This movement coordination permits to recycle the potential energy of the knee springs to push the robot forward. Furthermore, an Adaptive Frequency Oscillator (AFO) tunes its frequency to the knee oscillations, and this frequency is used for the hip oscillators. This controller has two feedback loops: one permits phase synchronization to proprioceptor signals and the second frequency adaptation.

In the rest of the article, we describe our implementation of this system for a simulated and a real Aibo robot (Section 2). Experiments in simulation (Section 3) and in the real world (Section 4) demonstrate how the system is capable of producing efficient walking gaits that are tuned to the resonant frequency of the robot, and that are continuously adjusted to changing body properties.

2   Adaptive Four Legged Locomotion Model

In this section we describe the main elements of our model. First, we explain what properties make a good locomotion controller and our approach to building one. Second, we describe the mechanical specifications the robot shall satisfy. Third, we introduce our CPG (Central Pattern Generator), a central component in our locomotion model. Eventually, we elaborate on the adaptive equations, which evolve the walking frequency parameter depending on the physical properties of the robot.

2.1   Mechanical Model

We propose a locomotion controller for four legged robots. Every limb has two joints: the upper one (hip) and the lower one (knee). Every hip is actively controlled and is composed of: a servo (actuator), and an encoder (angle sensor). Every lower joint is passively controlled and is composed of: a spring-damper system (actuator), and an encoder (angle sensor). The controller receives information from the knee and only controls the hip angle to produce locomotion. The body behaves like an inverted pendulum. Our controller does not explicitly deal with: inertial forces, moments of inertia, static and dynamic forces, and contact ground areas. We will call this mechanical framework "elastic locomotion framework." The ideas introduced by Blickhan [2] and by Fukuoka [8] inspired this framework.

Our controller reproduces a walking gait. Walking is easy to implement in this framework, and seems to be the most energy efficient gait compared to trot and gallop [12]. For reasons of stability the terminal limbs are oriented in forward direction, as in Figure 1. In fact, if the limbs were turned backwards, the four legged robot would tumble and fall during the swing phase. A spring-damper rotational system produces in the knee torques proportional to two components: the angle (γ, i.e. the spring torque) and the the angular velocity ($\dot{\gamma}$, i.e. the viscous damping force): $T = k\gamma - d\dot{\gamma}$, where k and d are positive spring and damper coefficients.

2.2 Approach to Learning

A good controller to be used in the elastic locomotion framework allows to recycle the potential energy stored in the knee spring during the step cycle, converting it into kinetic energy. Our challenge was to build and tune a controller for a general four legged robot without specific body information, that satisfies the following properties: (1a) knees and hips should move at the same frequency and with a constant phase shift; (1b) the four hips move with a fixed phase shift depending on the gait; and (1c) the knee resonant frequency depends on the weight of the robot as well as on the constant of the knee spring (k). The weight can change during the robot life, and the controller should be able to adapt to it. Our approach is based on nonlinear dynamical systems. There are two kinds of dynamical systems of interest to us: (2a) Hopf oscillators [11,16], which are interesting because of their limit cycle behavior with the possibility of phase synchronization; and (2b) "Adaptive Frequency Oscillators" (AFOs) [5,15], which are capable of synchronizing their frequency and phase to an external oscillating signal. In earlier contributions [6] it has been shown that such systems in a feedback loop with the mechanical system can indeed adapt to the resonant frequency of the body. Thus, our controller should satisfy constraints (1a), (1b), and (1c) using dynamical system (2a) and (2b) as building blocks.

2.3 CPG with Feedback

Our controller is composed of a fully connected network of four oscillators inspired by animal CPGs [4] (see Figure 1). A continuous arrow means that the signal of the source oscillator at time t is rotated by means of a rotational matrix (R) and summed to the differential equations of the target oscillator. The phase shifts (ρ_{ji}) introduced by the rotation matrix are constant and correspond to the one specified by the walking gait. The coupling values are as in Table 1. Each connection adds a perturbation to the target oscillator that contributes to the amplification of the signal. We have used a fully connected network because it permits to have a very stable CPG system. In such an architecture, any perturbation is quickly absorbed by the system, and, moreover, we have a reduced influence of the noise. A discontinuous arrow also adds a perturbation, but this time the source is the knee angle value, and it is rotated by a constant angle (ξ). Eqs. 1–2 describe each oscillator in the network. x, y are the state variables describing the oscillator, μ is a parameter which determines the amplitude of oscillations, k is a damping constant, ω is the intrinsic frequency of the oscillator, and a is a global coupling constant, finally to keep the expressions shorter we use $r_i^2 = x_i^2 + y_i^2$.

$$\dot{x}_i = (\mu - kr_i^2)x_i + \omega y_i + a \sum_{\forall j \in I \wedge j \neq i} R_x(\rho_{ji}, x_j(t), y_j(t)) + cR_x(\xi, s_i(t), 0) \quad (1)$$

$$\dot{y}_i = (\mu - kr_i^2)y_i - \omega x_i + a \sum_{\forall j \in I \wedge j \neq i} R_y(\rho_{ji}, x_j(t), y_j(t)) + cR_y(\xi, s_i(t), 0) \quad (2)$$

$$\begin{bmatrix} R_x(\alpha, x, y) \\ R_y(\alpha, x, y) \end{bmatrix} = \begin{bmatrix} x\cos(\alpha) - y\sin(\alpha) \\ x\sin(\alpha) + y\cos(\alpha) \end{bmatrix}$$

$$I = \{LF, RF, LH, RH\}$$

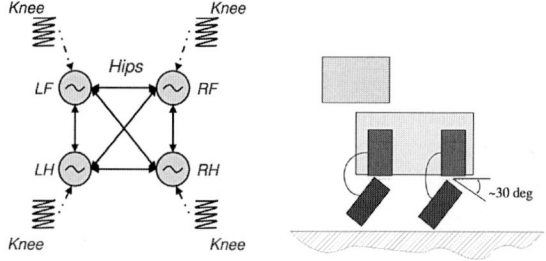

Fig. 1. (Left) Feedback system. In the figure each oscillator corresponds to one of the four hips. The arrows identify the phase coupling. **(Right)** Aibo limbs orientation in the absence of external forces, we have chosen an angle of 30 deg for the knee, and an angle of 0 deg for the hips.

2.4 Adaptive CPG

We have defined a mechanical framework and a walking controller in the previous section. Now we want to reduce the number of parameters to make the CPG frequency adaptive. As explained in Section 2.2, we aim at making the walking frequency adaptive to the resonant properties of the body. As input we use the knee angles. Hence, we compute the signal of the knee angle for every leg. The signal is periodic (but not sinusoidal) and has a similar shape in all legs. The four legs create similar oscillatory signals shifted by a constant phase difference defined by the hip oscillator connections. However the signals slightly differ from the specified phase and shape of the signal. We have to find a means to extract

Table 1. Default parameter values, inter-limbs coupling phase difference, upper limb joints moving range on Aibo platform

parameter	value	
μ	0	(bifurcation point)
k	0.15	
ω	8	[rad/s]
a	1	
c	5	
ξ	0.2	[rad]

ρ	LF	RF	LH	RH
LF	0	$-\pi$	$-\frac{3}{2}\pi$	$-\frac{\pi}{2}$
RF	π	0	$-\frac{\pi}{2}$	$\frac{\pi}{2}$
LH	$\frac{3}{2}\pi$	$\frac{\pi}{2}$	0	π
RH	$\frac{\pi}{2}$	$-\frac{\pi}{2}$	$-\pi$	0

	MIN ANGLE	MAX ANGLE
Front limbs	0.0 [rad]	0.6 [rad]
Hind limbs	-0.1 [rad]	0.3 [rad]

the frequency information of these signals despite these differences. This can be achieved with an adaptive frequency oscillator (AFO) [5,15]:

$$\dot{x} = (\mu - kr^2)x + \omega_l y + c\sum_{\forall j \in I} R_x(\beta_j, s_j(t), 0)$$
$$\dot{y} = (\mu - kr^2)y - \omega_l x$$
$$\dot{\omega}_l = c\eta\frac{y}{r}\sum_{\forall j \in I} R_x(\beta_j, s_j(t), 0)$$

(3)

The AFO (Eq. 3) has three state variables. Compared to a Hopf oscillator, it has one additional state variable used for frequency adaptation and one additional parameter used for learning. The state variable x will synchronize to the input signal. Again, the variable ω_l stands for the frequency [in rad/s]. Due to the additional differential equation the frequency will adapt to one of the frequencies of the input signal (see [15] for further discussion). The parameter η represents the learning rate: a too high learning rate influences the stability of variable ω_l, and a too low learning rate does not permit the adaptation process. In our experiment, we have varied η between 1 and 10.

As a first step, we have tested an intermediate solution where the controller "learns" the main knee oscillation frequency (**"Open Loop" solution**). Our system rotates ($\beta = \{0, -\pi, -\frac{3}{2}\pi, -\frac{\pi}{2}\}$) and sums the knee signals (s_j) so that the walking frequency becomes the most powerful frequency component[1]. This signal is used to perturb a frequency adaptive oscillator (Eq. 3). As a result, the oscillator smoothly learns the input frequency and maintains it, reaching a steady state. After a few seconds of transition the dynamical system adjusts its frequency and phase to the one coming from the perturbation signal. We will discuss results showing this properties in Section 3.1.

---

[1] If we sum the signals without rotating them, as the shapes of the signals are comparable and shifted among each other by about 90 deg, they are going to annihilate each other (see Figures 3 and 5). In the following section we propose the experimental results obtained with both "un-rotated" ($\beta = \{0,0,0,0\}$) and "rotated" ($\beta = \{0, -\pi, -\frac{3}{2}\pi, -\frac{\pi}{2}\}$) solutions, to support the choice of the "rotated" solution.

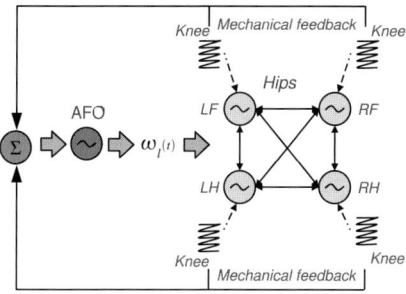

Fig. 2. Closed loop adaptive walking feedback system. In the figure the CPG presented in Section 2.3 is integrated with the AFO and the CPG becomes adaptive.

Eventually, we have rendered the controller adaptive (**"Closed Loop" solution**), as one can see in Figure 2. We have used the value of the state variable ω_l instead of the parameter ω in the CPG of the walking controller (explained in Section 2.3) making the system adaptive. These ideas permit a smooth adaptation of the walking frequency to the frequencies of the knee. Furthermore, the dynamic formulation of the adaptation by means of dynamical systems allows to adapt the frequency in the case of changes of the body properties or changes in the environment (e.g. ground friction) during the robot life. The online adaptation process also introduces a new feedback loop in the dynamical system. This feedback loop is nonlinear and it is not clear from the outset that it will work. It can however be expected from previous results in simulation [5,6] and has recently been treated analytically [3]. Even more, the presented experimental results show how well and stable this solution works.

3   Experimental Results (Simulation)

Webots [13] is an integrated environment for robot simulation, and the physics is simulated using the ODE Library [14]. The robot platform chosen is a Sony Aibo[2]. Aibo has advantages and drawbacks. The advantages are that one can find a detailed model of Aibo in the Webots environment, and can test the controller on a real Aibo robot. However, Aibo is not an optimal platform for our "Elastic Locomotion Framework" because the robot legs do not have a real spring-damper system. Two different strategies have been applied to solve this problem: in the simulation, a knee is controlled by a simulated spring, and, hence, it is the most accurate model of a spring using the Webots simulator. In the real Aibo robot springs are simulated using a PID and an active spring law control to simulate the spring behavior.

All differential equations in our controller are numerically integrated using the Runge-Kutta method with a fixed time integration step. The simulator integrates

---

[2] Sony AIBO by Sony Corporation. "Aibo" is a registered trademark of Sony Corporation.

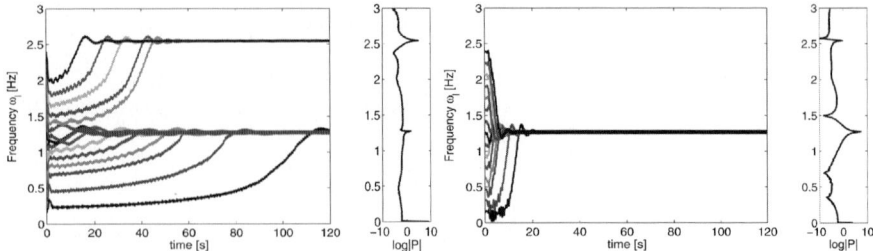

Fig. 3. In the first figure, a set of open-loop experiments (over the same signal) shows the adaptivity properties of an AFO dynamical system ($\eta = 1$). Here the perturbing signal is un-rotated in the left figure and rotated in the right one (Webots). The small graphs are the power spectral density of the un-rotated and rotated signals.

all the equations at every iteration, using a time step of 0.008 [s] of the virtual simulation time.

3.1   Open Loop Results

First, we have successfully tested our CPG network, using the default parameters shown in Table 1. Then, we have applied our adaptive component, Equation 3, and plotted the learning curve of the knee signal starting from a range of initial frequency values. In Figure 3, we plot an example of frequency learning using as a perturbing signal the un-rotated (left) and the rotated (right) sum of the knee angles. Comparing the graphs in Figure 3 with the Fourier spectrum (on the right), one can see that the AFO converges to the higher power frequency component and reaches a steady state. Furthermore, rotation provides a wider basin of attraction.

3.2   Closed Loop Results

Second, we have substituted the parameter ω of the hip oscillators with the AFO variable ω_l. Consequently, the CPG parameter ω (walking frequency) has become adaptive and equal to ω_l. This new feature introduces new feedback in the walking controller, as described in Section 2.4. In this system, eventually, knee angles are used for tuning the walking phase and frequency. Moreover, the Hopf oscillators as well as the AFO work at the bifurcation point ($\mu = 0$), which means that in the absence of a stimulus they will not oscillate. Hence, in order to create a locomotion process, at the begin the spring system must be stressed enough in order to create a chain reaction of oscillating stimuli, and must be strong enough to maintain the oscillation without stopping it. To make this possible, it is necessary that one chooses spring-damping parameters that allow the signal to oscillate a few times (in experiments 3 or 4) before being damped out. In other words, the damping parameter d of the knee must be high enough to avoid instability in the dynamics of the spring-mass system, but small enough to permit oscillations. It is expected that every instance implementing the "elastic locomotion framework" has this basic behavior.

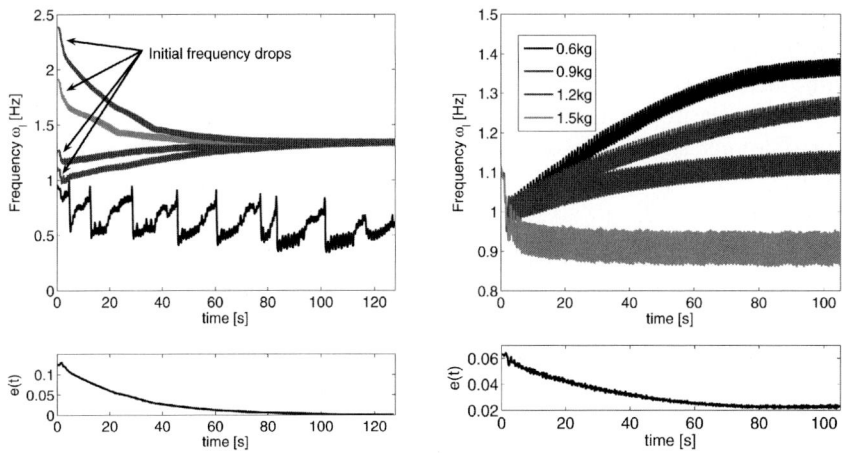

Fig. 4. (Left) closed loop frequency adaptation with different initial frequency but same initial condition ($\eta = 1$). The controller converges if the initial frequency is close to the asymptotic one (Webots). **(Right)** closed loop frequency adaptation with different robot weights but same initial conditions ($\eta = 1$)(Webots). The two bottom small plot show the average error (see text for definition) between the solution of our controller and the spring-mass oscillating law $\sqrt{\frac{k}{m}}$.

We present two experiments: the first is a set of adaptation processes, and the second shows the relation between four different experiments with the same initial conditions but different robot weights.

The first plot (Figure 4 left) shows a series of experiments with the same initial conditions (environment, robot) but different initial frequency $\omega_0=\{0.95, 1.11, 1.27, 1.90, 2.38\}$, in order to demonstrate how the final frequency purely comes from the adaptive process. Furthermore, in the plot the variable ω_l converges to the same asymptotic value. In the plot, the first experiment ($\omega_0 = 0.95Hz$) shows that an initial condition too far from the one at the steady state will never converge. Moreover, one can see how in every experiment in the first few seconds the frequency drops because while the CPG synchronizes the movements of the four legs, the AFO is perturbed by the signal coming from the knees, and this signal is not yet stabilized. After these few seconds the variable ω_l converges. Moreover, the AFO initial conditions give to the dynamical system an initial moving input. As the oscillators parameter μ is equal to zero, without an initial input the system remains inactive. Hence, the initial conditions must bring the AFO to oscillate enough in order to maintain the initial oscillation of the overall system. The two small graphs at the bottom of Figure 4 show the reduction of the difference between the walking frequency and the frequency of the spring-mass oscillating law $\sqrt{\frac{k}{m}}$. Where the error is defined as $e(t) = \sigma(S(t)); S(t) = \sum_i \sum_j a_{ij}; a_{ij} = \frac{\omega_i(t)}{\omega_j(t)} - \sqrt{\frac{m_j}{m_i}}$ where m_i or m_j stands for the mass, ω_i or ω_j stands for the frequency of experiment i or j, and σ stands for standard deviation.

The second plot (Figure 4 right) shows four experiments using the same controller as in the previous experiment (Figure 4 left), with the following initial conditions: (1) the robot has an initial frequency of 1.1 Hz. (2) In each of the four experiments the robot has a different weight (a) 0.6kg, (b) 0.9kg, (c) 1.2kg, (d) 1.5kg. The plot shows how a higher weight leads to a lower walking frequency. This behavior respects the spring-mass law where the resonant frequency can be calculated as $\sqrt{\frac{k}{m}}$. These results lead to two conclusions: the robot behaves like a spring-mass system, and it adapts its walking frequency to the resonant frequency of the knee. This second experiment shows an interesting property of our system: online adaptation to the physical properties of the robot. In other words, the robot is able to adapt its walking following the weight change during life (for ex. payload change).

Integrating the torque on the angle of the four hips and summing the four values, we have computed the energy consumed by the robot. Then, to obtain the efficiency, we have divided the energy over the distance covered by the robot. We have proved in simulation that this adaptation permits to save energy (in comparison to non-adaptive CPG) when loading a payload of 0.6 Kg of about 15% (data not shown). Our adaptive system does not maximize the walking speed but seems to find a more efficient walking frequency. The frequency found seems to optimize the conversion of the spring potential energy into kinetic energy to propel the robot forward. Our adaptive system can help make the walking locomotion more efficient. This is in line with earlier findings in simulations[6].

4   Experimental Results (Real World)

In this section, we show the experimental results on the Aibo robot. As outlined in Section 3, the Aibo is not the optimal platform to implement the "elastic locomotion framework," since it has activated knee joints, it does not have passive springs in the knee joints. In order to simulate the spring law in the knee joints, we have added controllers for the those joints such that the to a large extent the knees behave like a spring, and thus the robot corresponds to the requirements as stated in the "elastic locomotion framework". As in the case of the simulation, the equations are numerically integrated with a Runge-Kutta algorithm with a fixed time step of 0.008 [s]. The design of Aibo controller also takes care of further implementation problems occurring when simulating a spring law such as encoder resolution and accuracy, mechanical gear backlash, digital system delay, system identification and other problems (cf. [1] for more details). In the following two sections we have repeated the two simulation experiments on real world Aibo.

4.1   Open Loop Results

As in the simulation, we have successfully tested our CPG network, using the default parameters shown in Table 1. Then, we have applied our adaptive component, Eq 3, and plotted the results in Figure 5. When the AFO is perturbed

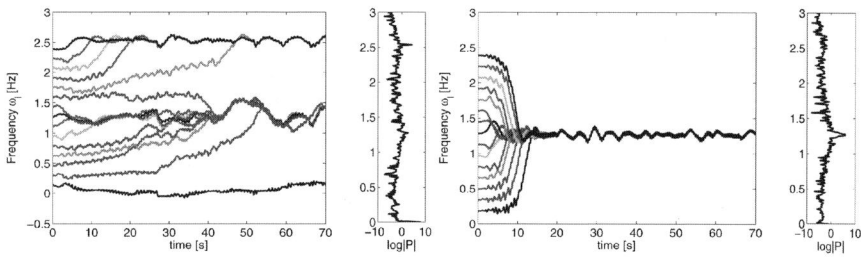

Fig. 5. This experiment is equal to the ones in Figure 3, but on a real AIBO robot. Here the perturbing signal is the un-rotated (left) sum of knees and the rotated one (right). The plot shows how the noise introduced by a real environment produces unstable solutions, in case of un-rotated signal, and very stable solutions in the case of rotated signal ($\eta = 1$). The small graphs are the power spectral density of the un-rotated and rotated signals.

using the rotated sum of the knee signals it quickly converges to a steady state, it also happened when the initial conditions are not close to the steady state. The rotated solution, using Aibo real world robot signal, seems to have maintained the convergence properties shown in the simulation results (Figure 3).

4.2   Closed Loop Results

In the first experiment the initial frequency is $\omega_0 = 0$. The plot, in Figure 6, shows how in this case using a higher learning parameter $\eta = 10$ the robot can learn to walk from scratch. In this case, a (randomly applied) hand-made stress on the knee joints provides the initial input to walking. Aibo quickly reaches a steady state frequency.

The second experiment, as described in Section 3.2, involves the parameter adaptation in case of different weights. In this case, we have simulated a real payload change application. The robot starts its life (weight 1.68 [Kg]), (1) adapts its

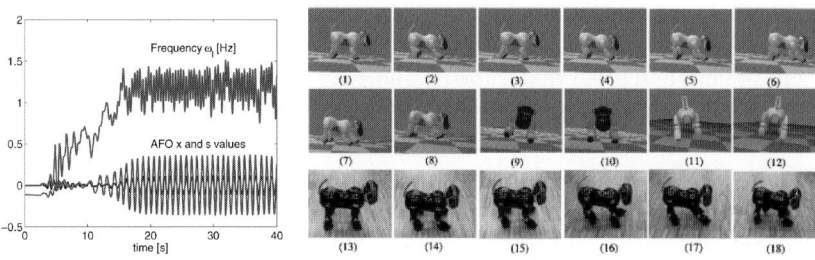

Fig. 6. The figure on the left shows how AIBO learns the walking frequency from scratch ($\omega_0 = 0$) because of the learning parameter $\eta = 10$. On the right there are a series of snapshot during the initial step of Aibo in simulation (from 1 to 12) and in the real world (from 13 to 18). (cf. movies [1])

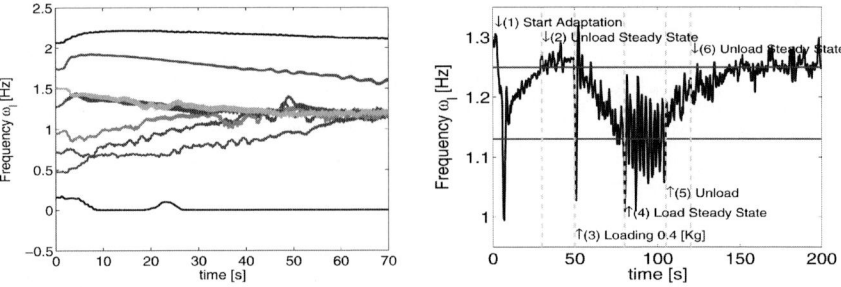

Fig. 7. (left) This figure is the same experiment of Figure 4 (left) but in this case using Aibo real robot. In this case AIBO uses a learning parameter $\eta = 3$. **(right)** This plot shows the variable ω_l and demonstrates the adaptability of our controller in a simple load transport application ($\eta = 10$, plotted is the running average of the frequency: $\omega_{l,p}(t) = \frac{1}{50} \sum_{i=0}^{50} (\omega_l(t-i))$ where $dt = 0.008[s]$, further details are given in the text).

walk, then (2) once a steady state has been reached, (3) one can load a payload weight (0.4 [Kg]) on the robot saddle, (4) the controller reaches a new steady state adapting the frequency to the new weight, (5) in the end we unload the payload, and (6) the robot returns to a frequency close to the initial one, see Figure 7. This experiment resumes well the interesting autonomous adaptation provided by our controller.

5   Conclusion

In biology, adaptation and memory are major properties of living systems. The main ideas presented in this article are to develop a walking controller where learning is embedded in the dynamics and not an offline optimization process.

The simulated and the real robot have different properties, but thanks to the feedback loop introduced, the controller adapts well its behavior. This demonstrates that the controller is flexible and not designed to work on a specific mechanical system. Moreover, since the robot is autonomous, the controller permits to adapt the walk using proprioceptive signals. Bio-mechanics suggests that the gait of dogs and other quadrupeds can be compared to our elastic locomotion framework. Moreover, the building blocks as well as the entire controller may possibly find implementations in neuro-biological models. We have not dealt with problems such as direction modulation, discontinuous terrain management, and other classical locomotion issues. But we believe that our model is open and flexible enough to be adapted to address these tasks.

We have tested the controller both in computer simulations and on a real robot with successful results in both cases. Furthermore, the frequency adaptation was shown to be useful to provide an efficient gait capable of recycling the energy. Hence, the controller was proved useful for quadruped transportation system applications.

Acknowledgments

G.B. wishes to especially thank his friend, Fabrizio Patuzzo, for his encouragement, support, and review during the entire work. This work was made possible thanks to a Young Professorship Award to Auke Ijspeert from the Swiss National Science Foundation.

References

1. Movies and more technical details of the implementation are available online at http://birg.epfl.ch/page57636.html.
2. R. Blickhan. The spring-mass model for running and hopping. *J. Biomechanics*, 22(11–12):1217–1227, 1989.
3. J. Buchli, F. Iida, and A.J. Ijspeert. Finding resonance: Adaptive frequency oscillators for dynamic legged locomotion. Submitted.
4. J. Buchli and A.J. Ijspeert. Distributed central pattern generator model for robotics application based on phase sensitivity analysis. In *Proceedings BioADIT 2004*, volume 3141 of *Lecture Notes in Computer Science*, pages 333–349. Springer Verlag Berlin Heidelberg, 2004.
5. J. Buchli and A.J. Ijspeert. A simple, adaptive locomotion toy-system. In S. Schaal, A.J. Ijspeert, A. Billard, S. Vijayakumar, J. Hallam, and J.A. Meyer, editors, *From Animals to Animats 8. Proceedings of the Eighth International Conference on the Simulation of Adaptive Behavior (SAB'04)*, pages 153–162. MIT Press, 2004.
6. J. Buchli, L. Righetti, and A.J. Ijspeert. A dynamical systems approach to learning: a frequency-adaptive hopper robot. In *Proceedings ECAL 2005*, Lecture Notes in Artificial Intelligence, pages 210–220. Springer Verlag, 2005.
7. A.H. Cohen and D.L. Boothe. Sensorimotor interactions during locomotion: principles derived from biological systems. *Autonomous Robots*, 7(3):239–245, 1999.
8. Y. Fukuoka, H. Kimura, and A.H. Cohen. Adaptive dynamic walking of a quadruped robot on irregular terrain based on biological concepts. *The International Journal of Robotics Research*, 3–4:187–202, 2003.
9. S. Grillner. Control of locomotion in bipeds, tetrapods and fish. In V.B. Brooks, editor, *Handbook of Physiology, The Nervous System, 2, Motor Control*, pages 1179–1236. American Physiology Society, Bethesda, 1981.
10. J. Honerkamp. The heart as a system of coupled nonlinear oscillators. *J Math Biol*, 18(1):69–88, 1983.
11. E. Hopf. Abzweigung einer periodischen Lösung von einer stationären Lösung eines Differentialsystems. *Ber. Math.-Phys., Sächs. Akad. d. Wissenschaften, Leipzig*, pages 1–22, 1942.
12. T. McGeer. Passive dynamic walking. *International Journal of Robotics Research*, 9:62–82, 1990.
13. O. Michel. Webots: Professional mobile robot simulation. *International Journal of Advanced Robotic Systems*, 1(1):39–42, 2004.
14. Russell Smith & others. *Open Dynamics Engine*. Available online at http://ode.org.
15. L. Righetti, J. Buchli, and A.J. Ijspeert. Dynamic hebbian learning in adaptive frequency oscillators. *Physica D*, 216(2):269–281, 2006.
16. S. Strogatz. *Nonlinear Dynamics and Chaos. With applications to Physics, Biology, Chemistry, and Engineering*. Addison Wesley Publishing Company, 1994.

The Control of Turning in Real and Simulated Stick Insects

Hugo Rosano and Barbara Webb

The University of Edinburgh, Edinburgh EH9 3JZ, UK
H.Rosano@ed.ac.uk

Abstract. This paper describes a model for six-legged robot turning control based upon the stick insect (*Carausius morosus*). Ethological observations were made on freely walking stick insects turning towards a visual target. It was found that there is a tendency for the prothorax to move directly towards the object while the rear of the body mainly rotates. The front legs are proposed to shape most of the body trajectory, affecting the other thoracic segments such that it is not necessary to calculate individual leg trajectories for the middle or rear legs. A 3D dynamical robot simulation proved able to replicate complicated insect leg trajectories by means of this simple principle.

1   Introduction

Implementing robotic models based on biological systems is a challenging goal. One characteristic of biological systems is redundancy — for example, stick insects can be modelled with three degrees of freedom (DOF) on each leg, for a total of 18 DOF, to move the body into a position represented by at most 6 DOF. This provides the system with a vast repertoire of movements, which could help solve complex navigational problems, but having so many degrees of freedom poses problems for designing a controller. Decentralising the walking controller into 6 coupled oscillators [1] following a small number of coordination rules [2] has been shown to reduce the dimensionality of the problem. Having only one type of leg controller copied across the body reduces the controller complexity. However, it is also known that the legs on different thoracic segments (i.e. the front, middle and hind legs) can behave differently in similar situations [3].

A non-trivial problem even for movement on flat surfaces is how insects manage to turn in such a way that almost all body trajectories are possible. Collaboration of all legs for turning has been suggested for the cockroach [4]; however, difference in walking dynamics and morphologies between insect species might result in different leg roles. Modulation of leg coordination and leg trajectories during turns has previously been analysed using the insect's response to a continuous visual flow [5]. In that experiment, insects were fixed on top of a movable ball in the middle of a rotating visual scene. The leg trajectories during turns varied considerably depending on the thoracic segment. It was suggested [5] that the front legs respond first to the visual stimulus, and this might trigger the response of the other legs. Furthermore, it is not clear to what extent the precise

S. Nolfi et al. (Eds.): SAB 2006, LNAI 4095, pp. 150–161, 2006.

trajectories have to be calculated by each leg and how each thoracic segment might interpret signals locally or from the brain. Neurophysiological information on leg control in the stick insect is largely limited to mechanisms in the middle leg [6].

We are particularly interested in the body trajectory and the different roles of legs on different thoracic segments when the insect decides to turn towards a fixed target. Hence, it was crucial to have experiments with insects walking freely. We observed the turning response of insects presented with an attractive visual target. We show that the front legs play an important role in turning and are responsible for shaping most of the body trajectory. This is demonstrated using a 3-dimensional dynamic simulation of insect walking. We found that if each front leg can follow trajectories at the desired angle, the only requirement for the middle and hind legs is to change the degree to which they respond to external forces. This approach is sufficient to replicate the observed insect behaviour for turning, while limiting the complexity of the walking controller.

2   Experiment

Adult stick insects (*Carausius morosus*) reared at our institute were placed in an arena (67cm by 177cm) with white walls (50 cm tall) around it to eliminate external visual stimuli. Stick insects are known to be attracted to bush-shaped objects and to be strongly stimulated by vertical edges [7]. The visual target consisted of a black bar, 4.5 cm wide and 60 cm tall. Insects were first allowed to walk continuously for about one minute before introducing the visual target. The target was placed within the insect's visual field in a different direction to its current heading. It would reliably respond by turning to walk in this direction. Just before the insect reached it, the target was quickly removed vertically and then placed in a different position, no more than 30 cm away, inducing another turn. This could be repeated around 10 times before the insect changed its attention to the walls or ceased walking.

Trajectories were recorded with a moving video camera[1] at a height of about 30 cm. Sequences were analysed with visual tracking software developed specifically for this task. A typical segment trajectory after processing a video sequence is shown in figure 1. Three marks on the body and one on each tarsus (foot) were followed.

Insects were sometimes attracted to the walls of the arena despite these being white and smooth. The analysis did not include sequences where insects hesitated between the black bar and walls, or when the response caused by the black bar was clearly weak. The first response to presentation of the object was also eliminated as these were less consistent. A total of 24 turns were combined for the analysis. The turns were normalised to start in direction zero, turn to a target angle of 'one', and to take the same time to complete. The average time to complete a turn was 60 frames (2.4 seconds).

---

[1] DCR-TRVHE with a resolution of 720x576 at 25fps.

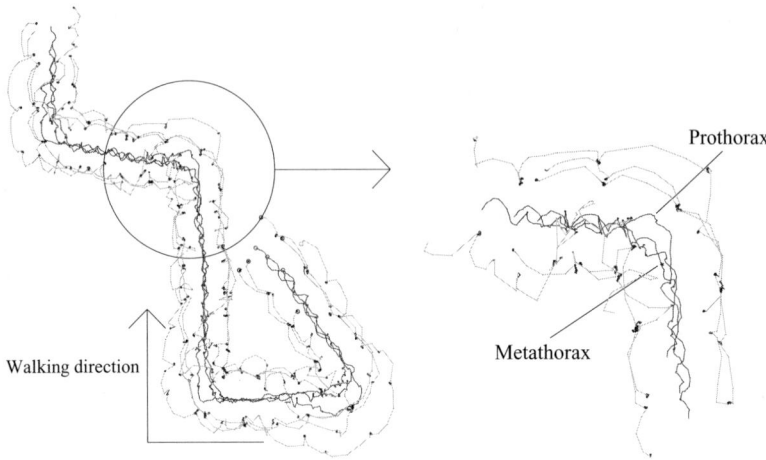

Prothorax

Metathorax

Walking direction

Fig. 1. Representative example of the paths stick insects follow when attracted to a black vertical bar. A zoomed section when turning is shown on the right, this represents a typical turn as analysed in section 2.1. Here, the prothorax direction suddenly changes almost 90 degrees.

A smoothing procedure had to be carried out for studying changes in body direction because of the limitations on camera movement compensation and image resolution, and the oscillation of the insect's body. The variables studied are shown in figure 4: the body angle θ_B at time t; and the direction of movement of the prothorax θ_P and metathorax θ_M at time t, calculated with respect to a point in time n frames ahead of the current position, typically 20 frames ahead. This lag removed noise and intrinsic oscillations of the stick insect, but also tended to smooth fast changes in angle, particularly affecting θ_P.

2.1 Results

Figure 2 shows, in the upper three plots, the direction followed by the prothorax, metathorax and body during a turn. For comparison, the means are plotted on one graph on the lower left. Despite the smoothing, it can still be clearly seen that at the beginning of the turn, the prothorax θ_P changes direction within just a few time steps, and very early during the turn is pointing towards the target. It tends to overshoot the target during the second phase of the turn, particularly in turns larger than 70 degrees, in which the speed of the back of the body is very slow and tends to rotate on the spot. It was also noticed that on certain occasions both front legs were lifted off the ground for a small period of time. During this time, the body's forward speed was almost zero, but rotation continued, resulting in a sharp curved trajectory at the beginning of the turn.

The metathorax, on the other hand, follows a smoother transition, similar to that of the body itself. The speed for this part of the body is low compared to the prothorax. In some cases, the prothorax can move twice the speed of the metathorax, as seen on the lower middle plot in figure 2. During the initial

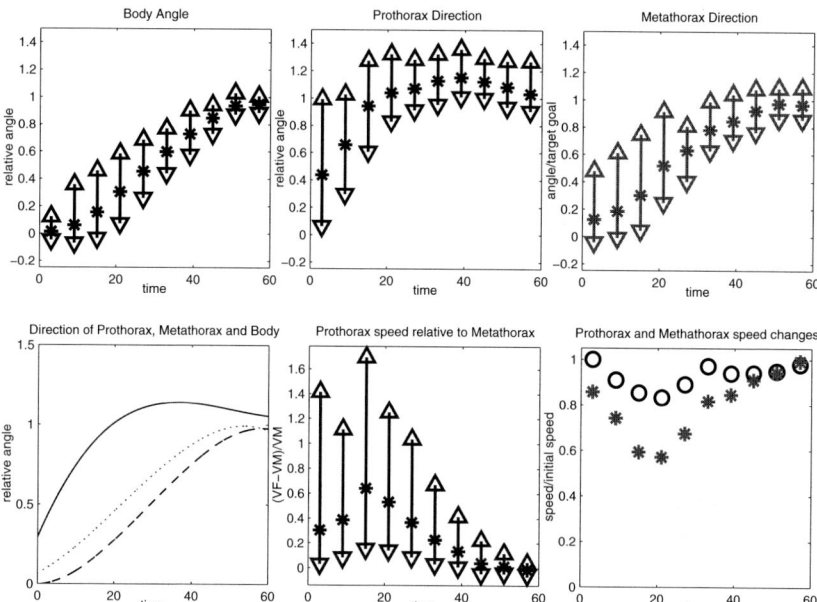

Fig. 2. Ethological results. Bottom left figure shows with a solid line the mean direction of the Prothorax θ_P; Metathorax θ_M with a dotted line; and the body θ_B with a dashed line. Top left shows progress of θ_B in more detail, deviation standard is shown by the arrows. The top middle is that of the prothorax θ_P and the top right is the metathorax direction θ_M. The bottom middle shows the relative speed of the prothorax with respect to the metathorax $(v_P - v_M)/(v_M)$. The bottom right shows changes in speed for the prothorax (o) and metathorax (*) relative to their velocity before the turn.

transition of the turn, the back legs are decelerated by the change in front leg direction; after one third of the turn they start accelerating again. On the bottom right plot in figure 2 it is shown that on average the speed of the metathorax is reduced by 40% at the slowest point, whereas the prothorax is only affected by a speed reduction of 15%.

The point of rotation for the body was calculated at each moment during the sequence, for rotations larger than 1 degree. Figure 3 shows a normalized graph of rotation positions weighted by the angle turned. Most rotations accumulated between the mesothorax and the metathorax, but there are trends to either side and towards the mesothorax. Having a rotation point close to the metathorax is in agreement with results shown for the speed of this segment, as $v_T = rd\theta$.

These data suggest that the specific movements of the stick insect's legs during turns results in the prothoracic segment following mostly straight lines, pointing most of the time towards the target, whereas the mesothorax and metathorax tend to follow curves, with a rotation point close to the metathorax. This can also be seen in figures 1 and 8. The individual leg trajectories for achieving the body trajectories described in this section vary considerably on different thoracic segments. Moreover, individual leg speeds on either side of the turn vary greatly.

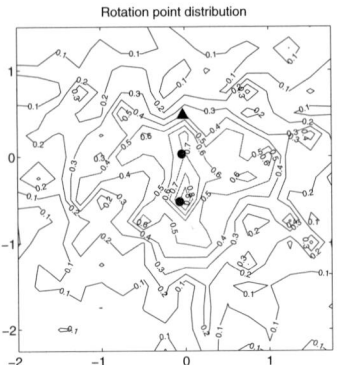

Fig. 3. A cumulative plot of the point of rotation during turns. The three black dots represent the three thoracic segments.

The problem we address in the following sections is how this complex pattern of leg behaviour might be achieved with a simple control model.

2.2   Body Trajectory Analysis

Typically, turning in six-legged robots is controlled by simple techniques, similar to wheeled robots, that cause a difference in speed on either side, which determines a point of rotation. If w is the distance between wheels, v_o is the outer speed to the turn and v_i is the inner speed, the point of rotation is given by $R = v_o w/(v_o - v_i)$. The closer this point of rotation is to the center of the robot, the sharper the turn. For legged robots there are alternative options to control this difference in speed, for instance, increasing frequency between step phases on one side. However, we are motivated to produce trajectories like those described in section 2.1, and describing this motion in terms of rotations does not fit naturally. Introducing the target angle error, ϕ, into body bearing directly $\theta_B = k\phi$ would produce curves in most cases even if the point of rotation is moved very close to the body. Therefore, the point of rotation for the stick insect is not a feasible variable to control; instead it is more likely to be a secondary effect of a controller causing the front of the body to follow straight lines.

Assuming the height of the body does not change, the insect body has only three DOF and moves in a 2D plane, i.e., body velocity $\dot{B}$ can be described as $\dot{B}(\dot{B}x, \dot{B}y, \dot{\theta}_B)$, or assuming constant speed in polar coordinates, $\dot{B}(\dot{\theta}_\eta, \dot{\theta}_B)$. If all leg positions were accurately calculated to control only these two variables, the problem of turning to visual targets would still have an infinite number of solutions. A point trajectory and rotation in the body will give us that of the rest of body. Therefore, we focused on describing the trajectory of the prothorax because it represents the most salient feature of stick insect turns; that is, tends to follow straight lines trajectories, i.e. $\dot{\theta}_\eta = 0$ for the prothorax.

Of the kinematical models describing straight lines we tested, the best describing body trajectories of section 2.1 is shown by equations (1), where η is the relative distance to the point of rotation R along the body, being 1 for the prothorax

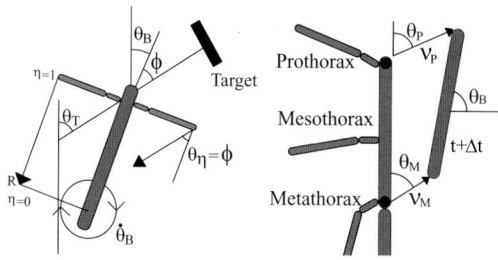

Fig. 4. The angle between the target and the body direction ϕ is shown on the left. As seen on the right hand side figure, prothorax and metathorax were considered just between coxae on that segment. All θ angles are in global coordinates for easy comparison, however, the simulation only requires to know ϕ, as $\theta_\eta = \theta_P - \theta_B = \phi$. Δt was 1/25 seconds.

and 0 for the point of rotation. The left hand side of figure (4) shows these variables with respect to the target. For the prothorax $\eta = 1$, therefore, the direction is always that of the target given its current position, $\theta_P = \theta_\eta + \theta_B = \phi + \theta_B = \theta_T$. This is in agreement with results shown in section 2.1 where the prothorax maintains direction towards the target most of the time.

$$\theta_\eta = \arctan\left(\tan\left(\phi\right)/\eta\right)$$
$$\dot{\theta}_B = |\boldsymbol{v_P}|\sin\left(\phi\right)/R \tag{1}$$

Transforming this model of body motion into leg trajectories $\boldsymbol{v}$, is fairly straightforward: if $\boldsymbol{L}$ is the distance from the coxa to the tarsus, then it follows that $\boldsymbol{v}(\dot{\theta}_B, \theta_\eta) = [-Ly, Lx]\dot{\theta}_B - \theta_\eta$. However, this equation for $\boldsymbol{v}$ implies that all legs in a kinematic model need to calculate at each point how much the body needs to rotate, and the rear legs need to know their relative position to the point of rotation. Furthermore, the rear leg direction will depend not only on ϕ, but on $\arctan\left(\tan\left(\phi\right)/\eta\right)$. Alternatively, implemented in a dynamic model, the equations in (1) could be executed independently. The prothorax could follow trajectories according to $\phi = \theta_\eta(\eta = 1)$, while the back of the body could account for the rotation $\dot{\theta}_B$. Moreover, given sufficient flexibility of different joints in the rear legs, rotation could be a passive result of the trajectory of the prothorax. This further simplifies calculations for leg trajectories in the prothorax, because if they no longer need to compute $\dot{\theta}_B$, the equation $\boldsymbol{v}$ for front leg trajectories becomes simply $\boldsymbol{v} = -\phi$ as seen in figure 4. In the next section we test the effectiveness of this simple method to control turning using a dynamic robot simulation.

3   Model

Simulation of the robot was programmed using ODE[2] libraries. Although the robot is based on the stick insect, the simulation does not exactly represent the

---

[2] Open Dynamic Engine. http://ode.org/

insect in scale or thorax morphology. Instead, the mass of body and leg segments were increased; and the centre of gravity was moved forward (by removing the back end of the body). This was because we wanted a controller that could be used on a real robot, and it is currently not plausible to replicate in any real robot the mass of the stick insect, nor the way the clawed tarsi are used by stick insects to grasp the ground to compensate for their centre of gravity being behind the metathorax. Figure 5 shows on the left the 3D model of the stick insect based robot and on the right the geometry of the leg. The angles controlled are α for rostral caudal movements; β for moving the leg up and down; and γ for movements towards and away from the body.

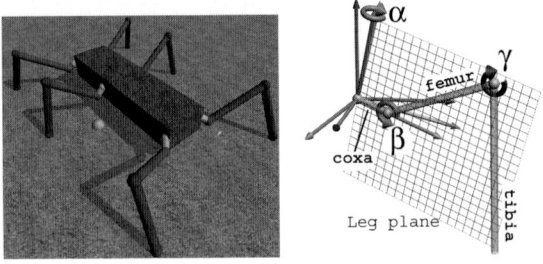

Fig. 5. The stick insect based robot created using ODE libraries is shown on the left. The right hand side shows the leg geometry.

Based on the results presented in section 2.1, it is assumed that the insect is able to sense the angular error between a visual target and its current heading. However, no visual processing was implemented, instead only the bearing of the body was controlled. This is equivalent to object targeting only if the body trajectory is exactly so as to have the prothorax follow a straight line. Fortunately, as previously shown in section 2.2, this is what the insect appears to be doing.

3.1   Inter–leg Co–ordination of Swing and Stance Phases

Body motion is determined by the complex interaction of forces produced by six legs alternating in stance. The problem is to control the sequence and direction of leg stance movements to get co-ordinated turns, but without using a centralised motion planner. Our walking controller is partly based on the WalkNet model[8], but with differences that implement the ideas presented in section 2.2. In WalkNet, there is no centralised control of gaits; instead transitions between leg phases (stance and swing) result from simple excitatory and inhibitory rules between adjacent legs. We used rules 1-3 and 5b from [8]: Rule 1, a leg swinging inhibits transition to swing in the anterior leg; Rule 2, a leg starting stance excites the anterior leg; Rule 3, a leg excites the caudal leg proportional to its distance travelled. Rules 2 and 3 are also implemented between contralateral legs. Rule 5b prolongs stance according to the load that leg is supporting. Additionally, we excited a leg when load was low.

3.2   Single Leg Stance Control

Using a decentralised architecture allows us to have legs following independent leg trajectories. We observed that the leg trajectories that change the most during turns are those of the middle and hind legs on the inner side. Their movement could be due to pulling from the front legs, but the variety of movements and directions for front legs could not be explained purely in terms of external forces. This implies they have to be responsible for their own movements. Therefore, in order to keep θ_η in equation (1), our model for single legs requires control of the movement direction during stance. However, having the legs moving independently can obviously lead to interaction problems. Furthermore, if we follow the approach suggested in section 2.2, it would be useful to have leg trajectories responding passively to other thoracic segments.

Thus, our model controls leg trajectories during stance by combining a velocity controller for each leg, which calculates joint velocities for a desired direction and speed, with a feedforward joint controller as proposed by [9]. The basic idea of the feedforward controller is that following the pulling of other legs will produce an emergent balance across all the joints. This method works particularly well for kinematic models, but is prone to follow all external forces, including gravity; the weight and internal forces to which the robot in our realistic dynamical simulation is subject to can cause further problems for this approach. By combining it with direct velocity control, these problems can be reduced.

For the velocity controller, equation 2 gives joint velocities, $\dot{A} = [\dot{\alpha}\,\dot{\beta}\,\dot{\gamma}]'$, where $T(\alpha, \beta, \gamma)$ is the tarsus position, J is the Jacobian matrix and $\boldsymbol{v}$ is the desired velocity of the tarsus. The height of the body (to compensate for gravity) was controlled locally as in [10] and the speed of all legs was kept constant for all experiments. Therefore the only free parameter was the direction of the leg, θ_L in the 'x–y' plane. For the front legs, θ_L was always that of the direction of the target, i.e. $\theta_L = \theta_\phi$. For the middle and hind legs this was always set to $\theta_L = 0$, i.e. without the front leg influence they would always walk forward. The three leg segments always stay in the same plane, as shown in figure (5), which allows us to solve equation (2) without further parameters.

$$\dot{A} = [\dot{\alpha}\,\dot{\beta}\,\dot{\gamma}]' = [J(\boldsymbol{T}(\alpha, \beta, \gamma))]^{-1}\boldsymbol{v}(\theta_L, speed). \tag{2}$$

The final angular joint velocity was a linear combination between the angular velocity calculated in equation (2) and a feedforward controller that adjusted the joint velocity proportionally to the position error caused by external forces. Therefore, this single leg controller was capable of following given directions, but at the same time it tended to follow external forces. How much it followed external forces was controlled by a subordination parameter $\boldsymbol{s}(s_\alpha, s_\beta, s_\gamma)$ for each thoracic segment. For the front legs $\boldsymbol{s} = 0$, i.e. they follow a straight line toward the target under velocity control alone. For the middle and rear legs, it was found (through empirical testing) necessary to have one set of values for target angles below 60 degrees ($\boldsymbol{s}_{Meta} = [0.10, 0.01, 0.15]$ and $\boldsymbol{s}_{Meso} = [0.40, 0.01, 0.20]$) and another for larger angles ($\boldsymbol{s}_{Meta} = [0.15, 0.01, 0.35]$ and $\boldsymbol{s}_{Meso} = [0.50, 0.01, 0.50]$).

3.3   Results

The robot model was made to turn at angles from 20 to 90 degrees by increments of 10 and due to the symmetry of the system all turns were made to the same side. Because gait coordination is probabilistic, and every time a different pattern was found, three runs were taken for each angle, for a total of 24 turns. Runs were stopped once the body angle was within 5 degrees of the target and the metathorax was aligned with the prothorax in the same direction.

Results from the simulation were analysed using the same approach as for the insect and are shown in figure 6. It can be seen that the prothorax, as for the insect results, tries to achieve the target orientation very early during the turn and maintains it until the metathorax and body are facing in the same direction. Overshooting of the prothorax is more evident for the simulation, meaning that it has more curvature in its turns than the insect. The behaviour of the metathorax direction and body angle are similar to the insect behaviour. However, note that the metathorax for the robot model takes longer to respond to the turn.

Speeds of the prothorax and metathorax are shown on the right hand side in figure 6. It can be seen that the velocity of the rear of the body is automati-

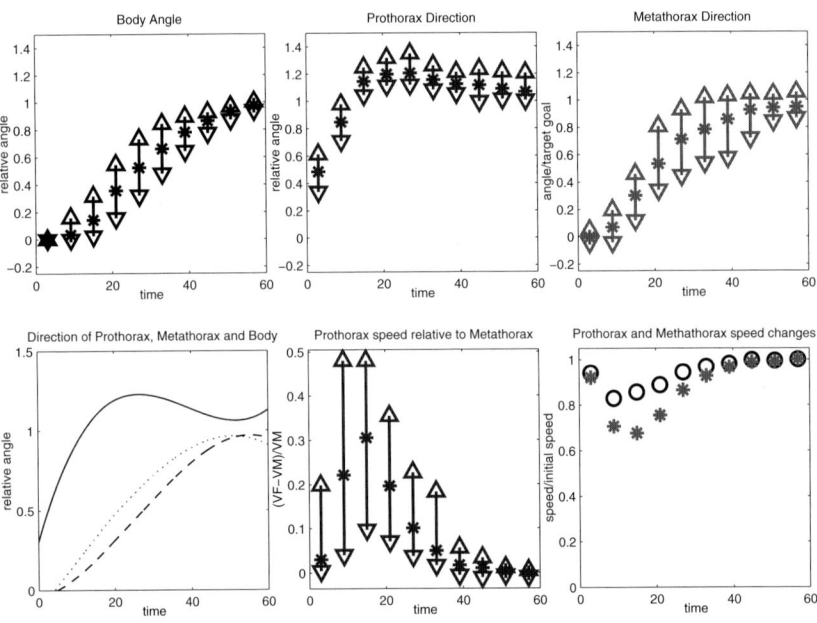

Fig. 6. Simulation results. Bottom left figure shows with a solid line the mean direction of the Prothorax θ_P; Metathorax θ_M with a dotted line; and the body θ_B with a dashed line. Top left shows progress of θ_B in more detail, deviation standard is shown by the arrows. The top middle is that of the prothorax θ_P and the top right is the metathorax direction θ_M. The bottom middle shows the relative speed of the prothorax with respect to the metathorax $(v_P - v_M)/(v_M)$. The bottom right shows changes in speed for the prothorax (○) and metathorax (∗) relative to their velocity before the turn.

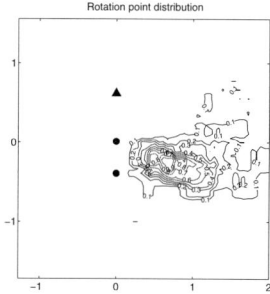

Fig. 7. Cumulative plot of the point of rotation for the simulated turns. The three black dots represent the main thoracic segments.

cally decreased with respect to the prothorax without explicitly controlling it. However, the maximum difference in speed is only 50%, whereas for the insect it can reach values higher than 100%. The simulation was found, like the insect, to recover its original speed before finishing the turn.

The accumulated rotation points of the simulation are shown in figure 7. The rotation peak for the simulation is also found between the hind and middle segment, however, it is clearly located away to one side of the body. If rotation is not so close to the metathorax, the speed of the segment does not decrease as much, as verified by figure 6. However, the opposite could also be the case, i.e., because the metathorax does not decrease in speed, rotation is moved away to one side.

4 Discussion

We were able to reproduce naturalistic insect body and leg trajectories during turns with a very simple control principal: that front leg stance trajectories are directed straight at the target. All the remaining complex pattern of leg movements, and consequent body rotation and speed changes, emerged through decentralised interaction between the legs. A typical simulation turn path is compared to one of the insect in figure 8. The prothorax follows a straight line just as the insect does for a similar angle. Note that calibration was different for each thoracic segment, but was identical for either side of the robot. In spite of this, the inner leg trajectories change abruptly, just as happens with the stick insect, whereas the outer leg trajectories are smoothly curved.

Statistics of the body while turning were also similar to those of the insect, as shown in section 3.3; particularly those of the prothorax, metathorax and body direction. However, there is still some underlying difference in the metathorax and mesothorax. The main difference is the point where the body tends to rotate most. As seen in figure 7, rotation is away from the body axis. Another difference is shown in figure 8, where the insect the hind leg seems to be arrested at that point, whereas for our simulation it is the middle leg [3]. Having the inner hind leg moving at a very low speed moves rotation further back. However, it is not

---

[3] Similar results for inset leg trajectories while turning are also shown in [5].

Fig. 8. Comparison between simulation, the 2 plots on the left, and an insect, the two plots on the right. Leg trajectories relative to the body show only the first third of the turn in both cases. Dark regions indicate stance and light is swing.

clear if this should be explicitly controlled or might, like body rotation, be a secondary effect that we have not entirely captured.

The role of front legs during horizontal forward walking is known to be less important than rear legs. For our simulation, there was no feedforward for front legs, $s_{Pro} = 0$; however, this value should be increased as turns finish. Similarly, only two sets of subordination values were used for the mesothorax and metathorax, depending on how big the angular error to the target was. This proved to be sufficient for replicating body trajectories. However, a more plausible approach might be to have this parameter modulated in a continuous fashion [11]. We found it was not necessary to change the speed and direction of the velocity control of these legs to get successful turning trajectories. However, allowing velocity to be influenced in some way by the turning information, for example, reducing the speed (as for the hind leg mentioned above) might create a closer match to the exact leg trajectories seen in the insect i.e. these legs could play a more active role in turning.

Future work could address the several ways in which the simulation differed from the stick insect. The direct introduction of the heading error angle was an oversimplification of the information the stick insect would receive from its visual system about the direction of the visual target as it turned. Not having tarsi in the simulation makes the robot slip when the feedforward controller is not properly set. This implies having very different values of subordination between systems with and without tarsi. The main difference is that our simulation tends to produce shorter stance trajectories. However, it is important to note that having tarsi is not a requirement for producing similar body trajectories.

Acknowledgement. This project was sponsored by CONACYT México.

References

1. Cruse, H., Bartling, G., Cymbalyuk, J., Dean, J., Dreifert, M.: Modular artificial neural net for controlling a six-legged walking system. Biological Cybernetics **72** (1995) 421–430
2. Cruse, H.: What mechanisms coordinate leg movement in walking arthropods. Trends in Neurosciences **13** (1990) 15–21

3. Bässler, U., Foth, E., Breutel, G.: The inherent walking direction differs for the prothoracic and metathoracic legs of stick insects. The Journal of Experimental Biology **116** (1985) 301–311

4. Jindrich, D.L., Full, R.J.: Many–legged maneuverability: Dynamics of turning in hexapods. The Journal of Experimental Biology **202** (1999) 1603–1623

5. Dürr, V., Ebeling, W.: The behavioural transition from straight to curve walking: kinetics of leg movement parameters and the initiation of turning. The Journal of Experimental Biology **208** (2005) 2237–2252

6. Ekeberg, Ö., Blümel, M., Büschges, A.: Dynamic simulation of insect walking. Arthropod structure and development **33** (2004) 287–300

7. Jander, R., Volk-Heinrichs, I.: Das strauch-spezifische visuelle perceptor-system der stabheuschrecke (*carausius morosus*). Zeithschrift für Vergleichende **70** (1970) 425–447

8. Cruse, H., Kindermann, T., Schumm, M., Dean, J., Schmitz, J.: Walknet-a biologically inspired network to control six-legged walking. neural networks **11** (1998) 1435–1447

9. Schmitz, J., Dean, J., Kindermann, T., Schumm, M., Cruse, H.: A biologically inspired controller for hexapod walking: Simple solutions by exploiting physical properties. The biological bulletin **200** (2001) 195–200

10. Cruse, H., Schmitz, J., Braun, U., Schweins, A.: Control of body height in a stick insect walking on a treadwheel. The Journal of Experimental Biology **181** (1993) 141–155

11. Dürr, V., Matheson, T.: Graded limb targeting in an insect is caused by the shift of a single movement pattern. Journal of Neurophysiology **90** (2003) 1754–1765

Kinematic Modeling and Dynamic Analysis of the Long-Based Undulation Fin of *Gymnarchus Niloticus*

Guangming Wang, Lincheng Shen, and Tianjiang Hu

College of Mechatronics Engineering and Automation,
National University of Defense Technology, Changsha 410073, China
wangguangming@sina.com

Abstract. Within median and/or paired fin (MPF) propulsion, many fish routinely use the long-based undulatory fins as the sole means of locomotion. In this paper, the long-based undulatory fin of an *Amiiform* fish"*G. niloticus*"was investigated. We brought forward a simplified physical model and a kinematic model to simulate the undulations of the long-based dorsal fin. Further, the equilibrium equations of the undulatory fin were obtained by applying the membrane theory of thin shells in which the geometrical non-linearity of the structure is taken into account. Last, we apply the derived kinematic model and equilibrium equations of the undulatory fin to analyze the thrust and propulsive efficiency varying with the aspect ratio of the fin and the maximum swing amplitude.

1 Introduction

Animal systems hold the promise of acting as models for robotic systems with improved performance in the aquatic realm. The morphology and behavior of animals have been copied for development of various technologies [1,2]. Among swimming modes of fish, an estimated 15% of the fish families use Median and/or Paired Fin (MPF) locomotion as their routine propulsive means. Within MPF propulsion, many fish such as knifefish, triggerfish and bowfin, routinely use the long-based undulatory fins as the sole means of locomotion, as well as for manuevering and stabilization. The elongate ribbon-like fins undulating has high performance swimming, precise maneuvering and low speed stability, which not only adapt to cruising swimming in calm waters, but also slow swimming, turning manoeuvres and rapid acceleration from stationary in structurally complex surroundings such as turbulent waters and seashore areas [3,4]. Therefore, the study of the propulsion mode concerning long-based undulatory fins may provide inspiration for the design of next generation autonomous underwater vehicles (AUVs) with remotely operated vehicles (ROVs) capabilities.

As one of the most characteristic representatives utilizing the long-based undulatory fins for swimming, *Amiiform* fish swim by undulations of a long-based dorsal fin, and hold the body axis straight in many cases when swimming. They are able to swim as well as backwards as they do forwards, by reversing the direction of waveform propagation on their long-based dorsal fin. The long-based fin consists of many fin-rays and a flexible membrane connecting them together. Fish have extensive

S. Nolfi et al. (Eds.): SAB 2006, LNAI 4095, pp. 162–173, 2006.

muscular control over fin-rays. The fish's precise maneuvering stems from the fact that the long-based fin is a propulsor having a high number of actively controlled inputs. By properly coordinating these inputs, exquisite control of the thrust vector is possible [5,6].

In order to investigate the undulatory dorsal fin propulsion and its potential for providing alternative approaches for future underwater vehicle design, the long-based undulatory fin of an Amiiform fish"*Gymnarchus niloticus*"was studied. We have quantificationally investigated the cruising swimming of *G. niloticus* by employing a high-speed digital video and the image-measure technology, and brought forward a simplified physical model and kinematic model to simulate the undulations of the long-based dorsal fin. The aim of the research is twofold. First, we aim at establishing the dynamic equations of the long-based fin motion and analyzing the propulsive performance of the locomotive means. Second, we aim at offering a kinematic model to develop control algorithms for undulatory motions.

2   The Simplified Physical Model of the Long-Based Fin

G. niloticus is a large aggressive freshwater electric eel distributing tropical Africa and Nile(fig.1). Maximum length of the fish is 1.5 m, usually below 0.9 m. The fish has a streamlined body, tapering to a posterior point. Swimming is accomplished by means of wavelike motion along the long-based dorsal fin, while the body axis is held straight in many cases when swimming. The anal and caudal fins have missed, while the dorsal fin extends along most of the body length and exhibits a large number of fin-rays (up to 183-230)[7]. The depth of the fin is almost constant except the original segment. Locomotor waves may pass in either direction along the dorsal fin, and may show widely varying amplitude and frequency, particularly during turning or braking.

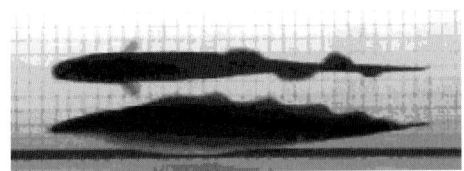

Fig. 1. Silhouette of *G. niloticus* during steadily swimming

The long-based fin system consists of the muscles, fin-rays and a flexible membrane connecting them together. A set of muscles for each fin-ray provide the later with two degrees-of-freedom movement capability. Fish can reduce the influence of the exterior fluid around the body and inner elasticity of the fin during swimming by actively modulating the amplitude and frequency of the fin-rays swaying [8]. For mathematical convenience, the long-based fin is simplified a system that makes up of N equal thin rods and a rectangular elasticmembrane connecting them together. Figure 2 shows the structure of the simplified physical model. Two consecutive rods and the membrane between them form a fin cell. All fin-rays are collocated at regular intervals along the spine, connecting with the spine by a simple

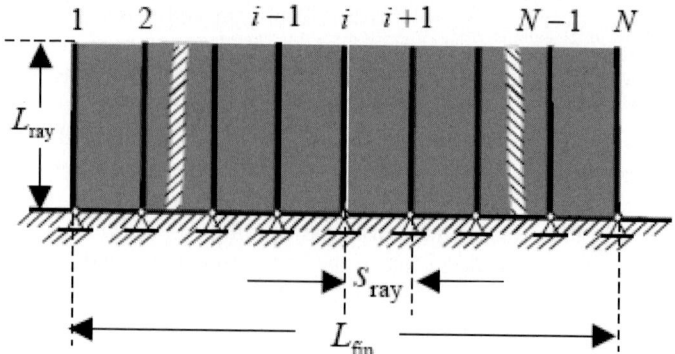

Fig. 2. The structure of the simplified physical model of the long-based fin. L_{fin} denotes the length of the fin, L_{ray} is the highness of the fin-ray, and S_{ray} is the space between consecutive fin-rays.

supported means. Fin-rays can swing to the lateral directions about the spine by the action of power forces. The bottom edge of the membrane is fixed on the spine, and the lateral edges are simply supported by fin-rays.

3   Analysis and Modeling of the Long-Based Fin Undulating

3.1   Analysis of the Long-Based Dorsal Fin Undulating

Figure 3 images show the body position and dorsal fin's waves at the same intervals during a specimen steadily swimming [9]. The fin-rays periodically wiggle around the spine, and drive the membrane waving with the same frequency during the long-based fin undulating. Since there is a phase lag between the consecutive fin-rays, the membrane twists and distorts along the chord of the membrane by the fin-rays powers and the fluid forces. The torsion angles and distortions periodically vary with the time, and augment along the spread of the membrane, 0 at the base and maximum at the edge of the membrane.

The shape along the spread of the membrane varies with the time during swimming. ①G. niloticus keeps the long-based fin a plane during suspending in the calm water. ②The membrane arches from the outer edge downwards to the spine during the fin cell swings downwards to the lateral. The arch curvature reaches its peak at the maximum angular deflection for the fin cell. ③During the fin cell swings from the lateral upwards to the middle, the membrane basically maintains arch, and the arch curvature gradually minishes until the membrane wiggles to the top. The half cycle of the membrane cell oscillating is done. ④The fin cell does the other half cycle during oscillating on the reversed side of the body. The process of oscillating and the shape of the fin cell are mirror-image symmetrical with the half cycle done. Then the membrane cell begins next a cycle, swings in cycles.

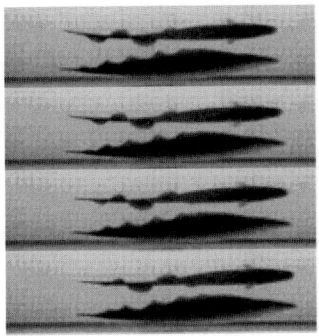

Fig. 3. Images show body position and dorsal fin's waves at regular intervals during steadily swimming (From [9])

3.2 Modeling for the Fin-Rays Swaying

During cruising, fish's muscular systems control the individual fin-rays periodically swaying around the spinal column, and the dorsal fin takes the shape of an analogous sine wave. Thus, a simple model imitating fin-rays locomotion is that the angular velocities of the fin-rays vary sinusiodally. Assume that every fin-ray swings with a common frequency and a constant phase lag between consecutive fin-rays, then the simplified model may be described as:

$$\phi^i = A_0^i \sin(2\pi ft + (i-1)\phi_{lag}) - \psi \quad , \quad i = 1,\dots,N \tag{1}$$

Where ϕ^i is the angular deflection for the ith fin-ray. A_0^i is the maximum angular deflection for the ith fin-ray. f is the swing frequency for the fin-rays. ϕ_{lag} is the phase lag angle between consecutive fin-rays. The angular ψ marks the offset of swimming paths for the body, and is set to $\psi = 0$ for locomotion in a straight line. Propagation direction for the wave depends on the sign of the phase lag parameter, and is from the body head to the tail for $\phi_{lag} < 0$. The condition $\phi_{lag} = \pm 2\pi \cdot n/N$ yields (exactly) n equal wavelength of the propulsive wave across the undulating fin.

3.3 Kinematic Modeling for the Fin Cell

We use a Cartesian coordinate system $o - xyz$ to describe undulatory moti-ons of the long-based fin (as Fig.4 shown). The coordinate origin o is at the tip of the body. The x-axis is along the base line and points to the head. The y-axis is parallel to the rest fin and points to the outer edge of the fin. The fin-rays rotate about the x-axis in the plane paralleled with the yz-plane and cause the membrane between consecutive fin-rays to make a lateral movement.

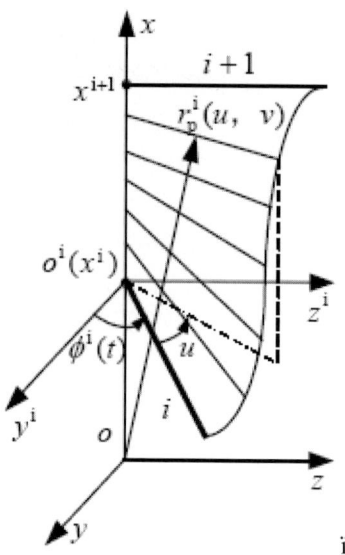

Fig. 4. Sketch of the curved surface of the fin cell at time t during the long-based fin undulating

The ith fin cell consists of the ith fin-ray, ($i+1$)th fin-ray and the membrane between them. The position of the ith fin-ray at the x-axis is x^i. The fin-rays oscillating and the fluid loading cause the membrane to form a curved surface. Assume that the fin cell forms a ruled surface at time t, and then the curved surface of the fin cell may be described as follows:

$$r^i = r^i(u, \ v) = \{b^i u + x^i, v\cos\theta(u), v\sin\theta(u)\}, \quad (u,v)\in D^i \qquad (2)$$

$$b^i = S_{ray}/(\phi^{i+1} - \phi^i), \theta(u) = \phi^i + u$$

$$D^i: \begin{cases} u = (x-x^i)/b^i, & x^i \le x \le x^{i+1} \\ 0 \le v \le L_{ray} \end{cases}$$

Where u and v are the parameter variables of the curved surface. r^i denotes the position vector concerning any surface point $P(u,v)$ on condition that the membrane is acted on by outside forces at time t. The rulings are parallel to the yz-plane and form an angle $\theta(u)$ with the xy-plane. the unit tangent vector of the ruling through the point u is $\{0, \cos\theta(u), \sin\theta(u)\}$.

We use $\tilde{r}_p^i$ and Δr_p^i to describe the curved surface deformation of the membrane during the long-based fin undulating. When the membrane surface is a plane, that is to say, when the membrane has no distortion at time t, the position vector concerning the point $P(u,v)$ is $\tilde{r}_p^i$. The deformation displacement with regard to the point $P(u,v)$ is Δr_p^i at time t. Expressions 3~4 correspond to $\tilde{r}_p^i$ and Δr_p^i, respectively.

$$\tilde{r}_p^i = \tilde{r}_p^i(u, v) = \{b^i u + c, \quad v\cos\phi^i, \quad v\sin\phi^i\} \tag{3}$$

$$\Delta r_p^i = r_p^i - \tilde{r}_p^i = \{0, \quad v\cos(\phi^i + u) - v\cos\phi^i, \quad v\sin(\phi^i + u) - v\sin\phi^i\} \tag{4}$$

In the orthogonal curvilinear coordinate system $o - uv$, the unit tangential vectors and unit surface normal vector are r_{eu}^i、 r_{ev}^i and r_{en}^i, respectively. The component parts of r_p^i and Δr_p^i along the curvilinear coordinates and normal direction of the surface can be obtained as follows:

$$(u_0^i \quad v_0^i \quad w_0^i)^{\mathrm{T}} = \begin{pmatrix} r_{eu}^i \\ r_{ev}^i \\ r_{en}^i \end{pmatrix}_{3\times3} \cdot (r_p^i)^T \tag{5}$$

$$(u_1^i \quad v_1^i \quad w_1^i)^{\mathrm{T}} = \begin{pmatrix} r_{eu}^i \\ r_{ev}^i \\ r_{en}^i \end{pmatrix}_{3\times3} \cdot (\Delta r_p^i)^T \tag{6}$$

Where $(u_0^i \quad v_0^i \quad w_0^i)$ is the component parts of the position vector at time t. $(u_1^i \quad v_1^i \quad w_1^i)$ is the component parts of the deformation displacement at time t.

4   Dynamic Analysis of the Long-Based Fin

4.1   The Stress-Strain Relations for a Membrane

The geometry of the membrane between consecutive fin-rays is such that its thickness dimension is much smaller than its in-surface dimensions. Thus, the kinematic deformation of the membrane in thickness direction can be omitted, and the kinematic deformation of the membrane is approximate to that of the middle surface. We apply the thin-shell theory to analyze the membrane dynamics by adopting some assumptions as follows [10]: ① points initially located along straight fibers in the through-the-thickness direction remain along straight fibers;② the principal stain in the direction vertical to middle surface is most small andïa negligible quantity in comparison with unity; ③ the principal stain in the surface parallel to the middle surface is far less than that in the vertical plane, and the deformation caused by the principal stain in the surface parallel to the middle surface may also be omitted; ④ body forces and surface forces can be transformed the loading acted on the middle surface. Since membranes have negligible bending stiffness and cannot sustain compressive stresses, a further assumption is that all cross sections of the membrane have no blending moment and twisting moment. According to above assumptions, components of the tangential strain with regard to point $P(u, v)$ at time t are given by:

$$\varepsilon_1^i = \frac{1}{A^i}\frac{\partial u_1^i}{\partial u} + \frac{1}{A^i B^i}\frac{\partial A^i}{\partial v}v_1^i - \frac{w_1^i}{R_1^i}$$

$$\varepsilon_2^i = \frac{1}{B^i}\frac{\partial v_1^i}{\partial v} + \frac{1}{A^i B^i}\frac{\partial B^i}{\partial u}u_1^i - \frac{w_1^i}{R_2^i} \tag{7}$$

$$\omega^i = \frac{A^i}{B^i}\frac{\partial}{\partial v}\left(\frac{u_1^i}{A^i}\right) + \frac{B^i}{A^i}\frac{\partial}{\partial u}\left(\frac{v_1^i}{B^i}\right) + \frac{2w_1^i}{R_{12}^i}$$

Where ε_1^i and ε_2^i are components of strain along u-line and v-line. ω^i is shear strain which corresponds to angle varieties between u-line and v-line. A^i and B^i are Lame modulus. R_1^i and R_2^i are curvature radius along curvilinear coordinates concerning the normal section. Suppose that the membrane is homogenous, isotropic and linearly elastic. The stress-strain relations on the membrane are therefore given by:

$$N_1^i = \frac{2Eh}{1-\mu^2}(\varepsilon_1^i + \mu\varepsilon_2^i)$$

$$N_2^i = \frac{2Eh}{1-\mu^2}(\varepsilon_2^i + \mu\varepsilon_1^i) \tag{8}$$

$$S^i = S_1^i = -S_2^i = \frac{Eh}{1+\mu}\omega^i$$

Where N_1^i and N_2^i are components of stress along curvilinear coordinates, S^i is shear stress in the middle surface, E is Yong's Elastic Modulus, μ is Poisson ratio, and h is the thickness of the membrane.

4.2 Differential Equations of Motion for the Membrane in Curvilinear Coordinates

We consider the dynamic balance for a differential element of the membrane using the theory of a flexible thin shell. The forces acting on a differential element include the elastic forces (N_1^i, N_2^i and S^i), fluid loading force (q^i) and turning inertia force ($\rho_m h A^i B^i du dv \frac{\partial^2 r_p^i}{\partial t^2}$). ρ_m is the density of the membrane. The equations of motion for a membrane in the curvilinear coordinates are written by:

$$\frac{\partial}{\partial u}(B^i N_1^i) + \frac{\partial A^i}{\partial v} S_1^i - \frac{\partial}{\partial v}(A^i S_2^i) - \frac{\partial B^i}{\partial u} N_2^i + A^i B^i q_u = A^i B^i \rho_m h \frac{\partial^2 u_0^i}{\partial t^2}$$

$$\frac{\partial}{\partial u}(B^i S_1^i) - \frac{\partial A^i}{\partial v} N_1^i + \frac{\partial}{\partial v}(A^i N_2^i) - \frac{\partial B^i}{\partial u} S_2^i + A^i B^i q_v = A^i B^i v \rho_m h \frac{\partial^2 v_0^i}{\partial t^2} \qquad (9)$$

$$\frac{N_1^i}{R_1^i} + \frac{N_2^i}{R_2^i} + \frac{S_2^i - S_1^i}{R_{12}^i} + q_w = \rho_m h \frac{\partial^2 w_0^i}{\partial t^2}$$

Where q_u, q_v and q_w are the components of q^i in the curvilinear coordinates. The integrated loading forces $Q^i (Q_x^i \quad Q_y^i \quad Q_z^i)$ with the Descarts coordinates for the ith fin cell membrane is given by:

$$Q^i = (Q_x^i \quad Q_y^i \quad Q_z^i) = \iint_{D^i}(q_u \ q_v \ q_w) \cdot (r_e^i)^T \, du \, dv \qquad (10)$$

4.3 Boundary Conditions on the Membrane

The stresses and displacement of deformation for the membrane should meet the boundary conditions as follows:

①Along the fixed edge (spine) of the membrane, $u_1^i = 0$ and $v_1^i = 0$ for $v = 0$.

②Along the simply-supported edges (fin-rays) of the membrane, $N_1^i = 0$ and $v_1^i = 0$ for $u = 0$ or $u = \phi^{i+1} - \phi^i$.

③Along the free edge of the membrane, $N_2^i = 0$ and $S^i = 0$ for $v = L_{ray}$.

4.4 Analysis of the Forces on the Fin-Rays

Assume that the fin-rays are rigid and have no elastic deformation during rotating. The forces acting on the ith fin-ray are membrane tensions (T^i and $\tilde{T}^i$), resultant force (q_0^i) of power force and fluid loading force, and turning inertia force ($\pi \rho_{0m} d^2 B^i \frac{\partial^2 r_p^i}{\partial t^2}$). T^i is caused by the ith fin cell membrane, $\tilde{T}^i$ is caused by the $(i-1)$th fin cell membrane, ρ_{0m} is the density of the fin-rays, and d is the radius of the cross section of the fin-rays. The differential equations of motion for fin-rays in curvilinear coordinates are given by:

$$T_u^i + \tilde{T}_u^i + B^i q_{0u} = B^i \pi \rho_{0m} d^2 \frac{\partial^2 u_0^i}{\partial t^2} \quad , \quad u = 0$$

$$T_v^i + \tilde{T}_v^i + B^i q_{0v} = B^i \pi \rho_{0m} d^2 \frac{\partial^2 v_0^i}{\partial t^2} \quad , \quad u = 0 \qquad (11)$$

$$T_w^i + \tilde{T}_w^i + q_{0w} = \pi \rho_{0m} d^2 \frac{\partial^2 w_0^i}{\partial t^2} \quad , \quad u = 0$$

Where q_{0u} , q_{0v} and q_{0w} are the components of q_0^i in the curvilinear coordinates, T_u^i , T_v^i and T_w^i the components of T^i in the curvilinear coordinates, and $\tilde{T}_u^i , \tilde{T}_v^i$ and $\tilde{T}_w^i$ the components of $\tilde{T}^i$ in the curvilinear coordinates. $\tilde{T}^i$ is 0 for $i=1$ and T^i is 0 for $i = N$.

The integrated loading forces $Q_0^i(Q_{0x}^i \quad Q_{0y}^i \quad Q_{0z}^i)$ for the ith fin-ray is given by:

$$Q_0^i = (Q_{0x}^i \quad Q_{0y}^i \quad Q_{0z}^i) = \int_{0 \le v \le L_{ray}} (q_{0u} \quad q_{0v} \quad q_{0w}) \cdot (r_e^i)^T \, d\,v \tag{12}$$

The integrated force Q acting on the long-based fin is therefore given by:

$$Q = (Q_x \quad Q_y \quad Q_z) = \sum_{i=1}^{N-1} Q^i + \sum_{i=1}^{N} Q_0^i \tag{13}$$

5   Numerical Calculation and Analysis

Configuration parameters of the long-based fin, characteristic parameters of fin material and movement parameters of the fin may influence the dynamic performance of the fin to a certain extent. In this section, we apply the derived the kinematic model and equilibrium equations of the undulatory fin to analyze the thrust and propulsive efficiency varying with the aspect ratio ($g1$) of the fin and the maximum swing amplitude ($A0$). Assume the parameters of the rectangle fin as follow: $E = 10^6 (\text{N/m}^2)$, $\mu = 0.4$, $h = 0.1\text{mm}$, $\rho_m = 950\text{kg/m}^3$, $\rho_{0m} = 7900\text{kg/m}^3$, $d = 0.2\text{mm}$, the length of the membrane is $L = 17\text{mm}$, the frequency of the oscillating membrane $f = 1\text{Hz}$, the deflexion angle of the spine $\alpha = 0$, the obliquity of the fin-ray $\beta = \pi/2$.

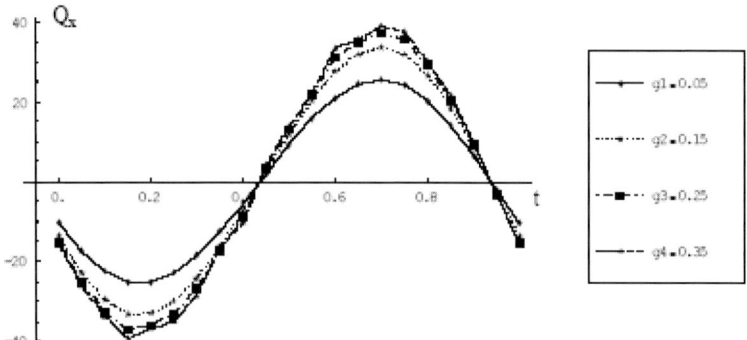

Fig. 5(a) Thrusts Q_x varying with different the aspect ratio ($g1$) of the fin and time t

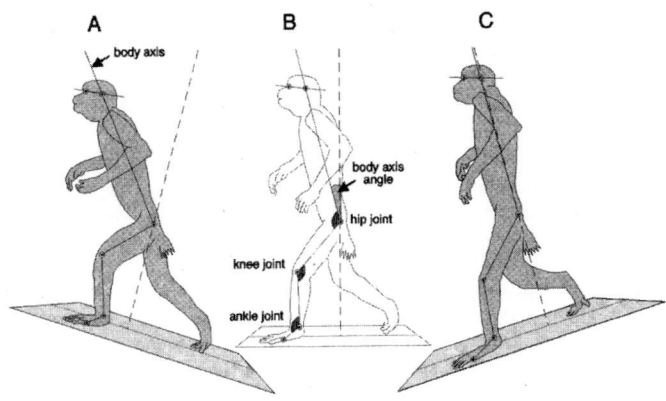

Fig. 1. Walking experiment of young monkey. Adaptation to inclination changes[11]. Reprinted with permission.

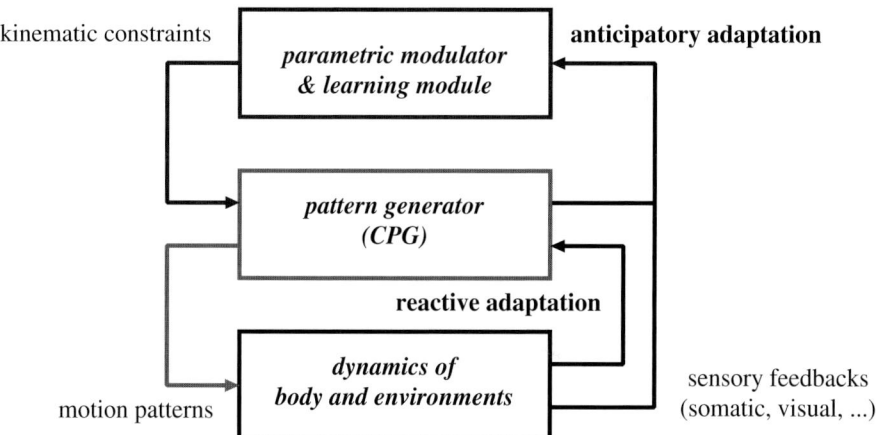

Fig. 2. Proposed dual adaptation loops model for motor control

dynamics of physical body and control inputs (i.e. outputs of CPG model) based on sensory feedbacks. This concept is sometimes called "global entrainment[6]".

In contrast, it has been generally clarified that there are a variety of adaptation loops in animals motor control systems, and thanks to the spatiotemporal redundancy for adaptation, we can adapt to a broad range of environments.

According to recent physiological study[11], well-trained monkey can walk on a moving treadmill even if its elevation angle are arbitrarily changed(Fig.1). It is considered that the monkey can generate suitable force to control its posture and walking pattern by using both of feedback and feedforward adaptation strategies.

From developmental system's point of view, a sophisticated motor control should be regulated based on feedforward adaptation since sensory feedback

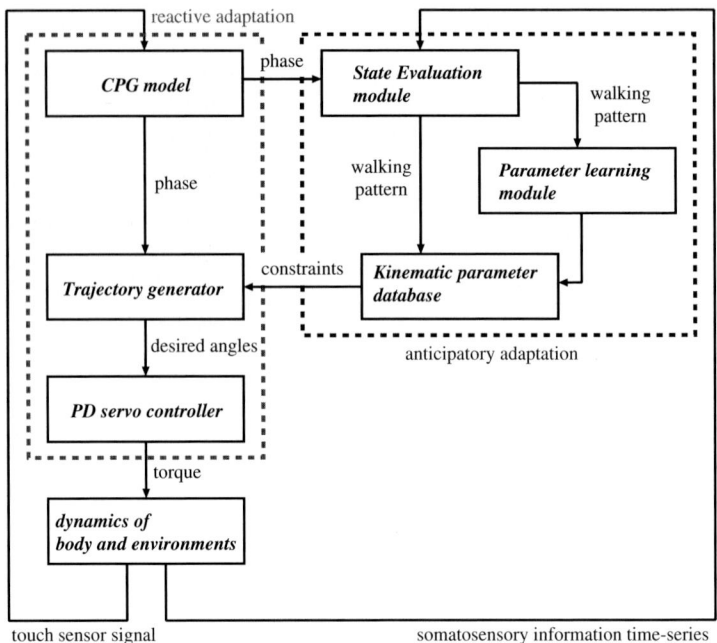

Fig. 3. An implementation of the proposed dual adaptation loops model

can not avoid dead time for observation. On the contrary, in the early stage of developmental process, feedback adaptation should be dominant. The roles of feedforward and feedback adaptation loops in motor control is well summarized in [12,13].

Based on the above, we had proposed a reactive and anticipatory adaptation model for environmental cognition and motor adaptation[14]. As shown in Fig.2, it basically consists of three components, parametric modulator, pattern generator, and the dynamics of body (e.g. musculo-skeletal system) and external environments.

2   Method

In this section, we explain the implementation of the proposed method. As shown in Fig.3, the proposed control system is basically composed of a coupled phase oscillators, trajectory generator, evaluation module, and learning module.

2.1   Phase Oscillator-Based Walking Control

As has been noted, the proposed method has dual adaptation loops for a biped walking control. First, we account for the basic control system denoted in the left hand side of Fig.3. The basic frequency of locomotion (i.e. walking speed) is governed by coupled phase oscillators depicted as **CPG** which are assigned

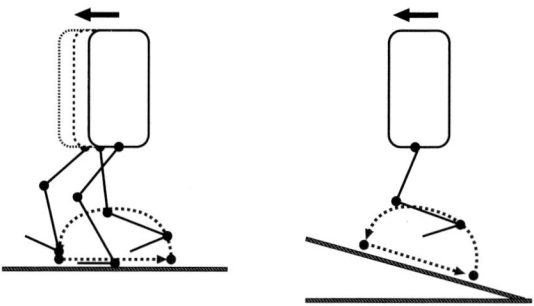

Fig. 4. Desired trajectory for a biped locomotion

to left and right leg, respectively. The dynamics of these oscillators are given by the following differential equation,

$$\dot{\phi}_i = \omega_i + K \sum_{k=1}^{N} \sin\left(\phi_k - \phi_i\right) + \Gamma_i \;, \tag{1}$$

where ϕ is the phase of the oscillator, ω is a angle velocity, and K is the interaction gain among neighbor oscillators. In addition, Γ represents the **phase resetting** term given by,

$$\Gamma_i = \left(\hat{\phi}_i^{touch}(t) - \phi_i^{touch}(t)\right) \delta\left(t - t^{touch}\right) \;. \tag{2}$$

where $\hat{\phi}^{touch}$ indicates desired phase of swinging leg in which it has just touch on the ground, and ϕ^{touch} corresponds to the actual phase of the leg. In addition, $\delta\left(\cdot\right)$ denotes the Dirac's delta function. The phase resetting method has been widely adopted in legged robots control studies for keeping stable locomotion[9,10].

According to the phase of each oscillator, subsequent **Trajectory generator** determines the position of the heel based on the kinematics (See Fig.4). Due to this, kinematic parameters for each leg should be constrained.

Based on the desired trajectories, each actuated joints are servo controlled with torque τ_i.

$$\tau_i = K_{P_i}\left(\hat{\theta}_i - \theta_i\right) + K_{D_i}\left(\dot{\hat{\theta}}_i - \dot{\theta}_i\right) \;. \tag{3}$$

where $\hat{\theta}$ and $\dot{\hat{\theta}}$ are desired, and θ and $\dot{\theta}$ are actual trajectories of angle and angle velocity, respectively. Furthermore, K_P and K_D are the feedback gain parameters.

2.2   Elicitation of Sensorimotor Constraints

Although adaptation by phase entrainment and phase resetting method is robust, these are available with respect to a limited perturbation. Thus in the outside of the available regions (i.e. unexperienced situations), other adaptation algorithm is necessary.

In this study, we propose actively changing the kinematic parameters such as "length of stride" and "inclination angle" based on the state evaluation module.

The elicitation of the previously-experienced sensorimotor constraints is executed by the case-based reasoning method[15]. For this aim, the once optimized kinematic parameters have to be stored by correlating with the corresponding sensorimotor information. As illustrated in the right hand side of Fig.3, state evaluation module monitors somatosensory, phase, and walking pattern information at any time. Thus in the module, code book vectors (i.e. reference vectors) for the case-based reasoning are maintained using LVQ (learning vector quantization) algorithm[15].

3   Simulations

3.1   Simulation Setting

To verify the feasibility of the proposed method, it was applied to a biped walking robot control in a physical simulator which is developed using an open source physics engine, ODE[16].

Fig.5 depicts schematic model of a biped walking robot and our developed simulator. As shown in the left figure, the biped only has lower body with seven DOF in pitch direction (i.e. hip, knee, ankle joints). Therefore it is constrained to move in sagittal plane. And the detail parameters of each body part are listed in Table. 1.

3.2   On-Line Optimization of Kinematic Parameters

In the proposed adaptation method for biped walking robot control, the two kinematic parameters, $\boldsymbol{p}_i = [l_i, \theta_i]^T$, here l and θ are the length of stride and

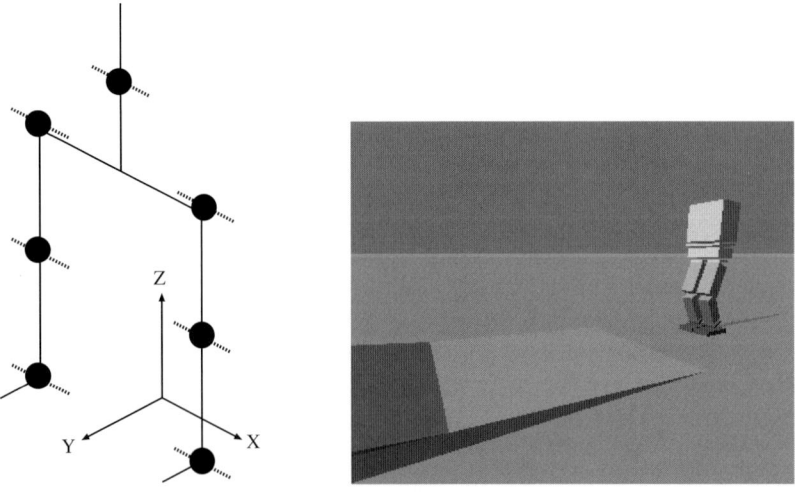

Fig. 5. Simulated biped walking robot and developed simulator

Table 1. Parameters of biped

Body Part	Mass [Kg]	Width [m]	Height [m]	Depth [m]
FOOT	0.10	0.07	0.02	0.13
SHANK	0.18	0.05	0.12	0.05
THIGH	0.30	0.06	0.14	0.06
HIP	0.36	0.15	0.04	0.10
BODY	1.84	0.16	0.12	0.12

estimated inclination angle respectively, have to be optimized for realizing natural (i.e. swing less) locomotion. In addition, the kinematic parameters that once
optimized have to be stored in dynamical memory by correlating with the corresponding sensorimotor information. These parameters are called **sensorimotor
constraints** in this paper.

Obviously, the sensorimotor constraints have to be optimized through dynamic interaction with the situated environments. Due to this, we tried to optimize these parameters in on-line under various kinds of environments using the
following updating laws.

$$E(t) = \int^{2T} x_{ZMP}(t)dt , \qquad (4)$$

$$l_i(t+1) = l_i(t) - K_c \cdot sgn\left(\frac{E(t) - E(t-1)}{l_i(t)}\right) , \qquad (5)$$

$$sgn(x) = \begin{cases} 1 & (x > 0) \\ -1 & (x < 0) \end{cases} ,$$

$$\theta_i(t+1) = \theta_i(t) + \left(K \cdot \frac{1}{T}\int^{T} \theta_{incl}(t)dt - \delta_{offset}\right) . \qquad (6)$$

where $E(t)$ indicates the performance of the biped locomotion, which is calculated from the x component of the ZMP (i.e. pitch angle). The equation (5)
represents the updating law of the stride $l(t)$, and $sgn(\cdot)$ denotes the a sign
function. Moreover, the equation (6) also represents the updating law for the
estimated inclination angle $\theta(t)$.

Fig.6 represents the transitions of evaluation $E(t)$ (dashed line) measured
from integration of swinging in a walking period (i.e. lower is better), and the
optimized length of stride $l(t)$ (solid line). The result shows that the stride can
be optimized in on-line.

On the other hand, Fig.7 demonstrates that transitions of body inclination
angle (solid line) and the estimated slope angle (dashed line). Owing to wrong
estimation, in early stage of iteration, the body inclination has been swinging.
As the estimation becomes optimal, the swinging range is converged.

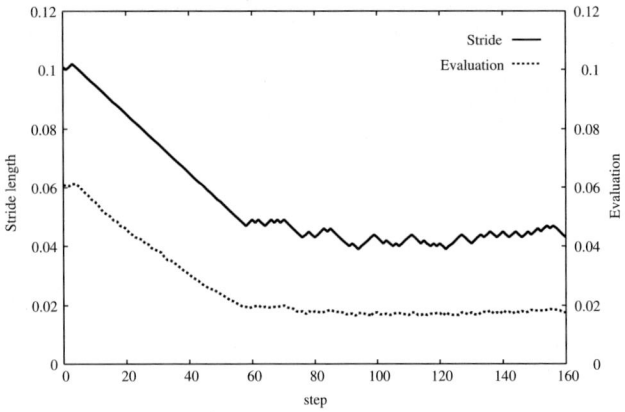

Fig. 6. Result of on-line optimization of a kinematic parameter (length of stride)

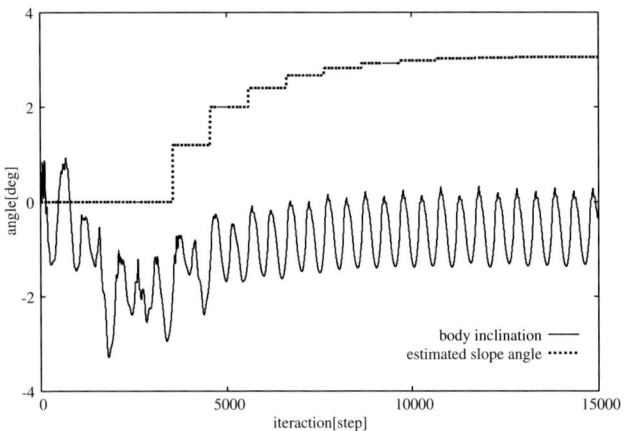

Fig. 7. Result of on-line estimation of slope angle

3.3 Adaptive Locomotion by Eliciting Sensorimotor Constraints

In order to evaluate the feasibility of the proposed feedforward adaptation method, i.e. eliciting sensorimotor constraints according to the situation, simulation has been executed.

Fig.8 shows (a) snapshots and (b) stick pictures of locomotion of biped walking robot on up slope, the elevation angle is gradually increased.

Fig.9 illustrates trajectory of COG (center of gravity) of the biped walking robot. As can be seen from the figure, walking pattern (i.e. dashed line) was changed when the biped walked into the 6.0[deg] slope. In other words, the biped robot can adapt the 3.0 [deg] slope only with the phase entrainment/resetting (i.e. feedback) adaptation. From the trajectory of Y component of the ZMP (Fig.10), we can see that the biped demonstrates backward inclining locomotion

(a) Snapshots

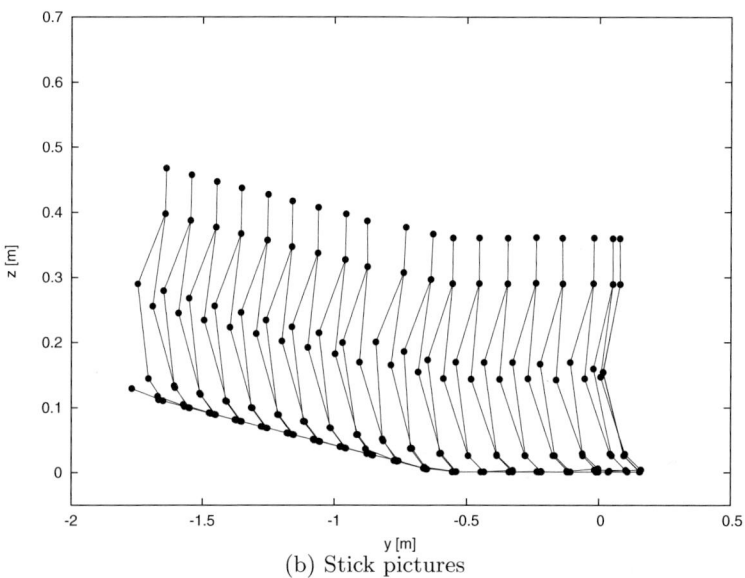

(b) Stick pictures

Fig. 8. Locomotion of biped walking robot on up slope with different angles. First, the robot is situated on a flat surface, the elevation angle of the slope gradually scarped (i.e. 0.0, 3.0, and 6.0[deg]).

on 3.0 [deg] slope, since it depends on phase resetting. In contrast, on 6.0 [deg] slope, its posture is recovered because appropriate kinematic constraints has been elicited.

3.4   Adaptive Locomotion on Down Slope

Similarly, walking experiment on down slope has been attempted. In this experiment, the slope angle was changed from flat surface to down slope (angle is -4.0[deg]).

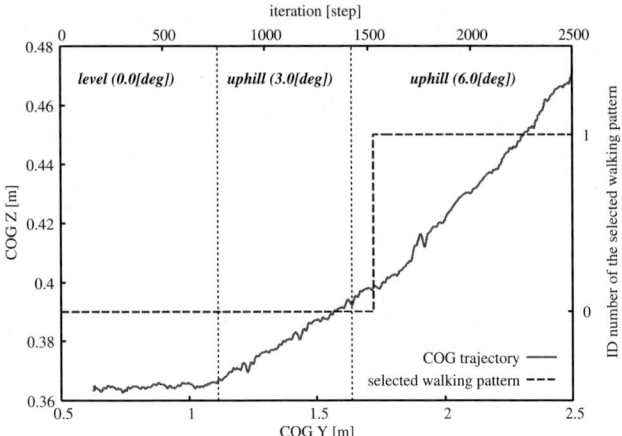

Fig. 9. Trajectory of COG

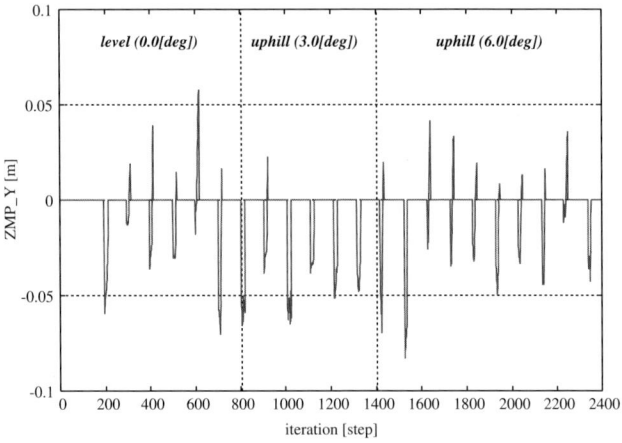

Fig. 10. Trajectory of ZMP

We can see (a) the snapshots and (b) stick pictures of locomotion on the down slope from the Fig.11.

4   Discussions

In this paper, we proposed an environmental adaptation method for a biped walking robot control. There are dual loops in the proposed adaptation mechanism, one is based on phase entrainment ability of pattern generators, and the other is feedforward elicitation of sensorimotor constraints.

As can be seen in Fig.8 and Fig.11, the robot with the proposed adaptation loops can walk continuously in spite of slope elevation angle are changed. In

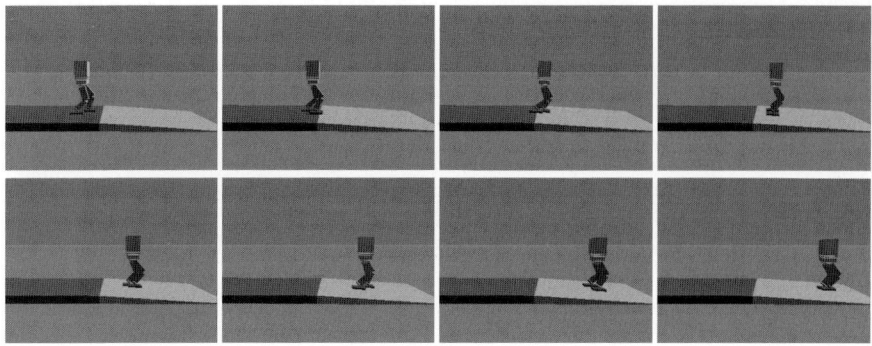

(a) Snapshots

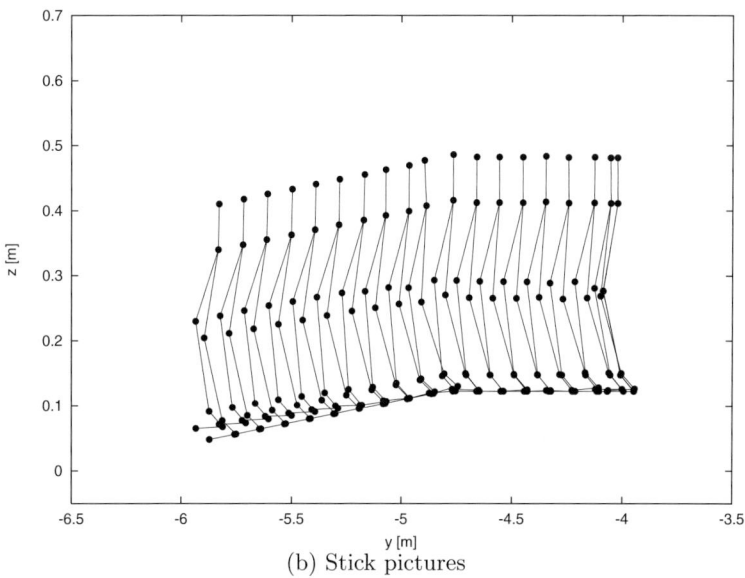

(b) Stick pictures

Fig. 11. Locomotion of biped walking robot on down slope

addition, we confirmed that the robot can continue to walk even though it is situated on the slope with unexperienced angles.

Acknowledgments

This research has been partially supported by a Grant-in-Aid for Scientific Research on Priority Areas (No.454, 2005–2010), and Young Scientists (B) (No.18700195) from the Japanese Ministry of Education, Culture, Sports, Science, and Technology.

References

1. Honda Worldwide – ASIMO at `http://world.honda.com/ASIMO/`
2. McGeer, T., Passive dynamic walking, *International Journal of Robotics Research*, **9**-2, 62–82, 1990.
3. Collins, S., Ruina, A. Tedrake, R., and Wisse, M., Efficient Bipedal Robots Based on Passive Dynamic Walkers, *Science Magazine*, **307**, 1082–1085, 2005.
4. Asano, F., Yamakita,M., Kamamichi, N. and Luo, Z.-W., A Novel Gait Generation for Biped Walking Robots Based on Mechanical Energy Constraint, *IEEE Tran. on Robotics and Automation*, **20**-3, 565–573, 2004.
5. Ogino, M., Hosoda, K. and Asada, M., Learning Energy-Efficient Walking with Ballistic Walking, *Adaptive Motion of Animals and Machines*, 155–164, Springer, 2005.
6. Taga, G., A model of the neuro-musculo-skeletal system for human locomotion. *Biological Cybernetics* **73**-1, 97–111, 1995.
7. Ijspeert, A. J., A connectionist central pattern generator for the aquatic and terrestrial gaits of a simulated salamander. *Biological Cybernetics*, **85**, 331–348, 2001.
8. Makino, Y., Akiyama, M., and Yano, M., Emergent mechanisms in multiple pattern generation of the lobster pyloric network, *Biological Cybernetics*, **82**, 443–454, 2000.
9. Tsuchiya, K., Aoi, S. and Tsujita, K., Locomotion Control of a Biped Locomotion Robot using Nonlinear Oscillators, *Proceedings of the IEEE/RSJ International Conference on Intelligent Robots and Systems*, 1745–1750, 2003.
10. Kimura, H., Fukuoka, Y. and Mimura, T., Dynamics based integration of motion adaptation for quadruped robot "tekken", *Proc. of the second International Symposium on Adaptive Motion of Animals and Machines*, 2003.
11. Mori, F., Nakajima, K. and Shigemi, M., Control of Bipedal Walking in the Japanese Monkey, *M. fuscata*: Reactive and Anticipatory Control Mechanisms, *Adaptive Motion of Animals and Machines*, 249–259, Springer, 2005.
12. Kuo, A.D., The relative roles of feedforward and feedback in the control of rhythmic movements, *Motor Control*, **6**, 129–145, 2002
13. Ito, S. and Kawasaki, H., Regularity in an environment produces an internal torque pattern for biped balance control, *Biological Cybernetics*, **92**, 241-251, 2005.
14. Kondo, T., Somei, T. and Ito, K., A predictive constraints selection model for periodic motion pattern generation. *Proc. of 2004 IEEE/RSJ International Conference on Intelligent Robots and Systems (IROS'04)*, 975–980, 2004.
15. Kohonen, T., Self-Organizing Maps, Springer, Berlin, Heidelberg, New York, 1995.
16. Smith, R.: Open dynamics engine v0.5 user guide, 2004. `http://ode.org/`

Dynamic Generation and Switching of Object Handling Behaviors by a Humanoid Robot Using a Recurrent Neural Network Model

Kuniaki Noda[1], Masato Ito[1], Yukiko Hoshino[1], and Jun Tani[2]

[1] Sony Intelligence Dynamics Laboratories, Inc.,
Takanawa Muse Building 4F, 3-14-13 Higashigotanda,
Shinagawa-ku, Tokyo, 141-0022, Japan
{noda, masato, yukiko}@idl.sony.co.jp
[2] Brain Science Institute, RIKEN,
2-1 Hirosawa, Wako-shi, Saitama, 351-0198, Japan
tani@brain.riken.go.jp
http://www.bdc.brain.riken.go.jp/~tani/

Abstract. The present study describes experiments on a ball handling behavior learning that is realized by a small humanoid robot with a dynamic neural network model, the recurrent neural network with parametric bias (RNNPB). The present experiments show that after the robot learned different types of behaviors through direct human teaching, the robot was able to switch between two types of behaviors based on the ball motion dynamics. We analyzed the parametric bias (PB) space to show that each of the multiple dynamic structures acquired in the RN-NPB corresponds with taught multiple behavior patterns and that the behaviors can be switched by adjusting the PB values.

1 Introduction

The learning of object handling behavior by robots is a difficult problem because the motor trajectories required to achieve adequate handling behaviors can be diverse as a result of the various types of situations that may be encountered. Even when manipulating the same object, the motor time-development differs widely depending on how the robot and the object are situated in the workspace. The present study shows that a dynamic neural network model is effective in learning and generating such diverse and situational behaviors for object handling.

The learning of object handling by robots has been investigated in a substantial number of studies. Recently, Bianco and Nolfi [1] showed that a simulated robot arm can acquire an object grasping behavior by evolving neural controllers. By evolving simple sensory-motor maps in layered networks, rather complex grasping behavior is generated dynamically, even with a significant range of perturbations in the position and direction of the object. However, their evolutionary approach may be difficult to apply to a real robot task because it

S. Nolfi et al. (Eds.): SAB 2006, LNAI 4095, pp. 185–196, 2006.
© Springer-Verlag Berlin Heidelberg 2006

requires a substantial number of trials, which real robot situations cannot easily accommodate.

In some reinforcement learning studies, behavior schemes are learned by combining predefined behavior primitives. For instance, for an object handling task, a robot learns to select from among predefined behavior primitives, such as approaching, grabbing, carrying and releasing an object, as appropriate for each step. However, this approach can hardly be applied to a dynamic object handling behavior such as object grasping [1] and juggling [2] because it is difficult to manually divide the dynamic behavior scheme into a set of discrete behavior primitives. On the other hand, some researchers [3,4] proposed models that can learn various behavioral skills from continuous sensory-motor flow without possessing any predefined behavior primitives. Recently, some of the authors proposed a neural network scheme, called the RNN with Parametric Bias (RNNPB) [5,6], and applied it to the task of object manipulation by an arm-type robot [5]. However, the task was quite simple because the object was manipulated only in a two-dimensional workspace and the interaction dynamics between the arm and the object were quite limited.

In the present study, complex tasks of ball manipulations utilizing a humanoid robot are considered. In order to let the robot acquire these task skills, an imitation learning framework is introduced in order to avoid an unrealistic number of trial and error instances, which is often observed when applying reinforcement learning and genetic algorithms to complex behavior tasks. In the present imitation learning method, the manipulation of objects is taught directly by humans who guide the movements of the robot by grasping its arms. After repeated guidance and corresponding neuronal learning, the robot becomes able to generate the taught behavioral patterns with generalization.

One specific goal of the present study is to show possible neuronal mechanisms that enable the robot to generate behavior adaptively corresponding to various situational changes of the robot and the object. For this purpose, reflex-type behavior generation for acquiring a simple sensory-motor mapping may not be sufficient because the recognition of situational changes in our task may require contextual information processing. In order to recognize the current situation in a contextual manner may require certain internal models. Here, the internal model does not refer to the global model of the task, but rather to the capability to anticipate encountering sensory flow in the future by regressing the current and past sensory-motor flows in a contextual manner.

In the present study, our previously described scheme of the RNNPB [5,7] is utilized as one possible neuronal network model to implement context switching. The ultimate challenge of the present study is to clarify the essential mechanism of context switching for the task of object handling from the dynamic systems perspectives [8,9]. The dynamic structures that appear in the tight coupling among the body, the object and the internal neuronal processes will be explained by means of attractor dynamics and their parameter bifurcation characteristics.

2 Model and Algorithm

2.1 Architecture

The RNNPB model has the same architecture as the conventional Jordan-type RNN model [10] except for the PB nodes in the input layer.

Figure 1 shows the network architecture. For the normal input and output nodes, both open-loop and closed-loop operations are performed simultaneously. In the open-loop operation, outputs of the network $(\hat{s}_{t+1}, \hat{m}_{t+1})$ are calculated as the result of prediction from the current inputs (s_t, m_t). In the closed-loop operation, copies of the previous prediction outputs $(\hat{s}_{t-1}, \hat{m}_{t-1})$ are copied to the current inputs, and outputs are then calculated based on the feedback. This feedback enables look-ahead prediction (rehearsal process) for an arbitrary number of future steps without perceiving the actual inputs.

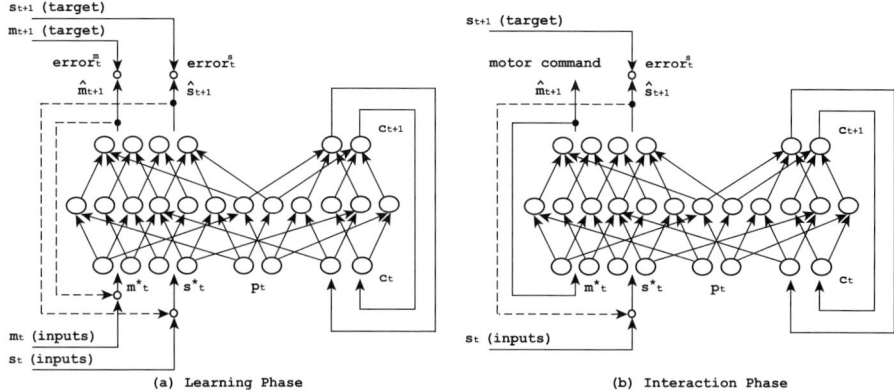

Fig. 1. Configuration of the RNNPB (a) in the learning phase and (b) in the interaction phase. In both phases, each normal input is calculated according to the weighted sum between current inputs and copies of previous prediction outputs. However, in the interaction phase in particular, motor inputs m_t^* are determined only from previous prediction outputs.

In the learning phase, the closed-loop operation is needed in order to enable the RNNPB to learn the memory structure by which the RNNPB can regenerate learned patterns autonomously. However, learning by the closed-loop operation only tends to be unstable. Therefore, we combined two types of operations for the learning of the RNNPB. In addition, in the interaction phase, both types of operations are combined to achieve both robustness against noise and flexibility to situational changes in the generation of learned behaviors. The combination of the open-loop and the closed-loop operations is calculated by the weighted sum for current inputs and previous prediction outputs, as follows:

$$(s_t^*, m_t^*) = \alpha(s_t, m_t) + (1 - \alpha)(\hat{s}_{t-1}, \hat{m}_{t-1}) \ . \tag{1}$$

The parameter α determines the influence ratio of each operation on the actual inputs (s_t^*, m_t^*).

Both the input and output layers contain context nodes, c_t. The output of the context nodes is copied to the context nodes in the input layer. The internal state is recursively computed for future steps utilizing the recurrent feedback loop for the context nodes. The input layer contains PB nodes, p_t. The PB nodes are additional network variables that can be manipulated to learn and generate diverse behavior patterns.

The common structural properties of the training data sequences are acquired as connection weights using the back propagation through time (BPTT) algorithm [11]. On the other hand, the specific properties of each individual time sequence are simultaneously encoded as PB values. Therefore, the modulation of the PB values shifts the modes of the behavior pattern. In the processes of learning and recognition, the PB values are iteratively computed utilizing the error between the target sensory-motor sequence and the predicted sequence.

2.2   Learning Process

The learning algorithm for the parametric bias vectors is a variant of the BPTT algorithm. The step length of a sequence is denoted by l. For each of the sensory-motor outputs, the back-propagated errors with respect to the PB nodes are accumulated and used to update the PB values. The update equations for the ith unit of the PB at t in the sequence are as follows:

$$\delta\rho_t = k_{bp} \sum_{t-l/2}^{t+l/2} \delta_t^{bp} + k_{nb}(\rho_{t+1} - 2\rho_t + \rho_{t-1}) \ . \tag{2}$$

$$\Delta\rho_t = \epsilon\delta\rho_t \ . \tag{3}$$

$$p_t = \text{sigmoid}(\rho_t/\zeta) \ . \tag{4}$$

The term $\delta\rho_t$ is the local gradient to update the interval values of the PB ρ_t, which is obtained from the sum of the following two terms, as shown in (2). The first term represents the delta error, δ_t^{bp}, back-propagated from the output nodes to the PB nodes and is integrated over the period from steps $t - l/2$ to $t + l/2$ ($0 \le t - l/2$ and $t - l/2 < l$). Integrating the delta error prevents local fluctuations in the temporal PB values. The second term is a low-pass filter that inhibits frequent and rapid change of the PB values. The terms k_{bp} and k_{nb} are the coefficient parameters of the above two terms.

The term $\Delta\rho_t$ is the differential of the internal value ρ_t, which is calculated using the local gradient $\delta\rho_t$ and the learning rate ϵ, as shown in (3). The current PB values p_t are then obtained from the sigmoidal outputs of the internal values ρ_t, as shown in (4), where ζ is the slope parameter of the sigmoid function.

2.3   Generation of Behaviors

The inverse computation of the PB values is performed during on-line operations of the robots even after the learning process is terminated. When the dynamic characteristics of ongoing sensory flow is alternated from one mode to another, the prediction error with the current PB values increases. The PB value is then updated in the direction of minimizing the prediction error. The obtained PB value sequences are used as the inputs of the PB nodes for the next sensory-motor sequence prediction process, which results in the update of generated motor patterns based on the current sensory inputs. This parallel mechanism of the inverse computation of the PB by means of the regression of the past sensory flow and the forward computation of motor flow in the future based on the current PB is essential for generating situated behaviors [5,7].

3 Robot System Configuration

The robot task of object handling involves two-arm ball handling. The ball is visually identified by color segmentation. The robot acquires an adequate behavior scheme by learning a set of sensory-motor sequences obtained through direct human teaching. During behavior learning, the target sequence to be learned is the paired trajectory of the sensory and motor values. The sensory information is obtained for both arm movement and object movement. The arm movement is the trajectory of the joint angles as measured by encoders in both arms. The object movement is obtained from the 3-D positions of the center of color-segmented regions of the objects. On the other hand, the motor values are the trajectories of the reference joint angles of the robot's arms. In the present experiments, the target reference trajectory is simply obtained as a copy of the measured arm movement during direct teaching by human supporters. A set of these paired sequences is learned by the RNNPB off-line. In the generation mode, the robot attempts to generate suitable behaviors depending upon the situation by predicting incoming sensory sequences. In this phase, the robot perceives the sensory input and generates its corresponding motor output while adapting the PB values by regression.

4 Dynamic Generation of Ball Handling Behaviors

4.1 Learning of Behaviors from Human Direct Teaching

In the present experiment, the robot learns two different types of ball handling behaviors. One type is 'rolling a ball', in which the robot swings both arms alternately to roll a ball on a table from left to right and vice versa. The other type is 'lifting a ball', which involves bringing the robot's hands together to grasp and lift vertically a ball that is located on a table and then releasing the ball.

This task is performed by a small humanoid robot that is seated on a chair and that handles the ball on a table. The ball is 6 cm in diameter. The table is 45 cm square and is equipped with guides (height: 1 cm, width: 3 cm) to prevent the ball from falling from the table. The table is inclined approximately 4 degrees on the near side so that the ball returns to the reachable area even if pushed away. The robot perceives the ball with a camera attached to its head and handles the ball with both of its arms. The RNNPB in the robot receives two types of information. One is the current ball position obtained by the robot vision system in Cartesian coordinates in the task workspace, where the size of the segmented color region of the ball indicates the distance from the camera. The other type of information is the encoder value for each of the joint angles (shoulder pitch, shoulder roll, shoulder yaw, and elbow pitch) of both arms at the current time step. The RNNPB outputs two types of information. One is the motor commands in terms of the reference values of all joint angles of the next time step, and the other is the prediction of the ball position in the next time step. (Note that the reference value and encoder value for each joint can be different because of the position error by the PID control in the robot.)

In the learning phase, the robot learns two different ball handling behaviors from human direct teaching. In the teaching process, a human user grasps the robot's arms and guides them to perform the target ball handling behaviors using an actual ball while the servo gain of the robot arms is set to approximately zero. In the present study, the reference trajectory is simply obtained as a copy of the measured arm movement in the direct teaching by the human trainer. The training data for the RNNPB were recorded with a time interval of 50 msec, which is as same frequency as the calculation interval of the RNNPB. For the ball rolling behavior, two cycles of the behavior, starting from the right and left sides (three samples each), were recorded. For the ball-lifting task, one cycle of the behavior (six samples) was recorded. The sequence length of the trajectories is approximately 120 steps (6 sec) for the ball rolling task and approximately 90 steps (4.5 sec) for the ball lifting task. It is important to note that during the teaching process these two behaviors are taught as separate sequences. Thus, the robot never learns the transition between these two behaviors.

A set of these paired sequences is learned by the RNNPB off-line. For the learning of the forward model of the behavior sequences, we employed an RNNPB having 19 input nodes and 19 prediction output nodes. In addition, the RNNPB has 50 hidden nodes, 70 context nodes, and two parametric bias nodes. The above numbers of hidden and context nodes were determined by a parametric study using a PC cluster system. In the present study, we examined only the RNNPB with two parametric bias nodes for analysis. The other parameters were heuristically determined as follows: k_{bp}: 0.4, k_{nb}:0.4, ϵ:0.1, and ζ: 0.03. In addition, in the learning phase, α is set to 0.16 for motor inputs and 0.3 for sensory inputs, and in the interaction phase, α is set to 0.0 for motor inputs, 0.08 for joint angle sensory inputs, and 1.0 for ball position inputs.

For the learning sample set, the learning is iterated for 50,000 steps, starting from an initial random set of synaptic weights. In order to avoid over-fitting to noisy data, we introduced a small artificial random noise to the output of the RNNPB in the learning process. The final root-mean-square error of the output nodes was less than 0.0003.

Subsequently, in the interaction phase, the RNNPB in the robot receives the current ball position and the current encoder values for all of the joint angles as inputs and generates its corresponding motor commands and the prediction of the ball position in the next time step as outputs in an online manner. For the online recognition process (PB regression by utilizing the prediction error), 50 instances of forward- and back-propagation iteration were conducted using a window length of 30 steps of the immediate past to determine the PB at each following time step. Along with this update of the PB, the motor references for the following step are also computed by means of forward computation using the window.

4.2 Dynamic Generation and Switching of Learned Behaviors

After the learning, we examined how well the robot with the trained RNNPB could generate two different learned ball handling behaviors. In addition, we observed how well the ongoing behavior could be switched depending on the situational differences between the robot and the ball.

Figure 2(a) shows snapshots of the ball rolling behavior generated by the robot. When the ball was rolling from the front of the robot to the left side, the robot hit the ball using its right hand. The ball then rolled to the left and the robot hit the ball with its left hand. This rolling ball behavior was stably repeated several times. Figure 2(b) shows snapshots of the ball lifting behavior generated by the robot after the ball rolling behavior. When the human trainer stopped the ball in front of the robot, after a short time, the robot began to grasp the ball with both arms without any irregular movements and then lifted the ball to a specified height. The robot then released the ball, which fell in front of the robot. The robot then began to grasp the ball again. This ball lifting behavior was also autonomously repeated several times.

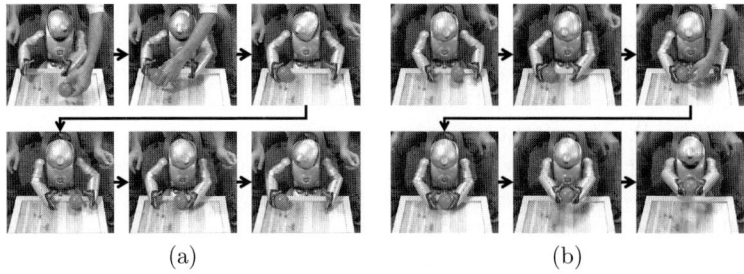

(a)   (b)

Fig. 2. Snapshots of (a) the 'rolling a ball' behavior and (b) the 'lifting a ball' behavior

Figure 3 shows the time course of the entire interaction and the parametric bias values of the RNNPB. In Fig. 3, the plots at the top and in the second row show the actual ball positions and those predicted by the RNNPB. The third-row plot shows the robot joint angle generated by the RNNPB. (Out of a total of eight DOFs, only two DOFs are plotted.) The plot at the bottom shows the parametric bias of the RNNPB.

We observed the transient status, in which two different behaviors were switched (from rolling to lifting) as a result of PB online adaptation according to the sensory input variation. In this case, the ball position was changed

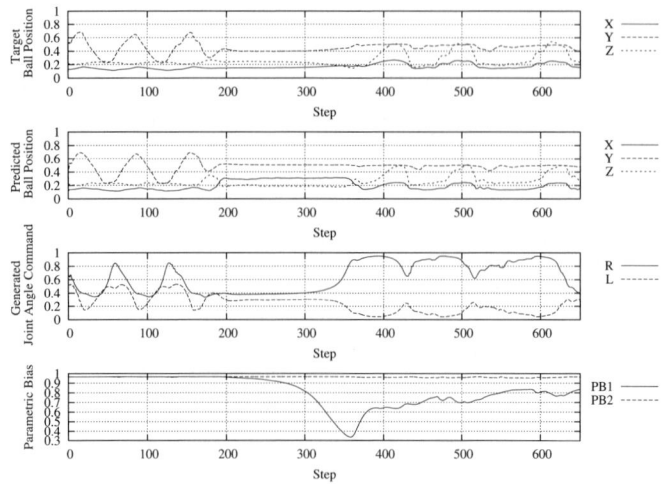

Fig. 3. Dynamic generation and switching of two learned behaviors for the ball handling task. The plot in the top row represents the measured position of the ball. The plots in the second and the third rows represent the predicted ball position and the robot joint angles generated by the RNNPB. The plot in the bottom row represents the parametric bias of the RNNPB.

intentionally by a human. At approximately 200 steps, the ball was stopped in front of the body. This resulted in the reduction of one of the PB values. From approximately 340 steps, the ball lifting behavior was generated according to the PB values.

In the present experiment, we observed that the learned ball handling behaviors were well generated through interaction between the RNNPB dynamics and the ball movement dynamics. Remember that the actual ball movement does not necessarily duplicate exactly the learned ball movement. Even under such noisy conditions, the learned behaviors were stably generated. The system seems to maintain a certain robustness against unknown irregularities. We speculate that such robustness originated from the characteristics of the attractor dynamics that emerged from the coupling between the RNNPB dynamics and the ball movement dynamics. The behavior switching, which the robot did not learn, was also observed to be performed smoothly. (Remember that in the learning process, these two behaviors were trained as separate patterns.) Novel behaviors in terms of behavior transitions were generated spontaneously utilizing emergent dynamic structures self-organized in the system.

4.3   Analysis of the Memory Structure of the RNNPB

To clarify the relationship between the memory structure organized in the RNNPB and the behavior appeared in the previous robot experiment, we analyzed the structure of the PB space and then examined the behavior dynamics embedded therein.

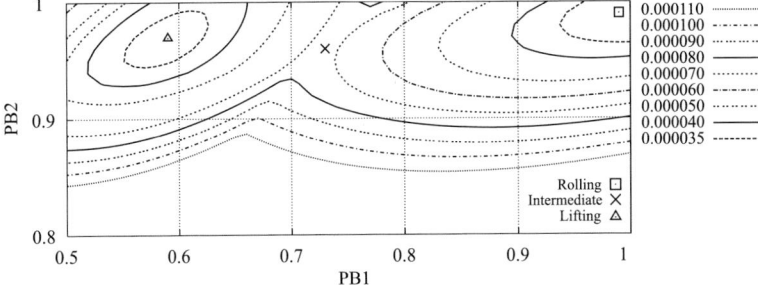

Fig. 4. Prediction error distribution in the PB space. Only the area in which the prediction error is less than 0.00011 is shown. The contour on the right-hand side indicates the area to realize ball-rolling behavior, and the contour on the left-hand side indicates the area to realize the ball lifting behavior.

First, to visualize the memory structure of learned behaviors in the RNNPB, we calculated the distribution of the prediction error for the ball movements of the two learned ball handling behaviors over the entire PB space. Each point in the PB space was set in the RNNPB as a constant value. As a result, the RNNPB predicted the ball movements without PB adaptation in the simulation. Both ball rolling and ball lifting were shown independently, and the prediction error was calculated for each sequence.

Figure 4 shows a contour map of the prediction error obtained by the above calculation over the two-dimensional PB space. Two of the error distributions were overlaid to show the distribution of the smaller prediction error value. Figure 4 shows that there are two continuous bowl-shaped error distributions and that two minimum values exist for each structure. In addition, there is a boundary of two structures in the region where PB1 is approximately 0.70 and PB2 is approximately 0.95. By relating the PB trajectory shown in Fig. 3 to this contour map, we speculate that continuous variation of the PB values between the two different regions in the PB space contributed to the smooth behavior switching.

Next, in order to examine the behavior dynamics embedded in the PB space, we chose three typical PB vectors from the distribution map and examined how well the RNNPB with these vectors could perceive the ball movements of each behavior and generate its prediction and its corresponding arm movements in the off-line simulation. The PB vectors chosen to correspond to each behaviors are as follows: (PB1: 0.99, PB2: 0.99) for the ball rolling task, (PB1: 0.59, PB2: 0.97) for the ball lifting task and (PB1: 0.73, PB2: 0.96) for the intermediate behavior. The ball movements of two learned behaviors were shown to the RNNPB with each PB vector, and the RNNPB then predicted the ball movement and generated its corresponding arm movement without PB adaptation. Figure 5 shows the ball movement predicted and the arm movement generated by the RNNPB with three typical PB vectors (a), (b), and (c) for ball rolling, and Fig. 6 shows that for ball lifting.

In these two figures, we found that the RNNPB with the intermediate PB value performed as well as the RNNPB with the proper PB, as compared to the

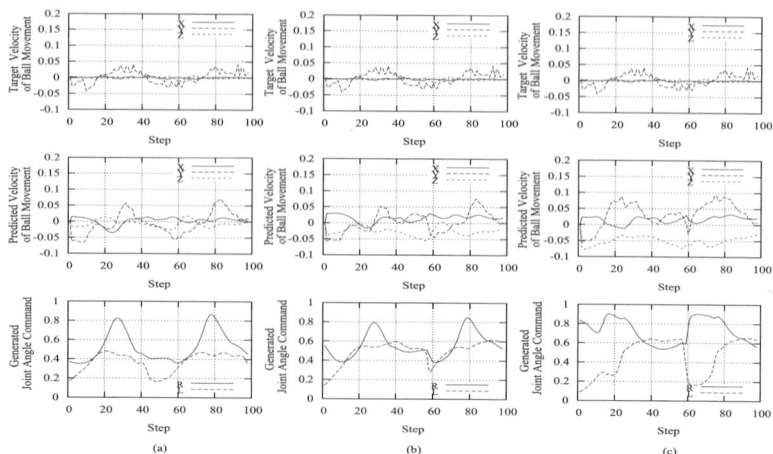

Fig. 5. Ball rolling behavior for constant PBs by the RNNPB that learned two different types of ball handling behaviors. Each of the plots labeled (a), (b) and (c) indicates the result that was acquired from the constant PB selected from the PB space, where the prediction error corresponds with rolling, intermediate, and lifting behaviors respectively. The plots in the top, second, and bottom rows represent the target velocity of the ball calculated from the teaching data, the predicted velocity of the ball calculated from the data generated by the RNNPB, and the joint angles generated by the RNNPB, respectively.

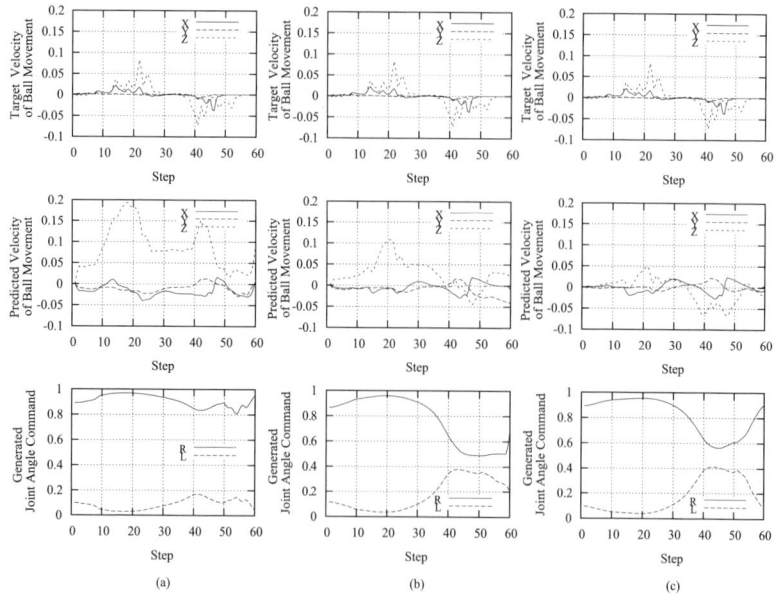

Fig. 6. Ball lifting behavior for constant PBs. The meanings of the labels under the plots are the same as those described in Fig. 5.

RNNPB with the opposite PB for both ball rolling and ball lifting. This indicates that two different behavior dynamics were incorporated in the single dynamic structure, which was represented by constant PB values at the boundary region.

5 Discussion

In order to learn multiple behaviors as switching dynamics, two distinct types of learning scheme have been proposed. One is the local representation scheme such as MOSAIC [3] and Mixture of RNN experts [4]. The other scheme is the distributed representation scheme such as RNNPB. RNNPB has remarkable generalization capability on learning multiple dynamic patterns since each pattern is embedded in shared structures of distributed representation rather than memorized independently [12,6,13].

However, RNNPB has a difficulty in increasing number of training patterns. Because the weight matrix which covers most of the memory resources in RNNPB participates in representing the common structures among memorized patterns and only a few PB values represent the difference in contrast to the fact that as the number of training patterns increases, the common structures among them tend to decrease and their difference tends to increase. We consider that the scalability problem of RNNPB is due to the inappropriate balance between the two types of memory resources.

One possible solution for this problem is to adjust each of the capacity among the two types of memory resources depending on a training set. However, it is difficult to estimate how much the common structures could be extracted before learning. The other possible direction is to extract the "local" common structures rather than the "global" common structures among the training patterns. In this direction, we consider how to combine both of the distributed and local representation scheme. One possibility is to extend the current model to have multiple RNNPB modules with a gating function. In this model, each of the modules affords distributed representation by having associated PB units. Therefore the "local" common structure can be represented in each module and it is easy to increase the number of memorized patterns by simply adding local modules as necessary, because there is little memory interference between the different independent modules. The characteristics of the local representation scheme can also contribute to the additional learning performance.

6 Summary

The present experiments involving a humanoid robot revealed that diverse ball handling behaviors can emerged as the result of mutual entrainment between the internal dynamics of the RNNPB and the external dynamics of the environment. As a result of the present experiments, the following four objectives were accomplished.

First, various behavior schemes were self-organized from continuous sensory-motor flow without any predefined behavior primitives. Second, learned behaviors were generated dynamically through the interaction between the internal memory dynamics and the external dynamics. Third, dynamic switching was achieved between multiple learned behaviors according to the external dynamics. Fourth, a novel behavior pattern was generated in the behavior transition phase without any explicit sensory-motor pattern teaching.

The present study reveals that the present learning mechanism enables the robot to learn new goal-directed behavior from the direct teaching of the trainer and to generate appropriate learned behavior according to environmental changes. Such learning function and adaptive behavior are essential for realizing entertainment robots that will provide long-term interaction with users. In the future, we expect these functions to be utilized to realize robots that share experiences with human users and that learn new tasks from their own experiences.

References

1. Bianco, R., Nolfi, S.: Evolving the neural controller for a robotic arm able to grasp objects on the basis of tactile sensors. Adaptive Behavior **12**(1) (2004) 37–45
2. Schaal, S., Sternad, D., Atkeson, C.G.: One-handed juggling: A dynamical approach to a rhythmic movement task. Journal of Motor Behavior **28**(2) (1996) 165–183
3. Wolpert, D.M., Kawato, M.: Multiple paired forward and inverse models for motor control. Neural Networks **11** (1998) 1317–1329
4. Tani, J., Nolfi, S.: Learning to perceive the world as articulated: an approach for hierarchical learning in sensory-motor systems. Neural Networks **12** (1999) 1131–1141
5. Tani, J.: Learning to generate articulated behavior through the bottom-up and the top-down interaction process. Neural Networks **16** (2003) 11–23
6. Ito, M., Tani, J.: Generalization in learning multiple temporal patterns using rnnpb. In: Proc. of the 11th International Conference on Neural Information Processing (ICONIP2004). (2004) 592–598
7. Ito, M., Tani, J.: On-line imitative interaction with a humanoid robot using a dynamic neural network model of a mirror system. Adaptive Behavior **12**(2) (2004) 93–115
8. Beer, R.: A dynamic systems perspective on agent-environment interaction. Artificial Intelligence **72**(1) (1995) 173–215
9. van Gelder, T.: The dynamical hypothesis in cognitive science. Behavior and Brain Sciences (1998)
10. Jordan, M.: Attractor dynamics and parallelism in a connectionist sequential machine. In: In L. Erlbaum, editor, Proc. 1986 Cognitive Science Conference. (1986) 531–546
11. D. E. Rumelhart, G.E.H., Williams, R.J.: Learning internal representations by backpropagating errors. Nature **332** (1986) 533–536
12. Tani, J., Ito, M.: Self-organization of behavioral primitives as multiple attractor dynamics: A robot experiment. IEEE Transactions on System, Man and Cybernetics Part B **33**(4) (2003) 481–488
13. Sugita, Y., Tani, J.: Learning semantic combinatoriality from the interaction between linguistic and behavioral processes. Adaptive Behavior **13**(1) (2005) 33–52

Action Selection and Behavioral Sequences

Distributed Action Selection by a Brainstem Neural Substrate: An Embodied Evaluation*

Mark Humphries and Tony Prescott

Adaptive Behaviour Research Group, Department of Psychology,
University of Sheffield, UK
{m.d.humphries, t.j.prescott}@shef.ac.uk

Abstract. Theoretical approaches to the problem of action selection in autonomous agents often contrast centralised and distributed selection schemes. Here we describe a neural substrate for distributed action selection in the vertebrate brain-stem, the medial reticular formation (mRF), which may form a evolutionary precursor to centralised schemes found in the higher brain. We evaluate its competence as a selection device for robot control in a simulated resource co-ordination task, and use a genetic algorithm to evolve the mRF's inputs and internal structure. Some configurations of the mRF could sufficiently co-ordinate actions to maximise the robot's energy, but this is critically dependent on a high rate of energy acquisition, which leaves an animal (or agent) susceptible to food shortages. Thus, the inflexibility of the mRF as a distributed selection mechanism may have provided impetus for the evolution of more complex, centralised, selection mechanisms in the brain.

1 Introduction

A generally effective strategy for designing controllers of autonomous agents is to reverse-engineer biological systems that have evolved as solutions to the control problems. One such problem is action selection: a mortal agent must continuously choose and co-ordinate behaviors appropriate to both its context and its current internal state if it is to survive. Animals necessarily embody successful solutions to the action selection problem. Thus, it is natural to look at what parts of the central nervous system — the neural substrate — have evolved to carry out the action selection process.

Recent proposals for the neural substrate of the vertebrate action selection system have focussed on the basal ganglia - a set of fore- and mid-brain nuclei whose input, output, and inter-connections seem to be consistent with a central (as opposed to distributed) resource switching device [1,2]. Decerebrate animals, altricial (helpless at birth) neonates, and lateral hypothalamic rats do not have fully intact or functioning basal ganglia, but are capable of expressing spontaneous behaviors and co-ordinated and appropriate responses to stimuli. During

---

* This research was supported by the EPSRC (GR/R95722/01), a Wellcome Trust VIP Award, and the European Union Framework 6 ICEA project.

S. Nolfi et al. (Eds.): SAB 2006, LNAI 4095, pp. 199–210, 2006.

decerebration the entire brain anterior to the superior colliculus is removed leaving only the hindbrain intact. Yet, the chronic decerebrate rat can, for example, spontaneously locomote, orient correctly to sounds, groom, perform co-ordinated feeding actions, and discriminate food types [3,4]. Such animals clearly have some form of intact system for simple action selection. We have recently argued that, of the potential candidate structures left intact in the brainstem of decerebrate animals, the medial reticular formation (mRF) is the most likely substrate of a generalised (if limited) action selection mechanism [5,6,7].

We first evaluated the single existing computational model of RF function — a landmark model proposed by Warren McCulloch and colleagues [8] — in both simulation and embodied form (on a version of the same task used here) [5]. However, inevitably, given its age, several aspects of the model were incorrect or implausible, or omitted features known from more modern studies of the mRF. We thus turned to reviewing the modern literature on the mRF, and found that the organisation of the mRF's inputs and outputs, and the functional properties of its cells, are all consistent with the action selection proposal [7]. We began addressing the question of how the mRF represents and resolves the competition between actions by synthesising the neurobiological data to determine the mRF's internal structure [9] — see Fig. 1. The mRF is made up of stacked cell clusters, each cluster a mix of projection and inter-neurons. The projection neurons are excitatory, project a long axon to the motor centres in the brainstem and spine, and contact cells in other clusters via collaterals from that long axon. The inter-neurons are inhibitory, and project only within their own cluster. We outlined a novel quantitative anatomical model that generated networks with this structure and we found that the networks had small-world properties [9].

Potential configurations of the mRF as an action selection system were explored by simulation of a new computational model whose connectivity was based on the anatomical model [6,7]. We found that a *sub-action* configuration most effectively supports selection: the projection neurons of each cluster represents a component of an action, and a coherent behavioural response is created by clusters recruiting other clusters which represent compatible components (in addition, incompatible components are suppressed by inhibition of their representing clusters, which occurs via activation of that cluster's inter-neurons).

The neurobiological and simulation data indicate that the mRF is a distributed selection mechanism, from which the selection of actions is an emergent phenomenon. This can be contrasted with the basal ganglia, which are a centralised control structure, selecting actions on the basis of inputs from multiple command systems. It thus appears that evolution has seen fit to produce both forms of selection structure that are often counter-posed in theoretical discussions [10]. In this paper we extend the assessment of the mRF's capabilities as an action selection mechanism by testing our new computational model of the mRF in an embodied form. The aim of this work was to determine whether or not the mRF was capable of carrying out action selection independently from other neural systems that may be involved in the intact animal, and to shed some light on the complexity of task that the mRF could cope with.

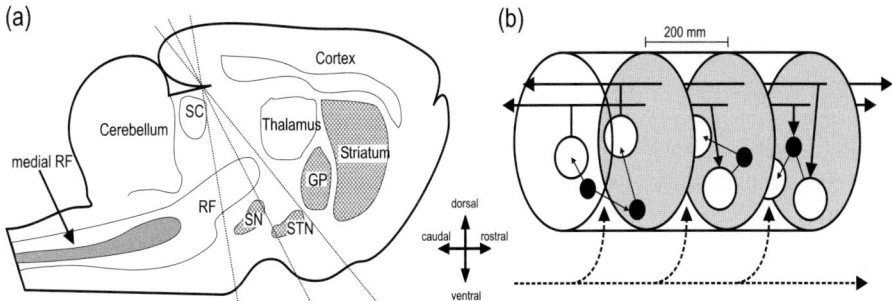

Fig. 1. Anatomy of the vertebrate medial reticular formation (mRF). Directional arrows apply to both panels. (a) The relative locations of major nuclei and structures including the basal ganglia (hashed) and the medial reticular formation (RF) shown on a cartoon sagittal section of rat brain. The dashed lines show the location of the common decerebration lines — all the brain rostral to the line is removed, leaving hindbrain and spine intact. GP: globus pallidus. SN: substantia nigra. STN: subthalamic nucleus. SC: superior colliculus. (b) The proposed mRF organisation: it comprises stacked clusters (3 shown) containing projection neurons (open circles) and inter-neurons (filled circles); cluster limits (grey ovals) are defined by the initial collaterals from the projection neuron axons. The projection neurons receive input from both other clusters (solid black lines) and passing fibre systems (dashed black line). The inter-neurons project within their parent cluster. Reproduced from [7].

2   Methods

We begin by describing the computational model of the mRF, which forms the basis for the robot controller, then describe the task on which the robot is evaluated, the input and output of the controller, and the form of the genetic algorithm used to evolve the mRF.

2.1   A Population-Level Model of the mRF

We do not have sufficient space to fully describe the anatomical model which provides the connectivity data for the computational model — see [9]. It is sufficient to note that the model has six parameters: the number of clusters N_c; the number of neurons per cluster n; a proportion ρ of those are projection neurons, the rest are inter-neurons; the probability of a projection neuron sending a connection to another cluster $P(c)$; the probability of that connection then contacting any neuron in that cluster $P(p)$; and the probability of an inter-neuron contacting any other neuron in its own cluster $P(l)$. Each of these parameters were limited to a range of values sourced from the neurobiological data: the specific values used in this paper are thought to be the most realistic. There are six sub-actions within the task to represent, hence $N_c = 6$; the other parameters were set to: $n = 50$, $\rho = 0.7$, $P(c) = 0.25$, $P(p) = 0.1$, and $P(l) = 0.1$ (see [9] for further details). The result of creating a particular instance of this model is a network of linked nodes.

We use a population-level computational model of the mRF. In this approach, populations of neurons are treated as a statistical ensemble, assuming functionally meaningful sub-groups of neurons cannot be further distinguished. Thus, the model is a set of simplified ordinary differential equations describing the change in the normalised mean firing rate of each population over time. Given the proposed cluster structure, and the hypothesis of projection neurons encoding the action representation, the most natural division of the mRF is into separate populations of projection and inter-neurons for each cluster. The computational model thus has two vectors encapsulating its behaviour: the projection neuron activity c and the inter-neuron activity i. Each vector element is a population: c_k is the normalised mean firing rate of the kth cluster's projection neuron population, and i_k is the normalised mean firing rate of the kth cluster's inter-neuron population. These activities evolve according to the differential equations given in [6,7]. Input to the model is described by vector u, where each element u_k is the scalar summation of all external input to cluster k from sensory and internal monitoring systems, and which thus represents the salience of that cluster's represented sub-action.

The connections between the populations are defined by the underlying network generated by the anatomical model: variables A_{jk} and C_{jk} are the mean number of contacts from cluster j to, respectively, the projection and interneurons of cluster k; b_k and d_k are the mean number of contacts from interneurons in the current cluster k to, respectively, the projection and inter-neurons in that same cluster. Figure 2 shows a schematic of the mRF population-level model, which further explains its structure, and details of the parameters which are optimised by a genetic algorithm (see below).

2.2   The Energy Task

We have previously evaluated bio-mimetic computational models on an energy-based task [11,5]. In those evaluations, fixed action patterns were selected by the models: for example, avoiding an obstacle was a complete pattern in which the robot reversed, turned, then moved ahead in a different direction. However, as our simulations have shown that the mRF is most likely to represent components of actions [6], we here decompose the action patterns into their constituent parts.

The form of the task is as described in [5]: a mobile robot explores an arena with a grey coloured floor (representing neutral) upon which are laid two white and two black tiles. The robot controller has six state variables: the states of the left and right bumpers, B_L and B_R; the values of the bright and dark infra-red floor sensors L_B and L_D; potential energy P_E (which is recharged on black tiles); and energy E (which is recharged on white tiles by consuming potential energy). Both the internal variables P_E and E were limited to the range $[0,1]$.

The change δP_E in potential energy when recharging on a black tile for T_{eat} seconds is

$$\delta P_E = E_{rate}\, T_{eat}\, L_D \ . \tag{1}$$

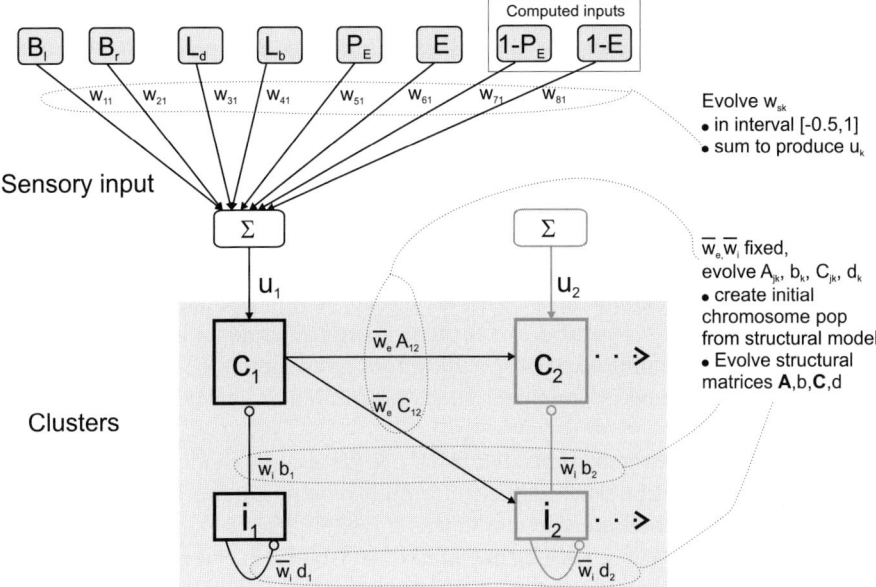

Fig. 2. Schematic of the population-level mRF model. Two clusters are shown; only the first cluster's (c_1) inputs and outputs are depicted. The state variables from the robot are weighted and summed to produce a scalar input signal that represents the urgency of request for the sub-action represented by the cluster. (In the GA, the weights are evolved to obtain the optimal balance of sensory input signals required for that sub-action.) The internal structural parameters $\boldsymbol{A}, \boldsymbol{b}, \boldsymbol{C}, \boldsymbol{d}$ are initially derived from the anatomical model — forming the initial population — and are then evolved as well. The mean excitatory weight $\bar{w}_e$ is constant, and the mean inhibitory weight $\bar{w}_i$ is computed for each model (see text). These values represent the average synaptic efficacy between neurons. Thus, the emphasis is on evolving the *structural* properties of the mRF.

The increase in energy δE and decrease in potential energy δP_E^- when recharging from stored potential energy on a white tile for T_{digest} seconds is

$$\delta E = E_{\text{rate}}\, T_{\text{digest}}\, L_{\text{B}}, \quad \delta P_E^- = -E_{\text{rate}}\, T_{\text{digest}}\, L_{\text{B}} \ . \tag{2}$$

The initial experiments set the acquisition (and conversion) rate $E_{\text{rate}} = 0.027$, following our prior work [11,5].

In the original version of the task, the robot had four actions available to it: *Wander*: a random walk in the environment, formed by forward movement at a fixed speed followed by a turn of a randomly selected angle; *Avoid Obstacles*: reverse movement, followed by a turn away from the object; *Reload On Dark*: stop on a black tile and charge potential energy; *Reload On Light*: stop on a white tile and charge energy by consuming potential energy.

We decomposed these into the following six sub-actions: move forward, move backward, turn left, turn right, recharge potential energy, and recharge energy.

Hence the mRF model used had six clusters, one per sub-action. This decomposition is based on studies of the mRF's control of movement: for example, in the lamprey there are separate mRF neuron sub-populations whose activity drives moving forward, moving backward, and turning [12]. The recharging sub-actions are distinct from the original fixed actions of reloading, as they do not include commands to stop movement: the model must co-ordinate the stopping of movement with the selection of recharging at the appropriate time.

All robot simulations were performed in Webots 4 (Cyberbotics). One robot simulation time-step is one second. At the beginning of each run the robot was initialised with $E = 1$ and $P_E = 0.5$ and placed at a particular location in the arena. Regardless of the action(s) selected, energy E is depleted at a constant rate of $E_{met} = 0.002$ unit/s, corresponding to a fixed metabolic rate. Therefore, if no recharging of energy occurred then the minimum survival time was 500 seconds.

At each time-step, the instantiated mRF computational model receives inputs computed as described below, and is run until it reaches equilibrium, or until $t = 0.5$s, whichever occurs first. The model was solved in discrete-time, using a zero-order hold approximation, with a time-step of 0.001s. Each run is initialised with the final state of the previous run, so that the model's dynamics are effectively continuous. Its output is then converted into the activity of the corresponding spinal motor centres. The behavioural vector of the robot is then computed by aggregating these signals. We describe each of these processes in turn: a schematic of the controller is given in Fig. 3b.

2.3   mRF Input

Our previous robot work based on the energy task used complex salience equations: computed levels of urgency of each action, based on functions of each sensory variable. However, mRF inputs are mostly directly from sensory and internal monitoring systems, so there is little scope for complex neural processing of those signals. We thus assume that the kth cluster's input u_k is given by a summed weighted input of the robot's sensory variables, energy variables, and the inverse of the latter (information on falling energy levels, or at least the volume of used gut capacity, is available to the mRF [4]). These inputs are shown in Fig. 2. The resulting total is re-scaled so that $u_k \in [-0.5, 1]$, following neurobiological data on the input to the mRF: see [6,7] for more detail.

2.4   Interpreting mRF Output

The mRF has direct control over vertebrate central pattern generators (CPGs), the circuits which drive limb and jaw movements [13]. Increasing activation of reticulo-spinal neurons causes the onset of and then increasingly rapid locomotion [12], which corresponds to the onset of oscillations and their increasing frequency within the locomotor CPG. Simulations have explained the seemingly paradoxical result that some increases in activation can terminate locomotion: given sufficient input, the oscillations terminate and a stable state is resumed

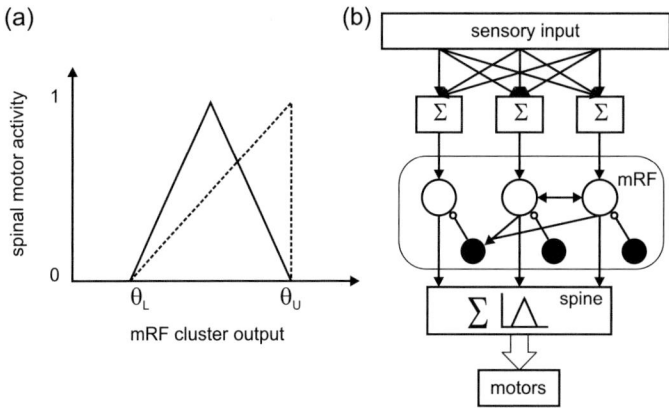

Fig. 3. (a) Activity in the spinal motor plants is a non-linear function of mRF drive. Two transfer functions, based on neurobiological data, are evaluated in this paper: "dual" (dashed line) and "triangle" (solid line), which give non-zero output between two thresholds (θ_L and θ_U). (b) Cartoon schematic of the robot controller. Sensory variables are weighted and summed, the total then input to the mRF computational model (projection neuron populations, white circles; inter-neuron populations, black circles). The output of each cluster is filtered through the spinal transfer function, then aggregated to produce a motor vector. This then drives the wheels and sets the rate of change for both energy and potential energy.

[14]. There thus appears to be separate thresholds for the onset and termination of activity in the CPG. It is not clear whether the increasing reticulo-spinal activity causes a continuous increase in CPG activity until a sudden, discontinuous, stop, or whether it causes an increase to some maximum followed by a decrease (see Fig 3a) — both will be examined.

The output vector c of the mRF computational model is thus converted to a spinal-command vector by $m_k = M(c_k)$, where M is one of the output transfer functions, "dual" or "triangle", shown in Fig. 3a. Lower and upper thresholds for M were set at $\theta_L = 0.1$ and $\theta_U = 0.8$ for the results reported below. The spinal-command vector then gates the contribution s^k of each sub-action to the final behaviour vector b: each sub-action contribution being either the requested drive of the robot's wheels, or the requested quantity of energy and/or potential energy to change, as detailed in Table 1. The final behaviour vector thus has four elements: the first and second elements are the motor speed sent to the left and right wheels respectively, the third element is the quantity to update E, and the fourth element is the quantity to update P_E.

2.5   Form of the Genetic Algorithm

We detail the features of the genetic algorithm (GA) used to search the space of mRF models. Following our previous work, we define our fitness function as the mean E over a fixed time window of 2500 seconds *after* the minimum survival period had elapsed [5]. (We do not use survival time as this is unbounded —

Table 1. Behavioural contributions of each sub-action: the first two elements are the requested motor commands sent to the left and right wheels of the robot; the second two are the requested changes in energy and potential energy, whose values are given by (1)-(2). The motor speeds for the turn sub-actions are the values necessary to turn the robot 180° in one time-step.

Sub-action	Vector s
Move forward	[80 80 0 0]
Move backward	[-40 -40 0 0]
Turn left	[10.5 -10.5 0 0]
Turn right	[-10.5 10.5 0 0]
Recharge potential energy	[0 0 0 δP_{E}]
Recharge energy	[0 0 δE $\delta P_{\mathrm{E}}^{-}$]

the robot may never expire). Our resulting fitness function naturally falls in the range [0,1], with 1 indicating maximum fitness.

An initial population of 200 chromosomes was created: 200 mRF anatomical models were instantiated with the parameters given above, and the connection matrices derived from them; the sensory input weights were randomly chosen from the interval [-0.5,1]. Every subsequent chromosome population had 50 members. Each chromosome of a population was converted into the set of connection matrices, and the resulting mRF model evaluated on the energy task. The population was then ranked by fitness level, and the best 20 chromosomes retained. From this remaining population, 30 pairs of chromosomes were randomly chosen for mating: from each pair, a new chromosome is created by conjoining the two chromosomes at a randomly chosen split point. Thus, a new population of 50 chromosomes results (20 parents, 30 offspring). The new population is subjected to mutation, where each element is changed to another value within its preset interval (defined below) with a probability of 0.05. The top chromosome of the parent population is never mutated, so that the most fit parent is always retained intact (elitism). Once all pairings and mutations have been carried out, the resulting population is again evaluated on the energy task. This process was iterated until the termination condition was reached, that the top chromosome was unchanged for 20 consecutive generations.

The models were encoded as a real-valued chromosome of 120 elements, broken down into: 48 input weights (8 sensory inputs × 6 clusters); 60 inter-cluster connections (total number of non-zero elements in A and C); and 12 intra-cluster elements (total number of elements in b and d). The intervals over which each element could be mutated were limited as follows. We used an interval of [-0.5,1] for the sensory weights. Each element of A, b, C, d was limited in the interval $[0, 3 \times E(x)]$ where $E(x)$ is the expected value of each element, given the connection probabilities used in the underlying anatomical model.

The mean connection weights $\bar{w}_e$ and $\bar{w}_i$ were not optimised, to reduce the number of free parameters. We set excitatory weight $\bar{w}_e = 0.2$, as before [7]. Previously, we set inhibitory weight by $\bar{w}_i = -\bar{w}_e \times N_e/N_i$, where N_e and N_i are the total number of excitatory and inhibitory connections in the network (see [7] for explanation): here we sum together the elements of A and C to compute N_e, and sum together the elements of b and d to compute N_i.

3   Results

3.1   Assessment of the Fitness Landscape

To assess the fitness landscape to be explored by the GA, we performed a series of Monte Carlo simulations: 1500 mRF models were instantiated, using the anatomical model parameter values given above, and each was assessed for its fitness as a controller for the robot. We thus hoped to gain some insight into the distribution of fitness over all possible models.

Three random searches were conducted: for the first we used the "dual" transfer function; for the second we increased the rate of energy acquisition by an order of magnitude so that $E_{rate} = 0.27$ (we ran this search because, when observing robot behaviour on individual model trials, we noted that the rates of recharging energy and potential energy were, consistently, roughly 10% of those used in the previous work [11,5], due to the gating of the sub-actions by the mRF output); for the third we retained the same increased E_{rate}, but used the "triangle" transfer function. For each search we computed the empirical cumulative distribution function (EDF) [1] of model fitness, shown in Fig. 4 as a function of fitness error (i.e. 1-fitness).

The first search found a maximum fitness of 0.165 and its EDF shows that most models had a fitness close to zero, and were therefore not surviving long beyond the minimum survival period. For the second search, the resulting fitness distribution had a markedly increased maximum of 0.518, but again the EDF shows that most models had low fitness. For the third random search, the fitness distribution had a further increased maximum of 0.897; however, the EDF again shows that the overwhelming dominance of low-fitness models remained. From these random search results we concluded that, though infrequent, there were forms of the model that a GA-based search could potentially find and optimise.

3.2   Increasing Energy Acquisition Rates Increases Fitness

We confirmed the first random-search results by conducting a series of GA-based searches with the original acquisition rate ($E_{rate} = 0.027$) and "dual" transfer function: altering numerous parameters of the GA search (the initial population, the retained population, the number of offspring), or features of the underlying model (using un-scaled salience inputs, using hand-coded sensory-input weights, changing the transfer function thresholds), never resulted in a maximum fitness that markedly exceeded that of the random-search by the time the search terminated.

---

[1] An empirical cumulative distribution function is an estimate of the underlying cumulative distribution function, each probability estimated by $P(x) =$ (number of observations $\leq x$) / (total number of observations).

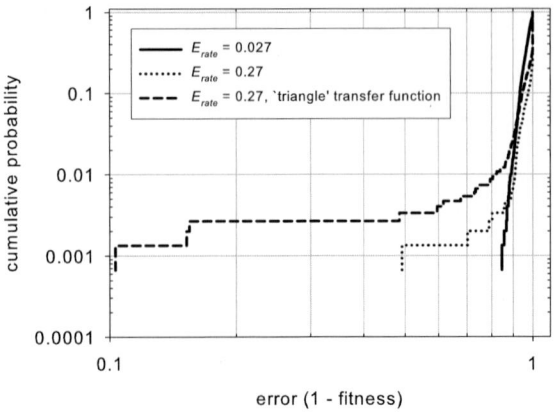

Fig. 4. Distributions of model fitness over a large random sample of all possible models in three different set-ups of the robot task. The distributions are shown as EDFs (see text), plotted on log-log axes as a function of fitness error (1 - fitness), to allow plotting of zero fitness on the log-scale. Increasing E_{rate} increased maximum fitness, but the distribution was always dominated by low fitness models.

Subsequent GA-based searches using the increased energy acquisition rate of $E_{rate} = 0.27$ confirmed the results of the other random-searches: for both "dual" and "triangle" transfer functions the maximum fitness was considerably increased following the increase in E_{rate}. Indeed, the maximum fitness achieved during the searches either exceeded ("dual", fitness $= 0.714$) or equalled ("triangle", fitness $= 0.883$) that found by the corresponding random search. The mean fitness of each population did not increase over the course of the generations (Fig. 5a), consistent with the dominance of low-fitness models in the random searches.

3.3   Dependence of Fitness on the Rate of Energy Acquisition

We then assessed the dependency of the performance of the GA-based search on the value of E_{rate}. A search was run using the GA set-up described in Sect. 2.5, and models with the "dual" transfer function, for each step increase of 0.027 from the initial value of $E_{rate} = 0.027$ up to $E_{rate} = 0.27$ — there were thus 10 searches. A linear regression showed a significant increase in maximum fitness ($r = 0.8426, p < 0.01, n = 9$; one outlier), and a significant increase in final generation mean fitness ($r = 0.725, p < 0.05, n = 10$) as a function of increasing E_{rate} (Fig. 5b). However, the increase in mean fitness was not of the same order of magnitude as the increase in maximum fitness, indicating that the majority of the population remained at low fitness regardless of E_{rate}.

4   Discussion

The complexity of the task seems to be a difficult one for the mRF models to solve, given the low fitness of the vast majority, but we have shown that mRF structures

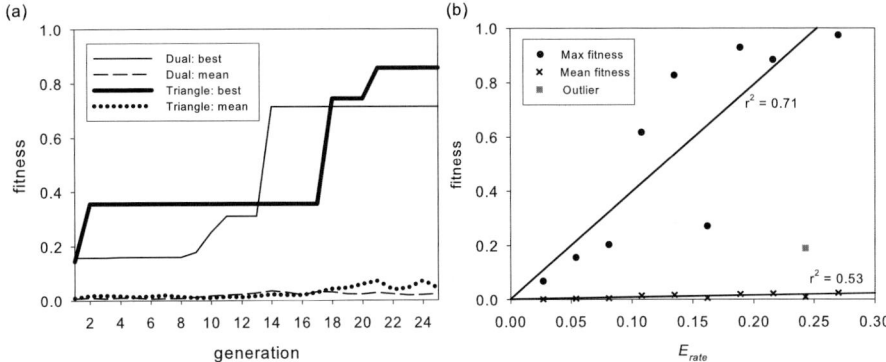

Fig. 5. (a) A small but high-fitness set of mRF models can evolve over time given a sufficiently high energy acquisition rate ($E_{\text{rate}} = 0.27$), using either the 'dual' or 'triangle' output functions. However, there is little increase in fitness of the majority of the population. (b) The effect of increasing E_{rate} on the maximum and mean fitness of the GA-based model search. Increasing E_{rate} is significantly correlated with an increase in both the maximum fitness and the final generation's mean fitness. Mean fitness did not show the same order of magnitude increase as maximum fitness.

can select action components sufficiently well to co-ordinate energy gathering and acquisition. We should not be surprised that mRF models with high evolutionary fitness were difficult to find: our previous similar evaluations of bio-mimetic models were based on fixed action patterns, whose salience was a complex function of sensory variables; the necessity of creating emergent actions from components, based only on direct sensory input, makes the task far more difficult for the mRF model. Indeed, if most structural configurations of the mRF could support efficient action selection, then why would more complex neural systems have evolved to deal with the same problem, rather than co-opting the existing solution? Nevertheless, it is testament to the potential computational power of even the most "basic" of brain structures that the mRF model was successful at all.

The success of the mRF as an action selection system is dependent on the energy acquisition rate (it may equivalently be dependent on the metabolic rate E_{met}, which will be explored in future work). Interestingly, an increase in E_{rate} did not result in a correspondingly large general increase in model fitness: it appears that it only promoted the models which had the potential to successfully co-ordinate the robot's behaviour (Fig. 5b).

The use of a fixed acquisition rate seems valid: data from studies of decerebrate rats suggests that they are unable to adaptively alter their rate of food acquisition during periods of food deprivation [4]. This, in turn, suggests that a brain-stem dominant animal may be susceptible to fluctuations in food supply, and indeed may be inflexible in its response to other environmental changes: future work will test this idea more rigorously.

The results of this work neatly parallel the evolution of the vertebrate brain: some ancient species, such as the lamprey, have their motor behaviour dominated by the reticulo-spinal system. Thus, the mRF seems to be a sufficient control system in some ecological niches. Yet most modern vertebrates have more complex neural systems that combine to control their behaviours. Studying the integration of these more complex, centralised, systems with the lower-level, distributed, mRF system may provide further insight into potential designs for control architectures of autonomous agents.

References

1. Redgrave, P., Prescott, T.J., Gurney, K.: The basal ganglia: A vertebrate solution to the selection problem? Neuroscience **89** (1999) 1009–1023
2. Prescott, T.J., Redgrave, P., Gurney, K.: Layered control architectures in robots and vertebrates. Adapt. Behav. **7** (1999) 99–127
3. Berntson, G.G., Micco, D.J.: Organization of brainstem behavioral systems. Brain Res. Bull. **1** (1976) 471–483
4. Grill, H.J., Kaplan, J.M.: The neuroanatomical axis for control of energy balance. Front. Neuroendocrin. **23** (2002) 2–40
5. Humphries, M.D., Gurney, K., Prescott, T.J.: Is there an integrative center in the vertebrate brainstem? A robotic evaluation of a model of the reticular formation viewed as an action selection device. Adapt. Behav. **13** (2005) 97–113
6. Humphries, M., Gurney, K., Prescott, T.: Action selection in a macroscopic model of the brainstem reticular formation. In Bryson, J.J., Prescott, T.J., Seth, A.K., eds.: Modelling Natural Action Selection. AISB Press, Brighton, UK (2005) 61–68
7. Humphries, M.D., Gurney, K., Prescott, T.J.: Is there a brainstem substrate for action selection? Phil. Trans. Roy. Soc. B. (2006) in press.
8. Kilmer, W.L., McCulloch, W.S., Blum, J.: A model of the vertebrate central command system. Int. J. Man-Mach. Stud. **1** (1969) 279–309
9. Humphries, M.D., Gurney, K., Prescott, T.J.: The brainstem reticular formation is a small-world, not scale-free, network. Proc. Roy. Soc. B. **273** (2006) 503–511
10. Maes, P.: Modeling adaptive autonomous agents. In Langton, C.G., ed.: Artificial Life, An Overview. MIT Press, Cambridge, MA (1995) 135–162
11. Girard, B., Cuzin, V., Guillot, A., Gurney, K.N., Prescott, T.J.: A basal ganglia inspired model of action selection evaluated in a robotic survival task. J. Integr. Neurosci. **2** (2003) 179–200
12. Deliagina, T.G., Zelenin, P.V., Orlovsky, G.N.: Encoding and decoding of reticulospinal commands. Brain Res. Brain Res. Rev. **40** (2002) 166–177
13. Noga, B.R., Kriellaars, D.J., Brownstone, R.M., Jordan, L.M.: Mechanism for activation of locomotor centers in the spinal cord by stimulation of the mesencephalic locomotor region. J. Neurophysiol. (2003)
14. Jung, R., Kiemel, T., Cohen, A.H.: Dynamic behavior of a neural network model of locomotor control in the lamprey. J. Neurophysiol. **75** (1996) 1074–1086

A Schema Based Model of the Praying Mantis

Giovanni Pezzulo and Gianguglielmo Calvi

Institute of Cognitive Science and Technology - CNR
Via S. Martino della Battaglia, 44 - 00185 Roma, Italy
giovanni.pezzulo@istc.cnr.it, gianguglielmo.calvi@noze.it

Abstract. We present a schema-based agent architecture which is inspired by an ethological model of the praying mantis. It includes an inner state, perceptual and motor schemas, several routines, a fovea and a motor. We describe the design and implementation of the architecture and we use it for comparing two models: the former uses reactive, stimulus-response schemas; the latter involves also forward models, which are used by the schemas for generating predictions. Our results show an advantage in using anticipatory components inside the schemas[1].

1 Introduction

Schemas [1] are basic functional units, permitting to investigate animal behavior without explicit assumptions about the physiological and neurophysiological realization and localization of the functions[2]. The model we propose is inspired by an ethological model of the praying mantis described in [2] but it is focused on anticipatory capabilities. It includes two kinds of schemas: **perceptual schemas** and **motor schemas**. Some schemas also are closely related (e.g. *detect predator* and *escape*): we call them **coupled perceptual-motor schemas**. In the rest of the paper we will call *schemas* the functional units, and *behaviors* the functions they realize, since many schemas can realize the same behavior.

Schema based design has three advantages: (1) it permits to integrate many competing behaviors in a coherent whole. While the animal has a large repertoire of behaviors (realized by its schemas), only few of them are useful in a given context. For this reason, *the activity level of the schema represents its relevance* [1,8,18]. The activity level of a perceptual schema represents a confidence level that a certain entity, encoded in the schema, is or is expected to be present. The activity level of a motor schema represents a confidence level that the behavior encoded in the schema is both applicable and useful in the current situation. (2) it affords distributed control: there is not a central executor, but the behavior of the animal emerges from the competition and cooperation of all the active schemas. (3) it permits to integrate in an unique framework data-driven,

---

[1] This work is supported by the EU project **MindRACES**, FP6-511931.

[2] For an hypothesis of implementation of the schemas in the nervous system, see the notion of *command neurons* in neuroethology [17] or [1].

S. Nolfi et al. (Eds.): SAB 2006, LNAI 4095, pp. 211–223, 2006.

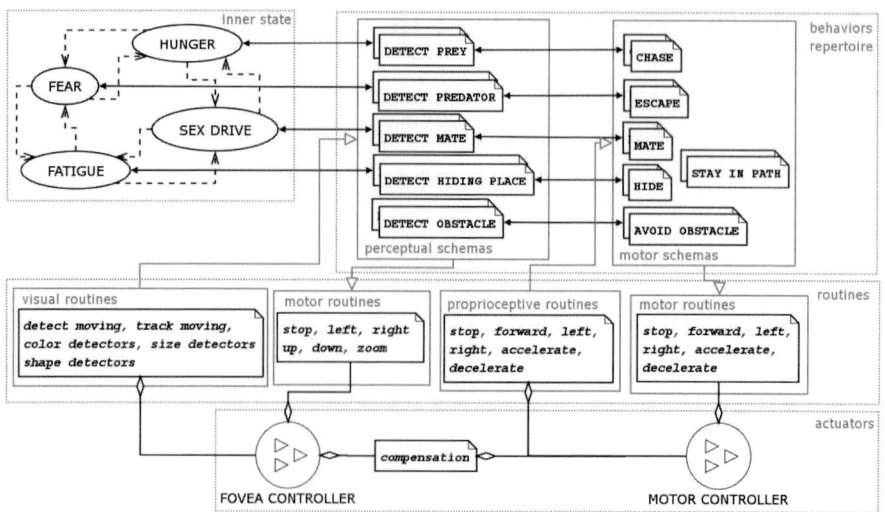

Fig. 1. The Components of the Mantis Architecture

bottom-up processes, such as the influence of stimuli on the behavior; and hypothesis-driven, top-down processes, such as forming, maintaining and testing a coherent interpretation of the stimuli.

In the rest of the paper we present our schema based agent model and we test it in a simulated environment, with two goals: (1) to evaluate its adaptivity in a dynamic environment, i.e. its capability to select the appropriate schemas for satisfying its drives; (2) to compare anticipatory vs. reactive strategies.

Fig. 1 shows the main components of the model: the **Inner State** (the drives); the **Behavior Repertoire** (Perceptual and Motor Schemas); the **Routines** (Visual, Motor and Proprioceptive Routines); the **Actuators** (the Fovea and Motor Controllers), that we introduce in the next Section.

1.1   Functional Principles of the Architecture

According to the previous definition, schemas are concurrent processes, each one encapsulating the procedures to realize a behavior. In order to be effective and adaptive, the agent has to adopt the most relevant schemas: in our implementation, this depends on schemas activity level. In our parallel architecture the activity level of each component (including schemas) determines the priority of its thread of execution; activity level thus represents the "power and influence" of a schema even in computational terms. As we will see, a more active perceptual schema can process the visual input more quickly and a more active motor schema can send more commands to the motor controller. Routines have a variable priority, too, reflecting their relevance for the schemas which exploit them: for example, in some situations color-detectors can be very relevant and size-detectors can be less. The activity level of schemas is set according to three parameters: absolute relevance, contextual relevance and predictive success.

The *absolute relevance* represents how much a schema is relevant by default; for example, *detect predator* is more relevant in absolute than *avoid obstacle* for a living creature, even if sometimes the latter is more contextually relevant. If the animal has a repertoire of schemas for the same behavior (e.g. many specialized *detect prey* schemas, e.g. for gray or red, big or small preys), some of them are more relevant in absolute, e.g. because gray preys are more common. Ceteris paribus, more absolutely relevant schemas have higher activation levels.

The *contextual relevance* represents how much the schema is relevant in the current situation. If a prey is detected and the mantis is hungry, *chase* is very relevant; but it is much less relevant if there is no prey or if the mantis is not hungry. The contextual relevance is not centrally calculated, but emerges from the dynamics of the components of the distributed architecture, in two ways. The first way is preconditions matching; for example, *chase* has as a precondition the presence of a prey (or, to be more precise, an high activity level of *detect prey*). Preconditions are not necessary but facilitating conditions: they have fuzzy values, so they can match even partially and provide graded reinforce (the more the match, the more activation is gained by the schema). The second way is exploiting the links between the components, affording spreading activation. Our design methodology includes both pre-designed links (stroke edges in Fig. 1) and evolved ones. Thanks to the pre-designed links, drives spread activation to related perceptual and motor schemas (e.g. *hunger* to *detect prey*), and coupled perceptual-motor schemas (e.g. *detect prey* and *chase*) spread activation to each other. As an emergent result of the dynamics involving both kinds of links, the most relevant schemas become more active in a context-sensitive way.

The *predictive success* also regulates schemas activation. As we will see, the schemas incorporate a predictive component (a forward model) which generates expectations; schemas generating accurate expectations gain activation. The rationale behind this principle is that schemas which predict well are "well attuned" with the current situation; for example, if *detect prey* is activated by error by a big, gray entity (assuming that some preys are gray in the environment), the schema will try to track the prey (by moving the fovea) according to its forward model, i.e. as a moving object. If the object is not a prey but, say, a stone, its tracking activity will fail (because the entity does not move). The *detect prey* schema is not well attuned with the environment, while the *detect obstacle* is: in fact, its forward model predicts a static object. While in the beginning *detect obstacle* could be not very active, as long as its forward model predicts well it becomes more and more active and overwhelms *detect prey*[3].

Schemas activation is also regulated by a general architectural principle. There is a *limited amount of activation* (i.e. computational resources) shared by all the components. All the components thus compete for limited resources and active schemas prevent other ones to gain more activation[4].

---

[3] In the current implementation expectations are treated as preconditions only in the next schema cycle; if partially matched, they provide it graded activation.

[4] This is similar to having lateral inhibition between the competing components, but the total amount of activation can be manipulated (for example by the drives).

2   The Four Components of the Mantis Model

Here we introduce the four components of the mantis model: the **Inner State**, the **Schemas**, the **Routines** and the **Actuators**.

2.1   The Inner State

The inner state includes four drives: *hunger*, *fear*, *sex drive* and *fatigue*, which have inhibitory links (dashed edges in Fig. 1). All the drives except *fear* are regulated by an endogenous factor, a "biological clock", creating an *habit* system: the mantis routinely needs food and repair and spreads activation to the related schemas for fulfilling these needs (stroke edges in Fig. 1); of course schemas can operate only if there are appropriate environmental conditions. The drives also receive exogenous influences, i.e. the activity level of the related schemas; for example, if *detect predator* is very active, *fear* grows up. There is thus an activation loop between internal drives and schemas. In the current implementation the mantis do not starve and is not really harmed by predators; on the contrary, fatigue has a real effect: it diminishes the overall amount of activation available.

2.2   The Schemas

The schemas (*perceptual* and *motor*) are the main components of the model. As shown in Fig. 1, many schemas realize the same behavior; as an example, there are *detect prey* schemas specialized for gray or red preys, or for big or small ones.

The Perceptual Schemas. The model includes five kinds of perceptual schemas: *detect prey*, *detect predator*, *detect mate*, *detect hiding place*, *detect obstacle*. Each perceptual schema has three components: a *detector*, a *controller* and a *forward model*. The main role of the detector is to acquire relevant input (preconditions) from the the fovea. The main role of the controller is to send motor commands to the fovea: in this way the mantis is able to orient its attention. Perceptual schemas are not passive data processing structure, but active ways for "navigating" the visual field [19]. The main role of the forward model is anticipate visual stimuli, i.e. the activation level of appropriate visual routines.

The perceptual schemas become more active if the kind of stimuli they process are indeed present in the environment. In our implementation, the detector has (graded) preconditions which are associated to visual routines; when the relevant visual routines are active, the schema gains activation. For example, in the mantis environment preys are gray; if the *gray-detector* visual routine is very active, the *detect prey* schema gains activation, too. The perceptual schemas also run their forward models: schemas which predict well gain activation.

The perceptual schemas receive activation from the inner states, too; a fearful mantis will search for predators even in absence of real danger. An "hallucinatory" phenomenon is in play: when a mantis is fearful, predators appear closer, moving entities appear to be predators (and get it even more fearful). In the long run hallucinations are ruled out: since the perceptual schemas also feedback on the inner states, the lack of dangerous signs (and the predictive errors of *detect predator*) will make the mantis less fearful.

Active perceptual schemas have two ways to induce top-down pressures. Firstly, they send motor commands to the fovea, orienting it toward relevant entities; more active schemas send commands with higher fire rate. By orienting the fovea, the schemas are able to partially determine their next input (they have an active vision, [19]). In an anticipatory framework, this functionality is mainly used to test the predictions of the forward models: for example, tracking a moving object is a way to acquire new stimuli in order to test the expectations. For this reason, the schemas orient the fovea towards the more informative points, i.e. those able to determine whether or not their predictions are correct. Secondly, they spread activation to the related visual routines. For example, *detect prey* schema activates the *gray-detector* visual routine, even in absence of real stimuli. This induces an "hallucinated" state (like a fake gray entity) which is close to **visual imagery** in [16]). As in the previous case, without real stimuli the hallucinated state lasts shortly.

As an effect of top-down pressures, the same stimulus is interpreted in different ways depending on the active perceptual schemas. For example, if a prey and an obstacle have the same color (say gray) and the gray-detector is very active, both *detect prey* and *detect obstacle* detect it. However, the more active schema detects it faster and takes controls of the fovea: if *detect prey* is more active, it is likely that the fovea will try to track it as a moving object (the *detect obstacle* schema, on the contrary, would have monitored it as a static object). Of course, more active perceptual schemas activate more their related motor schema, too.

The Motor Schemas. The model includes six motor schemas: *stay in path* (the default behavior), *chase, escape, mate, hide, avoid obstacle*. They have three components: a *detector*, which sets the value of the preconditions by monitoring the state of the perceptual schemas (e.g. *detect prey* is *very_active*); a *controller* (an inverse model), which send commands to the motor (e.g. *move left*); and a *forward model*. The motor schemas receive activation from the related perceptual schemas in the form of matched preconditions: a very active *detect prey* activates *chase* (which learns to interpret it as: "there is a prey"). The motor schemas receive also activation from the inner states: a fearful mantis activates its motor routines for escaping even in absence of real danger; as in the case of perceptual routines, they can only remain active if the right stimuli are in place. The main role of the controller is to send commands to the motor. The main role of the forward model is to produce expectations about perceptual stimuli (to be matched with sensed stimuli, including vision and proprioception).

Coupled Perceptual-Motor Schemas. Fig. 2 shows the pseudo-closed loop between controllers and forward models in a coupled perceptual-motor schema. The controllers send a control signal to the actuators, which integrate them and act accordingly; on the same time, an efference copy of the (final) command signal is sent to the forward models of all the schemas, which compute the next expected input. The dashed lines indicate that a feedback signal is received (via visual or proprioceptive routines); the dashed circles indicate that there is a

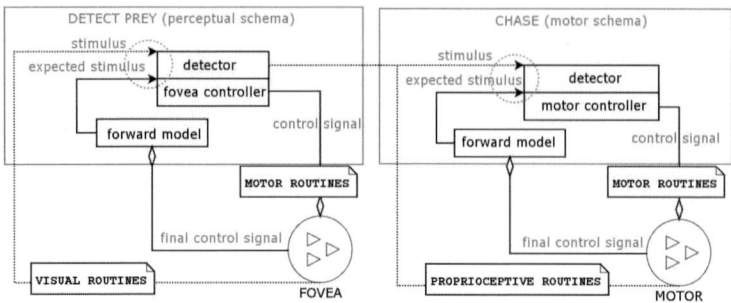

Fig. 2. A coupled perceptual-motor schema: *Detect Prey* and *Chase*

comparison between the actual input stimulus and the expected stimulus. The degree of (mis)match between actual and expected stimulus has two functions: (1) *Adjustment of Control*: the predicted signal can compensate time delays, filter or replace missing or unreliable stimuli; see [7] for a comparison with Kalman filtering; (2) *Schema Selection*: schemas which predict well gain activation.

Patterns of Actions. Some schemas include many concatenated actions (in a way similar to [4]); e.g. *detect prey* and *chase* have the following structure:

```
DETECT PREY:   IF red AND moving THEN find_prey
(loop)         ELSE IF prey_found THEN maintain_prey
               ELSE IF prey_lost THEN re_find_prey
               ELSE IF prey_maintained THEN maintain_prey


CHASE:         IF prey_found THEN approach_prey
(loop)         ELSE IF prey_close THEN grab_prey
               ELSE IF prey_in_contact THEN eat_prey
```

Schemas are always-looping procedures; for each cycle, depending on preconditions, an action is selected. This means that actions can be executed in different sequences, in parallel and also skipped: for example, *re_find_prey* is only needed when a prey is lost. Coupled perceptual-motor schema can realize complex strategies by coordinating their patterns of actions. Fig. 3 illustrates the example of a chasing behavior involving two schemas, *detect prey* and *chase*.

In the beginning, only *detect prey* is active, because some of its preconditions are true (e.g. the visual routines *red-detector* and *movement-detector* are highly active). *Detect prey* both activates its first action (*find prey*, which sends commands to the fovea) and spreads activation to *chase*. The first applicable action of *chase* (*approach prey*) can only start when *find prey* succeeds; subsequently, the actions of the two schemas continue in a coordinated way: as long as the perceptual schema succeeds in finding and maintaining the prey, the motor schema tries to reach, grab and eat it.

Interaction-Oriented Representations. It is worth noting that inside the forward models schemas process information in an interaction-oriented format,

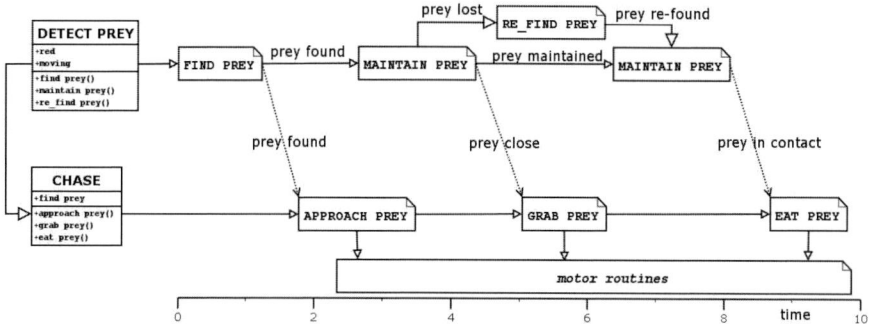

Fig. 3. Complex patterns of actions within a coupled perceptual-motor schema

and the mantis only has deictic representations. For example, *detect prey* is only able to represent "the prey I am looking at". However, even without storing extra information, the architecture implements a certain kind of object permanence. Since schemas activation decay gracefully, it is highly probable that schemas which where very active remain quite active even if the stimulus disappears for a while. This effect is magnified by the presence of drives, which have a stabilizing effect on behavior: since they continue to fuel schemas for a given span of time, drives can be seen as **task-specific memories**, introducing commitment without central control. Moreover, since schemas act according to their predictive models, they remain attuned with relevant entities by actively searching them. For example, during a successful chase *detect prey* and *chase* gain activation thanks to the success of their predictions. On the contrary, failure in finding, maintaining or reaching a prey weakens the schemas and eventually the chase ends, if another behavior becomes more active. This example shows how, in stable enough environments, deictic representations and agent-environment engagement based on predictions can realize (at least a limited version of) complex functionalities such as maintaining objects permanence.

2.3   The Routines

The perceptual schemas do not receive raw input from the fovea: a number of preprocessing units, the *visual routines*, filter fovea information (although with different priority). In the current implementation there are several routines of each kind, such as color-detectors specialized for detecting different colors, as well as for colors, sizes, shapes and for detecting and tracking moving entities. The activation level of the visual routine directly encodes the presence of absence of associate entities; for example, an active red-detector encodes directly the presence of red entities as provided by the 3D engine, without learning. In a similar way there are *motor routines*, commanding the fovea and the motors, and *proprioceptive routines*, providing feedback information from the motors.

2.4   The Actuators: Motor and Fovea Controllers

The actuators receive commands from the motor routines and perform command fusion. Differently from many systems in literature (see [5] for a review), in which many schemas can be partially active at once but only one is selected for commanding the actuators, in this model each active schema sends its motor command. Since we adopted a parallel architecture in which schemas can have different priority, commands are sent asynchronously and with different fire rates. *Fire rate encodes relevance*: more active (and thus more relevant) schemas are able to send more commands to the actuators and to influence it more.

Command Fusion. As already discussed, selection is needed both for adopting the most appropriate schema(s) for realizing the same behavior, and for adopting the most appropriate behavior. As an example of the first case, consider that there can be many *detect prey* schemas which are specialized e.g. for small and quick ones or for very big and red ones; in order to realize prey detection, often many *detect prey* schemas are needed, as in the "mixture of experts" model [13]. The case is similar for motor schemas. As an example of the second case, consider that the agent has a repertoire of behaviors and has to arbitrate between them (e.g. *chase* vs. *escape*), as long as it can not fulfill all them together.

In both cases, the fuzzy based command fusion mechanism we adopted [15] produces the course of actions accounting for more drives and stimuli. Strictly speaking, there is no actual "selection" since all the active schemas send their commands to the actuators, although with different fire rate; the course of action results from the graded contribute of all the active schemas. For example, a detect prey behavior is often realized not by a single *detect prey* schema, but integrating the graded contribute of many ones. The rationale is that exemplars of preys do not fall into clear cut categories (which prototypes are encoded in the schemas), such as "quick" or "slow" so it is often necessary to fuse the commands of the two schemas specialized for quick and slow preys. As an example, consider that a moving prey can match the preconditions of both schemas, although with different degrees of matching; thus, the degree influence they have on the fovea depends on the degree of membership of the prey to their prototype (expressed in fuzzy terms in the current implementation).

Mixed courses of actions can also emerge from the contribute of schemas realizing different behaviors. As an example, Fig. 4 shows the activation levels of two schemas, *stay in path* (black boxes) and *avoid obstacle* (white boxes), during obstacle avoidance. Both schemas are involved, with different priorities over time, as long as they can both be satisfied together. Note that the trajectory and the turning points are not preplanned but dynamically emerge depending on the size of the obstacle and the initial direction of the agent.

Exploitation also happens when the results of a schema are exploited by another behavior. For example, a mantis which is *escaping* can activate an *hide* schema as a part of the escaping strategy; the latter schema is not selected per se, but activated and exploited by the former behavior.

Constructive Perception. As discussed in the introduction, schemas permit to model both bottom-up and top-down phenomena, for example in perception. A classic experiment [25] shows that human attention varies with the nature of the task. When there is not an explicit task specification, bottom-up processes

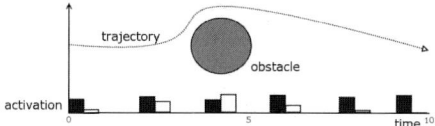

Fig. 4. Evolution over time of the activation of the schemas during obstacle avoidance

are mainly responsible for determining the salience of objects in the scene [12]. On the contrary, if there is an explicit task specification, top-down and volitional processes frame the scene and drive attention to the task-specific relevant objects in the scene [9]. We call this process *constructive perception*.

In this framework perception is distinct from sensing: *the active perceptual schemas represent multiple concurrent perceptual hypotheses which compete for being accepted*; they are prioritized according to the accuracy of their preconditions and predictions, i.e. how much their requirements are compatible with the actual perception. Schemas also actively drive perceptual exploration of the environment by orienting the fovea. The constructive process does not only influence stimuli categorization (such as prey vs. obstacle), but also behavior selection. An example of goal oriented constructive perception can clarify the point: if the mantis is not hungry and is escaping, it can approach a prey as an obstacle and activate *avoid obstacle*. Constructive perception is thus the abductive process of producing and testing hypotheses: the most active schemas drive subsequent actions (i.e. chase a prey or avoid an obstacle), active perception (where to orient the fovea) and visual imagery (to which visual routines give priority). Indeed, information is selected and it serves to confirm or disconfirm the running hypotheses, not to mirror the environment.

As discussed above, the behavioral and perceptual spaces of the mantis are also shaped by its internal drives. Drives provide activation to the behaviors, which can thus perform more epistemic actions, predict more often and influence more the fovea. An hungry mantis is much more likely to interpret ambiguous evidences as food; more precisely, an hungry mantis puts much more resources in classifying an evidence either as food or not food (instead of e.g. shelter or not shelter) even if the affordances of the object are the same.

3   Implementation and Testing

We implemented the mantis model by using the cognitive modeling framework AKIRA [11,21], and the 3-D engine Irrlicht [10], having realistic physics. Our aim is to investigate if our architecture fulfills adaptively its drives in a dynamic environment. We followed the above described architectural design, inspired by the ethological model reported in [2]; we also set up two learning phases. In the former the components of each schema, controllers (inverse models) and forward models, and the parameters such as the schemas absolute relevance or the weight of the edges, were first learned individually in a simple environment having a limited number of features (e.g. only preys or predators). In the latter all the schemas were integrated in an unique architecture. Since the framework puts schemas in competition, adding new behaviors did not waste the performance of old ones; the challenge is now to coordinate them in a complex environment. In this phase the inverse and forward models did not learn any more, but schemas which were active in the same span of time evolved energetic links (in addition to those shown in Fig. 1) with hebbian learning [15]. Schemas which were not active in a given context learned to spread energy to more successful ones, too, with a mechanism described in [20]. The rationale is that the energy has to be conveyed in a context-sensitive way from less to more relevant schemas[5].

Each schema is implemented by using a single thread which activation is set according to the principles explained above: absolute relevance (learned in the first phase), spreading activation (via edges learned in both phases) and degree of match of its preconditions and expectations. All the representational elements (activation, preconditions and expectations) have fuzzy values: in this way it is possible to compare all them and obtain graded results. For example, a *very_active* internal drive (hunger) provides high match with the *quite_hungry* precondition of *detect prey*. Or a poorly active *detect_movement* routine provides low match for the expectation *prey_moving* produced by the forward model of *detect prey*. Drives, inverse and forward models were implemented by using both Fuzzy Cognitive Maps and Neural Networks [15], with minor differences. Drives values vary according to their links, their "biological clock" and the input they receive from the active schemas. In turn, the values of the drives become input for the schemas, as described above. Even the motor commands have the form of fuzzy statements such as *turn_left* and a fuzzy controller is responsible for command fusion, as in [20]. A motor routine (*compensation*) compensates the movements of the agent, permitting to maintain the right orientation of the fovea during movement.

Related Literature. Similar models in literature are MOSAIC [24] and HAMMER [6], implementing coupled inverse and forward models for motor control and basing schema selection on predictive success; schema architectures [1,2] and the "mixture of experts" model [13]. However, we use a parallel architecture in which computational resources (such as speed) encode "responsibility"; command fusion is asynchronous and based on fuzzy logic. Our model also includes hebbian learning and spreading activation between the schemas.

Anticipatory vs. Reactive Systems. We compared the performance of two variants of the mantis model (MANTIS and MANTIS-R) in a complex environment including obstacles, preys, predators and hiding places. The first model (MANTIS) is the one we have described throughout the paper. The second model (MANTIS-R) lacks the forward models. Obviously, in this case *predictive success* can not be used for allocating activation; only absolute and contextual

---

[5] The main reason of having two phases is that it is very complex to learn many behaviors together. For example, the prediction error of the forward model can be interpreted either as scarce relevance of the schema or as poorly accurate forward model. Learning each forward model individually permits to disambiguate this signal.

relevance are used. Some anticipatory capabilities are however *implicitly* present in MANTIS-R, too: for example, a perceptual schema spreading activation to a visual routine implicitly assumes that its results will be useful; and this prediction is grounded on the history of their past interactions. However, endowing schemas with a predictive component (a forward model) which produces *explicit* expectations (as in MANTIS) permits also to use predictive error for action control and schema selection.

The two agents dwell separately in the same environment with the same inner states. **Drives satisfaction** was used as success metric: each agent had to satisfy its drives, i.e. to keep their values close to zero. Moving increases *fatigue*, while resting in hiding places lowers it; eating preys lowers *hunger*; the presence of close predators increases *fear*, while their absence lowers it; mating lowers *sex drive*. Four analysis of variance (ANOVA) with mean *fatigue, fear, hunger* and *sex drive* satisfaction (calculated as $1 - mean\_drive\_value$, in 100 real-time, 3-minutes simulations) as dependent variables were carried out.

Table 1. MANTIS vs. MANTIS-R (mean satisfaction)

DRIVE	MANTIS	MANTIS-R
Fatigue	0,845	0,657
Fear	0,812	0,665
Hunger	0,758	0,703
Sex Drive	0,891	0,793
Average	0,826	0,704

Tab. 1 shows the results. The main effect is significant for all drives ($F(1, 99) = 130, 53; p <, 00001$ for *fatigue*; $F(1, 99) = 68, 01; p <, 00001$ for *fear*; $F(1, 99) = 24, 82; p <, 00001$ for *hunger*; $F(1, 99) = 50, 65; p <, 00001$ for *sex drive*): in all cases, MANTIS satisfies its drives better than MANTIS-R. Our results indicate that in a scenario involving multiple entities and drives it is advantageous to exploit anticipatory representations, and this advantage overwhelms the cost to maintain and to run them in real time. However, our results can be considered preliminary: further investigation is needed for understanding at which level of environmental complexity anticipatory strategies become advantageous.

4 Conclusions and Further Work

We have presented a schema based agent architecture; illustrated its components and its action selection strategy; tested its behavior in a complex environment, also comparing it with a simpler model only having *implicit* anticipatory capabilities in an adaptive drives satisfaction task. Our results show that there is a significant advantages in using *explicit* expectations, produced online by forward models, for action control and schemas selection. Recently many authors [3,7,24] have provided evidences for a crucial role of anticipations and forward models in these and other cognitive functions, which we are now investigating.

In our architecture the top-down influences introduced by drives follow the **ideomotor principle** [14], arguing that action planning takes place in terms of anticipated features of the intended goal. For example, the drive *hunger* endogenously activates related perceptual and motor schemas (*detect prey* and *chase*), which in turn pre-activate visual and motor routines related to preys and pray-chasing. In [20] we have also show that this mechanism can also realize more complex goal states such as *search the red prey*. We are now investigating how to extend this principle to realize decoupled, off-line processing, such as planning by off-line producing, evaluating and comparing hypothetical, alternative goal states and courses of events, even if new and never experimented before.

In our architecture an high activation level of a perceptual schema represents the (actual or expected) presence of related entities, and motor schemas can exploit this information; in a similar way, active visual routines are exploited as preconditions by the perceptual schemas. We are now investigating how to extend this principle to schemas organized hierarchically; [20] describes the preliminary implementation of a layered architecture including feature-specific and increasingly complex concept-specific schemas, in which the activity of schemas in the lower layers *is interpreted as an information* by schemas in the higher layers, and in which *complex schemas exploit simpler ones* which realize some of their preconditions or expectations. Schemas in the higher layers are specialized for satisfying the drives of the agent; on the contrary, [22], also based on [13], shows that if hierarchical systems are evolved for a single task they do not specialize in a feature- or concept-specific way.

References

1. Arbib, M.A. (1992) Schema Theory. in the *Encyclopedia of Artificial Intelligence*, 2nd Edition, Editor Stuart Shapiro, 2:1427-1443, Wiley.
2. Arkin, R.C., Ali, K., Weitzenfeld, A., Cervantes-Prez, F. (2000) Behavioral models of the praying mantis as a basis for robotic behavior. *Robotics and Autonomous Systems*, Vol. 32, No. 1, pp. 39-60.
3. Barsalou, L. W. (1999) Perceptual symbol systems. *Behavioral and Brain Sciences*, 22, 577-600.
4. Bryson, J. (2000) *The Study of Sequential and Hierarchical Organisation of Behaviour via Artificial Mechanisms of Action Selection*. M.Phil Thesis. University of Edinburgh.
5. Crabbe, F.L. (2004) Optimal and non-optimal compromise strategies in action selection. In *Proceedings of SAB 2004*.
6. Demiris, Y., Khadhouri, B. (2005). Hierarchical Attentive Multiple Models for Execution and Recognition (HAMMER). *Robotics and Autonomous Systems Journal*.
7. Grush, R. (2004) The emulation theory of representation: Motor control, imagery, and perception. *Behavioral and Brain Sciences*, 27 (3): 377-396.
8. Gurney, K.N., Prescott, T.J., Redgrave P. (2001) A computational model of action selection in the basal ganglia, I. A new functional anatomy, *Biological Cybernetics* 84, 401-410.
9. Hopfinger, J. B., Buonocore, M. H., Mangun, G. R. (2000) The neural mechanisms of top-down attentional control. *Nature Neuroscience*, 3(3):284291.

10. http://irrlicht.sourceforge.net/
11. http://www.akira-project.org/
12. Itti, L., Koch, C. (2001) Computational modeling of visual attention. *Nature Reviews Neuroscience*, 2(3):194-203
13. Jacobs, R.A., Jordan, M.I., Nowlan, S.J., Hinton, G.E. (1991). Adaptive mixtures of local experts. *Neural Computation*, 3:79–87.
14. James, W. (1890) *The Principles of Psychology*. Dover Publications, New York.
15. Kosko, B. (1992) *Neural Networks and Fuzzy Systems*, Prentice Hall International Inc, Singapore.
16. Kosslyn, S.M., & Sussman, A.L. (1994). Roles of imagery in perception: Or, there is no such thing as immaculate perception. In M. Gazzaniga (Ed.), The cognitive neurosciences (pp. 1035-1042). Cambridge, MA: MIT Press.
17. Kupfermann, I., Weiss, K. (1978) The command neuron concept. *Behavioral and Brain Sciences* 1:3-39.
18. Maes, P. (1990) A Bottom-Up Mechanism for Behavior Selection in an Artificial Creature. In Jean- Arcady Meyer and Stewart W. Wilson, editors, *Proceedings of SAB 1990*, pages 238-246, Cambridge, Mass. The MIT Press.
19. O'Regan J.K., Noe, A. (2001) A sensorimotor account of vision and visual consciousness. *Behavioral and Brain Sciences*, 24(5):883–917
20. Pezzulo, G., Calvi, G., Lalia D., Ognibene D. (2005) Fuzzy-based Schema Mechanisms in AKIRA. *Proceedings of CIMCA'2005*, M. Mohammadian (ed.). Vienna.
21. Pezzulo, G., Calvi, G. (2005). Dynamic Computation and Context Effects in the Hybrid Architecture AKIRA . In Anind Dey, Boicho Kokinov, David Leake, Roy Turner, *Modeling and Using Context: 5th International and Interdisciplinary Conference CONTEXT 2005*. Springer LNAI 3554.
22. Tani, J., Nolfi, S. (1999) Learning to perceive the world as articulated: An approach for hierarchical learning in sensory-motor systems. Neural Networks 12:1131–1141.
23. Tyrrell, T. (1993) *Computational Mechanisms for Action Selection*. PhD thesis, University of Edinburgh.
24. Wolpert, D. M., Kawato, M. (1998) Multiple paired forward and inverse models for motor control. *Neural Networks* 11(7-8):1317-1329
25. Yarbus, A. (1967) *Eye Movements and Vision*. Plenum Press, New York.

Perceptual-Motor Sequence Learning
Via Human-Robot Interaction

Jean-David Boucher and Peter Ford Dominey

CNRS, Institut des Sciences Cognitives, 67 Bd Pinel, 69675 France
{jdboucher, dominey}@isc.cnrs.fr
http://www.isc.cnrs.fr/dom/dommenu-en.htm

Abstract. The current research provides results from three experiments
on the ability of a mobile robot to acquire new behaviors based on the
integration of guidance from a human user and its own internal representation of the resulting perceptual and motor events. The robot learns
to associate perceptual state changes with the conditional initiation and
cessation of primitive motor behaviors. After several training trials, the
system learns to ignore irrelevant perceptual factors, resulting in a robust representation of complex behaviors that require conditional execution based on dynamically changing perceptual states. Three experiments demonstrate the robustness of this approach in learning composite
perceptual-motor behavioral sequences of varying complexity.

1   Introduction

The current research explores mechanisms that allow autonomous systems to
acquire complex composed behaviors through a combination of interaction with
the sensory-motor environment, and a human teacher. We have previously developed a system that allows the user to use spoken language to teach the AIBO
ERS[1] robot the association between a name and a single behavior in the robot's
repertoire (Dominey et al. 2005 [3]). More recently, we have extended this so
that the system can associate a sequence of commands with a new name in a
macro-like capability.

The limitations of this approach result from the fact that all of the motor
events in the sequence are self contained events whose terminations are not
directly linked to perceptual states of the system. We can thus teach the robot
to walk to the ball and stop, but if we then test the system with different initial
conditions the system will mechanistically reproduce the exact motor sequence,
and thus fail to generalize to the new conditions.

Nicolescu & Mataric (2001, 2003) [5] [6] developed a method for accommo-
dating these problems with a formalized representation of the relations between
pre-conditions and post-conditions of different behaviors. They demonstrate how
pre- and post-condition relations between the successive behaviors can be extracted, generalized over multiple training trials, and finally used by the robot

---

[1] http://www.sony.net/Products/aibo

S. Nolfi et al. (Eds.): SAB 2006, LNAI 4095, pp. 224–235, 2006.

to autonomously execute the acquired behavior. A similar approach to learning generalized behavioral sequences by a bimanual humanoid robot has been developed by Zöllner et al. (2004) [10]. Billard and Mataric have also demonstrated how related approaches can be used for teaching through demonstration and imitation (2001) [1]. Kaplan et al. (2002) [4] used a progressive shaping approach, and Saunders et al. (2006) [7] use a related method in which behavior components are successively developed in a hierarchy of increasing complexity.

The current research builds upon these approaches in several important ways. First we enrich the set of sensory and motor primitives, that are available to be used in defining new behaviors (defined in Tab. 1). Second, we enrich the human-robot interaction domain via spoken language and thus allow for guiding the training demonstrations with spoken language commands, as well as naming multiple newly acquired behaviors in an ever increasing repertoire. Third, we ensure real-time processing for both the parsing of the continuous valued sensor readings into discrete parameterized form, as well as the generalization of the most recent history record with the previously generalized sequence. This ensures that the demonstration, test, correction cycle takes place in a smooth manner with no off-line processing required.

Before going into the technical details we provide a simple example scenario with AIBO entertainment robot (Sony) that is our platform for these studies. In this case the user will teach the robot a form of collision avoidance through demonstration.

1. The user initiates the learning by commanding the robot with a spoken command "turn around" that does not correspond to a primitive command nor to a previously learned command.
2. The robot thus has no knowledge of what to do, and awaits further instructions.
3. The user commands the robot to "march forward" and the robot starts walking.
4. The user sees that the robot is approaching a wall, and tells the robot to stop.
5. He then tells the robot to turn right. Behind and to the right of the robot is the red ball.
6. When the robot has turned away from the wall and is facing the ball the user tells it to stop turning, and then tells it that the learned behavior demonstration is over.

Now let us consider the demonstration in terms of the commands that were issued by the user, and executed by the robot, and the preconditions that could subsequently be used to trigger these commands. The robot was commanded to "turn around." Because it had no representation for this action, it awaited further commands. The robot was then commanded to walk. Before it collided with the wall the robot was commanded to stop walking. It was then commanded to turn right, and to stop when it was in front of the red ball. Now consider the perceptual conditions that preceded each of these commands, which could be used in a future automatic execution phase to successively trigger the successive commands. The pertinent precondition to start walking was that the command to "turn around" had been issued. The pertinent precondition to stop walking was the detection of an obstacle in the "near" range by the distance sensor in the robots face. The pertinent preconditions for subsequently turning right are that the robot is near something, and that it has stopped walking.

The goal then is for the system to encode the temporal sequence of all relations (which include user commands, exteroception and proprioception values) in a demonstration run, and then to determine what are the pertinent preconditions for each commanded action relation. Likewise, it may be the case that perceptual relations were observed during the demonstration that were not pertinent to the behavior that the human intended to teach the robot. The system must thus also be able to identify such "distractor" perceptions that occurred in a demonstration, and to eliminate these relations from the generalized representation of the behavioral sequence. The following sections will define the system architecture and its functioning that meet these requirements.

2   Perceptual Motor Learning Architecture - PML

2.1   Platform

The robot platform that we employ is the Sony AIBO ERS7, running the URBI[2] operating and control system. URBI provides a systematic access to both the entire set of onboard sensors (including vision of the red ball and other objects, sensitivity to presses of the several buttons on the robot's back and head, joint angles, position-orientation sensors etc.), and to movement commands for walking, turning, backing up etc.

A central aspect of the PML system (Fig. 1) is that there is a single coherent temporal representation of all perceptual and motor relations. That is, both the commands issued by the user, as well as the sensory values from the proprioceptors (e.g. selected joint angles) and exteroceptors (vision, distance sensors) are to be represented in a single temporal sequence of Boolean values. This transformation of continuous sensor values from a world model into a symbolic representation of logical predicates in a situation model corresponds to a form of conceptualization (Siskind 2001 [8], Steels & Baillie 2003 [9], Dominey & Boucher 2005 [2]).

Concretely, during the experiments, the "situation modeler" sends messages to the model whenever the robot detects a new event, i.e. a sensorimotor relation change (RC) (fig. 1.B). These Boolean relation change RC values (see Tab. 1) are communicated in an XML format. Based on the received RCs the model generates the world vectors (WVs) that are the vectors of all relation values. The WVs and RCs are stored in a chronological History that can be compared with a sensori-motor memory. For greater clarity, the History and its subsequences can be represented as directed acyclic graphs (Sect. 2.2) or as chronongrams (Sect. 2.3). During the first learning trial of a complex behavior, the user guides

---

[2] http://www.urbiforge.com

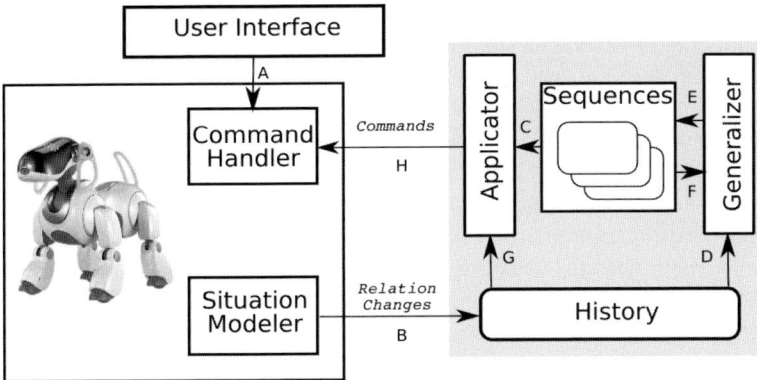

Fig. 1. System Architecture: The User Interface sends action command messages to the robot (A). The primitive actions are of the type command, and are implemented in the robot. Actions to be learned are of the type complex and are represented as sequences. The Situation Modeler sends RCs to the History (B) whenever the robot observes any change in a relation. When the behavior to be learned is completed (complex relation deactivated) the Generalizer extracts the last sequence from the History (D) and compares it with the corresponding sequence in the repertoire (F). The resulting generalized sequence replaces the previous one in the repertoire (E). When the user executes a learned behavior (complex relation activated) the Applicator executes (G). It retrieves the corresponding sequence in the repertoire (C) and sends the corresponding commands (H) if their preconditions are met.

Table 1. Actionnal and perceptual relations

Relation type	Meaning	
Action: When the relation is true, the command or the complex is going on. E.g. in Fig. 9, the robot is walking WV(3-4), is tracking WV(2-5) and is approaching WV(1-6).	**Command**	
	March(robot,front)	Walking forward
	March(robot,back)	Walking backward
	Rotate(robot,left)	Turning left
	Rotate(robot,right)	Turning right
	Track(robot,ball)	Tracking ball with head
	Complex	
	Approach(robot,ball)	Approaching the ball
	Align_right(robot,ball)	Aligning to the ball
	Turn_around(robot)	Turning around an obstacle
Perception: When the relation is true, the robot is perceiving. E.g. in Fig. 9, the robot is near something WV(4-6) and always see the ball.	**Exteroception**	
	See(robot,ball)	The robot is seeing the ball
	Near(robot,thing)	The robot is close to an object
	Proprioception	
	Neck(robot,left)	neck/head turned left ($\pm$ 10°)
	Neck(robot,right)	neck/head turned right ($\pm$ 10°)
	Neck(robot,center)	neck/head turned center ($\pm$ 10°)
	Touch(robot,shoulders)	shoulder button pressed

the robot through each step (fig. 1.A). When the complex command is issued, the Applicator is activated (fig. 1.G) but has no sequence to apply (fig. 1.C). Once the first learning trial is completed, the Generalizer extracts the sequence (fig. 1.D) from the History to include it in the sequence repertoire (fig. 1.E). For the next learning trial, the Applicator now finds the corresponding sequence in the repertoire (fig. 1.C) and attempts to apply it. Thus, while reading the sequence, when the preconditions for an action are met, the Applicator issues the appropriate command (fig. 1.H). After the second learning trial, the Generalizer extracts the just executed sequence from the History and compares it with the corresponding sequence in the repertoire (fig. 1.E). Sect. 2.2 and Sect. 2.3 respectively explain the roles of the Applicator and the Generalizer.

2.2   Generalization

The Generalizer performs generalization on a complex behavior that has just been executed and the existing generalized sequence for that behavior (respectively the sequence extracted from the History Fig. 1.D and the existing sequence

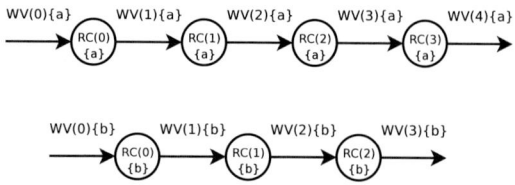

Fig. 2. Two complex behavior learning trials. The first trial yields 4 RC nodes and 5 WV arcs, RC(0-3){a} and WV(0-4){a} respectively. The second trial is identified by {b}.

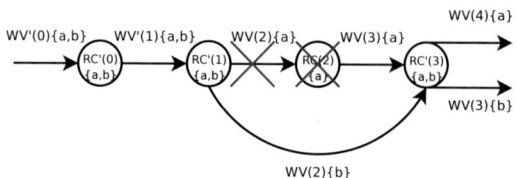

Fig. 3.   WV-RC   Fusion   &   deletion: RC'(0){a,b}=RC(0){a}=RC(0){b}
RC'(1){a,b}=RC(1){a}=RC(1){b}   RC'(2){a,b}=RC(2){a}=RC(2){b}
RC'(3){a,b}=RC(3){a}=RC(2){b}   WV'(0){a,b}=WV(0){a}=WV(0){b}
WV'(1){a,b}=WV(1){a}=WV(1){b}

in the repertoire Fig. 1.F). Fig. 2 represents these as graphs. Generalization serves to determine which are the pertinent components in this command sequence pair. For this we apply two operations. The first is to merge identical WVs and RCs and remove superfluous loops (Fig. 3). The second operation determines the sufficient conditions for sending a command or receiving a perception (Fig. 4 and Fig. 5). This requires comaring the WVs preceding a given RC and identifying relations whose values can indifferently be true or false.

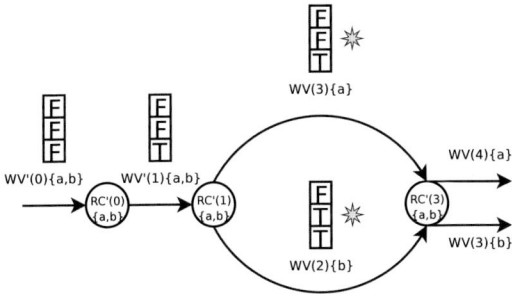

Fig. 4. Relation variability: The rectangles of 3 slots correspond to three arbitrary relations making up a WV. F and T correspond to False and True relations, respectively. We see that WV(3){a} and WV(2){b} lead to the same RC'(3){a,b}.

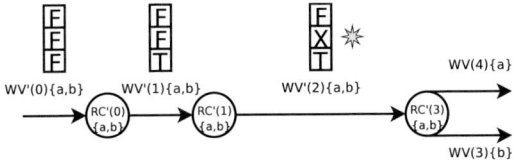

Fig. 5. Vectorial generalization: The new generalized vectorWV'(2){a,b} is equal to vect_gene(WV(3){a},WV(2){b}). If the value of a given relation in the WV can be either True or False, this is designated by an X.

2.3   Application

After an initial nave learning (where the user leads the robot through each stage) the user can then ask the robot to perform the learned behavior. This process of re-execution of a learned sequence (stored in the generalized sequence repertoire) is called "application". During generalized sequence application, if the robot is

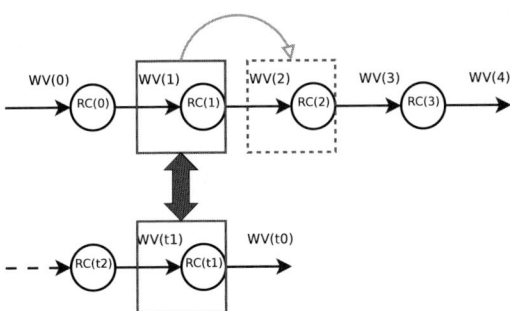

Fig. 6. Perceptive RC reception: If the current (WV(1), RC(1)) pair of the generalized sequence matches with the last (WV(t1), RC(t1)) pair of the History then the reading index is incremented. Concretely, the robot waits to be in the same state (WV(1) matches (WV(t1)) and to perceive the same state change (RC(1) = RC(t1)) in order to proceed to the next step.

in identical conditions (same WVs) then it can execute the next action in the sequence (sending a command Fig. 8). However, if not, the robot waits for these conditions (receipt of an appropriate RC Fig. 6) and thus remains static. The user can thus force the execution by explicitly commanding the robot if necessary (reception of an action RC Fig. 7).

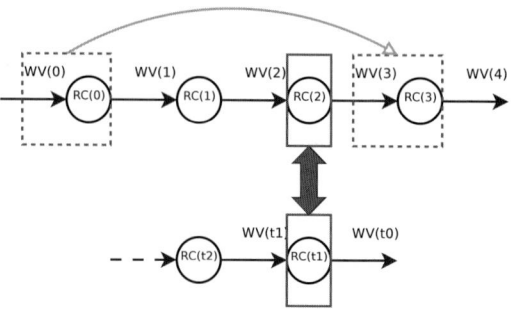

Fig. 7. Command RC reception: The reading head is positioned on the pair (WV(0), RC(0)) though the received command RC(t1) matches the next command in the sequence RC(2). The user has thus forced the robot. To continue the sequence execution correctly, the reading head is repositioned after the command matching that in the History.

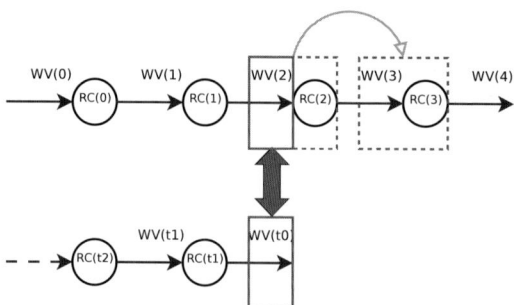

Fig. 8. Command RC emission: If the last History vector (WV(t0)) matches the current sequence vector (WV(2)) and RC(2) is a command, then the reading head is advanced. This incrementing occurs, of course, after the command is sent to the robot by the model.

3   Experimental Results

We can now present generalization results from three experiments that test the ability of the system to learn generalized representation of three distinct complex sensory-motor behaviors. In all cases, the human instructs the robot the first run through, and then the robot begins to generalize and attempts to execute the generalized sequence, with the intervention of the user when necessary.

3.1 "Approach" Behavior

The first experiment tests the ability of the system to learn to terminate an ongoing motor behavior based on a change in a perceptual state that occurs as a function of the execution of that behavior. This is a basic function of the generalized learning capability that allows the system to acquire behaviors that depend on generalized sensory-motor correlations, rather than on exact identical initial conditions.

For the behavior in question, the user places the red ball in front of the AIBO. The user then commands the AIBO to perform the complex action approach. At this time there is no Generalized Sequence stored for this behavior and so the AIBO does nothing. The user then commands the AIBO to visually track or look at the ball, and then to start walking. When the AIBO gets close to the ball, the user commands the AIBO to stop walking.

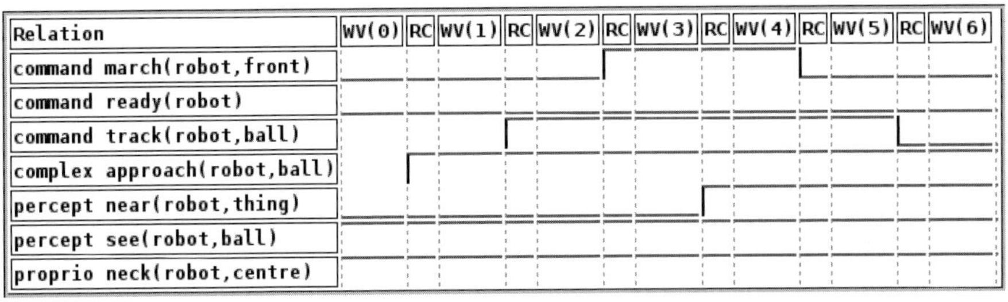

Relation	WV(0)	RC	WV(1)	RC	WV(2)	RC	WV(3)	RC	WV(4)	RC	WV(5)	RC	WV(6)
command march(robot,front)													
command ready(robot)													
command track(robot,ball)													
complex approach(robot,ball)													
percept near(robot,thing)													
percept see(robot,ball)													
proprio neck(robot,centre)													

Fig. 9. Chronogram of World Vectors and Relation Change states generated after two learning trials of the complex behavior approach(robot, ball). Note that distances along the horizontal time axis do not correspond to actual durations, though temporal order is accurately indicated. The user issues the approach command. This establishes the necessary preconditions to automatically initiate tracking the ball, which then establishes the preconditions to initiate the march command. The robot proceeds in this manner until the perceptual relation near changes to true. This establishes the preconditions for issuing the command to stop marching, and to stop tracking. At this point the learned behavior has successfully been performed in a fully autonomous mode by the system.

The crucial point is that there is a direct relation between the initial distance from the AIBO to the ball, and the distance that the AIBO walks before the user commands it to stop. The desired generalization is that whatever this distance is, on future trials, the AIBO will learn when to stop. Inspection of Fig. 9 reveals that prior to the command to stop walking (top trace command march(robot, front) goes to False or zero) the value of the perceptual relation near(robot, thing) has changed to true. That is, the proximity sensor value has reached a threshold indicating that the robot is now in physical proximity with an object. This physical proximity detection - which occurred because the robot walked sufficiently close to the ball independent of how far it was initially - can thus serve as the perceptual signal to stop walking. Thus, the human subject puts the robot through the action, and the robot learns to link what it perceives with the required issuing of commands to initiate and terminate motor functions. A crucial aspect of this adaptive behavior mechanism is that the robot is continuously testing its "hypothesis" - the current Generalized Sequence for the command in question - and relying both on regularities in the stream of relations coming from the perceptual environment, and relations coming from commands issued by the human in order to correct or shape the ongoing behavior that is being learned.

3.2   "Align" Behavior

The second experiment also tests the ability of the system to learn to terminate an ongoing motor behavior based on a change in a perceptual state that occurs as a function of the execution of that behavior in a different context. In this case, the AIBO is to search for the ball, orient its head to the ball, and then - maintaining this orientation -turn its body so that the neck is straight, and the AIBO body is aligned with the ball. In this configuration, when the AIBO starts to walk, it will be pointed in the direction of the ball.

For this behavior the ball is placed a few meters away from the AIBO, away from the direction in which it is currently looking. In the current example this will be to the right of the AIBO. The user starts by invoking the command align_right. In the first demonstration, there is no Generalized sequence, and so the system waits for instructions. The user then commands the robot to track or locate and look at the ball. Once this is achieved, the user then commands the robot to rotate its body towards the right, while continuing to fixate the ball, thus bringing the body into alignment with the head. When this alignment is achieved, the user commands the robot to stop turning. The goal here is that in the general case, independent of how far the ball initially is to the right, the robot will turn, and stop turning when it is aligned with the ball.

Examination of Fig. 10 indicates that just prior to the changing of command rotate(robot, right) to false, there is a pair of state changes in which the value of the proprioceptive relation neck(robot, center) becomes true, and then the value of the proprioceptive relation neck(robot, right) becomes false.

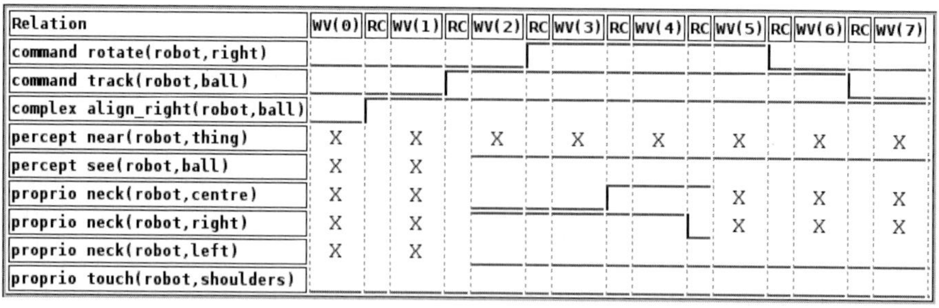

Relation	WV(0)	RC	WV(1)	RC	WV(2)	RC	WV(3)	RC	WV(4)	RC	WV(5)	RC	WV(6)	RC	WV(7)
command rotate(robot,right)															
command track(robot,ball)															
complex align_right(robot,ball)															
percept near(robot,thing)	X		X		X		X		X		X		X		X
percept see(robot,ball)	X		X												
proprio neck(robot,centre)	X		X								X		X		X
proprio neck(robot,right)	X		X										X		X
proprio neck(robot,left)	X		X												
proprio touch(robot,shoulders)															

Fig. 10. Chronogram of World Vectors and Relation Change states in learning and generalizing the complex behavior align_right(robot, ball). generalized sequence after 6 learning trials. See text.

Though not clearly indicated on the chronogram these two events occurred with a very short separating interval (<1 second). The next event is the user issuing the command to stop turning. Again, the crucial issue will be to automatically establish the link between the sending of the command to stop turning and the perceptual events corresponding to the centering of the head and neck with the body. If we compare the first history with the last generalied sequence, we can observe that there is a substantial shortening of the sequence is due to the fusion and deletion generalization (Fig. 3). The "X"s distributed throughout correspond to vectorial generalization illustrated in Fig. 5. The presence of an "X" indicates that the particular relation value can be either true or false at that point, and the generalized sequence can still proceed. This corresponds to relations whose values are not pertinent at the given time for the passage from the previous to the subsequent State.

Experiments 3.1 and 3.2 thus demonstrate that the generalization method is capable of determining the pertinent perceptual relation changes that trigger the onset and offset of motor commands. This provides generalized sequences that are sensory-motor programs that can be autonomously executed by the robot with a good deal of invariance to modifications in initial conditions.

3.3 "Turn Around" Behavior

The third and final experiment tests the ability of the system to learn to coordinate the initiation and termination of two distinct behaviors based on a succession of changes in perceptual states that occur as a function of the execution of these behaviors. In this case, the AIBO is to start walking and then stop when it detects a potential collision. It should then begin turning to the right, and stop turning when it sees the red ball.

Fig. 11 illustrates the generalized sequence for this behavior after 5 interactive training demonstrations. Examination of this generalized sequence provides a clear way to understand the successive elements of the behavior that were used to each this complex behavior. To begin training this behavior, the AIBO is placed a few meters from a wall, with the red ball placed about 1 meter to the right of the future collision point. The user starts by invoking the command turn_around. In the first demonstration, there is no Generalized sequence, and so the system waits for instructions. The user then commands the robot to center its head/neck in the forward direction. Once this is achieved, the user then commands the robot to begin to march forward. When the user sees that the robot is getting close to the wall, the user commands the robot to stop marching. Again, the goal here is that in the general case, independent of how far the robot initially is from an obstacle, the robot will stop walking when it comes within some threshold distance of that object.

In Fig. 11 this corresponds to the State change in which the perceptual relation near(robot, thing) becomes true. Indeed, this occurs just before the user-issued command to stop marching. Thus in the generalized sequence, this state change must precede the self generated command to stop marching. After this command, the next event is the user issuing the command to begin turning right.

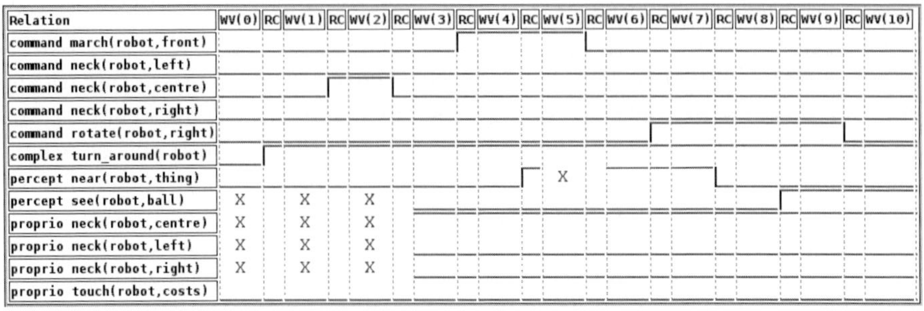

Relation	WV(0)	RC	WV(1)	RC	WV(2)	RC	WV(3)	RC	WV(4)	RC	WV(5)	RC	WV(6)	RC	WV(7)	RC	WV(8)	RC	WV(9)	RC	WV(10)
command march(robot,front)																					
command neck(robot,left)																					
command neck(robot,centre)																					
command neck(robot,right)																					
command rotate(robot,right)																					
complex turn_around(robot)																					
percept near(robot,thing)										X											
percept see(robot,ball)	X		X		X																
proprio neck(robot,centre)	X		X		X																
proprio neck(robot,left)	X		X		X																
proprio neck(robot,right)	X		X		X																
proprio touch(robot,costs)																					

Fig. 11. Chronogram of World Vectors and Relation Change states after generalizing the complex behavior "turn_around"; generalized sequence after 5 learning trials

Again, a crucial issue will be to automatically establish the link between the sending of this command to turn right, and the enabling pre-conditions. The robot continues turning right until the next sequence of relation events which are the value of near(robot, thing) changing to false, followed by the relation see(robot, ball) changing to true. Fig. 11 thus represents the combined sequence of perceptual events and command events that define the Generalized Sequence for the complex behavior turn_around. This generalized sequence is used by the system as a behavioral program that allows the successive selection of the appropriate motor behaviors based on the combination of perceptual and motor relations that define the preconditions for these successive motor commands.

4   Discussion

The current research provides new results from experiments with a robotic system that can acquire an open set of new behaviors through interaction with the sensory-motor milieu and a human teacher. Part of the novelty of the system is that it combines perception and action into a coherent representation in terms of a temporal sequence of perception and action state changes. This combined sequence is then the object of a generalization process whereby, through multiple demonstrations of a given behavior to be learned, the system extracts the minimal pertinent description of the relevant actions and their required preconditions. This resulting generalized sequence is thus an integrated sensory-motor program that the system can execute autonomously. In three experiments based on complex behaviors of varying complexity, we have demonstrated that the system indeed learns to generalize the behavioral sensory-motor sequences, and that it can perform this for arbitrary combinations of its base set of sensory and motor relations (illustrated in Tab. 1). Another novel aspect of the current approach is that it is not at all tied to the specifics of the robot platform used. It only requires that the robot has the equivalent of the situation modeler which provides logical predicate values for different sensor and motor command status values. Indeed, part of the long term aim of this research is to demonstrate that this Generalized Sequencing Model can easily adapt to a variety of robot platforms.

This provides the basis for an interesting set of extensions. In this context we are currently extending the system to allow embedding of these complex behaviors to create hierarchical complex sequences. A second line of extension is related to the command of the system via spoken language. In the current context, user commands can be issued via a console interface and by spoken language using the CSLU RAD system in which single spoken commands are recognized and used to send the appropriate URBI command to the robot. Our future work in this area will allow the use of predicate-argument commands (such as "get the X" where X can take different arguments like ball, bone, etc.) and corresponding grammatical constructions, rather than single words. We have begun to investigate this use of grammatical constructions in Dominey & Boucher 2005 [2], and Dominey et al. 2005 [3].

Acknowledgements

J-DB is supported by a Doctoral Fellowship from the Rhone Alps Region as part of the Cluster Project "Presence." PFD is supported in part by funding from LAFMI, ACI TTT and the ESF ECRP. We thank Gerard Bailly, James Crowley and Jean Caelen for constructive comments on an earlier version of this research.

References

1. Billard A, Mataric MJ (2001) Learning human arm movements by imitation: Evaluation of a biologically inspired connectionist architecture. Robotics and Autonomous Systems, 37, 145-160.
2. Dominey PF, Boucher JD (2005) Learning To Talk About Events From Narrated Video in the Construction Grammar Framework, Artificial Intelligence, 167 (2005) 31-61
3. Dominey PF, Alvarez M, Gao B, Jeambrun M, Weitzenfeld A, Medrano A (2005): Robot Command, Interrogation and Teaching via Social Interaction, Proc. IEEE Conf. On Humanoid Robotics 2005.
4. Kaplan F, Oudeyer P-Y, Kubunyi E, Miklosi A (2005) Robot clicker training, Robotics and Autonomous SystemsMarch 2002, pp. 197-206(10)
5. Nicolescu M.N., Matari M.J. (2003): Natural Methods for Robot Task Learning, Proc. AAMAS'03.
6. Nicolescu M.N., Mataric M.J. (2001): Learning and Interacting in Human-Robot Domains, IEEE Trans. Sys. Man Cybernetics B, 31(5) 419-430.
7. Saunders J, Nehaniv CL, Dautenhahn K (2006) Teaching robots by moulding behavior and scaffolding the environment, Proceedings of HRI'06, March 2-4, Salt Lake City, Utah USA
8. Siskind JM (2001): Grounding the lexical semantics of verbs in visual perception using force dynamics and event logic. Journal of AI Research (15) 31-90
9. Steels, L. and Baillie, JC. (2003): Shared Grounding of Event Descriptions by Autonomous Robots. Robotics and Autonomous Systems, 43(2-3):163–173. 2002
10. Zöllner R., Asfour T., Dillman R.: Programming by Demonstration: Dual-Arm Manipulation Tasks for Humanoid Robots. Proc IEEE/RSJ Intern. Conf on Intelligent Robots and systems (IROS 2004).

Navigation and Internal World Models

POTBUG: A Mind's Eye Approach to Providing BUG-Like Guarantees for Adaptive Obstacle Navigation Using Dynamic Potential Fields

Michael Weir, Anthony Buck, and Jon Lewis

School of Computer Science, University of St Andrews
North Haugh, St Andrews, Fife, Scotland
{mkw, arb11, jpl3}@st-and.ac.uk
http://www.dcs.st-and.ac.uk

Abstract. The problem we address is adaptive obstacle navigation for autonomous robotic agents in an unknown or dynamically changing environment with a 2-D travel surface without the use of a global map. Two well known but hitherto apparently antithetical approaches to the problem, potential fields and BUG algorithms, are synthesised here. The best of both approaches is attempted by combining a Mind's Eye with dynamic potential fields and BUG-like travel modes. The resulting approach, using only sensed goal directions and obstacle distances relative to the robot, is compatible with a wide variety of robots and provides robust BUG-like guarantees for successful navigation of obstacles. Simulation experiments are reported for both near-sighted (POTBUG) and far-sighted (POTSMOOTH) robots. The results are shown to support the theoretical design's intentions that the guarantees persist in the face of significant sensor perturbation and that they may also be attained with smoother paths than existing BUG paths.

1 Introduction

1.1 The Problem

The general problem we address is how to get a robot to navigate from A to B where there are intervening obstacles.

In static and familiar environments, a path between a particular A and B that circumvents obstacles may be known before travel commences. If not, a global map showing the location of intervening obstacles may allow a path between A and B to be computed [1]. Knowledge of a connecting path coupled with knowledge of a reliable mapping of robotic action into motion then enables a sequence of actions moving the robot from A to B to be readily computed.

However, such knowledge may not be available for autonomous robots working in unknown or dynamically changing environments. An important consequence is that the unknown throws up unexpected aspects of the environment resulting in perturbation of the robot motion. The motion task then is a challenging one of finding a route based on more limited knowledge through adaptive behaviour. A common

S. Nolfi et al. (Eds.): SAB 2006, LNAI 4095, pp. 239–250, 2006.

scenario is where knowledge is restricted to goal direction and local obstacle data provided through sensors. This scenario in autonomous obstacle navigation is addressed here for a 2-D travel surface.

1.2   Existing Approaches

There are numerous techniques for robotic navigation in 2-D space that do not rely on the existence of a global map providing a priori knowledge of the environment. They feature varying levels of pre-planning, reactivity and world modelling and differ with respect to their computational expense and success. Major general approaches as described in [2] are:

- BUG based algorithms (e.g. [3], [4]) which provide geometrical paths connecting to the goal. For other non-heuristic algorithms, see [5].
- Methods which develop a discrete model of the environment that may, for example, be searched via an A* algorithm to establish an optimal connection from start to goal. Overviews can be found in [6] and [7].
- Potential field based systems in which the motion of a robotic agent is directed by a combination of repulsive obstacle potential and attractive goal potential [8]. Overviews can be found in [6] and [7].

We focus here on the two approaches in the above that employ a direct travel path without search, i.e. potential fields and BUG algorithms.

1.2.1   Potential Field Based Navigation

The potential field approach views the robot as analogous to a charged particle attracted to the goal and repelled by the obstacles [8]. These virtual forces are used to guide the robot in navigating its environment. By converting sensor information found in the field into a combination of attractive and repulsive potential, motion may be generated towards the goal through the direction of the potential field gradients. The goal distance and the inverse of the local obstacle distance are commonly taken to represent the degree of attractive goal potential and repulsive obstacle potential respectively at each location. Provided the goal is reasonably clear of obstacles, it is identifiable as a unique location of lowest combined potential. The combined potentials may be visualised abstractly as a surface of varying height over the travel environment. Standardly, the locally steepest gradient downwards on the potential surface points the way to the goal so that travel proceeds like a ball rolling to the bottom of a hill. Such an approach offers the possibility of robust travel since increasing error between actual and intended motion will result in steeper gradients coming into play to guide the motion back on course. The underlying potential field framework also offers compatibility with a wide range of robot designs and a relatively seamless integration with robot features such as finite size and sensor-based perception.

While there has been substantial enthusiasm for the potential field approach in robotics [7], [9], progress has been blocked by the commonality of features such as the well known Local Minimum Problem [10]. In the latter case, obstacles containing common shapes of many varieties such as C shapes cause the robot to become stuck upon entering the convex inside of a region bordering the obstacle, e.g. the inside of the C. The robot cannot escape the region except by temporarily going further away from the goal, which would entail going upwards in potential against the downward flow in the static field. The behaviour is equivalent to a ball getting stuck in a local pit part way down the hill.

1.2.2 BUG Algorithms

There are many algorithm variants belonging to the BUG family of algorithms beyond those initially developed in [3], [4], that also provide a guarantee for their geometric paths to reach a realisable goal. The family commonly assumes a point robot analogous to a bug which proceeds directly towards the goal when it can, and steadily goes forwards along intervening obstacle boundaries until the boundaries can be left to carry on directly towards the goal.

The guarantee comes about because, in 2-D, going forwards along an obstacle boundary after hitting the obstacle always eventually results in a point being reached where the obstacle can be left *safely*. That is, the boundary traversed is left permanently with no danger of an infinite loop developing in the path through returning to the point. If leaving takes place at such a point, the obstacle is then only a temporary diversion. As the travel between obstacles always reduces the distance to the goal and the number and size of obstacles are finite, reaching a realisable goal is guaranteed. There are many BUG algorithms, each with different conditions for leaving an obstacle safely enough to preserve the guarantee, but all have the above basic modus operandi. Our own method has its own improvements for leaving safely.

The BUG algorithms also have their drawbacks though. The earlier versions were relatively inefficient in the degree to which they went round obstacles before leaving them. Subsequent variants attempt to reduce the inefficiency of their predecessors' paths so they leave an obstacle earlier, e.g. [11]. More recently DistBUG [12] and 3D Bug [13] feature a more prevalent use of real sensor data.

However, a central issue is that while providing a theoretically sound geometric path for a point robot, they require extra support, that is unspecified in detail, to realise their paths. There is no specification in BUG's path planner for how to follow an obstacle edge or how to cope with a finite sized robot or with perturbation of the robot's perception and action. In robotics, it is not trivial to provide such support given the uncertain nature of a robot's environment and idiosyncrasies of the robot itself. BUG paths are also inherently inefficient by closely clinging to the obstacle boundaries. There is no opportunity to curve a smoother course to go round obstacles at a greater distance from the obstacles, in the way natural agents with extended sensing such as vision can do.

In this paper, we aim to provide the best of both approaches, i.e. to provide robust and smooth paths through a potential field approach that has a more integral specification than BUG, but also to provide BUG-like guarantees of reaching realisable goals.

2   Previous Work on Overcoming Local Minima Using Potential Fields

The local minimum problem has been a serious problem for potential field methods. While there have been attempts to overcome this problem in a variety of ways [6], [7], [14], [15], none of the attempts have been seen to be an outright solution. Obstacle avoidance on its own using potential fields does not guarantee convergence to the goal. Franceschini et al [16] show that robots become stuck in local minima or wander aimlessly without a target. As described earlier, the use of a target induces local minima in common convex shapes for a static field. Varying a potential field dynamically during the behaviour may move the process on from local minima in a static field [17], but then introduces cyclic paths for various common convex obstacle shapes due to being attracted back onto a previously visited obstacle edge. BUG algorithms provide theoretical guarantees that obstacle edges are not returned to as described above. A potential field approach thus needs to provide more practical guarantees equivalent to BUG if success is to be assured on a universal basis.

Forward chaining is a technique recently developed by one of the authors [2] that aims to provide smooth plasticity and persistence for robotic agent navigation towards a goal. It is a relatively inexpensive dynamic potential field based approach capable in principle of traversing static local minima in combined obstacle and goal potential. Smooth adaptation of the robot's travel path while maintaining persistence towards a goal is provided by the use of intermediate subgoal attractors that dynamically form temporary stepping stones connecting to the goal. Other work which has made use of subgoals in order to address the local minimum problem for potential fields can be found in [18], [19], [20] and more recently in robotics in [21].

A subgoal as an individual acts as an attractor on a combined subgoal and obstacle potential surface just like the goal acts as an attractor on the combined goal and obstacle potential surface in traditional static potential field methods. The difference is to repeatedly replace the subgoal with one further forwards to generate a simple and dynamically changing potential surface. The dynamics carries the robot position within a moving *dip* that corresponds to the moving attractor basin at each stage of the process. That is, each new subgoal generates a different potential surface containing a single global minimum forwards of the previous location that is readily approachable through gradient descent from the current robot position. A suitable sequence of subgoals, set to track the goal and nearby obstacle edges in a forward direction as they are sensed, allows the robot to travel from start to goal without getting stuck in static local minima short of the goal.

Forward chaining has been shown in [2] to reach the goal successfully for a variety of obstacle courses including local minima for the traditional static method. However, the design was limited in being for a point robot, in its obstacle leaving condition restricting the type of obstacle course navigable, and by not using direct or perturbable sensor information. We now cater for these features in the present paper.

3   A Mind's Eye Approach to Providing a Goal Reachability Guarantee Using Dynamic Potential Fields

In this section we present POTBUG and POTSMOOTH flavours for Forward Chaining robots that have evolved to remove the limitations outlined above, and extend the goal reachability guarantee to arbitrary obstacle shapes by chaining subgoals along BUG-like paths.

3.1   Sensor and Subgoal Based Potential Evaluation

In the following sections, range sensors are used to repeatedly establish obstacle distances and directions relative to the robot throughout the behaviour. The direction to the goal is also repeatedly established through sensor readings and is not blocked by obstacles at any time. Repeated estimates of the distance to the goal are established by triangulation.

The sensor readings are used to construct a varying potential field relative to the robot's current position and orientation. The obstacle potential function has an infinite value on the edge of an obstacle and falls off to a value of 0 over a finite *fall-off* range from the obstacle. The obstacle potential function is given by

$$U_{\text{Obstacle}}(d_i) = \begin{cases} 1/d_i \cdot e^{-1/(s^2 - d_i^2)} & , 0 \le d_i \le s \\ 0 & , d_i > s \end{cases} \tag{1}$$

where d_i is the sensed distance of the robot from an edge point i of an obstacle, and s is the *fall-off* range which is generally set to be $3r$, where r is the robot radius. [For the given fall-off function to have a monotonic decrease in gradient size, $s \le 1$. Consequently, as s = 3r, the robot radius needs to be expressed in terms of units that result in such a fall-off range.] Setting the fall-off range in terms of the robot radius will be shown to be important for making the robot navigation suitable for its size.

Sensor readings of obstacle distances d_i are taken in various forwards linear directions that impact on obstacles within sensor range. The potential function used to model subgoals as attractors for the robot is a quadratic bowl in terms of the distance d_a to that attractor

$$U_{\text{Attractor}}(d_a) = d_a^2 \tag{2}$$

The combined obstacle and attractor potential may then be calculated as:

$$U_{\text{Combined}} = U_{\text{Attractor}}(d_a) + \sum_i U_{\text{Obstacles}}(d_i) \tag{3}$$

The robot uses the equivalent of a mind's eye to focus attention on a circle set concentrically around itself with a radius that is the distance of the nearest sensed obstacle point, or the sensor range itself if no obstacles are sensed, or the goal if this is within sensor range with no obstacles inbetween. The mind's eye is used by the robot to monitor and position the subgoal on the forwards semi-circle either relative to the

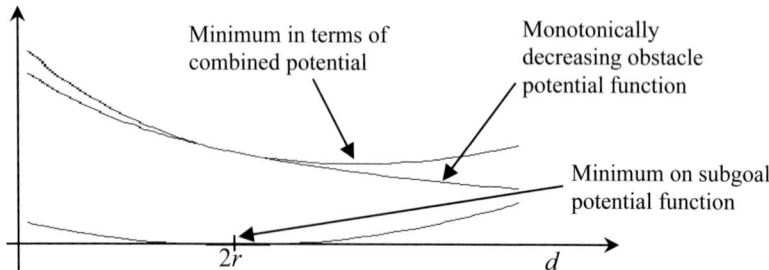

Fig. 1. Summation of obstacle potential and subgoal potential to form combined potential. The distance d from the obstacle increases from left to right.

goal or to any intervening obstacles. The robot performs steepest descent down the dip of the combined obstacle and subgoal potential surface at each stage.

3.2   Subgoal Setting and Reachability Guarantee

Each subgoal is set forward in free space using the sensors to ensure there are no obstacles between the current robot position and the subgoal. In the absence of any sensed obstructions, subgoals are set on the forward semi-circle in the goal direction so as to pull the robot along a straight line to the goal. In the presence of sensed obstructions, subgoals are set on the forward semi circle at a safe yet close distance from the obstacle edge to pull the robot around the obstacle. By selecting positions that have a summed obstacle potential value of $B=U_{Obst}(2r)$, where r is the robot radius, we effectively set a safe distance corresponding to $2r$ from a single obstacle sensor sample, and slightly more for multiple sensed obstacle points. The robot uses its mind's eye to set subgoals targeting an obstacle potential contour of this B-value, a *B-contour*, surrounding obstacles.

For each subgoal the combined subgoal and obstacle potential surface is the summation of monotonically changing functions and a quadratic bowl. Starting at an obstacle edge, the combined potential decreases monotonically until the minimum of the subgoal function is reached (Fig. 1). After this point, the obstacle function gradients continue to decrease monotonically in size towards zero while the subgoal function gradient now becomes positive and increases monotonically. The combined potential will therefore eventually start to turn and rise (monotonically). A single global minimum attractor on the combined surface occurs when the subgoal's positive function gradient matches the obstacle functions' negative total gradient, see Fig. 1. This always occurs between $2r$ and the edge of the fall-off range, $3r$, from a single obstacle sample point. Consequently the simple nature of the obstacle and subgoal surfaces ensure that monotonic descent towards the minimum is always possible with the global minimum basin containing the current robot position at each stage.

3.3   Subgoal Chaining Around Obstacles

The targeting of a B-contour will set a subgoal steadily to the left or right of the current robot trajectory when approaching an obstacle to make the robot turn to fall

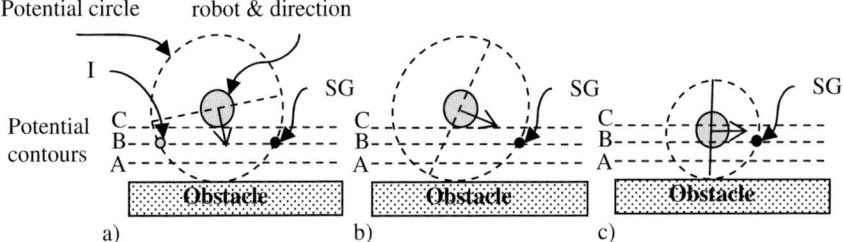

Fig. 2. (a) A subgoal setting SG on the forward half of potential circle (forward semi-circle) at the most forward intersection with a potential value of B (compare with I). (b) The robot body (inner circle) turns when targeting SG for subsequent travel towards the minimum on subgoal and obstacle surface. (c) Continued movement towards successive subgoals set on the B-contour pulls robot centre into and along the BC-band of obstacle potential.

into line along the obstacle edge as depicted in Fig. 2. The most forward B-contour intersection with the forwards semi-circle is chosen as the subgoal. In the pathological case when multiple intersections are exactly equally forward, one of them is chosen at random. Subgoals are continually set and replaced forwards of the robot along the B-contour of potential thus pulling the robot forwards towards the minimum on the combined subgoal and obstacle potential surface. This minimum lies in a BC-band of potential where $C=U_{Obst}(3r)$ as described earlier. The A-contour is associated with a potential value of $A=U_{Obst}(r)$ and represents a boundary of safety which is not crossed by the edge robot during travel.

We deem passages between obstacles to be too narrow for safe navigation when there is significantly less than a robot width either side of the robot. The setting of the obstacle potential *fall-off* range to 3r is sufficient to achieve occlusion of such passages without losing the ability to navigate other passages. Occlusion will occur for some passages that are less than 6r, and all those less than 4r, wide when the B-contour flows past the passage. The sets of obstacles surrounded by a BC-band can be treated as though they are just one virtual obstacle since the passages between them are unsafe to navigate.

The safe leaving of a BC-band to go to the goal or another obstacle requires satisfaction of the condition of being clear to leave with no immediate obstruction in the goal direction. If a BC-band has had to be followed to any extent, it is also required that finite progress has been made towards the goal since the first point on the band, the *hit point*, was reached by the robot. The progress ensures that in theory each traversed part of a BC-band is left permanently to go to another one nearer the goal. In this way, the chain connects to the goal with a BUG-like guarantee.

3.4   Modes of Operation

The robots chain their subgoals to reach the goal through three main modes of operation, two of which are much like those found in BUG. The two similar modes are *free* which entails moving directly towards the goal when there are no obstructions in the way, and *engaged*, when having to circumnavigate an obstacle to any extent until *safe to leave*. The third mode is a transitional *approach* mode following *free* mode.

Once an obstacle has been detected and deemed to be an obstruction on the way to the goal, the robot enters the *approach* mode. In this mode, it approaches the *B*-contour surrounding the obstacle to establish a hit point in the BC-band, turning while it does so. When the robot achieves this hit point, if it is then at a corner of the obstacle edge and is clear to leave immediately towards the goal it does so. Otherwise it enters the *engaged* mode in which it follows the BC-band further round the obstacle boundaries. This carries on until it is deemed safe to leave the BC-band and recommence travel towards the goal again in *free* mode. The process terminates at any stage if the robot centre comes within $2r$ of the goal where r is the robot's radius.

3.5   Travel Algorithm

The top level algorithm may be described as follows:

1.   Initialise
        1.1   set the sensor range and the radius of the robot

2.   **while** (not reached goal)

        2.1   scan for goal, rotate robot to face goal, scan for obstacles

        2.2   **while** (no *obstruction* in goal direction on forwards semi-circle arc of
                        width $2r$ and not reached goal, *free mode*)
                2.2.1   make a move (towards goal)
                end while

        2.3   **while** (hit point not flagged and not reached goal, *approach mode*)
                2.3.1 if the robot's position w.r.t. obstacle is deemed a hit point,
                        then flag hit point reached and record robot's current distance to
                                goal as hit point distance
                        else make a move (towards B-contour)

                2.3.2 if (hit point flagged and not reached goal and
                        clear to leave obstacle)
                        then safe to leave obstacle, make a move (towards goal)
                end while

        2.4   **while** (not *safe to leave* obstacle and not reached goal, *engaged mode*)
                2.4.1 make a move (along B-contour)
                2.4.2 if (not reached goal and progress attained and clear to leave
                        obstacle)
                        then safe to leave obstacle
                end while
        end while

make a move procedure
1. set the radius of the forwards semi-circle to minimum of
        (sensed distance to obstacles, sensor range, distance to goal)
2. set subgoal on semi-circle towards goal or towards or along B-contour
3. make a step down the subgoal obstacle potential surface through
        robot motion
4. test for goal reached

3.6 POTBUG and POTSMOOTH

The POTBUG robot has a near-sighted mind's eye due to the sensor range being $4r$. In the POTSMOOTH robot, the only difference is that the sensor range is extended. With a more far-sighted mind's eye, the robot is able to detect obstacles at a greater distance and turn earlier. The earlier turning makes for a smoother path around the obstacle with less of an abrupt turn near to the obstacle. This generates paths qualitatively different to straight BUG-like paths in their approach to obstacles. For both flavours, the algorithm's targeting of the desired B-contour ensures that, while turning, the robot continues to move closer to the desired sensed obstacle edge to register a hit point.

3.7 Example Run

Fig. 3. shows the POTBUG and POTSMOOTH robots on an obstacle course containing a variety of shapes that present local minima for static potential field methods. The local minima are overcome by the robots and the goal is reached.

The jagged edged rectangle demonstrates how, in contrast to BUG, summation of potential is able to smooth the robot's perception of an obstacle edge to allow for a smooth plastic path around an obstacle and persistent travel further towards the goal. The POTSMOOTH robot has a less BUG-like travel path due to turning earlier than POTBUG using its extended sensor range.

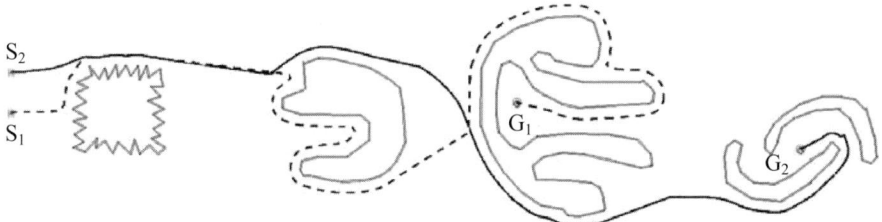

Fig. 3. POTBUG (dashed path from start S_1 to goal G_1) and POTSMOOTH (continuous path from start S_2 to goal G_2) on a difficult obstacle course with multiple local minima and a jagged edged rectangle

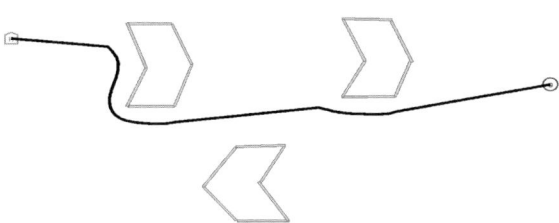

Fig. 4. Test arena for simulation experiments containing 3 chevron shapes with a typical POTBUG path (without sensor perturbation)

4   Simulation Experiments

In the following we present simulation experiments conducted on POTBUG and POTSMOOTH robots. The test arena for the experiments is an obstacle course containing 3 chevron-shaped obstacles (see Fig. 4). All the above tests are run on a set of 241 pairs of initial and goal positions thoroughly distributed throughout the test arena and with intervening obstacles. The robot radius and the sensor range are the only variables used to parameterise the robot, besides the degrees of perturbation experimented with. All robots have 30 sensors.

Table 1. Average ratio of unperturbed to perturbed path curvature (RPC) and length (RPL)

	gp											
%	**0**		**5**		**10**		**20**		**30**		**40**	
0	1.00	1.00	2.30	1.01	3.66	1.01	6.88	1.08	10.92	1.17	16.41	1.35
5	1.07	1.00	2.43	1.02	3.73	1.01	7.03	1.08	10.93	1.19	16.43	1.37
10	1.21	1.00	2.51	1.01	3.84	1.02	7.22	1.09	10.90	1.16	16.20	1.35
20	1.67	1.01	2.94	1.01	4.31	1.03	7.60	1.09	11.44	1.18	17.11	1.35
30	2.18	1.02	3.34	1.03	4.65	1.04	7.81	1.07	11.79	1.19	17.65	1.37
40	2.75	1.03	3.87	1.04	5.08	1.04	8.26	1.08	12.57	1.21	18.78	1.45
	RPC	RPL	RPC	RPL	RPC	RPL	RPC	RPL	RPC	RPL	RPC	RPL

op labels the rows.

4.1   Robust Smoothness (POTBUG)

Robustness here consists of smoothly maintaining plasticity and persistence of the travel path in the face of unreliable sensor readings. BUG on its own has no integrated method for addressing this issue. By contrast, adaptation of robot motion through the corrective potential field gradients allows the POTBUG robot to smoothly get back on course if the intended path is not realised at any stage.

The extent of successful adaptation possible may be tested through introducing perturbation repeatedly into the sensed distance to any detected object and sensed direction to the goal. The experimental aim is to show the extent to which relatively inaccurately sensed obstacles and goals leads to failures or departures from the path produced with relatively accurate sensing. Such robust smoothness may be quantified by comparing the ratio of overall curvature (RPC) and path length (RPL) of a POTBUG forward chaining path under various degrees of perturbation to the unperturbed counterparts for the tested pairs of initial and goal positions. The sensor range is small ($4r$).

A random percentage of the sensed obstacle distance and the sensed goal direction and distance is added to or subtracted from unperturbed sensor readings within a maximum given by each value of *op* and *gp* respectively in Table 1. In the unperturbed case there were no failures. In the perturbed cases, there was at most 1% failure. Table 1 shows that there is a broad trend of RPC and RPL increasing with increasing perturbation.   RPL increases by a relatively much smaller amount compared to RPC for up to 40% random perturbation in both individual obstacle and goal sensors. This is in line with our visual inspection of typical travel paths that shows a slightly drunken walk around, but close to, the true goal directions and B-C bands. There is considerable individual fluctuation about the mean curvature in the perturbed cases, most probably due to each transition moving the robot from side to side. The standard deviations are on average 86% of the mean. The path length for individual cases is again much less variable with standard deviations that are on average 13% of the mean.

Beyond the 40% random perturbation shown in Table 1, the goal is still achieved with the path length and curvature continuing to increase relative to the unperturbed case. The high degree of success in attaining the goal in the face of perturbation is quite possibly due to the recalibrating nature of the potential used. More specifically, the random effects of the goal and obstacle sensor perturbation tend to cancel out with repeated triangulation and polling of multiple obstacle sensors respectively.

4.2 Extended Smoothness (POTSMOOTH)

The empirical aim in this section is to investigate the extent to which the undulatory smoothness of travel paths may be enhanced by extending the sensor range. We extend the range from $4r$ to $12r$ in 4 steps of $2r$ to try out various types of POTSMOOTH robot and evaluate the ratios of path length and curvature for each range relative to those for the $4r$ counterpart set the same initial and goal positions. The sensors are unperturbed for this experiment.

Table 2 shows the path length and curvature decreasing as the sensor range is extended. There were no failures. This is consistent with our visual inspection of typical travel paths that shows the travel path curving earlier but still reliably joining the B-contour further to the right or left of the last goal direction taken in the *free* travel mode. There is consequently much less time spent in closely following the obstacle boundaries as was shown in Fig 3.

Table 2. Average ratio of path curvature (RPC) and path length (RPL) for sensor range (SR) compared to path curvature and path length for sensor range $4r$

SR	4r	6r	8r	10r	12r
RPC	1.000	0.975	0.955	0.946	0.932
RPL	1.000	0.995	0.992	0.989	0.986

5 Conclusion

A dynamic potential field method has been shown by design and empirically to achieve effective guarantees for goal realisability in the face of intervening obstacles. This was done by setting achievable and continually replaced subgoal attractors in the robot mind's eye as targets and descending towards them using potential field gradients. BUG-like travel modes were found to guarantee dynamically emerging connections of the subgoal chains to the goal based solely on significantly perturbable sensor readings without the aid of a global map or other prior knowledge.

The POTBUG and POTSMOOTH flavours of robot were developed with small and extended sensor ranges respectively. POTBUG's adaptive resilience to unreliable sensor readings and POTSMOOTH's adaptive smoothing of the global travel path have been empirically tested in simulations and found to be reliable to significant degrees. The initial results are promising for offering a relatively seamless integration with real world robotics and the implementation of the algorithms in physical robots is currently underway.

The authors thank G. W. Lucas for support through the Rossum Project Simulator.

References

1. Russell S. J., Norvig P., Artificial Intelligence, a Modern Approach, Prentice Hall. 2002.
2. Bell, G., Weir, M., Forward chaining for robot and agent navigation using potential fields, Twenty-seventh Australasian Computer Science Conference (ACSC2004), Vol 26, 2004.
3. Lumelsky, V. & Stepanov, A., Path planning strategies for a point mobile automaton moving amidst unknown obstacles of arbitrary shape, Algorithmica, 2(4):403-440, 1987.
4. Lumelsky, V., Mukhopadhyay, S. & Sun, K., Dynamic path planning in sensor-based terrain acquisition. IEEE Trans. Robotics and Automation, 6(4):462-472, 1990.
5. Rao, N.S.V. Kareti, S. Shi, W. and Iyenagar, S.S. Robot Navigation in Unknown Terrains: Introductory Survey of Non-Heuristic Algorithms. Oak Ridge National Laboratory Technical Report, ORNL/TM-12410:1--58, July 1993.
6. Latombe, J., Robot Motion Planning, Kluyver Academic Publishers, 1991.
7. Dudek, G., Jenkin, M., Computational Principles of Mobile Robotics, Cambridge University Press, 2000.
8. Khatib, O., Real-time obstacle avoidance for manipulators and mobile robots, The International Journal of Robotics Research, 5(1), 1986.
9. Arkin, R., Behavior-based robotics, MIT Press, 1998.
10. Koren, Y., Borenstein, J., Potential field methods and their inherent limitation for mobile robot navigation, IEEE Conference on Robotics and Automation, pp. 1398-1404, 1991.
11. V.J.Lumelsky and T.Skewis, Incorporating range sensing in the robot navigation function, IEEE Transactions on Systems, Man, and Cybernetics, vol. 20, no. 5, pp. 1058-1068, 1990.
12. Kamon, I., Rivlin, E., Sensory-based motion planning with global proofs, IEEE Transactions on Robotics and Automation, 13(6), 1997.
13. Kamon, I., Rivlin, E., Range-sensor-based navigation in three-dimensional polyhedral environments, The International Journal of Robotics Research, 20(1), 2001.
14. Koditschek, D., Exact robot navigation by means of potential functions: Some topological consideration, IEEE Conference on Robotics and Automation, pp. 1-6, 1987.
15. Alvarez, D., Online motion planning using Laplace potential fields, IEEE International Conference on Robotics and Automation, pp. 3347-3352, 2003.
16. 16.Franceschini, N., Pichon, J. M., and Blanes, C., From insect vision to robot vision, Philosophical Transactions of the Royal Society B, 337, 283-294, 1992.
17. 17.Huang, W.H., Fajen, B.R., Fink, J.R., and Warren, W.H., Visual navigation and obstacle avoidance using a steering potential function, Robotics and Autonomous Systems, 54(4), 288-299, 2006.
18. Gorse, D. et al., The new era in supervised learning, Neural Networks, 10(2):343-352, 1987.
19. Lewis, J., Weir M., Subgoal chaining and the local minimum problem, IEEE International Joint Conference on Neural Networks, 1999.
20. Lewis, J., Weir M., Using subgoal chaining to address the local minimum problem, International ICSC Symposium on Neural Computation, 2000.
21. Xi-yong, Z., Jing, Z., Virtual local target method for avoiding local minimum in potential fields based navigation, Journal of Zhejiang University SCIENCE, 4(3):264-269, 2002.

Navigation in Large-Scale Environments Using an Augmented Model of Visual Homing

Lincoln Smith, Andrew Philippides, and Phil Husbands

Centre for Computational Neuroscience and Robotics
University of Sussex BN1 9QG, United Kingdom
{lincolns, andrewop, philh}@sussex.ac.uk

Abstract. Several models have been proposed for visual homing in insects. These work well in small-scale environments but performance usually degrades significantly when the scale of the environment is increased. We address this problem by extending one such algorithm, the average landmark vector (ALV) model, by using a novel approach to waypoint selection during the construction of multi-leg routes for visual homing. The algorithm, guided by observations of insect behaviour, identifies locations on the boundaries between visual locales and uses them as waypoints. Using this approach, a simulated agent is shown to be capable of significantly better autonomous exploration and navigation through large-scale environments than the standard ALV homing algorithm.

1 Introduction

Many models of insect navigation have been devised which reproduce the animal's visual homing capabilities in simple environments [1]. However, they do not cope well with large-scale environments containing several visual locales, or areas, without significantly increasing computational and storage demands. In this context, large-scale environments are defined as those in which the visual input at any location does not define the entire environment - as in most natural settings. Whether a scene is cluttered (e.g. dense woodland) or sparse (e.g. desert) there will be objects at a variety of spatial frequencies and distances which will not all be visible at any one time. This means that a single room with no features of a size less than the agent's visual acuity is not a large-scale environment, whereas an environment containing objects far enough away that they cannot be resolved by the agent is large-scale. The scale of the environment can therefore be varied by changing the size of objects whilst keeping the size of the environment and the visual acuity of the agent fixed.

Importantly, as an agent passes though large-scale environments landmarks will enter and exit its visual field. They therefore present difficulties for navigation as visual landmarks which define the goal location may not be visible from other locations. If a subset of visual landmarks remains visible throughout the environment, navigation methods relying on this subset can be used. In large-scale environments, however, this feature set will change as landmarks exit or enter view and a new visual locale is entered. The navigational information

S. Nolfi et al. (Eds.): SAB 2006, LNAI 4095, pp. 251–262, 2006.

from the previous locale will very likely be useless in navigating from the current locale to the goal and navigation methods relying on this information will fail. In addition, navigation strategies will fail if distinct locations are visually indistinguishable, with the probability of such perceptual aliasing increasing in large-scale environments.

If a navigation algorithm is to function robustly in the real world, it must account for the problems arising from large-scale environments. To date, a major focus of biomimetic strategies capable of dealing with such environments has been on constructing databases or graphs that associate a given location with a navigational action (reviewed in [2]). On recognising the current location, an agent selects the associated action that leads to the next goal location. This approach replaces local navigation with a recognition-triggered response and enables navigation through large-scale environments. As visual input is only used to evaluate whether the agent has reached a goal location, landmarks which exit and enter the agent's perceptual field have no effect on navigation.

While they work in certain situations, associative databases have several serious shortcomings. If a location is incorrectly identified the error is not evident until the agent fails to find the next goal location. To compound matters, the agent does not know whether its failure to reach the goal was due to incorrect identification of the previous location or for some other reason. Thus, for clarification, the agent must attempt to return to the last known location. These problems arise from the ballistic nature of the navigational action stored in associative databases. A compass direction or vector (compass direction and distance) provide no feedback until followed to completion - the cost of ignoring available navigation cues between goals.

We have attempted to overcome the problems presented by large-scale environments by augmenting the average landmark vector (ALV) model [3] in two ways. We use a novel approach to waypoint selection during the construction of multi-leg routes for visual homing. The method recognises entry into a new visual locale and incorporates this information in the automatic creation of a series of intermediate goals, or waypoints, between which local navigation methods can be used. Navigation along the whole route is accomplished by navigating to each waypoint in turn. In this way, the larger visual environment becomes segmented into areas where a subset of visual features remain in view as the individual moves. Waypoints are only selected when required by the agent, resulting in the *minimum* number for successful navigation being used. The model also makes use of path integration information to 'scaffold' the visual learning of the route. The resulting algorithm achieves high performance despite very low computational and storage requirements. A comparative study of the ALV and augmented models was carried out in a series of environments of gradually increasing scale with the augmented model demonstrably superior in all instances.

Our aim was to produce a robust visual navigation system for autonomous robotics which is based on biological observation and attempts biological plausibility. The advantage of this constraint is twofold: it provides a successful and efficient model on which to base our algorithm and it retains the possibility of generating biological hypotheses from the work. To this end, in the algorithm presented we have reduced the level of computation, duration of route learning, and storage requirement found in current biomimetic models of navigation in large-scale environments. We have also taken note of the fact that ants and bees integrate several navigational cues (vision, compass, odometery, olfactory) *during* the trajectory between two locations of interest [4,5,6,7].

The paper proceeds as follows. Section 2 details key components of current insect visual navigation models and describes the ALV model. Our augmented algorithm is described in Section 3. The experimental setup used to evaluate the algorithm is described in Section 4 and Section 5 presents the experimental results. Lastly future directions for the work and conclusions are discussed.

2   Models of Insect Navigation

The visual navigation algorithm presented in Section 3 incorporates two extensively researched aspects of insect navigation - path integration and visual homing. Brief details of these processes are given here as background before the ALV model is described.

Path integration (PI), or dead-reckoning [8], is a pervasive strategy in nature, occurring in both vertebrate and invertebrate species. PI enables an individual, who may have travelled a tortuous outbound journey, to return to the starting location via a direct route. In its simplest form, direction and distance to home are continually integrated as the individual moves, creating a 'home vector' which can be followed to return to the starting location. PI is a continuous iterative estimate, and is therefore susceptible to cumulative error. To mitigate this, both bees [9] and ants [6,10] use visual information for homing within an area local to a target location. PI can provide an initial means of navigating a path, enabling the 'scaffolding' of more reliable visual learning of the route [5]. It is used in this context in the algorithm presented here.

Several models of insect visual homing use a comparative process to match the current view to that expected at the target location (see [1] for an overview). The basic model works by storing a (possibly parameterised) view at a location of interest. As an individual returns to that location the stored image is repeatedly compared with the current visual input, with a close match indicating that the individual has returned to the location where the view was stored. Several models extend this simple comparison procedure to produce the direction of movement that will increase similarity between views and thus guide the individual to the location of the original view. One such example is the ALV model [3], a highly parsimonious simplification of the 'snapshot' model proposed by Cartwright and Collett to describe navigation behaviour in bees [9].

The primary difference between the snapshot model and the ALV model is the visual information stored at the location of interest. The snapshot model stores a rather unprocessed image which is later compared to the current retinal image to produce a movement direction. The ALV model, however, processes the visual image into an abstract representation of the view - a single two-component vector - before it is stored. To calculate this vector, features (landmarks) are selected from a 360 degree panoramic view[1], each represented as the unit vector from the individual to the landmark. By averaging these individual landmark vectors a single vector (the ALV) that characterises the visual scene at a particular location is derived.

As an individual moves, landmark positions change relative to the individual and the ALV changes accordingly. To return to a location of interest, an individual compares the stored ALV from that location to the current ALV. It then moves so that the subsequent ALV is closer to the stored one. Since the difference between the ALVs gives the approximate direction of goal, iteration of this process brings the individual to the goal [3].

The ALV model is very cheap in terms of computation and memory, and has been shown to be effective for visual navigation in both computer simulation [12,3] and on autonomous mobile robots [13]. It works well in simplified small-scale environments in which the task is to home to a single location following a displacement. In such environments, however, all visual features are within the visual field of the agent. This is not the case in large-scale environments and we later show that the ALV fails in such situations.

3   The Augmented Navigational Algorithm

The augmented navigational algorithm developed in this research preserves the attractive qualities of the ALV method (very low computational and memory requirements) while adding capabilities that allow large-scale environments to be handled. The behavioural scenario used to evaluate the algorithm is food foraging from a nest (home) to which the agent must return. The algorithm contains three components: foraging, visual route learning and visual navigation. The foraging component is little more than random search. Initially, the agent is set to foraging from the nest location where the agent wanders randomly in directions sequentially chosen from a gaussian distribution about its current heading ($\sigma = 0.2$). While foraging, a global vector to the nest is maintained by the agent through path integration, as proposed for *Cataglyphid* ants [5]. This home vector is later used to return to the nest after food is found.

On locating food, the agent starts the process of visually learning the nest-bound route. The agent first stores its global home vector and starts on its nest-bound trajectory. During the inbound journey, the ALV (discussed in Section 2) is calculated at each time-step and compared to that of the previous time-step. When the ALV changes significantly, defined as an object appearing or disappearing from the visual array, the agent is deemed to have moved into a new visual locale and the previous ALV is stored. In this manner an ordered series of ALVs is accumulated - each representing a significant discontinuity in the ALV space. When the agent reaches the nest, the series is terminated with the current ALV.

---

[1] Prerequisites for the ALV are a 360° visual system and an ability to align views with an external reference (eg a compass direction). Ants and bees have near spherical vision and both gain compass information from (at least) celestial cues [11].

Outbound trajectories proceed in much the same manner as inbound route learning. The global vector previously stored at the food location is now used to determine the goal direction back to the food object. En route, continuity of the ALV is monitored and the ordered series of vectors is terminated with the ALV at the food object. After completion of one direct inbound and one direct outbound trip, inbound and outbound routes are stored as two series of waypoints (represented by the ALVs). The agent then enters the final behavioural component of visual navigation between nest and food regions. When leaving the food for the second time, the agent uses the standard ALV algorithm to home to the first location represented in the series of ALV's compiled in the previous inbound trajectory. As the agent approaches this location the difference between the current ALV and goal ALV approaches zero. At this point, the agent sets the next ALV in the series as the goal and repeats the process until it reaches home. The algorithm is summarised in Figure 1.

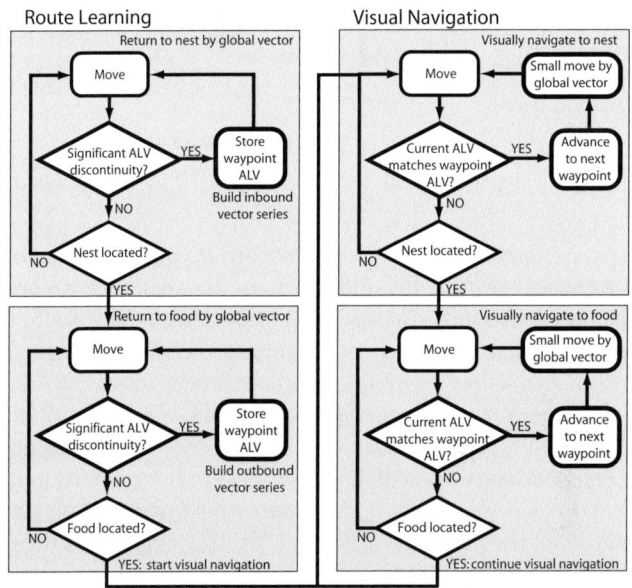

Fig. 1. Simplified flow diagram of the navigation algorithm. There are three primary behavioural components: foraging, route learning and visual navigation.

4   Computer Simulation: The Experimental Setup

Navigational runs were performed in a two dimensional computer simulated environment. The environment contained three object types - landmarks, nest and food. The landmarks are black cylinders as used in both simulated and real ant visual navigation experiments [6,4,7]. Nest and food regions were circular areas detectable only upon entry and therefore not used in navigation. The environment was 1000×1000 units and unless stated otherwise, object diameters were randomly selected in the range 30-95 units while nest and food are 20 units in diameter (Figure 2). 25 objects are placed randomly within each environment together with a randomly positioned nest and food item. However, to avoid the biologically implausible scenario of nest or food being on the edge of the environment and thus having no objects 'behind' them, they are constrained to lie within the central 500×500 area of the arena. We also ensured that nest and food do not overlap and that no object is placed over nest or food.

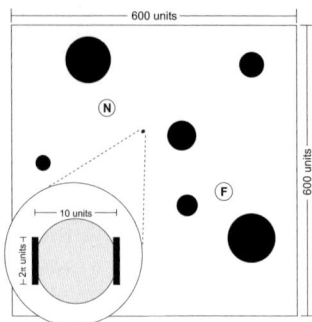

Fig. 2. Plan view of the computer simulated environment with nest and food regions denoted by **N** and **F** respectively. Inset: an enlargement to indicate agent dimensions.

The agent is circular with wheels tangential to its body. Movement is resolved into translation and rotation dependent upon the distance travelled by each wheel. A wheelbase of 10 units, and wheel circumference of 4 units ($diameter = 4/\pi$) characterise both maximum speed and rate of turn. Motor output is in the range $[-1.0, 1.0]$ and represents the percentage of wheel rotation for a given time-step, with a maximum wheel travel of $\pm100\%$ of wheel circumference (i.e. ±4 units). Motor output is calculated as the cosine of the angular difference between current heading and goal heading - skewed by 0.25π and -0.25π for right and left motors respectively [2]. The motor output equation produces turning proportional to the angular difference of the current and goal headings (Figure 3). Large angular differences produce a three point turn; with smaller angular differences, output converges to produce straight line travel.

Agents cannot move through landmarks and must perform rudimentary obstacle avoidance. This is implemented by a ring of simulated infra-red proximity sensors which go high when either a landmark or the edge of the simulation area is within 5 units. A vector opposite to the direction of the detected object is then added to the movement vector calculated from visual input. As well as obstacle avoidance, this procedure constrains the agent to the simulation area.

To produce a large-scale environment for the agent to navigate within, visual acuity of the agent is limited to 4° per sensor, with visual information collected by 90 sensors, creating a 360 degree panoramic view. If an object covers more

---

[2] The skew value (±0.25) was arbitrarily chosen to produce a reasonable turning rate proportional to angular difference of the current and goal headings.

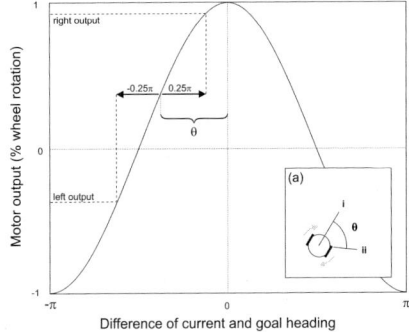

Fig. 3. Graph of the angular difference between the agent's current heading and its goal heading (θ) versus motor output values. The right motor output is $\cos(\theta + 0.25\pi)$ and left motor output is $\cos(\theta - 0.25\pi)$. **(a)** Shows an agent with initial heading **(i)**, the goal heading **(ii)**, and their angular difference (θ). Forward motion of the right wheel, and backward motion of the left results from the situation depicted.

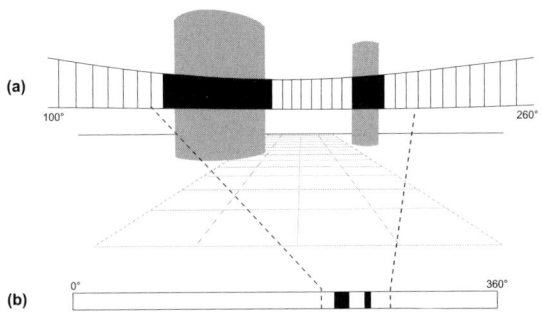

Fig. 4. Translation of environment to visual input: **a** Three dimensional view from agent's centre. Here a 40 degree subset of sensors is represented. Sensor arcs containing (visually) any portion of a landmark are set high (black). **b** The one dimensional visual array corresponding to the above view. Sensor states are transferred to a one-dimensional array with objects appearing as black segments.

than 50% of a sensor, the corresponding portion of the visual array is set to high (Figure 4). This means that the visual range of the agent is determined by the size of objects. The resulting visual input represents landmarks with black segments within a one dimensional array. At any one time, the agent was unable to see all landmarks within the environment. Landmarks would exit and enter from view as the agent moved through the environment.

5 Experimental Results

In this section we present the results of a comparison of the ALV and augmented ALV algorithms over a range of large-scale environments. A typical run of the navigational algorithm in the simulated environment is then looked at in detail before discussing situations in which it fails.

5.1 Comparison of Algorithms in Large-Scale Environments

For a thorough exploration of the effect of scale on the performance of the algorithms, we ran the ALV and augmented ALV on environments where all objects had equal diameter and then incrementally decreased this parameter. As discussed in the introduction, reducing the size of all objects while keeping other factors fixed increases the 'scale' of the environment. Positions of objects, nest and food were randomly selected and fixed while experiments were performed for each object radius in the range 10-40. This procedure was repeated 30 times, giving a total of 30 random environments for each object diameter value. As the environments were entirely random, the results include pathological examples where the task was impossible, such as when objects significantly obstructed movement towards nest or food or when there were sections of the environment where no objects are visible. The algorithm performance would be improved if such environments were excluded or if the algorithm was augmented to deal with these occurrences. The ALV homing algorithm was tested by providing the agent with a single ALV (stored at the nest location) to navigate through a large-scale environment. The augmented algorithm was run as described in Section 3.

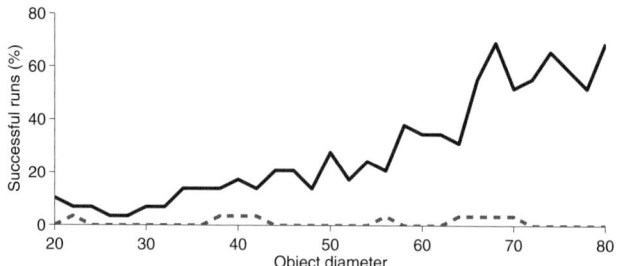

Fig. 5. Comparison of ALV (dashed line) and augmented ALV (solid line) on random large-scale environments with objects of fixed diameter

The percentage of trials in which the agent successfully returned to the nest using each algorithm are shown in Figure 5. As can be seen, the standard ALV model fails to cope with almost all environments and only returns home 9 times. The cases where it succeeds are not dependent on object diameter, but rather are fortunate cases where food and nest fall within the same visual locale. The augmented ALV performs significantly better over the entire range of diameters tested. For larger objects this translates into fairly consistent success with the agent returning to the nest in the majority of the trials. Failures are due to the problem of mistaken context discussed in the next section, and encountering visual locales not experienced during the single learning journey due to noise in the agent's trajectory. This occurs more frequently when there are many small visual locales.

5.2   Single Run

In this section we examine a single run of the augmented ALV algorithm in detail. The run begins by placing the agent at the nest location. The agent starts its outbound foraging journey on a random bearing. Foraging continues until a food location is discovered, at which point the agent navigates nestward by path integration. The agent's current ALV is monitored as it proceeds toward the nest. When a significant difference between the current and previous ALV is detected, the location is identified as a boundary between visual locales and selected as a route waypoint. In Figure 6(a) centre, four peaks clearly differentiate from the baseline and indicate visual locale boundaries. The causes of significant differences in ALV are illustrated by the traces of visual input over time (Figure 6(a) right) which depict landmarks entering and exiting perceptible range. Once the agent has reached the nest, it starts the outbound journey to the food by path integration. In the same way as in the nestbound journey, waypoints are selected (Figure 6(b)).

On returning to the food for the second time, the agent begins the nestbound journey by visually navigating to the first waypoint along the route to the nest. In comparing agent and waypoint ALVs, the agent is provided with both the direction to the goal, and a measure of current visual similarity to the goal location. When a significant difference between the agent ALV and waypoint ALV is detected, the agent moves a short distance in the direction of its global vector, visual navigation recommences, and the next waypoint is sought (Figure 7 centre). When the new goal becomes the next waypoint location, a jump in difference between the agent and waypoint ALVs is observed. This difference is again reduced as the agent approaches the waypoint.

Following the selection of waypoints during the inaugural inbound and outbound journeys, the sequence of waypoint seek-advance-seek continues until interrupted. The agent shuttles between nest and food, via the agent selected waypoints.

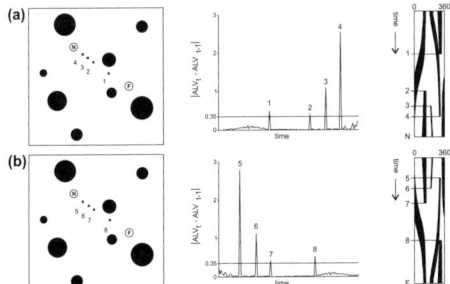

Fig. 6. Initial trips during which the route was learnt and visual navigation waypoints selected; **(a)** inbound trajectory, **(b)** outbound trajectory. **left:** Overhead view of environment with agent selected waypoints (numbers 1 to 8). **centre:** Magnitude of the difference between ALV at **t** and **t-1**. **right:** Trace of the 360 degree retinal image over the journey. Landmark disappearance and appearance are depicted by an abrupt end or start of a trace, marking the crossing of the boundary between two visual locales.

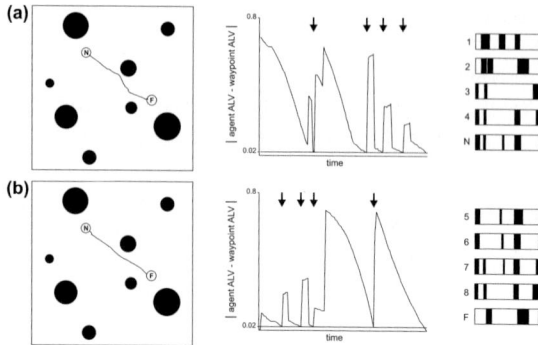

Fig. 7. Visually navigated routes; **(a)** inbound, and **(b)** outbound. **left:** Overhead view of environment and paths generated during visual navigation. **centre:** Magnitude of the difference between agent and waypoint ALVs. When the magnitude falls under the set threshold, and the goal is advanced to the next waypoint (arrows). **right:** The retinal image used to produce ALVs at selected waypoints.

Agents did mistake context if they failed to recognise they were sufficiently close to the waypoint to advance to the next ALV in the series. Although infrequent, these situations cause the algorithm to fail. This confusion is caused by the angle at which the agent approaches the visual locale boundary. In Section 6 we discuss ways of addressing this problem.

While the augmented algorithm performs well in may large-scale environments, it can be seen from Figure 5 that it does degrade as the object size decreases, although it is still much superior to the performance of the ALV. The reason for this degradation is that with smaller objects, there is an increased probability of the biologically implausible situation of areas where there are no visual landmarks. This necessarily results in failure of the algorithms. In addition, when there are fewer objects in view, the possibility of perceptual aliasing is greatly increased.

Such failures are to be expected; our intention was not to produce perfect navigation in these simulated environments, but rather to develop a strategy which addressed the shortcomings of the ALV. We are satisfied that our results demonstrate that we have a basic strategy that can be applied to navigation in large-scale environments. From this point, rather than tweaking this particular model to tune it to the environments used, improved algorithms will be built from this general strategy, via modifications as discussed in the next section.

6   Conclusions and Future Directions

Our navigational algorithm enables successful autonomous exploration of large-scale environments, and efficient selection of waypoints to connect separate visual locales. Several advantages are gained over current navigation models proposed for robotic platforms: multi-legged routes can be traversed entirely by non-ballistic navigation methods; continuous sensory feedback guides the agent to the next waypoint; and unlike ballistic navigation methods, errors can be detected as soon as they occur.

We included amongst the goals for this work the possibility of biological hypothesis generation. To this end, we have taken a model proposed for insect visual navigation, and extended it to a form that remains biologically plausible in its complexity. Although we have purposefully developed a model that operates at a low level of computation, higher level processes such as topological mapping and route planning could be added to the navigational algorithm presented here.

While the results reported here were achieved in simulation, we are currently transferring the algorithm onto a real robot platform. The transfer requires the algorithm to cope with sensory input noise not present in the simulated environment. The main problem is in the noise in visual input and in object segmentation in particular, though compass readings and physical interaction also contribute to the noise experienced by the robot. While the algorithm performs well in hand-picked 'un-noisy' conditions, we are currently augmenting the visual processing of the robot to aid object segmentation and recognition. In particular, we will use higher order object features such as centre-of-mass as well as edge information.

A second area of development is the identification and correction of mistaken context. As mentioned in Section 5.2, the algorithm will fail when an agent does not recognise it has crossed into an adjacent visual locale. The agent continues to use the visual information for a locale it is not in, and spurious movement vectors are produced. By comparing the ALV movement vector (Section 2.3) to the movement vector suggested by path integration, an indication of error can be obtained. If the movement vectors diverge substantially, a navigational error has occurred and corrective behaviour can take place at the point of divergence[3]. In some situations the conflict could be caused by detours around obstructions, or changes in a dynamic visual environment. In these situations, replacing, adding or deleting waypoints could adapt a previously learned route. Recent work in these directions has been successful. Finally, future plans include extension of the navigational algorithm to accept visual input in three-dimensions. Landmarks would not be restricted to a two-dimensional plane about the robot, which should improve accuracy and efficiency as well as freeing the robot to traverse uneven ground.

Acknowledgments. This work was partly supported by EPSRC grant GR T08753 01. Thanks to Tom Collett and Paul Graham for useful discussion and to the anonymous reviewers.

References

1. Vardy, A., Moller, R.: Biologically plausible visual homing methods based on optical flow techniques. Connection Science **17**(1-2) (2005) 47–89
2. Franz, M.O., Mallot, H.A.: Biomimetic robot navigation. Robotics and Autonomous Systems **30** (2000) 133–153

---

[3] Desert ants are shown to abandon landmark navigation when it takes them away from the direction that path integration indicates the goal should be [14].

3. Lambrinos, D., Möller, R., Pfeifer, R., Wehner, R., Labhart, T.: A mobile robot employing insect strategies for navigation. Robotics and Autonomous Systems **30** (2000) 39–64
4. Collett, T.S., Dillmann, E., Giger, A., Wehner, R.: Visual landmarks and route following in desert ants. Journal of Comparative Physiology A **170** (1992) 435–442
5. Collett, M., Collett, T.S.: How do insects use path integration for their navigation? Biological Cypernetics **83** (2000) 245–259
6. Wehner, R., Räber, F.: Visual spatial memory in desert ants. *Cataglyphis bicolor* (hymenoptera: Formicidae). Experientia **35** (1979) 1569–1571
7. Nicholson, D.J., Judd, S.P.D., Cartwright, B.A., Collett, T.S.: Learning walks and landmark guidance in wood ants (*Formica rufa*). Journal of Experimental Biology **202** (1999) 1831–1838
8. Mittelstaedt, H.: The role of multimodal convergence in homing by path integration. Fortschritte der Zoologie **28** (1983) 197–212
9. Cartwright, B., Collett, T.S.: Landmark learning in bees. Journal of Comparative Physiology A **151** (1983) 521–543
10. Wehner, R., Michel, B., Antonsen, P.: Visual navigation in insects: Coupling of egocentric and geocentric information. Journal of Experimental Biology **199** (1996) 129–140
11. Wehner, R.: The polarization-vision project: Championing organismic biology. In Schildberger, K., Elsner, N., eds.: Neural Basis of Behavioural Adaptation. Volume 30. Gustav Fischer Verlag, Stuttgart, Jena, New York (1994) 103–143
12. Lambrinos, D., Möller, R., Pfeifer, R., Wehner, R.: Landmark navigation without snapshots: the average landmark vector model. In Elsner, N., Wehner, R., eds.: 26th Goettingen Neurobiology Conference. Volume 1. (1998) 221
13. Möller, R., Lambrinos, D., Roggendorf, T., Pfeifer, R., Wehner, R.: Insect strategies of visual homing in mobile robots. Technical Report IFI-AI-99.06, University of Zurich, Artificial Intelligence Laboratory, Department of Computer Science (1999)
14. Collett, M., Collett, T.S., Bisch, S., Wehner, R.: Local and global vectors in desert ant navigation. Nature **394** (1998) 269–272

Evolutionary Active Vision Toward Three Dimensional Landmark-Navigation

Mototaka Suzuki and Dario Floreano

Ecole Polytechnique Fédérale de Lausanne (EPFL)
Laboratory of Intelligent Systems, CH-1015 Lausanne, Switzerland
`Mototaka.Suzuki@epfl.ch`, `Dario.Floreano@epfl.ch`
`http://lis.epfl.ch/activevision`

Abstract. Active vision may be useful to perform landmark-based navigation where landmark relationship requires active scanning of the environment. In this article we explore this hypothesis by evolving the neural system controlling vision and behavior of a mobile robot equipped with a pan/tilt camera so that it can discriminate visual patterns and arrive at the goal zone. The experimental setup employed in this article requires the robot to actively move its gaze direction and integrate information over time in order to accomplish the task. We show that the evolved robot can detect separate features in a sequential manner and discriminate the spatial relationships. An intriguing hypothesis on landmark-based navigation in insects derives from the present results.

1 Introduction

Active vision emphasizes the role of vision as a sense for robots and other real-time perception-action systems [1,2,3,4]. It picks out the properties of images which are necessary to perform its assigned tasks, and ignores the rest. In this context, there is no clear need for the sort of detailed reconstructions of the visible world that have been an accepted, traditional goal of machine vision [5].

Active vision may be useful to perform landmark-based navigation where landmark relationship requires active scanning of the environment. In this article we explore this hypothesis by evolving the neural system controlling vision and behavior of a mobile robot equipped with a pan/tilt camera so that it can discriminate visual patterns and arrive at the goal zone.

The experimental setup employed in this article has a notable characteristic: the visual landmarks are identical if the elevation of a robot's camera is fixed with the body. In that case, the robot could be unable to discriminate one from the other[1]. It needs to actively move its gaze direction and integrate information over time in order to differentiate these patterns. The sequential detection of spatially separate visual landmarks has been largely neglected in the literature. Instead most machine vision systems process an entire image of their large visual field every time step.

---

[1] The use of a panoramic camera which provides larger field of view is discussed in section 4.

S. Nolfi et al. (Eds.): SAB 2006, LNAI 4095, pp. 263–273, 2006.

We show that the best evolved robots successfully perform the task by using an effective scanning strategy. The evolved active scanning trajectory covers only a small region of the entire visual field and, more importantly, consists of a sequence of feature-driven, anticipatory, and context-dependent gaze movements. We address the advantages of the present method and neural architecture in terms of algorithmic, computational and memory resources.

The rest of this paper is organized as follows: the next section details the experimental setup, i.e. the environment, the simulated robot and the task for the robot. The neural network embedded in the robot and the genetic algorithm for developing the synaptic weights in the neural network are also described. Results and the analysis of the best evolved individual are described in Section 3. Finally an intriguing hypothesis on landmark-based navigation deriving from the present results and conclusions are discussed in Section 4 and 5 respectively.

2   Methods

The neural control system of a mobile robot equipped with a pan/tilt camera is evolved by means of a genetic algorithm to perform goal-directed navigation in an enclosed space using only visual information (Fig. 1). The evolutionary algorithm evaluates each neural controller with random mutations until an evolutionary stable control strategy is found [6]. In order to collect data from several independent runs and perform rigorous statistical analysis, we used fast, physics based simulations of the robot and its environment (Fig. 1).

Fig. 1. Left: The original six-wheeled robot Koala equipped with a pan/tilt camera. Right: The robot's perspective in a simulated environment. The robot can access the world with 5 by 5 retina at the center of the image.

We simulated the robot and the environment using physics-based Vortex libraries[2]. The robot has six wheels, but only the central wheel on each side is motorized. The robot base is 30cm(W)×32cm(L)×20cm(H). The pan and tilt angles of the camera are controlled by two separate and independent motors.

---

[2] http://www.cm-labs.com

2.1   Experiment and Task

Figure 2 shows the experimental setup where each of two facing white walls has two squares placed at different heights. The task of the robot is to visually discriminate one wall from the other in order to arrive at the goal zone at the end of each trial. There is no other identification of the goal than the visual patterns. Importantly this experimental setup is designed such that it does not allow the visual field of the robot to cover both of the two black squares at any given moment. Therefore the robot cannot discriminate the two walls by keeping the

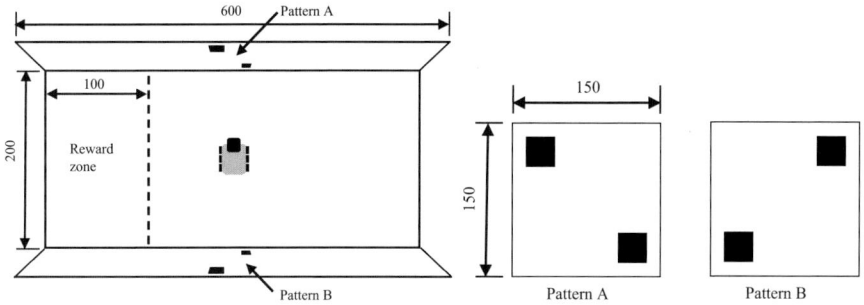

Fig. 2. The arena (200cm×600cm) and two visual patterns used in the simulation. The visual field of the robot can not cover both of the two black squares at any given moment. The difference of the two walls resides in the spatial relationship of the two squares (right). The position and direction of the robot are randomized at the beginning of each test.

vertical angle of the camera constant because both walls have an identical black square in the same height. The difference of the two walls resides in the spatial relationship of the two squares (Fig. 2, right). The robot needs to discriminate one pattern from the other by using active, sequential scanning of the two black squares of each pattern and integrating the information over time.

2.2   Neural Architecture and Genetic Algorithm

The neural network is characterized by a feedforward architecture with evolvable thresholds and discrete-time, fully recurrent connections at the associative layer (Fig. 3). A set of visual neurons, arranged on a grid, with non-overlapping receptive fields receives information about the gray level of the corresponding pixels in the image provided by the camera on the robot. The receptive field of each unit covers a square area of 48 by 48 pixels in the image. We can think of the total area spanned by all receptive fields (240 by 240 pixels) as the surface of an artificial retina. The activation of a visual neuron, scaled between 0 and 1, is given by the average gray level of all pixels spanned by its own receptive field or by the gray level of a single pixel located within the receptive field. The choice between these two activation methods, or filtering strategies, can be dynamically changed by one output neuron at each time step. An object detector unit

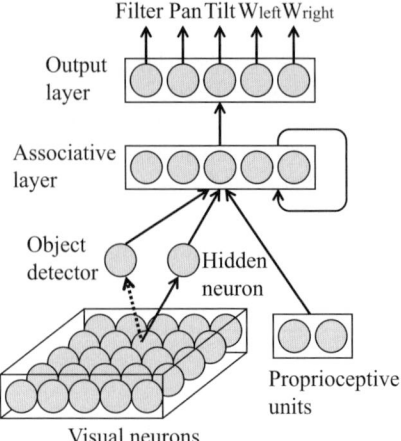

Fig. 3. The neural architecture is composed of a grid of visual neurons with non-overlapping receptive fields whose activation is given by the gray level of the corresponding pixels in the image; an object detector unit that is activated when any visual neuron is strongly activated; a hidden unit with incoming synapses from visual neurons; a set of proprioceptive neurons that provide information about the movement of the camera with respect to the chassis of the robot; a set of output neurons that determine at each sensory motor cycle the filtering used by visual neurons, the new pan and tilt speeds of the camera and the rotational speeds of the two wheels of the robot; a set of associative neurons with recurrent connections. Solid arrows between layers represent fully connected synaptic weights. Dashed arrow represents a predetermined (non-evolvable) OR filter (see main text for more detail).

is activated when any visual neuron is strongly activated. Therefore the synaptic weights incoming into this unit can be seen as a predetermined (non-evolvable) OR filter. Two proprioceptive units provide input information about the measured horizontal (pan) and vertical (tilt) angles of the camera. These values are in the interval $[-100, 100]$ and $[-25, 25]$ degrees for pan and tilt, respectively. Each value is scaled in the interval $[-1, 1]$ so that activation 0 corresponds to 0 degrees (camera pointing forward parallel to the floor). A set of memory units store the values of the associative neurons at the previous sensory motor cycle step and send them back to the associative units through a set of connections, which effectively act as recurrent connections among associative units [7]. The bias unit has a constant value of -1 and its outgoing connections represent the adaptive thresholds of associative, hidden and output neurons [8].

Associative, hidden and output neurons use the sigmoid activation function $f(x) = 1/(1 + exp(-x))$ in the range $[0, 1]$, where x is the weighted sum of all inputs. Output neurons encode the motor commands of the active vision system and of the robot for each sensory motor cycle. One neuron determines the filtering strategy used to set the activation values of visual neurons for the next sensory motor cycle. Two neurons control the movement of the camera, encoded as speeds relative to the current position. The remaining two neurons encode the direction and rotational speeds of the left and right motored wheels of the robot. Activation values above and below 0.5 stand for forward and backward rotational speeds respectively.

The present neural architecture has been incrementally developed based on our previous investigations [9,10]. The object detector unit incorporated in the architecture is explicitly designed to simplify the biological visual system capable of monitoring for change in the visual environment[3]. The hidden neuron is incorporated to equalize the contributions of the visual neurons, the object detector unit and the proprioceptive units to the activations of the associative neurons. The roles of the hidden and object detector units are further analyzed in section 3.

The neural network has 106 evolvable connections that are individually encoded on five bits in the genetic string (total length=530 bits). A population of 100 genomes is randomly initialized by the computer. Each individual genome is then decoded into the connection weights of the neural network and tested on the robot while its fitness is computed. The best 20% of the population (those with the highest fitness values) are reproduced, while the remaining 80% are discarded. Equal number of copies of the selected individuals are made to create a new population of the same size. The new genomes are randomly paired, crossed over with probability 0.1 per pair and mutated with probability 0.01 per bit. Crossover consists in swapping genetic material between two strings around a randomly chosen point. Mutation consists in toggling the value of a bit. Finally two copies of the best genomes of the previous generation are inserted in the new population at the places of the randomly chosen genomes (elitism) in order to improve the stability of the evolutionary process.

The fitness function was designed to select robots for their ability to arrive at the goal zone at the end of each life. Each individual is tested for six trials, each trial lasting for 300 sensory motor cycles. A trial can be truncated earlier if the operating system detects an imminent collision into the walls. The fitness criterion F is composed as follows:

$$F = F_{speed}(S_{left}, S_{right}) + F_{goal} \tag{1}$$

where $F_{speed}(S_{left}, S_{right})$ is a function of the measured speeds of the left S_{left} and right S_{right} wheels and F_{goal} is a reward given if the robot reaches the goal at the end of its life[4]. More specifically $F_{speed}(S_{left}, S_{right})$ is defined as follows:

$$F_{speed}(S_{left}, S_{right}) = \frac{1}{ET} \sum_{e=0}^{E} \sum_{t=0}^{T'} f(S_{left}, S_{right}, t) \tag{2}$$

$$f(S_{left}, S_{right}, t) = (S_{left}^t + S_{right}^t)(1 - \sqrt{|S_{left}^t - S_{right}^t|/2S_{max}}) \tag{3}$$

---

[3] In our preliminary studies it seemed difficult to develop the visual system capable of significantly responding to the black squares detected at any location of the retina.

[4] One might think that the first term $F_{speed}(S_{left}, S_{right})$ is not necessary, but in our preliminary study the fitness value remained zero without $F_{speed}(S_{left}, S_{right})$, meaning that evolution could not find the solution.

where S_{left} and S_{right} are in the range $[-8, 8]$ cm/sec and $f(S_{left}, S_{right}, t) = 0$ if S_{left} or S_{right} is smaller than 0 (backward rotation); E is the number of trials (six in these experiments), T is the maximum number of sensory motor cycles per trial (300 in these experiments), T' is the observed number of sensory motor cycles; F_{goal} is 10 if the robot reaches the goal area at the end of the test, otherwise it is 0. The reward is given only if at least one lower square and one upper square are detected before reaching the goal. This stronger constraint on F_{goal} is to prevent selecting 'blind' individuals which arrive at the goal by chance without using visual patterns.

At the beginning of each trial the position and orientation of the robot are randomized in the interval $[-50, 50]$ and $[-20, 20]$ for the longitudinal and short axes respectively.

3   Results and Analysis

We performed six replications of the evolutionary run starting with different initial populations. In all cases the fitness reached stable values in less than 30 generations (Fig. 4), and the fitness value of the best evolved individual ranged from 40 to 60.

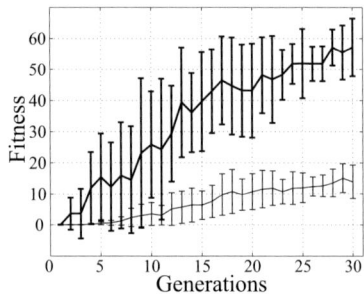

Fig. 4. Evolution of neural controllers for the simple three dimensional landmark navigation. Fitness values of the population average (thin line) and the best individual (thick line) across 30 generations. Vertical bars show the standard deviation. The results are averaged over six evolutionary runs.

We analyzed the behavior of the best evolved individual which arrived at the goal six times out of six trials. Figure 5 shows the scanning strategy, the trajectory of the robot, the camera movement with respect to the chassis of the robot and the activation of neuron 5 in the associative layer when the robot started in the face of pattern A and B. For clarity we show only the activation of neuron 5 because we found that it played the most significant role in the pattern discrimination.

The behavioral strategy of the best evolved robot can be illustrated as follows: 1. The robot searches for a lower square by moving the camera left-downward and turning its chassis counter-clockwise until it finds one; 2. Once it finds one,

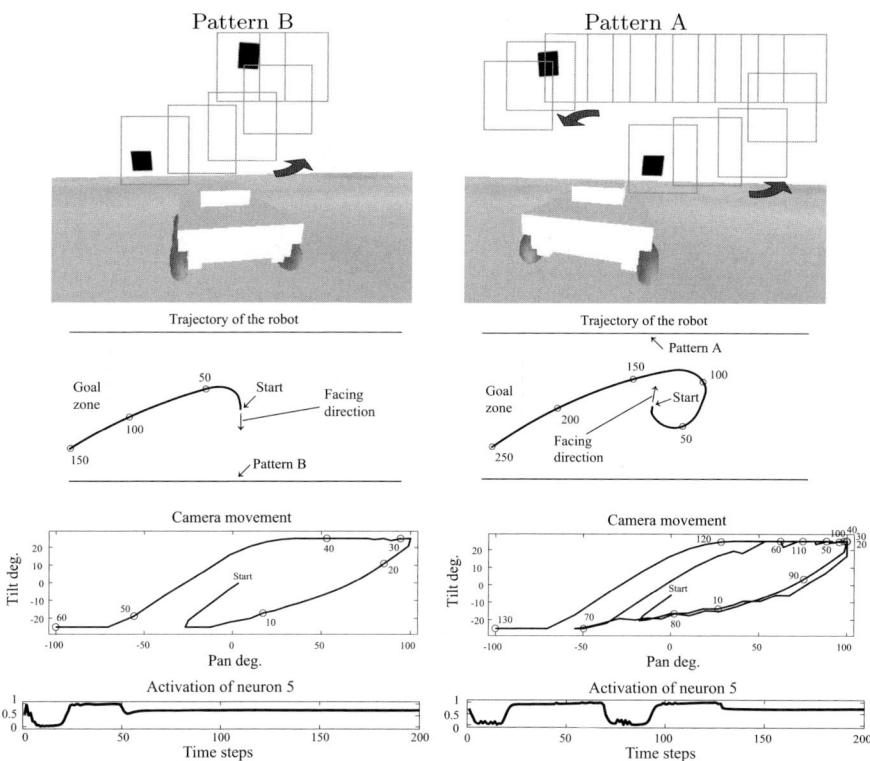

Fig. 5. From top to bottom, the best evolved robot scanning two black squares of each pattern sequentially (gray squares depicting the trajectory of the retinal perimeter), the trajectory of the robot, the camera movement with respect to the chassis of the robot and the activation of neuron 5 in the associative layer during the behavior (shown only for the first 200 sensory motor cycles) when the robot started in the face of pattern B (left column) and A (right column).

it points the camera right-upward to find an upper square; 3. If it finds an upper square after a short delay, it goes toward the goal while moving the camera left-downward. If not, it moves the camera left-downward again while turning its chassis counter-clockwise until it finds another lower square, and then goes back to step 2. Thus the robot always searches for pattern B to go toward the goal.

We studied the role of the hidden and object detector neurons by lesioning one at a time. Their operation was disrupted by clamping the activation value of the neuron to a constant value of 0.5 during behavior. Figure 6 (left) shows that both neurons significantly contribute to the successful performance. The best evolved individual while the object detector neuron was lesioned arrived at the goal zone five times out of 20 trials (10 in the face of pattern A plus 10 in the face of pattern B, at the beginning). However these successes were achieved only when the robot started facing pattern B. If the robot was facing pattern A it went in the opposite direction of the goal. In other words, the robot always goes

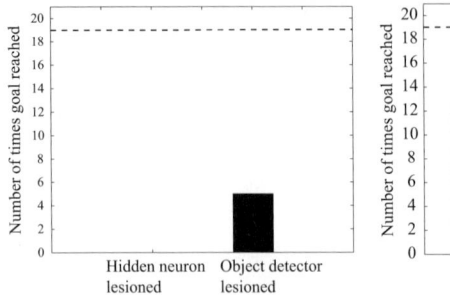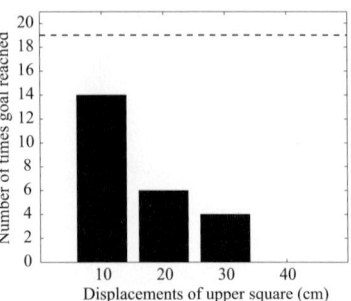

Fig. 6. The number of successful arrivals at the goal is counted in each condition out of 20 trials. Left: Lesion test of the best evolved individual. Horizontal dotted line shows the score of the intact best evolved individual. Right: Test of the best evolved individual when upper squares are displaced. Horizontal dotted line shows the score when upper squares are *not* displaced.

right in the face of both pattern A and B. This result suggests the crucial role of the object detector neuron in the behavior selection or decision making. That neuron significantly contributes to measuring the time interval between looking right-upward and subsequent detection of an upper square. Without the object detector neuron the robot can not measure the time interval and therefore can not discriminate the patterns.

While the hidden neuron was lesioned, the best evolved individual never arrived at the goal. This result suggests that the individual uses not solely the temporal information given by the object detector neuron, but also the visual information given through the hidden neuron.

One might think that the scanning strategy is reactive, i.e. the detection of a lower or an upper square always activates a particular behavior, but it is not. For example, in Fig. 5 (right) the lower square of pattern A was detected for the second time in the left side of the robot around the 130th time step, but this event did not affect the behavior of the robot going toward the goal. Therefore it seems that the behavior had been 'switched on' before the event[5]. The decision might be made when the upper square of pattern B was detected shortly after looking right-upward. If an upper square is detected late after looking right-upward, the robot does not go toward the goal, but resumes searching for a lower square.

This hypothesis was supported by another set of analyses where the upper square in each pattern was horizontally shifted toward the center (Fig. 6, right). The robot can not discriminate the two patterns any more if the upper square is shifted more than 20 cm.

The importance of the proprioceptive inputs is validated by another set of evolutionary runs without proprioceptive inputs (Fig. 7, left). Despite the shorter length of the genetic string (total length=480 bits), the best evolved individuals in all six evolutionary runs could reach the goal only three times out of six trials at max-

---

[5] The stable activation of the neuron 5 around 0.7 (see Fig 5, bottom) seems to reflect such a fixed behavior after the decision making.

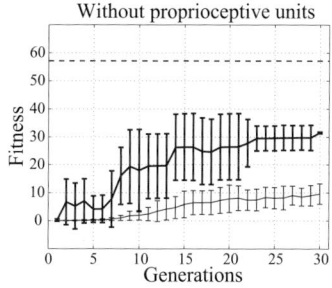
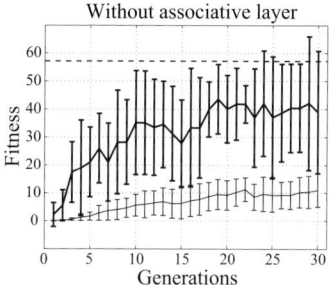

Fig. 7. Left: Evolution *without* proprioceptive inputs encoding pan and tilt movements. Right: Evolution *without* the associative layer. Fitness values of the population average (thin line) and the best individual (thick line) across 30 generations. Vertical bars show the standard deviation. Averaged over six evolutionary runs. Horizontal dotted line shows the averaged fitness value of the best evolved individual with the original neural architecture (see Fig. 4).

imum. Their behavioral analysis shows that these individuals always go left (or right depending on the evolutionary run) in the face of both pattern A and B. In other words they do not differentiate one pattern from the other.

One more set of evolutionary runs with another neural architecture which has fully recurrent connections at the output layer and does not have the associative layer shows worse fitness values than those with the original neural architecture (Fig. 7, right).

4   Discussion

We have shown that the evolved robot can detect two separate features in a sequential manner and discriminate the spatial relationships. If the system can perform active vision and sequentially store the events of visual feature detection, we do not need expensive computational power nor large memory storage capacity which would be required to resort to image memorization and matching. Although it has been shown that insects may indeed adopt such an image memorization and matching strategy [11], it is tempting to speculate that their tiny brain with restricted memory capacity may favor a more economical strategy as shown in this paper.

The evolved robot was able to effectively scan small regions of the broad visual field in an anticipatory manner in order to sequentially detect separate features. Such a characteristic of the evolved scanning strategy is in agreement with the evidence shown in [12,13] that people direct their gaze to points of the scene where information is to be extracted. Land et al. recorded human eye movements while playing cricket and table tennis. The eyes are very active and their activity takes roughly the same path as the ball. Contrary to popular belief, they do not follow the ball, but work in an anticipatory way. For example, eyes anticipate the position of the ball before it bounces, making saccades to positions where there is, as yet, no visible stimulus. In this article we have shown a computational model capable of an anticipatory "eye" movement in a freely moving behavioral system.

In order to detect the spatially separate features, the evolved robot executes a particular scanning sequence in front of the visual patterns. That is, after detecting a lower square, the robot routinely directs its gaze right-upward. Such a scanning sequence might be reminiscent of the human 'scanpath' during facial recognition [14,15]: Noton and Stark claimed that when a particular visual pattern is viewed, a particular sequence of eye movements is executed and furthermore that this sequence is important in accessing the visual memory for the pattern. The evolved scanning strategy presented in this article is similar to the 'scanpath' in that the moving sequence is crucial to identify a particular pattern. However, notice that the evolved scanning strategy is not for accessing the visual memory, but rather is tightly coordinated with the behavior of the robot.

From an engineering point of view one may argue that a panoramic camera could allow the robot to cover the entire visual field and discriminate the two patterns. However this approach would be computationally expensive if the entire image is to be uniformly processed in high resolution to extract tiny features out of a vast visual field as we have shown in this article. Active vision applied to an omnidirectional image is studied in a separate article [16].

Although the present neural architecture shown in Fig. 3 was investigated in the lesion test and additional evolutionary runs with modified neural architectures, further investigations must be done. We intend to identify the minimum components necessary for the neural controller of the robot to detect spatially separate features in the three dimensional visual environment.

5   Conclusions

In this paper we have shown that active vision may help not only to locate important features of the environment, but also to capture spatial relationships between those features that could provide behaviorally relevant information.

From these results it can be hypothesized that landmark-based navigation in insects and robots could be mediated by similar mechanisms instead of resorting to image memorization and matching [11]. We are currently exploring this hypothesis with simulated and physical robots.

Acknowledgments

Thanks to Danesh Tarapore and Claudio Mattiussi for enhancing the readability of this article. Two anonymous reviewers also provided helpful comments on the draft of this paper.

References

1. R. Bajcsy. Active Perception. *Proceedings of the IEEE*, 76:996–1005, 1988.
2. J. Aloimonos, I. Weiss, and A. Bandopadhay. Active Vision. *International Journal of Computer Vision*, 1(4):333–356, 1987.
3. J. Aloimonos. Purposive and Qualitative Active Vision. In *Proceedings of International Conference on Pattern Recognition*, volume 1, pages 346–360, 1990.

4. D. H. Ballard. Animate Vision. *Artificial Intelligence*, 48(1):57–86, 1991.

5. B. Horn. *Robot Vision.* McGraw-Hill, New York, 1986.

6. S. Nolfi and D. Floreano. *Evolutionary Robotics: Biology, Intelligence, and Technology of Self-Organizing Machines.* MIT Press, Cambridge, MA, 2000.

7. J. L. Elman. Finding Structure in Time. *Cognitive Science*, 14:179–211, 1990.

8. G. E. Hinton and T. J. Sejnowski, editors. *Unsupervised Learning: Foundations of Neural Computation.* MIT Press, Cambridge, MA, 1999.

9. D. Floreano, T. Kato, D. Marocco, and E. Sauser. Coevolution of Active Vision and Feature Selection. *Biological Cybernetics*, 90(3):218–228, 2004.

10. D. Floreano, M. Suzuki, and C. Mattiussi. Active Vision and Receptive Field Development in Evolutionary Robots. *Evolutionary Computation*, 13(4):527–544, 2005.

11. S. P. D. Judd and T. S. Collett. Multiple Stored Views and Landmark Guidance in Ants. *Nature*, 392:710–714, 1998.

12. M. F. Land and S. Furneaux. The Knowledge Base of The Oculomotor System. *Philosophical Transactions of the Royal Society of London, Series B*, 352:1231–1239, 1997.

13. M. F. Land and P. McLeod. From Eye Movements to Actions: How Batsmen Hit The Ball. *Nature Neuroscience*, 3(12):1340–1345, 2000.

14. D. Noton and L. Stark. Scanpaths in Saccadic Eye Movements while Viewing and Recognizing Patterns. *Vision Research*, 11:929–942, 1971.

15. D. Noton and L. Stark. Scanpaths in Eye Movements during Pattern Perception. *Science*, 171:308–311, 1971.

16. M. Suzuki, J. van der Blij, and D. Floreano. Omnidirectional Active Vision for Evolutionary Car Driving. In T. Arai, R. Pfeifer, T. Balch, and H. Yokoi, editors, *Proceedings of the 9th International Conference on Intelligent Autonomous Systems*, pages 153–161, March 7 - 9, 2006, Tokyo, Japan, 2006. IOS Press.

Global Navigation Through Local Reference Frames

John Pisokas

Computer Science Department, University of Essex,
Wivenhoe Park Colchester, CO4 3SQ, UK
john.pisokas@gmail.com

Abstract. The contribution of this paper is that illustrates the use of funneling actions in combination with local deictic reference frames for forming consistent and useful large scale maps. These maps do not rely on any geodetic sensors. Indications for the feasibility of such representations in humans, and other species, can be found in studies of spatial cognition. However, such implementations or applications in robotics have not been illustrated until now.

1 Introduction

The motivation for the research, which lead to this article, is the ability which humans and other species posses to develop internal spatial representations of their environment, and to use them for spatial reasoning and planning activity. The existence of such internal maps was established by O'Keefe and Nadel [13] with later developments leading to the work of Wills et al. [16]. This research established that the hippocampi of rats, dogs, monkeys and humans are the basis of spatial maps and reasoning.

After several attempts, neurophysiological studies have failed to identify any form of organization based on topology, perceptual similarity, or metric information in the spatial maps in the brain structures of rats, dogs and monkeys. Therefore, it is now believed that no topological, perceptual similarity, or metric organization is rendered on the maps' anatomy. Nevertheless, as previously mentioned, the existence of spatial maps has been well established. Furthermore, the same subjects – rats, monkeys, dogs and humans – are known not to posses any sensors of geodetic[1] orientation or position. It is also well known that the navigation skills of using printed maps and magnetic compasses (and later means such as astrolabe, GPS etc) are not innate competencies but rather a later development in human history, which presumes presence of multitude of competencies and tools: drawing maps, measuring distances, using magnetic compass, and logical reasoning abilities. It is fair to assume that earlier humans were able to find their way around before such developments; adequate proof for

---

[1] The term geodetic is used throughout this article for referring to a point of space in respect to a global geodetic reference frame such as used by GPS, magnetic compasses, and printed maps.

S. Nolfi et al. (Eds.): SAB 2006, LNAI 4095, pp. 274–285, 2006.

this is that rats, dogs, monkeys and other species can develop internal maps and they use them to find their way around, even though they do not possess skills for creating or using compasses and printed maps.

This line of reasoning brings us to the conclusion that, in our quest to understand human spatial reasoning, we must depart from a starting point where no global geodetic reference instruments (such as magnetic compass and printed maps) are available. We know that it is possible to build reliable and useful large scale maps without such tools because many species can do it; our problem is that we do not understand the necessary mechanisms. This paper puts together the necessary concepts and representations for achieving this end.

2   Funneling Actions

Funneling actions are particular types of activity which produce highly repeatable behavior and effects; that is, behavior that can be considered stereotypic. Such actions are implemented as groups of competencies which converge the robot's state variables of interest to specific ranges of values. For instance if the state variables of interest are the Cartesian coordinates (x, y, z) of the robot's position, in three-dimensional space, a corresponding funneling action will converge these three variables to specific values within appropriate ranges. These specified final values and ranges of the state variables essentially define what the effect of the particular funneling action is. The trajectory by which the goal values of the state variables are approached can vary and it is stipulated by the behavior and the environment. Our everyday lives are abundant of such funneling actions, which simplify our operation because they produce repeatable and convergent behavior.

To find examples where funneling actions are employed, you can observe people's paths in places where people aggregate. For instance people entering a building or a bus. In those cases people are coming from variety of initial conditions and they finally have similar spatial coordinates, bounded by well specified ranges. This process requires a method for moving from the initial conditions to the final ones. We argue that the employed method is behavioral funneling, or in other words use of funneling actions.

Other not so conspicuous examples of behavioral funneling can be found throughout various domains of human activity. For example, driving a bicycle

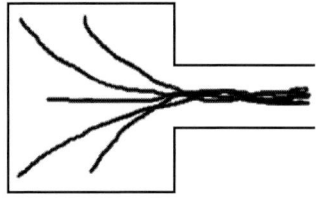

Fig. 1. Trajectory of five runs of a robot using a 'corridor transversing' competence. The robot starts in the room and drives in the corridor.

requires use of a multitude of behavioral funnels achieving motion within safe ranges of speed, inclination, acceleration and posture. But you should not think of funneling actions only as trajectories that resemble the shape of a funnel; funneling action can also be walking straight on the pavement. Keeping oneself on the pavement and walking on a constant direction is achieved with combination of multiple cooperating funneling behaviors.

Behavioral funneling is achieved by the interaction of environmental constrains with the agent's behavior. The environmental constraints allow only particular trajectories and positions of the agent, likewise the agent's structure and behavior allow only particular trajectories and activities. The interaction of those constrains stipulates the possible actions, positions and trajectories of the agent in the specific environment. Environment, positions and trajectories need not be referring only to spatial domains; it is possible for those parameters to refer in general to any group of substrate, states, and processes respectively.

The same principles can be employed in the domain of robotics. For example, in industrial assembly robotics, a variant of funneling actions has been used in order to limit the possible positions of an object on the workbench before grasping it (see [5]). Equivalent funneling actions are exhibited in mobile robot navigation when the trajectory of a robot converges to a certain region. Such results can be obtained with competencies such as 'wall following' or 'corridor transversing'. An example of the spatial convergence as result of a 'corridor transversing' competence can be seen in Figure 1. The concept of funneling actions will be proved to be significant to spatial reasoning and navigation in forthcoming sections.

As we defined previously, for an action to be a funneling one, the action needs to have a highly repeatable effect and thus produce routine behavior. Common characteristic in our implementations of funneling actions is that they use exteroceptive feedback control for adjusting the robot's trajectory or activity. It is of significant importance that any ambiguities and uncertainties of operating in the target environments are resolved within those funneling actions, because the mapping and spatial reasoning mechanisms, which will be presented in section 4.2, do not accommodate stochastic behaviors (the interested reader can refer to [11] for other work following the same philosophy). This is a deliberate choice, strongly supported by neurophysiological research which has shown that when animals are deprived of their reasoning faculties, they can still successfully walk, eat, drink etc [13]. Extrapolating this, it is reasonable to assume that humans would also be able to walk, run, eat etc if deprived of their reasoning faculties. Another reason for this choice is the fact that several areas of the human brain (cerebellum, M1, SMA, hippocampus and other areas depending on the task) are involved in motor planning, which controls balance, grasping, walking etc. For these reasons, we believe that significant effort should be devoted on building robust funneling actions.

2.1   How to Build Funneling Actions

Funneling actions can be preprogrammed or learned by experience. Given an initial preprogrammed robot competence there are three learning methods which can be used for deciding on whether this competence operates as a funneling action or what needs to be adapted in order to turn it into a funneling action. These three approaches can be followed individually or in combination:

- A competence can become a funneling one by adapting its parameters in order to produce increasingly funneling behavior. For instance reinforcement-learning methods can be used for learning a corridor transversing competence.
- By identifying a sensory condition which reliably predicts whether the agent will end up to the expected place or somewhere else, the agent can learn conditional changes in its behavior. For instance if the sensory condition indicates divergence from the desired trajectory additional corrective behaviors can be activated.
- Another technique is to identify the temporal part of a behavior which operates as funnel and place a stopping condition before the robot's trajectory starts diverging. Then this first segment can be reliably used as a funnel.

These three methods can be applied during a development period, but after that any destructive adaptation of the funneling actions must be stopped before starting using them for learning maps and making plans based on them. In this article we will not elaborate on those learning methods.

The controller which is going to employ these funneling actions — in our case the mapping and planning mechanism described in section 4.2 — must have some learning capacity for identifying if an action operates as a funneling action in each specific circumstance. This is because we do not have a method to warranty that an action, which has been proven to operate as funneling action in a wide range of circumstances, will operate as funneling action in every possible situation. Therefore, the controller should make sure that an action will operate as a funneling one in the particular situation. For this reason the controller must try the competence as it is, several times in the particular situation, and then decide whether is repeatable enough so that it can be used as a funneling action in the current situation.

For the implementation of the funneling actions — either being implemented by hand or using learning — some principles must be followed, which have been experimentally proven to lead to funneling behavior, as described in the sequel. We shall mention that the method presented in this paper is in concord with the views of behavior-based robotics and situated action research ([3], [14]). It is required that activity is situated in the agent's environment and skill decomposition and bottom-up building of the agent's competencies have been empirically proved to be successful in achieving situated behavior. However, having achieved situated behavior the next step is to discern the next level of behavioral achievement; that is funneling behavior and how to achieve it. We present here some principles for building funneling actions which have been derived through personal experience and thus are prone to augmentation and improvement; however, they constitute a working methodology. Funneling behavior can be achieved by building competencies in one of the behavior-based methodologies but taking into account few more constraints. These are:

- Build competence with stopping condition(s)
    - develop multiple alternative sensing pathways
    - develop multiple alternative reaction pathways
    - consider as many as possible environment conditions
- Experimentally find and exploit environmental dynamics, which apply forces to the agent. For example if it is a catamaran robot do not try to go against rough water, but rather exploit existing forces.
- Try a particular behavior in a specific environment to conclude if it operates as funneling action or not.

In section 4.2 the practical application of funneling actions is illustrated.

3 Local Deictic Frames of Reference

In this section, the concepts of deictic references and egocentric frames of reference will be introduced. '*Deictic*' means pointing and '*deictic references*' are pointers to objects. An example of *deictic reference* is referring to an object by pointing to it with your index finger, instead of referring to the object with a unique name. Deictic expressions are highly dependant on context, unlike proper nouns which refer to entities regardless of context. Examples of deictic references include, "I am trying to kill *the fly*", "...*the bee* that stung me", "I kicked *the ball* outside *the field*". The introduction of deictic references in Artificial Intelligence was made by Agre [1]. He used deictic operators for referring to objects. The work of Ballard and Kaelbling on deictic references has explored the use of deictic references as pointers, practically only as computational pointers (see for instance [2] and [6]).

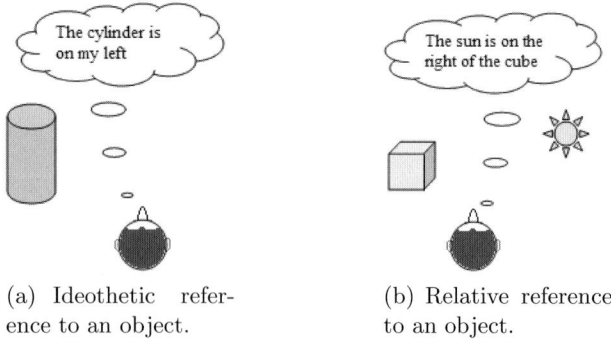

(a) Ideothetic reference to an object.

(b) Relative reference to an object.

Fig. 2. Two kinds of egocentric reference frame

Reasoning or talking about spatial relations between entities, requires the use of appropriate deictic references to entities; these are *ideothetic* and *relative* references (Figure 2). Both of them are deictic. *Ideothetic* references allow the speaker to use spatial relational terms locating an object in respect to his current position and orientation. *Relative* references allow the speaker to use spatial relational terms locating an object in respect to another and his current position and orientation. The next section will show how global map representations can be developed using these local egocentric reference frames and the funneling actions.

4   Developing Global Maps

In their research, Gillner and Mallot [8], discovered that while humans are capable of navigating around environments, when asked to draw a map of a familiar environment – encompassing topological and metric information — their drawings were largely inaccurate. Wang [15] was interested in the question "How people learn the directional relationship between places that share no common landmarks?" and found that people fail to learn the directional relationship between a room and the outside world, when no common landmarks are visible from both places. It seems that either humans do not use accurate global coordinate system and/or our communication mechanisms are incapable of accessing such information, if such information is stored in the human brain at all. The research stream of hippocampal and brain research found no internal maps which are organized based on topological or metric information (see for example [16]), rendering the second hypothesis unlikely. This leads us to take the most promising hypothesis that humans do not use an accurate global geodetic coordinate system, and therefore it is not necessary for rats, dogs, monkeys, or for robots. This leaves us with the question "How can we build maps without explicitly preserving topological and metric information?" and "Why should we? Are there any advantages?" The answer to the second question is that it is highly likely that there are advantages in such an implementation since several species are known to have evolved to this solution. The answer to the former question will follow.

We should add here that humans and other animals do not possess sensors which provide them with geodetic position and orientation information. Such geodetic information can be possibly extracted by virtual sensors using as source, for instance, dead reckoning. However, several experiments of Wang [15] vividly indicate that this is not the case in humans.

4.1   Experiments on Building Maps

A simple learning mechanism has been developed for illustrating how funneling actions together with local deictic frames of reference can be used for building global maps. The task of the robot was to move around in a three-roomed environment (Fig. 4) and build an internal representation of its interaction with the environment. Then this representation was used for planning paths in this environment. Figure 3 depicts a multigraph representation which was constru-

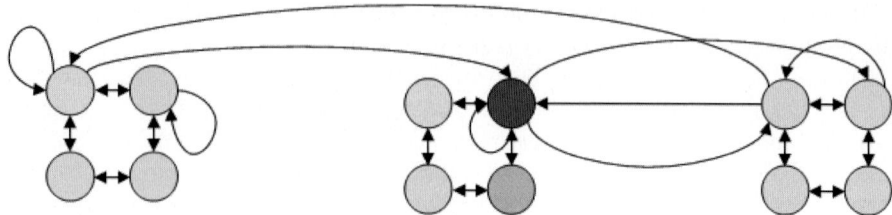

Fig. 3. Example of multigraph created while exploring a flat with three rooms. Each disc represents one view of the environment. The different colors (and intensity) represent the recency of the view's encounter. Only the first thirty-two actions are depicted, due to the cluttered nature of drawing such data in 2D.

cted by the robot. The sensory perceptions of the robot are represented as discs; each disc corresponding to one view in the environment. Each group of four discs corresponds to a single place, where the robot rotates on the spot and captures four snapshots, 90 degrees apart. Each of these four snapshots is connected to its neighbors with the actions 'turn left' and 'turn right', which enable the robot to change its point of view. The combination of views with these two actions forms *relative* references on a frame of reference local to the place.

When the robot activates one of its actions interacts with the environment and moves from one place to another or from one direction to another. If the robot moves, its sensory perception (view) will commonly change. In Figure 3, the arrows indicate which funneling actions can be initiated from the current place and direction and what will be their result to the sensory perception (view) of the robot. Essentially, *funneling actions* link the local frames of reference together to a global frame of reference. The very nature of funneling activity is exploited to connect those local frames of reference.

Experimental Setup. The experiments were conducted in an indoor experimental arena (Fig. 4), which consists of three rooms connected with corridors. The walls of each room are painted with a different color, the corridor entrances are made of two blue circular pillars placed on the sides and the charging station is a corner which is painted green in room of different color. Color and geometry are the means for allowing distinguishing among different places due to the perceptual capabilities that the robot is equipped with. A MagellanPro mobile robot has been used during the experiments. A color camera and a LASER scanner mounted on the robot are used, including ultrasonic SONAR distance sensors and bumpers for navigation purposes.

Sensor Signals Processing. Sensor signals are the only means for the robot autonomously to know the result of its actions. Therefore, special attention was paid to the implementation of reliable sensing modules. To achieve this we use redundant processing units estimating the same features with alternative methods.

First, we will describe the feature extraction units for the LASER scanner (Fig. 5). Each scan dataset is a sequence of 180 radial distance measurements. Initially, each scan dataset is preprocessed by three different units. The one is

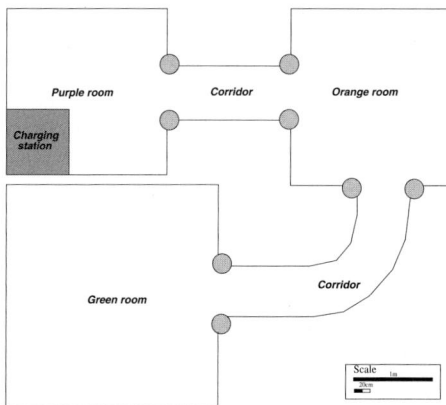

Fig. 4. Top view of the real world environment used for the experiments, in scale

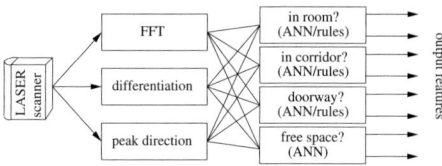

Fig. 5. Graphical representation of the features extraction units. The features extractors estimate the output values using two independent methods. The first is ANNs and the second rule-based estimators, providing two independent values for each feature.

calculating a 180 points FFT so that we get a frequency distribution measure. The second is differentiating the data sequence by calculating: $\forall i \in \{1..n-1\}, D_i = d_i - d_{i+1}$, where $n = 180$ elements of the dataset, d_i the ith element of the dataset and D the differentiated sequence. The third preprocessing unit finds the position of the two maximum and the two minimum elements of the dataset. The position of these elements indicates the angular directions of the corresponding measurements which are crucial for recognizing environmental features. In the sequel the outputs of those three preprocessing units are used by eight other processing units, four neural networks (ANN) and three rule-based processors. For training those processing units training datasets have been constructed corresponding to the three high level questions: in room?, in corridor?, doorway? and free space? The output of each processing unit is binary, signaling yes or no. The classification accuracy of each of these classifiers is between 70% and 90%. For obtaining more accurate classification we recorded the behavior of each of the processing units in variety of environmental conditions and the results were used for combining each pair of outputs with AND and OR logic operators to obtain more accurate outputs. For example, it was observed that in the majority of cases the robot was in a corridor only when both the ANN and the rule-based classifiers were giving positive output therefore a conjunction of the two outputs gave a better accuracy output.

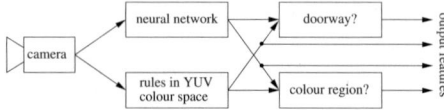

Fig. 6. Graphical representation of the features extraction units. The features extractors estimate the output values using two independent methods. The first is ANNs and the second rule-based estimators, providing two independent values for each feature.

Alternative pathways for calculating features of the input data in different ways provide different outputs due to the different approaches and performance characteristics. This has the advantage of allowing to extract information that would be elusive or ambiguous if a single method had been used.

The second sensor was a color camera, Fig. 6 depicts the processing of the camera output signal. Initially, the camera output is digitized and takes the form of a two dimensional matrix with each element having three values corresponding to the red, green and blue percentages of that pixel. In the sequel this digitized form is received from the two preprocessing units, one neural network and one YUV-colourspace rule-based classifier. Both units read and classify each pixel of the image matrix based on its color. Each classifier identifies six different color classes, so each pixel is classified to one of these six colors. The color classified images are then received by two other processing units, a doorway detector and color-region detector. The doorway detector is searching in the image for two parallel and vertical bars of the same color. This is done because in the environments we use all doorways are made of two blue pillars. The second detector searches for continuous regions of the same color. If an image region with single color is larger than a threshold then it appears in the output.

Behavioral Modules. Motor actions are produced by behavioral modules. Unreliable motor actions, of the kind 'move forward 10*cm*', that are usually employed in robotics research (see for example [9]) consign the mechanical and electrical uncertainties to the reasoning architectural layers. At the reasoning layers uncertainty is usually addressed with statistical methods which are computationally expensive and limit the scaling capabilities of such mechanisms.

Carefully designed funneling actions with high repeatability will be reliably resulting to the same effect whenever invoked and when failing will not cause damage to the robot but safely will fail and report (cognizant failure [7]). After all it is not disastrous to fail if the result is not fatal; how fatal the result might be depends on how well the robot has been designed. Both failure and success are useful sources of learning.

Special attention was paid for every target competence to have at least two motor pathways driving the robot. This achieved behavioral redundancy so that if one method was failing, the other still drives the robot towards the right direction. Behavioral redundancy was used in order to increase the probability that at least one pathway will drive the robot to an appropriate direction.

The behavioral repertoire of the robot was composed by five goal-directed funneling actions (i.e. 'go to charger', 'go through the deepest opening until you reach a room', 'find corridor and transverse it', 'turn to doorway' and 'turn to blue color').

4.2 Experimental Results

Initially, the robot was left to explore the environment by performing 100 actions randomly selected from those of its behavioral repertoire. Subsequently, the map representation was built off-line, using the acquired data.

Once trained, the mapping and planning mechanism was assessed by planning for five different, randomly selected tasks. For each task, the robot had to move from a starting to a goal position and take a picture with its camera. We performed 20 runs for each of the tasks. For planning paths, reaction–diffusion dynamics are used to spread a marker through the representation until a complete path from start to goal view is found.

The testing criterion was the *success rate* of the planner, which was defined as the percentage of successful runs over the number of trials for each task. An unsuccessful run occurs when the robot stops, not being able to make a plan from the current position to the goal, or if the time used for accomplishing the task exceeded *five* minutes. The success rate of each planner for five different tasks is depicted in Fig. 7. The measured overall success rate is 96%.

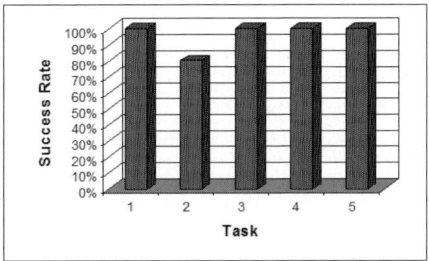

Fig. 7. Experimental results for the mapping and planning mechanism operating in the environment depicted in Fig. 4

These results confirm the hypotheses that: (1) deliberate planned actions for navigation in global scale can result by the synergy of funneling actions, pursued in respect to local egocentric frames of reference. And (2) deliberate planned activity can be achieved using only egocentric short-range sensors.

Further experiments have been performed in an unmodified furnished flat with three rooms, but the available space does not allow inclusion of these results which will be published in separate article. In that experiment different sensory processing and funneling actions have been used. The achieved overall performance was 93%.

5   Conclusion

Several researchers including Bugmann [4], Kuipers and Byun [10], Gillner and Mallot [8], Matarić [12] have explored the use of graph representations for representing topological and metric information acquired during exploration of the robot's environment. To this end they employed magnetic compasses and odometry for measuring the necessary quantities, in respect to a global geodetic frame of reference.

The novelty of the presented article is that it explores an alternative way to represent maps by using only local reference frames connected with funneling actions. Humans and animals have been proved capable of spatial mapping, navigating, planning, and reasoning without use of instruments that provide geodetic information and without need for common landmarks being visible from the related places. The question that was raised was "How do they achieve it?" This article provides an answer. The presented method was capable of successfully planning and navigating among rooms which share no common landmarks even though no explicit information about the topological or metric relation of the places was stored. This provided an answer to the questions raised by the research of Wang [15] and Wills et al. [16].

The mapping method which has been implemented in this article is designed for illustrating that by using funneling actions even such a primitive mapping mechanism can be successful. The reason for its success is that funneling actions produce highly repeatable behavior; therefore, the map does not need to employ statistical models in order to accommodate behavioral uncertainty. The presented mapping method is only an example; one can envisage that more advanced mechanisms, which might employ additional kinds of information, such as topological and metric, will perform at least as well.

In conclusion, we emphasize the importance of funneling actions in navigation and planning of paths. The results of Gillner and Mallot [8] illustrate that the conscious mental representations of maps are often largely inaccurate. However, they commonly suffice for planning and reaching one's goal place. We argue that this is not because these mental representations capture accurately enough the whole information needed, but because they capture adequate amount of information, while the rest of it is embedded in the funneling actions. For deciphering this we must remind the reader that funneling actions are carefully designed (or learned) to have highly repeatable effect, thus when the agent starts a funneling action from a certain place and direction will quasi-deterministically end at a specific place. Humans use funneling actions; therefore, even though their mental representation captures only approximate and distorted directional and distance relations among places; these pieces of information are adequate for initiating a funneling action towards the appropriate direction. The initiated funneling action will reliably lead the agent to the same place regardless of the fact that the exact spatial relationships on our mental representation are mistaken, since we do not move on the exact direction that the mental representation suggests but on that manifested by the funneling action we use. The distorted mental representation will persist as far as our distorted perception of our path remains consistent with that of the mental representation. We therefore conclude that the conscious mental representations are only tools for thought, for planning ahead; they are not full and accurate representations of the environment. They are good enough for what is their role and their role is to plan routes for reaching to places given constraints on the possible routes. These mental representations alone do not suffice for driving the agent to his goal. They are the funneling actions which the agent employs that enable him to reach his goals. The feasibility of this line of thinking was illustrated in this article by building robots which can successfully drive to goals.

References

1. P.E. Agre. *The Dynamic Structure of Everyday Life*. Technical report AI-TR-1085, MIT Artificial Intelligence Laboratory, 1988.
2. D. Ballard, M. Hayhoe, P. Pook, and R. Rao. Deictic codes for the embodiment of cognition. *Behavioral And Brain Sciences*, 20:723–767, 1997.
3. R. A. Brooks. A robust layered control system for a mobile robot. Technical Report 864, MIT AI Lab, September 1985.
4. G. Bugmann. *A connectionist approach to spatial memory and planning*, chapter 5, pages 109–146. Springer, London, 1997.
5. G.E. Deacon, P.L. Low, and C. Malcolm. Orienting objects in a minimum number of robot sweeping motions. Technical Report DAI Research Paper No. 619, Division of Informatics, University of Edinburgh, 2001.
6. S. Finney, P. Wakker, L. Kaelbling, and T. Oates. The thing that we tried didn't work very well: Deictic representation in reinforcement learning. In *Proceedings of the 18th Annual Conference on Uncertainty in Artificial Intelligence*, 2002.
7. E. Gat. Integrating planning and reacting in a heterogeneous asynchronous architecture for controlling real-world mobile robots. In *Proceedings AAAI-92*, volume 4(4), pages 809–815, 1992.
8. S. Gillner and H. Mallot. Navigation and acquisition of spatial knowledge in a virtual maze. *Journal of Cognitive Neuroscience*, 10:445–463, 1998.
9. Sven Koenig and Reid G. Simmons. *Artificial Intelligence and Mobile Robots*, chapter Xavier: A Robot Navigation Architecture Based on Partially Observable Markov Decision Process Models, pages 91–122. AAAI and MIT Press, 1998.
10. B. Kuipers and Y. T. Byun. A robot exploration and mapping strategy based on a semantic hierarchy of spatial representations. *Robotics and Autonomous Systems*, 8:46–63, 1991.
11. Chris Malcolm and Tim Smithers. Symbol grounding via a hybrid architecture in an autonomous assembly system. *Robotics and Autonomous Systems*, 6(1&2) (Special Issue – Designing Autonomous Agents), June 1990.
12. M. J. Matarić. Environment learning using a distributed representation. In *Proceedings of IEEE International Conference on Robotics and Automation*, pages 402–406, 1990.
13. J. O'Keefe and L. Nadel. *The Hippocampus as Cognitive Map*. Oxford University Press, 1978.
14. Lucy Suchman. *Plans and Situated Actions: The Problem of Human-Machine Communication*. Cambridge: Cambridge Press, 1987.
15. R. Wang. Learning and unlearning spatial relationships during navigation. *Journal of Vision*, 3, 2003.
16. T. Wills, C. Lever, F. Cacucci, N. Burgess, and J. O'Keefe. Experiencedependent attractors in the hippocampal representation of the local environment. *Science*, 308:873–876, 2005.

Transition Cells for Navigation and Planning in an Unknown Environment

N. Cuperlier, M. Quoy, C. Giovannangeli, P. Gaussier, and P. Laroque

ETIS-UMR 8051, Universite de Cergy-Pontoise - ENSEA
6, Avenue du Ponceau, 95014 Cergy-Pontoise, France
cuperlier@ensea.fr

Abstract. We present a navigation and planning system using vision for extracting non predefined landmarks, a dead-reckoning system generating the integrated movement and a topological map. Localisation and planning remain possible even if the map is partially unknown. An omnidirectional camera gives a panoramic images from which unpredefined landmarks are extracted. The set of landmarks and their azimuths relative to a fixed orientation defines a particular location without any need of an external environment map. Transitions between two locations recognized at time t and t-1 are explicitly coded, and define spatiotemporal transitions. These transitions are the sensory-motor unit chosen to support planning. During exploration, a topological map (our cognitive map) is learned on-line from these transitions without any cartesian coordinates nor occupancy grids. The edges of this map may be modified in order to take into account dynamical changes of the environment. The transitions are linked with the integrated movement used for moving from one place to the others. When planning is required, the activities of transitions coding for the required goal in the cognitive map are enough to bias predicted transitions and to obtain the required movement.

1 Introduction

Several biomimetic models allow to perform navigation tasks even without relying on localisation neither on maps (see [1] for a review of several insects like strategies). Nevertheless these models are constrained to use different "routes" for each goal to reach and can not exhibit some interesting behaviors like shortcut etc... Hence, in most of bio-inspired models, like in [2,3], localisation is based on particular neurons found in the rat hippocampus (particularly CA3, CA1 and dentate gyrus (DG), regions and also in the entorhinal cortex (EC)) named "place cells" (PC). A map of the environment may be built by linking these PC. One can refer to [4,5] for a comparative review of localisation and mapping models. In our "rodent like" model, we also use place cells (layer modelling EC see section 4) that learn patterns specific of a given location (spatial landmarks constellation, see section 3), but we do not directly use them to plan or construct a map. We rather use neurons ("transition cells") that explicitly code for these spatio-temporal transitions (in the layer modelling CA3/CA1). Details of their creation and arguments

S. Nolfi et al. (Eds.): SAB 2006, LNAI 4095, pp. 286–297, 2006.

in favor of such a coding are given in section 5. During exploration, these transition cells are created and allow to learn a cognitive map whose construction is explained in section 6. When a plan is needed, transitions are predicted and are then biased via top-down information from the cognitive map (section 7).

Hence we propose here a unified neuronal framework based on an hippocampal and prefrontal model where vision, place recognition and dead-reckoning are fully integrated (see Fig. 2 for an overview of the architecture). All neurons activity are analogous. There is no symbolic programming nor predefined object of high cognitive level. No assumption are made about the structure of the environment. We will conclude with improvements that may be proposed in our model.

2   Material and Methods

The robot is a koala platform (40*30cm) with six wheels. It has infrared sensors for obstacle detection. A low level obstacle avoidance mechanism is implemented (not described here). Images are taken by a panoramic camera at low resolution. A rectangular image (1500×240 pixels) is obtained from the panoramic image which is originally circular (640×480 pixels).

Since our robotic model is inspired from the animat approach [6], we use three contradictory animal like motivations (eating, drinking, and resting). Each one associated with a satisfaction level that decreases over time and increases when the robot is on the proper source according to coupled differential equations [7]. When a level of satisfaction falls bellow a given threshold, the corresponding motivation is triggered so that the robot has to reach a place allowing to satisfy this need. Hence this place becomes the goal to reach. More sources can be added and one can increase the number of sources associated with a given motivation.

3   Autonomous Landmark Extraction and Recognition Based on Characteristic Points

In order to reduce problems induced by luminance variability, we only use the gradient image as input of the system. Next, curvature points (corresponding to robust focal points) are detected by filtering this gradient image with a Difference Of Gaussian. Two processes then occur in parallel: first a log-polar transform of the local area extracted around each focal point is computed. Connection's weights of neurons are then modified to learn these small images. This allows to improve the pattern recognition when small rotations and/or scale variations on these small images occur [8,9]. These images are landmarks, and by extension, we also name the coding neurons landmarks. Second, for each landmark, an angular position relative to the north given by a compass is computed [10,11]. Thus, this visual system provides both a *what* and a *where* information: the recognition of a 32×32 pixels small images in log-polar coordinates, and the azimuth of the corresponding focal point. *What* and *where* informations are then merged in a product space leading to a spatial landmark constellation. The number of

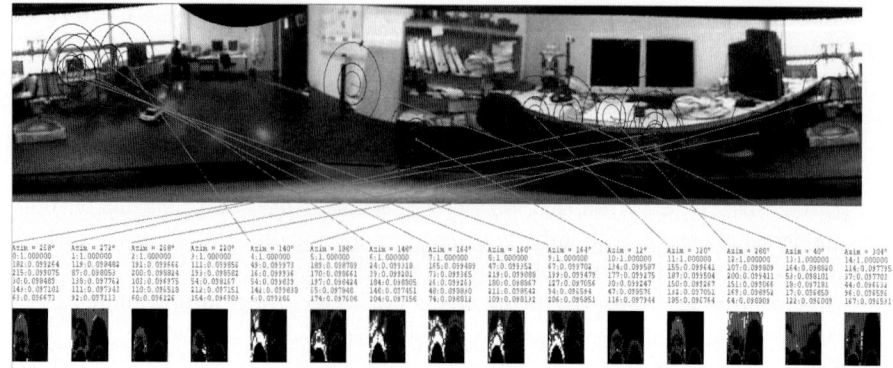

Fig. 1. Image taken from a panoramic camera. Below are 15 examples of 32×32 log-polar transforms taken as landmarks and their corresponding position in the image.

landmarks needed is a tradeoff between the robustness of the algorithm and the speed of the process. If all landmarks were fully recognized, only three of them would be needed. But as some of them may not be recognized in case of changing conditions like luminance or occlusion, taking a greater number is enough to guarantee the robustness.

4   Autonomous Place Building

The spatial landmarks constellation resulting from the visual input treatement characterizes one location. This constellation can thus be learned on a neuron of EC (place recognition at time t see fig. 2). The neuron coding for this location is called a "place cell" as the one found in the rat's hippocampus [11] since these cells fire when a rat is at a particular location in its environment. The activity of a PC results from the computation of the distance between the learned and the current local view. Thus, the activity of the kth PC can be expressed as follows:

$$P_k = \frac{1}{l_k} \left(\sum_{i=1}^{N_L} \omega_{ik}.f_s(L_i).g_d(\theta_{ik}^L - \theta_i) \right) \tag{1}$$

with $l_k = \sum_{i=1}^{N_L} \omega_{ik}$ the number of landmarks used for the kth PC, where $\omega_{ik} = \{0,1\}$ expresses the fact that landmark i has been used to encode PC k, with N_L the number of learned landmarks, L_i the activity of the landmark i, $f_s(x)$ the activation function of the neurons in the landmark recognition group, θ_{ik}^L the learnt azimuth of the ith landmark for the kth PC, θ_i the azimuth of the current local view interpreted as the landmark i. d is the angular diffusion parameter which defines the shape of the function $g_d(x)$. The purpose of $f_s(x)$ and $g_d(x)$ is to adapt respectively the dynamics of *what* and *where* groups of neurons. They are defined as follow :

$$g_d(x) = \left[1 - \frac{|x|}{d.\pi}\right]^+$$
$$f_s(x) = \frac{1}{1-s}\left[x - s\right]^+$$

where $[x]^+ = x$ *if* $x > 0$, *and* 0 *otherwise*.

The s parameter rescales the activity of the landmark neuron over s between 0 and 1. The d parameter modulates the weight of the angular displacement.

Experimental place cell formation has also been tested in outdoor environments [12]. The result confirmed the mathematical model which predicts that the size of the place field grows proportionaly with the landmark distance.

If the robot is at the exact position where the PC has learned, its activity is maximal (equal to one). When the robot moves from this position, the activity of this PC decreases. Hence the PC keeps a certain amount of activity around the learned position that is named the *place field* of a PC. Consequently, we have to use a rule that controls the recruitment of a new neuron to encode a new location. This mechanism is performed autonomously, without any external signal, relying only on the PC's population activity. If the activities of all previously learned place cells are below a *given* recognition threshold (R.T), then a new neuron is recruited. At a given place, every existing place cell responds with an analog recognition value that may be seen as the robot position probability. If at a given place several PC respond with activities greater than the R.T, a competition takes place so that the most activated one wins and codes the current location. The density of locations learned depends on the level of this threshold, but also on the robot position in the environment. Namely, more locations are learned near walls or doors due to the fast changes in the angular position that can occur near landmarks, or in the (dis)appearance of landmarks caused by these obstacles. In other locations, small changes produce a small variation in the place cell activity. When the environment has been entirely explored, and thus fully

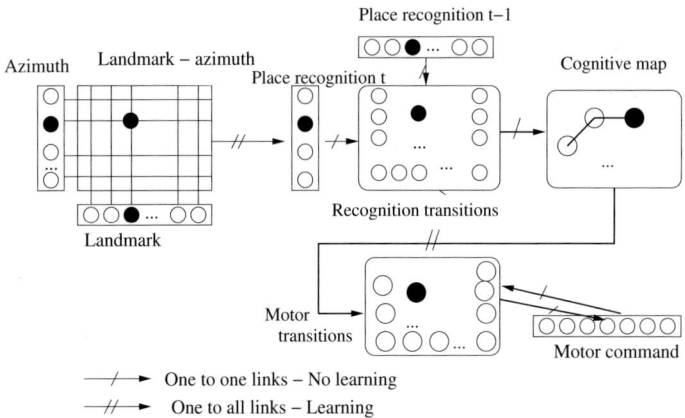

Fig. 2. Sketch of the model. From left to the right: merging landmarks and their azimuth, then learning of the corresponding set on a place cell. Two successive place cells define a transition cell. They are used to build up the cognitive map and are also linked with the integrated movement performed.

covered by place cells, a PC responds specifically for each location (see Fig. 6). Consequently the PC neural layer gives our robot a way to localize itself inside the environment it has explored.

5   Autonomous Building of Transition Cells

A natural question is "why using transitions instead of places"? In order to briefly answer this question, we have to focus first on how to plan using place cells. Several bio-inspired approaches rely on place cells, but to better illustrate our approach we will only describe briefly our past-model which allows to easily underline the problem. First a place cell may be linked with the movement needed to reach a goal without any map. This sensory-motor association may be generalized to the whole environment [7]. However, this simple reactive mechanism is not enough in environments composed of several rooms, or when there are contradictory motivations. A cognitive map will solve these drawbacks (see section 6). Two different approaches of this cognitive map exploitation have been proposed. First, the selection of the action in a place cell based model can be realized by an external mechanism applied to the cognitive map: the gradient algorithm. But, if this solution is enough for a navigation task, it might be more difficult to find an external mechanism for more complex tasks like robot arm control. Moreover from a biological point of view, using an external algorithm "looking for" the gradient of activity leads to the famous problem of the homonculus: "who is looking ?" Second, as a consequence, the action selection mechanism has to be integrated. This can be performed by associating an action with a place, thus defining a sensori-motor unit. But then, the choice of the direction to follow may be ambiguous. Indeed, in some place, several actions can be associated with the same place (see fig. 3) like in the T-maze example. In this case, which movement should select the robot if it must go to C?

In order to solve this problem, we do not directly use PC for planning in our model, but rather transitions between the two PC winning the recognition competition: respectively at time t (in EC) and time $t-1$ (in DG). Such spatio-temporal transitions are explicitly coded on neurons called *transition cells*. The idea of this coding has been inspired by a neurobiological model of timing and

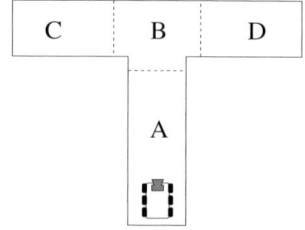

Fig. 3. In this example, from place B the robot had learned during exploration that it can go either to C by turning right or D by turning left. Both movements are thus linked with place B.

temporal sequences learning in the hippocampus [13]. Motivation from such a coding comes from the fact that transitions are better suited for sensori-motor association than places since only one direction can be linked with a transition: the movement used to go from A to B with the transition cell AB (see fig. 2).

Before going further about transitions, note that as transitions link two succesively recognized PC, transitions like AA are also coded. These kind of transitions are the equivalent of PC in transition coding. No movements are linked with these transition cells. We only associate a movement to a transition linking two different PC. An internal signal is computed from the automatic detection of a new wining PC at time t by temporal differences on EC. This signal is used to trigger the sensori-motor association.

A relevant question is about the growth of the number of transition cells created while exploring the environment. This number is intimately linked with the number of place cells. This number of place cells created for a fixed R.T value depends on the complexity of the environment. The degree of complexity of an environment relies mainly on two factors: the number and the location of its landmarks and the number of obstacles.

Thus we have performed several tests setting one of these parameters to underline the impact of the second upon the ratio between created transition cells over created place cells. Each simulation lasts 50000 cycles. This number has been chosen high enough to ensure that the robot has learned a complete cognitive map of the environment [1]. The results of both tests shown here are the average on ten simulation results. We have first studied the impact of obstacles configuration in three environments of increasing complexity. Tests have been performed for a single, a two and a four rooms environment. The number of landmarks have been fixed at a high value. The ratio remains stable around the mean value 5.45 for all environments once the cognitive map of the environment is complete (see table 1). The second study shows how this ratio evolves for an environment with the same complexity of obstacles but with simple to half landmark number. These tests have been done on the two and four rooms environment with the same experimental set-up than the previous study. The number of landmarks increases the number of particular cases in which a landmark previously visible becomes invisible (or the reverse) and consequently decreases the activity of all previously known place cells. This finally results in the creation of a higher number of place cells. The results (see Table 2) show a stable ratio of a mean value still around 5.35 for simple environments. This ratio does not depend on the value of R.T (but the number of place cells increases with increasing R.T). The stability of this ratio can be explained as follows: since the number of a place cell's neighbours is necessary limited and that a transition is a link between "adjacent" place cells, only a few transitions can be created from a given place cell. To conclude, there is no combinatorial explosion of the number of created transitions. Thus, they can be memorized and used for planning purpose.

---

[1] We consider the cognitive map is complete when the robot becomes unable to detect new places or new transitions.

Table 1. Results of the experiments on the ratio of the number of place cells (nbPC) created over the number of transitions created (nbT) according to the number of rooms in the environment. Standard deviation is given into brackets. This ratio remains stable. There are at most six times more transition cells than place cells. R.T is set at 0.97.

Param / Env	One room	Two rooms	Four rooms
nbPC	133.8(2.85)	606.2(6.89)	643.7(9,88)
nbT	735.8(19.80)	3389.2(56.38)	3281,2(48,80)
ratio	5.49(0.06)	5.59(0.08)	5.09(0,04)

Table 2. Results of the experiments on the ratio of the number of place cells (nbPC) created over the number of transitions created (nbT) according to the number and configuration of landmarks in the environment: with two rooms (first column for many landmarks and second column for few landmarks) and with four rooms (third column for many landmarks and fourth column for few landmarks). Standard deviation is given into brackets. This ratio remains stable. There are at most six times more transition cells than place cells. R.T is set to 0.97.

Param / Env	Two, many land.	Two, few land.	Four, many land.	Four, few land.
nbPC	606.2(6.89)	364.3(5.75)	643.7(9,88)	295.5(4.94)
nbT	3389.2(56.38)	1951.2(35.30)	3281,2(48,80)	1591.8(26.64)
ratio	5.59(0.08)	5.35(0.03)	5.09(0,04)	5.38(0,05)

Now that we know the number of possible transitions starting from a given place cell, we can use this information for modelling the transition layer. Transition cells building does not rely on a full "matrix" coding the relationships between successively reached places. This would be too memory consuming. Instead, we exploit the fact that a place cell has around 5 neighbours on average to compress the structure merging these informations (see fig. 4). In order to cope with extreme cases, we allow for a maximal number of 10 neighbours. Consequently the number of neurons of this structure has decreased since we only take

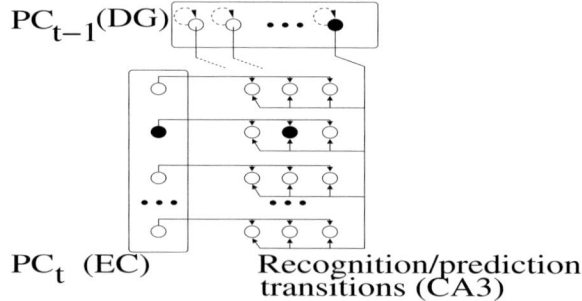

Fig. 4. Transition cells population inputs from population of place cells at time t and at time $t-1$. In order to have a clear figure, only 3 possible transitions are shown on the merging and compression group. For the same reason, connections from only one neuron of PC_{t-1} are drawn.

into account real possible transitions and not all the combination of place cells. Each neuron of a given line receives projections both from population coding for place cell at time t named PC_t and from population coding place cell at time $t-1$ named PC_{t-1}. Each transition neuron belongs to a particular neighbourhood supervised by a single PC_t neuron (a line in the figure 4). No learning is allowed on those links and their weights are not sufficient to trigger any activity on the associated transition neurons. Conversely, each transition neuron is connected to all the PC_{t-1} neurons through conditional links. The activation of PC_t neurons increases the weights coming from the activated neuron in PC_{t-1}, when no transition neuron already corresponds to this conjunction. Once those weigths are learned, in a prediction mode, the single activity of the corresponding PC_{t-1} neuron allows the activity of the transition neuron even if no signal comes from PC_t.

6   Autonomous Cognitive Map Building

Experiments carried out on rats have led to the definition of cognitive maps used for path planning [14]. Most of cognitive maps models are based on graphs showing how to go from one place to an other [15,16,17,18,19,20,21]. They mainly differ in the way they use the map in order to find the shortest path, in the way they react to dynamical environment changes, and in the way they achieve contradictory goal satisfactions. Other works use ruled-based algorithms, a classical functional approach, that can exhibit the desired behaviors, we will not discuss them in this paper, but one can refer to [22].

In our model, learning the cognitive map is performed continuously during the exploration of the unknown environment (latent learning) by linking transition cells successively reached if no link was yet created between these two transitions. Equation 2 shows the learning rule applied to the value of edge $W_{i,j}$ linking vertice j to i. $G(j)$ is the activity of transition j. $\overline{G(i)}$ is the memory term of G(i) that decreases with time. λ is a decay term that allows to forget erroneous transition due to an uncomplete exploration. $\frac{dR}{dt}$ is the variation of the reinforcement. The edge value is increased if the edge is used, and decreased if it is not. After some time, some edges are reinforced. These edges correspond to paths that are often used. In particular, this is the case when some particular locations have to be reached more often than others (see section 7) [7].

$$\frac{dW_{i,j}}{dt} = -\lambda.W_{i,j} + (1 + \frac{dR}{dt}).(1 - W_{i,j}).\overline{G(i)}.G(j) \qquad (2)$$

In the same time, if a source is present at the destination place the corresponding transition is associated with a motivation neuron. After some time, exploring the environment leads to the creation of the cognitive map. The prefrontal cortex is the place in our model where this cognitive map is coded. This seems to be coherent with neurobiological data [23]. This topological map may be seen as a graph where each vertices is a transition and where the edges code for a path between two transitions. No position in a fixed reference is assigned to the vertices of the graph and edges code for adjacence relation only.

7   Autonomous Planning Using the Cognitive Map

Some places are more important because they are goals that have to be reached
when necessary. When a goal has to be reached, the transitions leading to it are
activated. This activation is then diffused on the cognitive map graph, each node
taking the maximal incoming value which is the product between the weight on
the link and the activity of the node sending the link. After stabilization, this
diffusion process gives the shortest path between all nodes and the goal node.
This is a neural version of the Bellman-Ford algorithm [2][24,25] (see fig. 5).

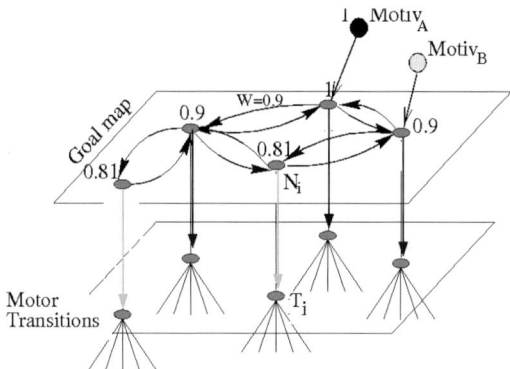

Fig. 5. Diffusion of the activity on the graph corresponding to the cognitive map.
Diffusion is starting from the goal. Each vertex keeps the maximal activity coming
from its neighbors. Corresponding motor transitions (integrated movement) are then
biased by this activity.

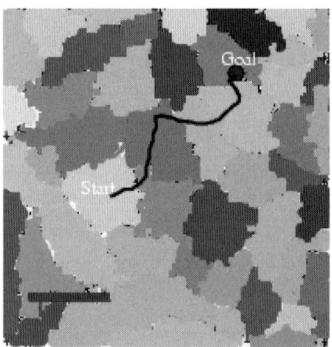

Fig. 6. A simulated environment fully explored. Each region represents the place field
of a particular place cell. After a full exploration, the entire environment is covered by
the place cell population. The curve is an example of a planned path to reach a goal
place (presence of a source).

---

[2] The Bellman-Ford algorithm allows to find the shortest path between any node and
a goal node of a weighted graph.

When the robot is in a particular location A, all possible transitions beginning with A are predicted and filtered from the n most activated place cells (similar to the multiple hypothesis position tracking, described in [5], where several position hypothesis can be used in constrast with a 'single' position following). The top-down effect of the cognitive map is to bias these predicted transitions such that the ones chosen by the cognitive map have a higher value. This small bias is enough to select/filter the appropriate transitions via a competition mechanism. This results in a unique movement vector to apply to the robot motor command. See fig. 6 for an illustration of a path followed.

8   Discussion

Exploration periods may be alternated with planning periods. The choice of the behavior is obtained through the self-regulation of two control variables: first the motivational information which allows to trigger a planning behavior, and second, a detection signal triggering a period of exploration. This signal is generated while a new transition is learned meaning that the planning behavior leads the robot in a place still unknown (case of an incomplete map). Planning then restarts as soon as the robot is able to predict transitions from the current place.

Our model currently running on robots (Koala robots and Labo3 robots) has interesting properties in terms of autonomous behavior. However, this autonomy has some drawbacks:

- we are not able to build a cartesian map of the environment because all locations learned are robot centered. However, the places in the cognitive map and the direction used give a skeleton of the environment.
- we have no information about the *exact* size of the rooms or corridors. Again, the cognitive map only gives a sketch of the environment.
- some parameters have to be set, in particular the recognition threshold (section 4). The higher the threshold, the more places are created.

The transitions used in this model may also be the elementary block of a sequence learning process. Thus, we are able to propose a unified vision of the spatial (navigation) and temporal (memory) functions of the hippocampus [26]. However how to go from a graph of transitions to a sequence of transitions of any length is still an open question. This will be part of the next step of the work. The same scaling problem appears when one wants to code several different maps. Each map should be linked with a kind of context signal (which floor or which room) that should be able to "reload" the previous learned map (or a part of it) into the different neural structures used here. Again, models are available and should be tested in simulation and on a robot.

Acknowledgements

This work is supported by two french ACI programs. The first one on the modelling of the interactions between hippocampus, prefrontal cortex and basal ganglia in collaboration with B. Poucet (CRNC, Marseille) JP. Banquet (INSERM U483) and R. Chatila (LAAS, Toulouse). The second one (neurosciences integratives et computationnelles) on the dynamics of biologically plausible neural networks in collaboration with M. Samuelides (SupAero, Toulouse), G. Beslon (INSA, Lyon), and E. Dauce (Perception et mouvement, Marseille). C. Giovannangeli is supported by a DGA Grant.

References

1. Franz, M.O., Mallot, H.A.: Biomimetic robot navigation. Robotics and Autonomous Systems **30** (2000) 133–153
2. Hafner, V.V.: Cognitive maps in rats and robots. Adaptive Behavior **13** (2005) 87–96
3. Arleo, A., Gerstner, W.: Spatial cognition and neuro-mimetic navigation: A model of hippocampal place cell activity. Biol. Cybern. **83** (2000) 287–299
4. Filliat, D., Meyer, J.A.: Map-based navigation in mobile robots - I. a review of localisation strategies. Journal of Cognitive Systems Research **4** (2003) 243–282
5. Meyer, J.A., Filliat, D.: Map-based navigation in mobile robots - II. a review of map-learning and path-planing strategies. Journal of Cognitive Systems Research **4** (2003) 283–317
6. Meyer, J.A., Wilson, S.W.: From animals to animats. In Books2-4, B., ed.: First International Conference on Simulation of Adaptive Behavior, MIT Press (1991)
7. Gaussier, P., Leprêtre, S., Quoy, M., Revel, A., Joulain, C., Banquet, J.: Experiments and models about cognitive map learning for motivated navigation. In: Interdisciplinary approaches to robot learning. Volume 24. Robotics and Intelligent Systems Series, World Scientific, ISBN 981-02-4320-0 (2000) 53–94
8. Schwartz, L.: Computational anatomy and functional architecture of striate cortex: a spatial mapping approach to perceptual coding. Vision Res. **20** (1980) 645–669
9. Joulain, C., Gaussier, P., Revel, A.: Learning to build categories from perception-action associations. In: International Conference on Intelligent Robots and Systems - IROS'97, Grenoble, France, IEEE/RSJ (1997) 857–864
10. Tinbergen, N.: The study of instinct. Oxford University Press, London (1951)
11. O'Keefe, J., Nadel, N.: The hyppocampus as a cognitive map. Clarenton Press, Oxford (1978)
12. Giovannangeli, C., Gaussier, P., Banquet, J.P.: Robot as a tool to study the robustness of visual place cells. In: I3M'2005: International Conference on Conceptual Modeling and Simulation (CMS 2005), Marseille (2005) 97–104
13. Banquet, J., Gaussier, P., Dreher, J., Joulain, C., Revel, A.: Space-Time, Order and Hierarchy in Fronto-Hippocampal System: A Neural Basis of Personality. In: Cognitive Science Perpectives on Personality and Emotion. Volume 124. Elsevier Science BV Amsterdam (1997)
14. Tolman, E.: Cognitive maps in rats and men. The Psychological Review **55** (1948)
15. Arbib, M., Lieblich, I.: Motivational learning of spatial behavior. In Metzler, J., ed.: Systems Neuroscience, Academic Press (1977) 221–239
16. Schmajuk, N., Blair, H.: Place learning and the dynamics of spatial navigation: a neural network approach. Adaptive Behavior **1** (1992) 353–385
17. Franz, M.O., Schölkopf, B., Mallot, H.A., Bülthoff, H.H.: Learning view graphs for robot navigation. Autonomous Robots **5** (1998) 111–125

18. Bachelder, I.A., Waxman, A.M.: Mobile robot visual mapping and localization: A view-based neurocomputationnal architecture that emulates hippocampal place learning. Neural Networks **7** (1994) 1083–1099

19. Trullier, O., Wiener, S.I., Berthoz, A., Meyer, J.A.: Biologically based artificial navigation systems: review and prospects. Progress in Neurobiology **51** (1997) 483–544

20. Schölkopf, B., Mallot, H.A.: View-based cognitive mapping and path-finding. Adaptive Behavior **3** (1995) 311–348

21. Bugmann, G., Taylor, J., Denham, M.: Route finding by neural nets. In Taylor, J., ed.: Neural Networks, Henley-on-Thames, Alfred Waller Ltd. (1995) 217–230

22. Donnart, J., Meyer, J.: Learning reactive and planning rules in a motivationnally autonomous animat. IEEE Transactions on Systems, Man and Cybernetics-Part B **26** (1996) 381–395

23. V. Hok, E. Save, P.L.S., Poucet, B.: Coding for spatial goals in prelimbic-infralimbic area of the rat frontal cortex. Proceedings of the National Academy of Sciences ((to appear in 2005))

24. Bellman, R.E.: On a routing problem. In: Quaterly of Applied Mathematics. Volume 16. (1958) 87–90

25. Revel, A., Gaussier, P., Leprêtre, S., Banquet, J.: Planification versus sensory-motor conditioning: what are the issues ? In: From Animals to Animats : Simulation of Adaptive Behavior SAB'98. (1998) 129–138

26. Banquet, J., Gaussier, P., Quoy, M., Revel, A., Burnod, Y.: A hierarchy of associations in hippocampo-cortical systems: cognitive maps and navigation strategies. Neural Computation **17** (2005)

Use Your Illusion: Sensorimotor Self-simulation Allows Complex Agents to Plan with Incomplete Self-knowledge

Richard Vaughan and Mauricio Zuluaga

Autonomy Lab, School of Computing Science,
Simon Fraser University, Burnaby, BC, Canada
{vaughan, mzuluaga}@sfu.ca

Abstract. We present a practical application of sensorimotor self-simulation for a mobile robot. Using its self-simulation, the robot can reason about its ability to perform tasks, despite having no model of many of its internal processes and thus no way to create an *a priori* configuration space in which to search. We suggest that this in-the-head rehearsal of tasks is particularly useful when the tasks carry a high risk of robot "death", as it provides a source of negative feedback in perfect safety. This approach is a useful complement to existing work using forward models for anticipatory behaviour. A minimal system is shown to be effective in simulation and real-world experiments. The virtues and limitations of the approach are discussed and future work suggested.

1 Introduction: Let Your Hypotheses Die in Your Stead

To solve some problems, autonomous agents must plan ahead. One common problem that requires planning is finding an efficient route between a set of places in the world: the family of problems that includes the classical Traveling Salesman Problem. Finding efficient routes between places of interest can clearly be seen to be adaptive; for example a squirrel visiting nut caches or a female lion patrolling and freshening her territorial urine marking sites can save time and energy for other tasks if a good route is chosen.

Formally, planning is the process of finding continuous trajectories through the agent's configuration space between the start and goal states. Configuration space is the set of all possible states that can be achieved by the system. Conventional planning techniques construct a model of the configuration space: either a static model such as a traversability map, or a generative model such as a production system. In either type of model, all possible state transitions are known.

Now suppose we have an intelligent agent, a robot, that contains some components of unknown function. By definition, the agent can not have an *a priori* model of its state evolution, due to the unknown internal states of the mysterious components, and their contribution to the system's output. Thus it can not construct an *a priori* model of configuration space. Such a system does not know what it can do, so how can it plan its future actions?

S. Nolfi et al. (Eds.): SAB 2006, LNAI 4095, pp. 298–309, 2006.

One solution is to learn a model of the mysterious systems by running them for a while and observing their inputs and outputs. Once a sufficiently good input/output mapping model of the mystery system is constructed, the model can be used to create a configuration space. The major flaw with this approach is that it suffers from the general learning problem of being critically dependent on the training data samples: the model can only be expected to be correct when it operates in situations similar to those seen during learning. This is a serious problem because there is a particular set of situations that are very important, and can never be experienced in training: the situations that cause the robot to be destroyed. Avoiding doom is a very important part of adaptive behaviour, and it can not be learned by negative experience.

This risky-learning problem can be solved by learning in simulation instead of the real world. It is often possible to construct a good *a priori* model of the outcome of a robot's motor actions in the world in terms of its new sensor readings: a *sensorimotor simulation*. It is commonplace for real adaptive systems to be usefully tested and trained in simulation. For example, commercial pilots spend a considerable part of their training time in flight simulators. In addition to routine flying, pilots can rehearse dangerous scenarios such as engine and instrument failures in complete safety. While we should be mindful of the advice of Brooks [1] about the limitations of world models, it is a fact that many robot control programs have been developed, learned, or evolved in simulation and successfully transferred to the real world with few or no changes, e.g.[2,3].

Thus a robot with unmodeled mystery components could employ a sensorimotor simulation, observe its simulated actions and build a model of the mystery components. The resulting model can be used to construct a configuration space in which to plan.

But with the sensorimotor simulation in place, we have a simpler alternative. Why model the mystery components at all? Instead we can just execute candidate plans in the simulation and evaluate the outcomes. The mystery components just run as they would in the real world, remaining an unmodeled mystery, but we can still observe their effects on the world. The only requirement is that the sensorimotor simulation is a usefully good approximation of the robot's interactions with the real world.

Once we have taken this step, an appealing simplification presents itself. Why have an explicit model of *any* part of the robot's control system? If we have an explicit model M of robot control code C that is intended to implement process P, there is always a possibility of discrepancies between M, C, and P. An unfortunately common situation is when the code C contains bugs which prevent it from implementing the programmer's intention P correctly. Model M derived from P will probably not contain the same bugs, but may contain different bugs of its own. Plans computed in a configuration space from M may not be executable using C. But plans observed to work in a sensorimotor simulation in which C runs directly are guaranteed to reflect the actual function of C, bugs and all. Again, this is limited by the fidelity of the simulation, but *only* by the fidelity of the simulation.

Brooks' aphorism "the world is its own best model" [1] is well known. We propose the complementary idea: *the agent is its own best model.* More strongly, we can say that an agent's control software (or a provably correct transcoding of that software) is the *only* reliable model of itself. Sometimes simulating the agent's interaction with the world is relatively easy if you choose the appropriate level of abstraction [2].

This idea is appealing from an intuitive point of view. We can consciously observe the imagined results of our actions in the world, but we often do not have conscious models of what we can do, or how we do it. Few people could write down a correct dynamical systems or neurophysiological model of themselves riding a bicycle, but most riders can imagine themselves cycling down the street - even a street they have never seen. Similarly, most people can imagine swinging a golf club to strike a ball, even if they have never held a club. This could be explained by the existence of a generalized model of our motor interactions with the world, which can be used to rehearse novel situations. Intriguingly, there is evidence that athletes can improve their performance at motor tasks by performing such in-the-head rehearsal. The effect is more pronounced among individuals who already have expert skills, suggesting that the fidelity of the in-the-head model may be important in the successful transfer of imagined performance to the real world [4,5].

In this way we relate sensorimotor self-simulation to the folk psychology notion of *imagination* as a mechanism to consider the outcomes of our behaviour without having to fully understand it, and without having to try everything out for real. The relationship between imagination and simulation is concisely expressed by Dawkins:

> "We all know, from the inside, what it is like to run a simulation of the world in our heads. We call it imagination and we use it all the time to steer our decisions in wise and prudent directions" [6].

This informal idea is consistent with *the emulation theory of representation*, developed by Rick Grush[7]. In this philosophical framework, phenomena external to an agent are represented internally by *processes* rather than the symbol systems of conventional cognitive science. This idea can be seen to underly this paper and much of the related work.

We can also consider an executable plan as a statement of truth about what the robot can do, and an untested plan as a hypothesis. In this sense we can view the robot as a Popperian scientist seeking truth by eliminating bad hypotheses through in-the-head experimentation:

> "The scientist can annihilate his theory by his critique, without perishing along with it. In science, we let our hypotheses die in our stead." [8]

Interpreting this famous statement rather more literally than originally intended, a robot can observe itself dying a thousand times in simulation as a result of bad plans, and thus eliminate those plans without risking its neck in the real world. The robot need have no model of itself beyond the immediate sensory outcomes of its motor behaviour.

In practical terms, the proposed model offers a method for allowing high-level strategic or "cognitive" function to reason about the actions of other behaviour-producing systems without understanding how they work. This could be a useful engineering strategy for adding strategic layers on to existing behaviour-based systems. It also may hint at how evolutionarily recent cognitive systems could come to usefully exploit the functions of more ancient control systems in the brains of animals. This approach may also be a useful principle for robustness: if all system components are treated as if they are unknown, then the results of internal failures will be immediately apparent in the simulation without needing to update any internal model.

1.1 Related Work

In the 1960s, Jewett placed lesions in the brains of cats which eliminated the inhibition of action commands during the REM sleep stages of dreaming, and (in Jewett's interpretation) allowed the cats to act out their dreams. The cats displayed such behaviours as fighting, grooming, exploring, running away and showing rage. Jewett concluded that dreams are rehearsal of vital survival activities that are likely to occur in real life [9]. In [10] rats were trained in a maze while awake, and could be observed rehearsing the maze experiments while sleeping.

Several authors have described systems in which sensorimotor forward models are able to predict how sensory information changes through sequences of motor commands [11,12,13,14]. In contrast, this paper shows how the outcomes of plans consisting of sequences of relatively high-level operations can be predicted. Our "motor commands" are goto() operations that abstract away a powerful and complex navigation system that is part of the agent, but completely unmodeled.

A framework that incorporates simulation to speed up learning in an evolutionary experiment is presented in [15]. Their proposed method combines into a single framework learning from reality and learning from simulation.

In this paper we present a practical application of sensorimotor self-simulation for a mobile robot. Using self-simulation, the robot can reason about its ability to perform tasks, despite having no model of many of its internal processes and thus no way to create an *a priori* configuration space. This approach is a useful complement to existing work using forward models for anticipatory behaviour. A minimal system is shown to be effective in simulation and real-world experiments. The virtues and limitations of the approach are discussed and future work suggested.

2 The Application of Imagination

2.1 Task

A robot lives in an office-like environment. At some moment it is asked to visit a set of places in the world. There is no preferred order of visits: the only requirement is that all of them are visited in the shortest possible time. This task is a variation of the Travelling Salesman Problem where the traversal cost on the arcs connecting nodes is initially unknown.

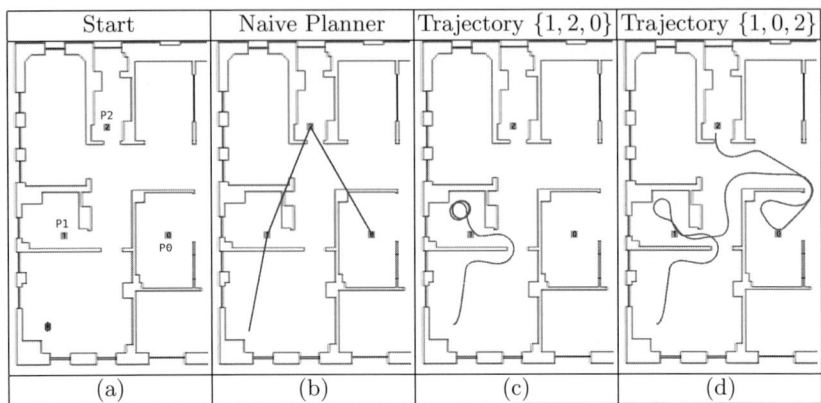

Fig. 1. (a) Starting state; (b) Naive shortest path, and two actual trajectories (c,d) of robots guided by VFH

Figure 1 shows such a scenario. Image (a) shows a map containing the robot start position and 3 goal locations. Image (b) shows the shortest path that visits all goal points. Image (c) shows the path taken by the robot using VFH navigation (described below) as it attempts to reach the locations in the order suggested by the shortest path: $\{1, 2, 0\}$. Due to the dynamics of VFH, the robot is fatally trapped. If the goal locations are submitted to the robot in the order $\{1, 0, 2\}$, the robot easily completes its task, reaching all locations by the path shown in Image (d). It is very difficult to characterise the configuration space created by VFH in a traversibility map. The system described below solves this problem.

2.2   Procedure

A typical lab robot R_0, operates in world W_0. At any time R_0 can run a sensori-motor simulator that models the interactions between a model robot R_i running controller C in model world W_i over time $S(R_i, W_i, t) \Rightarrow S(R_i, W_i, t+1)|i > 0$. The real and simulated robots use the identical controller C. To decide which order to visit the goal locations in reality, R_0 uses its simulator to internally rehearse all possible visit orders. If there are n possible plans, R_0 runs n simulations $S(R_m, W_m, t)|m = 1, ..., n$.

The simulation results are evaluated and the plan that caused the fastest traversal in any simulation is selected as the plan for real-world execution.

A naive implementation will not scale well to large values of n, but the difficulty of scaling is not unique to our approach as the Traveling Salesman Problem is known to be NP complete, i.e. no scalable solution is known. A sophisticated implementation could prune the space of simulations to improve performance, and this method is ridiculously parallel (i.e. it parallelizes perfectly to n processors).

One useful optimization is immediately apparent. If the simulations are real-time, or a constant multiple of real time, we can evaluate plans trivially: run all the simulations in parallel, the first to finish must be the route that takes least time to traverse, and all the others can be aborted without loss.

2.3   Control Architecture

Our control system makes extensive use of Player, a well-known Free Software system for robot control over a network interface [16]. To greatly simplify the robot controller, we use the *VFH* obstacle-avoidance algorithm [17] and the Adaptive Monte Carlo Localization map-based localization algorithm [18] provided with Player. The Player server and its VFH and AMCL modules are treated as "black boxes" of mysterious internal construction.

The robot controller C receives as input an ordered vector of places to visit $P = \{p_1, \ldots, p_n\}$ and an initialized index $i = 0$ marking the current goal, the ith member of P. $p_i = (x_i, y_i, r)$, where where x_i and y_i are Cartesian coordinates in the plane. If the Cartesian distance from the robot to place (x_i, y_i) is less than r then the robot is considered to have visited (x_i, y_i). The controller takes the initial goal location p_0 and submits it to Player's VFH implementation as a goal location. Player then attempts to drive the robot to that position while avoiding obstacles. All locations are specified in the robot's localization coordinate system, as determined by the AMCL implementation. Player reports the current robot pose back to the controller. When the robot visits its goal location, the location index is incremented $i = i+1$ and the new goal location p_i is submitted to VFH. When the last goal location is reached, the task is complete and the robot stops. A schematic of the controller is given in Figure 2.

The output from the "known", i.e. non-blackbox part of the controller is a sequence of commands of the form goto(x, y). The blackbox parts of the

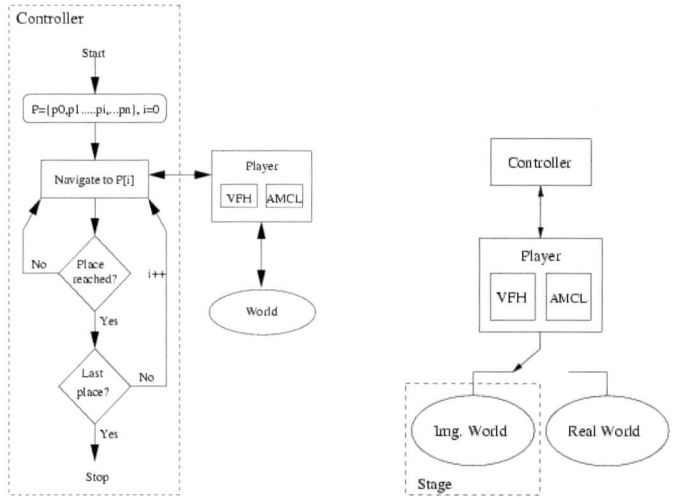

Fig. 2. Control Architecture (left) and Imagination Engine (right)

Fig. 3. Pioneer robot in the Real World (a) and VFH trap: The robot's goal is to reach the small square outside the room (b)

controller attempt to achieve the current goal without crashing into obstacles by reading from the range sensors and sending a stream of motor commands of the form move(v, ω), where v is the desired forward velocity and ω is the desired angular velocity. Though we as designers can state the purpose of these modules, there is no way for the system to know what they do, and how they would effect the robot's configuration space.

In fact VFH has some dynamic properties that are crucial to a planning system. It is an excellent local planner and obstacle-avoider, but as such it suffers from local minima problems that cause it to make bad decisions under certain conditions. Consider the situation presented in Figure 3: the robot is blocked from reaching its goal by walls to the front and sides. If the side walls are long enough, VFH is unable to escape from the "trap" and will instead make small loops indefinitely, eventually exhausting the robot's energy supply. Depending on the task and the availability of human assistance, this may be a fatal error that the robot should never experience for real.

We desire our robot system to be able to take into account the complex dynamic properties of VFH when choosing the best route to take, with no *a priori* model of VFH. The same argument applies to the dynamics of the AMCL localization system, and the Player TCP server, the details of which we omit for lack of space.

2.4   Sensorimotor Simulation Implementation

In the real robot, the move(v, ω) commands are converted by the robot's embedded computer into pulse-width modulated signals that drive amplifiers that power the wheel motors. The physical motion of the wheels is reflected in subsequent measurements taken by the robot's physical sensors.

We can replace the physical part of the system with the well-known Stage robot simulation engine [19]. Player using simulated Stage devices is known as *Player/Stage*, and has been used in published experiments by many authors. For convenience, we introduce the term *Player/real* to indicate a Player server connected to real robot hardware. If reasonably careful in the assumptions made in the implementation of a robot controller, one can expect grossly similar robot behaviour in Player/Stage and in Player/real. We have anecdotal evidence from the community that Stage models real robots reasonably well. We present experimental evidence below that backs this up (Section 2.6).

2.5   Experiment 1: Simulation Proof of Concept

In this proof of concept we work entirely in simulation, i.e. R_0 and W_0 are simulated and model the real robot and world, but R_0 is still unique in that only it can spawn child simulations. R_0 must choose the best order in which to visit n locations, by observing the behaviour of $R_1 \ldots R_{n!}$ simulated robots, one for each possible route.

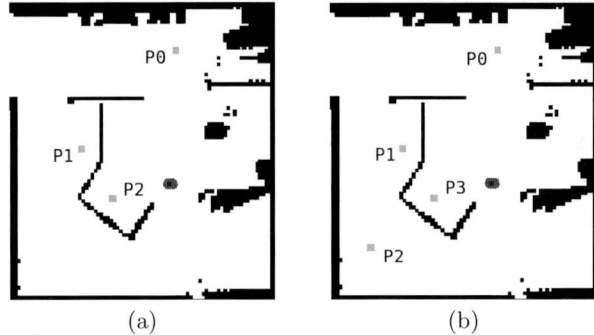

(a)   (b)

Fig. 4. Simulated world maps showing robot start location and goal locations for (a) Exp.1: with 3 locations, and (b) Exp.2 with 4 locations

Figure 4(a) shows the scenario of the first experiment. There are $n = 3$ places to visit, so $P = \{p_0, p_1, p_2\}$. To visit a place, the robot must come within $r = 0.5m$ of the place. The world map is an approximation of our real robot arena, and was automatically created with the pmap mapping utility[1].

R_0 spawns $m = n! = 6$ threads, each containing a complete simulation and robot controller. The first thread in which a robot visits all locations is the winner R_{win}. The other threads are stopped and R_0 executes the route taken by R_{win}.

With small values of m, we can run all the simulation threads in real time on a modest workstation. A more scalable solution would allow threads to be spawned on multiple computers. On a machine with a single CPU there is little advantage to be gained from multiple threads and instead we could explicitly run each simulation in turn for a short time in a single thread, thus avoiding the thread switching overhead.

Results. Figure 5 shows the progress in time of all 6 threads that execute all possible routes. Plots of the route travelled by each simulated robot at 20-second

---

[1] pmap was written by Andrew Howard and available from http://playerstage.sourceforge.net.

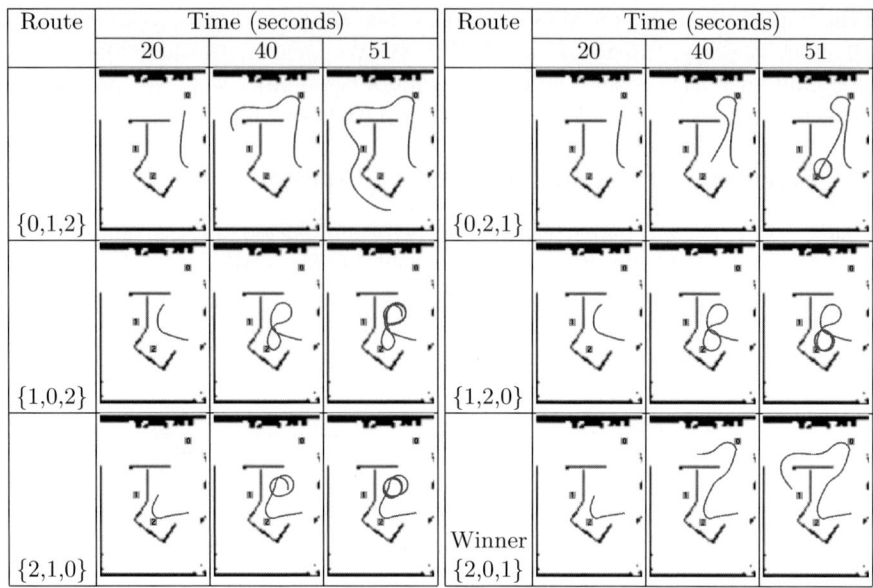

Fig. 5. Multi-threaded example: 6 threads are generated, one for each possible route. Route {2,0,1} is completed in 51 seconds, when this happens all the other threads are stopped.

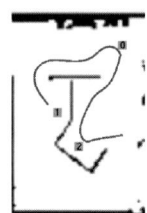

Fig. 6. The path taken by R_0, having selected the winning route $\{2, 0, 1\}$

intervals. At 50 seconds the robot executing route $\{2, 0, 1\}$ finishes and all the other simulations are stopped. All routes except the winner and Route $\{0, 1, 2\}$ were in looping states (the fatal situations we want the real robot to avoid) and were not chosen for execution. Route $\{0, 1, 2\}$ appears to be on its way to accomplishing the task, but it is taking a longer path than the winning route. R_0 now executes the winning route, and its path is shown in Figure 6.

2.6   Experiment 2: R_0 in the Real World

In this experiment we use the real-world Pioneer 3DX robot shown in Figure 3 for R_0 and increase the number of places to visit $n = 4$, located as shown in Figure 4(b). There are now $m = n! = 24$ possible routes, and the robot's onboard computer could not run 24 simulations in real time, so we implemented a single-

Route $\{3,0,1,2\}$	Route $\{3,0,2,1\}$	Route $\{0,1,2,3\}$	Route $\{3,2,1,0\}$	Route $\{3,1,2,0\}$

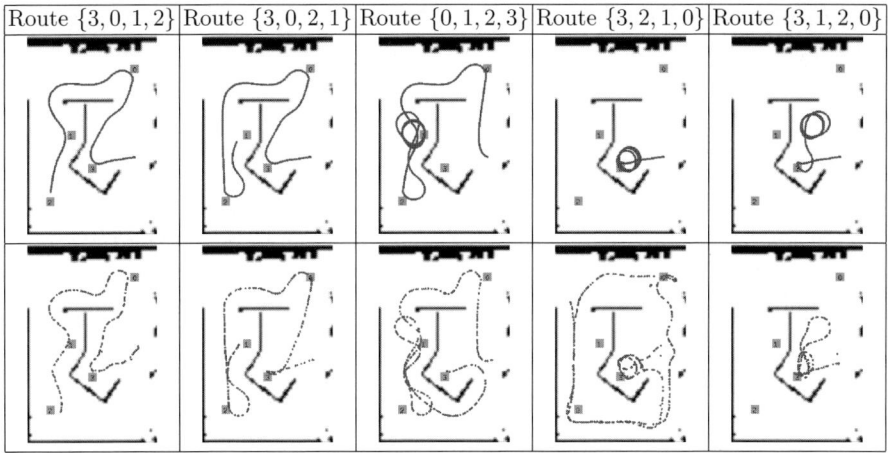

Fig. 7. Simulation (top row) vs. Reality (bottom row)

threaded imagination. Figure 7 (top-row) shows the simulated robot path in 5 of the 24 possible routes.

Results. Route $\{3,0,1,2\}$ 7 (top-row, left-most plot) was the first to finish and was selected for execution by the real robot. The path taken by the real-world robot (as estimated by the AMCL localization system) is shown directly below, in Figure 7(bottom-row, left-most plot). The real robot path is qualitatively similar to the simulation path, and completes the task successfully.

In order to examine how closely the behvaiour of the Stage-simulated robots predicts the real-world behaviour, we run four other possible routes on the real-world R_0. Compare the paths taken by the simulated robots in the top row of Figure 7 with their real-world executions in the bottom row.

Route $\{3,0,1,2\}$ and Route $\{3,0,2,1\}$ are qualitatively similar in simulation and reality. Unfortunately there is no standard metric for quantifying the similarity of robot trajectories. Route $\{3,1,2,0\}$ did not finish but shows the same behaviour in simulation and reality. Routes $\{0,1,2,3\}$ and $\{3,2,1,0\}$ finished in reality but stayed in a cycle during simulation. Though the eventual outcome in simulation and reality was different for these routes, they showed similar dynamics in that they spent most of their time stuck in a VFH trap. Note that because neither of these routes was the winner, the divergence of simulated and real behaviour did not effect the performance of the real-world robot.

3   Future Work

In the short term we plan to create more complex experiments that explore the possibilites of this approach. Some ideas include:

- **Optimized simulations:** We can explore the use of heuristics and the reuse of previous simulatons to speed up exploration of imagined worlds.

- **On-line world model aquisition:** Requiring a complete world model for simulation is a serious constraint. We seek to aquire world models on-line and simulate using the latest models.
- **Dynamic worlds:** The worlds in this paper have been static apart from the motion of the single robot. We seek to include models of other agents that can effect the environment and behaviour of our robot.
- **Concurrent imagination:** While acting in the real world, a robot can still imagine alternative scenarios, to anticipate its reaction to unexpected events.
- **Recursive imagination:** Currently only R_0 can spawn simulations. There may be utility in allowing simulated robots to spawn their own simulations, creating a tree of imaginary robots, rooted at R_0.
- **Multiple objective optimization:** In this paper we chose a simple task with a single objective function. A natural progression of this problem is to allow for multiple goals.

4   Conclusion

We have described a novel framework loosely analagous to imagination, in which agents can use sensorimotor self-simulation to reason about their ability to perform tasks, despite having no model of most of their internal processes and thus no way to create an *a priori* state-evolution model with which to search. This in-the-head rehearsal of tasks is particularly useful when the tasks carry a high risk of agent robot "death", as it provides a source of negative feedback in perfect safety. This approach is a useful complement to existing work using forward models for anticipatory behaviour. A simple but useful implementation was shown to be effective in simulation and real-world experiments, and future directions outlined.

References

1. Brooks, R.A.: Intelligence without representation. Artificial Intelligence Journal **47** (1991) 139–159
2. Jakobi, N., Husbands, P., Harvey, I.: Noise and the reality gap: The use of simulation in evolutionary robotics. Lecture Notes in Computer Science **929** (1995)
3. Zuluaga, M., Vaughan, R.T.: Reducing spatial interference in robot teams by local-investment aggression. IEEE/RSJ International Conference on Intelligent Robots and Systems IROS2005 (2005)
4. Damasio, A.R.: The Feeling of What Happens: Body and Emotion in the Making of Consciousness. Harcourt (1999)
5. Ramachandran, V.: Phantoms in the Brain. HarperCollins (1998)
6. Dawkins, R.: The evolved imagination: Animals as models of their world. Natural History magazine **104** (1995) 8–11,22–23
7. Grush, R.: The emulation theory of representation: Motor control, imagery, and perception. Behavioral and Brian Sciences (2004) 377–442
8. Popper, K.: Alles Leben ist Problemlsen. ber Erkenntnis, Geschichte und Politik. Piper (1996)

9. Holley, J.: First investigations of dream-like cognitive processing using the antic- ipatory classifier system. Technical Report UWELCSG04-002, Clares MHE Ltd, Wells, Somerset UK (2004)

10. Wilson, M.A., Louie, K.: Temporally structured replay of awake hippocampal ensemble activity during rapid eye movement sleep. In: Neuron. Volume 29. (2001) 145–156

11. Hoffmann, H., Moller, R.: Action selection and mental transformation based on a chain of forward models. In: Proceedings of the Eight International Conference on Simulation of Adaptive Behaviour, SAB 2004. (2004)

12. Möller, R.: Perception through anticipation. a behavior-based approach to visual perception. In Riegler, A., Peschl, M., von Stein, A., eds.: Understanding Repre- sentation in the Cognitive Sciences. Plenum Academic / Kluwer Publishers. (1999) 169–176

13. Ziemke, T.: Cybernetics and embodied cognition: On the construction of realities in organisms and robots. Kybernetes **1/2** (2005) 118–128

14. Jirenhed, D.A., Hesslow, G., Ziemke, T.: Exploring internal simulation of percep- tion in mobile robots. Neurocomputing **68** (2005) 85–104

15. Zagal, J.C., del Solar, J.R., Vallejos, P.: Back to reality: Crossing the reality gap in evolutionary robotics. In: IAV 2004 the 5th IFAC Symposium on Intelligent Autonomous Vehicles, Lisbon, Portugal. (2004)

16. Gerkey, B.P., Vaughan, R.T., Stoy, K., Howard, A., Sukhatme, G.S., Matarić, M.J.: Most valuable player: A robot device server for distributed control. In: IEEE/RSJ International Conference on Intelligent Robots and Systems. (2001) 1226–1231

17. Ulrich, I., Borenstein, J.: Vfh+: Local obstacle avoidance with look-ahead verifi- cation. In: IEEE International Conference on Robotics and Automation (ICRA). (2000)

18. Fox, D., Burgard, W., Dellaert, F., Thrun, S.: Monte carlo localization: Efficient position estimation for mobile robots. In: National Conference on Artificial Intel- ligence (AAAI). (1999)

19. Vaughan, R.T., Gerkey, B.P., Howard, A.: On device abstractions for portable, reusable robot code. In: IEEE/RSJ International Conference on Intelligent Robots and Systems, Las Vegas, Nevada, U.S.A (2003) 2121–2427 (Also Technical Report CRES-03-009).

Learning and Adaptation

Stabilising Hebbian Learning with a Third Factor in a Food Retrieval Task

Adedoyin Maria Thompson[1], Bernd Porr[1], and Florentin Wörgötter[2]

[1] Department of Electronics & Electrical Engineering, University of Glasgow,
Glasgow, G12 8LT, Scotland, United Kingdom
{mariat, b.porr}@elec.gla.ac.uk
[2] Bernstein Center of Computational Neuroscience,
University Göttingen, Germany
worgott@chaos.gwdg.de

Abstract. When neurons fire together they wire together. This is Donald Hebb's famous postulate. However, Hebbian learning is inherently unstable because synaptic weights will self amplify themselves: the more a synapse is able to drive a postsynaptic cell the more the synaptic weight will grow. We present a new biologically realistic way how to stabilise synaptic weights by introducing a third factor which switches on or off learning so that self amplification is minimised. The third factor can be identified by the activity of dopaminergic neurons in VTA which fire when a reward has been encountered. This leads to a new interpretation of the dopamine signal which goes beyond the classical prediction error hypothesis. The model is tested by a real world task where a robot has to find "food disks" in an environment.

1 Introduction

Hebbian learning [1] is the most prominent paradigm in correlation based learning: If pre- and postsynaptic activity coincides the weight of the synapse is strengthened. However, Hebbian learning is inherently unstable because of its *autocorrelation* term: Briefly, a changing weight will alter the output which will lead to further weight change, and so on. In this study we present a novel learning rule which is an extension of our differential Hebbian learning [2] rule ISO-learning [3] which minimises the destabilising autocorrelation term by switching learning on when the autocorrelation term is minimal. This switching is performed by a third factor which acts like a neuromodulator [4]. Consequently we call this learning rule ISO3 learning because it is ISO learning with a third factor. We will demonstrate the applicability of the rule with a robot that learns to retrieve "food disks".

2 Three Factor Learning

We are going to demonstrate in the open loop case how to minimise the destabilising autocorrelation term of Hebbian learning. Fig. 1A shows the basic components of the neural circuit. The learner consists of three inputs x_0, x_1 and r which

S. Nolfi et al. (Eds.): SAB 2006, LNAI 4095, pp. 313–322, 2006.
© Springer-Verlag Berlin Heidelberg 2006

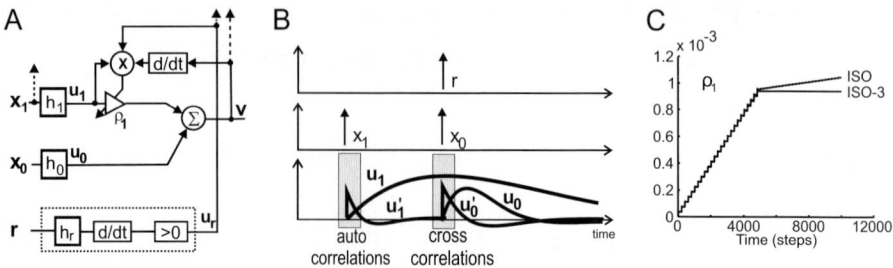

Fig. 1. A) General form of the neural circuit in a generic environment. The inputs x_0, x_1, r are filtered by standard resonators (h_0, h_1, h_r which have frequency f and quality Q as parameters) which smear out an input signal for about $1/f$ samples. u_0 and u_1 are summed at v with weights ρ_0 and ρ_1. The number of filters in the x_1 pathway can be extended to a filterbank with different resonators h_k and corresponding weights ρ_k which is indicated by the dotted lines. From the output of the filter h_r the derivative d/dt is taken and then rectified (> 0). The symbol $\otimes$ is a correlator and $\sum$ is a summation node. B) Signals u_0, u_1 and their derivatives which illustrate how learning works (see text for explanation). C) Comparing ISO and ISO3 learning rules. System parameters: $f_{h_0, h_1, h_r} = 0.1$ and damping parameter $Q = 0.51$ was used to filter inputs x_0, x_1 and relevance signal r. Learning rate was $\mu = 0.005$ for ISO learning rule and $\mu = 0.07$ for ISO3 rule. Time difference between x_1 and x_0 was $T = 10$ (x_1 always precedes x_0).

are filtered by low pass filters: $u_0 = x_0 * h_0$, $u_1 = x_1 * h_1$ and $u_r = \Theta((r * h_r)')$ where Θ is a threshold for > 0 as depicted in Fig. 1. The low pass filters smear out the input signals in time to generate appropriate motor responses. The circuit can easily be extended to a bank of filters with different resonators $h_j, j > 0$ and individual weights $\rho_j, j > 0$ to generate complex shaped responses [5]. The learning rule for the weight change $\frac{d}{dt}\rho_j$ is given as:

$$\rho_j' = \mu u_r u_j v', \; j > 0 \tag{1}$$

which is essentially ISO learning where we have added a third factor u_r.

To get a better understanding how the third factor u_r influences learning we split Eq. 1 into a superposition of a cross-correlation cc_j and an auto-correlation ac_j, multiplied by the third-factor u_r:

$$\rho_j' = \left(\underbrace{\rho_0 u_j u_0'}_{cc_j} + \underbrace{u_j \sum_{k=1}^{N} u_k' \rho_k}_{ac_j} \right) u_r \tag{2}$$

$$= (cc_j + ac_j) u_r \tag{3}$$

The cross-correlation cc_j drives learning by relating *different* inputs with each other so that, for example, in the case of simple conditioning the correlation of the conditioned stimulus (CS) and the unconditioned stimulus (US). Here the unconditioned input is x_0 which is smeared out in time by the filter h_0 and enters Eq. 2 in form of the signal u_0. The conditioned input x_1 enters Eq. 2 via a filterbank $h_j, j > 0$ and generates a number of different temporal traces u_j which are then correlated with u_0. Hence, the signals u_j are cross-correlated with the signal u_0. The autocorrelation term ac_j is the unwanted contribution to learning because it is correlating the conditioned responses (x_1) with themselves which lead to self-amplification of the corresponding weights.

To demonstrate how the third factor stabilises learning we generate artificially input signals x_0, x_1, r to our open loop circuit which are delta pulses (pulses that last for one unit step) that trigger damped filter responses (see Fig. 1B). It can be clearly seen that the autocorrelation ac and cross correlation terms cc happen at *different* moments in time. Consequently we can switch on learning when the autocorrelation is minimal and the cross correlation is maximal. This can be achieved by switching on the third factor u_r at the same time as the signal x_0 is triggered.

Fig. 1C shows the behaviour of ISO3 learning as compared to ISO-learning for a relatively high learning rate. To test the effect of the autocorrelation we switched off the signal x_0 after step 4000 which effectively removes the cross correlation. As shown in [3], at least for low learning rates in ISO-learning, the weights should stabilise after x_0 has been switched off. Instead, clearly one sees that ISO-learning contains an instability, which leads to an upward bend. This is different for ISO3 learning which does not contain this instability because learning is switched off when self amplifying autocorrelation terms would desta-bilise learning. ISO3 learning is also stable when there is a bank of filters in the x_1 pathway and/or when the filter functions are not orthogonal to each other because the autocorrelation is zero at the moment the third factor u_r is triggered.

In summary ISO3 learning uses the fact that auto- and cross correlation happen at different moments in time. Consequently we can stabilise differential Hebbian learning by switching learning on at the moment when the autocorrelation term is minimal.

3   Closed Loop

The behavioural experiments of this section have two purposes: They will give the signals x_0, x_1 and r a behavioural meaning as well as demonstrate the superiority of ISO3 compared to ISO learning. We will present a task where a robot has to learn to retrieve "food disks" [6,7]. This task will first be used for benchmarking and will then be demonstrated in a real robot. The robot has to find "food disks" from the distance. Initially the robot has only a pre-wired reflex which enables it to react to "food disks" at close range only. During learning this reflex reaction is correlated with distant stimuli which enable the robot to target "food disks" from the distance. In the simulation, we use sound and vision for distant and proximal stimuli which respectively replace the artificial input signals x_1, x_0 originally used in our open loop circuit. In the real robot experiment these two signals x_1, x_0 will be generated from two different scanlines from a video camera attached to the robot.

3.1   Benchmark

Fig. 2A,B presents the task and circuit diagram where the simulated robot had to learn to retrieve "food disks". The reflex x_0 is established by two light detectors (LD) which draws the robot into the centre of the "food disks" (Fig. 2A1). Learning uses the sound detectors (SD, Fig. 2A2) which feed into x_1 to generate an anticipatory reaction towards the "food disk". The reflex reaction is established by the *difference* of two light dependent resistors which cause a steering reaction towards the white disk (Fig. 2B). Hence x_0 is equal to zero if both LDs are not stimulated or when they are stimulated *at the same time* which happens during a straight encounter with a disk. The latter situation occurs after successful learning. The reflex has a constant weight ρ_0 which always guarantees stable behaviour. The

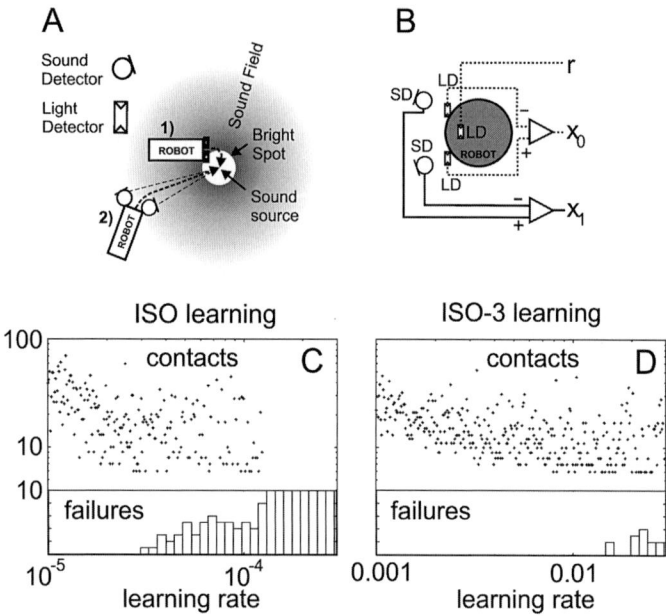

Fig. 2. The robot simulation. A) The robot has two pairs of sensors: It has two light sensors which detect the "food disk" only in their direct proximity. In addition it has two sound detectors which are able to "hear" the food source from a distance. B) The output v is the steering angle of the robot. The two light detectors (LD) establish the reflex reaction (x_0). The sound detectors (SD) establish the predictive loop (x_1). The weights $\rho_1 \ldots \rho_N$ are variable and are changed either by ISO or ISO3 learning. The signal r is generated by a third light sensor and is triggered as soon as the robot enters the "food disk". The robot has also a simple retraction mechanism when it collides with a wall ("retraction") which is not used for learning. The output v is the steering angle of the robot. Filters are set to $f_0 = 0.01$ for the reflex, $f_j = 0.1/j, j = 1 \ldots 5$ for the filter bank where $Q = 0.51$. Reflex gain was $\rho_0 = 0.005$. C) and D) plot the number of contacts for both learning rules needed for successful learning against the learning rate. In addition the number of failures against the learning rate are plotted.

predictive signal x_1 is generated by using two signals coming from the sound detectors (SD). The signal is simply assumed to give the Euclidean distance from the sound source. The difference of the signals from the left and the right microphone is a measure of the azimuth of the sound source to the robot. Successful learning leads to a turning reaction which balances both sound signals and results ideally in a straight trajectory towards the target disk ending in a head-on contact.

We quantify successful and unsuccessful learning for increasing learning rates μ. The learning rates have been chosen in a way that in both cases the contacts for successful learning are the same to make the failures comparable. Learning was considered successful when we received a sequence of five contacts with the disk at a sub-threshold value of $|x_0| < 1.1$. We recorded the actual number of contacts until this criterion was reached. The simulations demonstrate clearly that ISO3 learning is much more stable than the Hebbian ISO learning. ISO3 learning can therefore operate at more than ten times higher learning rates than ISO learning.

3.2 Real Robot

In this section we will demonstrate that ISO3-learning is also able to master the task with the "food-disk" in a physically embodied agent [8]. It will also

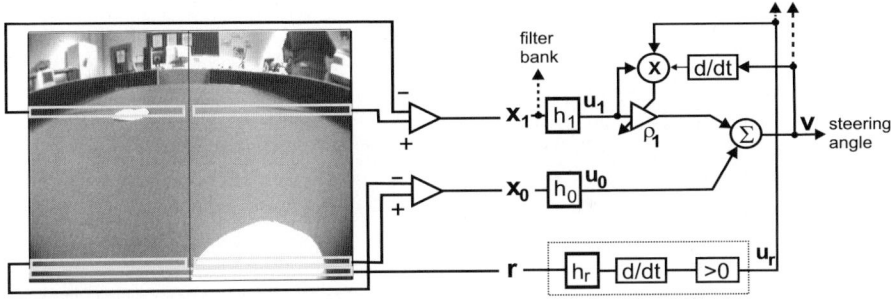

Fig. 3. The real robot's perspective showing two instances where the "food disks", represented by the white spheres lie in all scanlines at which the x_1, x_0 and the relevance input signals are established to produce the input signals for the learning circuit. The x_1 and x_0 signals are respectively triggered when the "food disk" appears in the upper scanline, where objects are further away from the robot's camera view and the lower scanline at bottom of the video image, where objects are closer to the robot's camera view. The relevance signal is obtained from the same scanline as the x_0 signal. When the "food disk" appears in either scanlines for the respective x_1, x_0 and relevance signal, a positive negative or zero value is generated depending on what side of the robots view the "food disks" lie. Parameters: frame rate was 25 frames/sec. The video image $f(x = [0 \ldots 95], y = [0 \ldots 64])$ was evaluated at $y = 53$ for the reflex x_0 and at $y = 24$ for the predictive signal x_1. Reflex and predictive signal were calculated as a thresholded (> 240) weighted sum: $x_{0,1} = \sum_{x=0}^{95} (x - 96/2)^2 \Theta(f(x, y))$. The reflex pathway was set to: $f_0 = 0.01, Q = 0.51$ with a reflex gain of $\rho_0 = 30$. The relevance filter was set to $f_r = 0.01, Q = 0.51$. The predictive filters were set to $f_1, k = 0.1/k, k \ldots 10, Q = 1$. The learning rate was $\mu = 0.0000035$.

be shown that ISO learning fails here completely because of its destabilising autocorrelation terms which drive the weights either very quickly to infinity or, alternatively, one has to run the robot for hours to see anticipatory behaviour which is impractical.

As before, the task of the robot is to target a white disk or "food disks" from a distance. As in the simulation the robot has a reflex reaction which pulls the robot into the white disk just at the moment the robot drives over the disk (Fig. 3). This reflex reaction is achieved by analysing the bottom scanline of a camera with a fisheye lens mounted on the robot. The predictive pathway is created in a similar way: A scanline which views the arena at a greater distance from the robot (hence "in its future") is fed into a bank of of ten filters. This enables the robot to learn to drive *towards* the "food disk" (Fig. 3).

The reflex behaviour of the robot before learning is shown in Fig. 4A, where the robot drives in a straight line and only makes a sharp bend when it encounters the "food disk" in very close proximity. i.e. when the "food disks" appears in the scanline that represents objects closest to the robot. Learning needs longer in these real robot runs than in the simulation. After about 5 minutes, the robot

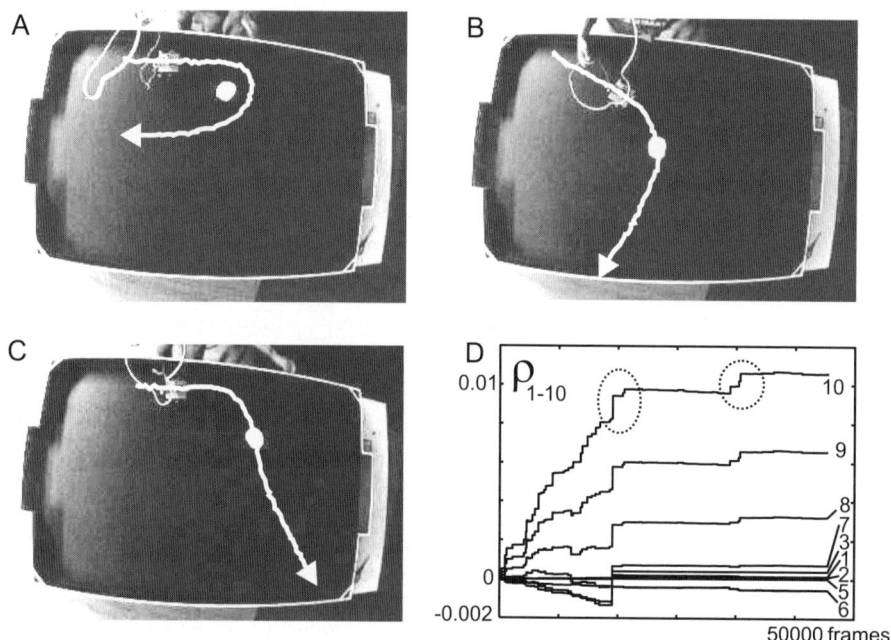

Fig. 4. Experiment with a real robot. A: start of the run at 00:12 mins, B: after 16:13 mins (92 contacts) weight change at a time step of approximately 24000, and C:after 24:10 mins (132 contacts) and the weight change at an approximate time step of 37000. The arrows at A and B show the trace of the robot while driving into "food disks" (white spheres). The weight development ($\rho_j, j = 1 \ldots 10$) is shown in D. The film can be viewed at http://www.berndporr.me.uk/films.

starts exhibiting a learned behaviour. Successful learning can be shown in Fig. 4B and C where the robot's turning reaction sets in from a distance of about 40cm. The robot has learned anticipatory behaviour.

The real robot is subject to complications which do not exist in the simulation. The inertia of the robot, imperfections of the motors and noise from the camera render learning more difficult than in the simulation. These elements contribute to the fluctuations in the weight change in Fig. 4D. The weights however remain stable. The two slightly large "jumps" (marked by circles) in the weight change between time steps $18000 \ldots 20000$ and $40000 \ldots 45000$ have been caused by typical problems which arise in real robots which have been mainly reflections on the floor and also the erroneous detection of the hand of the operator which caused weight changes. However, learning does not diverge and further learning makes the weights decrease again which points to the fact that the reflex reaction kicks in and corrects the slightly too strong steering reactions.

In order to fully appreciate the overall effects of the third factor, we have ran a real robot experiment implementing ISO learning without the third-factor by setting $u_r = 1$ all the time in Eq. 1. The learning rate has been reduced so that the weight development under ISO learning is comparable with ISO3 (see Fig. 4D). The weight change generated from this experiment is shown in Fig. 5. It can clearly be seen that ISO learning becomes unstable very quickly. Only after 2500 frames the weights diverge which leads to random behaviour of the robot so that the experiment was aborted.

In summary it can be concluded that ISO3 learning is much more stable than ISO learning: While ISO3 learning learns fast and remains stable, ISO learning

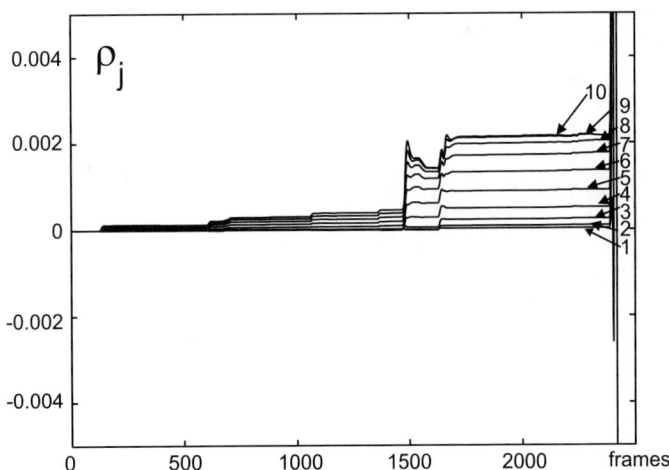

Fig. 5. The weight development of the real robot experiment implementing ISO learning. Parameters: The reflex pathway was set to: $f_0 = 0.01, Q = 0.51$ with a reflex gain of $\rho_0 = 30$. The predictive filters were set to $f_k, k = 0.1/k, k = 1 \ldots 10, Q = 1$ and the learning rate was $\mu = 0.0000035$.

diverges very quickly. This shows that the elimination of the autocorrelation term in ISO3 creates fast and reliable learning.

4 Discussion

In this work we have shown that a third factor is able to stabilise differential Hebbian learning by switching it on when its autocorrelation term is minimal.

Our ISO3-learning rule seems to have similarities with reinforcement learning (RL) which also employs a modulatory signal to select actions [9,10]. However, there are important differences. RL is usually implemented as an actor/critic architecture where the critic generates a delta error which tells the actor *what* to do. In other words the delta error is a teaching signal which actively reinforces or penalises actions. However, in ISO3 the signal u_r does not evaluate actions. ISO3 just switches learning on or off but does not force the system towards a certain behaviour. This is an important difference between our ISO3 and RL: The latter uses a global error signal to drive learning which tells the actor *what* to learn whereas our ISO3 tells the actor *when* to learn and leaves the "what" aspect to the actor itself. Learning of the actor in ISO3 is related to spike timing dependent plasticity [11,12].

Dopamine as a crucial factor for long term potentiation (LTP) has been suggested, for example, in [13,4] and been reviewed in [14,15]. Evidence suggests that LTP not only needs coinciding pre- and postsynaptic activity [11,16] but also dopamine transients as a third factor. Without dopamine no long term potentiation seems to be possible [4]. The third factor of ISO3 can be related to the dopaminergic neurons in the VTA (see Fig. 6) which respond strongly to primary rewards [17]. The VTA in turn is driven by the lateral hypothalamus (LH) which is the primary nucleus which becomes active while eating food. The circuit of LH and VTA could have the task to switch on learning in a number of brain areas like the prefrontal cortex, the hippocampus and the nucleus accum-

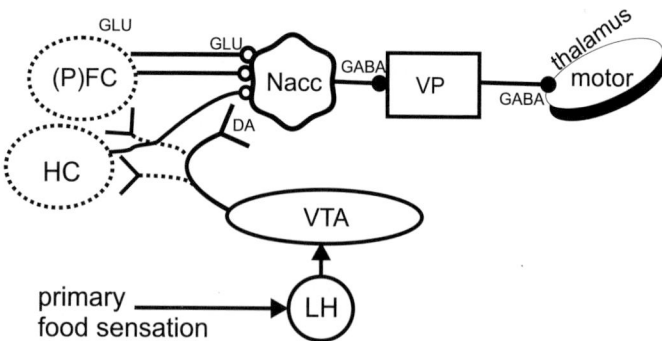

Fig. 6. Simplified diagram of the limbic system. NAcc=Nucleus Accumbens core, HC=Hippocampus, PFC=prefrontal cortex, VP=ventral pallidum, VTA=ventral tegmental area, LH=lateral hypothalamus.

bens which could act as a global signal for learning. In terms of behaviour the nucleus accumbens plays here a central role because it transforms information from the cortex and the hippocampus into motor commands. In our model the learner in Fig 1 can be directly associated with the NAcc: Initially the NAcc is pre-wired with certain behaviours which are then modified and superseded by learned inputs from the cortex and hippocampus. Thus, learning takes place on top of pre-wired behaviours. Consequently, models like the one developed by Prescott et al. [18] which work with pre-wired behaviour could be upgraded to accommodate learning so that anticipatory behaviour is generated. For the actual learning this means that the dopamine signal does not choose the actions in the striatum but that it rather tells it to learn at a certain moment in time. The striatal neurons would learn locally by themselves with the help of spike timing dependent plasticity and not by a dopaminergic error signal [19]

It is known that dopaminergic activity decreases at the primary reward and builds up at the location of the conditioned stimulus [17]. This behaviour can be re-interpreted if we accept that dopamine is telling the target structure when to learn rather than what to learn: it helps to stabilise behaviour associated with the primary reward because learning is switched off when the signal u_r is no longer happening at the moment the primary reward is experienced. Switching on learning at the first conditioned stimulus preserves the behaviour which is associated with the primary reward.

The applicability of ISO3 learning to other tasks depends on the availability of a third factor. In the food retrieval task it is obvious that the third factor is generated from the contact with food because it is a relevant event. Similarly a relevant event can be found in avoidance tasks, for example collision avoidance. However, the relevance is not tied to the primary trigger, for example food or pain. In more general terms relevance could be derived from novelty which triggers learning when unexpected events have happened which in turn switch on learning to reduce uncertainty.

References

1. Hebb, D.O.: The organization of behavior: A neurophychological study. Wiley-Interscience, New York (1949)
2. Kosco, B.: Differential hebbian learning. In Denker, J.S., ed.: Neural Networks for computing: Snowbird, Utah. Volume 151 of AIP conference proceedings., New York, American Institute of Physics (1986) 277–282
3. Porr, B., Wörgötter, F.: Isotropic Sequence Order learning. Neural Comp. **15** (2003) 831–864
4. Bailey, C.H., Giustetto, M., Huang, Y.Y., Hawkins, R.D., Kandel, E.R.: Is heterosynaptic modulation essential for stabilizing Hebbian plasticity and memory? Nat Rev Neurosci **1**(1) (2000) 11–20
5. Grossberg, S., Schmajuk, N.: Neural dynamics of adaptive timing and temporal discrimination during associative learning. Neural Networks **2** (1989) 79–102
6. Verschure, P.F.M.J., Voegtlin, T., Douglas, R.J.: Environmentally mediated synergy between perception and behaviour in mobile robots. Nature **425** (2003) 620–624

7. Porr, B., Wörgötter, F.: Isotropic sequence order learning in a closed loop behavioural system. Roy. Soc. Phil. Trans. Math., Phys. & Eng. Sciences **361**(1811) (2003) 2225–2244

8. Ziemke, T.: Are robots embodied? In: First international workshop on epigenetic robotics Modeling Cognitive Development in Robotic Systems. Volume 85., Lund (2001)

9. Sutton, R.: Learning to predict by method of temporal differences. Machine Learning **3**(1) (1988) 9–44

10. Sutton, R.S., Barto, A.G.: Reinforcement Learning: An Introduction. 2002 edn. Bradford Books, MIT Press, Cambridge, MA (1998)

11. Markram, H., Lübke, J., Frotscher, M., Sakman, B.: Regulation of synaptic efficacy by coincidence of postsynaptic aps and epsps. Science **275** (1997) 213–215

12. Saudargiene, A., Porr, B., Wörgötter, F.: How the shape of pre- and postsynaptic signals can influence stdp: A biophysical model. Neural Comp. **16** (2004) 595–626

13. Centonze, D., Picconi, B., Gubellini, P., Bernardi, G., Calabresi, P.: Dopaminergic control of synaptic plasticity in the dorsal striatum. Eur J Neurosci **13**(6) (2001) 1071–1077

14. Reynolds, J.N., Wickens, J.R.: Dopamine dependent plasticity of corticostriatal synapses. Neural Networks **15** (2002) 507–521

15. Wörgötter, F., Porr, B.: Temporal sequence learning, prediction and control - a review of different models and their relation to biological mechanisms. Neural Comp **17** (2005) 245–319

16. Zhang, L.I., Tao, H.W., Holt, C.E., Harris, W.A., Poo, M.m.: A critical window for cooperation and competition among developing retinotectal synapses. Nature **395** (1998) 37–44

17. Schultz, W.: Dopamine neurons and their role in reward mechanisms. Curr Opin Neurobiol **7**(2) (1997) 191–197

18. Prescott, T.J., González, F.M.M., Gurney, K., Humpries, M.D., Redgrave, P.: A robot model of the basal ganglia: Behaviour and intrinsic processing. Neural Networks, In Press (2006)

19. Dayan, P., Balleine, B.W.: Reward, motivation and reinforcement learning. Neuron **36** (2002) 285–298

Investigating STDP and LTP in a Spiking Neural Network

Daniel Bush, Andrew Philippides, Phil Husbands, and Michael O'Shea

Centre for Computational Neuroscience and Robotics
University of Sussex, Brighton, BN1 9QG
{daniel.bush, andrewop, philh, m.o-shea}@sussex.ac.uk
www.cogs.susx.ac.uk/ccnr

Abstract. The idea that synaptic plasticity holds the key to the neural basis of learning and memory is now widely accepted in neuroscience. The precise mechanism of changes in synaptic strength has, however, remained elusive. Neurobiological research has led to the postulation of many models of plasticity, and among the most contemporary are spike-timing dependent plasticity (STDP) and long-term potentiation (LTP). The STDP model is based on the observation of single, distinct pairs of pre- and post- synaptic spikes, but it is less clear how it evolves dynamically under the input of long trains of spikes, which characterise normal brain activity. This research explores the emergent properties of a spiking artificial neural network which incorporates both STDP and LTP. Previous findings are replicated in most instances, and some interesting additional observations are made. These highlight the profound influence which initial conditions and synaptic input have on the evolution of synaptic weights.

1 Introduction

The ability of the brain to translate ephemeral experience into enduring memories has long been attributed by neuroscientists to activity-dependent changes in synaptic efficacy. One of the first to suggest a mechanism that could govern this plasticity was Donald Hebb, who hypothesised that 'when an axon of cell A is near enough to excite a cell B, and repeatedly or persistently takes part in firing it, some growth process or metabolic change takes place ... such that A's efficiency as one of the cells firing B, is increased' (Hebb, 1949). This concept of 'Hebbian' learning has become a mainstay of neural theories of memory, but more precise rules of synaptic change have been difficult to elucidate.

It has become clear, however, that there are certain features which are crucial to a successful model of plasticity (Roberts and Bell, 2002 ; Song, Miller and Abbott, 2000 ; van Rossum, Bi and Turrigiano, 2000). It must generate a stable distribution of synaptic weights, and stimulate competition between inputs to a neuron, in order to account for the processes of activity-dependent development and forgetting, and to maximize the capacity for information storage (Miller, 1996). Pure Hebbian learning cannot achieve this, not least because it fails to make any mention of synaptic weakening processes, but also because those inputs which correlate with post-synaptic firing are repeatedly strengthened, thus growing to infinitely high values. This creates an inherently unstable, bimodal distribution of synaptic weights. Earlier plasticity models

S. Nolfi et al. (Eds.): SAB 2006, LNAI 4095, pp. 323–334, 2006.

have had to resort to a variety of means in order to solve this problem. Often these promoted competition through the use of global signalling mechanisms, such as limiting the sum of strengths of pre-synaptic inputs to a cell, but the biophysical realism of such protocols can be questioned. The exact nature of the additional constraints used can also strongly influence the behaviour of the model (Miller and McKay, 1994).

In considering the neural basis of memory, it is long-lasting alterations in synaptic strength that are of most interest. Experimental evidence for such changes was first found in the hippocampus – a region of the brain long identified with learning – when it was shown that repeated activation of excitatory synapses by high frequency spike trains caused an increase in synaptic strength which lasted for hours, or even days (Lomo and Bliss, 1973). This phenomena - known as long-term potentiation (LTP) - has since been the subject of a great deal of investigation, because it exhibits several features which make it an attractive candidate as a neural learning mechanism (see Malenka and Nicol, 1999, for a review). It is synapse specific, vastly increasing the potential storage capacity of individual neurons. It is also associative, in that the repeated stimulation of one set of synapses can simultaneously facilitate LTP at adjacent sets of synapses. This has often been viewed as analogous to the process of classical conditioning.

The wealth of research into LTP has helped to inform and inspire new plasticity models which are more easily reconcilable with the tenets outlined earlier. The 'BCM' model, named after its creators (Bienenstock, Cooper and Munro, 1982) and based on their consideration of input selectivity in the visual cortex, is a good example. It is Hebbian, but achieves stability through the existence of a 'threshold' firing rate, a crossover point between depression and potentiation which is itself slowly modulated by post-synaptic activity. This makes the strengthening of a synapse more likely when average activity is low, and vice versa, thus generating competition between inputs.

Another contemporary plasticity model, based on the more straightforward empirical observation of distinct pairs of pre- and post- synaptic action potentials (Roberts and Bell, 2002 ; Bi and Poo, 1998), has also generated a great deal of interest. It is known as spike timing dependent plasticity (STDP), because it dictates that the direction and degree of changes in synaptic efficacy are determined by the relative timing of pre- and post- synaptic spiking. Only pre-synaptic spikes which provoke post-synaptic firing within a short temporal window potentiate a synapse, while those which arrive after post-synaptic firing cause depression. Those inputs with shorter latencies or strong mutual correlations are thus favoured, at the expense of others.

The most pertinent feature of STDP is that it implicitly generates competition between synapses, and experiments with artificial neural networks (ANNs) have shown that this precipitates inherently stable weight distributions. The shape of the resulting distribution is dependent on the exact nature of the STDP implementation, and the values of parameters used. Some researchers, for example, include the experimental observation that stronger synapses seem to undergo relatively less potentiation than weaker synapses, or an activity dependent scaling mechanism such as that outlined by the BCM model (van Rossum, Bi and Turrigiano, 2000). These features help to generate a weight distribution that more closely resembles the stable, unimodal, and positively skewed distribution found *in vivo* (see *fig 1*). Their omission tends to produce a

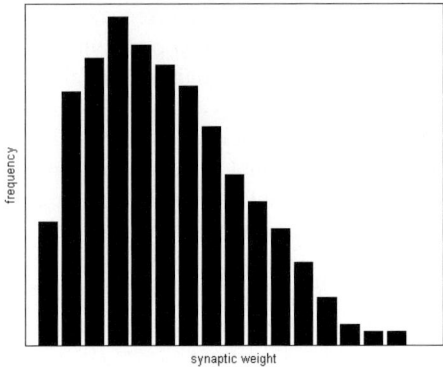

Fig. 1. Synaptic weight distribution found in vivo, taken from Bekkers et al., 1990

bimodal distribution (Song, Miller and Abbott, 2000; Iglesias et al. 2005) more similar to that produced by pure Hebbian learning, but stabilised by innate competition and the inclusion of hard limits on the maximum achievable strength of a synapse.

The analysis of STDP is based on isolated pairs of pre- and post- synaptic action potentials, while observations of LTP are mediated by the application of prolonged spike trains more characteristic of normal brain activity. It is not clear how the STDP model causes synaptic weights to develop with such input, which involves many possible spike pairings. We can presume that both forms of plasticity arise from the same underlying biophysical mechanisms, and some recent work has attempted to reconcile both models within a single theoretical framework (Izhikevich and Desai, 2003). By making a few biologically plausible assumptions, this research has demonstrated that the parameters of STDP can be linked directly with the sliding threshold of the BCM model.

This paper explores the emergent properties of an artificial neural network which implements spike timing dependent plasticity. The form of STDP used is compatible with the BCM model of long-term potentiation, and thus the value of the threshold firing rate can be directly manipulated. The effects this has on synaptic weight distributions and dynamics are examined. Size-dependent potentiation is also introduced into the model, and results obtained from the input of random uncorrelated or partially correlated spike trains are compared with those generated by the performance of two simple, embodied, sensorimotor tasks. The latter will have temporal patterns that are perhaps more representative of firing regimes found *in vivo*, and which STDP has previously been shown to make use of (Izhikevich, Gally and Edelman, 2004).

2   Methods

2.1   Neural Controller

In the majority of tests, the neural network consists of *20* neurons, which are divided into *9* sensory, *9* intermediate and *2* motor neurons. During the phototaxis task, however, the network has only *2* sensory neurons, and thus a total of *13* neurons. The network is realistic of the mammalian cortex in that these are *80%* excitatory and *20%*

inhibitory, and that each has a randomly chosen axonal delay in the range *[1ms, 20ms]*. Each neuron has *5* randomly assigned post-synaptic connections. Motor neurons have no post-synaptic connections, and sensory neurons have no pre-synaptic connections.

The neurons operate using the Izhikevich (2004) spiking model, which dynamically calculates the membrane potential (v) and a membrane recovery variable (u), based on the values of four dimensionless constants (a,b,c and d) and a dimensionless applied current (I), according to the equations below.

$$v' = 0.04v^2 + 5v + 140 - u + I$$
$$u' = a(bv - u)$$

$$\text{if } v \geq +30\,\text{mV} \quad \text{then} \quad \begin{cases} v \leftarrow c \\ u \leftarrow u + d \end{cases} \tag{1}$$

This model was chosen for two main reasons. Firstly, it uses very few floating point operations, and so is computationally advantageous. Secondly, it can exhibit firing patterns of all known types of cortical neurons, by variation of the parameters a - d. The values used for a standard excitatory neuron are *[0.02,0.2,-65,6]* respectively, and those for an inhibitory neuron are *[0.02,0.25,-65,2]*.

In order to introduce neural noise into the system, one neuron is selected at random each time step, and a small current applied to it. A value of *I=10* was used in most tests, although this was varied to assess the effects of neural noise. When distributed randomly over *20* neurons, an applied current of this size produces a spiking rate of approximately *3Hz* per neuron.

2.2 STDP

Mathematically, with $s = t_{post} - t_{pre}$ being the time difference between pre- and post-synaptic spiking, the change in the weight of a synapse (Δw) due to spike timing dependent plasticity can be expressed as:-

$$\Delta w = \begin{cases} A^+ \exp(-s/\tau^+) & \text{if } s > 0 \\ A^- \exp(s/\tau^-) & \text{if } s < 0 \end{cases} \tag{2}$$

The method of implementing this plasticity is outlined by Song et al (2000) and Di Paolo (2003). Two recording functions (P_+ and P_-) are kept for each synapse. These values decay exponentially according to the time constants of potentiation and depression, except when pre-synaptic spikes arrive or post-synaptic spikes are fired, in which case the values are reset to A_+ or A_- respectively. This means that only those spikes which are temporally adjacent affect the degree of synaptic weight change, and hence this is known as the 'nearest neighbour' model of STDP. Research has shown that this implementation allows the reconciliation of the BCM model with STDP (Izhikevich and Desai, 2003). It also outlines a formula for the calculation of the threshold firing rate, which is given by eqn. 3 below. The expressions $A_+ > |A_-|$ and $|A_- \tau_-| > |A_+ \tau_+|$ must be satisfied during experiments, to ensure that the threshold has a positive value at all times.

$$v = -\frac{A_+ / \tau_- + A_- / \tau_+}{A_+ + A_-} \tag{3}$$

Previous research (Bi and Poo, 1998) has shown that an inverse exponential relationship between the level of potentiation and initial synaptic weight may also exist *in vivo*. The modified formula which governs increases in synaptic weight when such 'size dependent potentiation' is examined is given in eqn. 4. It should be noted that there is no evidence for any such size-dependent effects in synaptic weakening.

$$w_{ij}(t) = w_{ij}(t) + P_+ e^{-k w_{ij}} \tag{4}$$

2.3 Tasks

The network was first examined using uncorrelated Poissonian spike trains of varying frequencies as input. In later experiments, correlated spike trains and two simple robotics tasks were used to assess how temporal patterns and more widely varying spike frequencies may affect the behaviour of the network. The tasks chosen were a simple phototaxis exercise similar to that used by Di Paolo (2003), and a falling block task which has been employed previously by Goldenberg et al (2004).

The correlated input was generated by creating a set number of Poissonian spike trains of a certain frequency, and distributing these amongst the 9 sensory neurons. Each time step, the spike trains were re-distributed amongst the inputs. The number of trains that exist thus determine the 'strength' of the correlation between inputs.

In the 'falling block' task, an agent of radius 15 moves horizontally in an arena which is 400 units wide. The agent has 9 sensory neurons with a range of 205 units, which are distributed evenly over a visual angle of $\pi/6$. These sensory neurons each have a randomly determined bias in the range [0.6:1.0] which is used to scale an applied current, relative to the distance of any object in their direct line of vision.

Two blocks of radius 13 fall from a height of 198 at randomly assigned angles and from randomly assigned horizontal start positions, constrained only by the criteria that it must be possible for the agent to catch them both. The first object has a random velocity in the range [0.03:0.04] and the second object in the range [0.01:0.02]. The agent's horizontal velocity is determined by the sum of the two opposing motors outputs, its maximum velocity being set at 0.05 units/ms. The two motor neurons are leaky integrators, operating according to eqn 5 below, where t° is the time at which a spike was last received. Each has a randomly assigned gain in the range [0.01:0.05] and a decay constant (τ) in the range [20ms : 40ms].

$$v = v e^{-(t-t^\circ)/\tau} \tag{5}$$

In the phototaxis task, an agent of radius 2 is placed at a random angle of orientation and randomly determined distance in the range [60 : 80] from a light source, in an arena of unlimited size. The agent has two sensory neurons, which are connected to light sensors separated by an angle of $2\pi/3$ on the agents body, plus or minus a random displacement of $\pi/36$. These light sensors have an angle of acceptance of π, and a randomly assigned bias in the range [10:50], which is used to scale the intensity of any light into an applied current. The intensity of the light source is assigned randomly in the range *[3000:5000]*. Two motors are placed diametrically opposite on the agents body, and driven in a forwards-only direction by the two motor neurons. The speed of each motor is limited at *0.5 units/ms*, and thus this is also the maximum achievable forwards velocity of the agent. As before, the motor neurons are leaky integrators with gains in the range *[0.01:0.5]* and decay constants in the range *[40ms:100ms]*. It is important to note that the capacity of the network to learn how to perform this task is not being tested in this paper. The embodiment is needed only to provide realistic sensorimotor input, which has correlated temporal properties that are considered important in assessing the properties of the plasticity model and network.

2.4 Stability

After each 100ms of experimental time, a histogram of synaptic weights is generated. If the values in each bin (which are of size 1) do not vary by more than ±1 for 10 of the 100ms steps (i.e. 1 second), then the network is considered to have achieved a stable synaptic weight distribution. In order to test that this criteria was adequate, 30 tests were performed, with random initial conditions and parameter values, and network operation was continued for 100 seconds of simulated time after stability was flagged. In all cases, no further discernible change in the synaptic weight distribution occurred. In each experiment, 30 random incarnations of the neural network are created, and each one is run twice, in each case until stability is achieved. Thus, the results presented in this paper are a conglomeration of 60 individual tests, or a total of 5400 synaptic weights (3300 in the case of the phototaxis task).

3 Results and Discussion

3.1 Manipulation of Threshold Firing Rate

Figure 2 represents a typical synaptic weight distribution generated when the network was operated with purely uncorrelated input at a rate of *30Hz*, and the results replicate previous research findings (Song et al., 2000). The values of STDP parameters used in this case correspond to a threshold firing rate of approximately $v=17Hz$. The effects of moving the threshold firing rate (by varying any of the four main STDP parameters) are intuitive, and demonstrated by figures 3 ($v=350Hz$) and 4 ($v=6.25\ Hz$). A higher threshold for long-term potentiation allows fewer synapses to reach the maximum possible strength, and a lower threshold has the reverse effect.

However, results suggest that the relationship between weight distribution and STDP parameters is dictated by more complex factors than simply the position of the BCM threshold. Figure 5 shows a weight distribution for an identical threshold firing rate as 2 ($v=17\ Hz$), but with different STDP values. The number of synapses which have been potentiated to saturation are fewer, and those which have been persistently depressed are larger, in frequency. The value of $|\ A_+\ \tau_+\ |$ is identical in both cases, but the longer temporal window for potentiation that existed in fig. 5 clearly had a lower overall strengthening effect on weight values, compared with the higher degree of synaptic strengthening per spike which was present in the results for 2. Further investigation demonstrated that the ratio of $A_+ : A_-$ is particularly important in determining the shape of the stable weight distribution. Results generated with identical values of this ratio are consistently very similar, more so than results with equal values of the modification threshold.

3.2 Firing Rates

It is clear that the key to a good plasticity model, and one of the reasons why STDP is so highly regarded, is that it regulates network output in the face of wide fluctuations in input. In previous research (Song, Miller and Abbott, 2000) an increase in input firing rate has been observed to cause a decrease in the number of synapses saturating at the uppermost weight values, a finding that was replicated in these experiments. One may expect that fewer strong synapses would correlate with lower post-synaptic activity, but previous work has shown that the STDP model actually exhibits a 'damping' effect – increasing the input firing rate precipitating a much smaller increase in post-synaptic firing rate. An analysis of firing rates in the intermediate, excitatory neurons during this investigation, however, led to a finding which, at least to some extent, contradicts this previous research (Song, Miller and Abbott, 2000). Figure 6 illustrates the correlation between input and intermediate firing rates for four different sets of STDP parameters.

The data demonstrates that, if any relationship exists between these two variables, then it is very complex, and could depend on many factors. As the figure shows, in some cases there seems to be an inverse relationship between the two firing rates, while in others previous research has been replicated and a simple damping effect can be seen. Once again, it seems that the ratio of $A_+ : A_-$ has a pronounced effect on post-synaptic firing rates. In the data presented here, similar values of this ratio do seem to produce similar relationships between input and intermediate firing rates. Further investigation will be required to elucidate the nature of this relationship.

3.3 Varying Network Input

It is useful to make a comparison between the weight distributions arising from uncorrelated input, correlated input, and those generated by input from closed-loop sensorimotor tasks. Figures 7 to 9 illustrate these results – in each case, identical parameter values to figure 2 were used, but in each case the stable synaptic weight distributions are markedly different.

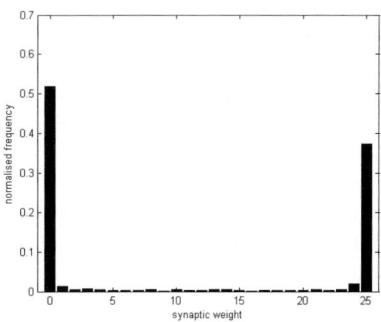
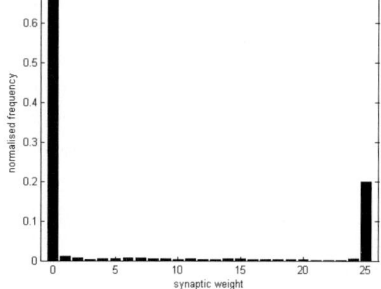

Fig. 2. A_+=0.16; A_-=-0.1; τ_+ = 20ms; τ_- = 40ms

Fig. 3. A_+=0.12; A_-=-0.; τ_+ = 10ms; τ_- = 40ms

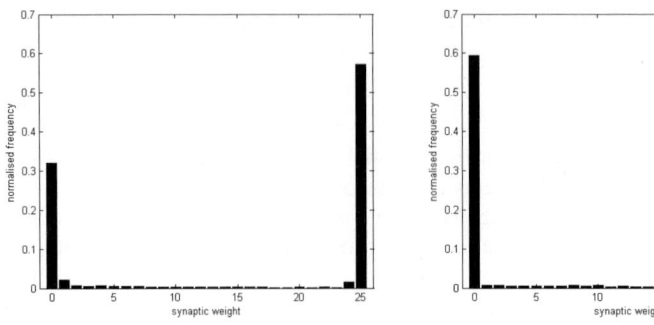

Fig. 4. A$_+$=0.18; A$_-$=-0.; τ$_+$ = 20ms; τ$_-$ = 40ms

Fig. 5. A$_+$=0.12; A$_-$=-0.1; τ$_+$ = 30ms; τ$_-$=40ms

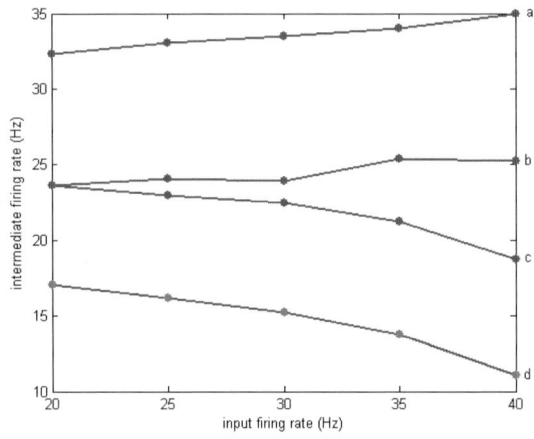

Fig. 6. – The relationship between input and intermediate firing rates

a - A$_+$=0.2; A$_-$=-0.1; τ$_+$ = 10ms; τ$_-$=40ms
b - A$_+$=0.2; A$_-$=-0.1; τ$_+$ = 20ms; τ$_-$=40ms
c - A$_+$=0.2; A$_-$=-0.15; τ$_+$ = 20ms; τ$_-$=40ms
d - A$_+$=0.2; A$_-$=-0.15; τ$_+$ = 10ms; τ$_-$=40ms

It seems that input in the sensorimotor tasks caused more synapses to adopt intermediate weight values, rather than be pushed to the bounds. In the phototaxis task there are also a much greater frequency of synapses at maximum strength and fewer at zero weight, in contrast to the falling block task. Other results showed that distributions generated by input from the robotics tasks are generally much more consistent in shape. The effects of manipulating the threshold rate can still be seen, but rather than simply altering the size of the bimodal peaks (as seen in figures 2 – 4), it is the frequency and distribution of the intermediate strength synapses that are most affected. Correlated input also produced a markedly different weight distribution, with much more similarly sized modal peaks and a more uniform intermediate distribution. As with uncorrelated input, the intermediate weight values are more sparsely populated.

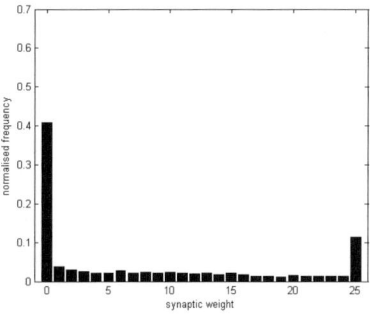

Fig. 7. The falling block task

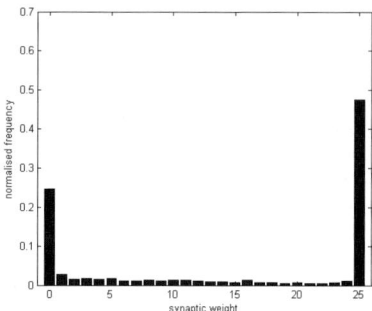

Fig. 8. The phototaxis task

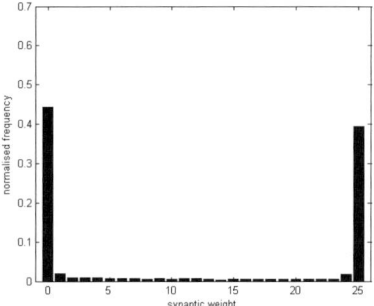

Fig. 9. Correlated input

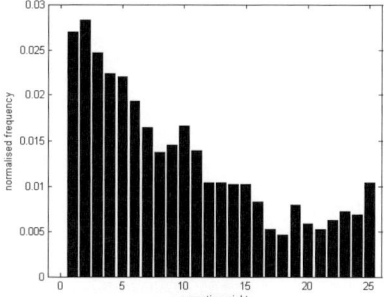

Fig. 10. Size dependent potentiation

The variations in the size and shape of the distributions seen can be loosely explained by the slight differences in the nature of the input presented to the neural network. However, the main issue is that these discrepancies support the intuitive hypothesis that the nature of input to an ANN has a pronounced effect on the evolution of synaptic weights in that network. Much of the previous research in this area has made exclusive use of uncorrelated input, but results found here show that care must be taken in generalising from these findings. The development of a network directed by any plasticity model is at least partially defined by the nature of the input it receives – and there are gross differences between uncorrelated and more realistic sensorimotor input.

3.4 Size-Dependent Potentiation

The introduction of size-dependent potentiation into the plasticity model also has a pronounced effect on synaptic weight distributions. Figure 10 (which was generated using a value of $k=50$ and identical parameter values to fig. 1) illustrates this, and more closely resembles results found *in vivo*. The peak at $w=0$ has been omitted, as these 'silent' synapses are not considered (and cannot be detected) when biological appraisals of weight distributions are made. It is interesting to note, however, that the

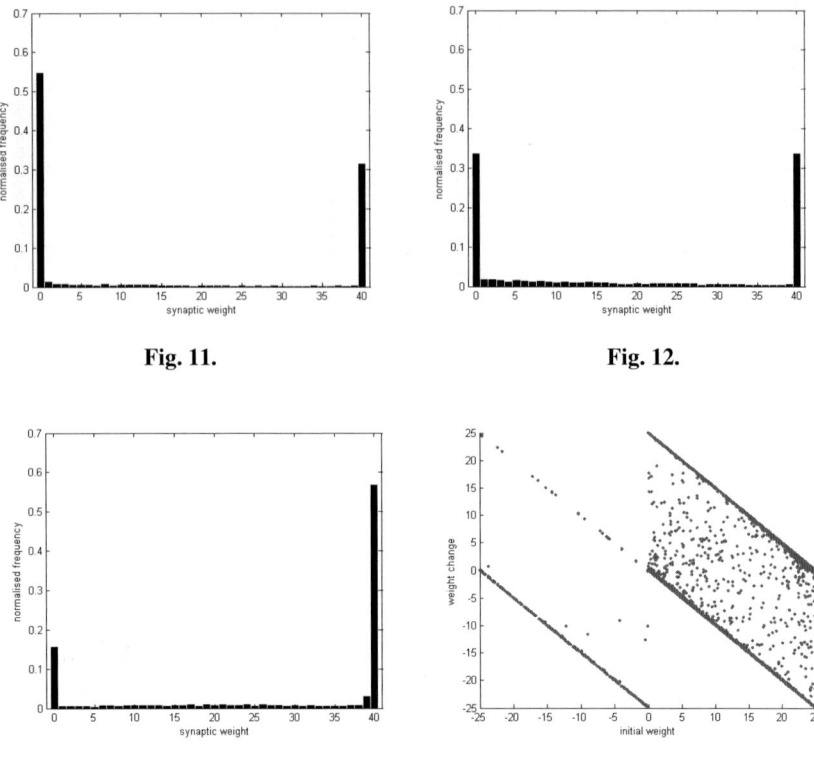

Fig. 11.

Fig. 12.

Fig. 13.

Fig. 14. Changes in synaptic weight

frequency of synapses found at the lower bound was generally consistent between experiments with and without size-dependent potentiation. This implies that the larger number of synapses adopting intermediate weights was simply a product of the fact that fewer synapses were able to saturate to the upper bounds. By tuning the value of k appropriately, the hard limit on synaptic weights can be completely removed, giving a more biophysically realistic plasticity model, which in turn will generate a more biophysically realistic weight distribution.

3.5 Effect of Initial Weight Values

The results obtained also demonstrated that the initial synaptic weight values have some considerable influence on the appearance of the stable weight distribution. The spread of initial weights has little effect on the shape of the distribution, although it does make the network slower to converge to stability. Experiments in which initial weights were uniformly distributed from 0 to w_{max} took the most time to reach stability, while those runs in which all weights were initially set at or near the maximum value were quick to converge. Uniform and Gaussian distributions produce very similar results, as do tests where all synapses begin with the exact same weight. The value of this weight, however, whether it be the point around which initial strengths are (relatively narrowly) distributed, or that which all synapses originate with, does have a considerable effect on the shape of the final distribution. This is illustrated by figures 11 – 13. In these instances, synaptic weights were given a Gaussian distribution around values of 20, 25 and 30 respectively, with a standard deviation of 5. The differences in the shape of the stable weight distributions are clear.

It is interesting to note that the initial weight of a synapse bears no indication as to what its final weight value will be. In more simple Hebbian plasticity models, synapses above a certain strength would immediately be correlated with post-synaptic firing, and thus persistently potentiated, while those below that strength were persistently weakened. This implies that final weights could be predicted based on the initial configuration. With STDP, however, the competition generated between synapses allows no such predictions to be made. Figure 14 illustrates the relationship between the initial and final weights of a synapse. Although there is a slight residual tendency for weights which are originally strong to remain so, and likewise, for those which begin as very weak to remain at zero, the only clear correlation is caused by the hard limits on synaptic weight.

3.6 Conclusions

The results obtained mostly support previous findings in this area. Manipulation of the BCM threshold firing rate directs synaptic weights in an intuitive manner. The STDP model has a strong regulatory effect on post-synaptic output, although in some cases it seems to reduce the intermediate firing rate in the face of an increased frequency of input. The initial conditions of the network underlying the plasticity model, and the nature of input used, seem to have a pronounced effect on the direction in which it develops, which is to be expected from any dynamical system. Although STDP implicitly generates competition between synapses, the weight distribution it creates is still not representative of that found *in vivo,* unless additional experimental observations such as size-dependent potentiation are included. It is left to future work to elaborate on the results presented here, and to assess how beneficial the phenomena identified by this paper are to developing simple learning behaviour in ANNs.

Acknowledgements

The authors would like to thank early reviewers for their input, and for the general help and support of the members of the CCNR at Sussex University.

References

1. Wickliffe Abraham et al. Heterosynaptic metaplasticity in the hippocampus *in vivo*: A BCM-like modifiable threshold for LTP. *PNAS,* 98 (19): 10924-10929, 2001.
2. J.M. Bekkers et al. Origin of variability in quantal size in cultured hippocampal neurons and hippocampal slices. *PNAS,* 87: 5359-5362, 1990.
3. Guo-qiang Bi and Mu-ming Poo. Synaptic modifications in cultured hippocampal neurons : dependence on spike timing, synaptic strength and post-synaptic cell type. *Journal of Neuroscience,* 18: 10464 – 10472, 1998.

4. Elie Bienenstock, Leon Cooper and Paul Munro. Theory for the development of neuron selectivity : Orientation specificity and binocular interaction in the visual cortex. *Journal of Neuroscience,* 2: 32 – 48, 1982.
5. Ezequiel Di Paolo. Evolving spike-timing-dependent plasticity for single-trial learning in robots. *Phil. Trans. R. Soc. London,* 361: 2299-2319, 2003.
6. Eldan Goldenberg, Jacob Garcowski and Randall Beer. May we have your attention : Analysis of a selective attention task. *Proc. Eighth Int. Conf. Sim. Adap. Behaviour,* 49-56, 2004.
7. Donald Hebb. The Organisation of Behaviour: A Neuropsychological theory. *Wiley, New York,* 1949.
8. Javier Iglesias et al.. Stimulus-driven unsupervised synaptic pruning in large neural networks. *Proceedings of BV & AI,* LNCS 3704: 59-68, 2005.
9. Eugene Izhikevich and Niraj Desai. Relating STDP to BCM. *Letters to Neural Computation,* 15: 1511 – 1523, 2003.
10. Eugene Izhikevich. Which model to use for Cortical spiking neurons? *IEEE Transactions on Neural Networks,* 15 (5): 1063 – 1070, 2004.
11. Eugene Izhikevich, Joseph Gally and Gerald Edelman. Spike timing dynamics of neuronal groups. *Cerebral Cortex,* 14: 933-944, 2004.
12. T. Lomo and T. Bliss. Long-lasting potentiation of synaptic transmission in the dentate area of the anesthetized rabbit following stimulation of the perforant path. *Journal Physioogy,* 232: 331-341, 1973.
13. Robert Malenka and Roger Nicoll. Long-term potentiation – A decade of progress? *Science,* 285: 1870 – 1874, 1999.
14. K.D. Miller and D.J. McKay. The role of constraints in Hebbian learning. *Neural Computation,* 6: 100-126, 1994.
15. K.D. Miller. Synaptic economics: competition and co-operation in synaptic plasticity. *Neuron,* 17: 371-374, 1996.
16. Patrick Roberts and Curtis Bell. Spike timing dependent plasticity in biological systems. *Biological Cybernetics,* 87: 392 – 403, 2002.
17. Sen Song, Kenneth Miller and L.F. Abbott. Competitive Hebbian learning through spike timing dependent synaptic plasticity. *Nature Neuroscience,* 3: 919 – 926 , 2000.
18. M.C.W. van Rossum, G.Q. Bi and G.G. Turrigiano. Stable Hebbian learning from spike timing dependent plasticity. *Journal of Neuroscience,* 20 (23): 8812 – 8821, 2000.
19. M.C.W. van Rossum and G.G. Turrigiano. Correlation based learning from spike timing dependent plasticity. *Neurocomputing,* 38-40: 409-415, 2001.

Spike-Timing Dependent Plasticity Learning for Visual-Based Obstacles Avoidance

Hédi Soula[1] and Guillaume Beslon[2]

[1] LBM/NIDDK, National Institutes of Health, Bethesda, MD, 20892, USA
[2] PRISMA, National Institute of Applied Sciences, 69621 Villeurbanne Cedex, France

Abstract. In this paper, we train a robot to learn online a task of obstacles avoidance. The robot has at its disposal only its visual input from a linear camera in an arena whose walls are composed of random black and white stripes. The robot is controlled by a recurrent spiking neural network (integrate and fire). The learning rule is the spike-time dependent plasticity (STDP) and its counterpart – the so-called anti-STDP. Since the task itself requires some temporal integration, the neural substrate is the network's own dynamics. The behaviors of avoidance we obtain are homogenous and elegant. In addition, we observe the emergence of a neural selectivity to the distance after the learning process.

1 Introduction

From a dynamical systems point of view, a behavior is a spatio-temporally structured relationship between an organism and its environment. The dynamical loops generated by the input/output flow and by the neural system are coupled together to produce a minimal cognition [1]. A non purely reactive architecture is then obtained through neural dynamics. Consequently, learning a behavior is a result from pairing (coupling) the dynamics of input/output loop with the dynamics generated by the artificial brain.

However, even from that point of view, the problem still needs to be solved so far. Many architectures proposed dynamical process for temporal series learning. Most of the time, the learning procedures are off-line and supervised (their performances rely solely on a good foreseeing from the designer). An alternative was found in the genetic algorithm [2,3,4]. Unfortunately, both approaches still lack of on-line adaptation methods and offer no obvious path for enabling them.

These facts will be our starting point. A neural controller must exhibit enough intrinsic features to encompass a wide range of dynamics. Learning will therefore be a plasticity mechanism that allows us to put constraints on particular (and interesting) dynamics depending on the experience of the agent. The mechanism of plasticity is inspired from biology: Spike-Time Dependent Plasticity. Using an over-simplistic scaling, we show that an agent can learn to avoid obstacles using only its visual flow. This approach provides an elegant solution to a non-trivial temporal task. We show also the emergence of a spiking selectivity to distance from obstacles when this distance is not provided as such to the agent.

S. Nolfi et al. (Eds.): SAB 2006, LNAI 4095, pp. 335–345, 2006.

2   Dynamics of Integrate and Fire Neurons

The following series of equations describe the discrete *leaky integrate and fire* (I&F) model we use throughout this paper [5]. Each time a given neuron fires, a synaptic pulse is transmitted to all the other neurons. This firing occurs whenever the neuron potential V crosses a threshold θ from below. Just after the firing occurs, the potential of the neuron is reset to 0. Between a reset and a spike, the dynamics of the potential of a neuron (labelled i) is given by the following (discrete) temporal equation :

$$V_i(t+1) = \gamma V_i(t) + \sum_{\substack{n>0 \\ j=1}}^{N} W_{ij}\delta(t - T_j^n) + I(t) \qquad (1)$$

The first part of the right hand side of the equation describes the leak current $-\gamma$ is the decay rate ($1 - \gamma$ is the leak $- 0 \leq \gamma \leq 1$). $I(t)$ is an external input current (up to some conductance constant). The W_{ij} are the synaptic influences (weights) and $\delta(x) = 1$ whenever $x = 0$ and 0 otherwise (Kronecker symbol). The T_j^n are the times of firing of a neuron j and is a multiple of the sample discretization time. The times of firing are formally defined for all neurons i as $V_i(T_i^n) \geq \theta$ and the n-th firing time recursively as :

$$T_i^n = \inf(t \mid t > T_i^{n-1} + r_i, V_i(t) \geq \theta) \qquad (2)$$

We set $T_i^0 = -\infty$ and r_i is the refractory period of the neuron (which imposes a maximal frequency). Once it has fired, the neuron's potential is reset to zero. Thus, when computing $V_i(T_i^n + 1)$ we set $V_i(T_i^n) = 0$ in equation (1).

Leaky integrate and fire neurons are known to be good approximates of biological neurons concerning spiking time distribution. Moreover, they are simple enough and easy to handle when embedded in a robot.

Populations coding and mean field techniques showed that spiking neural networks can display a broad variety of dynamics [6,7,8]. In simple case, sufficient conditions for phase synchronization and its stability were proposed in homogeneous networks [9,10]. In the precise case of integrate and fire neurons, equilibrium criteria have been calculated for networks of irregular firing neurons [11,12] and VLSI neurons [13,14].

Although not applicable directly to our problem, these works showed that, in case of random networks, the parameters of the distribution of synaptic weights can determine in which case the firing activity is conducted by the input (input or noise drift) or by the internal coupling (internal drift) of the neurons. To quantify this range of parameters, we described analytically the bifurcation map of totally connected networks at the limit of no input drift [15] – the so–called spontaneous mode. This bifurcation map allowed us to estimate rather precise conditions for a purely internal regime to occur.

Consequently, these two types of drifts can impose quite different behaviors to an agent controlled by such networks. Indeed, a network whose dynamics rely solely on the input drift gives an agent with an input-led behavior: A reactive

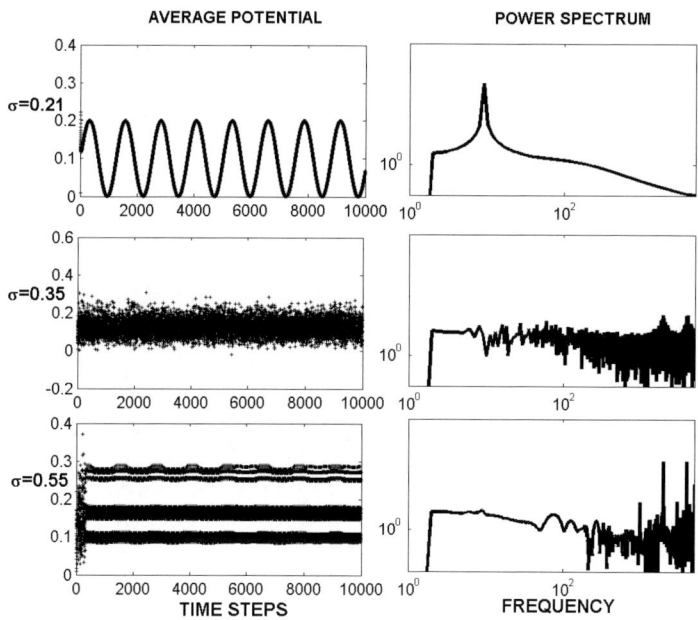

Fig. 1. Differences of behavior when given a sinusoidal input according to the size of the coupling. The figures on the left are the temporal evolution of the average potential (over all the neurons) of the network. On the right, there are the corresponding power spectra in a log-log plot. Parameters: $N = 100$, $\theta = 1.0$, $\gamma = 0.99$, $V_{rest} = V_{reset} = 0.0$, $r = 4$. The input signal is $I(t) = 0.2 \times \sin(t/1200)$ giving a frequency around 8.5 Hz.

architecture. The network serves as a temporally stable filter between sensors and actuators. On the other hand, networks with an internal drifting mode use only its own dynamical properties to compute the output. Then, the agent experiences a stereo-typical ("autistic") behavior. This is called an automatic architecture. In that case, the internal dynamics dominates the flow of input and is able to provide time-dependent responses.

This evolution from external to internal drift-led activity is shown on figure 1. We compute the average potential of an all-to-all coupled network of I&F neurons when all neurons are submitted to a sinusoidal input. The weights are chosen randomly following a centered normal law. In that case, the internal drift increases with the standard deviation of the weight distribution [15].

The network response thus ranges from a passive filter ($\sigma = 0.21$ – carrying the input frequency) to a signal that completely ignores the intrinsic frequencies of the input ($\sigma - 0.55$). In the former case, the output will react on both the spectral and tonic part of the input signal. In the latter case, however, the network acknowledges only the tonic part. The intermediate value of coupling seems to be a combination of both extreme cases. The network output depends critically on both internal and external component. In robot control terms, this kind of coupling allows both adaptivity to input variation and memory of past.

Since we obviously cannot simply increase the weight connectivity, a learning algorithm has therefore the task to make this coupling relevant for the situations experience by the robot.

3   The Learning Rule

We describe in this section the learning algorithm we used as well as the methodology for applying this learning rule.

Recent neurobiology experiments have suggested that the relative timing of pre- and post-synaptic potentials plays an important role in determining the intensity as well as the sign of change of a synapse strength [16,17]. The intensity of this Long Term Potentiation (and Depression) is directly dependent of the relative timing – the spike delay between the post-synaptic and pre-synaptic neurons. In addition, if this delay is high enough (order of tens of milliseconds) no modification occurs. On the other hand, the modification is maximal when the post-synaptic neuron fires just after (or just before) the pre-synaptic does.

As [16] put it, one can extract quite straightforwardly a very simple rule that rests upon inter-spikes delays. This "rule" is known as Spike-Time Dependent Plasticity (STDP). It has become a widespread implementation of Hebb's initial intuition on memory formation in the brain [18]. Therefore many STDP rules emerged from experiments (see [17]), we decide to use a simple one – namely an additive rule expressed as :

$$\Delta W_{ij} = \alpha(W_{max} - |W_{ij}|)h(\Delta_{ij})$$

where Δ_{ij} is the difference between the last firing dates of post-synaptic neuron i and pre-synaptic j. $h(t)$ is a function which depends on the axonal delay between j and i. α is a learning parameter. The "delay" function is chosen as :

$$h(t) = \begin{cases} \frac{T-t}{T} & 0 < t < T \\ 0 & t = 0 \\ 0 & t > T \\ -h(-t) & t < 0 \end{cases}$$

Here T is the time-out constant (i.e. the relative timing above which no modification occurs). Figure 2 shows the learning function $\alpha h(\Delta_{ij})$ (we chose $T = 50$ time steps).

In addition to classical Hebbian learning properties, STDP relies on a precise temporal frame. Neurons trained with STDP act as coincidence detectors and synapses of neurons that fire in a precise temporal manner will be modified in order to reinforce that order. Moreover, STDP introduces competition between Hebbian plasticity. As such, STDP seems a good candidate for temporal learning and dynamical coupling [19].

However, a real application of this rule *per se* does not seem straightforward. In order to see this, let's evaluate the effect of the rule on two neurons and introduce their crosscorrelation function :

$$C_{ij}(\tau) = \frac{1}{\mathcal{T}} \sum_{k=0}^{\mathcal{T}} S_i(k)S_j(k + \tau) \qquad (3)$$

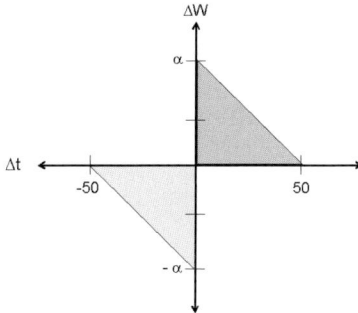

Fig. 2. Schematic description of the learning rule

where $\mathcal{T}$ is a window time and $S_i(t)$ is 1 whenever neuron i fires at time t and zero otherwise. We can then compute the evolution of the weights between the two neurons according to the learning law. It yields :

$$\Delta W_{ij} = \mathcal{T} \sum_{k \geq 1} [h(k)\,(C_{ij}(k) - C_{ij}(-k))] \tag{4}$$

Due to the shape of h, the sum is, in fact, finite and there is no modification for neurons strictly correlated ($C_{ij}(\tau) = 0$ for all $\tau \neq 0$). Moreover, for uncorrelated neurons $\mathbb{E}(C_{ij}(-\tau)) = \mathbb{E}(C_{ij}(-\tau))$ (i.e the correlation forward is equal to the correlation backward), the average modification of weight is zero.

Finally, periodic neurons with a phase ϕ of half the common period ($C_{ij}(\tau) = 0$ for all $\tau \neq \phi$ and $C_{ij}(\phi) = C_{ij}(-\phi)$) will have their weights unmodified. These are the two fixed points of the learning rule. The speed of convergence depends of $|\alpha|$ and the slope of h (that is $\mathcal{T}$). The sign of α determines the stability of the fixed points. For $\alpha > 0$, the strict correlation is stable and the dephasing of half the period is unstable. This is strictly the opposite for $\alpha < 0$ – the so–called "anti–STDP" also observed in real (fish) brains [17].

However, in any case, if an unmodified learning process is maintained for too long, synapses will eventually saturate, leading to a bimodal weight distribution [20]. This situation is illustrated on figure 3–top where the evolution of the average potential for various values of α is displayed .

Therefore, we are able to figure out the effect of the learning rule upon the coupling of the network and consequently on the overall behavior of the robot. Indeed, applying straightforwardly the learning rule with a constant learning parameter α will increase the coupling of the network. In that case, the resulting behavior for the robot will be ultimately a pure "autistic" one, ignoring any change in the input pattern.

This is not wanted. We need to introduce a mechanism that allows a regulation in the weight modification. The idea will consist of using a combination of both STDP and anti–STDP. Since the phase synchronization differs for the laws we expect to maintain the network in the range of intermediate coupling. As displayed on figure 3–bottom–left, we perform a flip experiment where the

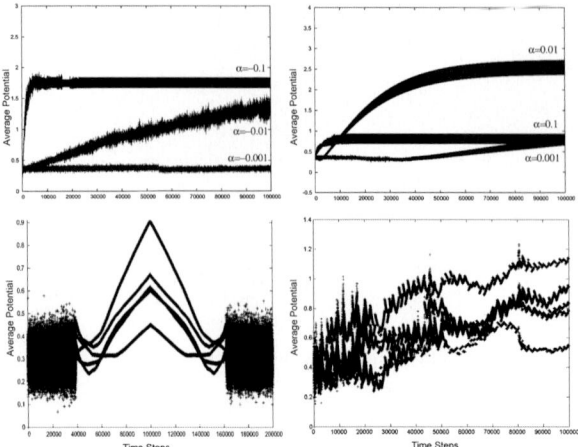

Fig. 3. Top: Evolution of the average potential for $\alpha \in \{-0.1, -0.01, -0.001\}$ (anti–stdp) on the left and $\alpha \in \{0.1, 0.01, 0.001\}$. **Bottom**: Evolution of the average potential using, on the left, first STDP then anti-STDP (flip experiment) and, on the right, switching between the two laws every 1000 time steps (switch experiment) $|\alpha| = 0.01$ in both experiments.

network is submitted first to STDP (for 100 000 time steps) then to anti-STDP (the remaining 100 000 time steps). This flipping of laws allows the network to remain in a safe zone. However, as figure 3–bottom–right shows it, the contribution of each of these two laws is not symmetrical. In the switch experiment (a higher frequency flipping in fact), the network increases its coupling. In other words, both laws couple the network and the STDP does it faster.

Such considerations will compel us to make careful decisions as when to apply both laws.

4   The Experiment

We tested our approach on a task of obstacle avoidance with visual flow. The robot has to avoid walls using only its camera information. More precisely, there will be neither proximity sensors nor positioning device available for it. In addition, the environment will not allow the robot to extract any simple rule to compute its exact position. Obviously in order to accomplish this difficult task, the network must exhibit important internal loops since no static input provides enough information by itself.

The simulated agent is round with a wheel at each side. It has two motors that control each wheel separately, allowing differential propulsion. It is also equipped with a linear camera of 64 pixels spanning 180 degrees (see figure 4–bottom–right). It is positioned in an arena with black and white vertical stripes of random size painted on the walls at irregular intervals (see figure 4–top). It is similar to the environment described in [4]. To simplify, we tested also the learning algorithm in an environment where the stripe sizes were equal.

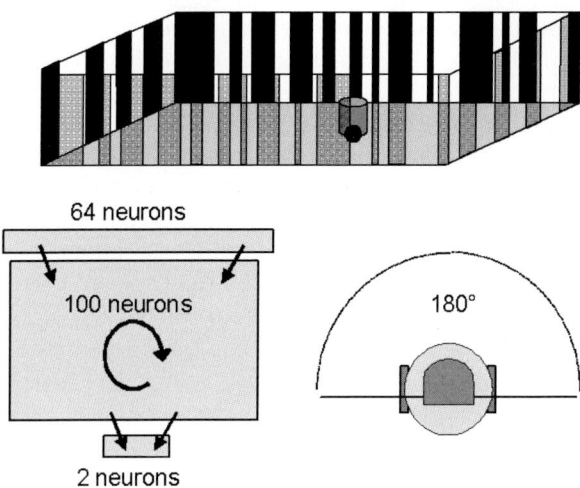

Fig. 4. Top: the environment. **Bottom left:** the neural architecture. **Bottom right:** top view of the robot.

The controller of the robot is a spiking neural network with three layers of neurons. The first and third serve as sensors and motor neurons respectively. More precisely, each of the 64 pixels of the linear camera is associated with a neuron. These neurons are fed with an input current computed to give either 20Hz or 200Hz for white and black pixel respectively. The current value I is calculated to provide the neuron a desired period P. This is done using the formula : $I = \theta \frac{1-\gamma}{1-\gamma^P}$. We recall that γ is the leak (decay rate) and θ the threshold. Both were constant throughout experiments and identical for all neurons (Parameters : $\gamma = 0.99$ and $\theta = 1.0$).

The 2 output neurons serve as motoneurons – one for each motor (left and right). The motor speeds are computed as a linear function of the corresponding neuron firing rate (over 20 time steps) in such a way that, if an output neuron does not fire in this time window, the motor goes rearward.

The intermediate layer – the hidden layer – consists of all–to–all connected 100 neurons. Each input neuron has a connection to each hidden neurons and all hidden neurons project to output neurons. There is no direct connection from input to output layer (see figure 4–bottom–left).

All the weights of the three layers are chosen randomly according to a normal distribution. We chose the distribution parameters in order to obtain, when given an average input (same number of black and white pixels), a balanced proportion of spiking activity coming from the input and from the internal (hidden) activity and an average behavior of near immobility. These parameters are the equivalent for the robot of the intermediate value of coupling described in section 2. (Parameters: $\mu_{input} = 0.0$, $\sigma_{input} = 0.05$, $\mu_{hidden} = 0.0$, $\sigma_{hidden} = 0.09$, $\mu_{output} = 0.04$ and $\sigma_{output} = 0.04$ where μ and σ are the mean and standard deviation of the normal laws).

The robot has two contradictory goals – moving forward while detecting and then avoiding the walls. In order to detect the walls, the robot has to move. It allows us to extract two "physiological" relevant states for the agent: Moving and colliding. We've decided to apply the learning rule only on those two situations. In addition, as put in the previous section, these laws must be antagonistic keeping in mind that one is stronger than the other. Consequently, we've decided that when the robot moves in line, whether forward or rearward (i.e. when both motors speed were equal) we apply anti–STDP. When the robot hit a wall, we apply STDP. Indeed, since hitting a wall is a rarer event for the robot, we apply the strongest law in that case. Note that both laws reinforce the internal drift of the neural controller but with different phases. They are both learning rules. The absolute value of the scaling factor was $|\alpha| = 0.001$.

5   Results

We drew ten random robots and compared the performance with or without learning for 100 000 time steps. The averaged number of collisions is shown in figure 5.

In the learning experiment, the number of collisions experienced decreased indicating an obstacle avoidance behavior. Moreover figures 6–left shows that moving forward is not impeded and is even increasing for regular size stripes. This means that the robots are not still and do not turn around themselves. The obstacle avoidance behavior is then non trivial. Indeed, figure 7 shows the trajectory of a typical individual after the learning process (with regular stripes). One notes that the obstacle avoidance is not one hundred per cent perfect since the agent actually collides with the wall (on the upper right). Nevertheless, the remainder of the run is collision-free. We can notice that the agent actually uses a temporal integration of the visual input to stay away from the walls. The avoidance behavior consists of small rear and forward movements when close to the obstacle. Away from the walls, however, the movement is faster and smoother.

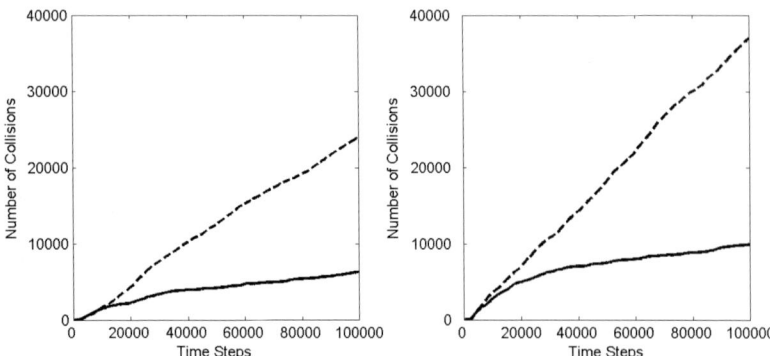

Fig. 5. Evolution of the number of collisions averaged over the ten robots. Experiment without learning is displayed with a dashed line while experiment with learning is displayed with a plain line. **Left:** with regular stripes. **Right:** with random stripes.

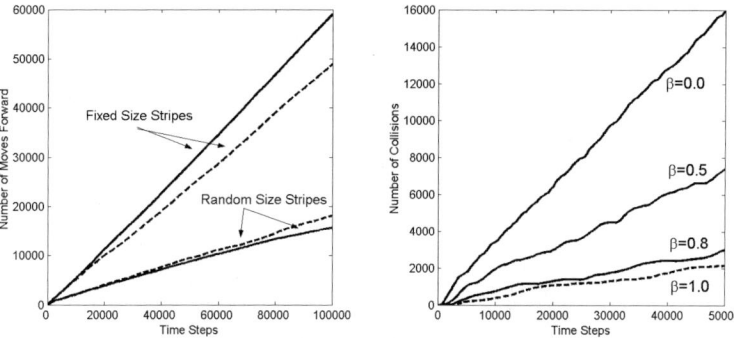

Fig. 6. Left: The average number of time the two motors speed are equal and > 0 for robots without learning (dashed line) and with learning (plain line). **Right:** average number of collisions for various noise value $\beta \in \{1.0, 0.8, 0.5, 0.0\}$.

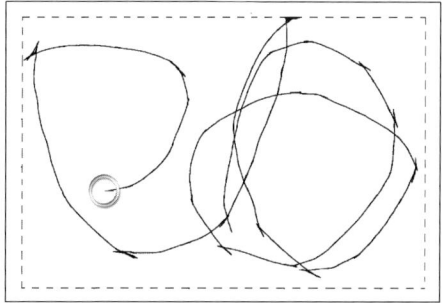

Fig. 7. The trajectory of a typical individual after the learning process. The black rectangle is the arena, the dashed rectangle indicates the radius of the agent.

In order to test the robustness of the learning, we introduce noise in the camera input. When computing the input current, let $I(t) = \beta I_{signal} + (1 - \beta)Z_t$ where β is a noise factor and Z_t a white noise. $\beta = 1$ corresponds to noise-free simulation while when $\beta = 0$ the input is made only of noise. We tested the ten robots after the learning process during 50 000 time steps for $\beta \in \{1.0, 0.8, 0.5, 0.0\}$. The figure 6(right) shows the average performance in terms of collisions. As already mentioned, even in the case $\beta = 1$, there are still some collisions but when $\beta = 0.8$ (that is 20% of the information is noise) the performance remains comparable.

These behaviors were homogeneous and depended only on the statistics of the neural controller's connectivity. They were not, moreover, dominated by either extreme regimes. Consequently, the robots was neither reactive neither automatic.

As a way to assess some properties of the resulting network, we recorded the spiking activity X_t (number of hidden neurons that fire at time t) during the evolution of the robot whose trajectory is shown in figure 7. We kept a trace of this activity by setting : $\bar{x}(t) = (1 - \omega)X_t + \omega\bar{x}(t - 1)$ (with $\omega = 0.99$)

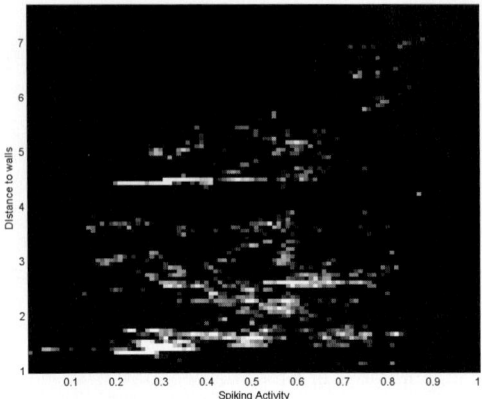

Fig. 8. Distance/Spiking activity heightmap. Distance is expressed as a factor of the agent radius and the spiking activity was scaled between maximum and minimum.

indicating that the higher the value of the trace the higher the past spiking activity. We recorded also the distance to the nearest obstacle and draw the bivariate heightmap of these two variables. This map is shown in figure 8. It shows that the distance is correlated to the spiking activity. It implies that the resulting neural network shows distance selectivity in its overall activity. It is an emergent result from the learning algorithm since we do not provide the robot with the distance and there is no such selectivity for untrained networks.

6   Discussion

In this article, we proposed that competing learning rules for competing dynamics can be a powerful way to a develop neural architecture that learns a temporal task. We are aware that, at first glance, this learning paradigm seems to be ad hoc or rather over tuned. Indeed, empirical use of this Hebbian rule may not be enough to extract more than simple behaviors. Orienting the learning toward an observed behavior corresponding with our wishes is probably a much more complicated task.

Nevertheless, to the best of our knowledge, this is the first scientific work where a spiking neural network learns a navigation task while the robot interacts with the environment. The results of this are homogeneous and depend only on the statistics of the network. Moreover, the collision-free navigation itself indicates non-trivial extraction *via the learning process* of the relevant features of the robot/environment relationship.

Acknowledgements

This work was supported by the ACI DYNN. We would like to thank Carson Chow for his critical comments on the paper.

References

1. I. Harvey, E. Di Paolo, R. Wood, and M. Quinn. Evolutionary robotics: A new scientific tool for studying cognition. *Artificial Life*, 11(1-2):79–98, 2005.
2. H. Soula, G.Beslon, and J. Favrel. Evolving spiking neurons nets to control an animat. In D. Pearson, N. Steel, and R. Albrecht, editors, *Proc. of ICANN-GA*, pages 193–197, Roanne, France, 2003.
3. E. di Paolo. Evolving spike-timing-dependent plasticity for single-trial learning in robots. *Phil. Trans. R. Soc. Lond. A.*, 361:2299–2319, 2003.
4. D. Floreano and C. Mattiusi. Evolution of spiking neural controllers for autonomous vision-based robots. In T. Gomi, editor, *Evolutionnary Robotics*, Berlin, Germany, 2001. Springer-Verlag.
5. H. C. Tuckwell. *Introduction to theoretical neurobiology: Vol.2:Non linear and stochastic theories*. Cambridge University Press, Cambridge, USA, 1988.
6. C. van Vreeswijk and H. Sompolinsky. Chaos in neuronal networks with balanced excitatory and inhibitory activity. *Science*, 274:1724–1726, 1996.
7. D.Q. Nykamp and D. Tranchina. A population density approach that facilitates large-scale modeling of neural networks: Analysis and an application to orientation tuning. *Journal of Computational Neuroscience*, 8:19–50, 2000.
8. C. Meyer and C. van Vreeswijk. Temporal correlations in stochastic networks of spiking neurons. *Neural Computation*, 14(2):369–404, 2002.
9. C. C. Chow. Phase-locking in weakly heterogeneous neural networks. *Physica D*, 118:343–370, 1998.
10. W. Gerstner. Population dynamics of spiking neurons: fast transients, asynchronous states and locking. *Neural Computation*, 12:43–89, 2000.
11. N. Brunel. Dynamics of sparsely connected networks of excitatory and inhibitory spiking neurons. *Journal of Computational Neuroscience*, 8:183–208, 2000.
12. D. J. Amit and N. Brunel. Model of global spontaneous activity and local structured delay activity during learning periods in the cerebral cortex. *Cerebral Cortex*, 7:237–252, 1997.
13. S. Fusi and M. Mattia. Collective behavior of networks with linear (vlsi) integrate-and-fire neurons. *Neural Computation*, 11:633–652, 1999.
14. M. Mattia and P. del Giudice. Population dynamics of interacting spiking neurons. *Physical Review E*, 66(5), 2002.
15. H. Soula, G. Beslon, and O. Mazet. Spontaneous dynamics of assymmetric random recurrent spiking neural networks. *Neural Computation*, 18(1), 2006.
16. G. Bi and M. Poo. Synaptic modifications in cultured hippocampal neurons: Dependence on spike timing, synaptic strength, and postsynaptic cell type. *The Journal of Neuroscience*, 18(24):10464–10472, December 1998.
17. L. F. Abbott and S. B. Nelson. Synaptic plasticity: taming the beast. *Nature America*, 3:1178–1182, December 2000.
18. D. O. Hebb. *The Organization of Behavior*. Wiley, New York, USA, 1949.
19. R. P. N. Rao and T. J. Sejnowski. Spike-timing-dependent hebbian plasticity as temporal difference learning. *Neural Computation*, 13:2221–2237, 2001.
20. S. Song, K.D Miller, and L.F Abbott. Competitive hebbian learning through spike-timing dependent plasticity. *Nature Neuroscience*, 3:919–926, 2000.

An Adaptive Robot Motivational System

George Konidaris and Andrew Barto

Autonomous Learning Laboratory
Department of Computer Science
University of Massachusetts at Amherst
{gdk, barto}@cs.umass.edu

Abstract. We present a robot motivational system design framework. The framework represents the underlying (possibly conflicting) goals of the robot as a set of drives, while ensuring comparable drive levels and providing a mechanism for drive priority adaptation during the robot's lifetime. The resulting drive reward signals are compatible with existing reinforcement learning methods for balancing multiple reward functions. We illustrate the framework with an experiment that demonstrates some of its benefits.

1 Introduction

Autonomy is central to intelligence—an intelligent system that requires the external specification of its goals is a tool, not an agent, because it fails the basic test of agency. To be autonomous, an agent requires an internal motivational system that appropriately values the actions available to it and generates its goals. In natural agents this system is evolved, but in artificial agents we must design it.

Reinforcement learning [17] is a learning, planning, and action selection paradigm based on maximising reward. Although it does not deal with the problem of designing the motivational system that generates those rewards, it is an intuitively appealing model of motivation-based learning. The importance of motivation to intelligent robot design was recognised early [7,3], and building motivational systems based on reinforcement learning is still an area of active research (e.g. [6]).

In this paper we attempt to bridge the gap between motivation and action selection by outlining the properties that a motivational system should have, and by introducing a design framework based on them. We illustrate our framework with an experiment demonstrating its benefits.

2 Background

Reinforcement learning relies on the existence of a reward function that penalises bad actions and reward good actions. If we are to use it as a method of action selection for an autonomous robot, we require a reward generating mechanism (or *motivational system*) that expresses the robot's internal goals and motivations [3]. This mechanism will likely consist of multiple parts—real animals want

S. Nolfi et al. (Eds.): SAB 2006, LNAI 4095, pp. 346–356, 2006.

more than one thing, and autonomous robots are likely to have multiple (possibly conflicting) simultaneously active concerns (at the very least, to keep running and simultaneously complete whatever task it is that we designed them for).

This leads to the concept of a drive as *the motivational unit underlying behavior*, a module that expresses one of the robot's purposes and produces motivational force as a *common currency* [12] for use in action selection.

Two types of drive are present in the artificial intelligence literature. Systems using homeostatic regulation [1] endow the agent with a set of internal physiological variables, each with an optimal range. Actions that move a physiological variable outside of this range are punished, and actions that move it toward this range are rewarded. Systems using Hullian drives (e.g. [10,11]) maintain a set of drives, each with a drive level that varies between totally unsatisfied and fully satiated. Reward is generated by drive level difference, so actions that raise the level are rewarded and those that lower it are punished.

In this paper we use Hullian rather than homeostatic drives because not all drives are homeostatic, and any homeostatic drive can be simulated using a pair of Hullian drives (one to penalise going above the ideal range, and one to penalise going below). This increases the number of drives, but also adds flexibility because the two directions can be handled separately.

3   Desirable Properties

An ideal robot motivational framework would possess the following properties:

1. A **drive interface specification** that provides a well defined and consistent way of specifying drives.
2. A **reward generating mechanism** that allocates drive-specific reward to actions given their effect on the drive, rooting action selection in the agent's motivational state.
3. A **drive priority** mechanism so that the agent can adjust the relative urgency of each of its drives during its lifetime.
4. **Numerically comparable rewards and priorities**, so that drive rewards can be used as a common currency when comparing and balancing the demands of various drives.
5. An efficient **action selection mechanism** that balances the demands and priorities of multiple drives.

The split between reward and priority is not self evident, but we treat them separately because such a split is adaptive. Drives and their associated reward mechanisms will in most cases be fixed by design or evolution, and be grounded in the agent's "physiology". Priority, however, should be a property of the environment the agent finds itself in, reflected in its own history. Agents in an environment where water is scarce but food is not should be able to learn to value water over food (and therefore have a higher water drive priority). These aspects of the environment are difficult to predict at design time and may change during the agent's lifetime.

4   Overview

Figure 1 shows an overview of our motivational framework. The framework employs a collection of drives $d_1, ..., d_n$, where each drive d_i maintains a satiation level σ_i and a priority level ρ_i. The priority level determines the shape of a priority curve, which translates satiation level to drive priority κ_i. For each time step, the reward generated by the drive is obtained by multiplying the difference in satiation between time steps by the drive priority, and the agent's aim at any given time is to maximize the sum of these rewards. A detailed description of each of these elements is given the following sections.

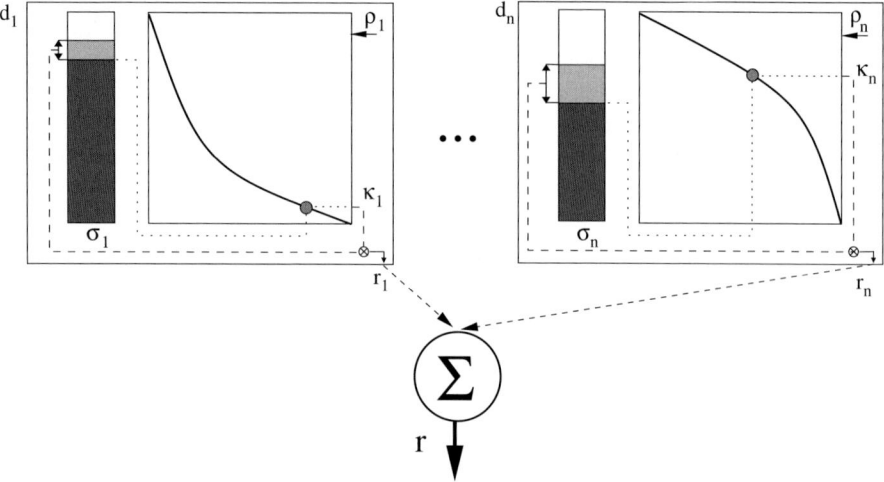

Fig. 1. An overview of our motivational framework. Each drive d_i maintains a satiation level σ_i, and the reward signal it generates is the difference in σ_i from one point in time to the next (shown in light gray) multiplied by a drive priority κ_i. Drive priority is determined by translating σ_i using a priority curve, the shape of which is determined by the drive's priority parameter ρ_i. The agent aims to maximise $r = \Sigma_i r_i$ for action selection.

5   Representing Individual Drives

Each individual drive d_i consists of the following components:

1. A *satiation level* $\sigma_i \in [0, 1]$, where at $\sigma_i = 0$ the drive is starved (and the agent may cease functioning), and at $\sigma_i = 1$ the drive is satiated and should have no effect on the agent's behavior.
2. A *drive process* that updates σ_i according to the drive's intended purpose.
3. A *priority parameter* $\rho_i \in (0, 1)$, which reflects the agent's long-term belief about the difficulty of raising σ_i. A high value for ρ_i indicates that d_i is difficult to satiate and should thus have a high priority, whereas a low value indicates that it is easy to satiate and should thus have a low priority.

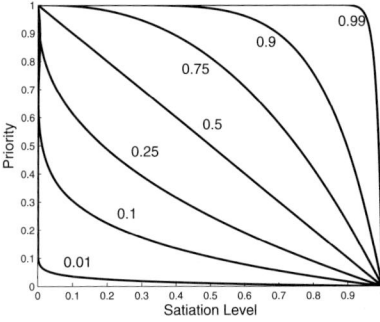

Fig. 2. Satiation level modulated by sample priority parameters

4. A *priority process* that monitors d_i's satiation history and slowly increases or decreases ρ_i to reflect the drive's long-term priority.

The priority parameter ρ_i is used to affect the shape of a priority curve that determines the drive's priority κ_i given its current satiation level, according to the following equation:

$$\kappa_i = 1 - \sigma_i^{\tan \frac{\rho_i \pi}{2}}.$$

Thus ρ_i allows the agent to adjust its drive priorities without changing the underlying drive process. Figure 2 shows sample priority curves for a few values of ρ_i. A very high value of ρ_i means that the drive attains a high priority even when σ_i is near 1 (satiation), but a low value of ρ_i means that the drive reaches a high priority only when it is near 0 (starvation). As we would expect, drive priority is 1 at $\sigma_i = 0$ and 0 at $\sigma_i = 1$ irrespective of ρ_i, and at $\rho_i = 0.5$ the priority curve is a straight line.

When the agent performs an action, it results in a change in σ_i for drive d_i, and that change multiplied by the priority level for the drive results in drive reward r_i:

$$r_i(t+1) = \kappa_i(t)[\sigma_i(t+1) - \sigma_i(t)].$$

This results in two design problems per drive: defining a drive process that expresses the drive's intended purpose and defining an appropriate priority process. The first will differ for each drive, but will in most cases be grounded in the robot's physical state. The second will likely be based on the drive's satiation level history over a long period of time. Section 7 employs a simple priority adjustment heuristic, but in cases with more drives we expect more complex, drive specific rules will be required.

6   Combining Drives for Action Selection

We are faced with an action selection problem over a state space comprising both the internal state of the robot and the external state of the environment. More specifically, we are trying to find a policy π:

$$\pi : (s_E, s_I) \mapsto a,$$

where s_E is an external state descriptor, $s_I = (\sigma_1, ..., \sigma_n)$ is an internal state descriptor, and a is an action. We omit the priority parameters, $\rho_1, ..., \rho_n$, from s_I because we can consider them constant as they are expected to change slowly over the lifetime of the agent. This means that π is nonstationary, but it will only change gradually.

Learning this policy directly may be difficult because the resulting state space may be much larger than s_E, the state space of the problem the robot is solving. Furthermore, this space has significant redundant structure: only (s_E, a) is useful in predicting s'_E (s_E at the next timestep), and since drive satiation levels do not interact, (s_E, a, σ_i) uniquely determines σ'_i and thus r_i (and each satiation level may only depend upon a subset of s_E).

Unfortunately, it is easy to construct examples where varying just one drive (e.g., by starving it) drastically changes the optimal policy for a given environmental state. Therefore any individual drive's value function or policy that is not a function of both s_E and all of s_I must be an approximation.

There are two possible ways to exploit the structure of this state space. The first is to attempt to learn π directly, using a function approximator specifically designed to take advantage of this structure. Although learning would take place over both s_E and all of s_I, if the function approximator is well designed then the extra dimensions may not make the problem significantly harder.

Alternatively, if such a solution is not feasible, we can learn a value function Q_i for each drive simultaneously and independently (as a function of (s_E, a, σ_i) only), and combine them to form an overall value function Q. Several solutions to this problem have been proposed in the literature. Of these, Sprague and Ballard [16] show that using Sarsa (an *on-policy* reinforcement learning algorithm) to learn each Q_i and then setting $Q(s, a) = \Sigma_i Q_i(s, a)$ performs best. Using an on-policy learning algorithm prevents each Q_i from overestimation, since an off-policy algorithm (like Q-learning) would compute each drive's action values assuming that their own optimal policies will be followed thereafter.

Using this method is equivalent to treating the values of the other drives as hidden state, so the resulting state values for each drive will be the same as the correct value state values for that drive at the expected value of each of the other satiation levels.

7   Experiments

In this section we use Spier's domain [14] to illustrate the benefits of the priority aspects of our framework. An agent (of width 60 units) is placed in a $10,000 \times 10,000$ toroidal grid containing two types of uniformly scattered resources (of width 60 units). There are 14 of the first (dark) type of resource, but only 8 of the second (lighter) type. The agent is able to perceive the proximity (scaled from 0 to 1) and angular distance (scaled from -1 to 1) of the nearest three of each type of resource within a perceptual range of 1500 units, resulting in a sensor space of twelve continuous variables. The agent can move forward ten units in the direction it is facing or rotate in either direction by ten degrees, and has a high-

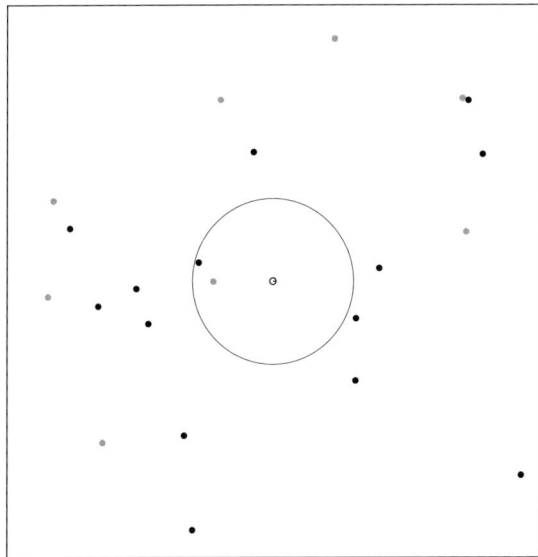

Fig. 3. An example domain instance. The large circle indicates the agent's perceptual radius, so it can see one of each resource type.

level behavior for approaching and consuming each visible target. Consumed resources are randomly replaced. Figure 3 shows an example instance.

The agents were given two drives (one for each resource type), each using a satiation level penalty of 0.00015 for a movement and a satiation level gain of 0.1 for each successful consumptive act. Each drive's priority parameter and satiation value was initially set to 0.5. Learning was by Sarsa(0) and gradient descent using a linear function approximator ($\alpha = 0.01, \gamma = 0.9$) representing separate value functions for each drive, each over the twelve continuous-valued sensory attributes plus one continuous satiation level. Actions were chosen from the available high-level approach and consume behaviors. We randomly generated 50 sample environments, and ran three types of agents for 250 episodes of 10 consumptive acts each. The first type of agent used drive reduction as a reward, the second used our framework but with fixed priority parameters ($\rho = 0.5$), and the third moved each drive's ρ value after each episode toward the ratio of the two drives' number of successful consumptive acts (using $\alpha = 0.1$).

The results are shown in Figure 4. The agent that does not take priority into consideration (Figure 4a) first satiates (at about episode 100) the drive associated with the plentiful resource, and only then (when further consumptive acts on this resource create less reward because it is near satiation) makes progress bringing its second drive towards satiation, although by the end of 250 episodes the second drive has not reached 80% satiation. The agent using priority curves (Figure 4b) allocates a higher reward to the lower drive—because it is higher on the priority curve by virtue of its low satiation value—and therefore increases both drives simultaneously, with the second drive reaching well over 80% satia-

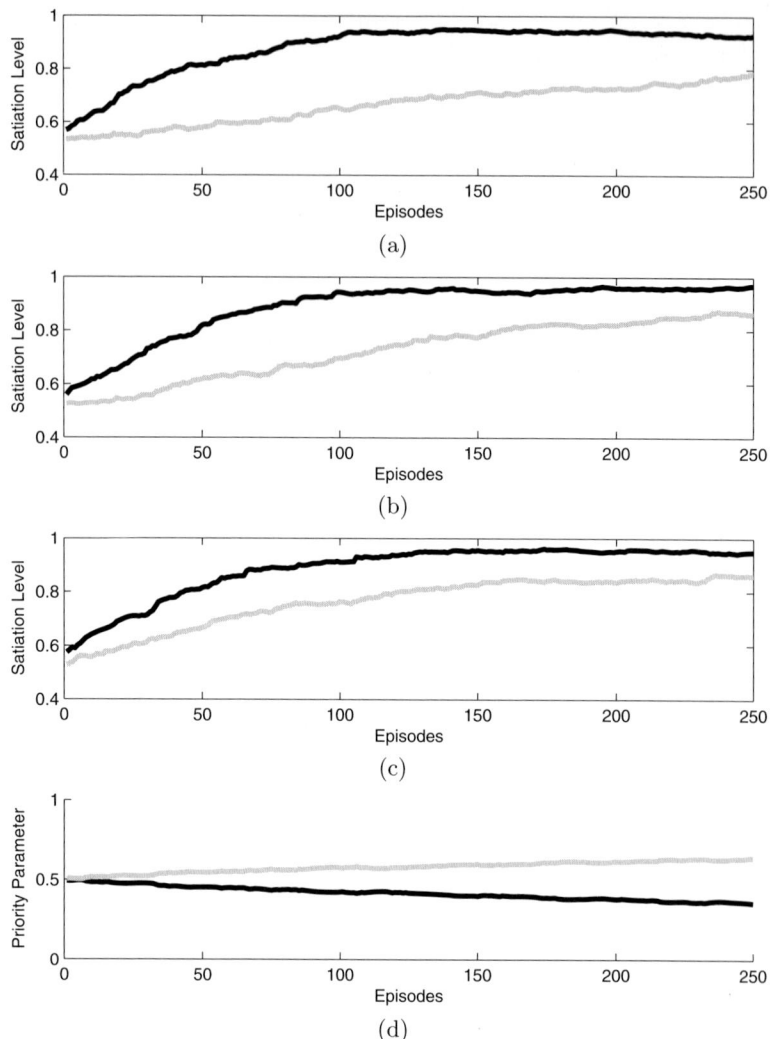

Fig. 4. Average (over 50 trials) drive satiation levels for two drives given an uneven distribution of resources. The first graph (a) shows satiation for an agent using drive difference directly as reward. The second (b) shows satiation for an agent with fixed priority parameters, and the third (c) shows the same for an agent that adjusts each drive's priority parameter to match observed resource frequency. The final figure (d) shows the third agent's priority parameters changing over time.

tion by 250 episodes. This occurs even though both drives have the same priority parameters. Finally, the agent with flexible ρ values (Figure 4c) is able to better balance the two satiation levels, even though they cannot be made to match because of differences in resource frequency. It also reaches well over 80% satiation by 250 episodes.

Note that (as can be seen in Figure 4d) the final ρ values inversely reflect the frequency of each resource, with the ρ value for the light (scarce) resource converging to approximately double that of the dark (plentiful) one.

8 Related Work

The relevant work in motivational system design[1] is primarily split into two threads: research on motivational models based on drives and research on balancing multiple reward functions.

8.1 Motivational Models Based on Drives

Cañamero [4] introduced a motivational model using homeostatic drives where each drive has an error signal (similar to a difference in satiation) which translated to an activation level (similar to a priority level), which could also be increased by an incentive stimulus (the presence of a goal object) or an activation modifier. The behavior attached to the drive with the highest activation is selected for execution. This system did not use learning for action-selection, but Cos Aguilera, Cañamero and Hayes [5] built a similar system using a simpler model and reinforcement learning with dominant-drive action selection.

Blumberg [2] described a behavior-based architecture that uses internal Hullian drives, where action selection uses the behavior with the highest activation (internal motivation multiplied by incentive stimulus). Reward generated by drive difference is used to associatively learn the value of incentive stimuli using reinforcement learning, but not for action selection.

Spier and MacFarland [15] empirically compared five different models of decision-making in a two-resource domain using Hullian drives. This work did not use reinforcement learning (although some of the decision models are similar), and the resources were available in equal quantities so the notion of drive priorities was absent.

Finally, Konidaris and Hayes [10,11] recently built a situated reinforcement learning system based on Hullian drives similar to ours (but without a priority mechanism), using a circadian switching mechanism for drive selection.

8.2 Balancing Multiple Reward Functions

To the best of our knowledge, Whitehead, Karlsson and Tenenberg [19] were the earliest to recognise that a reinforcement learning robot might have multiple goals expressed as separate reward functions and need to balance them. They introduced the idea of reward functions that could be "switched off" and proposed both setting Q to the highest individual Q_i value and setting it to the sum of the Q_i values as modular action selection mechanisms. They pointed out that these methods must respectively understimate and overestimate the monolithic

---

[1] We note that a great deal of research exists on computational models of natural motivational systems. Since this paper is concerned with motivational system *design* rather than modelling, we refer the interested reader to Savage's review [13].

(true) value function, and that using the maximum Q_i does not perform drive balancing, while summing the Q_i values may lead to actions where no drive receives a reward at all. However, their empirical results suggest that the difference between either modular method and the monolithic method is small, and may be more than made up for by the resulting increase in learning speed.

Sprague and Ballard [16] pointed out that these problems arise from using off-policy methods, which compute each drive's action values assuming that its own optimal policies will be followed thereafter, and show improved performance using Sarsa(0) (an on-policy learning algorithm). However, unlike Whitehead et al. [19] they assume that each value function is always active, in which case their method is not an approximation.

Humphrys [8] surveys several methods for balancing multiple reward functions, placing them on a continuum from single-minded to cooperative, and introduced W-learning, where agents explicitly learn a weight for each drive in each state expressing the extent to which that drive would suffer if its preference is not taken.

Although all of these papers are strongly related to this research, none of them employed drive mechanisms. Thus they did not include satiation (beyond on or off) or priority levels, and could not guarantee numerically comparable rewards.

9   Discussion

9.1   The Multiplicity of Drives

Even the most intuitive drives may not be atomic upon closer inspection. For example, sodium-depleted rats displays an enhanced appetite for food containing sodium [18], which suggests that they have a separate sodium seeking drive and possibly other nutrient seeking drives, instead of a single hunger drive. However, a system approaching the complexity of an animal will need to maintain so many internal variables that creating and balancing separate drives for each of them is not likely to be feasible.

One way to get around this would be for each drive to represent many internal variables likely to be systematically reduced by the same activity (e.g., nutrient levels are all modified by eating). The drive process could then modify what changes its satiation level according to the system's current needs (e.g., fruit becomes more rewarding when the agent needs sugar). We could even view each agent-level drive as composed of several subdrives, so that a hunger drive is composed of a salt drive, a sugar drive, etc., each with a very small state space.

9.2   Motivational Systems as a Basis for Further Learning

Once the motivational system described in this paper is present in an agent, it could also be used a way of focusing other types of learning. For example, Cos Aguilera, Cañamero and Hayes [5] learn object affordances based on changes in motivational state, where the effect of a behavior is quantified in terms of its effect on the agent's drives. Another example is provided by Konidaris and Hayes [10], where a robot learns associations between reward and the sensations present at reward states to speed up reinforcement learning in novel environments. This results in guided (as opposed to blind) searches in new environments [9], using a form of heuristic that can be learned autonomously. We expect that robot control architectures based on a motivational system will provide further opportunities for motivationally grounded learning.

9.3 Non-physiological Drives

When designing a sufficiently complex autonomous robot, we may wish to include motivational aspects that do not involve "physiological" attributes. For example, we may wish to motivate the robot to seek social interaction, or to avoid verbal reprimand. Such motivations can be represented in our framework, and thus integrated and balanced against physiological factors, although their implementation may not be as natural as physiological drives. In such cases the drive satiation level or priorities may be kept within a smaller range than usual so that the robot's "physiological" needs are never completely overridden for less immediate motivations. Alternatively, constraining drive priorities to a particular ordering may be a useful way to build safety measures or sanity checks into the robot.

10   Summary

We have presented a robot motivational system framework that provides a simple interface specification for drives, a mechanism for reward generation that guarantees numerically comparable rewards, and a natural method for adjusting drive priorities. The resulting reward structure is compatible with existing reinforcement learning methods for balancing multiple drives.

Acknowledgements

We would like to thank Gillian Hayes, John Hallam and Ignasi Cos Aguilera for their useful input. Andrew Barto and George Konidaris were supported by the National Science Foundation under Grant No. CCF-0432143. Any opinions, findings and conclusions or recommendations expressed in this material are those of the authors and do not necessarily reflect the views of the National Science Foundation.

References

1. H. Bersini. Reinforcement learning for homeostatic endogenous variables. In *From Animals to Animats 3: Proceedings of the Third International Conference on the Simulation of Adaptive Behavior*, pages 325–333, 1994.
2. B.M. Blumberg, P.M. Todd, and P. Maes. No bad dogs: Ethological lessons for learning in hamsterdam. In *From Animals to Animats 4: Proceedings of the Fourth International Conference on the Simulation of Adaptive Behavior*, pages 295–304, 1996.

3. R.A. Brooks. The role of learning in autonomous robots. In *Proceedings of the Fourth Annual Workshop on Computational Learning Theory (COLT '91)*, pages 5–10, 1991.

4. L. Cañamero. Modeling motivations and emotions as a basis for intelligent behavior. In W. Lewis Johnson, editor, *Proceedings of the First International Conference on Autonomous Agents*, pages 148–155, New York, NY, 1997. ACM Press.

5. I. Cos-Aguilera, L. Cañamero, and G.M. Hayes. Motivation-driven learning of object affordances: First experiments using a simulated khepera robot. In F. Detjer, D. Dörner, and H. Schaub, editors, *The Logic of Cognitive Systems: Proceedings of the Fifth International Conference on Cognitive Modeling (ICCM'03)*, pages 57–62, 2003.

6. K. Doya and E. Uchibe. The cyber rodent project: Exploration of adaptive mechanisms for self-preservation and self-reproduction. *Adaptive Behavior*, 13(2): 149–160, 2005.

7. J.R.P Halperin. Machine motivation. In *From Animals to Animats: Proceedings of the First International Conference on the Simulation of Adaptive Behavior*, pages 213–221, 1990.

8. M. Humphrys. Action selection methods using reinforcement learning. In *From Animats to Animats 4: Proceedings of the Fourth International Conference on the Simulation of Adaptive Behavior*, pages 135–144, 1996.

9. G.D. Konidaris and G.M. Hayes. Anticipatory learning for focusing search in reinforcement learning agents. In *The Second Workshop on Anticipatory Behavior in Adaptive Learning Systems*, July 2004.

10. G.D. Konidaris and G.M. Hayes. Estimating future reward in reinforcement learning animats using associative learning. In *From Animals to Animats 8: Proceedings of the 8th International Conference on the Simulation of Adaptive Behavior*, pages 297–304, 2004.

11. G.D. Konidaris and G.M. Hayes. An architecture of behavior-based reinforcement learning. *Adaptive Behavior*, 13(1):5–32, 2005.

12. D. McFarland and T. Bosser. *Intelligent Behavior in Animals and Robots*. MIT Press, Cambridge, MA., 1994.

13. T. Savage. Artificial motives: a review of motivation in artificial creatures. *Connection Science*, 12(3/4):211–277, 2000.

14. E. Spier. *From Reactive Behaviour to Adaptive Behaviour: Motivational Models for Behaviour in Animals and Robots*. PhD thesis, Balliol College, University of Oxford, 1997.

15. E. Spier and D. McFarland. Possibly optimal decision-making under self-sufficiency and autonomy. *Journal of Theoretical Biology*, 189(3):317–331, 1997.

16. N. Sprague and D.H. Ballard. Multiple-goal reinforcement learning with modular Sarsa(0). In *Proceedings of the Eighteenth International Joint Conference on Artificial Intelligence (IJCAI 03)*, pages 1445–1447, 2003.

17. R.S. Sutton and A.G. Barto. *Reinforcement Learning: An Introduction*. MIT Press, Cambridge, MA, 1998.

18. F. Toates. *Motivational Systems*, chapter 4. Cambridge University Press, Cambridge, UK, 1986.

19. S. Whitehead, J. Karlsson, and J. Tenenberg. Learning multiple goal behavior via task decomposition and dynamic policy merging. In J.H. Connell and S. Mahadevan, editors, *Robot Learning*, pages 45–78. Kluwer Academic Publishers, 1992.

Incremental Skill Acquisition
for Self-motivated Learning Animats

Andrea Bonarini, Alessandro Lazaric, and Marcello Restelli

Department of Electronics and Informatics
Politecnico di Milano
piazza Leonardo da Vinci 32, I-20133 Milan, Italy
{bonarini, lazaric, restelli}@elet.polimi.it
http://www.elet.polimi.it

Abstract. A central role in the development process of children is played by self-exploratory activities. Through a playful interaction with the surrounding environment, they test their own capabilities, explore novel situations, and understand how their actions affect the world. During this kind of exploration, interesting situations may be discovered. By learning to reach these situations, a child incrementally develops more and more complex skills. Inspired by studies from psychology, neuroscience, and machine learning, we designed SMILe (Self-Motivated Incremental Learning), a learning framework that allows artificial agents to autonomously identify and learn a set of abilities useful to face several different tasks, through an iterated three phase process: by means of a random exploration of the environment (*babbling phase*), the agent identifies interesting situations and generates an intrinsic motivation (*motivating phase*) aimed at learning how to get into these situations (*skill acquisition phase*). This process incrementally increases the skills of the agent, so that new interesting configurations can be experienced. We present results on two gridworld environments to show how SMILe makes it possible to learn skills that enable the agent to perform well and robustly in many different tasks.

1 Introduction

In this paper, we describe SMILe (Self-Motivated Incremental Learning), a learning framework leading an agent to incrementally learn general abilities through a direct interaction with the environment guided by self-generated interest. This approach integrates ideas coming from *cognitive sciences* and *intrinsically motivated reinforcement learning* and defines a self-development process that enables animats to autonomously operate in complex environments.

In recent years, studies on the inner mechanisms of human development, pursued in many different areas (such as neuroscience, psychology, developmental sciences, robotics, machine learning) converged to a new field, commonly referred to as *developmental robotics*[7,20]. Traditionally, a designer must specifically program the set of skills needed for an animat to accomplish a given task. Often, these skills are tuned to perform a predefined task on a specific environment,

S. Nolfi et al. (Eds.): SAB 2006, LNAI 4095, pp. 357–368, 2006.

and the learned abilities can hardly be reused if the task or the environment changes. On the other hand, developmental robotics tries to reproduce the basic mechanisms at the basis of human and animal development processes so as to propose frameworks in which the agent does not directly address any specific problem, but develops a set of basic skills up to very general abilities that can be used to solve many different tasks.

Because of the complexity of its goal, developmental robotics has many different facets [7]. In this paper, we focus on a subset of them and we will consider the developmental process as an *incremental process* where an agent *organizes* its initial skills through *spontaneous exploratory phases* and *self-motivated learning activities*. Self-motivated learning proved to be one of the most challenging aspects of development processes, as shown in [19,2,8]. One of the most promising approaches is *intrinsically motivated reinforcement learning* [2], that enables an agent to autonomously develop a hierarchy of skills through a process guided by an intrinsic motivation, without any commitment to achieve a specific task.

The SMILe framework extends the intrinsically motivated reinforcement learning model to a more general development process, in which the notion of interest is not hardwired, but autonomously extracted from characteristics of the environment. The learning process of each skill has been decomposed into three phases (*babbling, motivating* and *skill acquisition*), that are endlessly iterated to develop a hierarchy of abilities that can be exploited by animats to better control the environment.

The rest of the paper is organized as follows. In the next section we give a general description of SMILe and we introduce a novel framework for self-motivated learning. Section 3 gives an overview of the implementation of the framework using Reinforcement Learning (RL) techniques, and we provide a general definition of the interest function. Section 4 provides some experimental results on two gridworlds that simulate simple robotic environments, showing how the acquired skills help the agent to reduce the learning time in many different tasks. Finally, in the last section we discuss the results and propose some possible future directions.

2   The Learning Process

As stated in [20], one of the most promising approaches to achieve the ambitious goal of autonomy in artificial systems, is the definition of a suitable lifelong development process. This consists of an open-ended learning process in which an agent pursues self-motivated goals and develops highly reusable skills. Developmental robotics has its main source of inspiration in studies from neuroscience and psychology [7], that show how similar mechanisms could be traced in the developmental process of children.

Many approaches in developmental robotics refer to the studies by Piaget [13], and to his research on children's early stages of development. Piaget showed that childish development can be considered as an incremental process of acquisition of new abilities in which children modulate the complexity of their activities in association with the increasing complexity of their cognitive and morphological

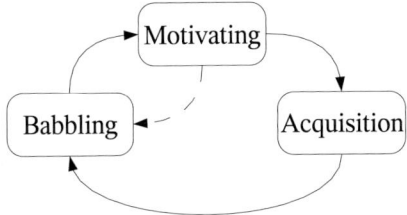

Fig. 1. The self-motivated developmental learning process of SMILe

structures. Another important contribution to the comprehension of the mechanisms attending human development comes from the research carried out by Berlyne [3] about the notion of curiosity and its influence on behavior and the rising of intrinsic motivation. Berlyne asserts that, in absence of a particular aim, human behavior is partly determined by an innate will of exploring what is perceived as interesting. Psychologists define curiosity as a form of motivation that promotes exploratory behavior to learn more about a source of uncertainty.

In summary, life-long learning seems to be characterized by a progressive, self-motivated *development* that leads to the *incremental* acquisition of more and more complex skills. SMILe implements this concept into a simplified learning process suitable for artificial systems, whose aim is to incrementally learn new skills that could be potentially useful to face different tasks. Each skill is learned by a self-motivated process that iterates on three main phases (see Fig. 1):

- *Babbling*: the agent playfully interacts with the environment to get aware of the relationships between its actions and the environment dynamics.
- *Motivating*: the agent evaluates which is the most interesting situation it has experienced during the exploration performed during the babbling phase.
- *Skill acquisition*: the agent learns the skill to reach the interesting situation.

2.1   Babbling Phase

One of the crucial activities in the development process of puppies and babies is *self-exploration* [7]. Through self-explorative acts, they become aware of their own capabilities with respect to the surrounding environment, understand the consequences of the actions they have autonomously selected, and learn to control and exploit the dynamics of their bodies. In analogy to vocal babbling, this experiential process has been called *body babbling* [10].

Moving from these observations, we introduced in SMILe, at the beginning of each iteration, a *babbling* phase. The acquisition of a new skill starts with a self-explorative phase in which the agent, for a certain time, randomly executes its admissible actions. The choice of taking actions according to a uniform probability distribution over the action space, even if it is not fully compliant to body babbling theory, is consistent with the fact that this exploration is completely goal free, without any external motivation leading the agent behavior. The goal of the babbling phase is to collect information about the environment dynamics that can be used in the next phase to determine whether there is any *interesting* skill that is worth learning.

2.2   Motivating Phase

There is a huge body of evidence about the central role played by *intrinsic motivation* in development process as the main driver of organisms behavior when no extrinsic motivation is available. In this way, they may increase their competence to control the environment, by acquiring a broad set of skills that can be reused for different goals. Studies from psychology, like those of Piaget [13] and Berlyne [3], and from neuroscience [5], suggest that intrinsic motivation may be generated by several factors: surprise, incongruity, and novelty. All these factors act together to determine an intrinsic *interest* associated to different situations.

Many studies [15,12] relate interest to the current knowledge of the observer and its capability to predict the outcome of its interaction with the environment, and propose particular quantitative definitions of novelty and surprise.

In SMILe, this second phase computes the *interest function*, which associates an interest value to each state visited during the babbling phase. Despite previous approaches, we propose a general methodology to define and compute interest values for each state obtained by the propagation (through the estimated transition model of the environment) of a given local measure of interest (a formal definition will be given in Section 3). Once the interest function has been computed, SMILe determines the next goal by searching for the state associated to the maximum interest. Learning to reach this state is the goal of the third phase.

It may happen that the agent has no strong motivation in learning to reach a state rather than another. In this case, it makes no sense to spend time and efforts in learning something that is not so interesting, but it is better to start a new babbling phase in order to collect more experience that could allow to discover new interesting situations (represented by the dashed line in Fig. 1).

2.3   Skill Acquisition Phase

During a development process an organism starts with very simple skills and acquires more and more complex abilities. Each time a new skill is learned, it may be used to simplify the learning of the following ones, thus progressively increasing the complexity of the tasks that could be successfully faced.

Recently, the idea of hierarchically decomposing complex problems into simpler sub-problems has been successfully exploited also in RL with the introduction of formalisms for managing temporally extended actions [1]. Several of these approaches work with fixed hand-coded decompositions, even if some proposals have been advanced to dynamically decompose a given goal into simpler sub-goals [9,11]. Barto et al. [2] have proposed an intrinsically motivated approach to generate the hierarchy of skills.

In SMILe, during the skill acquisition phase the agent learns, through an intrinsic reward function, a skill that leads to the most interesting state identified by the motivating phase. While learning the skill, the agent, in addition to its basic actions, may benefit also from other previously acquired skills.

After the acquisition of a new skill, the development process of SMILe starts a new iteration activating a new babbling phase, in order to experience how the new skill modifies the agent interaction with the environment. This leads to the computation of a new interest function that defines a new learning goal, thus obtaining an incremental learning process that continuously increases the agent capabilities of controlling its environment.

3 SMILe

In this section we propose an implementation of the learning framework described in Section 2. As already proved in many studies [2,21,19], Reinforcement Learning (RL) is one of the most suitable frameworks to deal with learning problems in developmental robotics. Furthermore, the incremental development of simple skills into complex activities can be efficiently described using Hierarchical Reinforcement Learning (HRL) [1], as suggested in [2].

3.1 Formal Representation of Skills: The Option Framework

HRL problems are generally formalized using Semi-Markov Decision Process (SMDP) models. In particular, in the *option framework* [1] an SMDP is defined by tuple $\langle \mathcal{S}, \mathcal{O}, \mathcal{P}, \mathcal{R} \rangle$, where $\mathcal{S}$ is the set of states (i.e. perceptions), $\mathcal{O}$ is the set of options (i.e. skills), $\mathcal{P}(s, o, s')$ is the transition model, that is the probability to get to state s' taking option o is state s, and $\mathcal{R}(s)$ is the reward in state s. The main difference between traditional RL approaches and intrinsically motivated learning, concerns the source of reinforcement. While in the usual interaction model the agent receives a reinforcement signal provided by an external critic, we consider the reward as the result of an intrinsic motivation of the agent that pursues self-generated goals according to the model proposed in [2].

Formally, a skill is represented as an option o, i.e. a tuple $\langle \pi_o, \mathcal{I}, \beta \rangle$, where $\pi_o : \mathcal{S} \times \mathcal{O} \to [0,1]$ is the control policy that describes the probability to execute an option when the agent is in a specific state, $\mathcal{I} \subset \mathcal{S}$ is the set of states where the option is defined and $\beta(s)$ is the probability for an option to terminate at state s. When the development process starts, the agent has an initial set of basic options $\mathcal{O}^0$, at the k-th iteration, the set of options is incrementally modified adding the option learned in the skill acquisition phase: $\mathcal{O}^k = \mathcal{O}^{k-1} \cup \{o^k\}$.

3.2 Incremental Learning of Reusable Skills

In the following, we give a brief description of the implementation of the development phases of SMILe, summarized in Algorithm 1 (for more details see [4]).

In the *babbling phase*, at each time instant the agent simply executes one skill at random, choosing among the set of admissible skills $\mathcal{O}^k$. The aim of this phase is to build, at each iteration k, an estimate (even partial) $\widehat{P}_{\pi_R^k}(s, s')$ of the state transition probabilities when the random policy π_R^k is used for a sufficient number of steps. Since the state transition probabilities do not depend only on characteristics of the environment, but also on the abilities of the agent, when a new skill is learned, the capabilities of the agent to control the environment dynamics change and the state transition probabilities must be recomputed.

Through the playful exploration performed in the babbling phase, the agent experiences several different situations. In the *motivating phase*, SMILe computes the interest associated to each state on the basis of the information contained in the estimated state transition probabilities $\widehat{P}_{\pi_R^k}(s, s')$.

Although there are several characteristics of a model that could be used to compute the local interest of a state, such as transition entropy and controllability (details can be found in [14]), here we will focus on the following definition:

$$\rho(s) = (1 - p_{in}(s)) - p_{in}(s)(1 - p_{out}(s)), \qquad (1)$$

where $p_{in}(s) = \frac{1}{|S|} \sum_{s' \in S} P_{\pi_R}(s', s)$ and $p_{out}(s) = \sum_{s' \neq s} P_{\pi_R}(s, s')$. The first term of Equation 1 is the probability of not moving into state s in one step following the policy π_R, given that the agent starts from a random state. To this term, we subtract a second term that represents the probability to reach s in one step starting from a random location and then to remain in s for another step.

Algorithm 1. The SMILe Algorithm

1: **repeat**
2:    **Babbling Phase**
3:    **for all** Babbling episodes **do**
4:       **for all** Steps **do**
5:          Given state s, choose action o at random over $\mathcal{O}^k$
6:          Take action o, observe state s'
7:          Update state transition probability estimation $\widehat{P}_{\pi_R^k}(s, s')$
8:       **end for**
9:    **end for**
10:   **Motivating Phase**
11:     Given model $\widehat{P}_{\pi_R^k}(s, s')$, compute local interest $\rho(s)$
12:     Compute interest function $I(s)$
13:     **if** no interesting state can be identified **then**
14:        step back to the Babbling Phase
15:     **else**
16:        Extract subgoal $s^* = arg \max_s I(s)$
17:        Create reward function $R(s)$
18:     **end if**
19:   **Skill Acquisition Phase**
20:    **for all** Skill Acquisition episodes **do**
21:       **for all** Steps **do**
22:          Given state s, choose action o according to ϵ-greedy
23:          Take action o, observe state s' and reward r
24:          Update state-action value function
25:       **end for**
26:    **end for**
27: **until** forever

The intuition behind Equation 1 is that states that, under a random policy, are difficult to be reached or that, once reached, can be easily left, are relevant as subgoals for many complex tasks whose solution needs the agent to pass through states that cannot be easily reached without a specific skill.

This measure defines the concept of interest of a state only on the basis of information about its input and output transition probabilities, without taking into account the characteristics of the surrounding states; for this reason we call it *local interest functions*. Using the estimated state transition probabilities and a local interest function, we define the global interest function with the following Bellman-like equation:

$$I^k(s) = \rho^k(s) + \gamma \sum_{s' \in \mathcal{S}} \widehat{P}_{\pi_R^k}(s, s') I^k(s'). \tag{2}$$

In this way, the interest of a state depends, not only on the characteristics of its local transitions, but also on the interests of those states that may be reached from it. The discount factor $\gamma \in [0, 1)$ determines how much distant states should influence the interest of the current state. To compute $I(s)$ we can use an iterative policy evaluation algorithm that uses Equation 2 as an update rule [17]. The formulation of the interest function $I(s)$ given in Equation 2 is such that it can represent a large set of the aspects of the concept of interest depending on the specific definition of local interest $\rho(s)$ that is used.

Once $I^k(s)$ has been computed, the agent self-determines its next goal by choosing the most interesting state $\bar{s}^k = arg \max_s I^k(s)$, and produces an intrinsically motivated reward function that simply returns a positive reward when the agent achieves state $\bar{s}^k$ and null otherwise. It is possible to show that, using the definition of local interest previously introduced, the acquisition of new skills decreases the interest in goal states (*boredom effect*), thus preventing the agent from choosing them again.

As stated in Section 2.2, after some iterations, the interest function tends to flatten until no state with relevant interest can be identified in the motivating phase. In this case, the agent has no advantage from learning to reach new useless goals and the babbling phase is started again in order to either refine the transition model estimation or adapt to changes in the environment dynamics [4].

After having identified the goal state $\bar{s}^k$ and generated the intrinsically motivated reward function $\mathcal{R}^k(s)$, the agent starts the *skill acquisition* phase in which it learns the policy of a new option o^k whose goal is $\bar{s}^k$. The policy of the new option is learned according to the option learning algorithm described in [1]. At each time step, the action value function $Q(s, o)$, that is the estimation of the amount of reward the agent can obtain by taking option o in state s, is updated according to the following update rule:

$$Q(s, o) \leftarrow (1 - \alpha)Q(s, o) + \alpha \left[\tilde{r} + \gamma^i \max_{o' \in O^{k-1}} Q(s', o') \right] \tag{3}$$

where α is a learning step size, i is the number of steps taken by option o to meet its termination condition and $\tilde{r}$ is the reward accumulated from s to s' in i steps according to the reward function $\mathcal{R}^k(s)$.

Once the skill acquisition is finished, the new option o^k is created and added to the set of options $\mathcal{O}^{k-1}$. This new option is characterized by a deterministic policy that can be directly derived from the action value function $Q(s, o)$ by choosing in each state s the option o that maximizes its value. The termination condition $\beta(s)$ is set to 1 for $s = \bar{s}^k$ and to 0 elsewhere. For what concerns the initial set, it can be limited to a subset of the state space $\mathcal{S}$ composed by the states that have been most visited in the skill acquisition phase.

The incremental generation of new options makes the agent able to develop a hierarchy of skills, where new options can reuse previously learned options to achieve the goals extracted in the motivating phase.

4   Experiments

In this section, we provide experimental results obtained by SMILe in two different environments. In the first problem, we show how SMILe learns general purpose options that may be effectively reused for learning to reach a large number of goals. The second experiment puts in evidence how the SMILe development process can significantly reduce the learning times.

4.1   Four-Room Gridworld

The *Four-Room* (Fig. 2) environment [18] is a 10x10 grid with a set of walls that delimit four rooms. The initial set of actions is $\mathcal{A} = \{down, right, up, left\}$ and the starting state is the upper left corner. To introduce stochasticity in the world dynamics, actions have a probability of 0.3 to fail. When an action fails the agent moves to one of the adjacent states at random.

The development process of SMILe led to the identification of interesting goals only in limited regions. The density plot in Fig. 3 shows the frequency of subgoal identification for each state: the lighter the region the higher the frequency of extraction. It is worth noting that, using the interest function described in Section 3, SMILe finds the states in the middle of the rooms as most interesting. Recalling the definition of local interest (Eq. 1), the explanation of this result is that from these states the agent can easily reach all

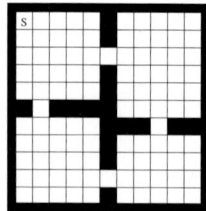

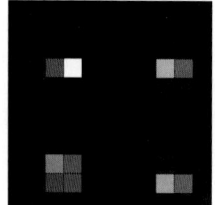

Table 1. Average and maximum number of learning steps in Four-Rooms with random goal

Algorithm	Mean	Max
Q-Learning	$1.293 \cdot 10^4$	$3.569 \cdot 10^4$
Random	$1.756 \cdot 10^4$	$6.777 \cdot 10^4$
SMILe	$0.957 \cdot 10^4$	$2.044 \cdot 10^4$

Fig. 2. Four Rooms world

Fig. 3. Density plot of subgoal distribution in Four Room world

the other states in the room. The usefulness of the learned skills can be measured only by imposing many different external goals to the agent and by evaluating the global learning performance. Therefore, we have performed a comparison among Q-learning [17], Q-learning with four skills whose goals have been chosen at random, and Q-learning with the four skills learned by SMILe, over 1000 randomly extracted external goal states. Then, we have recorded the sum of learning steps for each goal over the first 100 episodes. Table 1 reports the number of steps in the average and in the worst case. As it can be noticed, both the average and the maximum number of steps needed by SMILe are less than those needed by the other two algorithms. This means that the skills acquired by SMILe produced a relevant advantage when facing different learning problems. Furthermore, since Q-Learning with skills for random goals obtained the worst performance, the result of SMILe is not simply determined by the use of the option framework, but it strongly depends on the identification of interesting states that lead to the acquisition of general-purpose skills.

4.2   Playworld Environment

The second experiment we discuss, is a version of the Playworld proposed in [16]. The Playworld is an abstraction of a real environment characterized by two rooms with a door in between, two panels and a charger (see Fig. 4). The panels are in the room at left: the light panel switches the light on and off, while the door panel opens and closes the door. The animat perceives the light intensity, whether the door is open or not, its charge level and its position (i.e., absolute coordinates and orientation). The animat is initially placed at random in the left room and the light is switched off. When in the dark, the animat may fail in taking the selected action with a probability of 0.2, it cannot perceive the status of the door, and the door panel is deactivated. Once the light is switched on, actions always succeed, the animat can open the door, move to the other room and charge. The animat can turn left, turn right, and move ahead.

The experiment consists of two main stages: intrinsically motivated incremental learning and extrinsically motivated learning. In the first stage the animat explores the environment and develops new skills according to the process described in Section 2. In the second stage, five different goals are imposed by an external designer by providing an extrinsic reward function.

In the first stage, the salient events we can expect the animat to find are: *light on, light off, open door, close door, charge*. The upper graph of Fig. 5 shows the events occurred in the babbling phase at first iteration, when the agent succeeds in switching the light on and off only a few times. The lower graph of Fig. 5 shows the changes in the babbling phase introduced by the skills learned after five iterations. As it can be noticed, the skills developed in the previous iterations bias the random exploration so that the animat succeeds in activating new and more complex events (e.g., open the door and charge). This shows how SMILe enables the animat to autonomously discover interesting configurations in the environment and to develop new skills for achieving them.

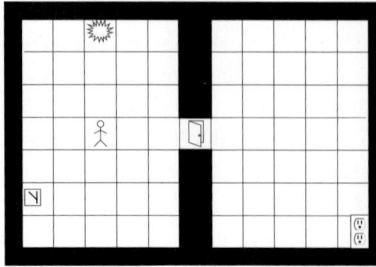

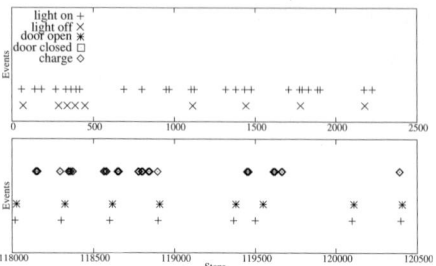

Fig. 4. The Playworld environment

Fig. 5. Sequence of events in the first (*upper*) and fifth (*lower*) babbling phase

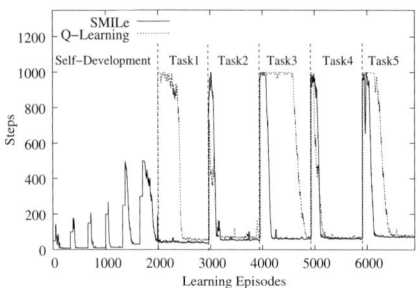

Fig. 6. Comparison of performance between Q-Learning and SMILe

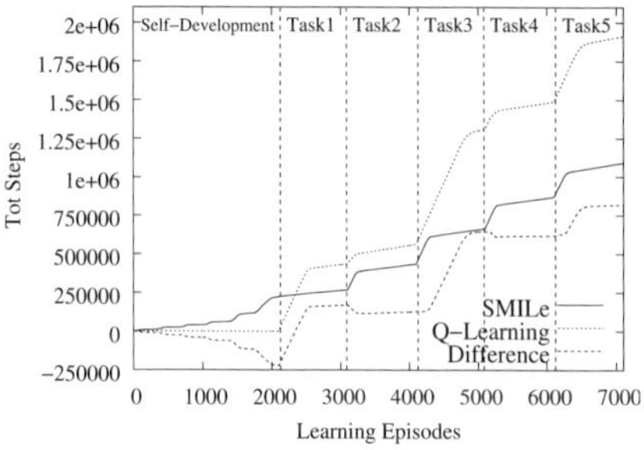

Fig. 7. Comparison of total number of steps between Q-Learning and SMILe

In the second stage, when the development process is over, we compare the performance of an animat that exploits the new skills, to that of an animat using Q-Learning with basic skills, on five different tasks:

Task1: charge
Task2: charge, move to upper left corner of right room
Task3: charge, move to upper left corner of left room
Task4: charge, move to left room and close the door
Task5: charge, move to left room, close the door, switch the light off

While *Task2* and *Task3* are not strictly related to any salient event, the other tasks require the animat to achieve configurations relevant in the Playworld environment. In the comparison, we adopted the same learning parameters for both Q-Learning and SMILe (learning rate $\alpha = 0.6$, ϵ-greedy exploration with $\epsilon = 0.2$, discount factor $\gamma = 0.95$). Each 1,000 learning episodes, the extrinsic reward function is changed according to the task that must be accomplished and the learning animat should be able to adapt its policy to the new task without restarting the learning from scratch.

Fig. 6 shows the number of steps per learning episode. The first 2,100 episodes, labeled as *Self-Development* in the graph, represent the first stage of the experiment in which the SMILe animat autonomously identifies six different interesting states, used as goals for learning six new skills. On the other hand, in the first stage the Q-Learning animat does nothing, since no extrinsic reward is provided. The second stage starts with the introduction of a positive extrinsic reward for achieving the charger. While the Q-Learning animat can only use the basic skills, the SMILe animat exploits the skills learned in the first stage and succeeds in finding the optimal policy to reach the charger in less episodes than those needed by Q-Learning. Similarly, SMILe succeeds in exploiting its skills even for changing tasks, while Q-Learning took more time to adapt to new extrinsic reward functions. Furthermore, in Fig. 7 we compare the total number of steps for both the algorithms and we report their difference. In the first stage, SMILe takes almost 250,000 steps to explore the environment and to learn the new skills, while no steps are taken by the Q-Learning robot. Notwithstanding the initial loss, the total number of steps needed by SMILe after the accomplishment of *Task1* is less than that of Q-Learning. The advantage of SMILe becomes even more relevant at the end of the second stage when Q-Learning took almost twice as many steps as SMILe. This comparison shows that SMILe, even though it requires potentially expensive exploration of the environment, leads to the development of useful skills that can be profitably reused in many different tasks. In particular, the number of steps saved during the extrinsically motivated learning stage is greater than those used in the first stage already in the first goal.

5   Conclusions

Hand-coded abilities, though useful in domains where tasks are fixed, proved to be inadequate to enable artificial systems to solve even slightly different tasks in uncertain environments. On the other hand, the capability to develop new skills from basic abilities without any imposed goal, is what makes human beings and animals able to reuse their skills in many complex tasks.

In this paper, we have presented SMILe, a self-development RL framework that incrementally acquires more and more complex skills through an iterative three phase learning process similar to those taken by children and animal puppies in their early development stages. Experimental results show the effectiveness of the skills learned by SMILe when operating in environments where different tasks may arise, thus developing agents with a good degree of autonomy.

Currently, we are investigating the use of function approximation techniques to scale to large, high dimensional domains. Future work includes the integration of SMILe with developmental robotics approaches in real robotic tasks [6].

References

1. A. G. Barto and S. Mahadevan. Recent advances in hierarchical reinforcement learning. *Discrete Event Dynamic Systems*, 13(4):341–379, 2003.
2. A.G. Barto, S. Singh, and N. Chentanez. Intrinsically motivated learning of hierarchical collections of skills. In *Proceedings of ICDL*, 2004.
3. D. E. Berlyne. *Conflict, Arousal, and Curiosity*. McGraw-Hill, 1960.

4. A. Bonarini, A. Lazaric, and M. Restelli. Smile: Self-motivated incremental learning. Technical report, Politecnico di Milano, http://www.airlab.elet.polimi.it/papers/bonarini06smile.pdf, 2006.
5. S. Kakade and P. Dayan. Dopamine: Generalization and bonuses. *Neural Networks*, 15:549–559, 2002.
6. G.D. Konidaris and G.M. Hayes. An architecture for behavior-based reinforcement learning. *Adaptive Behavior*, 13(1):5–32, 2005.
7. M. Lungarella, G. Metta, R. Pfeifer, and C. Sandini. Developmental robotics: a survey. *Connection Science*, 15(4):151–190, 2003.
8. J. Marshall, D. Blank, and L. Meeden. An emergent framework for self-motivation in developmental robotics. In *Proceedings of ICDL*, 2004.
9. A. McGovern and A. G. Barto. Automatic discovery of subgoals in reinforcement learning using diverse density. In *Proceedings of ICML*, 2001.
10. A. Meltzoff and M. Moore. Explaining facial imitation: a theoretical model. *Early Development and Parenting*, 6:179–192, 1997.
11. I. Menache, S. Mannor, and N. Shimkin. Q-cut - dynamic discovery of sub-goals in reinforcement learning. In *Proceedings of ECML*, 2002.
12. P-Y. Oudeyer, F. Kaplan, V. Hafner, and Whyte A. The playground experiment: Task-independent development of a curious robot. In *AAAI Spring Symposium Workshop on Developmental Robotics*, 2005.
13. J. Piaget. *The Origins of Intelligence in Children*. Norton, N.Y., 1952.
14. B. Ratitch and D. Precup. Using mdp characteristics to guide exploration in reinforcement learning. In *European Conference on Reinforcement Learning*, 2003.
15. J. Schmidhuber. Self-motivated development through rewards for predictor errors / improvements. In *AAAI Spring Symposium on Developmental Robotics*, 2005.
16. A. Stout, G. Konidaris, and A. Barto. Intrinsically motivated reinforcement learning: A promising framework for developmental robot learning. In *AAAI Spring Symposium on Developmental Robotics*, 2005.
17. R. S. Sutton and A. G. Barto. *Reinforcement Learning: An Introduction*. MIT Press, Cambridge, MA, 1998.
18. R. S. Sutton, D. Precup, and S.P. Singh. Between mdps and semi-mdps: a framework for temporal abstraction in reinforcement learning. *Artificial Intelligence*, 112:181–211, 1999.
19. E. Uchibe and K. Doya. Reinforcement learning with multiple heterogeneous modules: A framework for developmental robot learning. In *Proceedings of ICDL*, 2005.
20. J. Weng, A. McClelland, O. Sporns, I. Stockman, M. Sur, and E. Thelen. Autonomous mental development by robots and animals. *Science*, 291:599–600, 2001.
21. J. Weng and Y. Zhang. Novelty and reinforcement learning in the value system of developmental robots. In *International Workshop on Epigenetic Robotics: Modeling Cognitive Development in Robotic Systems*, 2002.

Piagetian Adaptation Meets Image Schemas: The Jean System

Yu-Han Chang[1], Paul R. Cohen[1], Clayton T. Morrison[1], Robert St. Amant[2], and Carole Beal[1]

[1] Information Sciences Institute, University of Southern California
4676 Admiralty Way, Marina del Rey, CA 90292
{cohen, ychang, clayton, cbeal}@isi.edu
[2] Department of Computer Science,
North Carolina State University
2268 Engineering Building II, 890 Oval Drive,
Raleigh, North Carolina 27695
stamant@ncsu.edu

Abstract. Jean is a model of early cognitive development based loosely on Piaget's theory of sensori-motor and pre-operational thought [1]. Like an infant, Jean repeatedly executes schemas, gradually extending its schemas to accommodate new experiences. Jean's environment is a simulated "playpen" in which Jean and other objects move about and interact. Jean's cognitive development depends on several integrated functions: a simple perceptual system, an action-selection system, a motivational system, a long-term memory, and learning methods. This paper provides an overview of Jean's architecture and schemas, and it focuses on how Jean learns schemas and transfers them to new situations.

1 Introduction

Jean is both a synthesis of ideas about cognitive development and the foundations of concepts, and an integrated software system that implements perception, action, learning and memory. Most of this paper is devoted to the Jean system, so let us begin with the underlying ideas. From Piaget we borrow the ideas that children learn some of what they know by repeatedly executing schemas, and executing schemas is in a sense rewarding, and some new schemas are modifications or amalgamations of old ones [1]. The Image Schema theorists [2,3,4,5,6] promote the ideas that primitive schemas are encodings or redescriptions of sensorimotor information; and these schemas are semantically rich, general, and extend or transfer to new situations, some of which have no salient sensorimotor aspects. Another idea, represented by various authors, is that semantic distinctions sometimes depend on dynamics — how things change over time — and so schemas should have a dynamical aspect [7,8,9,10].

Jean will test several conjectures about developmental AI: First, it will be possible to provide a relatively small, core set of schemas and a general algorithm to learn others as they are needed or indicated by experience. We are betting on a compositional account of knowledge, in which newly learned schemas are assembled from previously learned and appropriately modified components. Second, schemas will have to be more than

S. Nolfi et al. (Eds.): SAB 2006, LNAI 4095, pp. 369–380, 2006.

the declarative, logical structures proposed by AI researchers over the decades; they will have to include behavior-generating controllers, dynamic maps, deictic variable bindings, and causal theories; these components will not all develop simultaneously. Third, the generality of these schemas provides a basis for the effective *transfer* of knowledge learned in one task to a new, related task.

2   The Image Schema Language

Image schemas are Jean's elementary and innate representations, and much of Jean's design follows from these representation commitments, so we begin our description of Jean with them. Image schemas are representations that are "close" to perceptual experience. Some authors present them as re-descriptions of experience [5]. Their popularity is due to their supposed generality and naturalness: Many situations are naturally described in terms of paths, up-down relations, part-whole relationships, bounded spaces, and so on. Even non-physical ideas, such as following an argument, containing political fallout, and feeling "up" or "down," seem only a short step from image-schematic foundations [2,3].

These ideas are attractive but vague, as we discovered when we tried to build a formal Image Schema Language (ISL) [11]. We found it necessary to distinguish three kinds of image schema. *Static* schemas describe unchanging arrangements of physical things; *dynamic* schemas describe how the environment changes; and *action* schemas describe intentional aspects of static and dynamic schemas. Thus, the action schema *approach-object* includes a path (a static schema) but gives it the intentional gloss that it is the path one intends to follow (or is following). *Moving* might be intentional or it might simply be the result of force acting on an object. Both cases involve a path, but the latter is described by a dynamic schema, not an action schema.

2.1   Static Schemas

As represented in ISL, static image schemas are objects, in the sense of the object-oriented data model. Each schema has a set of operations that determine its capabilities. For example, operations for a basic container schema include putting material into a container and taking material out. Each schema also has a set of internal slots that function as roles in a case grammar sense [12]. Slots permit image schemas to be related to each other through their slot values. For example, the contents of a container can be other image schemas.

To illustrate static schemas in ISL, it will be helpful to walk through an example, which we take from our work on representing chess patterns. Consider a chess board in which the Black queen has the White king in check. In image schema terms, we say that there exists a path from the queen to the king. In ISL, we generate a path schema, which contains a set of locations. Through a mechanism called *interpretation*, we can substitute one kind of schema for (part of) another in ISL, which allows us to make the locations along the paths be *container* schemas. Then, by setting the capacity of these containers to one piece, we capture the idea that the location (interpreted as a container) is "full" if it is occupied by a piece. A blocked path is then a path with at least one full container/location.

2.2 Dynamic and Action Schemas

If Jean lived in a chess game, we might choose to represent the dynamics of the environment as a sequence of static schemas, like fluents in the situation calculus: Some schemas describe the environment until an instantaneous chess move happens, then others do. Jean's environment changes continuously, and there is much structure in *how* it changes, so we have adopted the following representation for dynamic schemas and action schemas: Both are finite state machines, the states of which are themselves composed of image schemas. For example, to catch a simulated cat, Jean must sneak up on the cat slowly and then, when it gets near the cat, Jean must pounce. As illustrated in Figure 1, the action schema for catching a cat must have at least these two states (and actually, a couple more). Each state contains several static and dynamic schemas. For instance, state s_3 of Figure 1 comprises an object schema that binds its deictic variable to the cat, and a near-far schema that binds its two deictic variables to the robot (i.e., Jean) and the cat, respectively. The near-far schema also asserts that the cat is more than six units away from the robot. s_3 is associated with two action schemas, fast-approach-object (F) and slow-approach-object (S), represented as arcs leaving s_3.

Composite action schemas like catching a cat are called *gists* because they involve story-like sequences of actions and states and they abstract away many of the details of particular instances.

In addition to the familiar, declarative components of schemas, dynamic and action schemas contain dynamic *maps*, which describe continuous, smooth changes in state variables. Dynamic maps have three functions: They let Jean estimate when changes will occur, they tell Jean when its actions are, or are not, likely to achieve its goals, and they help Jean find the boundaries between states. These functions are elaborated in Section 4.

Action schemas also contain *controllers* that control behavior. For instance, Jean's schema for *approaching* binds its variables to the approaching object (typically Jean) and the approached object, a map that shows how distance between the objects changes over time, and a controller that makes Jean move toward the location of the object

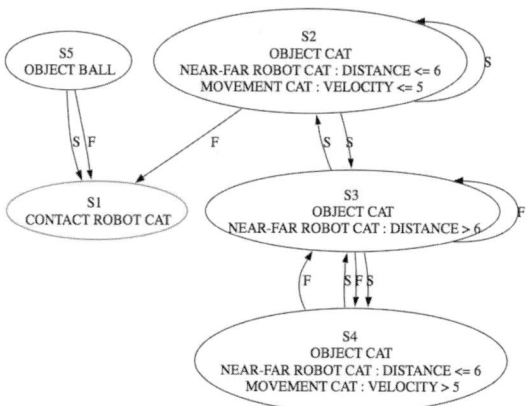

Fig. 1. A learned composite action schema for catching a cat

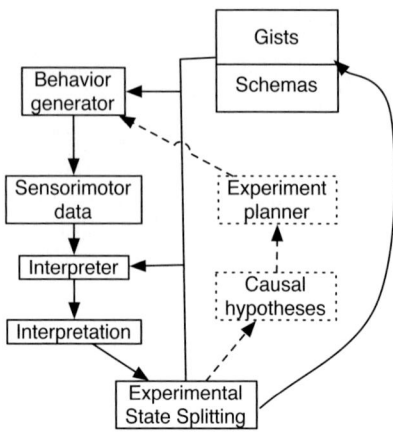

Fig. 2. The main functional components of Jean

it is approaching. Eventually, schemas will also be augmented with causal relations (see Sec. 7).

3   The Architecture of Jean

The main functional components of Jean are illustrated in Figure 2. Over time, Jean learns new schemas and gists, all represented in the Image Schema Language (ISL). When Jean has a goal, such as catching a simulated cat, it retrieves an appropriate gist and runs its controller, which means taking the actions that produce transitions between states. The component of Jean that runs schemas is called the *behavior generator*. Exercising schemas produces sensory data, and lots of it. The job of the *interpreter* is to retrieve and instantiate (i.e., bind the deictic variables of) schemas from the repository. Interpretation also updates conditional probabilities associated with state transitions within gists. Jean will try to interpret the sensory data in terms of the schemas in the gist it is running; for example, if it is trying to catch a cat, then it will prefer to interpret the sensory data as matching the states in Figure 1. Sometimes, though, the fit is poor, and another schema in the repository does a better job of explaining the sensory data. And sometimes, it is necessary to construct a novel static, dynamic or action schema, as described in the next section.

4   Learning: Experimental State Splitting

Jean learns new schemas in two ways, by *composing* schemas into gists and by *differentiating* states in schemas. Both are accomplished by the Experimental State Splitting (ESS) algorithm. The basic idea behind Experimental State Splitting (ESS) is simple. The algorithm starts with a minimal state model of the world, in which it has only one all-encompassing state. This model is modified as the agent explores its world, so that it becomes more predictive of some measure of observed action outcomes.

For a general developmental account we want a general measure, not a task-specific one. To accord with the idea that learning is itself rewarding, this measure might have something to do with the informativeness or novelty or predictability of states. In Jean, the ESS algorithm uses a measure we call *boundary entropy*, which is the entropy of the next state given the current state and an intended action. Initially, when there is only one state in the model, the entropy will always be zero. One way to drive Jean toward more states, and, thus toward states that have boundary entropy, is to have a goal state in addition to the initial (non-goal) state. At any moment in time, the agent is in one of these two states, and each state-action pair generates some probability distribution over the set of possible next states. ESS calculates the entropy of this distribution and uses it as a state splitting criterion.

In general, Jean is driven by ESS to modify its world model by augmenting existing states with new states that reduce the boundary entropies of state-action pairs. This augmentation is achieved by splitting an old state into two (or more) new states based on distinguishing characteristics. For example, if an agent is navigating a city intersection, it might decide that "green light" is an important characteristic because, where it observes a green light, it is much less likely to be involved in crashes. Thus, its non-goal state would be split into two states: (1) a green light is observed and its goal has not yet been attained, and (2) neither a green light is observed nor its goal has been obtained. The state machine thus grows over time as the agent adds more attributes to its state descriptions.

As Jean interacts with the world, it counts the transitions between the states via particular actions. If the developing state machine model is Markov, then the model is a Markov Decision Process (MDP), which Jean can solve for the optimal policy to reach its goal state. In this way, the developing model can be used for planning. However, it is worth pointing out that since Jean operates in a continuous environment using fairly general action schemas, its world model is more like a semi-Markov Decision Process (SMDP), where each action results in a transition between states after a certain amount of time, and this time interval is drawn from some probability distribution.

If Jean lives always in one environment with one set of goals, then ESS will eventually produce optimal policies for the environment. However, the purpose of the Jean project is not to produce optimal policies for each task and variant of Jean's environment, but, rather, to explain how a relatively small set of policies may quickly *accommodate* (as Piaget called it) or *transfer* to new tasks and environments. Our approach to this problem is to extract *gists* from policies. Gists are like policies, in that they tell Jean what to do in different situations, but they are more general because they extract the "usual storyline" or essential aspects of policies. We claim that these essential aspects are typically the *causal relationships* that govern actions and effects in the environment. If an agent can identify and learn these causal relationships, then it should have a very good idea of how its actions affect the world and how act to achieve its goals, *even in novel situations*. We will discuss gists and transfer further in Section 5.

4.1 An Example

Figure 1 illustrates a gist for approaching and contacting a simulated cat. In the scenario in which this gist was learned, the cat is animate, capable of sitting still, walking or running away. The cat responds to Jean. In particular, if Jean moves toward the cat rapidly, the cat will run away; if Jean approaches slowly, the cat will tend to keep doing what it is doing. Because of these programmed behaviors, there is uncertainty in Jean's representation of what the cat will do, but there is a general rule about how to catch the cat, and it can be represented in a gist: The only way to catch the cat is to first get into state s_2 (Fig. 1), where the cat is nearby and not moving quickly, and then to move fast toward the cat, reaching state s_1. All other patterns of movement leave Jean in states s_3 or s_4. This corresponds to the strategy of slowly sneaking up to the cat and then quickly pouncing on it to catch it.

To learn a gist like the one in Figure 1, Jean repeatedly retrieves action schemas from its memory, runs the associated controllers, producing actions, specifically slow and fast movement to a location; assesses the resulting states, and, if the transitions between states are highly unpredictable, Jean splits states to make the resulting states more predictable.

In fact, the three states, s_2, s_3 and s_4 were all originally one undifferentiated state in which Jean moved either fast or slowly toward the cat. Jean's learning history — the distinctions it makes when it splits states — begins with a single, undifferentiated non-goal state. Then, Jean learns that the type of object is an important predictor of whether or not it can catch the object. Balls are easy to catch, whereas cats are hard to catch. From here, ESS recognizes that distance also influences whether or not it can catch a cat. Starting near the cat, a fast-approach-object (F) action will often catch the cat, whereas this action will not usually catch the cat from further away. Thus, ESS splits on distance with a threshold of 6, where $<= 6$ is considered near, and > 6 is far. Finally, ESS may notice that even when Jean is near the cat, sometimes it does not succeed in catching the cat. This might be because the cat is already moving away from the agent with some speed. Thus, ESS may do a final split based on the velocity of the cat. This process leads to the states s_2, s_3, s_4 and s_5 that we see in Figure 1.

4.2   The Algorithm

We give a formal outline of the ESS algorithm in this section. We assume that agent receives a set of schema features $F^t = \{f_1, \ldots, f_n\}$ from the environment at every time tick t; these features could be schema slots that represent sensor readings, for example. We also assume that Jean is initialized with a goal state s_g and a non-goal state s_0. S_t is the entire state space at time t. A is the set of all actions, and $A(s) \subset A$ are the actions that are valid for state $s \in S$. Typically $A(s)$ should be much smaller than A. $H(s_i, a_j)$ is the boundary entropy of a state-action pair (s_i, a_j), where the next observation is one of states in S_t. A small boundary entropy corresponds to a situation where executing action a_j from state s_i is highly predictive of the next observed state. Finally, $p(s_i, a_j, s_k)$ is the probability that taking action a_j from state s_i will lead to state s_k.

For simplicity, we will focus on the version of ESS that only splits states; an alternative version of ESS is also capable of splitting actions and learning specializations of parametrized actions. The ESS algorithm follows:

- Initialize state space with two states, $S_0 = \{s_0, s_g\}$.
- While ϵ-optimal policy not found:
  - Gather experience for some time interval τ to estimate the transition probabilities $p(s_i, a_j, s_k)$.
  - Find a schema feature $f \in F$, a threshold $\theta \in \Theta$, and a state s_i to split that maximizes the boundary entropy score reduction of the split: $\max_{S,A,F,\Theta} H(s_i, a_i)$

$-\min(H(s_{k_1}, a_i), H(s_{k_2}, a_i))$, where s_{k_1} and s_{k_2} result from splitting s_i using feature f and threshold θ: $s_{k_1} = \{s \in s_i | f < \theta\}$ and $s_{k_2} = \{s \in s_i | f \geq \theta\}$.

- Split $s_i \in S_t$ into s_{k_1} and s_{k_2}, and replace s_i with new states in S_{t+1}.
- Re-solve for optimal plan according to p and S_{t+1}

The splitting procedure iterates through all state-action pairs, all of the schema features F, and all possible thresholds in Θ and tests each such potential split by calculating the reduction in boundary entropy that results from that split. This is clearly an expensive procedure. In future we will speed it up with heuristics that limit the Jean's attention to relevant features and state-action pairs. Heuristics to find potential thresholds are discussed next.

4.3   Finding Splitting Thresholds, Learning New Dynamic Schemas, and Other Criteria for Splitting

Given a candidate schema feature f for splitting, ESS must find a threshold on which to split the state using that schema. To do this, Jean uses a simple heuristic: States change when several state variables change more or less simultaneously. This heuristic is illustrated in Figure 3. The upper two graphs show time series of five state variables: headings for the robot and the cat (in radians), distance between the robot and the cat, and their respective velocities. The bottom graph shows the number of state variables that change value (by a set amount) at each tick. When more than a set number of state variables change, Jean concludes that the state has changed. The value of the schema f at the moment of the state change is likely to be a good threshold for splitting f. For example, between time period 6.2 and 8, Jean is approaching the cat, and the heuristic identifies this period as one state. Then, at time period 8, several indicators change at once, and the heuristic indicates Jean is in a new state, one that corresponds to the cat moving away from Jean. The regions between these changes become the dynamic maps associated with dynamic and action schemas, and the active schemas in these regions

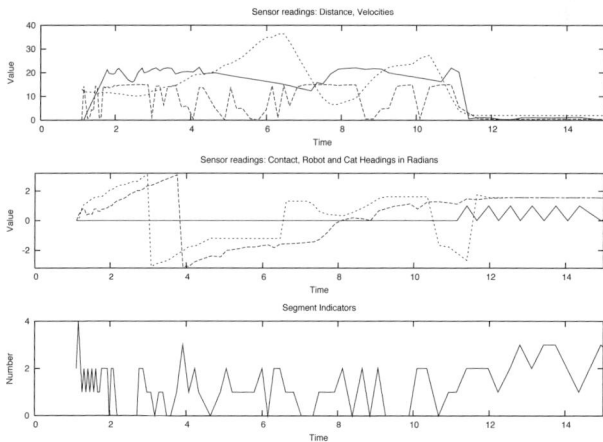

Fig. 3. New states are extracted by cutting multiple time series at places where multiple state variables change simultaneously

are bundled together into composite dynamic and action schemas such as "s_2: Object : Cat ; Near-Far : Robot, Cat :Distance $\leq$ 6 ; Movement Cat : Velocity $\leq$ 5."

This segmentation of the time series helps Jean learn new dynamic schemas. Segments correspond to dynamic *maps* in dynamic and action schemas. As long as Jean is safely within a schema, state variables will change as the schema's maps prescribe. Over time, Jean builds up a statistical model of how state variables are expected to change in a state. If the variance of this model is high, then it suggests the state is a candidate for splitting.

Other useful indicators that states need to be split include rewards (and costs), and repeatability or closure. Reward is fairly straightforward: we wish to split states so that new states will better predict future rewards (rather than simply predicting future states). Repeatability or closure refers to whether actions are easily repeated once executed. For example, picking up and dropping a block onto a table is easily repeated because the action leads to a state where it can immediately be repeated. Such "closed" actions are more easily learned by children, thus we may want closure to be an indicator for splitting actions.

5   Transferring Learning

Although ESS can learn policies for new situations from scratch, we are much more interested in how previously learned policies can accommodate or transfer to new situations. We call transferrable policies *gists*, or generalized summaries of policies. Gists capture the most relevant states and actions for accomplishing past goals. It follows that gists may be transferred to situations where Jean has similar goals. Jean could, of course, try to execute gists without modification in these situations. However, in situations that are not very similar, it may be more effective to execute a gist from which state transition probabilities have been excised. The extra cost of relearning the probabilities may be offset by the cost of *negative transfer*. Sometimes a learned gist can actually inhibit learning of a new gist. For instance, having learned to drive in the U.S. can interfere with learning to drive in the U.K. In cases like this, one wants to keep the general structure of a gist because it describes the states and transitions that occur in both situations, but learn new transition probabilities. By dropping transition probabilities, we may have an easier time finding a policy in the new situation without being burdened by the weight of past experiences.

6   Experiments

We tested Jean's transfer of schemas between situations in a simulated 3-D physical environment. Jean's task is to catch a given target object as quickly as possible. The targets, which we refer to as a "ball" and a "cat," have different dynamics. The ball moves ballistically when Jean makes contact with it (and may be moving at the start of a learning episode), whereas the cat is self-moving and has a preference to not be caught by Jean. In these experiments, the cat runs away if Jean is far away but approaching quickly, or when Jean is very near. Thus, the best way to catch the cat is to sneak up slowly and then quickly pounce upon it. In these experiments, Jean has four innate (not learned) action schemas: fast-approach, slow-approach, stop, and wander. In some experiments there are obstacles, such as walls.

In each trial, Jean has a chance to complete its given task — catch a ball or catch a cat — within a time limit. The time Jean requires to perform its task is the dependent variable, and is expected to decrease as Jean learns to catch its targets. An experimental condition includes 100 trials. The value of the dependent variable, time to catch the target, is smoothed over these trials using a smoothing window of 15 trials. Good learning performance should corresponds to a line that slopes down from left (early trials) to right (later trials).

To evaluate the effect of transfer, we follow an "B vs. A+B" protocol. A and B refer to tasks across which we might observe transfer; for instance, A might be catching a ball, and B, catching a cat. The protocol involves two conditions: The "B" condition involves learning to perform a task, B, whereas the "A+B" condition involves learning task B after learning to perform task A. If the gist for task A transfers, then it will improve some aspect of performance on B, either the initial level of competence on B (before learning) or the rate of learning to perform B.

Jean was tested on three tasks: **A:** Catch a ball in an unobstructed room; **B:** Catch a cat in an unobstructed room; **C:** Catch a cat in a room with obstacles such as additional walls. Here Jean must learn to avoid bumping into the walls while chasing the cat. We constructed five experimental conditions:

B. Catch the cat without any prior training.

A+B. First learn to catch a ball, then learn to catch a cat.

C. Catch the cat in the obstructed room without any prior training.

B+C. First learn to catch a cat in an unobstructed room, then learn to do this in a room with additional interior walls.

B+A. First learn to catch a cat in an unobstructed room, then learn to catch a ball in the same room.

The data presented here are a representative sample of system performance in the conditions in which we expect positive or negative transfer.[1]

Figure 4 shows Jean learning to catch a cat in an unobstructed room. The dotted line represents Jean learning to catch a cat without any prior knowledge (i.e., condition **B**). We see little improvement in the performance measure as Jean acquires more experience. The bold line represents Jean's performance on the catch-a-cat task after learning to catch a ball. Clearly Jean is able to transfer some of its knowledge from catching a ball to the task of catching a cat. In particular, it has already established a preference ordering on its action schemas, whereby it prefers fast-approach and slow-approach over stop and wander, since these result in shorter ball-catching times. Thus, in the **A+B**, condition, Jean does not explore the stop and wander actions, but directly accommodates its ball-catching gist to cat-catching. It is quickly able to identify the proper state to split (i.e., split based on the NEAR-FAR schema), and learns the optimal policy for catching a cat. This is clearly represented by the large drop in catching time seen around time 20.

---

[1] We are currently conducting tests for statistical significance, based on the methods of Piater et al. for comparing learning curves [13]; these will be ready in time for the camera-ready submission. These are particularly important since there is a high degree of variability inherent in the simulation domain.

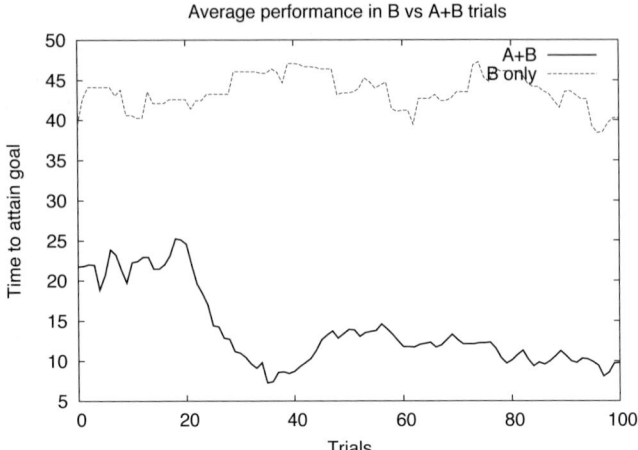

Fig. 4. Graph showing average performance over repeated trials in the B versus A+B regimes. Performance is measured as the time taken for the robot to catch the target object. If in some trial the object is not caught within the time limit, then the time limit value is used as that trial's performance.

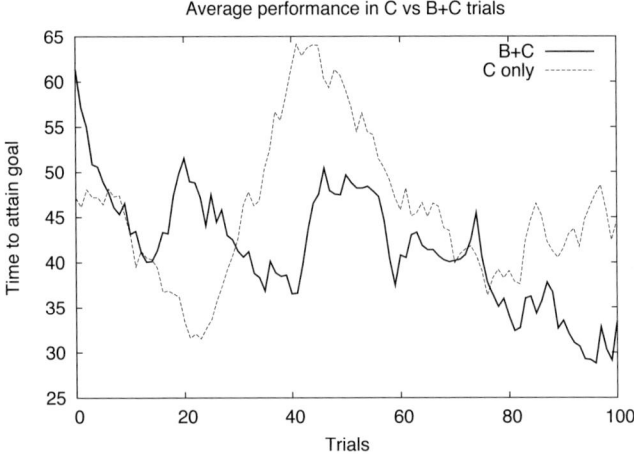

Fig. 5. Graph showing average performance over repeated trials in the C versus B+C regimes

A similar benefit of prior learning is seen in Figure 5. Here, the dotted line shows Jean's performance while attempting to learn from scratch how to catch a cat in an obstructed room (condition **C**). The bold-face line shows the performance given that Jean has already learned how to catch a cat in an open room (condition **B+C**). The transfer premium is due to Jean using its learned Task B gist to accomplish Task C when the cat is not hiding behind a wall. Jean then only needs to learn to amend this gist to deal with the situation in which the cat is hiding. There is more variability in the **C** condition because random exploration, which is more prevalent in the **C** condition, may cause the cat to run far away.

Finally, in the **B+A** condition, we found an example of negative transfer (graph not shown). Recall that transfer is not always beneficial when applied blindly. Here, Jean first learns task **B**. It learns to sneak up to, and then pounce on, the cat. When Jean directly applies this gist to task A, it results in a sub-optimal behavior, since one does not need to sneak up to a ball. Depending on its exploration behavior, Jean might take a long time to realize that its behavior is suboptimal. However, we can ameliorate this negative transfer by modifying the knowledge being transferred. Instead of transferring a gist that includes all states, actions, and transition probabilities, we can transfer only the state descriptions, leaving out the transition probabilities that were observed when learning the old task. This removes most of the negative transfer effect. Any remaining decrease in performance simply results from the fact that the agent has a slightly larger state space to explore. This transfer mechanism, where transition probabilities are dropped but state description are retained, is used in the **A+B** and **B+C** conditions as well.

7   Future Work

We are currently extending this work in a variety of ways. We have implemented a different learning domain in a real-time 3-D strategy game, complete with terrain, natural obstacles such as trees and water, and enemy units. We have observed some interesting transfer between the simple robot-and-cat domain and this military domain. Sneaking up to the cat is analogous to sneaking up on the enemy, for example. Future experiments will explore learning in this new domain, and cross-domain transfer.

Another line of development for Jean is suggested by the word "experimental" in Experimental State Splitting, and by the boxes labeled *causal hypotheses* and *experiment planner* in Figure 2. State splitting finds factors that reduce the entropy of state transitions, or conversely, increase the predictability of these transitions. Not all predictive relations are causal. The *counterfactual theory of causality* says X causes Y iff X precedes Y, and X and Y covary, and X is necessary to affect Y. The necessity condition is framed as a counterfactual: $\neg X \rightarrow \neg Y$. The problem with this theory is that it does not distinguish true causes from mere conditions; for instance, a wire is necessary for electricity to travel from a light switch to a light bulb, but we would not call a wire the cause when we turn on the light. A heuristic to get around this is to assign counterfactually necessary and proximal actions to X's in causal models. Thus flipping a light switch, being the most proximal action to illumination, and counterfactually necessary, is a candidate cause. It is easy to find actions that are proximal to effects, and to formulate counterfactuals relating these actions to effects. These counterfactuals serve as causal hypotheses for Jean to try to refute. Jean will develop causal models of its action schemas, and will learn not only what works, but why it works.

References

1. Piaget, J.: The Construction of Reality in the Child. New York: Basic (1954)
2. Lakoff, G.: Women, Fire and Dangerous Things. University of Chicago Press, Chicago, IL (1987)
3. Johnson, M.: The Body in the Mind: The Bodily Basis of Meaning, Imagination, and Reason. University of Chicago Press, Chicago, IL (1987)

4. Gibbs, R.W., Colston, H.L.: The cognitive psychological reality of image schemas and their transformations. Cognitive Linguistics **6** (1995) 347–378
5. Mandler, J.: The Foundations of Mind: Origins of Conceptual Thought. Oxford University Press (2004)
6. Oakley, T.: Image schema. In Geeraerts, D., Cuyckens, H., eds.: Handbook of Cognitive Linguistics. Osford University Press (2006)
7. Thelen, E., Smith, L.: A Dynamic Systems Approach to the Development of Cognition and Action. The MIT Press, Cambridge, MA (1994)
8. Regier, T.: The Human Semantic Potential: Spatial Language and Constrained Connectionism. The MIT Press (1996)
9. Cohen, P.R.: Maps for verbs. In: Proceedings of the Information and Technology Systems Conference, Fifteenth IFIP World Computer Conference. (1998)
10. Talmy, L.: Toward a Cognitive Semantics. Volume 1: Conceptual Structuring Systems (Language, Speech and Communication). The MIT Press, Cambridge, MA (2003)
11. St. Amant, R., Morrison, C.T., Chang, Y., Cohen, P.R., Beal, C.: An image schema language. To appear in the Proceedings of The 7th International Conference on Cognitive Modelling (ICCM 2006) (2006)
12. Fillmore, C.: The case for case. In Bach, E., Harms, R.T., eds.: Universals in Linguistic Theory. Holt, Rinehart & Winston, London (1968)
13. Piater, J.H., Cohen, P.R., Zhang, X., Atighetchi, M.: A randomized anova proceedure for comparing performance curves. In: Proceedings of the Fifteenth International Conference on Machine Learning (ICML 1998). (1998) 430–438

A Model of Reaching that Integrates Reinforcement Learning and Population Encoding of Postures[*]

Dimitri Ognibene, Angelo Rega, and Gianluca Baldassarre

Laboratory of Autonomous Robotics and Artificial Life,
Istituto di Scienze e Tecnologie della Cognizione,
Consiglio Nazionale delle Ricerche (LARAL-ISTC-CNR),
Via San Martino della Battaglia 44, 00185 Roma, Italy
{dimitri.ognibene, angelo.rega,
gianluca.baldassarre}@istc.cnr.it
http://laral.istc.cnr.it/

Abstract. When monkeys tackle novel complex behavioral tasks by trial-and-error they select actions from repertoires of sensorimotor primitives that allow them to search solutions in a space which is coarser than the space of fine movements. Neuroscientific findings suggested that upper-limb sensorimotor primitives might be encoded, in terms of the final goal-postures they pursue, in premotor cortex. A previous work by the authors reproduced these results in a model based on the idea that cortical pathways learn sensorimotor primitives while basal ganglia learn to assemble and trigger them to pursue complex reward-based goals. This paper extends that model in several directions: a) it uses a Kohonen network to create a neural map with population encoding of postural primitives; b) it proposes an actor-critic reinforcement learning algorithm capable of learning to select those primitives in a biologically plausible fashion (i.e., through a dynamic competition between postures); c) it proposes a procedure to pre-train the actor to select promising primitives when tackling novel reinforcement learning tasks. Some tests (obtained with a task used for studying monkeys engaged in learning reaching-action sequences) show that the model is computationally sound and capable of learning to select sensorimotor primitives from the postures' continuous space on the basis of their population encoding.

1 Introduction

This research is motivated by the idea that when humans and monkeys learn to solve complex tasks by trial-and-error they select and execute *sensorimotor primitives* (that is behavioral chunks that tend to achieve whole goals, cf. [2, 6, 7]) that have a coarse granularity with respect to the detailed commands sent to muscles. By using these primitives, they can learn to tackle complex tasks by assembling relatively few "behavioral chucks" instead of a multitude of fine muscular movements that would make the problems' search space huge. The computational advantages of this strategy have been explored in reinforcement learning literature (see [4] for a review; note that

---

[*] This research has been supported by the project "MindRACES - From Reactive to Anticipatory Cognitive Embodied Systems", European Commission's grant FP6-511931.

S. Nolfi et al. (Eds.): SAB 2006, LNAI 4095, pp. 381–393, 2006.

within this context sensorimotor primitives are called "macro actions" or "options"). This work is part of a research program directed to design, implement and test computational models that not only mimic animal's behaviors organized on sensorimotor-primitive repertoires, but also account for the neuroscientific evidence related to the brain's mechanisms underlying them. With this regards, an increasing amount of empirical evidence is giving specific indications on how vertebrates' brains *encode repertoires* of sensorimotor primitives and *select* and *assemble* them to flexibly produce complex behaviors. For example, it has been shown that when different areas of frogs' spinal cord are electrically stimulated, their lower limbs tend to assume a discrete number of particular postures in space independently of the initial configuration [6]. Moreover, recordings of neurons' activity in *premotor areas* controlling arms in monkeys that freely move in ecological conditions showed that the biggest amount of variance of the neurons' firing rate is explained by the final postures achieved by the limbs [1, 8]. Remarkably, other aspects of movement previously hypothesized to be encoded in premotor cortex, such as direction of movement, hand position, torques, and speed of motion, explained much less or none of the remaining variance.

A general hypothesis on the brain's architecture that might underlie reinforcement learning and behavior based on sensorimotor primitives has been proposed in [10] and has been used for building a modular reinforcement-learning model in [3]. According to this hypothesis sensorimotor primitives are acquired and executed by *cortical pathways* that involve sensory, associative, premotor, and motor cortex. These primitives are then *assembled, selected* and *triggered* to produce reinforcement-based complex behavior by *basal ganglia* (deep nuclei of vertebrates' brain that receive input signals from virtually the whole cortex, send output signals mainly to pre-frontal, premotor and motor cortex [12], and play an important role in chunking and assembling motor primitives in order to accomplish complex reinforcement-based behaviors [7, 8, 11]). This hypothesis has been further investigated in [16] by building a biomimetic model that explicitly incorporates the aforementioned biological evidence reported in [1, 8].

As the model presented shares many features with the model reported in [16], first these features are reviewed and then the main novelties introduced here are highlighted. In both models sensorimotor primitives are neural schemes that allow the system to produce sequences of fine movements that lead the arm to assume particular *final postures*. Both models learn the primitives through a *direct inverse modeling* process [14] based on *spontaneous random movements* performed by the system. The latter aspect of the process is interesting as it is very similar to *motor babbling* observed in infants [15] and might have functions similar to it. In both models, random movements are used for learning to associate limbs' final postures with the movements that led to them. Final postures are represented in a 2D neural map that mimics the function of premotor cortex reported in [1, 8]. Note that such *final postures* can be considered as the *goals* of the corresponding primitives, in fact: (a) the activations of the map's units correlate with the final postures of primitives, but not with other aspects of them (e.g., initial and intermediate postures); (b) the activations take place before the corresponding final-posture states are achieved; (c) the activations drive the system to act in order to get in the states that they encode. The representation of primitives in terms of their goals in the map has the computational advantage of being (almost) local: this eases the selection of them by reinforcement-learning systems (see section 2). Both models assume that basal ganglia select primitives by fueling a *dynamic competition* between their representations in the map: the representation that wins the competition triggers the execution of the corresponding primitive. In both models, the functionalities of basal ganglia are reproduced with an actor-critic rein-forcement-learning model [23]. This model captures several anatomical and physio-logical properties of basal ganglia [3, 10, 11]. The dynamic competition between goals is simulated through an *accumulator model* [24]. Accumulator models are among the best behavioral models of decision making and reaction times; moreover, the activation patterns of their units are similar to those of neurons of premotor cortex of monkeys engaged in action selection tasks [21, 22].

The first novelty of the model presented here is that, while in [16] the representa-tions of the sensorimotor primitives' goals in the 2D map were hand coded, they are now developed through a Kohonen network [13] which takes the arm's angles as input. This has the advantage of leading the map's units to cover the space of "legal" postures in a uniform fashion. Moreover, contrary to [16], the model is now capable of representing all possible postures of the arm in the continuous space of postures by representing them through a *population encoding* [18]. To this purpose, the previously used winner-take-all dynamic competition taking places within the accumulator model has been substituted with a many-winner dynamic competition. A second nov-elty is the proposal of a modified version of the actor-critic reinforcement-learning algorithm capable of selecting postures on the basis of such population encoding (to the best of the authors' knowledge, the learning rule used for training the actor is new). A third novelty is that the system performs a "pre-training" of the actor on the basis of the same motor babbling used for training the sensorimotor primitives. This pre-training allows the actor to learn to associate the *perceived hand's position* with the posture that produces it, and so biases the actor to select sensorimotor primitives that drive the hand on *salient* points in space such as those occupied by objects. This greatly speeds up learning when the system tackles new reinforcement-learning tasks. The whole architecture is tested through a task similar to the one used in [19] to con-duct physiological studies in monkeys engaged in reinforcement-learning action-sequence tasks.

The paper is organized as follows. Section 2 illustrates the architecture and func-tioning of the model, and the task used to test it. Section 3 presents the results of the tests. Section 4 illustrates the strengths of the model, its limitations, and future work.

2 Methods

The Task. The model has been tested with a task similar to the one used by Hikosaka and coworkers [19] to carry out physiological studies of various brain's districts (e.g., frontal cortex, basal ganglia, and cerebellum) of monkeys engaged in learning to perform sequences of reaching actions. In this task a monkey is set in front of a panel containing 16 LED buttons. These buttons are contained in 16 squares organized in a 4×4 grid, each with sides measuring 5 cm (see Fig. 7). The task (see figure Fig. 1) is formed by "hypersets", each composed of five "sets" organized in sequence. In each set, two buttons turn on and the monkey has to press each of them in a precise se-quence, which has to be discovered by trial-and-error, in order to obtain a reward. In case of error, the task re-starts from the first set, while in case of success the task continues with the second set, and so on, until it terminates with the fifth set. Here for simplicity: (a) the test is composed of only one particular hyperset (see Fig. 1) presented to the system several times; (b) the buttons involved in different sets are different; (c) the first LED to be "pressed" in each set is turned off when reached.

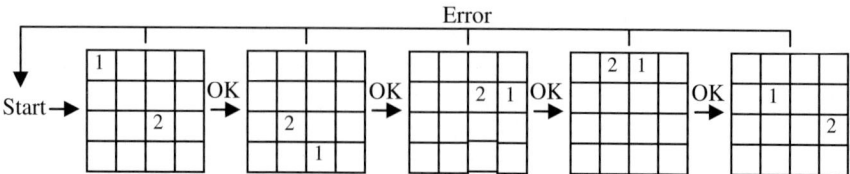

Fig. 1. The "hyperset" of Hikosaka's task used for testing the architecture. Each grid represents a "set": numbers "1" and "2" represent the two LEDs to be reached in sequence within the set.

The system's "body". The system is composed of a two-segment arm that moves on a 2D plane (Fig. 7, left), and a 2D retina. The *retina* is formed by 20×20 units and is supposed to correspond to an "eye" that watches the whole area that the arm can reach from above. The retina's visual field has a size of 40×40 cm and is centered on the arm's shoulder joint (so as to cover the whole area that the arm can reach). The centers of the retina's units are organized in a 20×20 grid that cover to whole visual field. The two segments of the *arm* measure 20 cm each. The arm has two degrees of freedom: the upper arm can move 180° with respect to the system's torso, by pivoting on the shoulder joint, while the forearm can move 180° with respect to the upper arm, by pivoting on the elbow joint (only simple kinematics of the arm were simulated).

The Architecture of the Model. The architecture of the model is shown in Fig. 2. The functioning and learning processes of its components will now be explained in detail (note: the corresponding brain parts will be indicated in *Italics* in brackets).

The *retina*'s units are activated by LEDs. Each LED is simulated as a point with coordinates (c_1, c_2) and when it is on, it activates the retina's units with an activation $x_i \in [0, 1]$ on the basis of Gaussian receptive fields having standard deviation σ (0.75 cm) and centers (c_{1i}, c_{2i}) that correspond to the positions of the units in the visual field:

$$x_i = \exp\left(-\frac{(c_{1i} - c_1)^2 + (c_{2i} - c_2)^2}{\sigma^2}\right) \tag{1}$$

The *actor-critic* components are a neural implementation of the actor-critic model [23]. The *actor* (*basal ganglia's matrix*, cf. [10]) is a two-layer feed-forward neural network with 20×20 input units, that correspond to the units of the retina, and 20×20 output units. The output units have a Sigmoid transfer function with activation y_j and each has a topological one-to-one connection (with weights equal to $\upsilon = +1$) with the posture controller's input units. The *critic* (*basal ganglia's striosomes* and *substantia nigra pars-compacta*, cf. [10]) is mainly composed of a neural network ("evaluator") having a linear output unit. At each step t this output unit produces evaluations V_t of perceived states, and the critic uses couples of successive evaluations, together with the reward signal R_t, to compute the *surprise* signal S_t (*dopamine*) (cf. [23]):

$$S_t = (R_t + \gamma V_t) - V_{t-1} \tag{2}$$

where γ is a discount factor ($\gamma = 0.3$). The surprise signal is used for training both the actor and the evaluator (see [10, 23] and the learning algorithms presented below).

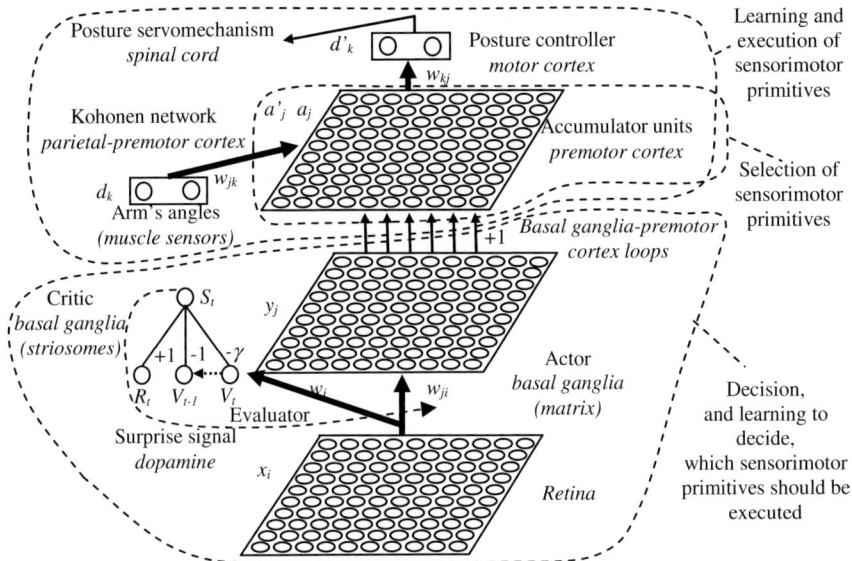

Fig. 2. The neural components of the architecture with the corresponding brain areas in *Italics*. Symbols: grouping: broad functionalities implemented by the architecture's main parts; bold arrows: all-to-all trained connections; thin arrows (only few of them are shown): one-to-one connections (weights = +1); dashed arrow: surprise learning signal; dotted arrow: delay connection; the weights of the critic's one-to-one connections are indicated in the figure.

The *accumulator units* (*premotor cortex*) form a 2D 20×20 map, have all-to-all lateral inhibitions, and have local excitations that decrease with distance on the map. The units engage in a many-winner competition on the basis of the signals ("votes") that they receive from the actor's output units via the one-to-one connections. In particular, they behave as *leaky-integrators* and have an activation a_j as follows:

$$a_{jt} = \max\left[\left(a_{jt-1} + \frac{dt}{\tau}(D)\right), 0\right]$$

$$D = \chi\left(-\delta a_{jt-1} - \iota \sum_{l,l\neq j} a_{lt-1} + \eta \sum_{l,l\neq j} e_{jl} a_{lt-1} + \upsilon y_j + \varepsilon_{jt} + \varepsilon_{jc}\right) \tag{3}$$

where τ is a time constant, corresponding to 1/10s, dt is the integration time step ($dt = 0.05$ 1/10s, so $dt/\tau = 0.05$; a_j is numerically updated every 0.005 s), χ regulates the speed of the dynamics ($\chi = 1$), δ is a decay coefficient ($\delta = 0.1$), ι regulates the all-to-all lateral inhibition ($\iota = 0.15$), η regulates the local lateral excitation ($\eta = 1$), e_k represents the fixed weights of the lateral excitatory connections (e_k is set to 0.4 for neighboring units along the x/y-axes directions, to 0.2 for neighboring units along the diagonals, and to zero for all other units), ε_{jt} is a noise component that ranges over [-0.1, +0.1] and varies in each cycle, ε_{jc} is a noise component that ranges over [-0.25, +0.25] and is constant for time intervals c randomly drawn from [0, 5] s (ε_{jc} is important for exploration of reinforcement learning as various ε_{jt} tends to sum to zero over many steps). When the activation a_j of one accumulator unit reaches a threshold T ($T = 1.9$), the total activation of accumulator units is *normalized to 1*, their dynamics is "frozen", and the execution of a reaching sensorimotor primitive is triggered.

The *posture controller* has an input-unit layer corresponding to the accumulator units and two Sigmoid output units, with activation d'_k, that range over [0, 1] (*motor cortex/spinal cord neurons*). The activations of these output units are remapped onto the arms' angles and form the commands issued to the posture servomechanism in terms of arms' desired angles (posture). It is important to notice that these desired angles are generated by the *cluster* of accumulator units that are active at the end of the many-winner competition. This implies that the target of the executed sensorimotor primitive is a *mixture* of the targets "suggested" by all active units: this *population encoding* allows the arm to cover the whole continuous space of postures.

The *posture servomechanism* is a hardwired closed-loop controller (*Golgi tendon-organs, muscle-fiber afferents,* and *spinal cord,* cf. [22]) that issues commands to the arm's actuators (*muscles*) on the basis of the desired-posture command received from the posture controller. In practice, this component simply changes the arm's current angles in the direction of the desired angles, with maximum changes of 10 degrees.

Learning Phases. The learning processes take place in two phases, the *childhood phase* (three processes) and the *adulthood phase* (one process). Now we first present an overview of these learning processes and then describe them in detail.

During the childhood phase the system performs motor babbling: in practice the arm randomly varies its joints' angles, with changes $\Delta d'_k$ belonging to [-10, +10] degrees, without violating the joints' constraints. Motor babbling is used for performing three learning processes. The first two processes allow the system to learn to perform sensorimotor primitives, in particular: (a) to train the 2D map of accumulator units, through a Kohonen algorithm [13], to represent the postures perceived by the proprioceptive units d_k (during the childhood phase the proprioceptive units, the accumulator units, and their connections, function as a Kohonen network); (b) to train the posture controller, through a Widrow-Hoff algorithm [20] (the generalized "delta rule"), to return as output the arm's angles corresponding to postures encoded in the Kohonen map. These two training processes lead the whole network formed by the Kohonen network and the posture controller to implement an "auto associative" function (i.e., the arm's angles encoded in the proprioceptive units are returned by the postural controller's output units). This whole network allows the system to recode postures, at the level of accumulator units, in an expanded format suitable to perform actor-critic reinforcement learning (cf. [23]). Notice that suitable *population encodings* at the level of the accumulator units allow the system to select *any posture* in the *continuous space* of postures: this is precisely what the actor-critic components learn to do while solving reinforcement-learning reaching tasks in the adulthood phase.

With the third learning process of the childhood phase the system's actor learns, through a Widrow-Hoff algorithm, to associate the point in space where the retina sees the arm's "hand" (i.e., the forearm segment's tip) with the activation pattern of the Kohonen map's units corresponding to such point (pattern caused by the arm's perceived angles). With this training, the actor acquires a bias to select sensorimotor primitives that drive the arm's hand to points in space corresponding to the retina's active units. This bias makes reinforcement learning performed during the adulthood phase quite fast notwithstanding the fact that the continuous space of postures is quite large. Note that two simplifying assumptions allow obtaining this result: (a) the retina does not perceive the arm and hand in the adulthood phase; (b) retina's units activated by the hand in the childhood phase are activated by the LEDs in the adulthood phase.

During the adulthood phase the system learns by trial-and-error to accomplish Hikosaka's task. The actor-critic model used to this purpose has been suitably modified to be capable of selecting "actions" represented with population encodings. The four learning processes are now illustrated in detail.

Childhood phase: training of the Kohonen network. During the childhood phase, while the system performs motor babbling, the accumulator units receive input signals from two input units, having activation d_k, that encode the arm's current angles (remapped in [-1, +1]: this information is thought to be returned by proprioceptive sensors located in the muscles, e.g. *Golgi tendon-organs* and *muscle-fiber afferents*, cf. [22]). An extra pseudo input unit is used to perform a "z-normalisation" of the input pattern: this is a normalization that preserves size information [13]. The accumulator units are trained with a Kohonen algorithm [13] that allows them to develop representations of the arm's angles in their weights. The output units give place to a winner-take-all competition: the unit with the highest activation potential activates with 1 ("winning unit"), while the other units activate at levels decreasing with their distance from the winning unit on the basis of a Gaussian function. In particular, the activation a'_j of the unit j and the rule to update its weights w_{jk} are as follows:

$$a'_j = \exp\left[-\frac{h_{fj}^2}{\sigma^2}\right] \qquad w_{jk\,t} = w_{jk\,t-1} + \phi\, a'_j \left(d_k - w_{jk\,t-1}\right) \qquad (4)$$

where h_{fj} is the distance on the map between the unit j and the winning unit f ($h_{fj} = 1$ for two contiguous units), σ is the standard deviation of the Gaussian function ($\sigma = 1$), ϕ is a learning coefficient ($\phi = 0.01$). Note that the Kohonen algorithm uses a *winner-take-all* competition to activate the accumulator units instead of the *dynamic competition* reported in equation 3, used in the adulthood phase: indeed, the former tends to lead to an activation of the accumulator units that approximates the steady state activation that the same units would get through the latter (cf. [13]).

Childhood phase: training of the posture controller. The posture controller is trained on the basis of a direct inverse modeling procedure [14] that exploits the random movements $\Delta d'_k$ produced by motor babbling as follows: (a) the arm's angles are perceived and categorized by the Kohonen net; (b) a Widrow-Hoff algorithm ([20], learning rate = 0.3) is used for training the posture controller's weights w_{kj} to associate the Kohonen-map units' activation (input pattern) with the angles d'_k caused by the random movements considered as desired output.

Childhood phase: pre-training of the actor. Through this pre-training, based on a Widrow-Hoff algorithm, the actor's weights w_{ji} are trained to associate the position of the hand perceived with the retina (input pattern x) with the corresponding posture (desired output a') encoded in the Kohonen map (learning rate 0.1).

Adulthood phase: actor-critic's reinforcement learning. During the adulthood phase, the actor-critic component is trained to solve the Hikosaka's task by reinforcement learning. During training, R_t is set to 1 when the arm reaches the two targets of any set of the hyperset in the correct order, and to 0 otherwise. The *evaluator* is trained after the selection and execution of a whole sensorimotor primitive (the primitive terminates when the arm reaches the desired posture selected by the posture controller). In particular its weights w_i are trained, on the basis of a Widrow-Hoff algorithm (learning rate $\psi =$ 0.6) and a *TD-rule* (cf. [23]), as follows:

$$w_{it} = w_{it-1} + \psi S_t x_{it-1} = w_{it-1} + \psi\big((R_t + \gamma V_t) - V_{t-1}\big)x_{it-1} \tag{5}$$

Through this learning process, the evaluator's evaluations V_t of the perceived states x_t tend to become higher for states corresponding to postures "closer" to reinforced states, and to form a gradient over the space of postures. The *actor* uses this gradient to learn to select highly rewarding sequences of primitives (cf. [23]). In particular the actor updates its weights w_{ji} with a Widrow-Hoff algorithm (learning rate $\zeta = 0.6$):

$$w_{jit} = w_{jit-1} + \zeta\big((y_{jt-1} + S_t a_{jt-1}) - y_{jt-1}\big(y_{jt-1}(1 - y_{jt-1})\big)x_{it-1} \tag{6}$$

where $(y_{jt-1}(1-y_{jt-1}))$ is the derivative of the Sigmoid function. The functioning of this learning rule is illustrated in Fig. 3. The rule tends to update only the weights of the units of the "winning cluster" because the activation a_j of other units tends to be zero at the end of the race. The votes of the winning units are decreased or increased in correspondence of respectively positive and negative surprises.

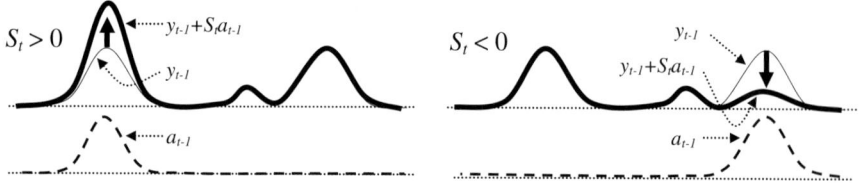

Fig. 3. Effects of the actor's learning rule of equation 6 illustrated with a scheme relative to a 1D layer of actor's output units (horizontal axis). Left: with a surprise $S_t > 0$, the actor's votes y_{t-1} (upper graph), that caused certain accumulator units' final activations a_{t-1} (lower graph), are moved toward the target $y_{t-1} + S_t a_{t-1}$ (upper graph): this causes the votes of the winning cluster of accumulator units to increase (bold arrow) while other votes are not changed. Right: with a surprise $S_t < 0$, actor's votes y_{t-1} are moved toward the target $y_{t-1} + S_t a_{t-1}$: this causes the votes of the winning cluster of accumulator units to decrease, while other votes are not changed.

3 Results

Now we present some tests that prove the computational soundness of the model, illustrate the functioning of its components, and show its capacity to learn sensorimotor primitives, by motor babbling, and to compose sequences of them, by reinforcement learning, on the basis of their population encoding.

During the first training of the childhood phase, the Kohonen network's error (measured as the average over 1,000 cycles of the square of the norm of the difference between the vector of weights and the vector of the input pattern) decreases from 0.411 to 0.034 after 600,000 random arm's movements. After this training the network learns to represent the whole perceived postural space by using its units in a statistically well-distributed fashion (Fig. 4, left graph). This representation is at the basis of the *population encoding* of postures used in the adulthood phase.

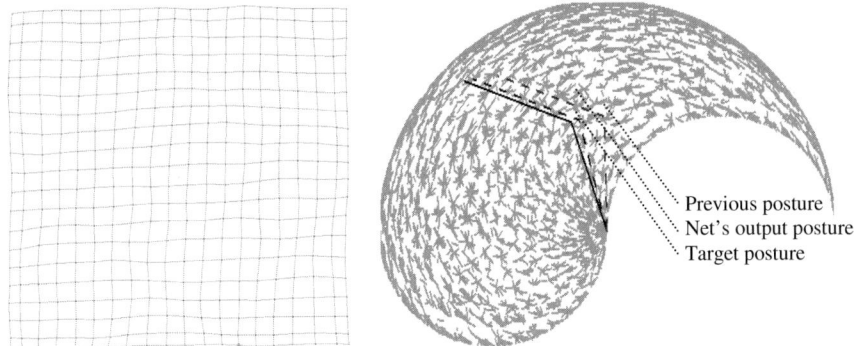

Fig. 4. Left: result of the training of the Kohonen network. Each vertex of the grid represents a node of the Kohonen map, and its x-y coordinates correspond to the node's two weights encoding the arm posture. Right: errors of the posture controller after training, collected while the arm produces several random movements; the graph represents the errors as gray segments plotted between the x-y positions of the hand corresponding to the target actual posture (e.g., black arm) and the position that the hand would have achieved on the basis of the posture controller's output pattern (e.g., dark gray dashed arm; the light gray dashed arm indicates the previous posture assumed by the arm during motor babbling).

During the second training of the childhood phase, the posture controller's error (measured as the average over 1,000 cycles of the distance between the point reached by the arm and the target point) decreases from 8.62 cm to 1.19 cm. Note that this error cannot become very low since the Kohonen network's units are activated on the basis of a Gaussian function *centered on the winner units*, that are in a *finite number*, while the desired output belongs to the whole *continuous* space of arm's postures. Indeed, the right graph of Fig. 4, which shows the residual errors after training, indicates that the hand tends to reach only few specific points corresponding to the vertex of a grid that covers the whole postural space (this grid is explicitly represented in Fig. 5, right graph). In the adulthood phase, this problem is overcome by the population encoding of postures resulting from the accumulator units' activation.

During the third training of the childhood phase, the actor's error (measured as the output units' mean error averaged over 1,000 cycles) decreases from 0.513 to 0.052. This training leads the system formed by the actor, accumulator units, and postural controller to acquire the capacity to perform fine reaching movements in the continuous space of postures even if the accumulator units cover such space at a gross granularity. This can be illustrated by showing the system a sequence of 100 targets positioned along a circumference having a ray of 10 cm and located near the arm's shoulder (see Fig. 5, right graph). The left graph of Fig. 5, which shows the errors between the targets and the points reached by the hand in the test, indicates that the errors are very small (mean: 3.2 mm). Moreover, and more importantly, the system succeeds in reaching virtually any point in the continuous space of postures even if the accumulator units cover such a space with a gross granularity. This skill depends on the mentioned accumulator units' capacity to represent postures by population encodings.

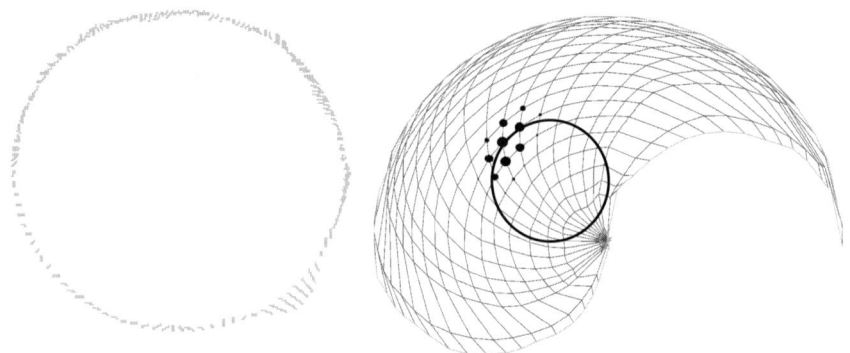

Fig. 5. Left: errors (indicated by the gray segments) between 100 target points positioned on a circumference (shown in the right graph) and the corresponding points reached by the hand. Right: activation (proportional to the size of the full dots) of the actor's output units caused by a target. The positions of the dots and vertexes of the grid plotted in the graph correspond to the positions of the hand related to the "postures" encoded in the accumulator units' weights of the posture controller.

During the adulthood phase, the system is tested with the Hikosaka's task illustrated in section 2. During 120,000 learning cycles, the performance of the system (measured as a 1000-step moving average of rewards) increases from 0.187 to the theoretical maximum of 0.500, when it successfully completes all the five sets of the task in sequence. The results show that the pre-training of the actor gives it a useful bias to reach the targets perceived by the retina. In particular the left graph of Fig. 6, reporting the activations of the actor's output units when the system sees two targets, shows that the units that "vote" for the two possible correct arm's postures form two clusters and have an activation higher than that of other units. The same figure (right graph) shows that the two clusters compete, at the level of the accumulator units, and only one of them "survives" and triggers the corresponding arm's posture when the activation of one of its units reaches the threshold. The left graph of Fig. 7 shows how the arm moves from one target to another, after target postures have been selected, on the basis of the postural servo controller. The same figure (right graph) also shows that the final points reached by the trained arm are quite accurate.

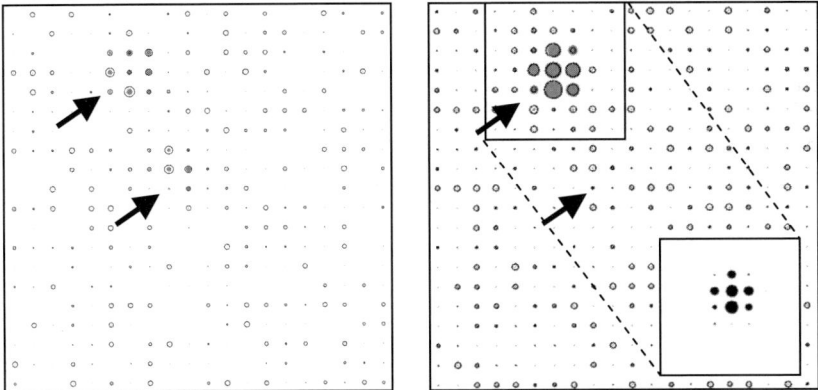

Fig. 6. Left: activations of the actor's output units before adulthood training caused by the perception of two targets in the Hikosaka's task (the area of the gray dots and black circumferences is proportional to the units' activations respectively before and after the addition of noise); the two arrows indicate two clusters of units with activation higher than that of the other units due to the actor's pre-training. Right: activation of the same units after training; notice how one of the two clusters has been strengthened while the other one has disappeared; the activation of the units of the strengthened cluster cause an activation of the accumulator units, at the end of the race, as plotted in the bottom right small graph.

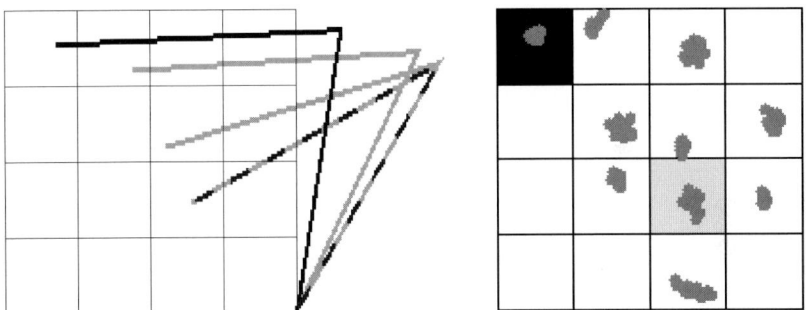

Fig. 7. Left: the trained arm that moves from the first to the second LED of "set 1" of Hikosaka's test under the control of the postural servomechanism (the two LEDs are represented by the black and light gray squares in the right graph). Right: panel with the LEDs, with gray dots indicating the positions reached by the hand of the trained arm in several trials of the hyperset.

4 Conclusions

This paper presented an architecture to solve reaching tasks by reinforcement learning. The architecture is based on the idea, suggested by recent neuroscientific research, according to which monkeys' sensorimotor behavior involving upper-limbs is organized on the basis of a repertoire of sensorimotor primitives that are represented in premotor cortex in terms of the limbs' final postures that they produce. The architecture uses motor babbling to learn sensorimotor primitives, develops a map of units that represent the corresponding postures on the basis of population encodings (so mimicking premotor cortex), and selects primitives on the basis of a biological-plausible accumulation model. Moreover, it proposes a novel learning rule which allows the actor of the actor-critic component (supposed to correspond to basal ganglia) to learn to select sensorimotor primitives on the basis of the population-encoding of their postural goals. The relevance of these novelties resides in the fact that population-encoding representations are widespread in real brains [18], so it is important to have reinforcement-learning models that can function on the basis of them.

The main limitations of the architecture that will be the starting point for future work. First, tests are needed to verify if the system can scale to arms with redundant degrees of freedom and/or to arms with a number of degrees of freedom higher than the number of the dimensions of the Kohonen network. Second, the Kohonen network functions on the basis of a winner-take-all competition: in the future this will be substituted with the same many-winner competition used while performing reinforcement learning. This improvement is relevant for the biological plausibility of the system. Third, although very detailed, the architecture takes into account only a part of the relevant available neuroscientific empirical evidence. For example, it does not model the different time courses of learning in basal ganglia and prefrontal cortex [17], the role of basal-ganglia direct and indirect pathways [10, 12], the possible separation of selection vs. control pathways [9], and the role of ventral and dorsal portions for appetitive and consummatory behaviors [5].

References

1. Aflalo, T.N., Graziano, M.S.A.: Partial Tuning of Motor Cortex Neurons to Final Posture in a Free-Moving Paradigm. Proceedings of the National Academy of Science 103(8) (2006) 2909-2914
2. Arbib, M.: Visuomotor Coordination: From Neural Nets to Schema Theory. Cognition and Brain Theory 4 (1981) 23-39
3. Baldassarre, G.: A Modular Neural-Network Model of the Basal Ganglia's Role in Learning and Selecting Motor Behaviours. Journal of Cognitive Systems Research 3 (2002) 5-13
4. Barto, A.G., Mahadevan S.: Recent Advances in Hierarchical Reinforcement Learning. Discrete Event Dynamic Systems, 13 (2003) 341-379
5. Girard, B., Filliat, D., Meyer, J.-A., Berthoz, A., and Guillot, A.: Integration of Navigation and Action Selection Functionalities in a Computational Model of Cortico-Basal Ganglia-Thalamo-Cortical Loops. Adaptive Behavior 13(2) (2005) 115-130
6. Giszter, S.F., Mussa-Ivaldi, F.A., Bizzi, E.: Convergent Force Fields Organised in the Frog's Spinal Cord. Journal of Neuroscience 13 (2) (1993) 467-491
7. Graybiel, A. M.: The Basal Ganglia and Chunking of Action Repertoires. Neurobiology of Learning and Memory 70 (1998) 119-136
8. Graziano, M.S., Taylor, C.S., Moore, T.: Complex Movements Evoked by Microstimulation of Precentral Cortex. Neuron 34 (2002) 841-851.
9. Gurney, K., Prescott, T. J., Redgrave, P.: A Computational Model of Action Selection in the Basal Ganglia I. A New Functional Anatomy. Biological Cybernetics, 84 (2001) 401-410.
10. Houk, J.C., Davis, J.L., Beiser, D.G. (eds.): Models of Information Processing in the Basal Ganglia. MIT Press, Cambridge MA (1995)

11. Joel, D.E.E., Niv, Y., Ruppin, E.: Actor-critic Models of the Basal Ganglia: New Anatomical and Computational Perspectives. Neural Networks 15 (2002) 535-547.
12. Kandel, E.R., Schwartz, J.H., Jessell, T.M.: Principles of Neural Science. McGraw-Hill, New York (2000)
13. Kohonen, T.: Self-Organizing Maps. 3rd edn. Springer-Verlag, Berlin Heidelberg New York (2001)
14. Kuperstein, M.: A Neural Model of Adaptive Hand-Eye Coordination for Single Postures. Science 239 (1988) 1308-1311
15. Meltzoff, A. N., Moore, M. K.: Explaining Facial Imitation: A Theoretical Model. Early Development and Parenting 6 (1997) 179-192
16. Ognibene, D., Mannella, F., Pezzulo, G., Baldassarre, G.: Integrating Reinforcement-Learning, Accumulator Models, and Motor-Primitives to Study Action Selection and Reaching in Monkeys. In: Fum, D., Del Missier, F., Stocco, A. (eds.): Proceedings of the 7th International Conference on Cognitive Modelling - ICCM06 (2006) 214-219
17. Pasupathy, A., Miller E.K.: Different Time Courses of Learning-Related Activity In the Prefrontal Cortex and Striatum. Nature 433 (2005) 873-876
18. Pouget A., Lathaam P. E.: Population Codes. In : Arbib, M. A. (ed.): The Handbook of Brain Theory and Neural Networks. MIT Press, Cambridge MA (2003) 893-897
19. Rand, M.K., Hikosaka, O., Miyachi, S., Lu, X., Miyashita, K.: Characteristics of a Long-Term Procedural Skill in the Monkey. Experimental Brain Research 118 (1998) 293-297
20. Widrow, B., Hoff, M.E.: Adaptive Switching Circuits. IRE WESCON Convention Record, Part 4 (1960) 96-104
21. Schall, J.D.: Neural Basis of Deciding, Choosing and Acting. Nature Reviews Neuroscience 2 (2001) 33-42
22. Shadmehr, R., Wise, S.: The Computational Neurobiology of Reaching and Pointing. MIT Press, Cambridge MA (2005)
23. Sutton, R.S., Barto, A.G.: Reinforcement Learning: An Introduction. MIT Press, Cambridge MA (1998)
24. Usher, M., McClelland, J.L.: On the Time Course of Perceptual Choice: The Leaky Competing Accumulator Model. Psychological Review 108 (2001) 550-592.

Combining Self-organizing Maps with Mixtures of Experts: Application to an Actor-Critic Model of Reinforcement Learning in the Basal Ganglia

Mehdi Khamassi[1,2], Louis-Emmanuel Martinet[1], and Agnès Guillot[1]

[1] Université Pierre et Marie Curie - Paris 6, UMR7606, AnimatLab - LIP6, F-75005 Paris, France ; CNRS, UMR7606, F-75005 Paris, France
[2] Laboratoire de Physiologie de la Perception et de l'Action, UMR7152 CNRS, Collège de France, F-75005 Paris, France
{mehdi.khamassi, louis-emmanuel.martinet, agnes.guillot}@lip6.fr
http://animatlab.lip6.fr

Abstract. In a reward-seeking task performed in a continuous environment, our previous work compared several Actor-Critic (AC) architectures implementing dopamine-like reinforcement learning mechanisms in the rat's basal ganglia. The task complexity imposes the coordination of several AC submodules, each module being an expert trained in a particular subset of the task. We showed that the classical method where the choice of the expert to train at a given time depends on each expert's performance suffered from strong limitations. We rather proposed to cluster the continuous state space by an *ad hoc* method that lacked autonomy and generalization abilities. In the present work we have combined the mixture of experts with self-organizing maps in order to cluster autonomously the experts' responsibility space. On the one hand, we find that classical *Kohonen maps* give very variable results: some task decompositions provide very good and stable reinforcement learning performances, whereas some others are unadapted to the task. Moreover, they require the number of experts to be set a priori. On the other hand, algorithms like *Growing Neural Gas* or *Growing When Required* have the property to choose autonomously and incrementally the number of experts to train. They lead to good performances, even if they are still weaker than our hand-tuned task decomposition and than the best Kohonen maps that we got. We finally discuss on propositions about what information to add to these algorithms, such as knowledge of current behavior, in order to make the task decomposition appropriate to the reinforcement learning process.

1 Introduction

In the frame of the Psikharpax project, which aims at building an artificial rat having to survive in complex and changing environments, and having to satisfy different needs and motivations [5][14], our work consists in providing a simulated robot with habit learning capabilities, in order to make it able to associate efficient behaviors to relevant stimuli located in an unknown environment.

S. Nolfi et al. (Eds.): SAB 2006, LNAI 4095, pp. 394–405, 2006.

The control architecture of Psikharpax is expected to be as close as possible to known anatomy and physiology of the rat brain, in order to unable comparison between functioning of the model with electrophysiological and behavioral recordings. As a consequence, our model of reinforcement learning is based on an Actor-Critic architecture inspired from basal ganglia circuits, following well established hypotheses asserting that this structure of the mammalian brain is responsible for driving action selection [16] and reinforcement learning of behaviors to select via substantia nigra dopaminergic neurons [17].

At this stage of the work, our model runs in 2D-simulation with a single need and a single motivation. However the issue at stake already has a certain complexity: it corresponds to a continuous state-space environment; the perceptions have non monotonic changes; an obstacle-avoidance reflex can interfere with actions selected by the model; the reward location provides a non instantaneous reward. In a previous paper [11], we demonstrated that this task complexity requires the use of multiple Actor-Critic modules, where each module is an expert trained in a particular subset of the environment. We compared different hypotheses concerning the management of such modules, concerning there more or less autonomously determined coordination, and found that the classical mixture of experts method - where the choice of the expert to train at a given time depends on each expert's performance [3][4] – cannot train more than one single expert in our reinforcement learning task. We rather proposed to cluster the continuous state space and to link each expert to a cluster by an ad hoc method that could indeed solve the task, but that lacked autonomy and generalization abilities.

The objective of the present work is to provide an autonomous categorization of the state space by combining the mixture of experts with self-organizing maps (SOM). This combination has already been implemented by Tang et al. [20] - these authors having criticized the undesirable effects of classical mixture of experts on boundaries of non disjoint regions. However, they did not test the method in a reinforcement learning task. When they were used in such tasks [18][13] - yet without mixture of experts –, SOM were applied to the discretization of the input space to the reinforcement learning model, which method suffers from generalization abilities. Moreover, the method has limited performance in high-dimensional spaces and remains to be tested robustly on delayed reward tasks.

In our case, we propose that the SOM algorithms have to produce a clustering of the responsibility space of the experts, in order to decide which Actor-Critic expert has to work in a given zone of the perceptual state space. In addition, the selected Actor-Critic expert of our model will receive the entire state space, in order to produce a non constant reward prediction inside the given zone.

After describing the task in the following section, we will report the test of three self-organizing maps combined with the mixture of Actor-Critic experts, for the comparison of their usefulness for a complex reinforcement learning task. It concerns the classical *Kohonen* algorithm [12], which requires the number of experts to be a priori set; the *Growing Neural Gas* algorithm [6], improved by [9], which adds a new expert when an existing expert has a important error of classification; and the *Growing When Required* algorithm [15], which creates a new expert when habituation of the map to visual inputs produces a too weak output signal when facing new visual data.

In the last section of the paper, we will discuss the possible modifications that could improve the performance of the model.

2   The Task

Figure 1 shows the simulated experimental setup, a simple 2D plus-maze. The dimensions are equivalent to a 5m * 5m environment with 1m large corridors. In this environment, walls are made of segments colored on a 256 grayscale. The effects of lighting conditions are not simulated. Every wall of the maze is colored in black (luminance = 0), except walls at the end of each arm and at the center of the maze, which are represented by specific colors: the cross at the center is gray (191), three of the arm ends are dark gray (127) and the fourth is white (255), indicating the reward location equivalent to a water trough delivering two drops (non instantaneous reward) – not a priori known by the animat.

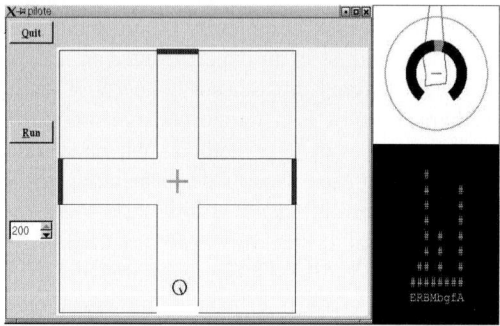

Fig. 1. Left: the robot in the plus-maze environment. Upper right: the robot's visual perceptions. Lower right: activation level of different channels in the model.

The plus-maze task reproduces the neurobiological and behavioral experiments that will serve as future validation for the model [1]. At the beginning of each trial, one arm end is randomly chosen to deliver reward. The associated wall becomes white whereas the other arm ends become dark gray. The animat has to learn that selecting the action *drinking* when it is near the white wall (distance < 30 cm) and faces it (angle < 45°) gives it two drops of water. Here we assume that reward = 1 for n iterations (n = 2) during which the action *drinking* is being executed. However, the robot's vision does not change between these two moments, since the robot is then facing the white wall. As visual information is the only sensory modality that will constitute the input space of the Actor-Critic model, this makes the problem to solve a Partially Observable Markov Decision Process [19]. This characteristic was set in order to fit the multiple consecutive rewards that are given to rats in the neurobiological plus-maze, enabling comparison between our algorithm with the learning process that takes place in the rat brain during the experiments.

We expect the animat to learn a sequence of context-specific behaviors, so that it can reach the reward site from any starting point in the maze:

- When not seeing the white wall, face the center of the maze and move forward
- As soon as arriving at the center (the animat can see the white wall), turn to the white stimulus

- Move forward until being close enough to reward location
- Drink

The trial ends when reward is consumed: the color of the wall at reward location is changed to dark gray, and a new arm end is randomly chosen to deliver reward. The animat has then to perform another trial from the current location. The criterion chosen to validate the model is the time – number of iterations of the algorithm - to goal, plotted along the experiment as the learning curve of the model.

3   The Animat

The animat is represented by a circle (30 cm diameter). Its translation and rotation speeds are 40 cm.s^{-1} and 10°.s^{-1}.
Its simulated sensors are:

- Eight sonars with a 5m range, an incertitude of ±5 degrees concerning the pointed direction and an additional ±10 cm measurement error. The sonars are used by a low level obstacle avoidance reflex which overrides any decision taken by the Actor-Critic model when the animat comes too close to obstacles.
- An omnidirectional linear camera providing every 10° the color of the nearest perceived segment. This results in a 36 colors table that constitute the animat's visual perception (see figure 1).

The animat is provided with a visual system that computes 12 input variables and a constant equal to 1 $(\forall i \in [1;13], 0 \le \text{var}_i \le 1)$ out of the 36 colors table at each time step. These sensory variables constitute the state space of the Actor-Critic and so will be taken as input to both the Actor and the Critic parts of the model (figure 3). Variables are computed as following:

- *seeWhite* (resp. *seeGray, seeDarkGray*) = 1 if the color table contains the value 255 (resp. 191, 127), else 0.
- *angleWhite, angleGray, angleDarkGray* = (number of boxes in the color table between the animat's head direction and the desired color) / 18.
- *distanceWhite, distanceGray, distanceDarkGray* = (maximum number of consecutive boxes in the color table containing the desired color) / 18.
- *nearWhite* (resp. *nearGray, nearDarkGray*) = 1 – *distanceWhite* (resp. *distanceGray, distanceDarkGray*).

The model permanently receives a flow of sensory information and has to learn autonomously the sensory contexts that can be relevant for the task resolution.

The animat has a repertoire of 6 actions: *drinking, moving forward, turning to white perception, turning to gray perception, turning to dark gray perception,* and *waiting*. These actions constitute the output of the Actor model (described below) and the input to a low-level model that translates it into appropriate orders to the animat's engines.

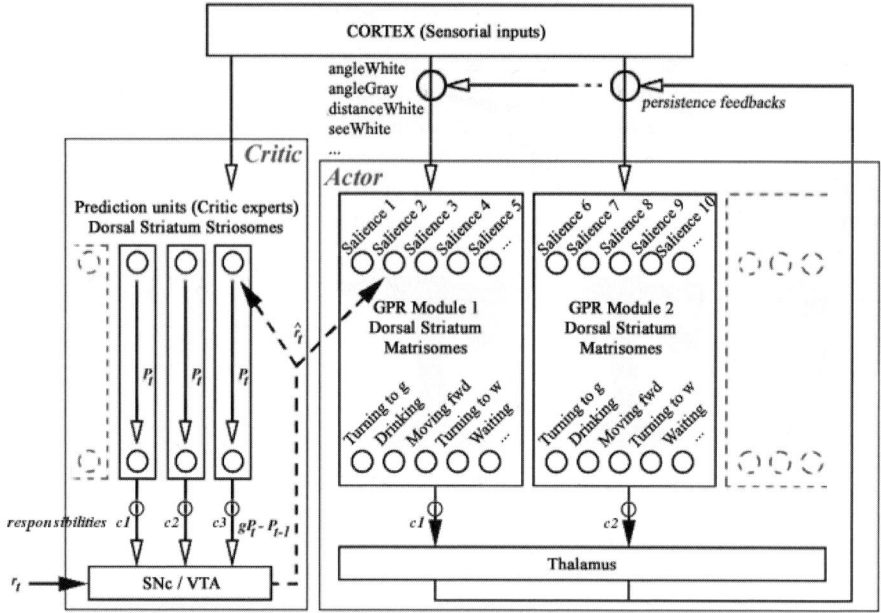

Fig. 2. General scheme of the model tested in this work. The Actor is a group of "GPR" modules [8] with saliences as inputs and actions as outputs. The Critic (involving striosomes in the dorsal striatum, and the substantia nigra compacta (SNc)) propagates towards the Actor an estimate ř of the instantaneous reinforcement triggered by the selected action. The particularity of this scheme is to combine several modules for both Actor and Critic, and to gate the Critic experts' predictions and the Actor modules' decisions with responsibility signals. These responsibilities can be either computed by a Kohonen, a GWR or a GNG map.

4 The Model

4.1 The Multi-module Actor-Critic

The model tested in this work has the same general scheme than described in [11]. It has two main components, an Actor which selects an action depending on the visual perceptions described above; and a Critic, having to compute predictions of reward based on these same perceptions (figure 2). Each of these two components is composed of N submodules or *experts*. At a given time, each submodule k ($k \in [1 ; N]$) has a responsibility $c_k(t)$ that determines its weight in the output of the overall model. In the context of this work, we restrict to the case where only one expert k has its responsibility equal to 1 at a given moment, and $\forall j \neq k, c_j(t) = 0$.

Inside the Critic component, each submodule is a single linear neuron that computes its own prediction of reward:

$$p_k(t) = \sum_{j=1}^{13} w'_{kj}(t) \cdot \text{var}_j(t) \tag{1}$$

where $w'_{kj}(t)$ is the synaptic weight of expert k representing the association strength with input variable j. Then the global prediction of the Critic is a weighted sum of experts' predictions:

$$P(t) = \sum_{k=1}^{N} c_k(t) \cdot p_k(t) \tag{2}$$

Concerning the learning rule, derived from the Temporal-Difference Learning algorithm [19], each expert has a specific reinforcement signal based on its own prediction error:

$$\hat{r}_k(t) = r(t) + gP(t) - p_k(t-1) \tag{3}$$

The synaptic weights of each expert k are updated according to the following formula:

$$w'_{kj}(t) \leftarrow w'_{kj}(t-1) + \kappa \cdot \hat{r}_k(t) \cdot \text{var}_j(t-1) \cdot c_k(t) \tag{4}$$

Actor submodules also have synaptic weights $w_{i,j}(t)$ that determine, inside each submodule k, the salience – i.e. the strength – of each action i according to the following equation:

$$sal_i(t) = \left[\sum_{j-1}^{13} \text{var}_j(t) \cdot w_{i,j}(t) \right] + persist_i(t) \cdot w_{i,14}(t) \tag{5}$$

The action selected by the Actor to be performed by the animat corresponds to the strongest output of the submodule with responsibility 1. If a reinforcement signal occurs, the synaptic weights of the latter submodule are updated following equation (4).

An exploration function is added that would allow the animat to try an action in a given context even if the weights of the Actor do not give a sufficient tendency to perform this action in the considered context.

To do so, we introduce a clock that triggers exploration in two different cases:

- When the animat has been stuck for a large number of timesteps (*time* superior to a fixed threshold α) in a situation that is evaluated negative by the model (when the prediction $P(t)$ of reward computed by the Critic is inferior to a fixed threshold).
- When the animat has remained for a long time in a situation where $P(t)$ is high but this prediction doesn't increase that much ($|P(t+n) - P(t)| < \varepsilon$) and no reward occurs.

If one of these two conditions is true, exploration is triggered: one of the 6 actions is chosen randomly. Its salience is being set to 1 (Note that: when exploration

= false, $sal_i(t)<1$, $\forall i,t,w_{i,j}(t)$) and is being maintained to 1 for a duration of 15 timesteps (time necessary for the animat to make a 180° turn or to run from the center of the maze until the end of one arm).

4.2 The Self-organizing Maps

In our previous work [11], we showed that the classical method used to determine the experts' responsibilities – a gating network, giving the highest responsibility to the expert that approximates the best the future reward value [3][4] – was not appropriate for the resolution of our reinforcement learning task. Indeed, we found that the method could only train one expert which would remain the more responsible in the entire state space without having a good performance. As our task is complex, we rather need the region of the state space where a given expert is the most responsible to be restricted, in order to have only limited information to learn there. As a consequence, we propose that the state space should be clustered independently from the performance of the model in learning the reward value function.

In this work, the responsibility space of the Actor-Critic experts is determined by one of the following self-organizing maps (SOMs): the Kohonen Algorithm, the Growing Neural Gas, or the Growing When Required. We will describe here only essential aspects necessary for the comprehension of the method maps. Each map has a certain number of nodes, receives as an input the state space constituted of the same perception variables than the Actor-Critic model, and will autonomously try to categorize this state space. Training of the SOMs is processed as following:

```
Begin
   Initialize a fixed number of nodes (for the Kohonen
   Map) or 2 nodes for GNG and GWR algorithms;
   While (iteration < 50000)
     Move the robot randomly; //Actor-Critic disabled
     Try to categorize the current robot's perception;
     If (GNG or GWR) and (classification-error > threshold)
        Add a new node to the map;
     End if;
     Adapt the map;
   End;
   // After that, the SOM won't be adapted anymore
   While (trial < 600)
     Move the robot with the Actor-Critic (AC) model;
     Get the current robot's perception;
     Find the SOM closest node (k) to this perception;
     Set expert k responsibility to 1 and others to 0;
     Compute the learning rule and adapt synaptic weights of the AC;
   End;
End;
```

Parameters used for the three SOM algorithms are given in the appendix table. Figure 3 shows some examples of categorization of the state space obtained with a GWR algorithm. Each category corresponds to a small region in the plus-maze, where its associated Actor-Critic expert will have to learn. Notice that we set the parameters so that regions are small enough to train at least several experts, and large enough to require that some experts learn to select different actions successively inside the region.

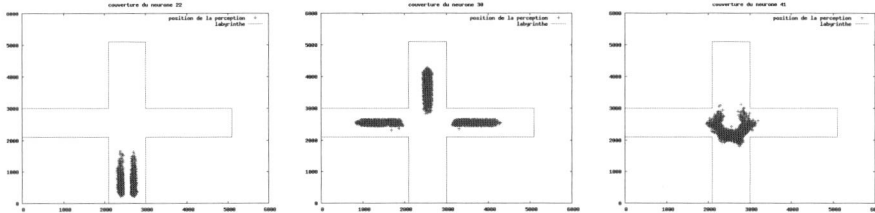

Fig. 3. Examples of clusterings found by the GWR self-organizing map. The pictures show, for three different AC experts, the positions of the robot for which the expert has the highest responsibility – thus, positions where the Actor-Critic expert is involved in the learning process.

5   Results

The results correspond to several experiments of 600 trials for each of the three different methods (11 with GWR, 11 with GNG, and 11 with Kohonen maps). Each experiment is run following the algorithmic procedure described in the previous section.

Table 1. Summarized performances of the methods applied to reinforcement learning

Method	Average performance during second half of the experiment (nb iterations per trials)	Standard error	Best map's average performance
Hand-tuned map	93.71	N/A	N/A
KOH (n=11)	548.30	307.11	87.87
GWR (n=11)	459.72	189.07	301.76
GNG (n=11)	403.73	162.92	193.39

Figure 4 shows the evolution with time of the learning process of each method. In each case, the smallest number of iterations occurs around the 250th trial and remains stabilized. Table 1 summarizes the global performances averaged over the second half of the experiment – e.g. after trial #300. Performances of the three methods are comparable (Kruskall-Wallis test reveals no significant differences: $p > 0.10$). When looking at the maps' categorizations precisely and independently from the reinforcement learning process, measure of the maps' errors of categorization highlights that Kohonen maps provide a slightly worst result in general, even while using more neurons than the GWR and GNG algorithms. However, this doesn't seem to have consequences on the reinforcement learning process, since performances are similar. So, the Kohonen algorithm, whose number of experts is a priori set, is not better than the two others which recruit new experts autonomously.

Performances with GNG and GWR algorithms are not very different either. In their study, Marsland et al. [15] conclude that GWR is slightly better than the GNG algorithm in its original version. Here, we used a modified version of GNG [9]. In our simulations, the GNG recruited on average less experts than the GWR but had a classification error a little bigger. However, when applied to reinforcement learning, the categorizations provided by the two algorithms did not show major differences.

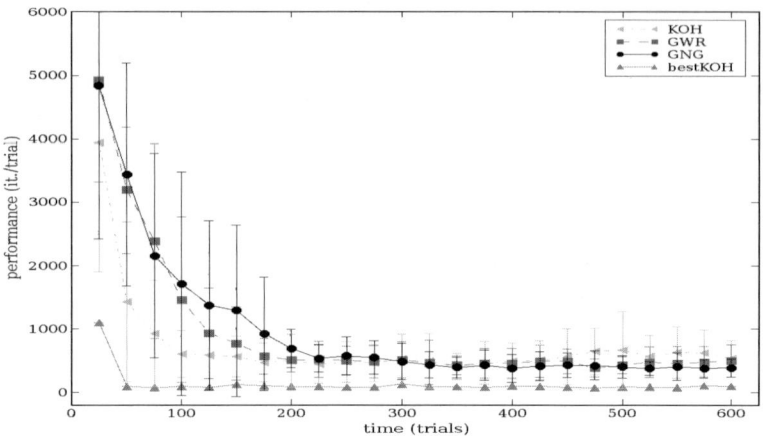

Fig. 4. Learning curves of the reinforcement learning experiments tested with different self-organizing maps

Qualitatively, the three algorithms have provided the multi-module Actor-Critic with quite good experts' responsibility space clustering, and the animat managed to learn an appropriate sequence of actions to the reward location. However, performances are still not as good as a version of the model with hand-tuned synaptic weights. The latter has an average performance of 93.71 iterations per trial, which is characterized by a nearly "optimal" behavior where the robot goes systematically straight to the reward location, without loosing any time (except the regular trajectory deviation produced by the exploration function of the algorithm). Some of the best Kohonen maps and GNG maps reached similar nearly optimal behavior. As shown in table 1, the best Kohonen map got an average performance of 87.87 iterations per trial. Indeed, it seems that the categorization process can produce very variable reinforcement learning depending on the map built during the first part of the experiment.

6   Discussion

In this work, we have combined three different self-organizing maps with a mixture of Actor-Critic experts. The method was designed to provide an Actor-Critic model with autonomous abilities to recruit new expert modules for the learning of a reward-seeking task in continuous state space. Provided with such a control architecture, the simulated robot can learn to perform a sequence of actions in order to reach the reward. Moreover, gating Actor-Critic experts with our method strongly ressembles neural activity observed in the striatum – e.g. the input structure of the basal ganglia – in rat performing habit learning tasks in an experimental maze [10]. Indeed, the latter study shows striatal neurons' responses that are restricted to localized chunks of the trajectory performed by the rat in the maze. This is comparable with the clusters of experts' responsibilities shown in figure 3.

However, the performance of the model presented here remains weaker than a hand-tuned behavior. Indeed, the method produces very variables results, from maps with nearly optimal performance to maps providing unsatisfying robotics behavior.

Analysis of the maps created with our method shows that some of them are more appropriate to the task than others, particularly when the boundaries between two experts' receptive fields corresponds to a region of the maze where the robot should switch from one action to another in order to get the reward. As an example, we noticed that the majority of the maps obtained in this work had their expert closer to the reward location with a too large field of responsibility. As a consequence, the trunk of the global value function that this expert has to approximate is more complex, and the behavior to learn is more variable. This results in selecting inappropriate behavior in the field of this expert – for example, the robot selects the action "drinking" too far from reward location to get a reward. Notice however that this is not a problem with selecting several different actions in the same region of the maze, since some experts managed to learn to alternate between two actions in their responsibility zone, for example in the area close to the center of the plus-maze. A given expert having limited computational capacities, its limitations occur when its region of responsibility is too large.

To improve the performance, one could suggest setting parameters of the SOM in order to increase the number of experts in the model. However, this would result in smaller experts' receptive field than those presented in figure 3. As a consequence, each expert would receive a nearly constant input signal inside its respective zone, and would need only to select one action. This would be computationally equivalent to the use of small fields place cells for the clustering of the state space of an Actor-Critic, which has been criticized by several authors [2], and would not be different than other algorithms where the winning node of a self-organizing map produces a discretization of the input space to a reinforcement learning process [18].

One could also propose to increase each expert-module's computational capacity. For instance, one could use a more complex neural network than the single linear neuron that we implemented for each expert. However, one cannot a priori know the task complexity, and no matter the number of neurons an expert possesses, there could still exist too complex situations. Moreover, "smart" experts having a small responsibility region could overlearn the data with poor generalization ability [7].

7 Perspective

In future work, we rather propose to enable the experts' gating to adapt slightly to the behavior of the robot. The management of experts should not be mainly dependent on the experts' performances in controlling behavior and estimating the reward value, as we have shown in previous work [11]. However, considering the categorization of the visual space as the main source of experts' specialization, it could be useful to add information about the behavior in order for boundaries between two experts' responsibility regions to flexibly adapt to areas where the animat needs to switch its behavior. In [21], the robot's behavior is a priori set and stabilized, and constitutes one of the inputs to a mixture of experts having to categorize the sensory-motor flow perceived by a robot. In our case, at the beginning of the reinforcement learning process, when behavior is not yet stable, visual information could be the main source of experts' specialization. Then, when the model starts to learn an appropriate sequence of actions, behavioral information could help adjusting the specialization. This would be similar to electrophysiological recordings of the striatum showing that, after extensive training of the rats, striatal neurons' responses tend to translate to particular "meaningful" portions of the behavioral sequences, such as the starting point and the goal location [10].

Acknowledgments

This research has been granted by the LIP6 and the European Project Integrating Cognition, Emotion and Autonomy (ICEA). The authors wish to thank Thomas Degris and Jean-Arcady Meyer for useful discussions.

References

1. Albertin, S. V., Mulder, A. B., Tabuchi, E., Zugaro, M. B., Wiener, S. I.: Lesions of the medial shell of the nucleus accumbens impair rats in finding larger rewards, but spare reward-seeking behavior. *Behavioral Brain Research.* 117(1-2) (2000) 173-83
2. Arleo, A., W. Gerstner,W.: Spatial cognition and neuro-mimetic navigation: a model of hippocampal place cell activity. *Biological Cybernetics*, 83(3) (2000) 287-99
3. Baldassarre, G.: A modular neural-network model of the basal ganglia's role in learning and selecting motor behaviors. *Journal of Cognitive Systems Research,* 3(1) (2002) 5-13
4. Doya, K., Samejima., K., Katagiri, K., Kawato, M.: Multiple model-based reinforcement learning. *Neural Computation,* 14(6) (2002) 1347-69
5. Filliat, D., Girard, B., Guillot, A., Khamassi, M., Lachèze, L., Meyer, J.-A. : State of the artificial rat Psikharpax. In: Schaal, S., Ijspeert, A., Billard, A., Vijayakumar, S., Hallam, J., Meyer, J.-A. (eds): *From Animals to Animats 8: Proceedings of the Seventh International Conference on Simulation of Adaptive Behavior,* Cambridge, MA. MIT Press (2004) 3-12
6. Fritzke, B.: A growing neural gas network learns topologies. In: Tesauro, G, Touretzkys, D.S., Leen, K.(eds): *Advances in Neural Information Processing Systems*, MIT Press, (1995) 625-32
7. Geman, S., Bienenstock, E., Doursat, R.: Neural networks and the bias/variance dilemma. *Neural Computation* 4 (1992) 1-58
8. Gurney, K.,Prescott, T.J., Redgrave, P.: A computational model of action selection in the basal ganglia. I. A new functional anatomy. *Biological Cybernetics,* 84 (2001) 401-10
9. Holmström, J.: Growing neural gas : Experiments with GNG, GNG with utility and supervised GNG. Master's thesis, Uppsala University (2002)
10. Jog, M.S., Kubota, Y., Connolly, C.I., Hillegaart, V., Graybiel, A.M.: Building neural representations of habits. *Science*, 286(5445) (1999) 1745-9
11. Khamassi, M., Lachèze, L., Girard, B., Berthoz, A., Guillot, A: Actor-critic models of reinforcement learning in the basal ganglia: From natural to artificial rats. *Adaptive Behavior, Special Issue Towards Artificial Rodents* 13(2) (2005) 131-48
12. Kohonen, T.: Self-organizing maps. Springer-Verlag, Berlin (1995)
13. Lee, J. K., Kim, I. H.: Reinforcement learning control using self-organizing map and multi-layer feed-forward neural network. In: *International Conference on Control Automation and Systems,* ICCAS 2003 (2003)

14. Meyer, J.-A., Guillot, A., Girard, B., Khamassi, M., Pirim, P., Berthoz, A.: The Psikharpax project: Towards building an artificial rat. *Robotics and Autonomous Systems* 50(4) (2005) 211-23
15. Marsland, S., Shapiro, J., Nehmzow, U.: A self-organising network that grows when required. *Neural Networks*, 15 (2002) 1041-58
16. Prescott, T.J., Redgrave, P., Gurney, K.: Layered control architectures in robots and vertebrates. *Adaptive Behavior, 7* (1999) 99-127
17. Schultz, W., Dayan, P., Montague, P. R. : A neural substrate of prediction and reward. *Science*, 275 (1997) 1593-9
18. Smith, A. J.: Applications of the self-organizing map to reinforcement learning. *Neural Networks* 15(8-9) (2002)1107-24
19. Sutton, R. S., Barto, A. G.: Reinforcement learning: An introduction. The MIT Press Cambridge, MA. (1998)
20. Tang, B., Heywood, M. I.: Shepherd, M.: Input Partitioning to Mixture of Experts. In: *IEEE/INNS International Joint Conference on Neural Networks*, Honolulu, Hawaii (2002) 227-32
21. Tani, J., Nolfi, S.: Learning to perceive the world as articulated: an approach for hierarchical learning in sensory-motor systems.*Neural Networks*12(1999)1131-41

Appendix: Parameters Table

Symbol	Value	Description
Δt	1 sec.	Time between two successive iterations of the model.
α	[50;100]	Time threshold to trigger the exploration function.
g	0.98	Discount factor of the Temporal Difference learning rule.
η	0.05 / 0.01	Learning rate of the Critic and the Actor respectively.
N	36	Number of nodes in Kohonen Maps.
$\eta\text{-}koh$	0.05	Learning rate in Kohonen Maps.
σ	3	Neighborhood radius in Kohonen Maps.
Ew, En	0.5, 0.005 / 0.1, 0.001	Learning rates in the GNG and GWR respectively.
$a\text{-}max$	100	Max. age in the GNG and GWR.
S		Threshold for nodes recruitment in the GNG.
$\alpha\text{-}gng, \beta\text{-}gng$	0.5, 0.0005	Error reduction factors in the GNG.
λ	1	Window size for nodes incrementation in the GNG.
$a\text{-}T$	0.8	Activity threshold in the GWR
$h\text{-}T$	0.05	Habituation threshold in the GWR.

From Motor Babbling to Purposive Actions: Emerging Self-exploration in a Dynamical Systems Approach to Early Robot Development

Ralf Der[1] and Georg Martius[2]

[1] University of Leipzig, Institute of Computer Science,
POB 920, D-04009 Leipzig, Germany
der@informatik.uni-leipzig.de
http://robot.informatik.uni-leizpig.de
[2] Bernstein Center for Computational Neuroscience, University of Goettingen,
Bunsenstr. 10, D-37073 Goettingen, Germany
georg.martius@ds.mpg.de

Abstract. Self-organization and the phenomenon of emergence play an essential role in living systems and form a challenge to artificial life systems. This is not only because systems become more lifelike, but also since self-organization may help in reducing the design efforts in creating complex behavior systems. The present paper studies self-exploration based on a general approach to the self-organization of behavior, which has been developed and tested in various examples in recent years. This is a step towards autonomous early robot development. We consider agents under the close sensorimotor coupling paradigm with a certain cognitive ability realized by an internal forward model. Starting from *tabula rasa* initial conditions we overcome the bootstrapping problem and show emerging self-exploration. Apart from that, we analyze the effect of limited actions, which lead to deprivation of the world model. We show that our paradigm explicitly avoids this by producing purposive actions in a natural way. Examples are given using a simulated simple wheeled robot and a spherical robot driven by shifting internal masses.

1 Introduction

Adaptation and survival in uncertain and ever changing environments are one of the key challenges in natural and artificial beings. The field has seen many impacts from life sciences, one of the directions being epigenetic and developmental robotics [11] trying to mimic natural ontogenesis. Moreover, the role of embodiment has become an important subject in the past decade under (*i*) the practical aspect of reducing computational efforts for control by exploiting the physical properties of the robot in its environment, see [12], [9], and (*ii*) the more conceptual aspect that embodied sensorimotor coordination is vital for the self-structuring of the sensor space necessary for categorization and higher level cognition, see [15], [10].

S. Nolfi et al. (Eds.): SAB 2006, LNAI 4095, pp. 406–421, 2006.

The approaches are very diversified and often oriented towards specific goals, leading to nice results up to the humanoid level. Our work aims more towards an approach from first principles. We consider agents under the close sensorimotor coupling paradigm, controlled by a neural network. Moreover, the robot disposes of a certain cognitive ability realized by an internal forward model (world model) predicting future observations on the basis of present observations and controls. Such models are speculated to play an important role for human motor control, cf. [17] as an example.

In the engineering sense the world model is learned by trying random actions. This is also assumed to take place in human development and is called motor babbling. However this approach is infeasible in high dimensional systems. This problem is called the curse of dimensionality in statistical learning theory and was realized to be a serious problem in learning sensorimotor tasks by Bernstein [1] long ago. Moreover, usually it is even not necessary to try all actions, but just those that contribute most to the information gain of the model. Our approach aims at the realization of self-exploration with emerging purposive actions instead of motor babbling.

The concomitant learning of both, the controller and the model, faces among others the cognitive bootstrapping problem. Starting at a "do nothing" and "know nothing" initialization of the controller and the internal model, respectively, the robot does not have any information on the structure and dynamics of its body so that the world model has to learn this from scratch. However, in order to learn effectively, the controls have to be informative or purposive so that the world model is provided with the sensorimotor patterns necessary for its improvement. On the other hand, these actions require a certain knowledge of the reactions of the body – information is acquired best by informed actions. This bootstrapping situation in principle reappears on all stages of the developmental process. We consider here a solution at a level, which is essentially based on the feed-back of proprioceptive sensors, i.e. self-exploration of the physical properties of the body. We understand this as early robot development, i.e. the first step of a self-organized development towards ever increasing behavioral competencies and understanding of the behavior of the body in its environment.

In recent years we have derived a systematic approach to the self-organization of behavior which has proven its practical applicability in a number of examples, see Refs. [4], [8] or the videos on [7]. This has been achieved not only for wheeled robots in a cluttered environment, see the video [2] and others on our video page, but also for high dimensional snake like robots, see the zoo videos on [7]. These creatures have no program (set of rules defining behavior), no aims, and no purpose. Yet they deploy activities by itself which are rooted in their bodies and related to the environment in which they "live". We will discuss in the present paper how this is related to the solution of the bootstrapping problem by emerging self-exploration.

The paper is organized as follows. In Sec. 2 we give a brief introduction to our general control paradigm. We demonstrate on a theoretical basis how purposive actions, necessary for self-exploration, emerge in a natural way in Sec. 3. These theoretical findings are verified in the following section with two examples, a wheeled robot and a spherical robot driven by internally shifting masses.

2   Controller Learning – Between Sensitivity and Predictability

The learning of the controller is based on the papers [3], [5]. We give here only the basic principles of our approach. We start from the information the robot gets by way of its sensor values.

2.1   The Sensorimotor Dynamics

Let us consider a robot which produces in each instant $t = 0, 1, 2, \ldots$ of time the vector of sensor values $x_t \in \mathbb{R}^n$. The controller is given by a function $K : \mathbb{R}^n \to \mathbb{R}^m$ mapping sensor values $x \in \mathbb{R}^n$ to motor values $y \in \mathbb{R}^m$

$$y = K(x)$$

all variables being at time t. In the example of a two-wheeled robot we have $y_t = (y_{t1}, y_{t2})^\top$, y_{ti} being the target wheel velocity of wheel i. In the cases considered explicitly below, the controller is realized by a one layer neural network defined by the pseudolinear expression (omitting the time index)

$$K_i(x) = g(z_i) \tag{1}$$

where $g(z) = \tanh(z)$ and

$$z_i = \sum_j C_{ij} x_j + h_i \tag{2}$$

This seems to be overly trivial concerning the set of behaviors which are observed in the experiments. Note, however, that in our case the behaviors are generated essentially also by an interplay of neuronal and synaptic dynamics (see Eq. 11 below), which makes the system highly nontrivial.

Our robot is equipped with a world model which is a function $F : \mathbb{R}^n \times \mathbb{R}^m \to \mathbb{R}^n$ predicting the current sensor values in terms of the earlier sensor and motor values, i.e.

$$x_t = F(x_{t-1}, y_{t-1}) + \xi_t \tag{3}$$

where ξ is the modeling error. In practical applications, F may be represented by a neural network with parameter vector w, which might be learnt by standard back propagation. The world model realizes the cognitive abilities of the robot. Cognition is understood on a very low level, meaning essentially the ability to predict the future consequences of the actions undertaken by the robot. This is actually what the world model does.

Introducing Eq. 1 into the equation for the world model, we get the dynamical system representing the dynamics of the SM loop as

$$x_t = \psi(x_{t-1}) + \xi_t \tag{4}$$

The dynamics for the parameters of the controller are derived from the following two objectives. We aim on the one hand, at a maximum sensitivity of the effects of the controls to the current sensor values. This induces a self-amplification of changes in the sensor values and thus is the source of activity. On the other hand, we require a maximum predictability of these effects, which are represented by future sensor values. This keeps the behaviour in "harmony" with the physics of the body and the environment.

The first objective is realized by requiring a high sensitivity of the map ψ of the sensorimotor loop towards small changes in its inputs. In more detail, we require that ψ realizes the new vector of sensor values x_t by applying a small shift to the inputs, i.e. we put

$$x_t = \psi\left(x_{t-1} + v_{t-1}\right) \tag{5}$$

or

$$\psi\left(x\right) + \xi = \psi\left(x + v\right) \tag{6}$$

where v is the input shift. This equation has a unique solution if ψ is invertible. If not, convenient approximations must be used. This question has to be solved in order to find a stable algorithm but we are not going into these details in the present paper.

At each time step we can find the value of v and define the error (omitting the time index)

$$E = \|v\|^2 = v^T v \tag{7}$$

where $\|\ldots\|$ means the Euclidean norm. The quantity $\hat{x}_{t-1} = x_{t-1} + v_{t-1}$ is the vector of previous sensor values as reconstructed from the current ones. We may therefore call E the reconstruction error. Moreover, from the point of view of time step $t-1$ the vector $\hat{x}_{t-1}$ is obtained by going one step forward in time by the true dynamics and then back to time $t-1$ by the inverse world model dynamics given by ψ. This is why we also call E the time loop error.

In order to get a more explicit expression we use Taylor expansion, which in leading order yields

$$\xi = L\left(x\right) v$$

where L is the Jacobian matrix defined as

$$L_{ij}\left(x\right) = \frac{\partial}{\partial x_j}\psi_i\left(x\right)$$

which is a direct measure of the stability of the dynamical system, see below for a discussion. If L exists we immediately find

$$v = L^{-1}\xi$$

so that

$$E = \left\|L^{-1}\xi\right\|^2 = \xi^T \left(LL^T\right)^{-1} \xi \tag{8}$$

which is the error function used in the algorithm for adapting C, see Eq. 9 below.

The Gradient Flow of the Parameters. The adaptation of the parameters of the controller can be realized by gradient descending the error function E as usual

$$\Delta C = -\varepsilon \frac{\partial}{\partial C} E \qquad (9)$$

The resutling dynamics of the parameters can, at a formal level, be argued to produce the desired properties of the system. In fact, since LL^T is symmetric we may decompose it as

$$LL^T = \sum_{i=1}^{n} \lambda_i P_i$$

where $P_i = n_i n_i^T$ is the projector on the eigenvector n_i with λ_i the corresponding eigenvalue. Then

$$E = \sum_i \lambda_i^{-2} \xi_i^2$$

with $\xi_i = n_i^T \xi$ being the projection of the model error into the subspace spanned by n_i, both λ_i and P_i depending on the parameters C of the controller in an intricate way. This expression is only valid if the $n \times n$ matrix $Q = LL^T$ is of full rank, so that none of the λ_i is equal to zero. However, we also note that, if we start with an L of full rank, the parameter dynamics will drive Q away from impending singularities due to the divergence of E for any $\lambda_i \to 0$.

In more detail, writing the gradient rule as

$$\varepsilon^{-1} \Delta C = \sum_{i=1}^{n} \left(\frac{\xi_i^2}{\lambda_i} \frac{\partial \lambda_i}{\partial C} - \xi_i \frac{\partial \xi_i}{\partial C} \right) \lambda_i^{-2} \qquad (10)$$

we see that the gradient flow is driven by two objectives. The first term on the right hand side obviously tends to increase each of the eigenvalues λ_i and hence the instability in the corresponding subspace. The interesting point is in the prefactors ξ_i^2/λ_i which mean that the update is strong where λ_i is small (high stability) and/or ξ_i^2 is large (high modeling error component in this subspace). This can be interpreted as the tendency of the parameter dynamics to produce in all directions the same degree of instability with subspaces of higher modeling error being destabilized even more strongly. Destability corresponds to a higher rate of noise amplification, such that one may say that those subspaces are explored more intensively, which are less well represented by the model. This is the effect which is relevant for the present paper and will be discussed by way of example in Sec. 3.2 below.

The second term in Eq. 10, the strength of which is modulated by ξ_i, essentially counteracts the overshooting destabilization of large error subspaces caused by the first term. It is to be noted, that the error components ξ_i not only depend on the quality of the model, but in an essential way on the behavior of the robot. Hence, both the ξ_i and eigenvalues λ_i change with changing parameters.

Altogether we may say, that our parameter dynamics generates an explorative behavior of the robot (by the first term), which however is related to the environmental reactions by the second term. This has been demonstrated in many applications realized in recent years, see for instance [5] and our video page.

Explicit Learning Rule. In the present paper we consider the one-layer neural network controller given by Eq. 1 so that the Jacobian is

$$L_{ij} = \sum_k A_{ik} g'(z_k) C_{kj}$$

where

$$A_{ik} = \frac{\partial}{\partial x_k} F_i(x, y)$$

and F is learnt concomitantly with the controller by supervised learning on the basis of the new sensor values. We use $g(z) = \tanh(z)$ where $g'' = -2gg'$ so that we get the explicit expressions (omitting the time indices everywhere)

$$\varepsilon^{-1} \Delta C_{ij} = \zeta_i v_j - 2\zeta_i \rho_i y_i x_j \tag{11}$$

$$\varepsilon^{-1} \Delta h_i = -2\zeta_i \rho_i y_i$$

where $v = L^{-1}\xi$, $\mu = A^\top Q^{-1}\xi$, $\zeta_i = g_i'\mu_i$, and $\rho = Cv$. The inversion of the matrix $Q = LL^\top$ is done by standard techniques, and has proven in many applications to be feasible and not time critical with up to 20 independent degrees of freedom.

Note that the parameter ε is chosen such that the parameters change at about the same time scale as the behavior. The interplay between synaptic and state dynamics of the controller induces a high dynamical complexity of the sensorimotor loop. The resulting robot behaviors are of a much larger complexity than the pseudolinear expression with fixed parameters might ever realize.

3   Model Learning – Problems and Challenges

Internal models are one of the prerequisites for a robot to become a cognitive system. In the case of human motor systems the role of internal models has in particular been emphasized by the work of Wolpert [18], [16]. In the present paper we are concerned with forward models as given by Eq. 3 which are learnt in a supervised way on a training set of sensorimotor patterns (x_{t+1}, y_t). However, in order to learn the relevant information about the world, the training instances must be guaranteed to sufficiently sample not only the sensor space, but also the action space. In practice it is complicated to ensure this sampling property. In case of on-line learning there is always only a part of the state action space covered in a restricted interval of time. This fact actually is widely recognized but we will demonstrate it in an extremely simple situations in order to work out explicitly the bootstrapping problem involved.

3.1   The Deprivation Effect

Let us consider a very simple example of a sensorimotor loop given by a robot with two wheels. The only sensor values are given by the current velocities which can be measured by a wheel counter, i.e. we have only proprioceptive sensors in this case. Assuming that the reactions of the wheels are largely independent of each other, we expect the model to be given as $x_{t+1} = Ay_t + \xi$ where $x \in \mathbb{R}^2$ and $y \in \mathbb{R}^2$ are the measured and target wheel velocities, respectively, and ξ is the modeling error. In the case given $A = \alpha\mathbf{E}$ is essentially the unit matrix. This is what one expects to be learnt by for instance gradient descending the error $E = \xi^\top\xi$ with learning rule

$$\Delta A_t = \varepsilon_A \xi_t y_{t-1}^\top - \beta A_t \tag{12}$$

where $\xi_t \in R^n$, $y_{t-1} \in R^m$, and $\left(\xi y^\top\right)_{ij} = \xi_i y_j$. The small damping term $-\beta A$ has to be introduced in order to damp away the influence of the initial conditions. The scaling factor α is a hardware constant.

However, convergence to the correct solution $A = \alpha\mathbf{E}$ is guaranteed only if the training instances (x_t, y_{t-1}) cover the full state-action space. Now let us assume that the behavior is restricted to a certain subspace of the action space. Under our closed loop control paradigm, behavior is parameterized by the matrix C of the controller. A restricted behavior is produced by assuming the C matrix of the controller as

$$C = \gamma pp^\top \tag{13}$$

where p is a normalized vector, $pp^\top$ is the projector onto p, and γ a constant with $\gamma > 0$ and $\gamma\alpha > 1$. The sensorimotor dynamics[1]

$$x_t = Ag\left(Cx_{t-1} + h\right) + \xi_t$$

converges towards a fixed point. The controller will produce the vector $y = g\left(sp + h\right)$ defining the wheel velocities, where s is obtained from the solution of the fixed point equation. In particular if ($h = 0$ for the moment)

$$p = \frac{1}{\sqrt{2}}\begin{pmatrix} 1 \\ 1 \end{pmatrix} \tag{14}$$

the robot will move either straight forward or backward if $s > 0$ or $s < 0$, respectively. Choosing instead $p = (1, -1)^T/\sqrt{2}$ the robot will rotate on site. The behavior can still be further modified by changing h.

The point now is, that instead of converging towards the unit matrix, A is learnt as

$$A = \alpha pp^\top \tag{15}$$

so that A is essentially the projector on the subspace given by the degenerate controller. This is a correct solution in the space covered by y_t which of course is completely wrong in the complementary subspace of the motions of the robot.

---

[1] Consider $g\left(z\right)$ as a vector function, i.e. $g_i\left(z\right) = g\left(z_i\right)$.

This effect makes the learning unstable if the controller changes the motor vector on a slow time scale since the matrix A will follow this change. Of course this result hinges on the time scales. In fact the problem will not arise so strongly if the time scales for the learning are much larger than the intervals of persistent directions of the robot.

3.2   The Bootstrapping Scenario

As explained in the introduction, the aim of our approach is the concomitant learning of the controller and the model from scratch. The toy example given above has demonstrated that, if the actions $y \in R^m$ (with $m = 2$ in the example above) are restricted to a certain subspace, the model will degenerate to a projector onto that subspace with the effect that it will be completely wrong in the orthogonal subspace. The challenge is that the controller needs to "feel" this deprivation of the world model and to issue motor commands which provide the world model with the state-action pairs (x_t, y_{t-1}) necessary for learning in the orthogonal subspace neglected so far.

This is exactly what happens in our approach for the learning of the controller. We will now demonstrate this theoretically in terms of the above model with degenerate $C = \gamma pp^\top$. With A from Eq. 15 we get in the linear (low z) case

$$L = \gamma \alpha pp^\top$$

so that the Jacobian matrix is singular, hence E has a singularity, and the degenerate C is seen to be an instable fixed point of the gradient dynamics. Without loss of generality we may use the specific form Eq. 14 for p. Now let us assume that C has a small deviation δC which corresponds to the projector into the orthogonal subspace, i.e. we put

$$C = \gamma pp^T + \mu p_\perp p_\perp^T$$

where $p_\perp$ is orthogonal to p, i.e. $p_\perp p_\perp^T$ is the projector onto the orthognal complement of p, and μ is arbitrarily small. With a noisy input or with a random motor event (motor babbling) the action may be

$$y = sp + \sigma p_\perp \tag{16}$$

where $|\sigma|$ is small. We are now going to show now that this small fluctuation leads to a strong amplification of the $p_\perp p_\perp^T$ component in C.

If the robot is executing this action, we get a model error (we assume the p subspace is already learnt correctly and neglect other noisy events)

$$\xi = \alpha \sigma p_\perp \tag{17}$$

since A is still the degenerate matrix $A = \alpha pp^T$. The learning step for A produces

$$\Delta A = \varepsilon \xi y^T = \varepsilon \alpha \sigma s \begin{pmatrix} 1 & 1 \\ -1 & -1 \end{pmatrix} + \varepsilon \alpha \sigma^2 pp_\perp^T$$

(we drop the damping term because it only contributes to the degenerate part of A). In the small z case considered we have now $L = (A + \Delta A) C$. In leading order (small μ and $\varepsilon\sigma$) we find with simple matrix algebra that

$$\left(LL^T\right)^{-1} = \frac{1}{4\alpha^2\varepsilon^2\sigma^4\mu^2} p_\perp p_\perp^T$$

which is the projector on just that subspace which is not well covered by the model so far, i.e. in this order the learning so to say concentrates fully onto the subspace not "understood" by the model. Using Eq. 17 the error is obtained as

$$E = \frac{1}{2\sigma^2\varepsilon^2\mu^2} \tag{18}$$

This shows that the learning will rapidly increase the strength μ of that part of C which projects into the orthogonal subspace. However, in this way also the contribution of the orthogonal actions is increasing so that the constant μ in Eq. 3.2 is increasing as long as the model is still wrong. We may interprete this by saying that the controller tries more and more actions which force the model to learn also the behavior in the orthogonal subspace. This is a kind of purposive behavior, the purpose being to feed the model with the necessary input-output pairs for complete learning. The process has to be started by some fluctuation in the output of the controller which may be called motor babbling.

The difference of this behavior to the usual strategy of issuing random motor commands consists in the fact that the novel motor commands are directed into the unknown regions of the state-action space. This of course is of relevance for high dimensional systems where random commands face the curse of dimensionality. This has been clearly demonstrated in our experiments with high dimensional (up to 20 independent motors) systems, see our videos of the snake robots, where collective modes are excited by this bootstrapping phenomenon. The background behind the high dimensional scenario is that the paradigm ensues spontaneous symmetry breaking and creating low dimensional searching modes in high dimensional search spaces, which will be demonstrated in a later paper.

4   Experiments

In the sections before we have studied deprivation of the world model and we have seen how purposive actions can efficiently eliminate this effect. In order to illustrate this in practice, we will consider different experiments. First, we consider the rather artificial setting as described in section 3.1 theoretically with a simulated two-wheeled robot. Second, the self-explorative character of you controlling paradigm is analyed using the same robot. Third, a simulated spherical robot is considered on a flat surface to show self-explorative behavior at a more complex system. Finally the spherical robot is considered in a basin like environment, where deprivation occurs naturally.

4.1 Experiment I – Two-Wheeled Robot: Deprivation and Bootstrapping

The idea of this experiment is to show, that first in the case of limited motor commands deprivation of the world model occurs, and second that our controller effectively produces purposive actions. We use the physics engine ODE (open dynamic engine [14]) for the computer simulation experiments. A simulated two-wheeled robot is controlled with motor commands within a subspace of the action space. Motor commands y are understood as wheel velocities and the sensors values x are the read back wheel velocities obtained from the wheel counters. Please recall the controller function $K(x) = y = \tanh(Cx + h)$ (Eq. 2). As decribed in section 3.1 the controller matrix is degenerated as $C_{ij} = 0.6 + \lambda u_{ij}$, where u_{ij} are random numbers and $0 < \lambda \ll 1$. We modulate h such that the robot drives backward and forward periodically.

As expected, we observe a degeneration of the world model, see Figure 1. After the model learning is basically converged, the learning of the controller according to Eq. 11 was switched on (at time 4550). After a short break down of the activity one observes the emergence of motor commands which live mainly in the orthogonal subspace. This means rotational behavior of the robot. Later on both, straight and rotational modes, are equally visited so that the model gets the necessary information. A is converging towards the unit matrix as it should be. The behavior and the parameter dynamics are displayed in Figure 1.

4.2 Experiment II – Two-Wheeled Robot: Frequency Wandering

Besides the effects discussed so far there is more to the self-exploration properties of our approach. In particular the fact that the error $E = v^\top v$ is invariant to rotations of v introduces a certain invariance of the state dynamics against frequency changes (in a linear approximation). This leads to the effect that the robot self-regulates the frequencies of its motor values.

In the experiments we use the simulated two-wheeled robot as in the previous section. Most of the time the robot moves by sequences of straight and rotational motion primitives. This corresponds to the exploration of the physical space and is what one would call an explorative behavior in the usual sense. However, occasionally the controller gradually increases the frequency of the dynamics in the sensorimotor loop so that rather complex trajectories emerge in the physical space. With even higher frequencies we observe a jiggling of the body where physical effects due to inertia, swing, and even gyro effects come into play. We may say that this is the phase of the self-exploration of these physical properties of the body. However, the simple world model does not understand the high frequency modes very well, so that they are left after some time. The robot returns to its "normal" behavior with a succession of rotational and straightforward driving modes. This play repeats more or less forever with a strong influence of the noise. In Figure 2 the short-time fourier transform of the motor values are displayed, which reflect the frequency in the sensorimotor dynamics.

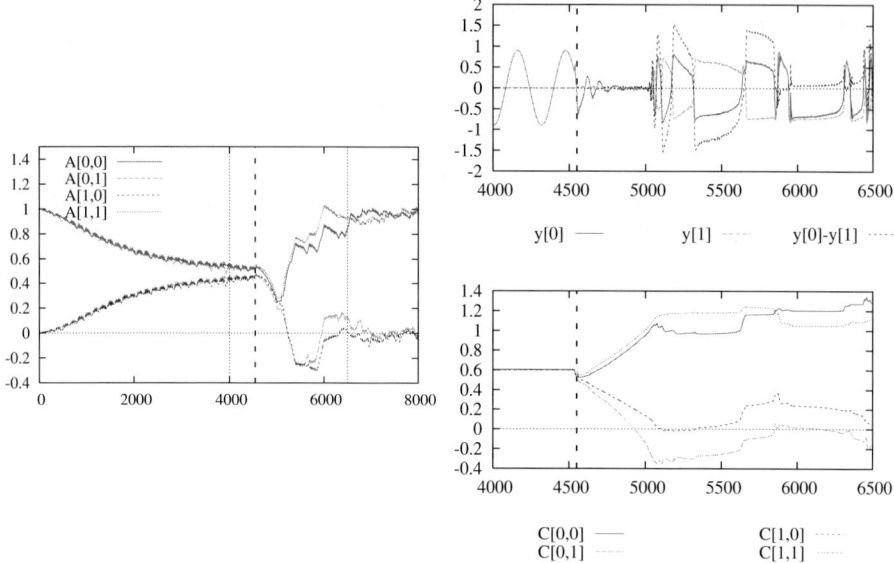

Fig. 1. Model deprivation and recovery in case of a two-wheeled robot. From time 0 – 4550 the robot was controlled by fixed C and modulated h (oscillating forward/backward), and after time 4550 the learning of the controller is enabled. The time scale is 1/50 sec, i.e. the whole run is 160 sec long. Left: Model matrix A. Degenerates until controller learning is enabled. After that it learns towards a unit matrix. Upper right: Motor commands y_i and steering $y_0 - y_1$. One can clearly see that the controller performs rotational actions in a dedicated manner after the activation of the learning (> 5000). Later on the motion consists of straight and rotational modes leading to the full deployment of the world model. Lower right: Controller matrix C.

This scenario actually reminds one of the fact that the controller with its learning dynamics does not know about the physical space so that everything it does is the exploration of the properties of the body and the exploration of the space is only in the eye of the beholder. A relation to the space would emerge if we include sensors informing about positions in space. The emergence of the concept of space will be the subject of a later paper.

4.3 Experiment III – Spherical Robot: Emerging Self-exploration on Flat Surface

The wheeled robot is a rather simple example of a sensorimotor loop. In order to show the emergence of sensorimotor coordination by self-exploration we demonstrate the above phenomena with a more complicated robotic object. The object of study is a simulated spherical robot see Figure 3, inspired by Julius Popp [13].

The motor commands y are the nominal positions of the masses along the axes. The sensor values are in this case the components of the vector of the z-axis of the robot in the world coordinate system. In this way, the controller has only very restricted information about the physical state of the sphere. Nevertheless, our

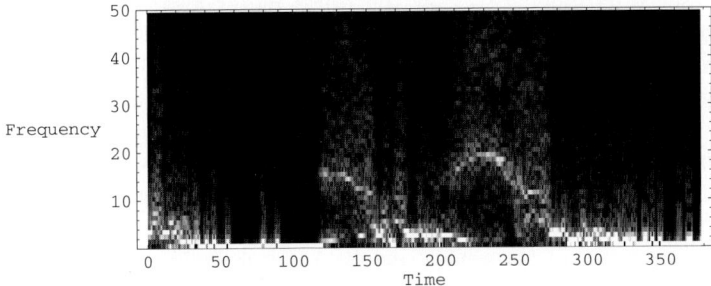

Fig. 2. Power spectra of the motor values (speeds) over time. Each column is the lower frequency part of the discrete Fourier transform of a 2 seconds time window of the motor values (wheel velocities). Subsequent lines are overlapping, so that the time scale is in units of 0.5 seconds. Dark pixels correspond to low energy and bright to high energy in the corresponding frequency band. High energy in a certain frequency band means that changes between forward and backward driving occur at about that frequency. Note that at times 120 and 200 a jump in the frequency occurs meaning that the robot suddenly changes to a highly complex motion pattern which then gradually decays towards the mentioned low frequency regime.

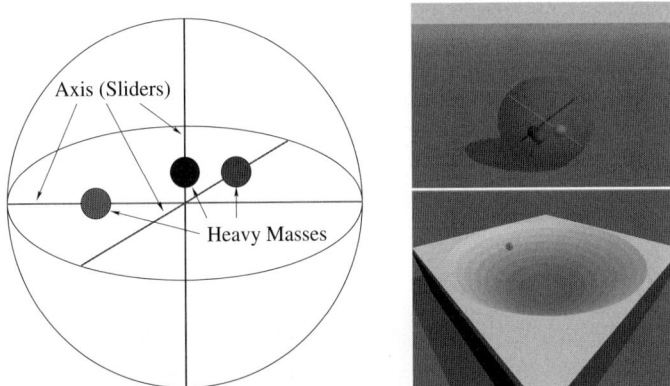

Fig. 3. Simulated spherical robot used in the experiment. Left: Sketch of a spherical robot. Inside the robot there are three orthogonal axes equipped with sliders. To each slider a heavy mass is attached which can be shifted along the axis. There is no collision or interaction of the masses at the intersection point of the axes. Upper right: Picture of a spherical robot on the ground. Lower right: Picture of a spherical robot in a basin.

learning algorithm manages to produce highly coordinated sensorimotor patterns corresponding to different rolling modes in the course of time.

Let us consider the case of the sphere on a flat surface, see the video [6]. In the beginning the controller and world model is initialized in the "do nothing" and "know nothing" situation (C and A are small random matrices). The parameter dynamics given by Eq. 11 drives C until noise amplification sets in and the inner masses start moving so that the world model also starts learning. After some time the sphere starts to roll slowly. Different movements are probed and later a nice and constant rolling mode emerges. Such periods stay for quite a long period of time. What happens in that time is that the world model gets restricted information and the deprivation sets in which leads to the sudden appearance of new controller actions which lead to a kind of explorative periods followed by a new stable rolling period about a different axis.

Besides of the rolling mode we also observe a kind of jumping mode and rolling modes with all three axes involved. These behaviors demonstrate the self-exploration of the body and show how our algorithm manages to close the sensorimotor loop in order to excite stable behavioral modes adequate to the physical properties of the body.

This effect is also demonstrated with a different sensor set. In this experiment we equipped the spherical robot with six infra-red sensors, which are installed in each point of intersection of the axes with the surface of the sphere with the direction along the axis and range of about two diameters of the sphere. The six sensors values are fed directly to the controller. No other sensors are in use. The sensor characteristic was chosen nonlinear as $x = s^{\alpha}$, where s is the primary sensor value (distance) and $\alpha = 1.5$. The effect is that the sensor characteristic is a smoother function of the angular position of the robot. Still the sensor information is extremely unreliable and related to the position of the sphere in a very complicated way. Nevertheless, starting with the "do nothing" and "know nothing" initialization, we observe many different rolling modes which are visited in the course of time. In Figure 4 the power spectra of the sensor values over time are displayed. High frequency means here high velocity. We observed different behavioral modes. For example rolling with different velocities around one of the slider axis or also the tumbling mode involving all three axis.

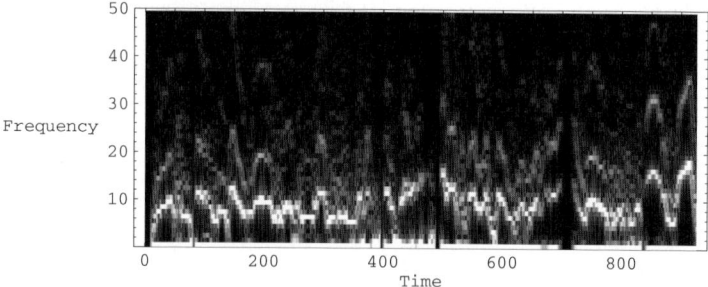

Fig. 4. Power spectra of the infra-red sensor values of the spherical robot on flat ground over time based on 10 s time windows. The bright pixels indicate that there is a dominating frequency of the sensor values which means that the robot is in a rolling mode, the rolling velocity being roughly proportional to the frequency. Periods of stable rolling modes of different velocities are seen to sometimes change rapidly into a resting mode (frequency zero) or to other velocities.

4.4   Experiment IV – Spherical Robot: Deprivation in a Basin

An interesting effect is produced if we put the sphere into a circular basin. We use as sensors the projections of the z-axis of the body coordinate system on the z-axis of the world coordinates. The motor commands are the nominal position of the mass points on the inner axis. In Fig. 5 the behavior of the robot in a basin is shown using the power spectra of the sensor values and the determinant of the world model. The initial phase is of the same nature as on the flat surface, i.e. from time 0 – 80 one can see self-explorational modes, where different frequencies are probed. Then a stable rotational mode emerged (time 80 – 120), which is the circulation in the basin at a constant height. The circulation mode is manifest in the power spectrum by the low frequency excitation, the high frequency excitations being the motions of the axes of the robot due to the rolling motion.

The circulation mode is a behavior which is not so easily realized with the internally shifting masses. Contrary to the rolling on the flat surface the circulation in the basin permanently changes the direction of the axes and hence of the sensor vector. Nevertheless, the controller finds a strategy, such that the circulation mode is stable over many laps. Interestingly this stable sensorimotor pattern is realized by a trajectory which directly reflects the specific geometry of the world. In a certain sense one might say that the robot by its behavior recognizes this geometry.

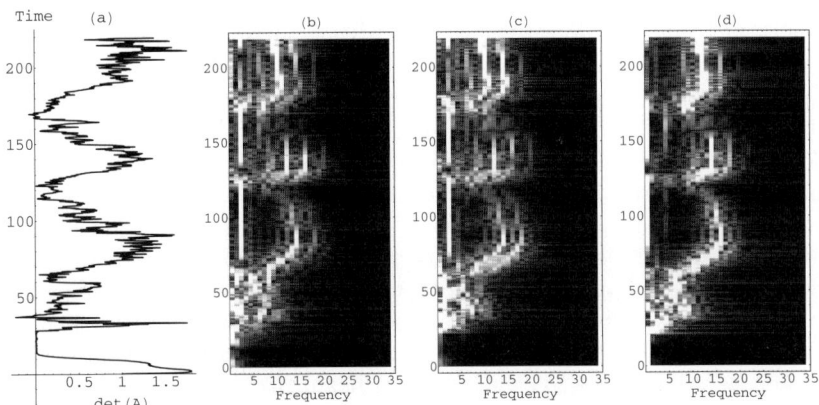

Fig. 5. Behavior of a spherical robot in a basin. (a): Determinant of the model matrix A, (b – d): Power spectra of the sensor values (x, y, z) over time. In the time intervals 80 – 110 and 140 – 170 there are components of stable low frequency in sensor 1 (x) and 2 (y), which correspond to the circulation in the basin at constant height. The higher frequencies reflect the rolling of the sphere as in the flat surface case. One can see that the value of the determinant of A decreases while the robot stays in one mode of behavior (80 – 110, 140 – 170). This is an indication for the deprivation of the model arising from the restriction to a specific mode of behavior. Once the deprivation reaches a certain measure the bootstrapping of new actions sets in which leads to the recovery of the model (increasing determinant).

The behavior in the basin also shows that staying in a stable mode for a longer time leads in general to a deprivation of the world model. This is seen from the time plot of $\det A$, which is taken as a crude measure for the degeneracy. In Fig. 5 the deprivation can be seen at the behavior of $\det A$. It is seen to be indeed decreasing until it reaches nearly 0 at time 115, which means A is closed to a singularity. An explorational period follows, which effectively explores the orthogonal subspace, and the determinant of A is seen to increase rapidly. The same starts again at time 140 and so forth.

5   Conclusions and Outlook

The results presented in the present paper can be considered as a step towards autonomous early robot development, meaning the scenario where an unbiased robot might learn the essential sensorimotor coordination by self-exploration. The important point of our approach is, that it is completely domain invariant, so that the emerging behaviors are dictated by the physical properties of the body and the environment. This has a direct bearing for embodied AI in the sense, that our controller learns to excite certain physical modes of the body, which are qualified by the fact that they can be understood by the world model in easy terms. Hence, we may understand these modes as behavioral primitives which may be used in more complex behavioral architectures.

We have given a theoretical approach to the deprivation problem which arises in the interplay between the world model and the controller. The system does not have any information on the structure and dynamics of the body, so that the world model has to learn this from scratch. This involves the so called bootstrapping problem, meaning that on the one hand the controls have to be such, that the world model is provided with the necessary information. On the other hand, these actions require a certain knowledge of the reactions of the body – information is acquired best by informed actions. The concerted manner by which both the controller and the world model evolve during the emergence of the behavioral modes seems to be a good example of this process.

We consider our approach as a novel contribution to the self-organization of complex robotic systems. At the present step of our development the behaviors, although related to the specific bodies and environments, are without goal. As a next step we will realize a so called behavior based reinforcement learning. When watching the behaving system one often observes behavioral sequences which might be helpful in reaching a specific goal. The idea is to endorse these with reinforcements in order to incrementally shape the system into a goal oriented behavior.

Acknowledgement

This study was supported by a grant from the german BMBF in the framework of the Bernstein Center for Computational Neuroscience Goettingen, grant number 01GQ0432.

References

1. N. A. Bernstein. *The Co-Ordination and Regulation of Movements.* Pergamon Press, 1967.
2. Ralf Der. Video of robot in cluttered environment. `http://robot.informatik.uni-leipzig.de/Videos/Pioneer/2004/maze2.wmv`, 2004.
3. Ralf Der, Frank Hesse, and René Liebscher. Contingent robot behavior generated by self-referential dynamical systems. *Autonomous robots*, 2005. submitted.
4. Ralf Der, Frank Hesse, and Georg Martius. Learning to feel the physics of a body. In *Proceedings CIMCA '2005*, 2005.
5. Ralf Der, Frank Hesse, and Georg Martius. Rocking stumper and jumping snake from a dynamical system approach to artificial life. *J. Adaptive Behavior, submitted*, 2005.
6. Ralf Der, Frank Hesse, and Georg Martius. Video of spherical robot in free space. `http://robot.informatik.unileipzig.de/Videos/SphericalRobot/2005/spherical_floor_trace.mpg`, 2005.
7. Ralf Der, Frank Hesse, and Georg Martius. Videos of self-organized creatures. `http://robot.informatik.uni-leipzig.de/Videos`, 2005.
8. Ralf Der, Georg Martius, and Frank Hesse. Let it roll – emerging sensorimotor coordination in a spherical robot. In *Proceedings AlifeX*, 2006.
9. F. Iida, G. J. Gomez, and R. Pfeifer. Exploiting body dynamics for controlling a running quadruped robot. In *International Conference of Advanced Robotics (ICAR 05)*, 2005.
10. Yasuo Kuniyoshi, Yasuaki Yorozu, Yoshiyuki Ohmura, Koji Terada, Takuya Otani, Akihiko Nagakubo, and Tomoyuki Yamamoto. From humanoid embodiment to theory of mind. In *Embodied Artificial Intelligence*, pages 202–218, Berlin, Heidelberg, New York, 2003. Springer.
11. Max Lungarella, Giorgio Metta, Rolf Pfeifer, and Giulio Sandini. Developmental robotics: a survey. *Connection Science*, 0(0):1–40, 2004.
12. Rolf Pfeifer and Fumiya Iida. Embodied artificial intelligence: Trends and challenges. In *Embodied Artificial Intelligence*, pages 1–26, 2003.
13. Julius Popp. Sphericalrobots. `http://www.sphericalrobots.com`, 2004.
14. Russel Smith. Open dynamics engine. `http://ode.org/`, 2005.
15. Rene te Boekhorst, Max Lungarella, and Rolf Pfeifer. Dimensionality reduction through sensory-motor coordination. In *Kaynak, Okyay (ed.) et al., Artificial neural networks and neural information processing – ICANN/ICONIP 2003. Joint international conference ICANN/ICONIP 2003, Istanbul, Turkey, 26-29, 2003. Proceedings. Berlin: Springer. Lect. Notes Comput. Sci. 2714, 496-503* .
16. D. Wolpert and M. Kawato. Multiple paired forward and inverse models for motor control. *Neural Networks*, 11:1317–1329, 1999.
17. D. M. Wolpert, R. C. Miall, and M. Kawato. Internal models in the cerebellum. *Trends in Cognitive Sciences*, 2(9), 1998.
18. Daniel M. Wolpert, R. Chris Miall, and Mitsuo Kawato. Internal models in the cerebellum. *Trends in Cognitive Sciences*, 2:338 – 347, 1998.

Modelling Multi-modal Learning in a Hawkmoth

Anna Balkenius[1,2], Almut Kelber[2], and Christian Balkenius[3]

[1] Chemical Ecology, SLU Alnarp, Box 44,
230 53 Alnarp, Sweden
[2] Vision Group, Department of Cell and Organism Biology,
Lund University,
Helgonavägen 3, 223 62 Lund, Sweden
anna.balkenius@cob.lu.se,
almut.kelber@cob.lu.se
[3] Lund University Cognitive Science,
Kungshuset, Lundagård,
222 22 Lund, Sweden
christian.balkenius@lucs.lu.se

Abstract. The moth *Macroglossum stellatarum* can learn the colour and sometimes the odour of a rewarding food source. We present data from 20 different experiments with different combinations of blue and yellow artificial flowers and the two odours honeysuckle and lavender. The experiments show that learning about the odours depends on the colour used. By training on different colour-odour combinations and testing on others, it becomes possible to investigate the exact relation between the two modalities during learning. Three computational models were tested in the same experimental situations as the real moths and their predictions were compared to the experimental data. The average error over all experiments as well as the largest deviation from the experimental data were calculated. Neither the Rescorla-Wagner model or a learning model with independent learning for each stimulus component were able to explain the experimental data. We present the new categorisation model, which assumes that the moth learns a template for the sensory attributes of the rewarding stimulus. This model produces behaviour that closely matches that of the real moth in all 20 experiments.

1 Introduction

Flowers attract pollinators mainly by colour and odour stimuli. For newly eclosed moths and butterflies, it is important to quickly recognise a rewarding flower, and innate colour and odour preferences contribute to this ability [10,33]. By their innate preference for blue, naive honeybees are guided to flowers with a large amount of nectar [15]. A preference for blue is shared by other insects but innate colour preferences can differ between species [34]. Rapid and flexible learning to associate colour or odour with a reward has been demonstrated in honeybees, butterflies and moths [1,32,31,20,24,30,33].

The diurnal hummingbird hawkmoth, *Macroglossum stellatarum*, uses colour vision in food-searching, and spontaneously forages from coloured artificial flowers without any odour [19,18]. *M. stellatarum* has a strong innate preference for

S. Nolfi et al. (Eds.): SAB 2006, LNAI 4095, pp. 422–433, 2006.

blue flowers as a food-source and a weaker preference for yellow [18], but it can easily and equally fast learn other colours including green which is not a colour of a typical flower [2,18].

M. stellatarum has most probably evolved from a nocturnal ancestor, and in nocturnal hawkmoths, odour is very important in food-searching [6,28]. It has recently been shown that the ability of *M. stellatarum* to learn an odour that accompanies a colour depends on the choice of colour [3]. When an innately preferred blue colour is learned together with an odour, the moth will not learn the odour. On the other hand, if the less preferred colour yellow is used in, the moths can readily learn the odour.

Stimuli of one sensory modality can influence learning of stimuli of another modality in different ways. In most cases, two stimuli are more effective than one and the advantages of multi-sensory integration are of great importance in many animals [22]. In honeybees, colours attract attention before odour, while odour attracts attention when the bees are very close to the food source [32,14]. There is also evidence for increased learning when two stimulus types are combined [29]. In bumblebees, it has been shown that the presence of odour enhances colour discrimination, and increases attention and memory formation [21]. In honeybees, the similarity between colours modulate odour learning [32,14]. The similarity between colours has also been shown to modulate place learning in a hawkmoth [4].

A special case of multi-modal learning is configural learning where an animal learns to respond to a configuration of stimuli, but not to the single stimulus modalities themselves [23]. The hawkmoth *Manduca sexta* needs both an odour and a visual stimulus to unroll the proboscis for feeding [28], which also might be a preference for a configuration of both cues.

In contrast, two different situations have been found where learning of one stimulus prevents the learning of another stimulus. First, animals trained to a stimulus compound consisting of, for instance, a colour and an odour, sometimes only learn one of the components. For example, they learn the colour but not the odour. This effect is called overshadowing [25]. Second, when animals are first trained to one stimulus component and later to the compound they will not learn the stimulus component that was initially absent. The first component already predicts the reward and blocks learning of the second component [17]. Blocking and overshadowing were originally defined for classical conditioning but have also been found in instrumental conditioning [8,9,23]. A possible reason for the lack of learning of the second stimulus may be that the animal directs its attention only to the first stimulus [35]. The existence of blocking and overshadowing in insects is controversial and experiments have given mixed results [8,9,7,11,13]. In particular, it has been disputed whether the learning of one stimulus modality depends on the other.

To test this, we collected data from 20 different learning experiments with *M. stellatarum* where multi-modal stimuli were used. Most of the animal data has been previously published [3], but experiments 8-12 are reported here for the first time. We also tested a number of learning models on these experiments using computer simulations. The models tested were: (1) the Rescorla-Wagner model, which assumes that learning depends on all stimuli present and the prediction error of the reward magnitude, (2) the independence model, which assumes that learning of each stimulus component is independent of the other, and (3) the new categorisation model, which assumes that the moth learns a template for the rewarded stimulus when there is a prediction error. As a base case, we also simulated random selection of stimuli.

2   Materials and Methods

M. stellatarum were bred in the laboratory throughout the year. The larvae were fed their natural food plant, and the pupae were kept at 20°C. On the day after eclosion, the naïve moths were released in the cage with two feeders [27]. The experimental cage measured 50 x 60 x 70 cm and was illuminated from above with four fluorescent tubes (Osram, Biolux). Two feeders were placed 35 cm above the cage floor and 30 cm apart from each other. To prevent place learning [4], the feeders were randomly shifted between four locations during learning. During training, the rewarded feeder was filled with sucrose solution and the unrewarded contained water. Groups of up to 25 moths were flying and feeding in the same cage. The tests occurred after four days of training. During tests, both feeders were filled with water and each moth was tested on its own. The first artificial flower the moth touched with its proboscis was recorded.

Two colours, blue (B), and yellow (Y) and two odours, artificial honeysuckle (H) and extract of lavender (L) (oil) were used in the experiments. 25 μl of the odour extract was distributed in 10 ml of water of sucrose solution in the feeders and refilled every second day. Both honeysuckle and lavender flowers are visited by *M. stellatarum* in the wild [16]. In electroantennograms, *M. stellatarum* responded strongly to both odours [5].

We run 20 different experiments with different combinations of colours and odours. Experiments 1-5 were different preference tests. Untrained moths were presented with two stimuli and their first choice was recorded. The stimulus combinations used were B/Y, YH/YL, BH/BL, BL/YH, and BH/YL. The results of the preference tests were used to set the initial weights of the different computational models. The number of animals tested in the first five experiments were 25, 38, 21, 25 and 10 respectively.

In experiment 6 and 7, we tested the ability of the moths to learn which colour was rewarded. The training used B+/Y and Y+/B respectively, where + indicates that this stimulus was rewarded. The tests used the same stimuli, but without any reward. There were 20 animals in each experiment.

In experiment 8-12, the moths were trained on one combination of colour and odour and tested on another. These combinations are shown in Fig. 2. The number of animals tested in these experiments were 50, 18, 21, 10 and 18 respectively.

We also used additional data from 8 experiment previously reported by Balkenius and Kelber [3] summarised in Fig. 1 and Fig. 3. The experiments shown in Fig. 3 started with a pre-training phase where the preference for a colour was changed. In two of the experiments, the weak preference for yellow was strengthened by a pre-training procedure, and in the the two other experiments, the strong preference for blue was weakened to see how this would influence subsequent learning (For details, see [3]).

3   Computational Models

The experiments run with the real moths were also tested with four computational models to see if these models were able to explain the behaviour of the animals. These models were a random selection model, the Rescorla-Wagner model, an independence model, and the new categorisation model which is described here for the first time.

For all models, each flower stimulus was coded as a vector $s = \langle s_0, s_1, s_2, s_3 \rangle$ with four components coding for blue colour (s_0), yellow colour (s_1), honeysuckle odour (s_2), and lavender odour (s_3). Each of these components were set to 1 when the corresponding stimulus component was available and 0 otherwise. For example, the stimulus BL was coded as $s = \langle 1, 0, 0, 1 \rangle$.

3.1   Random Selection

In the random selection model, a stimulus was always selected with probability $p(s(t)) = 1$. This model serves as a base case against which the other models could be compared. This model tests the assumption that no learning takes place at all. The result of the random model is not shown in the figures since it is always the same.

3.2   The Rescorla-Wagner Model

Since it appears that learning of one stimulus component can block learning of another in the moth experiments, it seems reasonable to test how well the Rescorla-Wagner model is able to reproduce the results of the experiments. Let $w(t)$ be the current weight vector and $R(t)$ the current reward at time t. When the moth attempts to forage, the weights are updated according to the equation

$$w_i(t+1) = w_i(t) + \gamma \delta(t) s_i(t) \tag{1}$$

for both rewarded and unrewarded trails, where γ is the learning rate and $\delta(t)$ is the difference between the actual and expected reward

$$\delta(t) = R(t) - \sum_{i=0}^{n} w_i(t) s_i(t), \tag{2}$$

where $n = 3$ since there were four different stimulus components. The stimulus $s(t)$ is selected according to the probability

$$p(s(t)) = \sum_{i=0}^{n} w_i(t) s_i(t). \tag{3}$$

3.3 The Independence Model

The independence model is similar to the Rescorla-Wagner model except that learning about one stimulus component is independent of the connection strengths of the other associations. In this model, one stimulus component is not able to block learning of another.

Let $w(t)$ be the current weight vector and $R(t)$ the current reward. When the moth attempts to forage, the weights are updated according to the equation

$$w_i(t+1) = w_i(t) + \gamma[2R(t) - 1]s_i(t). \tag{4}$$

A stimulus is selected according to the same probability as in the Rescorla-Wagner model (Eq. 3).

3.4 The Categorisation Model

The categorisation model assumes that the animal learns a template for the sensory attributes of the rewarding flower during foraging. When the moth is rewarded, its template for the flower will change towards the current flower. The moth has only one preferred stimulus, which is only updated when the moth is rewarded. Let $w(t)$ be the current weight vector coding for the flower template and $R(t)$ the current reward. When the moth is rewarded, the weights are updated according to the equation

$$w_i(t+1) = \frac{u_i(t+1)}{\sum_{i=0}^n u_i(t+1)} \tag{5}$$

where

$$u_i(t+1) = \begin{cases} w_i(t) + \gamma\delta(t) & \text{when } s_i = 1 \\ w_i(t) - \epsilon & \text{otherwise} \end{cases} \tag{6}$$

and $\delta(t)$ is calculated as for the Rescorla-Wagner model (Eq. 2). To function as a template it is necessary that the weight vector w is normalized as described by Eq. 5. The match between the learned stimulus and the current external stimulus is thus used to predict the magnitude of the reward. Unlike the Rescorla-Wagner model, however, the learning attempts to move the learned template towards the rewarded stimulus instead of directly decreasing the prediction error. As a consequence, the prediction error will still decrease as the template reaches the rewarded stimulus.

Because of the δ in Eq. 6, learning only occurs when there is a prediction error, which makes blocking possible when the reward is already predicted by the stimulus. Since the template is normalized, but the stimulus input is not, It is possible for a stimulus component to completely block learning even if the stimulus and the template are not identical.

The probability of selecting a stimulus is set to

$$p(s(t)) = \left[\sum_{i=0}^n w_i(t)s_i(t)\right]^q. \tag{7}$$

The sum describes the matching process and the exponent q is a parameter which is used to derive selection probabilities from the matching. This parameter was set to $q = 2.00$ to quantitatively fit the experimental data.

3.5 Simulations

The four models were tested on the 20 experiments describe above. During each simulation, the simulated moth was randomly presented with one of two stimuli and was allowed to select it with the probability given by the selection functions described above. Data from 1000 simulated animals were recorded for each experiment and each model. The simulated moths were rewarded 50 times during each learning phase.

In the simulations, the initial weights were set to parallel the stimulus preferences of the moth as closely as possible. The weights were optimized to two decimal places. The constants for each model were subsequently set to minimise the average error over all experiments. These constants and initial weight values are given in table 1.

Table 1. Optimal parameter for each of the models. Note that the sum of the weight for the categorisation models equals 1.

Model	γ	ϵ	w_0	w_1	w_2	w_3
Rescorla-Wagner	0.05	-	0.10	1.00	0.15	0.00
Independence	0.04	-	0.10	1.00	0.15	0.00
Categorisation Model	0.05	0.10	0.18	0.74	0.08	0.00

4   Results

The results of experiments 1-5 showed that the moth had a marked preference for blue, but no clear preference for any of the odours (data not shown). Since the parameters of each model were optimised to reflect these preferences, all models behaved as the real moth in these preference tests. Experiments 6 and 7 verified that the moth could be trained to select either a blue or yellow flower. The real moth selected yellow in 80% of the trials after being trained on yellow, and blue in 95% of the trials after being trained on blue. All models, except the random selection, were able to learn these discriminations.

Fig. 1 shows the result of experiments 8-11 with the same colour but different odours [3]. In the real moth, the blue colour prevents odour learning from occurring, but with the yellow colour, the moth is able to learn which odour is rewarded. The categorisation model gives almost the same result as the real moth on all experiments. Contrary to the real moth, the Rescorla-Wagner model learns the odour in all experiments. The same is true about the independence model, although the learning is less pronounced for this model regardless of which colour was used.

The behaviour of the real moth and the different models differ even more in experiments 12-16 shown in Fig. 2. Here, it is again evident that the real moth

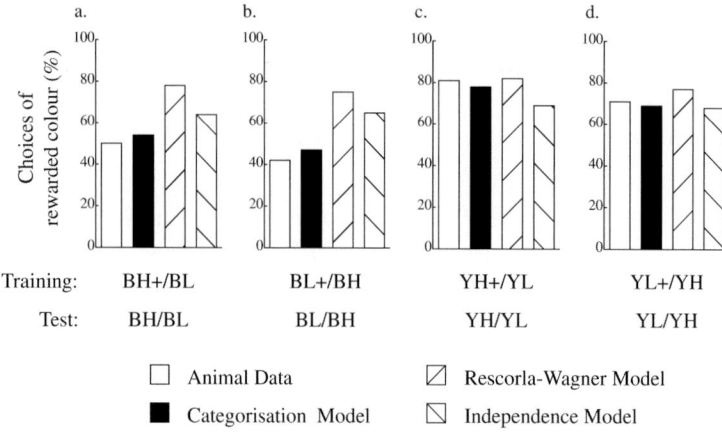

Fig. 1. Choices of the stimulus with the rewarded odour after discrimination training in experiments 8-11 for the real moth (data from [3]) and the three models. B: blue; Y: yellow; H: honeysuckle; L: lavender. With the yellow colour, the moths learn the odours, but with blue colour, they do not. This is predicted by the categorisation model, but not by the Rescorla-Wagner or independence models.

learns odour when it is presented together with yellow (Fig. 2a, b and e). Since the test with colour and odour and the test with only colour differs, the animals must have learned the odour. With the blue colour, the animals did not learn the odour and the result is the same with and without odour (Fig. 2c and d).

Again, the predictions of the categorisation model were very close to the actual data, but the other two models differed in different ways. In experiment 12 and 16 (Fig. 2a and e), the Rescorla-Wagner model did not make the correct discrimination and appears to select the correct odour and ignore the colour. The independence model does not take the colour into account when learning odour and learns the colour in experiment 10 (Fig. 2c), when the other models and the real moth does not.

For the real moth, the preference for the colour could be changed by pre-training [3]. In the experiments shown in Fig. 3a and b, the innate preference for blue is extinguished during pre-training. As a result, the moth can later learn odours together with a blue artificial flower. The opposite situation is shown in Fig. 3c and d where the less preferred yellow is made more attractive during pre-training. As a consequence, the real moth no longer learns the odour together with yellow.

Like the real moth, the categorisation model behaves differently depending on which colour is used and whether it was pre-trained or not. This is also true of the independence model, although the difference in the two cases is not as large. For the Rescorla-Wagner model, however, the learning is almost the same regardless of the colour or pre-training.

Fig. 4 shows the overall results of the simulations for the different models. The average error of the new categorisation model is clearly much lower than that

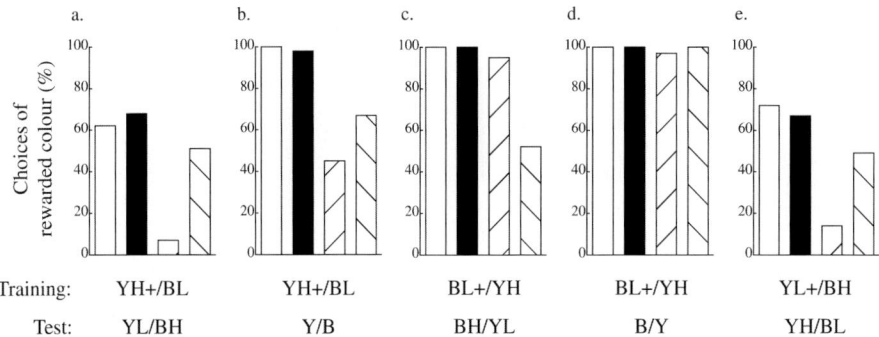

Fig. 2. Results of experiment 12-16. Choices of the stimulus with the rewarded colour after discrimination training in five experiments for the real moth and the three models. In the experiments, the moths (and models) were first trained on one combination of colour and odour and later tested on another combination to see how much of the learning that involved colour and odour respectively. See Fig. 1 for further explanations.

of the other models. Both the Rescorla-Wagner and the independence model are much better than random. Looking at the maximal error, the categorisation model reproduces the data much more closely than the other models. Surprisingly both the Rescorla-Wagner and the independence models perform at close to the random model in the worst case. The Rescorla-Wagner is even worse than the random model on some experiments.

5   Discussion

We have reported the results of 20 experiments with moths in different discrimination tasks involving multi-modal stimuli with colour and odour. Three computational models were tested on the data to try to determine the mechanisms behind this type of learning in hawkmoths. This is the first time multi-modal learning in sphingids has been modelled and the results shows that the learning mechanisms in insects can be far from trivial.

We observed behaviours that are reminiscent of overshadowing (Fig. 1a and b) and blocking (Fig. 3c and d). In naive moths, the degree of odour learning depended on the colour used during training (Fig. 1). Although the Rescorla-Wagner model is often proposed as an explanation for these phenomena, it was not able to reproduce our experimental results without changing the parameters for each individual experiment. Since the parameters were set to minimise the overall error on all experiments, the model failed in some instances. In fact, in one case, this model performed worse than random selection (Fig. 4). This parameter sensitivity is a well known problem with this model [12]. For example, in the experiments shown in Fig. 1a-b, the blue colour is not able to block odour learning since the colour undergoes extinction on the non-rewarded trials and thus looses it ability to block odour learning. For the same reason, the Rescorla-Wagner model fails to reproduce the blocking like situation in Fig. 3c-d. It is clear

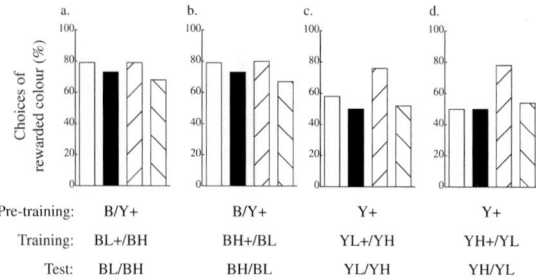

Fig. 3. Choices of the rewarded colour in experiments 17-20 for the real moth (data from [3]) and the three models. By pre-training the moths, their learning could be changed. (a-b). When the innate preference for blue was extinguished through discrimination learning, the moths could learn to discriminate between the two odours. This behaviour was predicted by all models. (c-d). When the moths were pre-trained to prefer yellow, they lost their ability to learn a discrimination between the two odours. The categorisation model as well as the independence model predicts this behaviour, while the Rescorla-Wagner model fails. See Fig. 1 for further explanations.

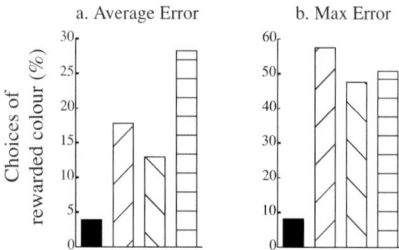

Fig. 4. Overall results of the three models and a random selection strategy (horizontal stripes). (a). Average error on all 20 experiments. The new model clearly outperforms the other models with an average error of 3.91%. (b). The maximum error for each of the models and the random selection strategy. Again, the new model is much better than the two alternatives.

that the influence of this extinction is highly dependent on the precise learning rate and the number of trials. In contrast, in the real moth, this phenomenon does not critically depend on the number of trials.

This was the motivation for the learning rule in the categorisation model, where learning only occurs during rewarded trials. This model is thus immune to extinction during non-rewarded trials and can accurately predict the behaviour of the moth in all the experiments in Fig. 1 and Fig. 3. In particular, the model will never learn the odour with a blue colour since this colour is never extinguished, and thus blocking remains intact throughout the experiment.

Surprisingly, the independence model was slightly better than the Rescorla-Wagner model both on average and in the worst case (Fig. 4). In particular, this was the case in the blocking like experiments in Fig. 3. The reason for this is the interaction between the initial preferences and the particular number of trials despite a fundamental disability to handle these learning situations. However, from the data, it is clear that the learning of one modality depends on the other.

These experiments are in line with colour preference tests that have shown that *M. stellatarum* prefers blue to yellow [18]. Blocking has been demonstrated with free-flying honeybees. Experiments have shown that the blocking effect depended on the salience of the different stimuli, that is, how easy it was for the animal to detect the stimulus [7]. In honeybees, the salience of a stimulus, e. g. the concentration of an odour, also influenced its ability to overshadow other stimuli [26].

In the future, it would be interesting to test more learning models available in the literature on the experimental data. In particular, we would like to test attentional models that may be able to explain the overshadowing like results in some of the experiments. This would possibly lead to alternative explanations for the results. We would also like to further study two of the assumptions of the model in experiments with real moths. One is that extinction never occurs which appears rather counterintuitive. The other is that the moth can only learn a single template.

In summary, we have presented experimental results from 20 different experiment with the hawkmoth *M. stellatarum*, which shows that the particular colour of an artificial flower determines whether the moth will learn its odour or not. Also, when the moth has learned a combination of colour and odour, colour is most important. By manipulating the preference for the colours, its effect on odour learning could be changed. Furthermore, we have shown that neither the Rescorla-Wagner model, nor the independence model are able to explain the experimental results. Instead, we have proposed a new model, the categorisation model, which is based on the idea that the moth learns a template for the rewarded multi-model stimulus when it is rewarded. This new model faithfully reproduces all the experimental data.

Acknowledgments

We would like to thank Michael Pfaff for help with breeding the *M. stellatarum*. We are grateful for the financial support by the Swedish Research Council. The code for the simulations is available at http://www.lucs.lu.se/ Christian.Balkenius/Downloads.

References

1. S. Andersson. Foraging responses in the butterflies *Inachis io, Aglais urticae* (nymphalidae), and *Gonepteryx rhamni* (pieridae) to floral scents. *J. Chem. Ecol*, 13:1–11, 2003.
2. A Balkenius and A. Kelber. Colour constancy in diurnal and nocturnal hawkmoths. *J. Exp. Biol.*, 207:3307–3316, 2004.
3. A. Balkenius and A. Kelber. Colour preferences influence odour learning in the hawkmoth, *Macroglossum stellatarum*. *Naturwissenschaften*, pages 1–4, 2006.

4. A. Balkenius, A. Kelber, and C. Balkenius. A model of selection between stimulus and place strategy in a hawkmoth. *Adaptive Behavior*, 12(1):21–35, 2004.

5. A. Balkenius, W. Rosén, and A. Kelber. The relative importance of olfaction and vision in a diurnal and a nocturnal hawkmoth. *Journal of Comparative Physiology A: Sensory, Neural, and Behavioral Physiology*, 192(4):431–437, 2006.

6. N. B. M. Brantjes. Sensory responses to flowers in night-flying moths. In A. J. Richards, editor, *The pollinaton of flowers by insects*, pages 13–19. The Dorset Press, Dorchester, 1978.

7. P. A. Couvillon, L. Arakaki, and M. E. Bitterman. Intramodal blocking in honeybees. *Anim. Learn. Behav.*, 25:277–282, 1997.

8. P. A. Couvillon, A. C. Campos, T. D. Bass, and M. E. Bitterman. Intermodal blocking in honeybees. *Exp. Psychol. Soc.*, 54B:369–381, 2001.

9. P. A. Couvillon, E. T. Mateo, and M. E. Bitterman. Reward and learning in honeybees: Analysis of an overshadowing effect. *Anim. Learn. Behav.*, 24:19–27, 1996.

10. J. P. Cunningham, C. J. Moore, M. P. Zalucki, and S. A. West. Learning, odour preference and flower foraging in moths. *J. Exp. Biol.*, 207:87–94, 2004.

11. E. S. Funayama, P. A. Couvillon, and M. E. Bitterman. Compound conditioning in honeybees: Blocking tests of the independence assumption. *Anim. Learn. Behav.*, 23:429–437, 1995.

12. R. C. Gallistel. *The organization of learning*. MIT Press, Cambridge, MA, 1990.

13. B. Gerber and J. Ullrich. No evidence for olfactory blocking in honeybee classical conditioning. *J. Exp. Biol.*, 202:1839–1854, 1999.

14. M. Giurfa, J. Núñez, and W. Backhaus. Odour and colour information in the foraging choice behaviour of the honeybee. *J. Comp. Physiol. A.*, 175:773–779, 1994.

15. M. Giurfa, J. Núñez, L. Chittka, and R. Menzel. Colour preferences of flower-naive honeybees. *J. Comp. Physiol. A*, 177:247–259, 1995.

16. C. M. Herrera. Activity pattern and thermal biology of a day-flying hawkmoth (*Macroglossum stellatarum*) under mediterranean summer conditions. *Ecol. Entomol.*, 17:52–56, 1992.

17. L. J. Kamin. Predictability, surprise, attention and conditioning. In B. A. Campbell and R. M. Church, editors, *Punishment and Aversive Behavior*, pages 279–296. Appleton-Century-Crofts, New York, 1969.

18. A. Kelber. Innate preferences for flower features in the hawkmoth *Macroglossum stellatarum*. *J. Exp. Biol.*, 200:826–835, 1997.

19. A. Kelber and U. Hénique. Trichromatic colour vision in the hummingbird hawkmoth, *Macroglossum stellatarum*. *J. Comp. Physiol. A*, 184:535–541, 1999.

20. A. Kelber, M. Vorobyev, and D. Osorio. Animal colour vision - behavioural tests and physiological concepts. *Biol. Rev.*, 78:81–118, 2003.

21. J. Kunze and A. Gumbert. The combined effect of color and odor on flower choice behavior of bumble bees in flower mimicry systems. *Behav. Ecol.*, 12:447–456, 2001.

22. R. C. Luo and M. G. Kay. Data fusion and sensor integration: state-of-the-art 1990s. In M. A. Abidi and R. C. Gonzalez, editors, *Data Fusion in Robotics and Machine Intelligence*. Academic Press, Boston, 1992.

23. N. J. Mackintosh. *The Physiology of Animal Learning*. Academic Press, London, 1974.

24. R. Menzel. Untersuchungen zum erlernen von spektralfarben durch die honigbiene (*Apis mellifica*). *Z. t vergl. Physiol.*, 56:25–37, 1967.

25. I. P. Pavlov. *Conditioned Reflexes*. Oxford University Press, Oxford, 1927.

26. C. Pelz, B. Gerber, and R. Menzel. Odorant intensity as a determinant for olfactory conditioning in honeybees: roles in discrimination, overshadowing and memory consolidation. *J. Exp. Biol.*, 200:837–847, 1997.
27. M. Pfaff and A. Kelber. Ein vielseitiger Futterspender für anthophile Insekten. *Entomol. Zt.*, 113:360–361, 2003.
28. R. A. Raguso and M. A. Willis. Synergy between visual and olfactory cues in nectar feeding by naive hawkmoths, *Manduca sexta. Anim. Behav.*, 64:685–695, 2002.
29. C. Rowe. Receiver psychology and the evolution of multicomponent signals. *Anim. Behav.*, 58(921-931), 1999.
30. M. V. Srinivasan, S. W. Zhang, and H. Zhu. Honeybees links sight to smell. *Nature*, 396:637–638, 1998.
31. K. von Frisch. Der Farbensinn und Formensinn der Biene. *Zool. Jb. Abt. Zool. Physiol.*, 15:193–260, 1914.
32. K. von Frisch. Über den Geruchssinn der Bienen und seine blütenbiologische Bedeutung. *Zool. Jb. Abt. Zool. Physiol.*, 37:2–238, 1919.
33. M. Weiss. Innate colour preferences and flexible colour learning in the pipevine swallowtail. *Anim. Behav.*, 53:1043–1052, 1997.
34. M. Weiss. Vision and learning in some neglected pollinators: beetles, flies, moths, and butterflies. In L. Chittka and J. D. Thomson, editors, *Cognitive Ecology of Pollination*, pages 171–190. Cambridge University Press, Cambridge, 2001.
35. T. R. Zentall and D. A. Riley. Selective attention in animal discrimination learning. *J. gen. Psychol.*, 127:45–66, 2000.

Adaptive Learning Application of the MDB Evolutionary Cognitive Architecture in Physical Agents*

F. Bellas[1], A. Faiña[2], A. Prieto[1], and R.J. Duro[1]

Integrated Group for Engineering Research,
Universidade da Coruña, Spain
[1]{fran, abprieto, richard}@udc.es, [2]iinafr00@ucv.udc.es
http://www.gii.udc.es

Abstract. This work is concerned with the study of the application of the MDB (Multilevel Darwinist Brain) evolution based Cognitive Architecture in real robots performing adaptive learning tasks. The experiments described here display the capabilities of this architecture when dealing with tasks that involve real time learning from a teacher and real time adaptation to changes in the goals provided or the communication pattern used by the teacher. One of the consequences of the interaction of the robot with the environment through the MDB is the generation of induced behaviors that allow the robot to continue its operation when no teacher is present. The experiments were carried out using a Sony AIBO robot and a Pioneer 2 robot with the same mechanism running on both just to demonstrate the robustness of the approach.

1 Introduction

In the field of autonomous robotics several approaches have been proposed to obtain controllers for physical agents through evolutionary processes [1]. Most of them use a stage of evolution in a simulated or real learning setting, but usually prior to the application in a real robot [2][3]. Thus, the acquisition of knowledge takes place in a controlled fashion and the problems of adaptation to the dynamics of the real environments are simplified. A relevant example of this is the system developed by Watson [4] which, starting from some pre-trained building blocks and behavior sequences, when released in a real environment adds its experiences to memory, building new behavior sequences, rules and procedures and deleting unused ones through a genetic algorithm. Nordin et al. [5] employ a memory based genetic programming mechanism in order to obtain a Cognitive Architecture for a Khepera robot that makes use of previous experience in its interaction with the world. Basically, a planning process incorporates a GP system that is used to evolve a suitable plan for the optimization of the outcome given the best current environment model. Walker [6] applies a training stage where a standard genetic algorithm is employed to obtain a robust and general controller through the presentation of many different situations. When operating on a real robot, they apply a minimal evolutionary strategy that adapts the controller in real time.

---

* This work was supported by the MEC of Spain through project CIT-370300-2005-24.

S. Nolfi et al. (Eds.): SAB 2006, LNAI 4095, pp. 434–445, 2006.

As mentioned before, the adaptive behavior in these cases is achieved starting from a controlled dynamic environment in the learning stage. However, if the objective is to obtain a controller in real time in a real agent operating in a real environment using evolutionary algorithms (called *Lifelong Adaptation by Evolution*, LAE), the number of examples found in literature decreases. In [7], the authors present an approach called "embodied evolution" where a group of robots can improve in real time a basic set of behaviours through evolution and use a system whereby they mate to transmit genetic information. Another relevant example that fits into the LAE approach is the one by Floreano and col. [8] where competitive coevolution is used to make two physical agents adapt their behaviours in real time. This work was continued by Ostergaard and Lund [9] applied to team competition.

After reviewing the literature, two main problems become evident in the LAE approaches: the real time adaptation to the dynamics of the real environment that must be solved by the Evolutionary Algorithm and the high computational cost of evolution. The solutions found in the literature, deal with these two problems in a theoretical way, but in real applications the dynamics of the environments and the real time adaptation of the systems are very limited. In this work, we present and apply a Cognitive Architecture (MDB) that follows this LAE approach. The objective of the MDB is to provide a robust evolutionary framework for the direct adaptation of the system to the dynamics of the environment and its interaction with it in such a way that computational cost can be reduced to make it efficient in real time applications. This is presented through the description of a set of experiments carried out on two different robotic platforms: an AIBO robot and a Pioneer 2 wheeled robot.

2 The Multilevel Darwinist Brain

The Multilevel Darwinist Brain (MDB) is a general Cognitive Architecture developed in our group and first presented in [10]. As usual in this kind of architectures, it has been designed to provide an autonomous agent with the capability of selecting the action (or sequence of actions) it must apply in its environment in order to achieve its goals. The main design objective was to automate the acquisition of knowledge in a real agent through the interaction with its environment so that it could autonomously adapt its behavior to fulfill its motivations. To carry out this requirement, we have resorted to classical bio-psychological theories by Changeaux [11], Conrad [12] and Edelman [13] in the field of cognitive science relating the brain and its operation through a Darwinist process. All of these theories lead to the same concept of cognitive structure based on the brain adapting its neural connections in real time through evolutionary or selectionist processes.

The MDB can be formalized through a cognitive model which is a particularization of the standard Abstract Architectures for agents [14]. In this case, a utilitarian cognitive model [15] is used which starts from the premise that to carry out any task, a *motivation* (defined as the need or desire that makes an agent act) must exist that guides the behavior as a function of its degree of satisfaction. We consider that the *external perception e(t)* of an agent is made up of the sensory information it is capable of acquiring through its sensors from the environment in which it operates. The environment can change due to the actions of the agent or to factors uncontrolled by the agent. Consequently, the external perception can be expressed as a function of the last action performed by the agent $A(t-1)$, the sensory perception it had of the external world in the previous time instant $e(t-1)$ and a description of the events occurring in the environment that are not due to its actions $X_e(t-1)$ through a function W:

$$e(t) = W [e(t-1), A(t-1), X_e (t-1)]$$

The internal perception $i(t)$ of an agent is made up of the sensory information provided by its internal sensors, its propioception. The internal perception can be written in terms of the last action performed by the agent, the sensory perception it had from the internal sensors in the previous time instant $i(t-1)$ and other internal events not caused by the agent $X_i (t-1)$ through a function I:

$$i(t) = I [i(t-1), A(t-1), X_i (t-1)]$$

The *satisfaction* $s(t)$ of the agent can be defined as a magnitude that represents the degree of fulfillment of the motivation of the agent and depends on the internal and external perceptions through a function S. As a first approximation we are going to ignore the events over which the agent has no control and reduce the problem to the interactions of the agent with the world and itself. Thus, generalizing:

$$s(t) = S [e(t), i(t)] = S [W [e(t-1), A(t-1)], I [i(t-1), A(t-1)]]$$

The main objective of the Cognitive Architecture is the satisfaction of the motivation of the agent, which, without any loss of generality, may be expressed as the maximization of the satisfaction s(t) in each instant of time. Thus:

$$\max\{s(t)\} = \max \{S [W [e(t-1), A(t-1)], I [i(t-1), A(t-1)]]\}$$

To solve this maximization problem, the only parameter the agent can modify is the action it performs, as the external and internal perceptions should not be manipulated. That is, the Cognitive Architecture must explore the possible action space in order to maximize the resulting satisfaction. To obtain a system that can be applied in real time, the optimization of the action must be carried out internally (without interaction with the environment) so W, I and S are theoretical functions that must be somehow obtained. These functions correspond to what are traditionally called:

- *World model (W):* function that relates the external perception before and after applying an action.
- *Internal model (I):* function that relates the internal perception before and after applying an action.
- *Satisfaction model (S):* function that provides a predicted satisfaction from predicted perceptions provided by the World and Internal models.

As commented before, the main starting point in the design of the MDB was that the acquisition of knowledge should be automatic, so we establish that these three models must be obtained in execution time as the agent interacts with the world. To develop this modeling process, information can be extracted from the real data the agent has after each interaction with the environment. These data will be called *action-perception pairs* and are made up of the *Sensorial Data* on instant t, the *Applied Action* on instant t, the *Sensorial Data* on instant t+1 and the *Satisfaction* in t+1.

Summarizing, for every interaction of the agent with its environment, two processes must be solved:

- The modeling of functions *W*, *I* and *S* using the information in the action perception pairs. As we will explain later, these are *learning* processes.
- The optimization of the action using the models available at that time.

2.1 Basic Operation

Fig. 1 displays a block diagram of the MDB with arrows indicating the flow of execution. The operation of the architecture can be summarized by considering that the selected action (represented by the *current action* block) is applied to the *environment* through an *acting* stage obtaining new *sensing* values. These acting and sensing values provide a new *action-perception pair* that is stored in the *Short-Term Memory (STM)*. Then, the *model learning processes* start (for world, internal and satisfaction models) trying to find functions that generalize the real samples stored in the STM.

The best models in a given instant of time are taken as *current world, internal and satisfaction models* and are used in the process of *optimizing the action* with regards to the predicted satisfaction of the motivation. After this process finishes, the best action obtained (*current action*) is applied again to the *environment* through an *acting* stage obtaining new *sensing* values.

These steps constitute the basic operation cycle of the MDB, and we will call it one *iteration*. As more iterations take place, the MDB acquires more information from the real environment (new action-perception pairs) and thus the learning model processes have more information and, consequently, the action chosen using these models is more appropriate.

The block labeled *Long-Term Memory* stores those models that have provided successful and stable results on their application to a given task in order to be reused directly in other problems or as seeds for new learning processes. We will later discuss the details of this memory due to its relevance in real applications.

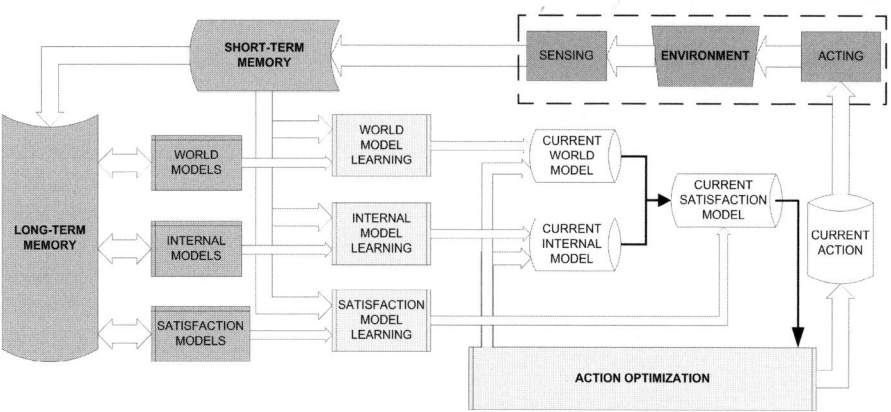

Fig. 1. Block diagram of the Multilevel Darwinist Brain

2.2 Darwinism

The main difference of the MDB with respect to other architectures lays in the way the process of modeling the functions *W, I* and *S* is carried out. According to the Darwinist theories that are the basis for this architecture, we have selected as modeling technique *Evolutionary Algorithms* and as the representation for the models *Artificial Neural Networks*, this is, the acquisition of knowledge in the MDB is basically a neuroevolutionary process. There is no design limitation in the type of Evolutionary technique to be used or in the type of neural structure to be applied in the MDB.

In this case, the modeling is not an optimization process but a learning process taking into account that we seek the best generalization for all times, or, at least, an extended period of time, which is different from minimizing an error function in a given instant *t* [16]. Consequently, the modeling technique selected must allow for gradual application, as the information is known progressively and in real time. Evolutionary techniques permit a gradual learning process by controlling the number of generations of evolution for a given content of the STM. Thus, if evolutions last just a few generations per iteration (interactions with the environment), gradual learning by all the individuals is achieved. To obtain a general model, the populations of the evolutionary algorithms are maintained between iterations (represented in Fig. 1 through the *world, internal and satisfaction models* blocks that are connected with the learning blocks), leading to a sort of inertia learning effect where what is being learnt is not the contents of the STM in a given instant of time, but of sets of STMs.

The MDB has been designed to be applied in real agents; this is, in real environments. The dynamics of such environments imply that the architecture must be intrinsically *adaptive*. The strategy of evolving for a few generations and maintaining populations between iterations permits a quick adaptation of models to the dynamics of the environment, as we have a collection of possible solutions in the populations that can be easily adapted to the new situation.

Obviously, the system must be able to respond in real time, and this is achieved through the use of current models to determine the action to be carried out, independently of what the modeling process is doing.

2.3 Short and Long Term Memory

The management of the Short Term Memory is critical in this real time learning process because the quality of the learned models depends on what is stored in this memory and the way it changes. The data stored in the STM are acquired in real time as the system interacts with the environment and, obviously, it is not practical or even useful, to store all the samples acquired in the agent's lifetime. A dynamic replacement strategy was designed that labels the samples using four basic features (distance, complexity, initial relevance and time) related to saliency of the data and temporal relevance. These four terms are weighted and, depending on the storage policy (depending on the motivation), the information stored in the STM may be different. For example, while the agent is exploring the environment or wandering (the motivation could be just *to explore*), we would like the STM to store the most general and salient information of the environment, and not necessarily the most recent. This can be achieved by simply adjusting the parameters of the replacement strategy. Details about the management strategy for the STM can be found in [17].

The Long Term Memory is a higher level memory element, because it stores information obtained after the analysis of the real data stored in the STM. From a psychological point of view, the LTM stores the knowledge acquired by the agent during its lifetime. This knowledge is represented in the MDB as models (world, internal and satisfaction models) and their context, so, the LTM stores the models that were classified by the agent as relevant in certain situations (context). In an initial approach we have considered that a model must be stored in the LTM if it predicts the contents of the STM with high accuracy during an extended period of time.

From a practical point of view, the introduction of the LTM in the MDB, avoids the need of re-learning the models in a problem with a real agent in a dynamic situation every time the agent changes into different states (different environments or different operation schemas). The models stored in the LTM in a given instant of time are introduced in the evolving populations of MDB models as seeds so that if the agent returns to a previously learnt situation, the model will be present in the population and the prediction will be accurate soon. In [18] there is formal presentation of the LTM in the MDB.

In addition, as explained before, the replacement strategy of the STM favors the storage of relevant samples. But, in dynamic environments, what is considered relevant could change during time, and consequently the information that is stored in the STM should also change so that the new models generated correspond to the new situation. If no regulation is introduced, when situations change, the STM will be polluted by information from previous situations (there is a mixture of information) and, consequently, the generated models will not correspond to any one of them. These intermediate situations can be detected by the replacement strategy of the LTM as it is continuously testing the models to be stored in the LTM. Thus, if it detects a model that suddenly and repeatedly fails in the predictions of the samples stored in the STM, it is possible to assume that a change of context has occurred. This detection will produce a regulation of the parameters controlling the replacement in the STM so that it will purge the older context. A more in depth explanation of the interaction between short and long term memory can be found in [19].

3   Adaptive Behavior Example

After presenting the basic operation of the MDB, in this section we describe an application example that uses the main features of the architecture: real time learning, real time operation, physical agent operation and adaptive behavior. We have carried out the same experiment using two different physical agents to show the robustness of the architecture: a Pioneer 2 wheeled robot and Sony's AIBO. Previous examples of the successful operation of the MDB in real problems can be found in [17].

3.1   Experimental Setup and Induced Behavior

The task the physical agent must carry out is simple: learn to obey the commands of a teacher that, initially, guides the robot towards an object located in its neighborhood.

In Fig. 4 we show the experimental setup for both agents. In the case of the Pioneer 2 robot (left image of Fig. 4), the object to reach is a black cylinder and in the case of the AIBO robot the object is a pink ball (right images of Fig. 4). The Pioneer 2 robot is a wheeled robot that has a sonar sensor array around its body and a laptop placed on its top platform. The laptop provides two more sensors, a microphone and the numerical keyboard, and the MDB runs on it. The AIBO robot is a dog-like robot with a higher range of sensors and actuators. In this example we use the digital camera, the microphones and the speaker. The MDB is executed remotely in a PC and communicates with the robot through a wireless connection.

Fig. 2 displays a schematic view of the current world and satisfaction models (with their respective numbers of inputs and outputs) that arise in this experiment in a given instant. The sensory meaning of the inputs and outputs of these models in both physical agents is summarized in Table 1. In this example, we do not take into account internal sensors in the agent and, consequently, internal models are not used. The flow of the learning process is as follows: the teacher observes the relative position of the robot with respect to the object and provides a command that guides it towards the object. Initially, the robot has no idea of what each command means in regards to the actions it applies. After sensing the command, the robot acts and, depending on the degree of obedience, the teacher provides a reward or a punishment as a pain or pleasure signal. The motivation of the physical agent in this experiment is to maximize pleasure, which basically means being rewarded by the teacher.

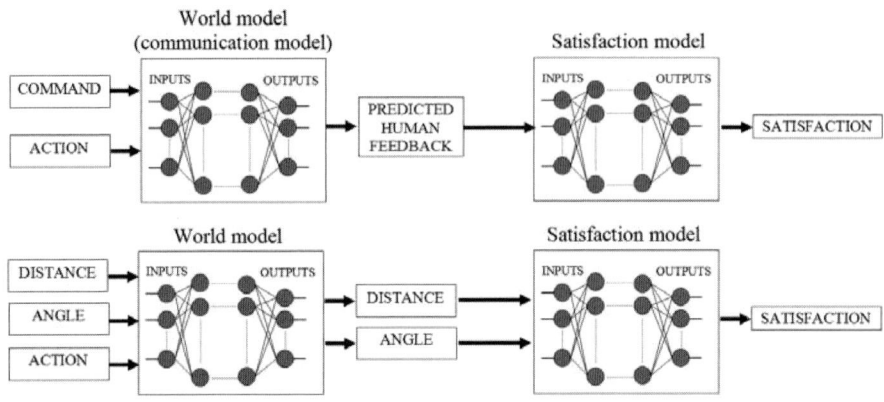

Fig. 2. World and satisfaction models involved in this example with their corresponding sensory inputs and outputs

To carry out this task, the robot just needs to follow the commands of the teacher, and a world model with that command as sensory input is obtained (top world model of Fig. 2) to select the action. From this point forward we will call this model *communication model*. The satisfaction model (top satisfaction model of Fig. 2) is trivial as the satisfaction is directly related to the output of the communication model, this is, the reward or punishment.

Table 1. Inputs and outputs involved in the models of the MDB for the two physical agents used in the experiment

	Pioneer 2 robot	AIBO robot
Command (1 input)	• Group of seven possible values according to the seven musical notes. • Provided by the teacher through a musical keyboard. • Sensed by the robot using the microphone of the laptop. • Translated to a discrete numerical range from -9 to 9.	• Group of seven possible values according to seven spoken words: hard right, medium right, right, straight, left, medium left and hard left. • The teacher speaks directly. • Sensed using the stereo microphones of the robot. • Speech recognition using Sphinx software translated into a discrete numerical range from -9 to 9.
Action (1 input)	• Group of seven possible actions: turn hard right, turn medium right, turn right, follow straight, turn left, turn medium left and turn hard left that are encoded with a discrete numerical range from -9 to 9. • The selected action is decoded as linear and angular speed.	• Group of seven possible actions: turn hard right, turn medium right, turn right, follow straight, turn left, turn medium left and turn hard left that are encoded with a discrete numerical range from -9 to 9. • The selected action is decoded as linear speed, angular speed and displacement.
Predicted human feedback (1 output /input)	• Discrete numerical range that depends on the degree of fulfillment of a command from 0 (disobey) to 5 (obey). • Provided by the teacher directly to the MDB using the numerical keyboard of the laptop.	• Group of five possible values according to five spoken words: well done, good dog, ok, pay attention, bad dog. • The teacher speaks directly • Sensed using stereo microphones of the robot. • Speech recognition using Sphinx software translated into a discrete numerical range from 0 to 5

Satisfaction (1 output)	• Continuous numerical range from 0 to 11 that is automatically calculated after applying an action. It depends on:
	  o The degree of fulfillment of a command from 0 (disobey) to 5 (obey).
	  o The distance increase from 0 (no increase) to 3 (max).
	  o The angle with respect to the object from 0 (back turned) to 3 (robot frontally to the object)
		• Continuous numerical range from 0 to 11 that is automatically calculated after applying an action. It depends on:
		  o The degree of fulfillment of a command from 0 (disobey) to 5 (obey).
		  o The distance increase from 0 (no increase) to 3 (max).
		  o The angle with respect to the object from 0 (back turned) to 3 (robot frontally to the object).

Distance and angle (2 outputs/ inputs)	• Sensed by the robot using the sonar array sensor. • Measured from the robot to the black cylinder and encoded directly in cm and degrees.	• Sensed by the robot using the images provided by the colour camera. • Colour segmentation process and area calculation taken from Tekkotsu software. • Encoded in cm and degrees. • Measured from the robot to the pink ball.

The interesting thing here is what happens to the models corresponding to other sensors. We assume that, in general, to design the models needed in the MDB for a particular task, no simplifications are made and world models are generated to cover the sensory capabilities of the physical agent. In this case, a second world model was simultaneously obtained (bottom world model of Fig. 2) that uses distance and angle to the object as sensory inputs. Obviously, this model is relating information other than the teacher's commands during the performance of the task. If the commands produce any regularities in the information provided by other sensors in regards to the satisfaction obtained, these models can be applied when operating without a teacher.

This is, if in a given instant of time, the teacher stops providing commands, the communication model will not have any sensory input and cannot be used to select the actions leaving this task in the hands of other models that do have inputs. For this second case, the satisfaction model is more complex relating the satisfaction value to the distance and angle, directly related with rewards or punishments.

In this particular experiment, the four models are represented by multilayer perceptron ANNs (with a number of neurons of 2-3-3-1 for the communication model, 3-6-6-2 for the world model and 2-3-3-1 for the second satisfaction model). They were obtained using the PBGA genetic algorithm [20] that automatically provides the appropriate size of the ANNs. Thus, in this case, the MDB executes four evolutionary processes over four different model populations every iteration. The STM has a size of 20 action-perception pairs in all the experiments.

Fig. 3 displays the evolution of the Mean Squared Error provided by the current models (communication, world and satisfaction) predicting the STM as iterations of the MDB take place in both physical agents. The error clearly decreases in all cases and in a very similar way for both agents (except at the beginning where the STM is being filled up). This means that the MDB works similarly in two very different real platforms and that the MDB is able to provide real modeling of the environment, the communication and the satisfaction of the physical agent. As the error values show in

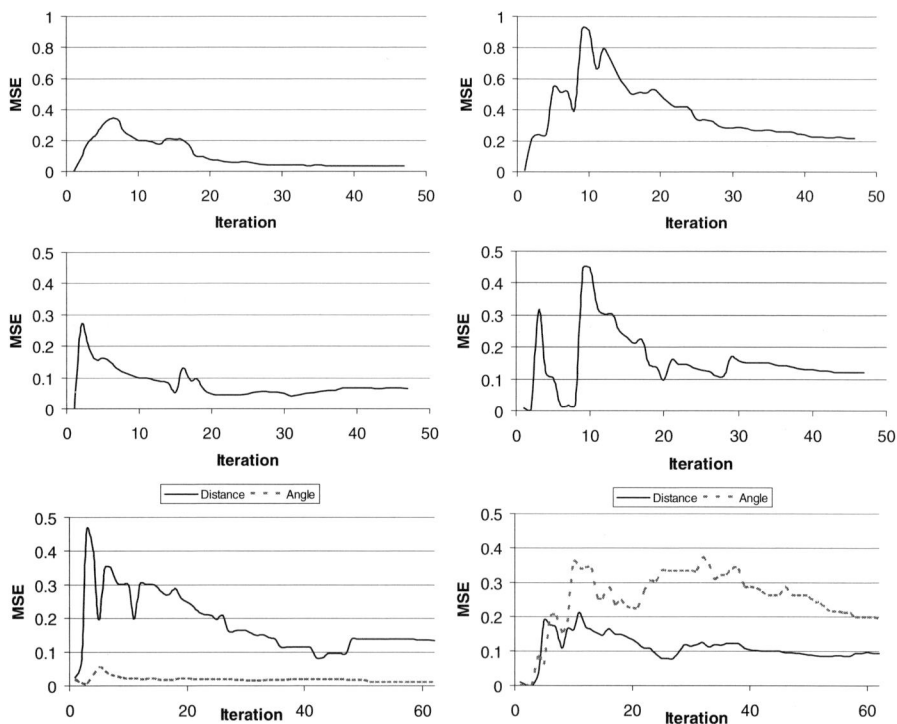

Fig. 3. Evolution of the Mean Squared Error of the current communication model (top), satisfaction model (middle) and world model (bottom) predicting the STM as iterations of the MDB take place in the AIBO robot (left column) and in the Pioneer 2 robot (right column)

Fig. 4. Left images show a sequence of real actions in the Pioneer 2 robot when the teacher is present (top) and the robot uses the induced models (bottom). Middle and right images correspond to the same situation but in the case of the AIBO robot.

Fig. 3, both robots learned to follow teacher commands in an accurate way in about 20 iterations (from a practical point of view this means about 10 minutes of real time) and, what is more relevant, the operation without teacher was successful using the world and satisfaction models. In this kind of real robot examples, the main measure we must take into account in order to decide the goodness of an experiment is the time consumed in the learning process to achieve a perfect obedience. Fig. 4 displays a real execution of actions in both robots. In the pictures with a teacher, the robot is following commands; otherwise it is performing the behavior without any commands, just using its induced models. It can be clearly seen that the behavior is basically the same although a little less efficient without teacher commands (as it has learnt to decrease its distance to the object and not the fastest way to do it).

Consequently, an induced behavior was obtained in the physical agents based on the fact that every time the robot applies the correct action according to the teacher's commands, the distance to the object decreases. This way, once the teacher disappears, the robot can continue with the task because it developed a satisfaction model related to the remaining sensors telling it to perform actions that reduce the distance.

3.2 Dynamic Adaptation

To show the adaptive capabilities of the MDB in a real application, in Fig. 5 (left) we have represented the evolution of the MSE provided by the current communication model during 200 iterations. In the first 70 iterations the teacher provides commands using the same encoding (language) applied in the previous experiment. This encoding is not pre-established and we want the teacher to make use of any correspondence it wants. From iteration 70 to iteration 160 another teacher appears using a different language (different and more complex relation between musical notes for the pioneer or words for the AIBO and commands) and, finally, from iteration 160 to iteration

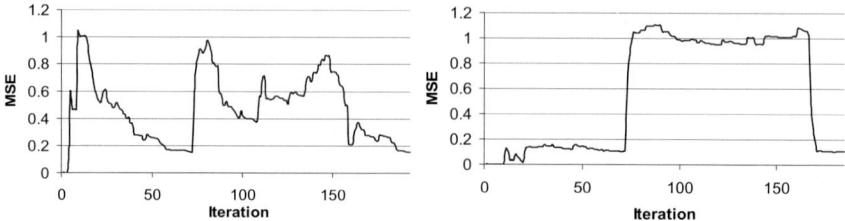

Fig. 5. Evolution of the MSE provided by the current communication model (left) and satisfaction model (right) predicting the STM as iterations of the MDB take place. The left graph corresponds to a dynamic language and the right graph to a dynamic satisfaction.

200 the original teacher returns. As shown in the figure, in the first 70 iterations the error decreases fast to a level (1.6%) which results in a very accurate prediction of the rewards. Consequently, the robot successfully follows the commands of the teacher. When the second teacher appears, the error level increases because the STM starts to store samples of the new language and the previous models fail in the prediction. At this point, as commented before, the LTM management system detects this mixed situation (detects an unstable model) and induces a change in the parameters of the STM replacement strategy to a FIFO strategy. As displayed in Fig. 5 (left), the increase in the value of the error stops in about 10 iterations and, once the STM has been purged of samples from the first teacher's language, the error decreases again (1.3% at iteration 160). The error level between iterations 70 and 160 is not as stable as in the first iterations. This happens because the language used by the second teacher is more complex than the previous one and, in addition, we must point out that the evolution graphs obtained from real robots oscillate, in general, much more than in simulated experiments due to the broad range of noise sources of the real environments. But the practical result is that about iteration 160 the robot follows the new teacher's commands successfully again, adapting itself to teacher characteristics. When the original teacher returns using the original language (iteration 160 of Fig. 5 left), the adaptation is very fast because the communication models stored in the LTM during the first iterations are introduced as seeds in the evolutionary processes.

In Fig. 5 (right) we have represented the evolution of the MSE for the current satisfaction model in another execution of this second experiment. In this case, the change occurs in the rewards provided by the teacher. From initial iteration to 70, the teacher rewards reaching the object and, as we can see in the graph, the error level is low (1.4%). From iteration 70 to 160, the teacher changes its behavior and punishes reaching the object, rewarding escaping from it. There is a clear increase of error level due to the complexity of the new situation (high ambiguity of possible solutions, that is, there are more directions of escaping than reaching the object). In iteration 160, the teacher returns to the first behavior and, as expected, the error level decreases to the original levels quickly obtaining a successful adaptive behavior.

4   Conclusions

This paper describes the application of the Multilevel Darwinist Brain to the control of two different robot platforms. The mechanism is robust with respect to the platform and sensor and actuator configuration. It allows the robot to learn on line while inter-acting with the environment and provides a way to do so redundantly in terms of world or satisfaction models. Another important characteristic is the capability of adapting in real time to changes in the environment, whether in terms of world or internal characteristics through world or internal models and task specifications through the satisfaction models obtained.

References

1. Walker, J, Garrett, A., Wilson, M.: Evolving Controllers for Real Robots: A Survey of the Literature, Adaptive Behavior vol 11, (2003) 179-203.
2. Floreano, D., Mondada, F.: Evolution on homing navigation in a real mobile robot. IEEE Transactions on Systems, Man and Cybernetics (1996), 396–407.
3. Marocco, D., Floreano, D.: Active vision and feature selection in evolutionary behavioural systems, From Animals to Animats: Proceeings SAB'02, (2002), 247–255.
4. Watson, J.B.: Behavior-Based Control for Autonomous Robotics. Proc. of the 3rd Annual Conf. on Evolutionary Programming, World Scientific, Singapore, (1994), 185-190.
5. Nordin, P., Banzhaf, W., and Brameier, M.: Evolution of a World Model for a Miniature Ro-bot Using Genetic Programming. Robotics and Auton. Systems, V. 25, (1998), 105-116.
6. Walker, J.: Experiments in evolutionary robotics: investigating the importance of training and lifelong adaptation by evolution. PhD thesis, University of Wales, (2003).
7. Nehmzow, U.: Physically embedded genetic algorithm learning in multi-robot scenarios: The PEGA algorithm. In 2nd International Workshop on Epigenetic Robotics: Modelling Cognitive Development in Robotic Systems, (2002).
8. Floreano, D., Nolfi, S., Mondada, F.: Co-evolution and ontogenetic change in competing robots. Advances in the Evolutionary Synthesis of Intelligent Agents, MIT Press (2001).
9. Ostergaard, E. & Lund, H.: Co-evolving complex robot behavior. In ICES'03, The 5th Int. Conference on Evolvable Systems: From Biology to Hardware. Springer, (20 03).
10. Duro, R. J., Santos, J., Bellas, F., Lamas, A.: On Line Darwinist Cognitive Mechanism for an Artificial Organism, Proceedings supplement book SAB2000, (2000), 215-224.
11. Changeux, J., Courrege, P., Danchin, A.: A Theory of the Epigenesis of Neural Networks by Selective Stabilization of Synapses, Proc. Nat. Acad. Sci. USA 70, (1973), 2974-2978
12. Conrad, M.,: Evolutionary Learning Circuits. Journal of Theoretical Biology, 46, (1974).
13. Edelman, G.: Neural Darwinism. The Theory of Neuronal Group Selection. Basic Books (1987), 167-188.
14. Genesereth, M.R., Nilsson, N.: Logical Foundations of Artificial Intelligence, Morgan Kauffman, (1987).
15. Mascaro, S., Korb, K., Nicholson, A.: Suicide as an Evolutionary Stable Strategy, Ad-vances in Artificial Life. 6th European Conference ECAL 2001, (2001), 120-132.
16. Yao, X.,: Automatic Acquisition of Strategies by co-evolutionary Learning, Proc. of the Int. Conf. on Computational Intelligence and Multimedia Applications, (1987), 23-29.
17. Bellas, F., Duro, R.J.: Multilevel Darwinist Brain in Robots: Initial Implementation, ICINCO2004 Proceedings Book (vol. 2), (2004), 25-32.
18. Bellas, F., Duro, R.J.: Introducing Long Term Memory in an ANN based Multilevel Dar-winist Brain, Lecture Notes in Computer Science, Vol 2686, (2003), 590-598.
19. Bellas, F., Becerra, J. A., Duro, R.J.: Construction of a Memory Management System in an On-line Learning Mechanism, ESANN 2006 Proceedings book (2006).
20. Bellas, F., Duro, R.J.: Statistically neutral promoter based GA for evolution with dynamic fitness functions. Proceedings of IASTED2002, (2002), 335-340.

Evolution

Why Are Evolved Developing Organisms Also Fault-Tolerant?

Diego Federici and Tom Ziemke

University of Skövde, Skövde, Sweden
{diego.federici, tom.ziemke}@his.se

Abstract. It has been suggested that evolving developmental programs instead of direct genotype-phenotype mappings may increase the scalability of Genetic Algorithms. Many of these Artificial Embryogeny (AE) models have been proposed and their evolutionary properties are being investigated. One of these properties concerns the fault-tolerance of at least a particular class of AE, which models the development of artificial multicellular organisms. It has been shown that such AE evolves designs capable of recovering phenotypic faults during development, even if fault-tolerance is not selected for during evolution. This type of adaptivity is clearly very interesting both for theoretical reasons and possible robotic applications.

In this paper we provide empirical evidence collected from a multi-cellular AE model showing a subtle relationship between evolution and development. These results explain why developmental fault-tolerance necessarily emerges during evolution.

1 Introduction

That biological organisms display various levels of robustness is a well known fact. Waddington referred to this tendency to suppress phenotypic variation as *canalization* [1,2]. Two types of canalization are distinguished. *Genetic canalization* describes the phenotypic resistance to alterations of the genotype (herein Mutational Robustness). *Environmental canalization* is instead the organism's capacity to suppress external influences. The latter comprises fault-tolerance as the ability to recover from transient phenotypic faults during development.

It is widely accepted that, when noise is present at the level of both the genotype and the phenotype, canalization emerges as an adaptive response under the influence of natural selection. This *stabilizing selection* captures the inclusive fitness advantage derived by robustness [3].

But is stabilizing selection strictly necessary for achieving robustness? (see also [4]).

This question is of great interest both for theoretical reasons and for engineering purposes. The possibility to develop designs/algorithms which can autonomously recover from faults during operation, much like living systems, is clearly very appealing. The classical engineering approach to fault-tolerance is

S. Nolfi et al. (Eds.): SAB 2006, LNAI 4095, pp. 449–460, 2006.

functional redundancy, but biological organisms can also recover faults by homeostatic processes, such as self-healing and regeneration. Theoretically a perfectly regenerating individual (e.g. a robot) could continue to operate ad infinitum without external maintenance.

A notable example is provided by the *Hydra Oligactis*. Hydras can regenerate any damaged or dead cell, and severed body parts can even reconstruct the complete organism [5]. Famous are also the limb of the salamander and the tail of the lizard, which can be regrown after being severed. Regeneration also takes place in the nervous system, as has been shown in recent studies [6,7].

But testing all possible sources of faults can be prohibitively expensive, if not impossible. This simple fact hinders the construction of artificial systems which are too complex for mathematical analysis. These unfortunately include most evolutionary designs.

Recently a few different models of multi-cellular development have been proposed as a possible solution to the scalability limitations of evolutionary computation [8,9,10]. An interesting property of these systems is that they have been shown to produce fault-tolerant designs in the absence of stabilizing selection [11,10], in other words "for free". We will refer to this property as *emergent fault-tolerance*.

Emergent fault-tolerance may be caused by some intrinsic property of the multi-cellular model. For example, due to their distributed nature, artificial neural networks are known to show a graceful functional degradation in response to the loss of a few units or connections. On the other hand, in [10] it was shown that this does not appear to be the case for multi-cellular development since recovery from faults was shown to be present in evolved individuals but not random ones. In [12] emergent fault-tolerance was shown to appear during evolution after a few generations. This fact is quite interesting since it is in agreement with biological findings in selection experiments, where canalization is shown to evolve in few generations [13,14].

It seems that the evolution of multi-cellular developing models shows a preference toward fault-tolerant individuals. Such a tendency can be exploited to produce designs with increased fault-tolerance even if during evolution only partial tests are carried on all possible sources of faults [15]. Similar results are also found in [10,16].

But if it is not caused by stabilizing selection, why are evolved individuals fault-tolerant? In [12] it was shown that individuals displaying high mutational robustness also proved particularly fault-tolerant. It was then hypothesized that fault-tolerance is the developmental counterpart of mutational robustness.

In fact with development phenotypes are constructed unfolding genotypes in time. A mutation can cause a phenotypic divergence in a phase, which can then be recovered in a following one. So, if with direct encoding mutational robustness can only be achieved by suppressing the phenotypic effects of mutations [17], with development variations can be expressed but still be neutral as long as they get corrected before the fitness test.

The correlation between mutational and phenotypic robustness appears because, if individuals can recover phenotypic perturbations caused by mutations,

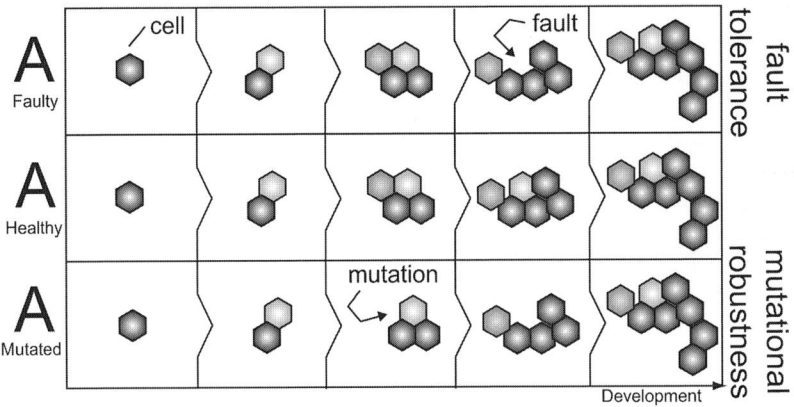

Fig. 1. Correlation between environmental and genetic canalization, $A_{healthy}$ is the original individual, $A_{mutated}$ its mutated offspring. Mutations that cause phenotypic effects can be recovered later (bottom row). In the same way faults occurring to A_{faulty} can be recovered as long as they are homologous to the typical divergence caused by mutations (top row). To be neutral to selection, divergence must be recovered before the fitness test.

they can also recover from similar phenotypic perturbations caused by faults (see Figure 1). In this paper, this hypothesis is put to the test and supporting is provided.

First developing artificial organisms are evolved to match specific targets. Then the robustness of the best individuals is checked offline. Results show that recovery from both genotypic and phenotypic perturbations becomes stronger as the fitness test gets closer in time, therefore validating the initial hypothesis: fault-tolerance is the ontogenetic homolog of mutational robustness.

These results highlight a subtle and indirect interaction between phylogeny and ontogeny. This interaction can both help to explain the evolutionary emergence of phenotypic robustness even in the absence of a direct evolutionary advantage (i.e. faults or noise), and be exploited to cheaply produce increasingly resistant artificial adaptive organisms (see also Conclusions).

2   Related Work

Typically introduced to increase the scalability and flexibility of evolutionary computation, several indirected encoding schemes have been proposed. These *artificial embryogeny* (AE, [18]) methods recursively construct the mature phenotype following the growth program defined in the genotype.

Since selection operates at the level of the phenotype, the relationship between the evolving genotype and its inclusive fitness is mediated by the development process. This indirect path may trigger complex gene-to-gene interactions, which are captured by the concept of the *gene regulatory network* (GRN).

Since phenotypic maturation in AE is de facto a rewriting process, early models were based on grammar-based approaches in which the genotype defines the substitution rules which are repeatedly applied to the phenotype. Examples include the *matrix rewriting* scheme [19] and the *cellular encoding* [20].

Some models introduced additional contextual information in each rule definition [21,22], so that phenotypic trait variations could be generated. Also, it is possible to implicitly define the grammar by means of an artificial GRN [23] and use the accumulated concentrations of simulated chemicals to modulate the characteristics of morphological constituents.

In this direction, and inspired by *cellular automata*, a second approach is to evolve the rules by which cells alter their metabolism and duplicate. Cells are usually capable of sensing the presence of neighboring cells [24], releasing chemicals which diffuse in simulated 2D or 3D environments [25,9], and moving and growing selective connections to neighboring cells [26].

Closely related to the one presented in this paper, the model proposed in [8] is based upon a fixed Cartesian 2D lattice, in which each cell occupies a given square. Artificial organisms are generated starting from a single cell. Every cell can replicate in the four cardinal directions taking the organism to maturation in a fixed number of development steps.

All cells share the same genotype encoding the cell growth program (its regulatory network). In [8] the growth program is structured as a sequence of rules. Rules are activated by matching the local neighborhood of a given cell and trigger specific cell responses: duplication, death and cell-state change.

In [9], the growth program is represented by a Boolean network. Cells belong to one of four different types and can release chemicals which undergo a simulated diffusion process. Specific evolutionary targets (2D patterns) were evolved and emergent self-healing dynamics were reported for the first time [11].

In [10] the previous model is extended with internal chemicals, which do not diffuse in the environment but are private to each cell. The growth program is encoded by a recursive neural network, and the organism's genotype can contain several chromosomes, each one specifying a complete growth program. Individuals are initialized with a single chromosome which controls the entire development process. During evolution, additional chromosomes can be introduced by duplication (i.e. gene duplication [27]), each one being associated to a specific stage of development. By allowing several independent *embryonal stages*, this method proved capable of increasing overall evolvability in the evolution of specific 2D patterns, also showing a higher scalability then direct encoding. In this case too, emergent fault-tolerance was reported.

In [8,9,10], fitness was based only on the topological displacement of cell in mature individuals. In [16] the AE model in [9] was used to produce a 2-bit multiplier capable of recovering transient phenotype faults. In [15] the AE model in [10] was used to evolve a regenerating spiking neuro-controller for simulated Khepera robots. These last results indicate the great potential that the evolution of complex fault-tolerant ontogenies can provide to the adaptive behavior community.

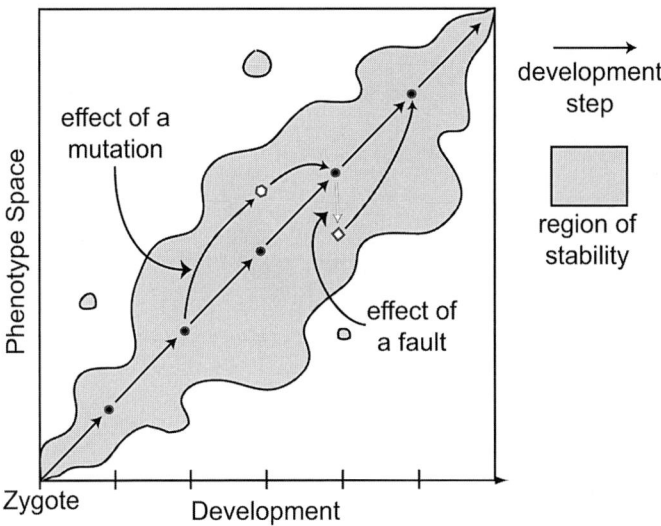

Fig. 2. Illustration of the relation between mutational robustness and fault-tolerance. If a development program has a region of stability, as long as they do not take phenotypes outside of the region perturbations can be recovered, whatever their cause.

3   Methods

It has been hypothesized that fault-tolerance is a side-effect of the genotype's mutational robustness (genetic canalization). Mutated genotypes could in fact produce identical mature organisms even if during development phenotypic divergence occurs. This as long as this phenotypic divergence is recovered/corrected before maturation, i.e. the fitness test, is reached (see Figure 2).

The central point is that once a genotype evolves a means to recover from phenotypic divergence, it will do so whatever the cause: either if it is caused by a mutation affecting the growth program or if it is produced by similar transient faults.

In either case, robustness should be stronger when the fitness test is closer in time, due to the fact that phenotypic divergence is only "apparent" to the selection mechanism at the time of the fitness test.

We can therefore test the level of mutational and phenotypic robustness and see how it varies during development. To do so we will perturb evolved genotypes and phenotypes for a single step and let the developing individual recover for an additional step. Divergence will be measured as the Hamming distance between the perturbed and non-perturbed individuals.

One problem, however, is created by the fact that multicellular development starts from a single cell (the zygote) to reach maturation in a certain number of steps. During the initial expansion phase perturbations will have a different impact on development.

To minimize this effect, it is possible to evolve individuals checking fitness at two different steps of development, in our case at step 12 and 17. It is reasonable

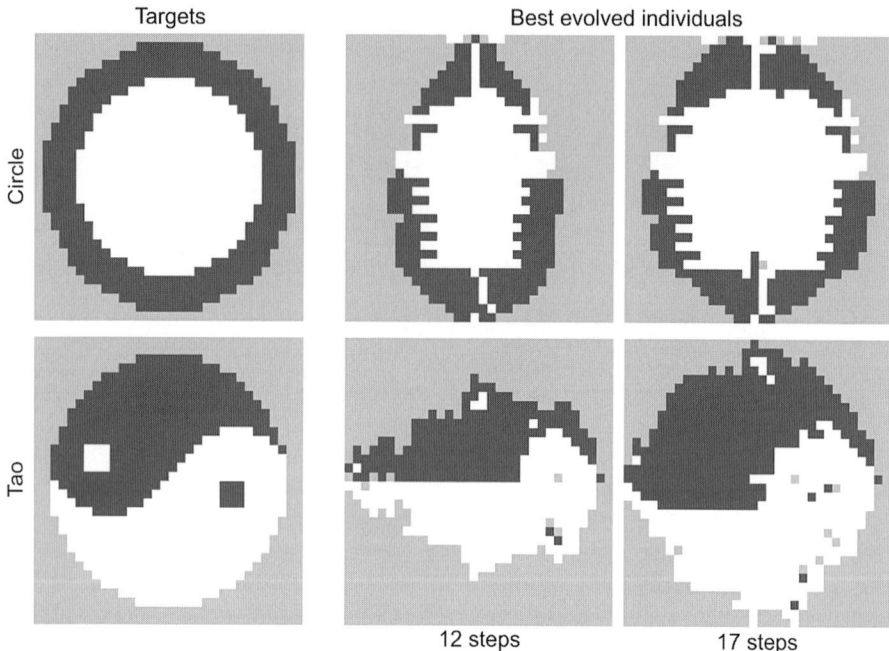

Fig. 3. Evolutionary targets and best evolved individuals at steps 12 and 17. Above the Circle and below the Tao target. Fitness is the average resemblance to the target computed at steps 12 and 17.

to assume that during the 5 growth steps from 12 to 17 the phenotypes will be more stable.

3.1   Development Model

The AE model used in this paper is explained in detail in [10]. For clarity, a short summary of the model's details follows.

Phenotypes develop starting from a single cell placed in the center of a fixed size 2D rectangular array. Multicellular organisms reach maturation in a precise number of developmental steps. Cells replicate and can release simulated chemicals in intra-cellular space (cell metabolism).

Unlike other possible approaches [8], no predefined chemical gradients are present. These in fact offer a global contextual information which biases the evolution of development toward trivial solutions.

Cell behavior in our model is governed by a growth program based on local variables, and represented by a simple recursive neural network (Morpher) with 4 hidden units. The Morpher input vector encodes the state of a particular cell (type and metabolism) and the types of the four neighboring cells in the North, West, South and East directions (NWSE).

At each developmental step, under the control of the Morpher, existing active cells can change their own type, alter their metabolism and produce new cells.

An active cell can also die or become passive. Each step, up to four new cells can be produced in any of the NWSE directions. In case of cell genesis, the mother cell specifies the daughter cells' internal variables (type and metabolism) and whether they are active or passive. If necessary, existing cells are pushed sideways to create space for the new cells. When a cell is pushed outside the boundaries of the grid, it is permanently lost.

Embryonal Stages. The regulatory system controls gene expression over two orthogonal dimensions: time and space. Development with Embryonal Stages (DES) implements a direct mechanism of Neutral Complexification for the temporal dimension [10].

As development spans over several consecutive steps, the idea is to start evolution with a single growth program (chromosome/Morpher) which controls all the development steps. As evolution proceeds, a new chromosome can be added by gene duplication.

The developmental steps are therefore partitioned into two groups/stages. The first, controlling the initial steps of embryogenesis, is associated with the old chromosome. The latter, completing growth, is associated with the new, identical, duplicated chromosome.

Being exact copies, new chromosomes do not alter development, and are therefore neutral. But possible mutations can independently affect each duplicated gene.

By unlocking the gene expression of different development phases, each chromosome can assume more specialized roles, de facto increasing the genotypic resolution around the area represented by the current mature phenotype. Overall, the effect is an increase in genotype-phenotype correlation leading to higher evolvability. In the simulations presented herein, only the chromosome controlling the latest stage is subjected to the evolutionary operators, while all other chromosomes remain fixed.

3.2   Evolutionary Details

Each cell in the mature phenotype is interpreted as a pixel, its color provided by the cell type (three possible). Fitness is proportional to the resemblance of an individual to the target pattern (see Figure 3) and is computed as the normalized Hamming distance to the target at steps 12 and 17. For fitness computation, dead cells are assigned the default type 0 (black color). Organisms grow in a 32x32 2D array starting from a single active cell in position (16,16), with type 1 and metabolism 0.

Results are obtained from 20 independent populations for each target. Every population is composed of 400 individuals. The best 50 individuals are copied to the next generation and reproduce (elitism). Evolution comprises 2000 generations. 10% of the offspring are produced by crossover.

The Morpher is modeled by a neural network with 7 inputs, 15 outputs and 4 hidden nodes. The genotype contains a floating point number for each of the 107 Morpher's weights. Mutation takes each weight of the Morpher with a .1 probability and adds to it Gaussian noise with 0 mean and .1 variance.

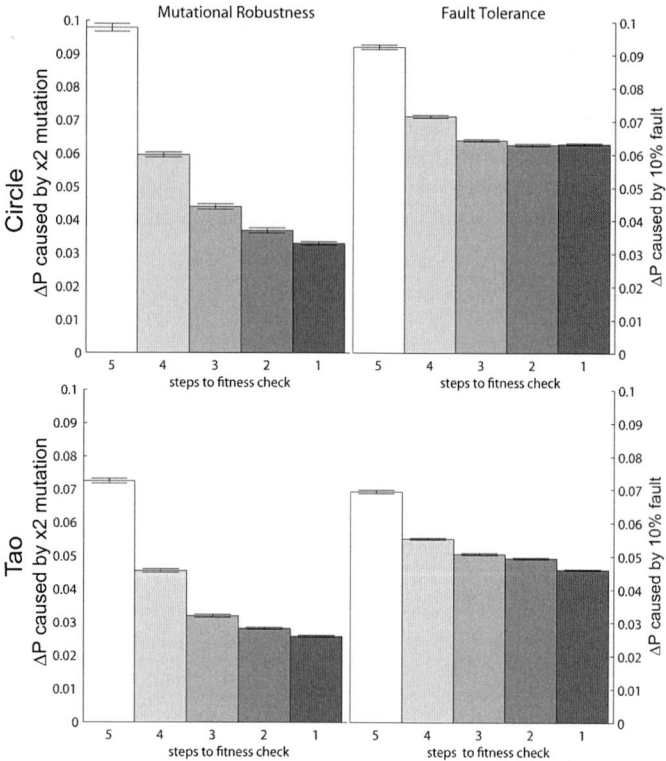

Fig. 4. Phenotypic divergence after the perturbation of a development step by means of two consecutive mutations (left) or the removal of each cell with a 10% probability (right). Circle (top) and Tao (bottom) individuals are allowed to recover one additional step. Robustness both to mutations and fault increases in temporal proximity of the fitness test. Averages and standard deviations over 100 perturbations, at each development step for each of the best individuals of the 40 populations.

4   Results

Fit individuals were produced in all the evolutionary runs. In the case of the Tao target the average fitness was $86 \pm 2\%$ with a maximum of 89%, while for the Circle it was $81 \pm 3\%$ with a maximum of 86% (see Figure 3).

In Figure 4 we show how genotypic and phenotypic perturbations are recovered in a single 'healing' step by the best evolved individuals of each population. Perturbations are generated either by two consecutive mutations to the tested genotypes, or by removing each cell in the phenotype with a 10% probability.

For both targets the phenotypic divergence ΔP decreases as the fitness test gets closer in time, both with genotype and phenotype perturbations. In the case of the Tao target, which is also the easier target to evolve, individuals appear more robust.

Since in all cases robustness measures the ability to dampen divergence with a single 'healing' step, it appears that phenotypic perturbations are taken care

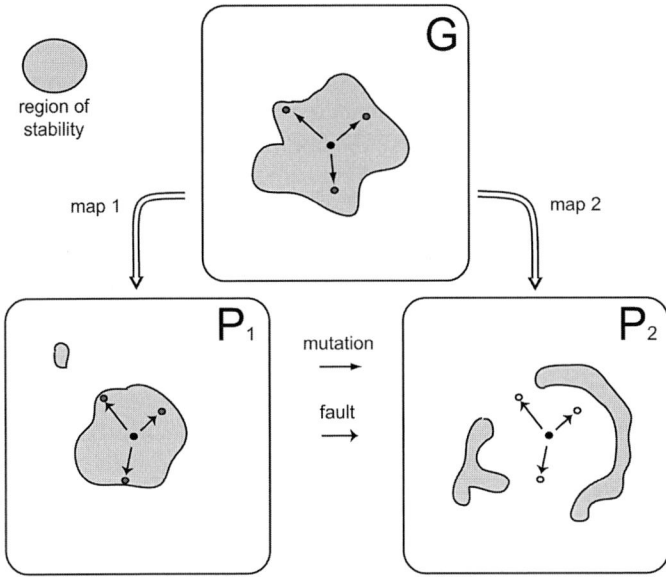

Fig. 5. An individual residing in a genotypic region of stability (top) will display a high degree of mutational robustness. Depending on the properties of the genotype-phenotype mapping, the genotypic region of stability can project in phenotype spaces in different ways. In P_1 small movements in genotype space produce small movements in phenotype space so that the region of stability is still, for the most, surrounding the individual's phenotype. The opposite in the case of P_2. Faults occurring in P_1 have a greater chance to be homologous to mutations in G and therefore be tolerated by development.

of more effectively in the temporal proximity of the fitness test. That means, the closer the fitness test in time, the more robust the organisms appear to be. This indicates that fault-tolerance is not a normal tendency of development but it is directed toward the dampening of the observable phenotypic deviations, i.e. from the point of view of selection and fitness.

5   Conclusions

We have argued that environmental canalization (fault-tolerance) is the developmental homolog of genetic canalization (mutational robustness).

Mutational robustness is the dampening of the observable[1] phenotypic consequences of mutations. It has been shown to emerge spontaneously as an adaptive response to the evolutionary dynamics [12], as regular regions of the fitness landscape are more stable under natural selection [28].

For direct mappings from genotypes to phenotypes, lacking a temporal dimension, robustness can only be achieved by means of epistatic interactions (see

---

[1] Observable by means of the fitness function.

for example [17]). With development on the other hand, phenotypic divergence can be recovered during growth.

When a good correlation between 'small genotypic changes' to 'small phenotypic changes' is present, mutational robust developing organisms will also display fault-tolerance. In fact, 'small genotypic changes' are dampened by the evolutionary preference for regular regions of the fitness landscape, while 'small phenotypic changes' are recovered because they are homologous to the those (see also Figure 5).

With this theory in place we can formulate a set of *predictions*:

Evolvability Tests. The presence of mutations pushes evolutionary-stable populations into genotype regions of stability. Since these regions are generated in genotype space by dampening the effects mutations, in general we cannot expect that the G-region of stability projects nicely into phenotype space.

For example, suppose you have a development system by which any small genotypic change can only cause big phenotypic consequences (a mapping based on a hashing function for example). In this case a mutationally robust individual may not display resistance to small faults, since small phenotypic variations are homologous to big leaps in genotype space, leaps which are unusual during evolution[2] and will probably take the genotype out of its region of stability.

In order for fault-tolerance to emerge, we must have a development system in which small phenotypic changes are homologous to the typical effects of mutations. The design of such systems is not a new problem in evolutionary computation since there is a wide consensus that they are associated with high levels of evolvability.

We can use the emergence of fault-tolerance as an indication of evolvability. Since fault-tolerance is shown to emerge in few generations, evolvability can be sampled rapidly on a wide range of parameters before more extensive searches are conducted.

Convergence and Robustness. As shown in selection experiments [13,14], populations undergoing a selective pressure for new characteristics also display less robustness. Evolution will in fact select those individuals capable of escaping the genotypic stability region, de facto pushing toward less robust genotypes. It might take several generations before robustness emerges again.

To boost fault-tolerance, it could be possible to first evolve a suitable individual. In a second evolutionary phase, the old population would now evolve not toward the old target, but toward the best individual found in the first phase.

By preferring younger individuals to old ones, or allowing selection with full replacement, the second phase will produce individuals with increased levels of canalization.

Mutation Levels and Fault-Tolerance. Since emergent fault-tolerance is a by-product of mutational robustness, larger regions of stability are to be expected as the mutation rate is increased.

---

[2] Using the mutation operator to define the genotype space metric.

This was in part validated by results contained in [12]. An increase in the mutation rate produced individuals with higher fault-resistance at the price of a decrease in overall fitness.

Acknowledgments. This work is part of the project "Integrating Cognition, Emotion and Autonomy" (IST-027819, www.his.se/icea), funded by the European Commission as part of the EC Cognitive Systems initiative.

References

1. Waddington, C.: Canalization of development and the inheritance of acquired characters. Nature **150** (1942) 563–565
2. Stearns, S.: Progress on canalization. Proc Natl Acad Sci USA (2002) 10229–30
3. Schmalhausen, I.: Factors of Evolution: The Theory of Stabilizing Selection. Univ. of Chicago Press; reprinted in 1986 (1949)
4. Siegal, M., Bergman, A.: Waddington's canalization revisited: developmental stability and evolution. Proc Natl Acad Sci USA **99(16)** (2002) 10528–32
5. Bode, P., Bode, H.: Formation of pattern in regenerating tissue pieces of hydra attenuata. i. head-body proportion regulation. Dev Biol **78**(2) (1990) 484–496
6. Zhao, M., Momma, S., Delfani, K., Calren, M., Cassidy, R., C.B.Johansson, Brismar, H., Shupliankov, O., Frisen, J., Janson, A.: Evidence for neurogenesis in the adult mammalian substantia nigra. Proc Natl Acad Sci USA **100(13)** (2003) 7925–30
7. Wu, D., Schneiderman, T., Burgett, J., Gokhale, P., Barthel, L., P.A.Raymond: Cones regenerate from retinal stem cells sequestered in the inner nuclear layer of adult goldfish retina. Invest Ophthalmol Vis Sci. **42(9)** (2001) 2115–24
8. Bentley, P., Kumar, S.: Three ways to grow designs: A comparison of embryogenies for an evolutionary design problem. In Banzhaf, W., al., eds.: Proc. of GECCO '99. (1999) 35–43
9. Miller, J.: Evolving developmental programs for adaptation, morphogenesys, and self-repair. In Banzhaf, W., Ziegler, J., Christaller, T., eds.: Proc. of the European Conference of Artificial Life, Lecture Notes in Artificial Intelligence, Vol. 2801. (2003) 256–265
10. Federici, D., K.Downing: Evolution and development of a multi-cellular organism: Scalability, resilience and neutral complexification. Artificial Life Journal (in press) **12:3** (2006)
11. Miller, J.: Evolving a self-repairing, self-regulating, french flag organism. In Deb, K., et al, eds.: Proc. of Genetic and Evolutionary Compuation, GECCO 2004. (2004) 129–139
12. Federici, D.: The evolutionary emergence of intrinsic regeneration in artificial developing organisms. In Ijspeert, A.J., Masuzawa, T., Kusumoto, S., eds.: Proc. of BioADIT 2006. (2006) 176–191
13. Kindred, B.: Selection for an invariant character, vibrissa number in the house mouse. v. selection on non-tabby segregants from tabby selection lines. Genetics **55(2)** (1966) 365–373
14. Maynard-Smith, J., Sondhi, K.: The genetics of a pattern. Genetics **45(8)** (1960) 1039–1050
15. Federici, D.: A regenerating spiking neural network. Neural Networks **18(5-6)** (2005) 746–754

16. Liu, H., Miller, J., Tyrrel, A.: Intrinsic evolvable hardware implementation of a robust biological development model for digital systems. In: Proc. of the 6th NASA Conference on Evolvable Hardware. (2005) 87–92
17. Wagner, A.: Robustness against mutations in genetic networks of yeast. Nature Genetics **24** (2000) 355–361
18. Stanley, K., Miikulainen, R.: A taxonomy for artificial embryogeny. Artificial Life **9(2)** (2003) 93–130
19. Kitano, H.: Designing neural networks using genetic algorithms with graph generation system. Complex Systems **4:4** (1990) 461–476
20. Gruau, F.: Neural Network Synthesis using Cellular Encoding and the Genetic Algorithm. PhD thesis, Ecole Normale Superieure de Lyon (1994)
21. Hornby, G., Pollack, J.: Body-brain co-evolution using L-systems as a generative encoding. In Spector, L., al., eds.: Proc. of the Genetic and Evolutionary Computation Conference, GECCO-2001, Morgan Kaufmann (2001) 868–875
22. Hornby, G., Pollack, J.: The advantages of generative grammatical encodings for physical design. In: Proc. of the 2001 Congress on Evolutionary Computation, CEC 2001, IEEE Press (2001) 600–607
23. Bongard, J.: Evolving modular genetic regulatory networks. In: Proc. of the 2002 Congress on Evolutionary Computation (CEC2002), IEEE Press, Piscataway, NJ, 2002 (2002) 1872–1877
24. Dellaert, F., Beer, R.: Toward an evolvable model of development for autonomous agent synthesis. In R.Brooks, Maes, P., eds.: Proc. of Artificial Life IV, MIT Press Cambridge (1994) 246–257
25. Eggenbergen-Hotz, P.: Evolving morphologies of simulated 3d organisms based on differential gene expression. In Husbands, P., Harvey, I., eds.: Proc. of the 4th European Conference on Artificial Life (ECAL97). (1997) 205–213
26. Cangelosi, A., Nolfi, S., Parisi, D.: Cell division and migration in a 'genotype' for neural networks. Network: Computation in Neural Systems **5** (1994) 497–515
27. Ohno, S.: Evolution by Gene Duplication. Springer (1970)
28. Nowak, M.: What is a quasi-species? Trends Ecol. Evol. **7** (1992) 118–121

GasNets and CTRNNs – A Comparison in Terms of Evolvability

Sven Magg[1,2] and Andrew Philippides[2]

[1] Department of Informatics
[2] Centre for Computational Neuroscience and Robotics
University of Sussex
S.Magg@herts.ac.uk

Abstract. In the last few years a lot of work has been done to discover why GasNets outperform other network types in terms of evolvability. In this work GasNets are again compared to CTRNNs on a shape discrimination task. This task is used as to solve it, or gain an advantage, a controller does not need timers or pattern generators. We show that GasNets are outperformed by CTRNNs in terms of evolvability on this task and possible reasons for the disadvantages of GasNets are investigated. It is shown that, on a simple task where there is no necessity for a timer or pattern generator, there may be other issues which are better tackled by CTRNNs.

1 Introduction

After GasNets, artificial neural networks inspired by gaseous signalling in biological neural systems, were introduced 1998 [4], they were used for evolution of controllers in many different tasks - from pattern generator tasks [12] to quadrupedal walking [5]. The findings in these experiments were that GasNets evolved faster than the same controller type without gas (e.g. [12] or [4]). Other studies also compared GasNet controllers to other controller types like continuous time recurrent neural networks (CTRNNs) or plastic neural networks (PNNs) [5][6]. In terms of evolvability (measured as the length of evolutionary runs till a successful controller was evolved or the robustness of evolved controllers), GasNets generally outperformed other network types or the GasNets without gas. After this higher performance in evolvability became evident, theoretical approaches were made to find the reason for this advantage. However, differences in fitness landscape properties between the GasNet and No-Gas classes which could explain the advantage for example, could not be found [12]. Other investigations focused on other properties of the GasNets, such as the coupling between the electrical and chemical signalling systems [7] and functional neutrality in evolution [14].

While some progress was made, none of these approaches found a definitive explanation of the improved evolvability of GasNets compared to other networks. This and other work did however lead to the current hypothesis that there are 3 possible reasons for improved evolvability [9]:

S. Nolfi et al. (Eds.): SAB 2006, LNAI 4095, pp. 461–472, 2006.

- modulatory effects
- different and separate temporal time scales of gas actions
- Flexible coupling of two different and interacting signalling systems through spatial embeddedness

In many of the tasks where GasNet controllers were analyzed, it turned out that the network had at least one sub network which produced a cyclic pattern or timing signal [11]. These timers or pattern generators (PG) are easy to realize in GasNets and it was suggested that this ability to tune pattern generators to given environments can be an advantage of GasNets [12 chapter 7.3.4]. Moreover, in almost all tasks used, the ability to tune pattern generators is helpful. Walking is clearly a cyclic process where timers/PGs can help. Also in the triangle-square discrimination task [4] the controller made use of pattern generators within the network [11]. But what happens if the task doesn't need a PG? If there are solutions which can be found without timers/PGs, are GasNets still better to evolve?

The aim of this work is to answer these questions. To do this, a shape discrimination task is introduced, which was previously used with CTRNNs [2]. Timer or pattern generator sub networks should not lead to an evolvability advantage in this task because no cyclic behaviour is needed. Different GasNet controllers are compared to different CTRNN controllers in terms of evolvability which is judged by the fitness of evolved solutions and the length of time to evolve them (if the word performance is used in this work, then always in terms of evolvability).

The hypothesis to be proven is that GasNets are good to evolve pattern generators and gain advantages in tasks where they are useful, but perform worse on tasks where these abilities are almost useless compared to different types of neural networks.

2 Experimental Setup

2.1 Shape Discrimination Task

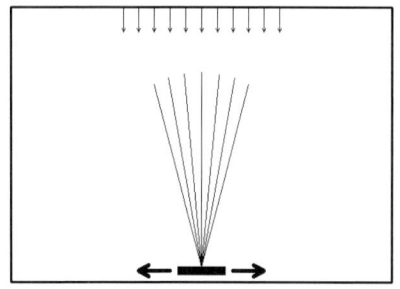

The shape discrimination task as used in this work was introduced by Randall Beer [2]. The robot has to discriminate between different falling objects and has to catch or avoid the objects, depending on the shape. The robot is represented by a line with a given length (30 Pixel) and is acting in a 2-dimensional, closed room (The room is 400 pixels wide and 275 pixels high, starting position is always (0,200)).

The robot has seven sensor rays which are uniformly distributed over a fixed angle (Π/6) starting from the centre of the robot and facing straight up. The rays act as proximity sensors with a maximum sensing range of 220 Pixel and a maximum output value if the object has reached the robot.

Each robot has two motor neurons for horizontal motion. They define the speed and direction of the robot. The speed is given by $Output_{Neuron0}-Output_{Neuron1}$. Negative speed stands for motion to the left, positive values for motion to the right. The maximum speed is 2 Pixel/time step.

Two shapes are used for the task: Circles (radius = 30) and diamonds (side length = 30). Reference points are always the centres of the objects or the robot. The robot has to catch circles and avoid diamonds. The falling speed for all objects is always one Pixel per time step. This leads to a simulation time of 275 discrete time steps. The horizontal offset of the object can be +/- 50 Pixel from the robots starting point.

After the falling object has reached the floor ($x_{Obejct} = 0$) the fitness of the robot is evaluated by the following formula:

$$\sum_{i=1}^{24} p_i \Big/ 24 \qquad\qquad (1)$$

where $p_i = d_i$ for diamonds and $p_i = 1 - d_i$ for circular objects. d_i is the distance between centre of robot and objects and is clipped to a maximum distance and normalized to [0,1]. The maximum distance is 1.5 * ($radius_{Object} + radius_{Robot}$) and has to be used, because it leads to a balance between avoidance distance for diamonds and accuracy in catching circles. 24 evaluation trials are performed for each robot with uniformly distributed dropping points and alternating object shapes.

2.2 Genetic Algorithm

In all experiments and for all network types, the same genetic algorithm is used. The competition is tournament based and the algorithm as follows:

The population is spatially distributed on a square plane and an individual randomly chosen. The tournament group consists of this individual and its 8 neighbours. The two fittest individuals are picked, recombined and the weakest individual in the group is then replaced by the mutated offspring.

As recombination, 1-point crossover is used with a randomly chosen crossing point between nodes (only whole nodes are transferred). For the shape discrimination task, the population size was always 324 and the genetic algorithm was running over a maximum of 200 pseudo generations or stopped if an individual with a fitness > 0.99 was found. One pseudo generation is equal to 324 selection and recombination processes.

Loci from the produced offspring are chosen for mutation with a fixed probability. This mutation rate is adjusted to the number of loci which can be mutated in the genotype (for some experiments, specific loci are locked) and therefore, always the same number of mutations are performed on average for each network type. If chosen, the locus is mutated using a Normal distribution with standard deviation of 1, which is scaled up (or down) to the range of the mutated value. The mean of the distribution is the original value of the locus. The maximum possible mutation is (upper limit – lower limit)/2. For example, the possibility to change a weight in a CTRNN ($\omega \in [-5,5]$) by less than 1.0 is 84%. The possibility for a change less than 2.0 is 98%.

2.3 Network Types and Characteristic Equations

2.3.1 CTRNN Network Types

All Continuous Time Recurrent Neural Networks used in this work have a fixed number of seven neurons. These neurons are divided into five inner neurons and two motor neurons. The inner neurons are fully connected. Each of them can have one non-weighted sensor input connection and is connected to the motor neurons. All connections between neurons have a weight value in each direction. The motor neurons are not connected to each other and can not receive input from sensors.

Three different CTRNN variations are used in this work, which are all based on this basic topology. Each network has a specific name in brackets for later references.

2.3.1.1 Standard CTRNN (C1). This is a conventional CTRNN [1] with the following characteristic equation:

$$y_i^{t+1} = y_i^t + \frac{\Delta t}{\tau_i}\left[-y_i^t + \sum_{j=1}^{N} \omega_{ji}\sigma(y_j^t + \theta_j) + I_i\right] \tag{2}$$

Where:

y_i^t is the activation of neuron i at time t

Δt is the time slice (Δt was 1.0 in all shape discrimination trials)

τ_i is the time constant of neuron i ($\tau \in [1,5]$)

ω_{ji} is the weight of the connection from node j to i ($\omega \in [-5,5]$)

θ_j is the bias term of neuron j ($\theta \in [-5,5]$)

I_i is the sensor input to the i'th neuron ($I \in [0,10]$, see experimental setup)

σ is the sigmoid function

$$\sigma(x) = \frac{1}{1+e^{-x}} \tag{3}$$

Biases, time constants, connection weights and the input source are under evolutionary control, i.e. for the input, the evolution can choose from which sensor ray the input is from.

2.3.1.2 CTRNN with no Temporal Dynamics (C2). This network has the same characteristic equation and topology as C1, but τ is set to 1.0 for all neurons and is not under evolutionary control. This effectively turns the neurons into reactive integrate-and-fire type neurons with no internal temporal dynamics.

2.3.1.3 CTRNN with Discrete Weights (C3). Same as C1, but only discrete weights are used. Connection weights in this network type can only be -5, 0 or 5.

2.3.2 GasNet Types

GasNets, introduced by Husbands et al. [4], have two different signalling mechanisms. They have electrical connections which can be compared to other ANN types and a gas signalling mechanism. The gases can be emitted by nodes and have modulatory effects on the transfer function of nodes in the vicinity of the emitting node. These gas connections work on different time scales than the electrical connections by build up and decay mechanisms. To model the gas diffusion the neurons of a GasNets are spatially distributed points on a square plane (in this work with side length 50 pixels). Electrical connections are based on neurons' positions in this plane with connections from a neuron being formed to all others within a genetically specified arc.

Detailed information on the GasNet model can be found in [12], [7]. The following chapters only repeat what is relevant for this work.

2.3.2.1 Standard GasNet (G1). This is a standard GasNet, already used in previous work [4-8,11-14] The characteristic equation is as follows:

$$y_i^t = \tanh\left[K_i^t \left(\sum_{j \in C_i} \omega_{ji} y_j^{t-1} + I_i^t \right) + \theta_i \right] \tag{4}$$

Where:

y_i^t is the output of neuron i at time t

K_i^t is the transfer function parameter

C is the set of nodes which have connections to node i

ω_{ji} is the weight of the electrical connection from node j to node i ($\omega \in [-1|1]$)

θ_j is the bias term of neuron j ($\theta \in [-1,1]$)

I_i is the sensor input to the i'th neuron ($I \in [0,1]$, see experimental setup)

In this work two gas types are used. Gas 1 increases the transfer function parameter K and gas 2 decreases it. This is done, dependent on the gas concentration at a given node. The gas concentration at a given node j with a distance d to the emitting node i at time t, is given by the following equations:

$$C(d,t) = \begin{cases} e^{-(d/r)^2} \cdot T(t), d < r \\ 0 \qquad\qquad , else \end{cases} \tag{5}$$

$$T(t) = \begin{cases} H\left(\dfrac{t - t_e}{s}\right) & emitting \\ H\left(H\left(\dfrac{t_s - t_e}{s}\right) - H\left(\dfrac{t - t_s}{s}\right)\right) & not\ emitting \end{cases} \tag{6}$$

$$H(x) = \begin{cases} 0 & x \le 0 \\ x & 0 < x < 1 \\ 1 & else \end{cases} \tag{7}$$

where r is the radius of the gas cloud, s is the parameter that controls the build up/decay rate of the gas, t_e is the time node i started emitting and t_s the time node i stopped emitting. The parameters r and s are genetically determined for each node.

The parameter K_i^t for node i at time step t is then given by equations 8 to 11:

$$K_i^t = \mathrm{P}[\,D_i^t\,] \tag{8}$$

$$\mathrm{P} = \{-4.0, -2.0, -1.0, -0.5, -0.25, -0.125, 0.0, 0.125, 0.25, 0.5, 1.0, 2.0, 4.0\} \tag{9}$$

$$D_i^t = f(D_i^0 + C_1^t(13 - D_i^0) - C_2^t D_i^0) \tag{10}$$

$$f(x) = \begin{cases} 0 & x \leq 0 \\ \lfloor x \rfloor & 0 < x < N \\ N-1 & else \end{cases} \tag{11}$$

2.3.2.2 GasNet without Gas (G2). This network type was also used in previous work and is a standard GasNet but using no gas (NoGasNet). This means no neuron can emit gas and only electrical connections can be used to connect neurons.

2.3.2.3   GasNet with Weight Table (G3). This is a fully connected GasNet. All neurons are connected and evolution can set the weights of these connections. As weight values, -1, 0 and 1 are used, as in the original GasNet. This means, that the network initially is never totally connected, because on average 1/3rd of the connections has weight = 0. However it is much easier for evolution to connect two nodes by mutating only one locus than when using the spatial connectivity scheme of the standard GasNet.

2.3.2.4 GasNet with Weight Table but Real Weights (G4). Same network type as G3, but using real weights. The initial connection ratio is much higher in this net, because it is very unlikely to have connections with weight=0.0.

2.3.2.5 GasNet with Weight Table, Real Weights but no Gas (G5). Same network as G4 but again no gas is used.

2.3.2.6 GasNet without Spatial Distribution (G7). In this GasNet the G5 type is changed a little bit. Because the G5 type does not use gas any more and all connections are given by a weight table, the neurons don't have to be spatially distributed any more. So, in the G7 type, the x and y loci are set to 0.0 and locked for mutation.

3 Results

As can be seen in Table 1, the CTRNN controlled robots clearly outperform the GasNet controlled robots. The best robots evolved over all runs were controlled by C1 and C2 networks (standard CTRNN and CTRNN with $\tau = 1$). They had a fitness value of 0.990 on the evaluation trials and an average of around 0.995 on 100 random trials. The best of the best GasNet controlled robot had an average fitness of .991 and an average of 0.959 on 100 random trials. It should be pointed out that the genetic algorithm stopped if a best individual was found (with a fitness of 0.990) or if the limit of 200 generations was reached. This explains the maximum of 0.990 on the CTRNN runs and the average generation number of around 200 on the GasNet runs.

The C1- and C2-robots only once reached the maximum of 200 generations. The best individual in this run already had a fitness value of 0.949. Only one G1 run stopped before the maximum number of generations was reached (177 generations). Two G2 runs stopped before the maximum was reached (151 and 182 generations).

Table 1. Results of 20 evolutionary runs. The table shows the average fitness of the best individual and the average of the whole population after 20 runs, the average fitness of the best individual on 100 random trials and the average number of generations needed to evolve the best individual. The values in brackets specify the standard deviation.

Network type:	Average of best individual:	Average fitness of population:	Average fitness on random trials:	Average Nr of generations:
C1	0.990 (0.010)	0.826 (0.030)	0.949 (0.043)	100.000 (42.7)
C2	0.990 (0.010)	0.798 (0.030)	0.953 (0.031)	108.000 (55.3)
G1	0.882 (0.099)	0.762 (0.075)	0.825 (0.119)	198.000 (5.1)
G2	0.775 (0.100)	0.683 (0.062)	0.723 (0.125)	196.000 (11.2)

Notice that the C2 variant, which has no internal temporal dynamics, performed as well as C1. This is evidence that different temporal scales are not needed in the nodes to complete the task and that the use of different temporal scales in the network does not lead to improved performance. Samples did not show evidence that the evolved solutions are different between both network types.

4   Why Are CTRNNs Better?

There are a lot of differences between CTRNNs and GasNets which could account for the CTRNN's better performance. In the following sections specific differences are highlighted and their influence on the advantage of CTRNNs evaluated.

4.1   Connection Scheme

The most eye-catching difference is the different connection scheme between both network types. CTRNN variants are almost fully connected whereas GasNets are only sparsely connected. This means that an evolving CTRNN only has to find a working set of weights, but an evolving GasNet also has to find the right sensor to motor connection mapping.

To find out if the connection scheme is accountable for a performance gain, a fully connected GasNet version (G3) was evaluated and compared to G1 and G2. For 7 neurons in a network, the gene length of G3 is the same as for G1. (In G1 seven loci are used to specify the connection scheme. In G3 the seven connection weights (-1, 0 or 1) are stored instead) so different mutation factors are no issue. No performance gain occurred by connecting all neurons in a GasNet (Table 2). In fact, the opposite occurred and the fully connected GasNet performs worse than the standard GasNet. This shows clearly that it is not the connection scheme on its own that leads to a better performance. It should be pointed out, however, that even in a G3 network the nets are never fully connected and the chance of a connection with weight = 0.0 is much higher than in a CTRNN with real weighted connections.

Table 2. Evolution of GasNet G3 compared to standard GasNet G1 and NoGasNet G2

Network type:	Average of best individual:	Average fitness of population:	Average fitness on random trials:	Average Nr of generations:
G1	0.882 (0.099)	0.762 (0.075)	0.825 (0.119)	198.000 (5.1)
G2	0.775 (0.100)	0.683 (0.062)	0.723 (0.125)	196.000 (11.2)
G3	0.743 (0.048)	0.591 (0.022)	0.699 (0.062)	200.000 (0.0)

4.2 Real Weights

Apart from the connection scheme, CTRNNs also use real weights for connections while GasNets only use inhibitory (weight = -1) or excitatory (weight = 1) connections. To find out if this influences evolution, a CTRNN variant (C3) with discrete weights (-5, 0 or 5) and GasNet variants (G4 and G5) with weight tables and real weights (with and without gas) were evaluated and compared to the standard CTRNN and GasNet versions.

Table 3. Comparison of evolution results for networks with varying weight type

Network type:	Average of best individual:	Average fitness of population:	Average fitness on random trials	Average Nr of generations:
C1	0.990 (0.010)	0.826 (0.030)	0.949 (0.043)	100.000 (42.7)
C2	0.990 (0.010)	0.798 (0.030)	0.953 (0.031)	108.000 (55.3)
G5	0.973 (0.041)	0.703 (0.035)	0.911 (0.075)	160.000 (36.4)
G4	0.954 (0.048)	0.766 (0.049)	0.920 (0.063)	185.000 (26.0)
G1	0.882 (0.099)	0.762 (0.075)	0.825 (0.119)	198.000 (5.1)
C3	0.874 (0.051)	0.631 (0.019)	0.835 (0.062)	200.000 (0.0)
G2	0.775 (0.100)	0.683 (0.062)	0.723 (0.125)	196.000 (11.2)

As one can see in Table 3, if the connections in a fully connected GasNet are combined with real weights, the results of the best individual almost reach the CTRNN results. Real weights for connections seem to be crucial for successfully evolved controllers in this task. As a crosscheck the network C3 was used which performed worse than all network types with real weights. Possible reasons for this are that it is easier for evolution to fine tune connections or to use the same neuron output as input with different strengths in different neurons.

As it turned out in the GasNet experiments, a successfully evolved solution doesn't need a lot of connections. It seems reasonable that if there are only a few connections, it is crucial to be able to fine tune them. Real weights provide this opportunity. More experiments have to be done to prove this and to find the specific reasons for the necessity of real weights. However, even with full connection scheme and real weights, the GasNets are still outperformed by both CTRNN variants in terms of evolution time. Thus, real weights are important, but not the only reason.

4.3 Neutrality

After examining many GasNet runs, we noticed that the best individual often does not change over long periods. Moreover, the fitness of the best individual plotted over generations shows long, flat regions: much longer than CTRNN runs. In [14], Tom Smith et al. show that GasNets have high functional neutrality, i.e. "many distinct neural network structures will produce the same functional mapping from sensory input to motor output" [14]. Perhaps this neutrality causes the long and flat regions?

To answer this question, a different kind of mutation operator is used. Every time a genotype is mutated, the corresponding phenotype is compared to the phenotype corresponding to the original genotype. If no change in the phenotype can be detected (i.e. phenotype has same electrical and gas connections), then the genotype is mutated again. This procedure continues till the mutation affects the phenotype. Although the mutation operator prevents neutral mutations, the picture does not change significantly. There are still long periods without change which leads to the conclusion that evolutionary search got stuck, but not for neutrality reasons. Also the overall results do not change (Table 4).

However, type G7 was also used which suggests that neutrality is not unimportant. Types G7 and G5 nn are functionally the same since changes to the coordinates in G5 are functionally neutral. Thus, in G7 mutation cannot change these coordinates and so less neutral mutations are made.

Table 4. Results from GasNet variants compared to the same variants using the "noNeutrality"-mutation operator (nn) and the results of the GasNet variant G7

Network type:	Average of best individual:	Average fitness of population:	Average fitness on random trials	Average Nr. of generations:
G5 nn	0.986 (0.024)	0.702 (0.044)	0.914 (0.048)	143.000 (49.3)
G7	0.984 (0.037)	0.722 (0.043)	0.913 (0.065)	119.000 (53.1)
G5	0.973 (0.041)	0.703 (0.035)	0.911 (0.075)	160.000 (36.4)
G4	0.954 (0.048)	0.766 (0.049)	0.920 (0.063)	185.000 (26.0)
G4 nn	0.948 (0.064)	0.710 (0.043)	0.894 (0.089)	163.000 (44.1)
G1	0.882 (0.099)	0.762 (0.075)	0.825 (0.119)	198.000 (5.1)
G1 nn	0.865 (0.057)	0.738 (0.050)	0.794 (0.093)	197.000 (9.4)

The time to evolve a successful G7 individual was significantly less than other GasNet types and is close to the result of the CTRNN results.

4.4 Fine Tuning Ability

The fitness value attained is strongly dependent on the exact position of the robot. If a robot is able to distinguish between circles and diamonds but cannot reach the exact position of the object at the end, the fitness value can be the same or even worse as the fitness of a robot that fails to distinguish shapes a few times, but has reached the exact position in all other trials. Exact positioning is rewarded. In 100 random trials (50% circles), the fitness of a robot that misses the exact position by one pixel every time while catching circles is 0.989 ((0.978 * 50 + 50) / 100). The average fitness of a robot that fails to distinguish a shape once, but positions exactly in 99 other trials is 0.990 (assuming that they all successfully avoid diamonds).

To find out if GasNets are likely to evolve controllers which can correctly distinguish shapes, but fail in finding the exact position (fine tuning), a different fitness function is used:

$$f = \begin{cases} 0.0 & , d < min \\ (d - min)/(max - min) & , min < d < max \\ 1.0 & , d > max \end{cases} \tag{12}$$

$$fitness = \begin{cases} f & , diamond \\ 1 - f & , circular\ object \end{cases} \tag{13}$$

Where d is the distance between the centres of robot and object, $min = radius_{Object}$ and $max = 1.5 * (radius_{Object} + radius_{Robot})$. This means that the robot does not have to find the exact position but only has to reach the area under the object.

Table 5. Results with new fitness (nf) compared to standard GasNet (G1) and NoGasNet (G2)

Network type:	Average of best individual:	Average fitness of population:	Average fitness on random trials	Average Nr. of generations:
G1	0.882 (0.099)	0.762 (0.075)	0.825 (0.119)	198.000 (5.1)
G1 nf	0.845 (0.096)	0.720 (0.049)	0.741 (0.150)	186.000 (37.5)
G2	0.775 (0.100)	0.683 (0.062)	0.723 (0.125)	196.000 (11.2)

The results in Table 5 show there is no difference between G1 and G1nf with new fitness. Hence, the problem of the GasNet evolution does not seem to be fine tuning of parameters governing the robot's final position.

5 Discussion

This work set out to give evidence that GasNets are outperformed by other neural network types if the solution to a given task does not need timer or pattern generator sub networks. To prove this, a task was chosen where the ability to use different time scales in the network gives no advantage. Samples over evolved controllers from both types showed, that successful solutions used active scanning and no examined GasNet had a pattern generator sub net. As shown in section 3, CTRNNs with or without evolvable time constants perform the same, which shows that timing is not necessarily needed for a successful solution. CTRNNs solve this task easily, while GasNets perform much worse. Further evidence that it is timing that is important is that GasNets have been shown to outperform other network types on tasks where pattern generation is needed [4] [5]. However, in a comparison with CTRNNs on a simple pattern generation task, while GasNets were superior to CTRNNs on one pattern, the converse was true for a second pattern [16]. It is possible that these differences are due to the range of temporal dynamics available to the two types of networks, but further work is needed to investigate this fully.

Different reasons for the disadvantage of GasNet controllers were examined. It was shown, that connectivity and fine tuning issues have no big impact on the results of the evolutionary runs. A fully connected GasNet with original GasNet weights (-1,0,1) using a weight table does not lead to a measurable performance gain. The crucial issue seems to be to have real weighted connections. As soon as a GasNet has real weighted connections, its performance is much better, while a CTRNN with discrete weights performs much worse. While the reason for the necessity of real value weights in this task is not known it is possible that the difference between the standard GasNet and its no-gas counterpart can be explained by the need for evolution to use different connection types where real weights are not available to enrich the connection scheme.

It was also shown that the time evolution needs to find a reasonable good individual decreases significantly for GasNets if loci with a high possibility of neutral mutations are taken out of evolutionary control. This is not surprising but can be still be outweighed by other issues for more complex tasks (e.g better robustness) and therefore worth accepting.

The results support the initial hypothesis, that while GasNets are good to evolve timers and pattern generators, they have disadvantages if other issues are more important. No successfully evolved GasNet controller that was analyzed during this work was using a timer/PG sub network or used a technique where timer/PG sub networks are useful. This work therefore provides evidence that for simple tasks which do not require timers or pattern generators, other issues which are not suited to GasNet type networks become more important. More research on the dynamics of GasNets is thus needed to further classify the type of tasks where they outperform other network types.

References

1. Beer, R.D. (1995): On the dynamics of small continuous-time recurrent neural networks. Adaptive Behavior 3(4):471-511.
2. Beer, R.D. (1996): Toward the evolution of dynamical neural networks for minimally cognitive behavior. In P. Maes, M. Mataric, J. Meyer, J. Pollack and S. Wilson (Eds.), "From animals to animats 4: Proceedings of the Fourth International Conference on Simulation of Adaptive Behavior" (pp. 421-429). MIT Press.
3. Funahashi, K. (1989): On the approximate realization of continuous mappings by neural networks. Neural Networks, 2, 183—192.

4. Husbands, P., Smith, T., Jakobi, N., and O'Shea, M. (1998): Better living through chemistry: Evolving GasNets for robot control. Connection Science, 10(3-4):185-210.
5. McHale, G. and Husbands, P. (2004): GasNets and other Evolvable Neural Networks applied to Bipedal Locomotion. From Animals to Animats 8: Proceedings of the Eigth International Conference on Simulation of Adaptive Behavior (SAB 2004)
6. McHale G., Husbands P. (2004): Quadrupedal Locomotion: GasNets, CTRNNs and Hybrid CTRNN/PNNs Compared. In Jordan Pollack, Mark Bedau, Phil Husbands, Takashi Ikegami and Richard A. Watson (Eds), "Artificial Life IX: Proceedings of the Ninth International Conference on the Simulation and Synthesis of Living Systems" , 106-113, MIT Press, 2004.
7. Philippides, A., Husbands P., Smith, T. and O'Shea, M.. (2005): Flexible couplings: diffusing neuromodulators and adaptive robotics. Artificial Life , 11(1&2), 139-160, 2005.
8. Philippides, A.O., Husbands, P., Smith, T. and O'Shea, M. (2002): Fast and Loose: Biologically Inspired Couplings. In Standish, R.K., Bedau, M.A. and Abbass, H.A. , editors, "Artificial life VIII. Proceedings of the 8th international conference on artificial life", pages 292-301. MIT Press.
9. Philippides, A. Personal communication
10. Slocum, A.C., Downey, D.C. and Beer, R.D. (2000): Further experiments in the evolution of minimally cognitive behavior: From perceiving affordances to selective attention. In J. Meyer, A. Berthoz, D. Floreano, H. Roitblat and S. Wilson (Eds.), "From Animals to Animats 6: Proceedings of the Sixth International Conference on Simulation of Adaptive Behavior" (pp. 430-439). MIT Press.
11. Smith, T., Husbands, P., Philippides, A., and O'Shea, M. (2003): Temporally adaptive networks: analysis of GasNet robot control networks. In R. K. Standish, M. A. Bedau, and H. A. Abbass, Eds. "Proceedings of the Eighth international Conference on Artificial Life" MIT Press, Cambridge, MA, 274-282.
12. Smith, T. (2002): The evolvability of artificial neural networks for robot control. PhD Thesis. School of Biological Sciences, University of Sussex.
13. Smith, T.M.C. and Philippides, A. (2000): Nitric Oxide Signalling in Real and Artificial Neural Networks. British Telecom Technology Journal, 18(4):140-149.
14. Smith, T.M.C., Philippides, A., Husbands, P. and O'Shea, M. (2002): Neutrality and Ruggedness in Robot Landscapes. In "Congress on Evolutionary Computation: CEC2002", pages 1348-1353. IEEE Press.
15. Yamauchi, B. and Beer, R.D. (1994): Sequential behavior and learning in evolved dynamical neural networks. Adaptive Behavior 2(3):219-246.
16. Magg, S. (2005): CTRNNs and GasNets : A comparison in terms of evolvability. Master Thesis, Department of Informatics, University of Sussex.

Incremental Evolution of Robot Controllers for a Highly Integrated Task

Anders Lyhne Christensen and Marco Dorigo

IRIDIA, CoDE, Université Libre de Bruxelles,
50, Av. Franklin. Roosevelt CP 194/6,
1050 Bruxelles, Belgium
alyhne@iridia.ulb.ac.be, mdorigo@ulb.ac.be

Abstract. In this paper we apply incremental evolution for automatic synthesis of neural network controllers for a group of physically connected mobile robots called *s-bots*. The robots should be able to safely and cooperatively perform phototaxis in an arena containing holes. We experiment with two approaches to incremental evolution, namely behavioral decomposition and environmental complexity increase. Our results are compared with results obtained in a previous study where several non-incremental evolutionary algorithms were tested and in which the evolved controllers were shown to transfer successfully to real robots. Surprisingly, none of the incremental evolutionary strategies performs any better than the non-incremental approach. We discuss the main reasons for this and why it can be difficult to apply incremental evolution successfully in highly integrated tasks.

1 Introduction

Automatic synthesis of robot controllers is an interesting field, which is likely to some day contribute significantly to the advancement and adoption of robots by industry and the general public. Techniques such as artificial evolution of controllers for autonomous robots can free us from having to understand every detail related to mapping sensory inputs to actuator outputs. Instead, we can focus on more high-level aspects in order to obtain a controller capable of solving a given task.

A robotics setup where artificial evolution can be applied usually starts off with one or more robots and some task. A fitness function is defined, which, given a behavior, assigns a number reflecting the goodness of that behavior with respect to the task. An evolutionary algorithm is then used to find an appropriate controller. The controllers themselves may consist of rule sets, decision trees or similar, but it has become common to use artificial neural networks (ANNs) due to their versatility and tolerance to noisy sensory input. If the controller is represented as an ANN, an evolutionary algorithm can be employed to optimize the weights, and possibly the morphology, of the network. Solutions found in this way can exploit subtle environmental features as they are perceived through the robot's sensors. Therefore, artificial evolution might not only be a time-saving

S. Nolfi et al. (Eds.): SAB 2006, LNAI 4095, pp. 473–484, 2006.

approach for synthesizing controllers: better controllers than those hand-crafted by human developers can be obtained in some cases [1].

The field in which evolutionary techniques are applied in order to develop robotics hardware and/or software is called *evolutionary robotics*. One direction of studies in this field is concerned with cognitive science and psychology [2], while another direction focuses on the use of evolutionary techniques as an engineering tool. Our interest falls in the latter category. We focus on the feasibility and efficiency of different approaches to automatic synthesis of controllers. Hence, our objective is to find evolutionary setups that frequently produce controllers capable of solving a given task.

The task we are concerned with is the evolution of controllers for a number of autonomous mobile robots called *s-bots* [3]. Each *s-bot* has a variety of sensors and actuators. Among these, particularly important is the gripper, which enables multiple robots to physically connect and form an artifact called a *swarm-bot*. In *swarm-bot* formation each *s-bot* maintains autonomous control. Our objective is to obtain controllers for a group of real *s-bots*, in *swarm-bot* formation, that allow them to safely navigate through an arena containing holes. The target location is indicated by a light source.

In our previous work we managed to evolve controllers for the combined phototaxis and hole-avoidance task in simulation, and we showed that the controllers could be transferred successfully to real robots [4]. In that work we also compared the performance of various non-incremental evolutionary algorithms: genetic algorithms [5,6], (μ, λ) evolutionary strategies [7], and cooperative coevolutionary genetic algorithms [8,9]. We found that the (μ, λ) evolutionary strategy in general out-performed the other evolutionary algorithms with respect to the number and quality of the successful solutions found. Furthermore, we tested a number of ANN structures and found that a multilayer perceptron with a hidden layer of two neurons is sufficient to represent successful solutions that can be transferred to real robots. For the study presented in this paper, we use the neural network topology, the fitness function components, and the (μ, λ) evolutionary strategy, which we previously found be the highest performing while resulting in transferable controllers [4], [10].

In the following section we discuss what incremental evolution is and provide examples of studies in which this technique has been applied in the field of autonomous robots. The task and the robotic hardware are explained in Section 3 and 4. Our approach and experimental setup is discussed in Section 5. In Section 6, our results are presented, discussed, and compared to results obtained in our previous work.

2   Incremental Evolution and Related Work

Incremental evolution, applied in order to obtain controllers for a given task, is a method in which evolution begins with a population that has already been trained for a simpler, but in some way related, task [11]. This is done by changing the fitness function during evolution in order to make the task progressively more complex. In this way, bootstrapping problems can possibly be overcome and evolution can be sped up. The use of incremental evolution can, however, require a substantial engineering effort, because the goal-task has to be organized into a number of sub-tasks of increasing complexity.

Note that some authors use the term *incremental evolution* for algorithms in which the morphology of ANNs is under evolutionary control. Such algorithms include SAGA where the morphology of ANNs is evolved in an incremental manner by the algorithm itself [12]. Another example is the SGOCE paradigm in which ANNs are constructed based on developmental programs that change size and composition during evolution [13]. We will not consider such algorithms here, but instead we use the term incremental evolution to denote evolutionary setups in which the fitness function and/or the environment in which the robots operate are modified during evolution.

A number of studies of incremental evolution of robot controllers has already been performed. Nolfi et al. [14] used incremental evolution to overcome some of the discrepancies between simulation and the real world. Controllers evolved in simulation were transferred to real robots on which evolution was continued. After a few generations, the performance of the controllers on real robots reached the same level as achieved in simulation. Thus, incremental evolution was used to adapt controllers, trained in simulation, to the sensory noise and behavior of the physical robot hardware, which are both impossible to simulate accurately [15].

Harvey et al. [11] evolved controllers incrementally to let a robot distinguish between white triangular and rectangular objects on a dark background. The goal was to evolve controllers that would move robots towards triangles only. The task was divided into sub-tasks where the robots would first learn to orient themselves to face a large rectangle easily detectable by their sensors, then to face and approach a smaller, moving rectangle, and finally to distinguish between rectangles and triangles, and only move towards triangles. Thus, controllers were first trained to follow white rectangles and then later trained not to follow them, but instead to follow triangles only. The authors divided the goal-task into sub-tasks in which recognition and pursuit were learnt in the first evolutionary phase, or *increment*, while discrimination between the two geometric shapes was learnt during later increments. The controllers obtained with incremental evolution were shown to be more robust than controllers trained on the complete task from an initial random population.

Gomez and Miikkulainen [16] used incremental evolution, combined with enforced sub-population and delta-coding, to evolve obstacle avoidance and predator evasion. Incremental evolution was performed by first evolving populations of neurons capable of avoiding a single enemy moving at low speed on a discrete 10x10 grid. The size of the grid was then increased to a 13x13 grid and another enemy was added. Two increments followed in which the speed of the two enemies was increased. The authors found that evolving controllers for the complete task directly was infeasible, while incremental evolution yielded satisfactory results.

In this paper we test different approaches to dividing the goal-task into sub-tasks, one inspired by Harvey et al. [11], which we denote *behavioral decomposition*, and one inspired by Gomez and Miikkulainen [16], which we call *envi-*

ronmental complexity increase. By performing incremental evolution we hope to guide the evolution search towards regions of the fitness landscape containing successful solutions. This should make the evolutionary process more efficient and should therefore increase the likelihood that an evolutionary run finds a good solution.

3   The Task

A group of physically connected *s-bots* should be able to navigate through each of the four arenas shown in Fig.1. The *s-bots* are physically connected using the rigid grippers mounted on the *s-bots*. The group is initially located in a starting zone and should navigate to the location of the light source without falling into any holes or over the side of the arena. Phototaxis and obstacle avoidance for physically connected robots has previously been studied in simulation by Baldassarre et al. [17].

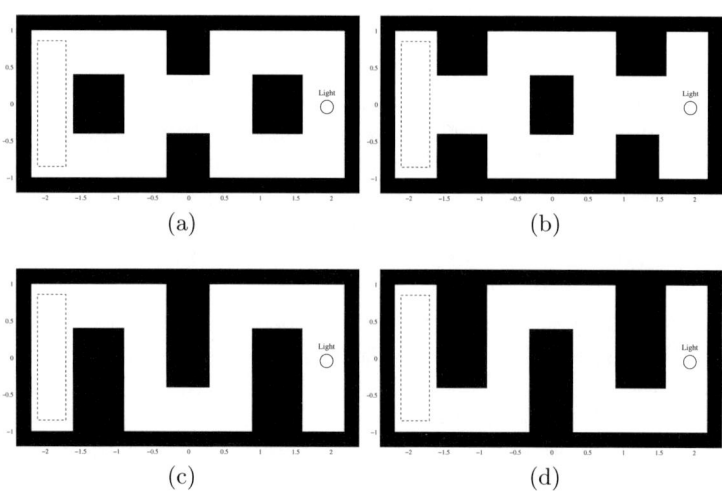

Fig. 1. The four arenas used to evolve controllers. Each of the arenas measures 480x240 cm. The dark areas denote holes, while the white patches denote the arena surface on which the robots can move. The *swarm-bot* must move from the initial location shown on the left-hand side of each arena to the light source on the right without falling into any of the holes or over the edge of the arena.

4   The Robots

In this study, we develop controllers for the SWARM-BOTS platform [3]. Fig. 2 shows photos of an *s-bot* and a *swarm-bot*. Each *s-bot* is equipped with four infrared ground sensors, one pointing 45 degrees forward, two pointing straight downward, and one pointing 45 degrees backward. The ground sensors are mounted between the differential treels© (a combination of tracks and wheels, see Fig. 2).

Fig. 2. Different views of an *s-bot* highlighting the location of the sensors used and a *swarm-bot*. An *s-bot* has a diameter of 120 mm and a height of 190 mm.

Microphones and speakers allow *s-bots* to emit and perceive sounds. An *s-bot* can sense forces exerted on its body in the horizontal plane via a traction sensor. These forces allow the *s-bot* to gauge the direction of motion of the rest of the group. Thus, each *s-bot* can align its own direction of motion to that of the other *s-bots*, allowing the *swarm-bot* to move coordinately. The traction sensor is mounted inside the robot between the bottom part (the chassis) and the top part (the turret). The turret can rotate independently of the chassis: up to 180 degrees in each direction from the neutral position. The result of an action in a given situation is likely to depend on the current rotational difference between the top and bottom parts of the *s-bot*. We therefore use two sensors that read the rotational difference in the clockwise and counter-clockwise directions, respectively, at every control step. The relative direction of the target, identified by a light source, is perceived via 8 light-sensors distributed evenly around the plastic ring on the chassis of the *s-bot* as shown in Fig. 2.

The ground sensors are located directly under the *s-bot*, which means that the *s-bot* will only detect the presence of a hole once it is already partly over it. If a single robot tries to navigate through an arena containing holes, it is very likely to fall into a hole unless it approaches the hole perpendicularly. In *swarm-bot* formation, however, the *s-bots* should be able to cooperate to safely navigate through the arena and reach the location of the light source.

We have preprogrammed the *s-bots* to emit a sound, which can be perceived by the other *s-bots* in the *swarm-bot*, when the presence of a hole is detected. This has previously been found to be an efficient aid when evolving hole-avoidance for a *swarm-bot* [18].

5 Methodology

5.1 Preliminary Fitness Function Components

In this section we present four components of the fitness function, which are used in the incremental evolution setups. In our previous studies, we found that these fitness components assist evolution in finding controllers capable of solving the combined phototaxis and hole-avoidance task on real robots [10], [4].

The first component scores controllers depending on how close they manage to get to the light source. In case they manage to get in the immediate vicinity (within 50 cm) of the light source they are scored based on how fast they do so:

$$f_{light} = \begin{cases} 1 - \dfrac{min\ distance}{initial\ distance} & \text{if the light source is not reached,} \\ 2 - \dfrac{time\ light\ is\ reached}{total\ time} & \text{otherwise,} \end{cases} \quad (1)$$

where *total time* is 240 seconds. A component penalizing controllers for falling into holes:

$$f_{stayalive} = \begin{cases} 0.5 \text{ if the } swarm\text{-}bot \text{ falls into a hole,} \\ 1.0 \text{ otherwise.} \end{cases} \quad (2)$$

Previous studies have shown that coordinated-motion in a group of connected *s-bots* can be obtained by minimizing the traction between the *s-bots* [19]. The traction forces are measured in the two dimensions of the horizontal plane with 0 corresponding to no traction perceived and 1 to the maximum traction force perceivable. At each control step i, we record the maximum traction τ_i^{max} perceived by any of the *s-bots* in the simulation:

$$f_{minimizetraction} = \frac{\sum_i (1 - \tau_i^{max})}{total\ number\ of\ control\ steps}. \quad (3)$$

The three components above are multiplied to form the function f_{final}:

$$f_{final} = f_{light} \cdot f_{stayalive} \cdot f_{minimizetraction}. \quad (4)$$

Finally, we introduce an additional fitness component f_{move} that is used in one of our proposed incremental evolutionary setups. This component rewards coordinated-motion and exploration by measuring the distance covered, measured in a straight line, during different time intervals. Initial experiments showed that measuring the distance moved during multiple time intervals is necessary in order to prevent circular paths and therefore three "good" intervals were found by trial-and-error. Every 7, 13 and 29 seconds, the position of the *swarm-bot* is compared to its position respectively 7, 13, and 29 seconds earlier. The controller achieves a fitness score based on the accumulated distances covered during these intervals divided by the maximum theoretical distance coverable.

5.2   Sub-task Divisions

We divide our goal-task in two different ways based on *behavioral decomposition* and *environmental complexity increase*, respectively.

Behavioral decomposition: Assuming that a successful overall behavior can be decomposed into the sub-behaviors: coordinated-motion, hole-avoidance, and phototaxis, it is possible that these behaviors can be learnt in an incremental fashion. That is, the first learning task is concerned with coordinated-motion in an arena without holes under a fitness function that rewards coordinated-motion ($f_{minimizetraction} \cdot f_{move}$). Once a satisfactory solution has been found, holes are added and the fitness function is extended with a component, which rewards controllers that avoid steering the *s-bots* into holes ($f_{stayalive}$). Finally, phototaxis too is rewarded and the evolved controllers should be able to solve the goal-task (f_{final}). Hence, we assume that the most fundamental task is coordinated-motion, since coordinated-motion is needed for performing both phototaxis and hole-avoidance, followed by combined coordinated-motion and hole-avoidance, and finally the goal task (including phototaxis) in an arena containing holes[1].

Environmental complexity increase: Evolution is started in one of two simple arenas, one containing no holes and the other containing a single hole. More holes and different arena layouts are added as evolution finds solutions. The purpose of applying incremental evolution in this manner is to shape the initial fitness landscapes in such a way that solutions are easier to find because the task is less difficult. When the complexity of the arena is increased, we expect the evolutionary search to resume in region(s) of the fitness landscapes closer to good solutions than a random population would cover. This way of performing incremental evolution could, for instance, prevent evolutionary runs in which the *s-bots* fail to coordinate and move, because in the first increments *swarm-bots* are less likely to encounter a hole. Evolutionary pressure is therefore towards controllers that cause the *swarm-bot* to move. In all increments fitness scores are computed by f_{final}.

For our experiments with environmental complexity increase we use the additional arenas shown in Fig. 3. Two different evolutionary setups are used to test incremental evolution based on environmental complexity increase. One setup consists of six increments while the other consists of three. In the setup comprising six increments, fitness scores of individuals are computed based on trials in the simplified arenas shown in Fig. 3a, 3b, 3c, and 3d, in the first four increments, respectively. In the 5th increment individuals are scored in two of the final arenas shown in Fig. 1a and 1b during the fitness evaluation, while in the 6th and final increment all four arenas shown in Fig. 1 are used. In the environmental complexity increase setup consisting of three increments, a population

---

[1] We experimented a different ordering of sub-tasks and with an initial increment in which only $f_{minimizetraction}$, as opposed to $f_{minimizetraction} \cdot f_{move}$, was used. However, the behavioral decomposition described above was the highest performing of those tested.

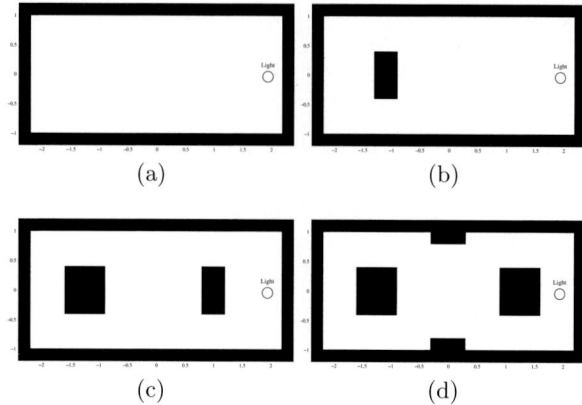

Fig. 3. Additional arenas used for incremental evolution based on environmental complexity increase. A population is first trained in (a) until an acceptable performance is achieved, then in (b) and so on.

starts in the arena shown in Fig. 3b containing a single hole. In the second increment the arena shown in Fig. 3d containing multiple holes is used. Finally, in the last increment, the goal-task is used and individuals are evolved based on fitness scores in all of the four arenas shown in Fig. 1.

We base the transition from one increment to the next on the performance of the population on the current sub-task. The fitness components we use are all relatively noisy. We have to take this into account to avoid that a noisy fitness function makes evolution move from one increment to the next before the current sub-task has been truly learnt. We therefore require the fitness score of the best performing individual to be above a certain threshold for 10 consecutive generations before a transition is made. The thresholds are different for each increment and they are determined based on the fitness function, stagnation of fitness scores during trial runs, and visual inspection of strategies found during the trial runs. Thus, the task of finding these thresholds is, like finding a suitable sub-division of the goal-task, an engineering effort.

For each of the evolutionary setups described above, we run 20 evolutions. Each evolutionary run comprises 1000 generations with a population size of 100 individuals. In all cases, we have used a (μ, λ) evolutionary strategy with $\mu = 20$ and a mutation rate of 15% on a chromosome of floating-point genes. This evolutionary algorithm was found to be the highest performing in our previous study [4]. All evolutionary runs are conducted in our software simulator TwoDee [10].

6   Results

In order to compare the performance of the controllers evolved in the different evolutionary setups, we took the highest scoring controller from each of the final generations, post-evaluated them 25 times in each of the four arenas shown in

Post–evaluation Results

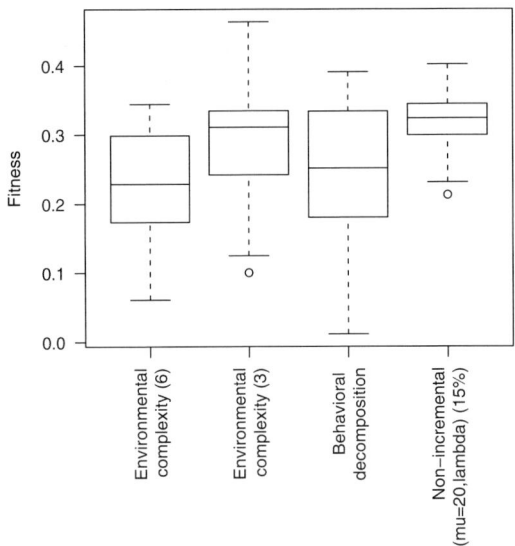

Fig. 4. Box-plot of post-evaluation results for incremental evolution through environmental complexity increase (results for both six and three increments, see text) and behavioral decomposition. We have included the results for non-incremental evolutionary runs using a (μ, λ) evolutionary strategy with a mutation rate of 15%. Each box comprises observations ranging from the first to the third quartile. The median is indicated by a bar, dividing the box into the upper and lower part. The whiskers extend to the farthest data points that are within 1.5 times the interquartile range. Outliers are shown as circles.

Fig. 1, and recorded their average fitness scores. The results for incremental evolution through behavioral decomposition and through environmental complexity increase are shown on the box-plot in Fig. 4. Each box represents post-evaluation results for 20 evolutionary runs. We have included results for the (μ, λ) evolutionary strategy with $\mu = 20$, obtained without the use of incremental evolution.

An example of a successful strategy can be seen in Fig. 5. All evolved controllers capable of performing integrated hole-avoidance and phototaxis displayed a similar strategy: The *swarm-bot* moves coordinately towards an edge of the arena and follows that edge in the direction of the light until the light source is reached.

As it can be seen in Fig. 4, on average the incremental evolutions did not produce better controllers than the evolutionary runs without increments. In the following, we discuss why this is the case.

In the evolutionary setup where integrated phototaxis and hole-avoidance behaviors were evolved incrementally based on behavioral decomposition, we assumed that a successful behavior can be decomposed into coordinated-motion,

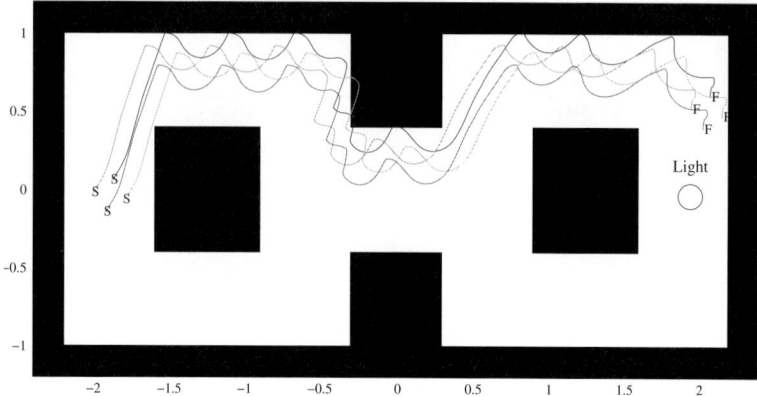

Fig. 5. Example of how a controller capable of solving the task behaves using a simple, but nonetheless general and successful, hole-following strategy

hole-avoidance, and phototaxis. If we take a closer look at the successful solutions found by artificial evolution (an example is shown in Fig.5), it is questionable if such a decomposition is valid and if a suitable decomposition can be found at all. Regardless of the initial position of the *swarm-bot*, the initial orientation of the *s-bots*, and the layout of the arena, the result we observe is always that the *swarm-bot* starts moving left (or right) with respect to the light. Therefore, it appears that the *s-bots* mainly coordinate based on the sensed direction of the light, which serves as a global point of reference, and not on the readings of the traction sensors. Hence, in a successful integrated behavior, coordinated-motion is partly a by-product of phototaxis and it is therefore not beneficial to evolve the two behaviors independently.

In the evolutionary setup based on environmental complexity increase with six increments, only 11 out of the 20 evolutionary runs reached all increments. In the remaining 9 setups the populations did not reach an adequate performance in one of the previous increments. If an evolutionary run does not reach the final increment, then the controllers produced by this run have not been evolved with respect to the goal-task but only with respect to a simpler task. We assume that such controllers obtain a lower post-evaluation score. In the evolutionary setups using only three increments, 17 of the 20 evolutionary runs reached the final increment before the 1000th generation.

No assumptions regarding decomposition of behaviors were made in the experiments for incremental evolution through environmental complexity increase. Nonetheless, the resulting controllers did not on average perform better than controllers evolved non-incrementally. According to the results in Fig. 4, it appears as if the results for incremental evolution through environmental complexity increase are not better than the results for the non-incremental approach. This is surprising given that we started evolution in a simplified arena and increased the complexity gradually in order to avoid bootstrapping issues and assist evolution in finding good solutions. We believe the major reason for the lack of superior performance is that some solutions found in the earlier increments are in fact not in regions of the fitness landscape that contain successful strategies for later increments. To illustrate this, take for instance the arena shown in Fig. 3b containing a single hole. In this arena, successful solutions include moving directly towards the light and then to move either left or right around the hole if/when it is encountered. However, this strategy does not work in the arenas shown in Fig. 1c and 1d, which both contain a number of turns and which cannot be solved by controllers that turn either only left or only right around holes. Given the reactive nature of the ANN controller, such strategies cause the *swarm-bot* to get trapped in one of the corners. In this way, incremental evolution through environmental complexity increase does in fact results in evolution taking a detour, because the highest scoring solutions in the initial increments are not simpler versions of successful behaviors in later increments.

The results of our study illustrate a fundamental issue related to applying artificial evolution in the field of robotics: The fact that evolution can find solutions, which we as human developers would not have anticipated, is a double-edged sword. On the one hand we can obtain novel solutions, while on the other it can complicate the applicability of techniques like incremental evolution for the very same reason.

Acknowledgements

Anders Christensen acknowledges support from COMP2SYS, a Marie Curie Early Stage Research Training Site funded by the European Community's Sixth Framework Programme (grant MEST-CT-2004-505079). The information provided is the sole responsibility of the authors and does not reflect the European Commission's opinion. The European Commission is not responsible for any use that might be made of data appearing in this publication. Marco Dorigo acknowledges support from the Belgian FNRS, of which he is a Research Director.

References

1. Nolfi, S., Floreano, D.: Evolutionary Robotics: The Biology, Intelligence, and Technology of Self-Organizing Machines. MIT Press/Bradford Books, Cambridge, MA (2000)
2. Harvey, I., Di Paolo, E.A., Wood, R., Quinn, M., Tuci, E.: Evolutionary robotics: A new scientific tool for studying cognition. Artificial Life 11(1-2) (2005) 79–98
3. Mondada, F., Gambardella, L.M., Floreano, D., Nolfi, S., Deneubourg, J.L., Dorigo, M.: The cooperation of swarm-bots: Physical interactions in collective robotics. IEEE Robotics and Automation Magazine 12(2) (2005) 21–28
4. Christensen, A.L., Dorigo, M.: Evolving an integrated phototaxis and hole-avoidance behavior for a swarm-bot. In: Proceedings of ALIFE X, MIT Press, Cambridge, MA (2006) In press
5. Goldberg, D.E.: The Design of Innovation: Lessons from and for Competent Genetic Algorithms. Kluwer Academic Publishers, Boston, MA (2002)
6. Mitchell, M.: An Introduction to Genetic Algorithms. MIT Press, Cambridge, MA (1996)

7. Schwefel, H.P.: Evolution and Optimum Seeking. Wiley & Sons, New York (1995)
8. Potter, M.A., De Jong, K.: A cooperative coevolutionary approach to function optimization. In: Proceeding of the Third Conference on Parallel Problem Solving from Nature – PPSN III. Volume 866 of LNCS, Springer Verlag, Berlin, Germany (1994) 249–257
9. Potter, M.A., De Jong, K.: Cooperative coevolution: An architecture for evolving coadapted subcomponents. Evolutionary Computation 8(1) (2000) 1–29
10. Christensen, A.L.: Efficient neuro-evolution of hole-avoidance and phototaxis for a swarm-bot. Technical Report TR/IRIDIA/2005-14, IRIDIA, Université Libre de Bruxelles, Belgium (2005) DEA Thesis.
11. Harvey, I., Husbands, P., Cliff, D.: Seeing the light: artificial evolution, real vision. In: Proceedings of the third international conference on Simulation of adaptive behavior: From animals to animats 3, Cambridge, MA, MIT Press (1994) 392–401
12. Harvey, I.: Species adaptation genetic algorithms: a basis for a continuing SAGA. In Varela, F.J., Bourgine, P., eds.: Proceedings of the First European Conference on Artificial Life. Toward a Practice of Autonomous Systems, Paris, France, MIT Press, Cambridge, MA (1992) 346–354
13. Kodjabachian, J., Meyer, J.A.: Evolution and development of neural controllers for locomotion, gradient-following, and obstacle-avoidance in artificial insects. IEEE Transactions on Neural Networks 9(5) (1998) 796–812
14. Nolfi, S., Floreano, D., Miglino, O., Mondada, F.: How to evolve autonomous robots: Different approaches in evolutionary robotics. In: Proceedings of Artificial Life IV, Cambridge, MA, MIT Press/Bradford Books (1994) 190–197
15. Jakobi, N.: Evolutionary robotics and the radical envelope-of-noise hypothesis. Adaptive Behavior 6(2) (1998) 325–368
16. Gomez, F., Miikkulainen, R.: Incremental evolution of complex general behavior. Adaptive Behavior 5 (1997) 317–342
17. Baldassarre, G., Parisi, D., Nolfi, S.: Distributed coordination of simulated robots based on self-organisation. Artificial Life (2006) In press
18. Trianni, V., Tuci, E., Dorigo, M.: Cooperative hole avoidance in a swarm-bot. Robotics and Autonomous Systems 54(2) (2006) 97–103
19. Trianni, V., Labella, T.H., Dorigo, M.: Evolution of direct communication for a swarm-bot performing hole avoidance. In: Ant Colony Optimization and Swarm Intelligence – Proc. of ANTS 2004 – 4th Int. Workshop. Volume 3172 of LNCS, Springer Verlag, Berlin, Germany (2004) 131–142

An Evolutionary Selection Model Based on a Biological Phenomenon: The Periodical Magicicadas

Marco Remondino[1,2] and Alessandro Cappellini[1,3]

[1] University of Torino
[2] Lagrange Project
[3] ISI Foundation

Abstract. *Magicicada* is the genus of periodical cicadas which display a unique combination of long life cycles, periodicity, and mass emergences. Their nymphs live underground and stay immobile before constructing an exit tunnel in the spring of their 13th or 17th year, depending on the species. Once out, the adult insects live only for a few weeks with one sole purpose: reproduction. Both 13 and 17 are prime numbers; why did the cicadas "choose" these lengths for their life cycles? Two are the most interesting hypotheses (limited resources and hybridization avoidance) drawn by biologists, both bringing to the conclusion that the prime number cycles were selected because they were least likely to emerge with other cycles. If that's the case, then these lengths would have been selected via a sort of "tacit coordination by evolution". In the agent based model presented here, it is shown how the two major hypotheses must be both present in order to cause the emergency of the prime numbers based life cycles. A very important point here is that the agents in the model are not endowed with a "calculating" ability, in particular they lack the capacity of determining divisors. In the model no form of learning is present; the emergency of prime numbers is then a fact of evolutionary biology, a natural selection by adaptation, a la Darwin's theory known as "survival of the fittest". Collective behaviour thus emerges from simple atomic reactions at agents' level and from their reactions to the constraints imposed by the environment. In order to explore the space of parameters and to understand their role in the evolutionary selection of the life cycles, the multi-run technique is used (i.e.: changing a value at a time, the others being the same).

1   Introduction

There are two species of cicada, called *Magicicada Septendecim* and *Magicicada Tredecim*, which have a life cyle of 17 and 13 years respectively. These are among the longest living insects in the world; they display a unique living behaviour, since they remain in the ground for all but their last few weeks of life, when they emerge *en masse* from the ground into the forest where they sing, mate, eat, lay eggs and then die. The nymphs of the periodical cicadas live underground, at

S. Nolfi et al. (Eds.): SAB 2006, LNAI 4095, pp. 485–497, 2006.

depths of 30 cm (one foot) or more, feeding on the juices of plant roots. They stay immobile and go through five development stages before constructing an exit tunnel in the spring of their 13th or 17th year. Adult periodical cicadas live only for a few weeks: by mid-July, they will all be gone. Their short life has one sole purpose: reproduction. After mating, the male weakens and dies. The female lives a little longer in order to lay eggs: it makes between six and 20 V-shaped slits in the bark of young twigs and deposits up to 600 eggs there. Shortly afterwards, the female also dies. After about six to ten weeks, the eggs hatch and the newborn nymphs drop to the ground, where they burrow and begin another 13 or 17 year cycle.

The fact that they have both evolved prime number life cycles is thought to be key to their survival. Many are the hypothesis about why these insects display these life cycle lengths. Two are the most interesting ones, and both of them require an adaptation by evolution of these species. The first one is about limited resources, that must be shared by the insects once out. By evolving life cycles of 17 and 13 years, the two species only have to share the forest floor every 221 years, that's 13 times 17. Resource boundedness can then be considered as an upper limit for the cicadas: no more than a threshold could survive with the available food, and then the fewer insects are out at the same time, the better.

The second hypothesis, actually interacting with the first one, is somewhat opposite to it; it's called predator satiation hypothesis and moves the focus from the insects to their main predators: dogs, cats, birds, squirrels, deer, raccoons, mice, ants, wasps, and even humans make a meal of the cicadas. Predator satiation is when a species can survive because its abundance is so great that predators do not have a large enough impact to effect the species' survival. In order to prove that predator satiation is occurring in a certain situation one must prove that above a certain prey density, the frequency of predation does not increase as the prey density increases (1). In the case of *magicicadas* this has been proved to occur. When the first cicadas emerge from the soil there is a very high predation rate, especially from avian predators. However, the predation rates decline over the next couple of days as predators have indulged in all the food they needed, or "satiated". Then, by the time that the satiation of the predators has worn off and foraging activities increase again, the density of adult cicada's has begun declining and they have already mated. This creates a situation where only a small portion of the adult population is consumed by predators. This is indeed a sort of lower bound for the number of cicadas that can be out at the same time; in few words, the more cicadas out at the same time, the least the possibility of being decimated by predators. Also according to this hypothesis, the prime numbers have a motivation: the prime number cycles were selected for because they were least likely to emerge with other cycles. For periodical cicadas emerging with other cycles of cicadas would mean hybridization, which would split up populations, shift adult emergences, and create lower densities below the critical size: (2). This would have made it harder for the hybridized broods to survive predation. Thus, prime number cycles which emerged with other cycles the least, would grow in population size over time because they would have the highest survival rates.

Considering these two hypothesis as real and founded, then we can assume that there has been a real selection among the species through many generations or, better a "tacit coordination by evolution". In this way, the prime number cycles can be seen as a very interesting emergent natural phenomenon, where the conclusions (results) were not embedded in any way into the initial data. For this reason, we chose to model this phenomenon using Agent Based Simulation (ABS) (3), a paradigm able to capture emergent behaviour arising from complex systems. Using an agent-based model of the cicadas' life cycle, we simplify the world in which they live and reduce it to just few parameters, essentially the limited resources and the predators. The cicadas have a reproduction rate, and so the predators and the food; we wonder if these parameters are enough for life cycles based on prime numbers to emerge as a result of adaptation and evolution. A very important thing is that the agents used in the model are reactive ones, i.e.: able to react to the stimuli coming from the environment, but not endowed with a mind. This means that they neither able to reason on the actions to perform at the individual level, nor to calculate which is the "optimal length" of their life cycle in order to optimize their perspectives. Every result in the model is thus to be considered as a collective evolutionary behaviour based on the interaction among the agents and the environment (4).

In the study of aggregate behaviour within Biology, it is more and more recognized that in addition to real experiments and field studies, also simulation experiments are a useful source of knowledge and verification. Using simulations for testing and validation of computational models could be seen as performing an experiment: since in the social sciences real experiments are in many cases not possible or only in a very restricted way, the use of computer simulations plays a decisive role: very often computer experiments have to play the part of real experiments in the laboratory sciences. By the way, that is also the case in those natural sciences where for similar reasons experiments are not (yet) possible, in particular in the those sciences like Entomology or evolutionary Biology.

ABS looks at agent behaviour at a decentralized level, at the level of the individual agent, in order to explain the dynamic behaviour of the system at the macro-level. Instead of creating a simple mathematical model, the underlying model is based on a system comprised of various interacting agents. For this reason the agent based paradigm has been chosen to model this biological phenomenon.

Emergent structures are patterns not created by a single event or rule. There is nothing that commands the system to form a pattern, but instead the interactions of each part to its immediate surroundings causes a complex process which leads to order. One might conclude that emergent structures are more than the sum of their parts because the emergent order will not arise if the various parts are simply coexisting; the interaction of these parts is central. Life is a major source of complexity, and evolution is the major principle or driving force behind life. In this view, evolution is the main reason for the growth of complexity in the natural world.

As stated before, we are facing a situation in which an emergent (possibly evolutionary) behaviour occurs: in the real world, life cycles based on prime numbers could have been a result of the environment in which the cicadas live. The agent-based model could give us an empirical answer to the following question: is the "predator satiation" hypothesis enough, along with the limited food quantity, to explain the emergence of life cycles based on prime numbers, through evolution?

2   The Model

An agent based model was created using the JAS library (`http://jaslibrary.sourceforge.net/`), in order to simulate different situations in which many species of cicadas compete for finite resources and are threatened by some predators. The only difference among the species, in the model, is the duration of their life cycle; population #1 will have a one year long life cycle, while population #20 a twenty years life cycle and so on. In this paragraph we describe the model as it's been implemented; words in italic identify the names of the variables.

We define a world with a fixed amount of resources (*resources*), that can be consumed by cicadas, a fixed amount of predators (*predatorsNumber*) and a probability to survive at wake up (*chanceToSurviveAtBirthRate*). Except for the constrain represented by food (*resources*), the other (*chanceToSurviveAtBirth* and *predators*) can be switched on and off, through the parameters window.

The population of cicadas is characterized by the number of members (*magicicadasNumber*), randomly distribuited among the different classes and by a growth rate (*reproductionRate*).

We use a fixed number of predators in order to underline that their population is not influenced by cicadas; In fact these insects represent only an alternative food, among the many present in nature, as they have to survive also in years without cicadas.

At the beginning of the simulation the cicadas are uniformly randomly distributed among C classes/sub-species. In our metaphorical world, each class of cicadas has a different life cycle length, so that they wait a different number of years underground, from a minimum of 1 to *maxSleepingTime*. We decided to use 20 as a maximum number of years a cicada can live, since in nature the *Magicicada Septendecim* is already the longest living insect, and we wanted to have a realistic setup. However, our model supports whatever number as a maximum life cycle.

At each simulation step, one year passes and the model inquires every cicada in order to update/reduce the number of years left for it to stay underground, or if this time is over, to wake it up.

The probability *chanceToSurviveAtBirthRate* also increases every year they spent underground. This is done according the biological theory that a longer life cycle is usually a good achievement. In particular, for the cicadas, (2) points out that during the climate cooling of the Glacial period, growth and development of cicada nymphs was slowed down by lowering soil temperature. He supports this by pointing out that it is well known that cumulative temperature is very critical in insect development. Also, the fact that the cooling climate slowed down the development of host plants from which cicadas get their nutrient may have also slowed their development. Because cicadas needed proper nutrients at each stage of their developmental cycle, and they were provided less because of the cooling temperatures, their life cycle was extended to larger range of years. So in the model we increase the *chanceToSurviveAtBirthRate* at any simulation step, to reproduce this natural phenomenon. When the cicadas have to go outside (i.e.: when their time underground is over), they perform a first check according to this probability.

As soon as they go outside, the cicadas must eat and reproduce themselves. Since the resources are limited, in our model only a certain number of cicadas can survive at each year/tic; this is a strong constraint existing also in the real world. Synthetically, if there are n resources in the world at time t, only n cicadas will survive.

Then predators can also eat cicadas, and reduce the population. This is the second constraint, present in the real world. In order to fulfill the "predator satiation" hypothesis we decided that after eating a fixed number of cicadas, the predators are satiated and let the other cicadas live and reproduce.

Finally the survived cicadas can reproduce and die; we clone the cicadas according to *reproductionRate*: the new cicadas have the same properties of their parents, and start a new life cycle underground, according to their duration.

Obviously, before each step we shuffle the list of alive cicadas, in order to randomize the process.

3   Results

Here we present some of the results from the simulation. Each run represent 20.000 years of time, in which we compute each step presented before, in order to define the population dynamics. For each run we show the graphs at times 20, 1000, 5000 and 20000.

Each graph has on x axis the classes, ordered by the number of years to be spent underground before emerging to the surface; on y axis the number of cicadas.

The first run we present features these parameters:

- *magicicadasNumber* 10000
- *reproductionRate* 6.0
- *chanceToSurviveAtBirth* true
- *chanceToSurviveAtBirthRate* 0.15
- *predators* true
- *predatorsNumber* 190
- *resources* 1000
- *maxSleepingTime* 20

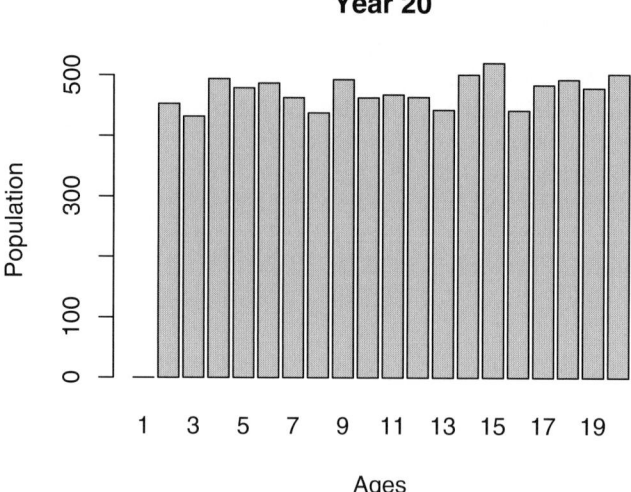

Fig. 1. Experiment 1 at 20 tics

During first 20 steps (figure 1) the cicadas with few years of sleeping died, according to their low *chanceToSurviveAtBirthRate*. At this time nothing can be said about the trend, yet, since most of the cicadas have been out for just one time and the food and predators dynamics have not yet influenced the results.

Normally the even classes are the least likely to survive; this could happen since they have many divisors. This causes many cicadas to be out at the same time, that starve to death according to the limited resources constraint. As you can observe in figure 2, after 1000 simulated years all the even classes are gone, except for class #20. This is because it's the longer living one (and in our model

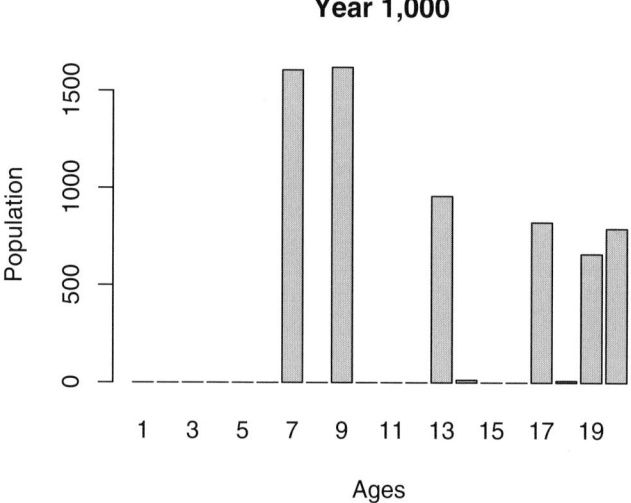

Fig. 2. Experiment 1 at 1000 tics

"longer is better" accorging to (2)) and because it's out less often then others. Notice that at this time (after 1000 years) four out of six classes which still exist have a prime number based life cycle.

Without the *chanceToSurviveAtBirthRate*, which increases at each time step the probability of a class to survive and go out, we'd obtain a different effect: in fact the populations with longest sleeping time are not competitive as the others, since they don't take advantage from the reproduction rate, being out less often. This would be highly unrealistic: the classes with a short life cycle (one to three years) will have much more chances to go out and reproduce themselves, when compared to those with longer cycles. As said before, nature often prefers, when possible, longer life cycles for cicadas, according to (2), so the model holds.

In figures 3 and 6 the typical trend of all the experiments can be observed, with a stable population concentrated in few classes. In particular, with the parameters we choose, after 5000 tics we have three high prime numbers left (13, 17 and 19) and the longest possible class (20), while after 20000 cycles (figure 3) the highest reproduction rate of the lower classes wins over the longest living one, and the only three remaining are represented by the three highest prime numbers lower than 20, which are 19, 17 and 13. While in nature a cicada with a life cycle of 19 years doesn't exist, probably because it wouldn't be possible for such insect to life so long, in our simulated world that is considered feasible. If we had classes just up to 18, we would have reproduced exactly the real situation, that is one in which two species emerge, the *Magicicada Tredecim* and the *Magicicada Septendecim*. In this experiment, then, you can observe that the populations of *magicicadas* confirm the "myth" of being biological prime number generator.

In order to investigate the individual trends of the single classes, some species have been analysed by considering the number of individuals belonging to them,

Year 5,000

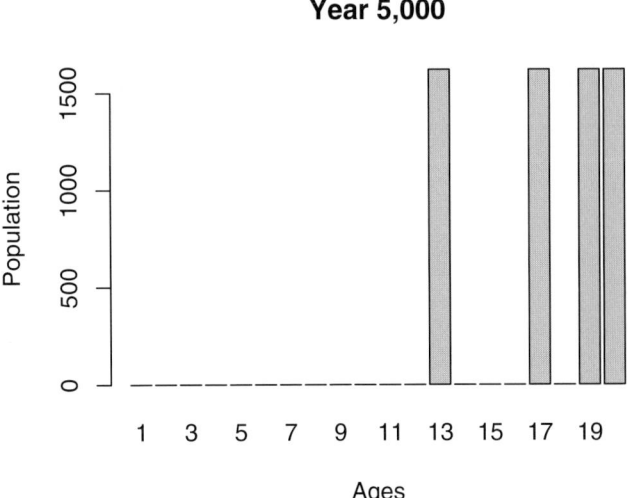

Fig. 3. Experiment 1 at 5000 tics

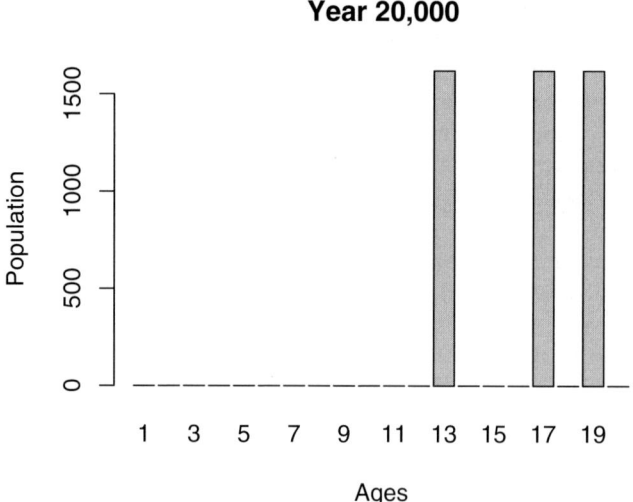

Fig. 4. Experiment 1 at 20000 tics

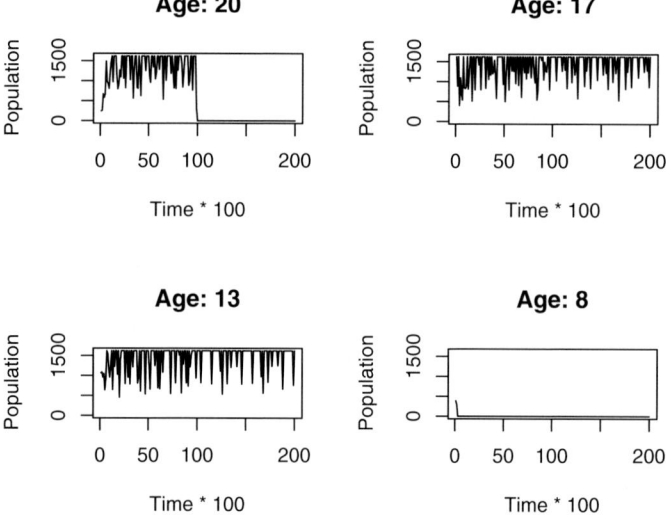

Fig. 5. Trends for ages 20, 17, 13, 8

over time. Class number 8 has been considered as representative for all the even classes, and, more generally, for all those that are not based on prime numbers and are low numbered. Classes number 13 and 17 are obviously very important, since in nature there are species with these life cycles. Class number 20 is then the "longest living" one among those in the simulation, and the last even class to "extinguish".

In the graphs in 5, the trends are examined with intervals of one hundred steps. The species with a life spam of 8 years is soon extinguished; this is probably due to the fact that when the cicadas belonging to this class step outside, they have to face an harsh competition for the limited resources. In fact, when they are "out", they have to share the resources with species number 1, 2, 4 and - once every two life cycles - with number 16. The same situation happens for species number 1, 2, 3, 4, 5, 6, 10, 12, 14, 16 and 18 soon to be followed by number 9 and, lastly, by number 7 (which was made weaker and weaker by the previous competition with number 14).

Species number 13 and 17 are the most interesting ones to analyze: they show a very similar trend and both of them survive after 20000 steps, exactly the same way as they survived in the real world. The trend is not linear at all, though, and both the species sometimes go all the way down to about 500 individuals, but they immediately reproduce and get back to about 1500, drawing a stable average of about 1000 individuals (these numbers are to be considered in function of the starting population of 10000 total cicadas for all the classes together).

A slightly different story is told by class number 20; this is an interest actor, since it exists till step 10000 and then suddenly extinguish. This is probably due to the previous competition with classes number 5, 10 and sometimes 15, which made it weaker than classes 13 and 17; class number 20 lives longer than other even classes thanks to the hypothesis by (2), implemented in the model.

3.1   MultiRun and Other Experiment

Our model, for its construction, is very "parameters sensitive", in the sense that a slightly different reproduction rate, or a negligible variation of the *cicadas*/predators ratio can lead to very different results in the long run. For this reason, we thought of a way to test many different parameters using a sort of "brute force" approach, i.e. by changing a value at a time, the others being the same (*ceteris paribus*). This is a sort of tuning, or even an empirical validation, useful to observe the aggregate results and to find dependencies among the parameters and the results.

Basically a MultiRun is a "super Model" class that launches sub-models, by changing a single parameter at each run, while the others remain the same. In the "result space" explored with this discrete method, zones without any cicadas surviving emerge, because of too many predators bundled with a low reproduction rate, or few chances to survive at birth.

A MultiRun simulation with these fixed parameters is performed:

− *magicicadasNumber* 10000
− *chanceToSurviveAtBirth* true
− *predators* true
− *resources* 1000
− *maxSleepingTime* 20

While those are fixed, *reproductionRate* iterates from 0.1 to 10.1 with steps equal to 0.1, *chanceToSurviveAtBirthRate* from 0.05 to 1.05 (step = 0.5) and *predatorsNumber* from 0 to 300 (step = 1).

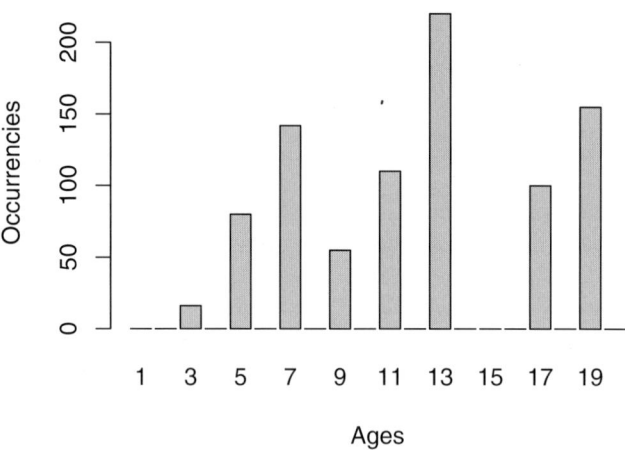

Fig. 6. Average of MultiRun simulations' results

The results of every run are collected after 20.000 years of evolution (tics), and represented in figure 6, where the aggregate average result of 1220 total runs are shown.

From these aggregate results a very interesting figure emerges; by exploring the parameter space, it's evident that prime numbers are the most likely results to appear. In fact, the first five positions are indeed prime numbers (13, 19, 7, 11 and 5 respectively). Number 9 follows, that's not prime, but after it there are other two prime numbers, # 17 and # 3. With the exception of # 2 (which

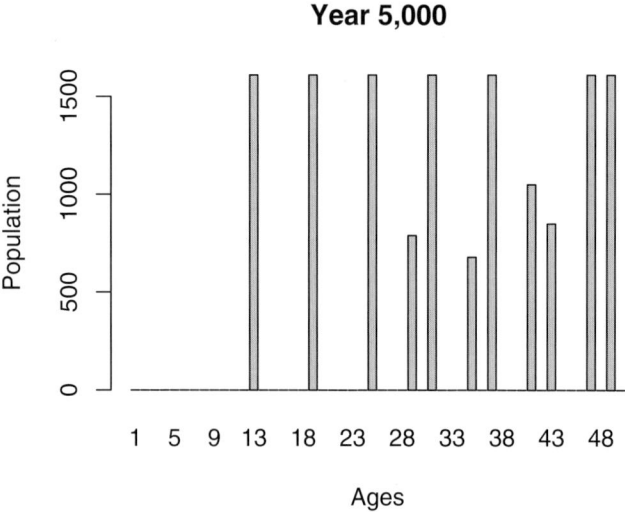

Fig. 7. 50 ages at 5000 tics

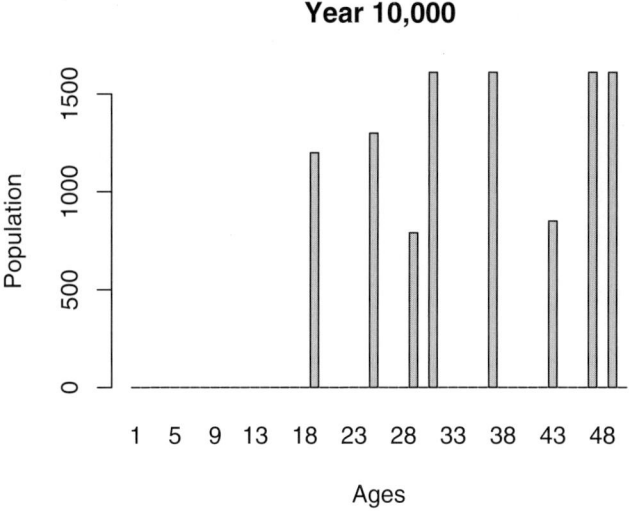

Fig. 8. 50 ages at 10000 tics

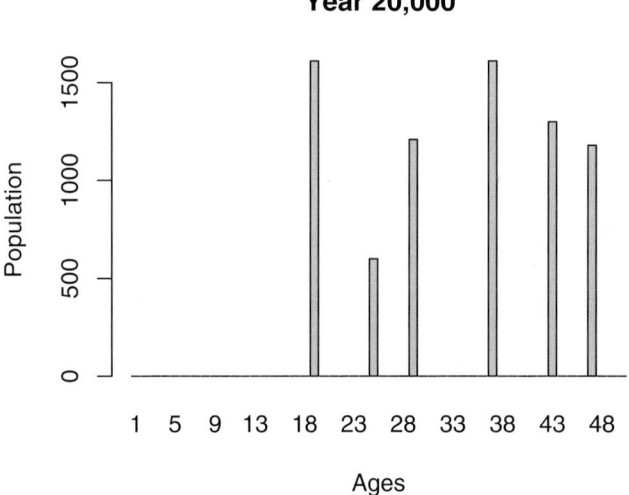

Fig. 9. 50 ages at 20000 tics

is probably too short a life cycle, to be selected), all the prime numbers among 3 and 20 were discovered by the model and, with the exception of # 9, all the selected numbers are prime.

The model allows "what-if" analysis; in particular the number of species can be higher than 20 (i.e.: with longer life cycles). Though impossible in nature, where a cicada living 17 years is already the longest living insect in the world, this could be very interesting to simulate, in order to see if the model can be a true (though inefficient) "biological prime numbers generator".

Fig. 10. *Magicicada Septendecim*

A simulation is now presented with the same parameters of the experiment in paragraph 2, except for the *maxSleepingTime*, now set at 50. This means that instead of 20 species of cicadas, we now have 50 species (unrealistic situation, of course, but useful for a "what-if" extrapolation).

In figure from 7 to 9 the distribution of the results after 5000, 10000 and 20000 simulated years can be observed, which seems, once more, to confirm the myth of periodical insects as "prime number generators". Even in an unrealistic biological situation, the mechanism inspired by *magicicadas* seems to work and after 20000 cycles the insects with a life cycle based on prime numbers are the most likely to survive (19, 29, 37, 43, 47, while 25 is the only number "surviving" not being prime).

4   Conclusion

Biology is an interesting field for agent based modelling, both as a source for examples/applications and as methodological inspiration. Many important results in Artificial Life and Computer Science are based on biological metaphors such as the ALife concept itself and, in detail Game of Life by John Conway, Holland's Genetic Algorithms, Neural Networks and many others. We also mention the simple ants or "vants" (virtual ants), with their simple rules and their simulated pheromones that are present in a great number of models and algorithms, since the Langton's ant.

In this paper agent based simulation was applied to a biological evolutionary phenomenon. Two species of cicadas living in North America show a unique behaviour: they remain in the ground for all but their last few weeks of life, when they emerge *en masse* from the ground into the forest where they sing, mate, eat, lay eggs and then die. The most interesting part is that their life cycle is, respectively, of 13 and 17 years, which are obviously prime numbers. Biologists tried to explain this phenomenon with some theories, among which we find the one stating that prime number were selected by evolution, since in this way the two broods would have had less chances to meet, avoiding to share resources.

An agent based evolutionary model was presented in which the agents represent the cicadas; there are different species of agents, who differ for the length of their life cycle. In the first experiment life cycles going from 1 to 20 were used, since beyond it wouldn't have been realistic, being *Magicicada Septendecim* the longest living insect in the world. In the proposed model there is a finite number of resources, and the agents must compete for them; when their time comes, the agents "come out" and must eat, before being able to reproduce themselves. Those who can't find food die, while the ones who find it live; also predators are present in the model. This is another threat for the insects, since some of them are caught and eaten by predators; the one who survive can reproduce themselves and then die of a natural death. The children inherit the life cycle length from their parents; of course, if no cicada of a brood is alive, then that class is extinct and disappears from the simulation. The model features a set of modifiable parameters, such as the reproduction rate, the number of predators, the maximum allowed life cycle and the probability to be alive after the period spent underground; this last parameter increases with the time, since for cicadas it's been proved that a longer life is a better option, according to (2).

The quantities of cicadas at four different times were examined: after 20, 1000, 5000 and 20000 years/tics. In the experiment, after 20000 years, the only cicadas alive are the ones with a life cycle of 13, 17 and 19 years. This is a great result, since it mimics the real world; 19 years is not applicable in the real world since it's too long a life cycle for an insect, but anyway it's a prime number itself, proving that the constraints coming from the environment lead to a selection of high prime numbers.

This model of evolutionary biology empirically validates, through an agent based technique, some different theories. A multi-run approach was then adopted, to show some robust results which partially confirm the "myth" of periodical cicadas being a biological prime number generator.

Bibliography

[1] Williams, K.S., Smith, K.G., Stephen, F.M.: Emergence of 13- yr periodical cicadas (cicadidae: Magicicada): Phenology, mortality, and predator satiation. Ecology **74**(4) (1993) 1143–1152

[2] Yoshimura, J.: The evolutionary origins of periodical cicadas during ice ages. The American Naturalist **149**(1) (1996) 112–124

[3] Ostrom, T.: Computer simulation: The third symbol system. Journal of Experimental Social Psychology (24) (1988) 381–392

[4] Remondino, M., Cappellini, A.: Agent based simulation in biology: the case of periodical insects as natural prime numbers generators. The International Journal of Veterinary Medicine – Vet On-Line (2006)

Evolving Reaction-Diffusion Controllers for Minimally Cognitive Animats

Kyran Dale

Centre for Computational Neuroscience and Robotics (CCNR)
Informatics, University of Sussex Brighton BN1 9QH, UK
Phone: +44 1273 678431; Fax: +44 1273 671320
kgd20@sussex.ac.uk
http://www.cogs.susx.ac.uk/ccnr/

Abstract. This paper describes work carried out to investigate whether a classic reaction-diffusion (RD) system could be used to control a 'minimally cognitive' animat. The reaction-diffusion system chosen was that first described by Gray and Scott (Gray-Scott) and the minimally cognitive behaviors those used by Beer et. al involving the fixation and discrimination of diamond and circle shapes by a whiskered animat. The parameters of this RD-controller were evolved using an evolutionary, or genetic, algorithm (GA).

1   Introduction

This paper describes work carried out to investigate whether a classic reaction-diffusion (RD) system could be used to control a 'minimally cognitive' animat. The reaction-diffusion system chosen was that first described by Gray and Scott (Gray-Scott) and the minimally cognitive behaviors those used by Beer et. al involving the fixation and discrimination of diamond and circle shapes by a whiskered animat. The parameters of this RD-controlller were evolved using an evolutionary, or genetic, algorithm (GA).

Within Evolutionary Robotics the prominent model dynamical system is the continuous time recurrent neural network (CTRNN) [7,8,3]. Many examples testify to the rich dynamics of which CTRNNs are capable [6,8], such as generating the patterns to regulate legged robot gaits, and controlling such simple cognitive tasks as navigation and shape-discrimination. Many classic RD systems also display rich dynamics, manifesting the full range of classic qualities such as Hopf bifurcation, stable and unstable limit-cycles, chaotic boundaries etc.. The main motivation for the work described in this paper was to see whether the tried and tested technique of evolving neural-network controllers for simple robotic behavior could be adapted to harnessing some of the rich dynamics displayed by these RD systems. In this sense the interest was both methodological, to show that evolutionary algorithms could be used successfully with a different class of non-linear system, but also focused on exploring the ability of RD controllers. For example the ability to sustain spatio-temporal patterns suggests a role in controlling gaited movement but can systems be tuned to particular requirements? Given the difference between the essentially 'spaceless' CTRNNs and

S. Nolfi et al. (Eds.): SAB 2006, LNAI 4095, pp. 498–509, 2006.

the necessarily spatial RD systems there is also the intriguing possibility that they might be able to compliment one another. By placing artificial neurons in an excitable medium with which they can interact, CTRNNs might be able to exploit the spatio-temporal properties of the medium. Given the dynamical potential of these RD systems there has been very little work dedicated to exploring it [1,2].

In place of the continuous time recurrent neural network used by Beer we used a one-dimensional ring of cells within which the concentration of two coupled chemicals changed according to two differential equations describing intra-cell reactions and inter-cell diffusion. Output from whisker-like proximity sensors was fed to the cells in the RD-ring via weighted links, perturbing the concentration of the two chemicals. Weighted links in turn allowed the concentration of particular chemicals in designated cells to specify motor activation, completing a sensor-motor loop. In some of the controllers described the RD-ring also received input from the motors. Links were made symmetrically about the animat's longitudinal axis. Parameters specifying the weighted links between cells, motors and sensors were evolved as were the values of a dimensionless feed rate and rate constant for the RD-system. A simple fitness-function encouraged animats to fixate circles and avoid diamonds.

1.1 Reaction-Diffusion Models

Perhaps the best known example of a reaction-diffusion model is that proposed by Alan Turing [5] as an attempt to explain cellular differentiation in early biological development. It is also one of the first examples of the use of a computer to solve differential equations. Turing was trying to understand how the chemicals in arrays, in this case one-dimensional, of identical cells could, by reacting within the cells and diffusing between them, form stable patterns. He was able to show that by constraining the chemical reactions within cells and the relative rate of diffusion between them one could guarantee a stable pattern. Subsequent work has shown analogous systems responsible for leopards' stripes, patternation of nautilus shells and many other natural patterns.

Within the class of model reaction-diffusion systems defined by two coupled chemicals (two rate equations) Turing was interested in those tending toward a stable configuration. But by altering the governing reactions and diffusion rates many other systems are possible, displaying a wide variety of spatio-temporal properties. One of the most intriguing is that proposed by Gray and Scott in their 1984 paper [9] and extensively analyzed by Pearson in his 1993 paper [4]. A variant of the autocatalytic Selkov model of glycolysis [4] the Gray-Scott model corresponds to the following reactions:

$$u + 2v \rightharpoonup 3v \tag{1}$$

$$v \rightharpoonup p \tag{2}$$

Both reactions are reversible so p is an inert product. A feed term for u introduces a non-equilibrium constraint with the feed process removing both u and v. This results in the following reaction-diffusion equations, expressed in dimensionless units:

$$\frac{\partial u}{\partial t} = d_u \nabla^2 u - uv^2 + F(1 - u) \tag{3}$$

$$\frac{\partial v}{\partial t} = d_v \nabla^2 v + uv^2 - (F + k)v \tag{4}$$

where k is a dimensionless rate constant and F a dimensionless feed constant. d_u and d_v are the diffusion rates for the two chemicals (see Sect. 2 for specific details). A trivial steady state of $u = 1$, $v = 0$ exists for all values of F and k. Gray-Scott proves a very robust simulation, showing no qualitative difference when implemented by forward Euler integration over a broad range of spatial and temporal scales [4].

When suitably perturbed Gray-Scott exhibits a large variety of spatio-temporal patterns that have to be seen to be appreciated. Pearson's paper is replete with beautiful images but the simulation is best appreciated in real-time with a two-dimensional simulation and a suitable colour-map. By fixing the diffusion rates of the chemicals and using F and k as control parameters Pearson was able to show that within suitable limits the two-dimensional phase-diagram described shows regions associated with specific spatio-temporal patterns, ranging from spot replication and stripes in a continuous transition to traveling waves and spatio-temporal chaos.

A disclaimer should be made at this point. The specific details of the Gray-Scott model are not central to this paper. It was chosen as a suitable candidate to provide an excitable medium, capable of rich dynamics and as likely as any to be exploited by evolution. There are many other classic two-chemical model-systems, with greater and lesser biological plausibility. Gray-Scott was the first tried and proved very capable.

1.2   Visually-Guided Agents

The choice of an evolved animat model, for example to demonstrate the potential of a novel reaction-diffusion controller, should be informed by two key considerations. The behavior in question must be cognitively 'interesting' and there should be a reasonable expectation that resultant controllers can be analyzed and understood.

> The term 'minimally cognitive behavior' is meant to connote the simplest behavior that raises cognitively interesting issues.
> Generally speaking, visually-guided behavior provides an excellent arena in which to explore the cognitive implications of of dynamical and adaptive behavior ideas, since it raises a host of issues of immediate interest. ([7] p.422)

In keeping with Beer's thesis we chose two of the visual-guidance tasks conforming to the requirements of 'minimal cognition'. In the first and easiest a whiskered animat, capable of moving along the floor of a two-dimensional arena (in the xz

plane), is required to orientate toward and track a circular object falling from the arena's ceiling with a large range of vertical and horizontal speeds (see Fig. 2 for details). The second, more challenging task took place in the same arena and required the animat to discriminate between diamond and circle objects dropped directly down from the arena's ceiling. The animat was rewarded for fixating the circles and avoiding the diamonds.

Beer evolved continuous time recurrent neural networks (CTRNNs) to control his animats, a control-system the author have some experience of [3]. His subsequent analysis [8] of the CTRNNs' dynamics makes them probably the best understood of all animat controllers, evolved or otherwise. This represents a useful benchmark and an obvious model to emulate. The use of such canonical models to provide a common point of reference would seem to be an efficient way to exploit the resources available. Broadly speaking this work preserves the details of Beer's model while replacing the CTRNN controller with a novel one using a reaction-diffusion medium.

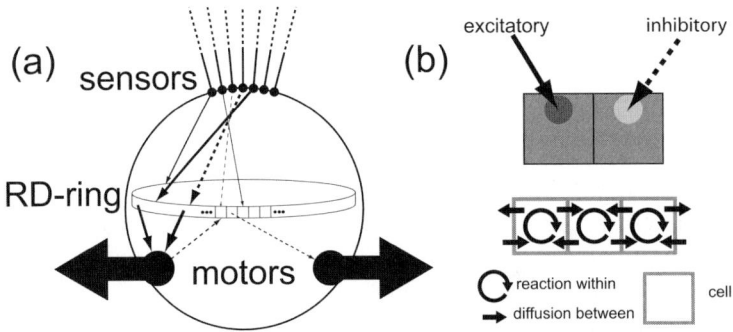

Fig. 1. (a) The animat model. Output from the proximity sensors (and occasionally motors) is fed, via weighted links, to the reaction-diffusion ring (*RD-Ring*) where it perturbs the cellular concentration of chemicals u and v. Solid links increase the chemical concentration in the cell while dashed links decrease it (*link-effects*). Following a number of reaction-diffusion cycles (*reaction and diffusion*), the chemical concentration levels in designated cells are in turn fed via weighted links to activate the animat's motors. Activation at a motors is summed and multiplied by a constant (10) to produce an output. The combined output of oppositional left and right motors is used to move the animat. (b) Excitatory links from the sensors increase chemical concentration while inhibitory links (dashed) decrease it.

1.3   Evolving Controllers

The Gray-Scott model, in keeping with most reaction-diffusion systems, is highly non-linear, at least unintuitive and often counter-intuitive [1]. It is not immedi-

---

[1] The speed of modern processors makes it possible to interact in real-time with 2D implementations of these reaction-diffusion systems. Having implemented and played with just such a model of, among others, Gray-Scott the we can attest to its counter-intuitiveness.

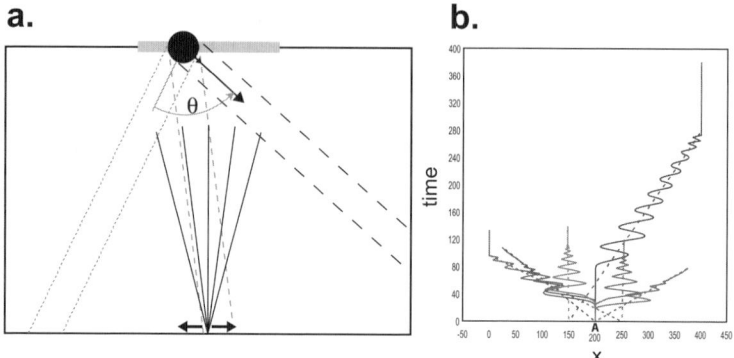

Fig. 2. (a) The fixation experiment (to scale). An animat with five whiskers spread over a 30° span is placed at the centre of the arena's floor. During a trial a circular object was placed in the starting zone (*grey-line*) on a trajectory within the limits defined by the animat's distal whiskers (θ). This was to ensure that the animat received some stimulation from the falling object. The object's speed is indicated by the relative position of large and small arrows, lying within 0.5 and 7 units per second. The end of the trial was signaled by the object reaching the arena floor, at which point the distance between object and animat was used, along with their relative start points, to calculate a fitness score for the animat. (b) Six animat-circle trajectory pairs with the circle's trajectory dashed.

ately clear how one could 'hand-wire' such a controller, but it would require an intuition about the rich dynamics of the system which escapes us. In cases such as this, where we require a controller capable of exploiting even a relatively simple dynamical system, it would seem that the need is pressing to leverage the increasing computer power at our disposal and automate the process of discovery. This approach is particularly appropriate to a robot that is intended to remain in-silico. The search algorithm employed here is a genetic algorithm (GA). A simplistic, but initially useful, way of understanding how a GA works is to picture the parameter space, describing in this case the details of our reaction-diffusion controller such as linkage points and weights, as a fitness landscape. Every point in this landscape describes an animat controller and height above ground corresponds to fitness. If the landscape is reasonably well-ordered it should be possible for the GA to find its way from low ground initially, corresponding to randomly-wired, poor performing controllers, to high, where the controllers are (much) better performing. This image leaves out important details, particularly the concept of neutral-networks [2], but the key detail is captured. From random parameters and allowing for a suitable encoding scheme, it should be possible to automatically produce good controllers by applying evolutionary pressure. The work described in this paper and elsewhere [7,8] is testament to that fact.

---

[2] A complex subject highlighting our poor intuition of movement in higher-dimensional space.

2   Method

To a large extent details from Beer's earlier simulations [7] were preserved and the required behaviors essentially the same. The arena was 400 units long by 275 units high (Figs. 2 and 3) in all the experiments. The animat's whisker sensors were 220 long and uniformly spaced over a 30° spread. The animat had five whiskers in the orientation experiment and seven in the discrimination experiment. Activation of the whiskers was a simple linear function with a minimal value of 0 when the whisker was unimpinged and 1 when it was intersected at base.

Fig. 1 shows a diagram of the animat. Activation from the sensors $\in [0, 1]$ was fed through weighted links $\in [-1, 1]$ to the one-dimensional reaction-diffusion ring (RD-ring) consisting of 128 cells subject to intra-cellular reaction and inter-cellular diffusion between near-neighbours (see the chemical reactions 1, 2 and rate equations 3 and 4). The weighted links were specific to either chemical u or v, this specificity being under evolutionary control.

The sensors, motors and input to the RD-Ring were updated using the forward Euler method with an integration step-size of 0.1. During this time-step each cell in the RD-ring was updated twice using the rate equations 3 and 4). Input via links to the cells perturbed the specified chemical's concentration by a simple multiple of time-step (0.1), sensor activation $\in [0, 1]$ and link weight $\in [-1, 1]$. The cellular concentration of u and v was bounded within the range $\in [0, 1]$.

The animat's motors received input from cells in the RD-ring. Input from a individual link was a product of link-weight $\in [-1, 1]$ and the concentration of the evolutionarily specified chemical in the cell. To update the animat's position, the activation of the oppositional motors was subtracted $(right - left)$ and the result multiplied by 10. This multiplier was fairly arbitrary, taking into account the need for the animat to move fast enough to catch objects with a maximal horizontal velocity around 5. It worked well enough but is probably too large. On reflection this value should probably have been an evolutionarily-specified parameter but given the fitness scores generated any gains could only have been very marginal.

Diffusion rates d_u and d_v were fixed at the standard values [4] of 2×10^{-5} and 10^{-5} respectively and the length of the RD-ring was 0.32. Each animat genotype specified a value for the rate constant k and feed constant F (equations 3 and 4) which were seeded at values 0.055 and 0.02 respectively in the otherwise randomly generated initial populations. By moving through this F, k parameter-space evolution had some control over the properties of the reaction-diffusion system (see section 1.1).

The GA consisted of a population of thirty animat genotypes which were updated generationally according to rank-based selection. The genotypes were essentially a list of weighted, chemically specific links, describing the wiring of an animat controller. As the animat controllers were symmetrical, each link on the list corresponded to two links on the controller. At each generation these lists were converted into their respective animat controllers and assigned a fitness value according how well the controller performed its task. It was neither practical or desirable to have the genotype describe a fully connected controller (1408 links in all) so the number of links was pre-set. The starting number for the orientation experiment was 8 sensor→RD-ring, 4 RD-ring→motor and 4 motor→RD-ring making 32 symmetrically arranged links in all. For the discrimination experiment it was 16 sensor→RD-ring, 8 RD-ring→motor and 4 motor→RD-ring making 56 links in all. These values seemed about right and worked well but subsequent analysis showed near optimal performance was possible with less than half this number of links (see Fig. 4.b for technical details).

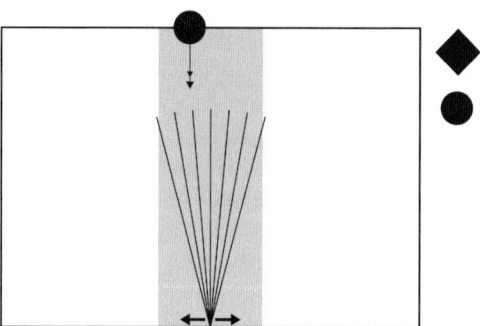

Fig. 3. The discrimination experiment (to scale). An animat with seven whiskers spread over a 30° span is placed at the centre of the arena's floor. During a trial an object was placed at the top of the arena within the grey drop zone on a straight downward trajectory of between 3 and 4 units per second. A pair of trials with the same starting position and speed was performed for diamond and circle objects. The animat was rewarded for it's ability to fixate the circle and avoid the diamond.

At the end of each generation a new generation was formed from the old and subjected to mutation operations. The numbers on the genotype were in the range $\in [0, 1]$, being mapped onto their respective controller parameters. Mutation consisted of the addition of a normally distributed random value with average 0 and standard-deviation 0.25. A second mutation operator was applied to each genotype with a probability of 10%, randomly deleting a link from or adding a link to the list. The link-addition operator allowed two links to share start and end points and chemical specificity.

Fitness was a function of the two values $dist_{start}$ and $dist_{end}$ specifying the absolute distance between animat and shape at the trial's start and end, signaled by the shape reaching the arena floor. For the circle object, required to be fixated in both experiments the fitness function f was:

$$f(dist_{start}, dist_{end}) = \begin{cases} 1 - \frac{dist_{start}}{dist_{end}} & \text{if } dist_{end} < dist_{start} \\ \max(\frac{dist_{start}-dist_{end}}{50}, -1) & \text{if } dist_{end} \geq dist_{start} \end{cases}$$

The fitness function for the diamond, required to be avoided, was the negative of that for the circle.

2.1   Training Protocol

A number of trials were conducted to assess the ability of individual controllers and allot their respective genotypes a fitness score. Fig. 2.a shows the trial set-up for the orientation experiment. The grey-zone at the top indicates potential starting points for the circle during a trial and theta the limits within which the trajectory line is set. In keeping with Beer's original model [7] the simulation was noiseless, meaning that the symmetrical animat controller was incapable of breaking symmetry without stimulus from the whiskers. The trajectories of falling objects were controlled to make sure the animat received some sensory stimulus during the course of a trial.

To test a controller for ability to orientate towards the circle, the grey starting-zone (± 70 from the animat's starting point $x = 200$) was broken into four equally spaced regions and 2 starting points chosen randomly from each region. Two random trajectories within the limits set by θ were chosen for each starting point and each of these tested at two random speeds, within the range $[0.5, 7]$. This makes a total of sixteen trials per assessment. Two assessments were carried out and the average returned as a fitness score.

Fig. 3 shows the set-up for the diamond-circle discrimination trials. Animat performance with dropped circles and diamonds was compared for eight random points starting within the drop zone (grey region ± 50 from the animat's starting point $x = 200$) and two random speeds $\in [3, 4]$ per point, making a total of 32 trials in all for a single assessment. Two assessments were made for each controller and the average returned as a fitness score for the genotype.

3   Results

3.1   Orientation

Animats capable of orientating toward and fixating a falling circle were easily evolved, almost invariably with close to optimal fitness for the best members. Perhaps not so surprising given the symmetrical nature of controller and task it is nevertheless remarkable how many randomly generated animat controllers were able to achieve respectable scores ab-initio. It is far too early to say but it appears that the symmetrically wired RD-rings have a natural tendency to resist perturbation and normalize left and right sensor input. It should be stressed that for the harder discrimination task no animat populations showed any ability ab-initio.

Fig. 2.b shows the tracking performance of a typical animat for three pairs of random circle trajectories, with evenly-spaced start points. The dashed-lines show circle trajectories and the solid the animat's response. Starting at $x = 200$ the animat remains stationary until its whiskers are stimulated by the falling object. It then oscillates back and forth below the falling object continually overshooting its position. These oscillations are damped as the object gets closer to the animat and sensor stimulus increases. The plots resemble attractor-cycles converging on the intersection of animat and falling shape at the ground. This animat controller had sixteen links in total. After an initial 200 generations, during which near optimal performance was achieved, the animat was evolved for a further 1000 generations with the link-addition mutator turned off but the link-deletion mutator still active. This effectively reduced the size of the controller while maintaining fitness.

3.2   Discrimination

Animats capable of distinguishing between circles and diamonds were readily evolved within 200 generations (see Fig. 3 for details of the task). From a typical batch of 20 at least a third achieved fitnesses of over 1.8 from a maximum 2 over 32 random trajectories. Much better results were achieved if the animats were trained initially to fixate a circle before being introduced to the discrimination task.

Fig. 4 shows the network of a typical near-optimal animat after the removal of most redundant links by an extended period of evolution with the addition mutator turned off(see Fig. 4.b for details). A total of 26 links are used with fairly even sampling of the seven available whisker sensors. Six links from the RD-ring to the motors dictate the animat's movement. The inner two links have converged on the same cell, allowing the cell to strongly influence the animat's movement by exciting one motor while inhibiting its opposite. Interestingly ,although allowed feedback from the motors to the RD-ring these links have been pruned away as unnecessary.

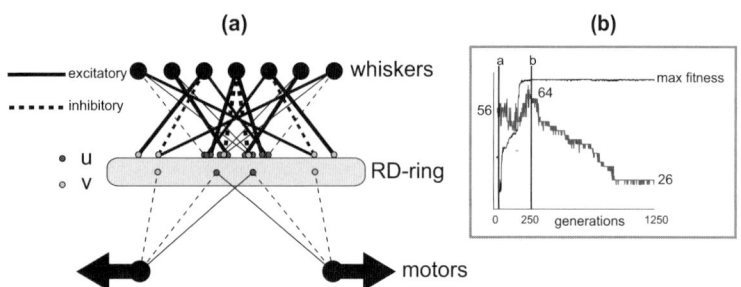

Fig. 4. (a) The complete reaction-diffusion controller for an animat capable of discriminating diamond and circle objects. Coloured nodes are used to indicate the chemicals affected by the whisker sensors and the chemicals affecting the motors respectively. This network has no feedback from motors to the reaction-diffusion ring. **(b)** Simplifying the network of a diamond-circle discriminator. The figure shows superimposed plots for the number of weighted links in the reaction-diffusion controller (*thick-line*) and the fitness of the best individual in the population (*thin-line*). After a short period during which the animat is evolved to fixate a circle the fitness drops as the animat is introduced to the diamond-circle discrimination task (*line a*). At this point the controller consists of 56 weighted links. Over 250 generations the GA is allowed to randomly insert and delete links from the controller genotype untill at *line b* the network has 64 links and the animat has achieved near-optimal fitness. At this point the GA's insertion mutator is turned off while continuing to allow evolution to delete links randomly. Over the next 1000 generations the animat maintains a close to optimal fitness while losing links, ending with 26 links, less than half the number it started with.

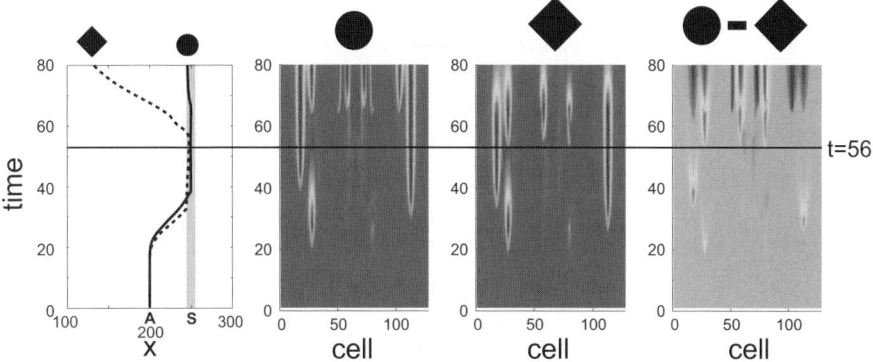

Fig. 5. The leftmost plot shows the x-position over time for two superimposed animat trajectories, both starting at **A**($x = 200$), in response to a shape dropped from **S** ($x = 250$)at speed 3.5. The *dashed* line shows the animat's response to a dropped diamond, the *solid* line to a dropped circle. The centre plots show activation of the reaction-diffusion cells during the circle and diamond trials. The right most plot shows the result of subtracting the activations of circle and diamond trials. Around time 56 a strong difference is seen in chemical concentrations in the reaction diffusion rings, corresponding to the animat's evasion of the diamond (*left-plot*). In this sense the right plot shows traces of the animat's active shape-detection.

Full analysis of the network will take some time but some provisional observations can be made. Fig. 5 shows two trajectories made by the animat controller shown in Fig. 4 in response to a diamond and circle dropped at the extreme end of its evolved range (see Fig. 3) at speed 3.5 and along the line **s**. The left plot shows the animat avoiding the diamond (*dashed-line*) while fixating the circle (*solid-line*). The central two plots show the concentration of chemical u in the RD-ring's cells over the course of the two trajectories and the plot on the right the result of subtracting the two central plots. This plot shows that following a period of relatively similar activity, during which time the animat is fixating both diamond and circle, there is a large difference in activity at time 56 around cells 27 and 79, these falling under the remit of the rightmost whisker. While the circle-plot concentrations remain symmetrical along the ring, an asymmetry can be seen in the diamond-plot. This corresponds to the initiation of a strong leftward avoidance tactic, taking the animat over a hundred units from the diamond at trial's end (maximum fitness is achieved for a distance of 50 units). Observing the diamond trajectory it is clear that the animat's attempt to fixate the diamond as it does the circle cause it to keep the diamond to the right of its central-line, introducing an asymmetry as the right whiskers stimulation is unbalanced by those on the left. This loss of balance initiates avoidance.

3.3   Generalization

The group of trajectories performed per animat assessment was not necessarily representative of the set of all possible trajectories. Random components in the construction of trial-sets introduced a noisy element into the fitness scores. This was reflected by the difference between average and maximum performance of the population. To test how well the animats generalized over all possible they were tested over batch of 200 randomly generated trajectories. This saw a marked reduction in the average score. Taking the animat from Fig 2 as a representative example, over the two-hundred trials its average was 0.82 with a standard-deviation of 0.24. On only eight of the two hundred trials did the animat respond inappropriately, scoring a negative fitness by ending the trial further from the shape at the start. On only 20 of the 200 did the animat fail to score less than 0.5. In other words for 90% of the trials the animat managed to at least halve horizontal distance between itself and the falling shape, on average covering over 80% of the optimal distance. Evolution under more stringent conditions saw the expected improvement in generalization confirming the limitation of the trial regime, not the controller model.

The range of possible trajectories was much smaller for the discrimination test (see Fig. 3) with a narrower range of velocities $\in [3, 4]$ and no horizontal movement of the shape. Generalization here was correspondingly better. Taking the animat discussed (Figs.4 and 5) as a representative example, it averaged 1.82 with standard-deviation 0.4 with 2 the maximum possible score. On only one out of the two hundred trials did the animat respond inappropriately, ending the trial closer to the diamond than to the circle. This network had been pruned of nearly all redundant links over a long period of evolution and would be expected to have fully exploited any deficiencies in the trial regime.

4   Conclusion

The challenge to emulate a classic, minimally cognitive animat control task has we believe been met by this novel control system. Given the much explored non-linearities in the Gray-Scott system it is perhaps surprising how readily the medium was exploited by evolution and some considerable work is needed to establish specific details of the successful controllers. As has been mentioned the Gray-Scott reaction-diffusion model was chosen because it is a classic of the literature with well documented properties. There is no reason to think it particularly well-suited for this task. It is conceivable that evolution be allowed to specify its own chemical equations but at this stage that is an unnecessary introduction of complexity. The results of using Gray-Scott are gratifyingly interesting.

The limitations of this class of controller remain to be seen, particularly in comparison to the performance of CTRNNs on similar problems. This work was in part prompted by the idea that hybrid systems, made by placing artificial neurons in an active medium, might allow CTRNNs to exploit spatio-temporal properties otherwise unavailable. As normally conceived CTRNNs are 'spaceless'. The one-dimensional RD-rings have two and three dimensional counterparts, another potential avenue of exploration.

To sum, a benchmark has been established and these RD-controllers have earned the right to further consideration.

Acknowledgments

Particular thanks must go to Phil Husbands for green-lighting the direction taken in this work, the paper's referees and to the EPSRC (grant GR/T11043/01).

References

1. Breyer, J., Ackerman, J., McCaskill, J.: Evolving Reaction-diffusion Ecosystems with Self-assembling Structures in Thin Films. Alife 4: 25–40. (1998)
2. Adamatzky, A., De Lacy Costello,B., Asai, T.: Reaction-diffusion Computers. Elsevier, (2006).
3. Dale, K., Collett, T.S.: Using Artificial Evolution and Selection to Model Insect Navigation. Current Biology 11: 51–62. (2001)
4. Pearson, John E.: Complex Patterns in a Simple System. eprint arXiv:patt-sol/9304003 (04/1993)
5. Turing, A.M.: The Chemical Basis of Morphogenesis. Philosophical Transactions of the Royal Society of London. Series B, Biological Sciences, Volume 237, Issue 641, pp. 37–72 (1952)
6. Beer, R.D.: On the Dynamics of Small Continuous-time Recurrent Nerual-networks. Adaptive Behaviour 3:469–510
7. Beer, R.D.: Toward the Evolution of Dynamical Neural Networks for Minimally Cognitive Behavior In P. Maes, M. Mataric, J. Meyer, J. Pollack and S. Wilson (Eds.), From animals to animats 4: Proceedings of the Fourth International Conference on Simulation of Adaptive Behavior (pp. 421–429). MIT Press.
8. Beer, R.D.: The Dynamics of Active Categorical Perception in an Evolved Model Agent. Adaptive Behavior 11(4):209–243
9. Gray, P., Scott, S.K.: Autocatalytic Reactions in the Isothermal Continuous Stirred Tank Reactor Oscillations and Instabilities in the System $A + 2B- > 3B, B- > C$. Chem. Eng. Sci., **39**, 1087–1097 (1984)

Emergence of Coherent Coordinated Behavior in a Network of Homogeneous Active Elements

Gentaro Morimoto and Takashi Ikegami

Graduate School of Arts and Sciences,
The University of Tokyo
3-8-1 Komaba, Tokyo 153-8902, Japan
{genta, ikeg}@sacral.c.u-tokyo.ac.jp

Abstract. We introduce a simulation model of coordinated adaptive behavior of agents with homogeneous elements connected by springs. The agents accomplish a hill climbing task by coordinating the dynamics of active elements and responding to the spatial pattern of surface convexity. Transition between discrete movement patterns is observed while moving in a rugged environment. The difference of dynamics between approaching hills and avoiding hollows is distinct and can be interpreted as a dynamic categorization of bumps on the surface.

1 Introduction

By moving one's fingers over the surface of objects, it is possible to obtain sensations of its shape, roughness and material and categorize these sensations to arrive at a perception of the object. The sensation of touching objects can not be deduced from the instantaneous sensory patterns on our fingers. These sensations exist inside the dynamics of our nervous system, which itself is modulated by the environment and our spontaneous movements.

In studies of haptic perception, the importance of self movement has been discussed and identified with a property called *Active Touch*[1]. Noë[2] concludes that all of our perceptions are grounded in our active body experiences.

However, in many computational models of embodied cognitive agents, the role of activeness has been neglected or underestimated. Perception has been considered as a transformation from a series of sensory inputs to internal representations that are used in selection of the next action. The effect of information about self action on perception is considered in order to calibrate the sensory input by using efferent copy of motor commands and proprioceptive feedback. The other aspects of activeness in perception that we want to emphasize in this work are as follows. Meaningful stimuli to sensory organs are not given passively but obtained actively by spontaneous exploration. The targets of perception, what we subjectively feel, are not attributed to sensations given from the object but induced internally by the sensory-motor coupling between agents and the environment. The correspondence between sensory-motor flow and internal representation can be more dynamic and complex than one would expect.

S. Nolfi et al. (Eds.): SAB 2006, LNAI 4095, pp. 510–521, 2006.

Agent's categorization by way of sensory-motor coordination is called *Dynamic Categorization*. Many examples of such categorization have been explored, see e.g., [3], [4], [5], [6], [7]. In this paper, we would like to introduce a novel simulation model, one in which artificial agents behave as if they have a categorization of the bumps on the surface yet without employing explicit, predefined sensory-motor coordination. The body of an agent is constituted from homogeneous active elements and plays mediating role between salience in the environment and coherent internal state dynamics.

2   Model Description

2.1   Body Dynamics

The body of each agent consists of seven active elements mutually connected by an elastic spring. The agent moves in a two dimensional space. Fig. 1(a) shows the connectivity of elements. A system in an equilibrium state has a hexagonal shape.

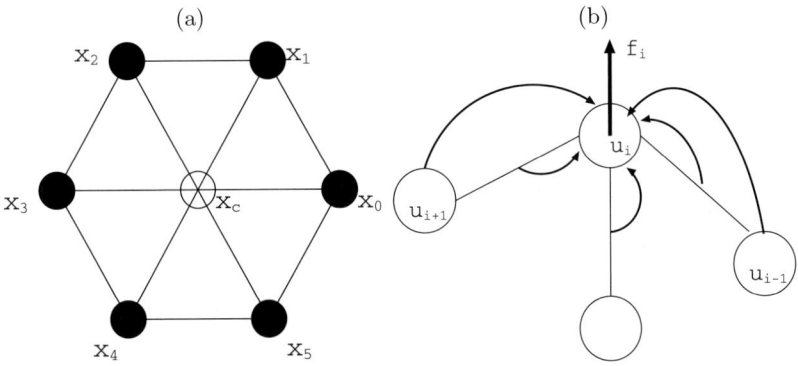

Fig. 1. (a) The body of an agent in an equilibrium state. Seven elements are connected by identical springs. Mass of all elements are the same. We denote the position and the velocity of the elements in a two dimensional space as $\mathbf{x}_i$ and $\mathbf{v}_i$, respectively. (b) Control system of the body. Every element has a variable u_i and produces force along the direction of the spring which connects it to the central element. u_i is a leaky integrator and its input is the force from connected springs and the signal from adjacent elements. Curved arrows indicate the flow of information to an element i.

The total energy of an agent's body is given as

$$E = \Sigma_i \frac{1}{2}m\mathbf{v}_i^2 + \Sigma_{(i,j)}\phi(\|\mathbf{x}_i - \mathbf{x}_j\| - L) + \Sigma_i V(\mathbf{x}_i), \tag{1}$$

where $\phi(r)$ is the potential energy of springs and $V(\mathbf{x})$ is the rest potential energy from the environmental field. We use exponential springs, which are easier to extend and harder to shrink than linear springs [1].

$$\phi(r) = ar + \frac{a}{b}\left(\exp(-br)\right) , \tag{2}$$

$$\phi'(r) = a\left(1 - \exp(-br)\right) . \tag{3}$$

The dynamics of the body is governed by the following equations.

$$\dot{\mathbf{x}}_i = \mathbf{v}_i , \tag{4}$$

$$m\dot{\mathbf{v}}_i = -\epsilon\mathbf{v}_i - \frac{\partial E}{\partial \mathbf{x}_i} + \mathbf{f}_i , \tag{5}$$

where $\mathbf{f}_i$ is the output of an element i, described in the next subsection.

2.2   Control System

Without any control system, the agent's body slides in a downward direction according to the gradient of the potential, reaching an equilibrium shape in a flat region of space or descending into a valley of the potential field. Each of the surrounding six elements has one variable u_i as internal state that can be activated. The active element is pulled or pushed from/to the central element by a force. The magnitude of the force acting on each element is computed from u_i as follows.

$$\mathbf{f}_i(u_i) = \tanh(\alpha u_i) \times \frac{\mathbf{x}_i - \mathbf{x}_c}{\|\mathbf{x}_i - \mathbf{x}_c\|} . \tag{6}$$

The central element does not receive any force and all of the surrounding six elements are homogeneous and symmetric. Therefore, to achieve spontaneous movement of the body, symmetry must be broken by coordinating the state of elements.

A state u_i is a continuous time leaky integrator that receives input from adjacent elements and connected springs. Fig. 1(b) shows the flow of information to an element i. The input from adjacent elements corresponds to signal transmission and the input from the connected springs corresponds to proprioception. The dynamics of u_i is governed by [2]

$$\begin{aligned}
\tau\dot{u}_i = \gamma - u_i \\
+ w^u g(\tanh(\alpha u_{i-1}) - \theta^u, \beta^u) + w^u g(\tanh(\alpha u_{i+1}) - \theta^u, \beta^u) \\
+ w^s g(\|\mathbf{x}_i - \mathbf{x}_{i-1}\| - L - \theta^s, \beta^s) + w^s g(\|\mathbf{x}_i - \mathbf{x}_{i+1}\| - L - \theta^s, \beta^s) \\
+ w^c g(\|\mathbf{x}_i - \mathbf{x}_c\| - L - \theta^c, \beta^c) ,
\end{aligned} \tag{7}$$

---

[1] We have also experimented with linear springs, $\phi(r) = \frac{1}{2}abr^2$. By simulating both cases two times each, we found that evolved agents with the exponential spring can climb a hill. Agents with linear springs can also break symmetry and move straight in flat regions of space, however they cannot climb hills. The reason is not yet clear, but we have a hypothesis. The effective elasticity of the exponential springs changes according to the length of the spring, which might be beneficial for the agent's reaction to the distortion of the space when the body shape is asymmetric.

[2] This equation is essentially the same as one introduced in CTRNN[3]. Differences between this model and a normal CTRNN implementation reside in the homogeneity of elements and the locality of connections.

where τ is the time constant and γ is the gain of the element. $w^{u,s,c}$ are the weights, i.e., the contribution to the activation of unit i from other elements or springs, and $\theta^{u,s,c}$ are the thresholds of these contributions. $g(x, \beta)$ is the sigmoid function with a nonlinear parameter β.

$$g(x, \beta) = (1 + \exp(-\beta x))^{-1} . \tag{8}$$

$\{\tau, \gamma, \alpha, w^{\{u,s,c\}}, \theta^{\{u,s,c\}}, \beta^{\{u,s,c\}}\}$ are the parameters of the control system and characterize the behavior of the agent.

2.3   Potential Field and Hill Climbing

Every element is affected by an external potential field $V(\mathbf{x})$. The strength of the potential is a function of its distance from a point source.

$$\psi(r) = h \exp\left(-\left(\frac{r}{d}\right)^2\right) , \tag{9}$$

where h is the strength of the source and d is the characteristic size. All effects from the positive and negative sources are linearly superposed. Hereafter, we consider the potential from the point sources as if the space is a surface in three dimensional space with gravity. A positive source is like a hill and a negative source is like a hollow.

By introducing an evolutionary algorithm, we enable agents to acquire adaptive behaviors. The first example of adaptive behavior will be hill climbing. The characteristic of biological adaptive behavior is decreasing entropy and maintenance of non-equilibrium states. Because in our model there is dissipation of energy as well as energy input resulting from the activation of elements, the system can enter into non-equilibrium states. It is in theory possible for agents to climb up a potential, behavior we would consider to be interesting.

2.4   Genetic Algorithm

Parameters of the control system, namely $\{\tau, \gamma, \alpha, w^{\{u,s,c\}}, \theta^{\{u,s,c\}}, \beta^{\{u,s,c\}}\}$ are encoded in genes as real values. It should be noted that all elements have the same parameter value, hence we refer to them homogeneous elements. Parameters in Table 1 are fixed in all the simulations in this paper. Behavior of individual agents is simulated in an environment in which two positive sources and two negative sources are randomly positioned. The boundary condition of the space is periodic.

Agents' performance at the hill climbing task is evaluated in terms of how long the central element is kept in the high potential position. The score of the agent is computed as

$$\int V(\mathbf{x}_c)dt . \tag{10}$$

The sum of the scores of the five runs from different initial positions, for 200 time periods each, totals the fitness of the agent. In each generation 100 individuals in the population are evaluated. The worst 30 individuals are removed from the population and 30 new individuals are produced from other agents by a mutation and a crossover operator. The strength, the size and the location of potential sources are updated at every generation.

Table 1. Parameters which are fixed in all the simulations in this paper

mass of an element	m	1.0
natural length of a spring	L	0.5
viscosity coefficient	ϵ	1.0
spring stiffness (constant part)	a	1.0
spring stiffness (exponential part)	b	5.0
strength of potential sources	h	$\pm[0.05, 0.1]$ (uniform distribution)
size of potential sources	d	$[0.5, 1.0]$ (uniform distribution)
size of the space		8.0×8.0 (periodic boundary)

3   Results

3.1   Behavior of an Evolved Agent

The control parameters of an evolved agent are shown in Table 2. Fig. 2 shows the shape and movement of the agent. The evolved agent can move straight in flat regions of space. Destabilization of symmetric shapes and stabilization of asymmetric shapes results from regulating the control system under no external force. The shape and the force from elements in the figure generates a fixed point attractor in flat regions of space.

Because the agents cannot sense the potential or the gradient of the potential directly, the only information that agents can use is the relative difference of the external forces between elements. The spatial difference of the external force from the potential, which is the second order differential of the potential, is the convexity of the space (Fig. 3). To climb a hill, the agents must be sensitive to information about convexity obtained through changes in body dynamics. Fig. 4 shows the time series of the output from all elements. A slight skew of the asymmetric shape according to the distortion of the space triggers changing of roles between three asymmetric elements, resulting in a change of direction.

Table 2. Control parameters of an evolved agent. The behavior of this agent is described in this section.

τ	γ	α	w^u	θ^u	β^u
0.02579	0.6631	2.486	-2.489	0.7149	173.2

w^s	θ^s	β^s	w^c	θ^c	β^c
5.683	0.6455	1.794	0.4211	2.129	35.38

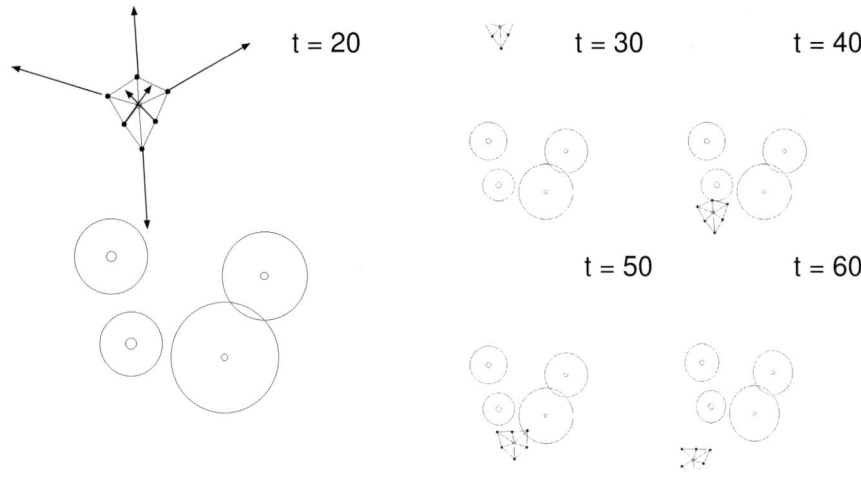

Fig. 2. Snapshots of the position and the shape of the evolved agent. Small circles are the position of point sources and large circles show the characteristic size of the potential area from the source. In this environment, top left and bottom right sources are positive and others are negative. The forces on the six surrounding elements are depicted in the figure from time 20. The force vector is $(-0.47, 1.0, 0.58, 1.0, -0.47, 1.0)$. This coordination corresponds to moving straight to the north in flat regions of space. The agent encountered a negative source at around time 40 and change direction to avoid the source. The time series of the force from all elements is shown in Fig. 4. The top left panel of the Fig. 5 shows the trajectory of the central element in this environment.

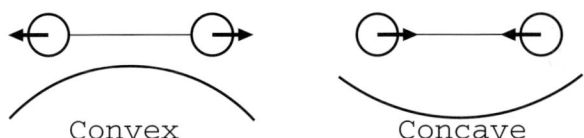

Fig. 3. Relative difference of external forces is interpreted as the convexity of the space. This is the only information about the potential field that agents can obtain through changes in body dynamics.

The top left panel of Fig. 5 corresponds to a trajectory of the central element in the same environment as that of Fig. 2. The agent explores the space, avoids hollows and tries to remain close to hills.

3.2   Internal Representation of Surface

It is possible to project the internal state dynamics of the agent in six dimensional space to the lower dimensional plane. We have chosen the mean and the variance of six outputs and traced the trajectory in two dimensional space. Fig. 5 and Fig. 6 show several trajectories (a) in the real space and (b) in the internal state space in various environments. In regular environments, four identical sources are

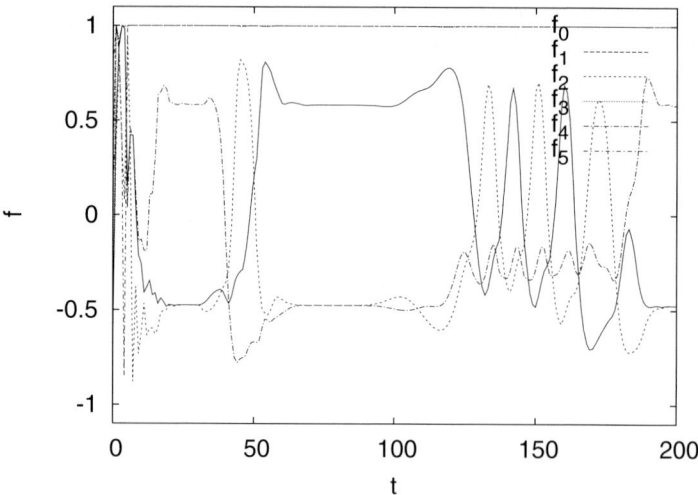

Fig. 4. Time series of output of the surrounding six elements. Output is the magnitude of the force by which the element pulls or pushes the central element. Three outputs (f_1, f_3, f_5) are almost fixed to one in this run, thus these three elements pull the central element as strong as possible. The differentiation of (f_0, f_2, f_4) is the cause of the movement. The role of these three elements change while moving in the environment when the body shape skews according to the distortion of the space. Direction of movement changes as a result.

positioned at regular intervals. Observed behaviors such as avoidance, approach and exploration correspond to different internal cycles.

Fig. 6 shows the dependency of the behavior of the agent on the shape of hills in the regular environments. The behavior and the internal cycle are distinct. In addition, the position of the cycles in the internal state space has a strong correlation with the value of the potential in the real space.

3.3   Evolution of Categorization

Fig. 7 shows the scoring profile in the regular environments as a function of the strength and size of the potential sources. The top left panel shows the initial generation, whereas the top right panel shows an earlier generation in the evolutionary process. Distinctive patterns appear in the evolution. The distinction is related to the behavioral categorization of hills in Fig. 6.

Although such categorization of bumps on the surface is not the objective of the agent, this functionality arises as a self-organized property through adaptive behavior. In earlier generations of evolution, agents begin to move straight and try to avoid hollows, but they cannot climb hills. Expected scores of such agents is close to zero. We have found no clear distinction of the bumps on the surface in these agents.

The difference of the potential shape is not explicit in the force on each element from the environment, because the change of the force is gradual. The distinction

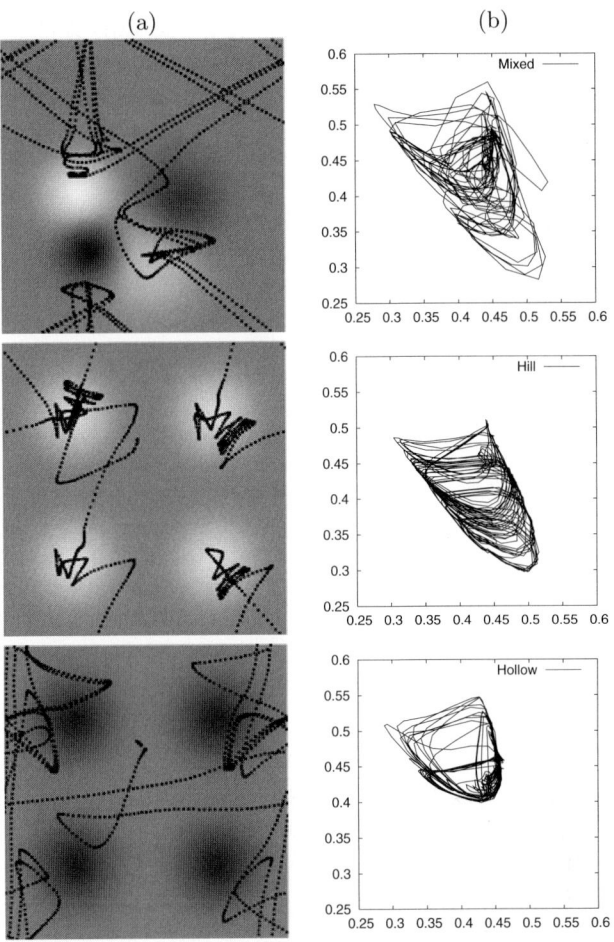

Fig. 5. Trajectories in various environments. Top: Mixture of positive and negative source randomly positioned. An example of an environment to which agents must adapt. Middle: Regularly positioned medium sized positive sources, whose strength is 0.075 and size is 0.75. Bottom: Regularly positioned medium sized negative sources, whose strength is -0.075 and size is 0.75. (a) Trajectory of the central element of the agent in the real space. The color is the potential at the point, the brighter the point the higher the potential. (b) Trajectory of the internal state as a two dimensional projection. Horizontal axis is the mean of the outputs of elements. Vertical axis is the variance. (0.44, 0.44) in the internal state space is the fixed point attractor corresponding to straight motion in flat region. Behaviors such as avoidance, approach and exploration correspond to different internal cycles.

is an outcome of the internally-driven sensory-motor coupling between agent and environment. Switching between several patterns of behavior is adaptive in the complex environment. The distinction simultaneously appears in both behavioral patterns and internal states.

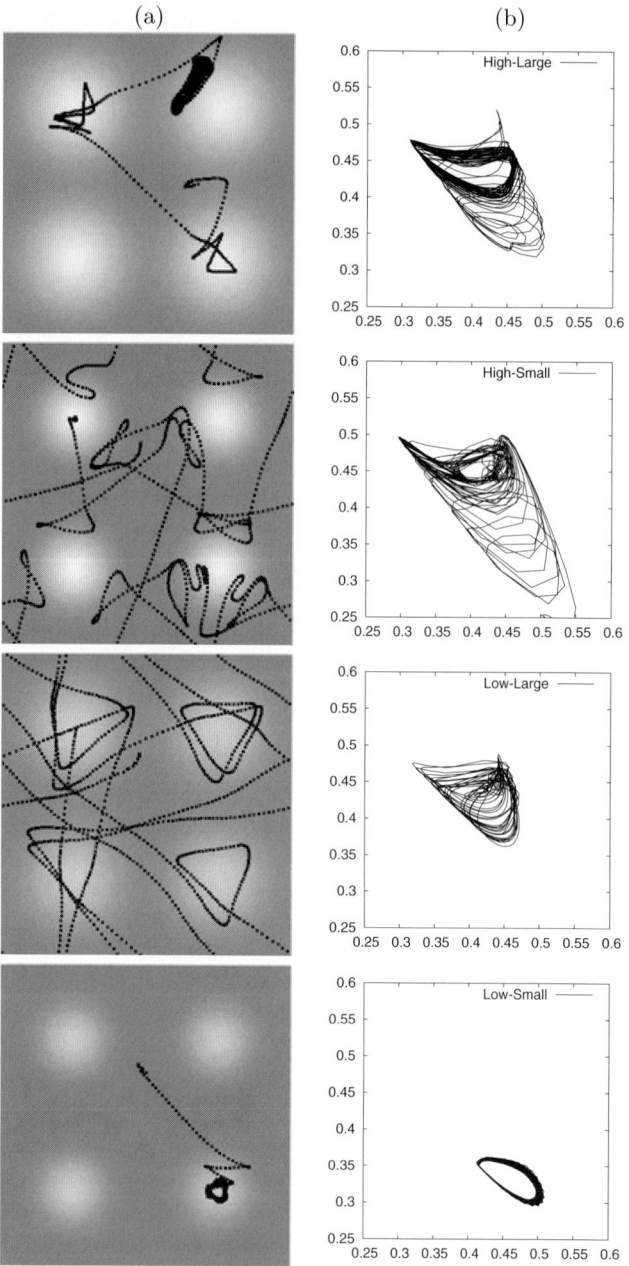

Fig. 6. Dependency of the behavior on the shape of hills. Trajectories of the agent in four regular environments, in which four identical hills are positioned at regular intervals, are shown. Strength: high – 0.09 low – 0.06; Size: large – 0.9, small – 0.6. The position of the internal cycles has a correlation with the value of the potential.

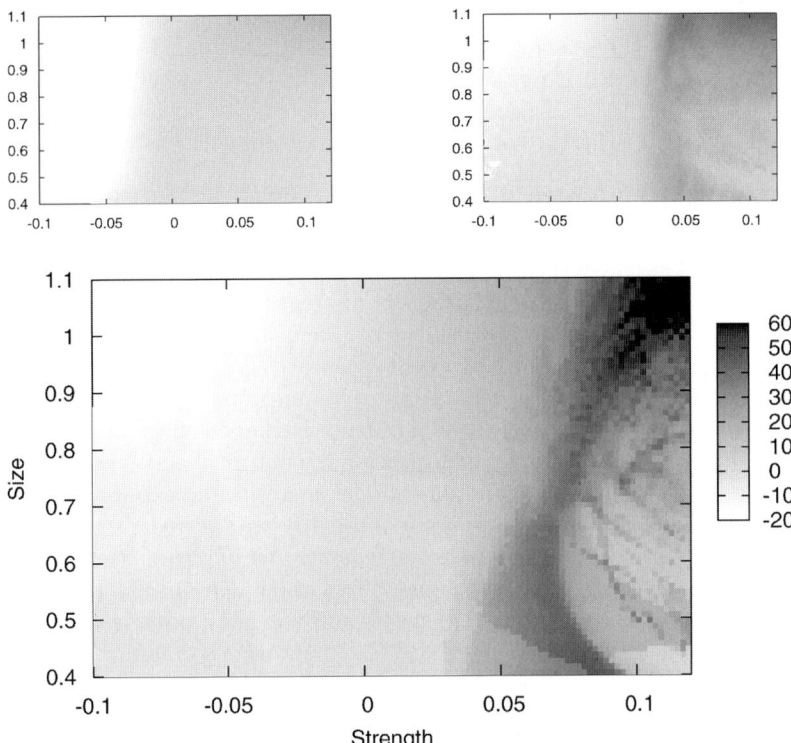

Fig. 7. Scoring profile of the evolved agent. Horizontal axis is the strength of the potential source (height of hills/hollows). Vertical axis is the characteristic size. The color shows the score of the agent in an environment in which four identical potential sources are positioned at regular intervals. The top left panel is the initial profile. The top right panel is the agent from an earlier generation. The agent breaks symmetry and moves straight in flat region, avoiding hollows. But still the agent cannot approach hills. There is no clear distinction between hills. The bottom panel is the profile of the evolved agent. The profile has a structure corresponding to the behavioral categorization in Fig. 6. In an environment with high hills that take up a large area, the agent can stay near the top of hills for a long time but sometimes fails to remain, and thus score is the highest but has a fluctuation. In an environment with high hill taking up a small area, hills are too steep to climb and the score is worse. Low and large hills are too vague to climb and remain close to. Low and small hills island in the profile corresponds to hills which the agent can climb and stay on top of. The oblique line with high scores shows the proportional hills, gradients of which are similar.

4   Discussion

To achieve sensory-motor coordination, neural networks are often employed. Neural networks with N elements generally have to tune N^2 parameters to achieve a given task. Here in this study, we only tune a single common parameter set shared by every element. Spontaneous symmetry breaking of the internal states is the key mechanism generating desired behaviors. The present model performs well and shows coordinated and coherent behavior. In flat region, the agent fixes the states of three elements and releases the other three elements (Fig. 4). This coordination generates a diamond shape and results in straight motion. To make a turn, the role of elements is alternated. The role of each element is both a cause and an effect of the body shape. This is the main difference between this study and other neural network studies.

The second point of this study is the origin of dynamic categorization by the evolved agent. The categorization is not the direct objective of the agent under the evolutionary pressure. The scoring profile is different from generalization of target categories. In other words, the evolved agent cannot help categorizing the object without being instructed in order to behave adaptively in the complex environment. In experimental studies, activations of hippocampal place cells of rats are known to diverge under the exposure of differently shaped environment without explicit reward[8]. Autonomous categorization is a nature of active perception.

Another interesting point is that the model agent has no explicit sensory inputs. Environmental information is only implicitly transferred to the agent by modifying the springs that connect the elements. We interpret this as proprioceptive flow producing "sensory" signals from the environment. This implicit coupling between sensory input and motor outputs may provide an origin of active perception. Iizuka[9] has discussed the spontaneous separation of sensors and motors in artificial cognitive agents.

As a prerequisite condition for having active perception, a flexible morphology is important. The change in the internal state of elements is reflected in the morphology of the agent, which enables coherent as well as exploratory motion. Pfeifer[10] has intensively discussed the relationship between control and morphology in adaptive behavior. For the moment, the agents have no memories and the morphology is restricted. Association of morphology with memory capacities is the next step.

Acknowledgments

This work is supported by The 21st Century COE (Center of Excellence) program (Research Center for Integrated Science) of the Ministry of Education, Culture, Sports, Science, and Technology, Japan.

References

1. Gibson, J. J.: Observations on Active Touch. Psychological Review **69** (1962) 477–491
2. Noë, A.: Action in Perception. MIT Press, Cambridge, MA. (2004)
3. Beer, R. D.: The Dynamics of Active Categorical Perception in an Evolved Model Agent, Adaptive Behavior **11**(4) (2003) 209–243
4. Morimoto, G., Ikegami, T.: Evolution of Plastic Sensory-motor Coupling and Dynamic Categorization. Proceedings of Artificial Life IX (2004) 188–193

5. Nolfi, S., Marocco, D.: Active Perception: A Sensorimotor Account of Object Categorization, From Animals to Animats VII; Proceedings of the International Conference on Simulation of Adaptive Behavior (2002) 266–271
6. Nolfi, S., Floreano, D.: Evolutionary Robotics, MIT Press, Cambridge, MA (2000)
7. Pfeifer, R., Scheier, C.: Understanding Intelligence, MIT Press, Cambridge, MA.
   (1999)
8. Lever, C., Wills, T., Cacucci, F., Burgess, N., O'keefe, J.: Long-term Plasticity in
   Hippocampal Place-cell Representation of Environmental Geometry, Nature **416**
   (2002) 90–94
9. Iizuka, H., Ikegami, T.: Emergence of Body Image and the Dichotomy of Sensory and Motor Activity. Proceedings of the Symposium on Next Generation Approaches to Machine Consciousness. Hatfield, UK (2005) 104–109
10. Pfeifer R.: On the Role of Morphology and Materials in Adaptive Behavior. Proceedings of the 6th International Conference on the Simulation of Adaptive Behavior. MIT Press, (2000) 23–32

Searching for Emergent Representations in Evolved Dynamical Systems

Thomas Hope, Ivilin Stoianov, and Marco Zorzi

Computational Cognitive Neuroscience Lab, University of Padova,
Via Venezia 12/2, Padova 35131, Italy
{thomas.hope, ivilin.stoianov, marco.zorzi}@unipd.it

Abstract. This paper reports an experiment in which artificial foraging agents with dynamic, recurrent neural network architectures, are "evolved" within a simulated ecosystem. The resultant agents can compare different food values to "go for more," and display similar comparison performance to that found in biological subjects. We propose and apply a novel methodology for analysing these networks, seeking to recover their quantity representations within an Approximationist framework. We focus on Localist representation, seeking to interpret single units as conveying representative information through their average activities. One unit is identified that passes our "representation test", representing quantity by inverse accumulation.

1 Introduction

In 1963, Feyerabend ([1] and [2]) claimed that improvements in our scientific understanding of the mind will eventually undermine our basic concepts of mental states. His position – Eliminative Materialism – stemmed from the intuition that "folk psychology" [3] is merely our current best "theory" of mind. Like any other theory, folk psychology may eventually be falsified, perhaps in favour of a neuro-biological account of cognition. Significantly, there is no requirement that this replacement must "explain" the theory that it replaces – like phlogiston and alchemy, mental states may simply disappear [4] in the face of scientific progress. In cognitive science, a debate has recently emerged that adds a practical dimension to Feyerabend's position.

Since the emergence of the digital computer during the 1940's, the Computer Metaphor (CM – [5]) has dominated the way in which scientists study intelligent behaviour. One of the principal methodological commitments of the CM is Functional Decomposition (FD), which implies that, like computers, cognitive systems can be understood as networks of functional "modules" [6]. The acceptance of this intuition is nearly ubiquitous in contemporary cognitive science; most experimental paradigms are designed explicitly to isolate and manipulate these putative modules. Seeking to account for experimental data, computational cognitive models have tended to be directed along similar lines. This division naturally emphasises the concept of "representation" in contemporary cognitive theories, since the specification of a module's interfaces (input and output representations) has a critical impact on its empirical behaviour (e.g. [7]).

S. Nolfi et al. (Eds.): SAB 2006, LNAI 4095, pp. 522–533, 2006.

Yet for all its evident utility, FD carries a heavy burden of explanation; functional modules must be *integrated* before they can reasonably be said to account for cognitive behaviour. Recognising the critical role that module interfaces play in designing these modules, some researchers (e.g. [8], [9]) have questioned whether this integration could ever be successful.

Dynamicism [10] offers an alternative. The Dynamicist programme construes cognition as intrinsically embodied, emerging from the interaction of adaptive behaviours. These behaviours are the Dynamicist equivalent of functional modules – the atomic components of intelligence [8]. To the extent that this framework addresses the "integration problem" mentioned previously, it must surely be welcomed even by researchers entrenched within the CM. The problem is that, at least as commonly construed [11], the Dynamicist programme is resolutely Eliminativist.

Dynamicist models are thought to be best understood in terms of the temporally situated causal processes that manage sensor-motor integration [11]. Traditionally critical concepts, like representation, simply do not appear to "fit" with how these systems work. Recognising this, many Dynamicist researchers have been moved to claim that the CM-inspired distinction between data and process – the very concept of representation itself – must go the way of phlogiston [8]-[11].

This proposition is antithetical to many neuroscientists because it simply does not appear to correspond with the observed structure of cognition in the brain. Selective neural disorders and brain imaging experiments [12] provide convincing evidence of functional specialisation in human cognition; though consistent with Dynamicism, a functionally specialised cognitive system naturally encourages analysis by FD. Concepts of representation are a powerful and intuitive tool in accounting for well-confirmed experimental data (e.g. [13], [14]); in some cases, such as [15], these accounts can even draw on "observations" of the representations themselves. In some respects at least, brains just do appear to represent and to compute. If Dynamicism must be accepted at the expense of the concept of representation, few cognitive neuroscientists will accept it.

To manage this tension, we propose an Approximationist response. Approximationism articulates the intuition that computational accounts of cognitive processes may be useful and approximately correct, without necessarily capturing every detail of the underlying causal processes [16]. One implication is that we might usefully search for – and discover – representations in neural systems, while at the same time accepting that the implied "computational story" will be at best a good approximation to the underlying "causal story".

Following the logic of [17], we evolve artificial agents to perform a "representation-heavy" task – a task for which some kind of representational structure appears to be required. Section 2 describes the artificial ecosystem and agents, as well as the representation-heavy task that they evolve to perform. The goal is then to recover the agents' evolved representations. Section 3 describes and applies a methodology designed to achieve this goal.

2   Through Foraging to Quantity Comparison

The focus for the current project is "quantity comparison", a common theme of study within the cognitive neuroscience of numeracy. Representation plays a critical role in contemporary accounts of the way in which subjects (humans and animals) manipulate numbers and numerosities [13] – indeed, there is great debate in this field concerning the precise format of that representation (e.g. [13], [18], [19]). Quantity comparison therefore meets our requirement for a representation-heavy process. Further, a growing body of evidence indicates that certain facets of the "number sense" may be inherited [20]. The implication is that evolution might engender a preparedness in humans and animals to represent quantity in a particular way, raising an independent question about what kinds of quantity representations can emerge "spontaneously" during evolution. The current work approaches quantity comparison as an evolutionary by-product of selection for quantity-sensitive foragers.

2.1   The Artificial Ecosystem

The environment is a 2-dimensional, toroidal grid, composed of 100x100 square cells. Agents navigate the grid by moving between neighbouring cells. Each cell can contain "food", construed as appearing in "bundles" of some specified numerosity (1-9). Any number of agents can co-exist in the same cell: the only upper limit is the size of the population itself (200), which remains constant throughout the run. Food can also "grow", in the sense that its numerosity can increase. A record is kept of the total depletion of food during the run, and this food is periodically reinserted by sharing it among randomly selected cells.

The ecosystem proceeds by iterative update. During each iteration, every individual is updated, with sensor activity propagated through the neural network and effector units interpreted to identify if any action has been made. The update order for agents is randomly specified at the beginning of each iteration.

The agents are recurrent, asymmetrically connected neural networks. In a network of N units, the activity u of the i-th unit (u_i) at time step t is calculated by

$$u_i(t) = S(\sum_{j=1}^{N} w_{ij}u_j(t-1)).(1-m) + u_i(t-1).m \qquad (1)$$

where w_{ij} is the weight of the connection from unit j to unit i, $S()$ is the sigmoid function and m is a fixed momentum term with a value of 0.5.

A subset of units act as sensors, which are clamped according to the salient features of the environment around the agent. Agents have a 3 cell field of view, and are also sensitive to food in the cell that they currently occupy (see Fig. 1), for a total of four sensor "fields". Each field represents its corresponding food quantity using a "Random Position Code"; this was used in [18], among others, to capture quantity information without biasing models in favour of specific representational strategies. To represent food numerosity N, the code requires that N (randomly chosen) sensor units (positions) should be active. The scheme is illustrated in Fig. 2, where N = 5.

Another subset of units are effectors, whose activity determines how the position / orientation of the agent's body is updated, as well as defining when agents try to eat. The remaining units are hidden and do not interact directly with the environment. Sensor units receive no input from the rest of the network, and have no direct connection to effector units, but the hidden layer is universally connected – every unit is connected to every other unit, and to itself.

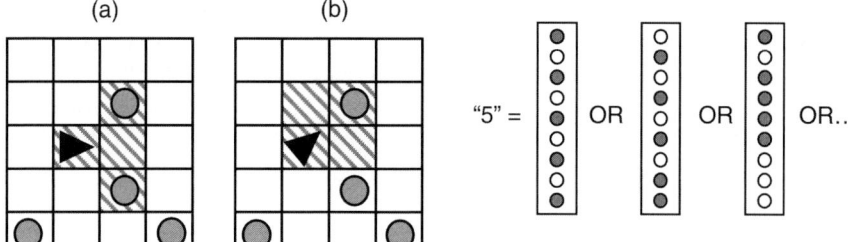

Fig. 1. An agent in its environment. (A) The agent – a black triangle – is facing right and can sense food (grey circles) in its right and left-most sensor fields. (B) The same agent, after making a single turn to the left. It can now sense only one cell containing food.

Fig. 2. The Random Position Code. For food = N, exactly N units are active (activity clamped to '1'), while the remainder are inactive (activity clamped to '0'). Each sensor field has a total of 9 units.

2.2 Evolution in the Ecosystem

All of the agents in the initial population are specified with random numbers of hidden units, and random weights. The "fitness" (F) of the i-th agent is just the rate at which it has consumed food,

$$F(i) = f_i / a_i \tag{2}$$

where f is the total food consumed and a is the agent's age (expressed as the number of iterations since the agent's creation).

At the end of each iteration, two "parents" are drawn at random from the population and their fitnesses compared. The structure of the "child" is defined by randomly mixing the parents' weight vectors (cross-over), followed by mutation. The mutation operator will usually increment or decrement a randomly selected weight value by a small constant (0.01), but may also add or remove a hidden unit. The resultant child replaces the least fit of the two parents.

The best signal that agents are discriminating quantity is high food collection efficiency: food collected per moves made in the environment. The agents' food collection efficiency rises above 5 after about 10 million iterations (~50,000 generations), indicating that genuinely "discriminating" foraging behaviour has evolved.

2.3 Quantity Comparison Performance

The evolved agents are not merely models of quantity comparison; to capture that facet of their behaviour, we need a methodology that can effectively isolate it. Fortunately, examples of the required kind of methodology already exist. In [21], the "subjects" (salamanders) were placed at the base of a clear Perspex T-maze. Two clear jars of *drosophilia* fruit flies, which the salamanders eat, were placed at the end of each branch of the maze – two flies in one jar and three flies in the other. The authors reported that twice as many salamanders "chose" the jar with more flies (signified by walking toward and touching that jar). Our methodology emulates this experiment.

The experimental environment is a 3x3 "mini-world". Two cells, the top left and top right of the world, contain food of varying quantity. In its initial position at the centre of the world, the agent can "see" both of these food quantities, though it may turn without constraint once the trial begins. Food "selection" occurs when the agent moves onto one or other of the filled cells – the only cells onto which the agent is allowed to move. A correct choice is defined as the selection of the larger of the two food groups. Every agent in the population was tested using this methodology, with 50 repetitions of every combination of food quantities (1-9, 72 combinations in all), for a total of 3,600 trials per agent. The results are displayed in fig. 3.

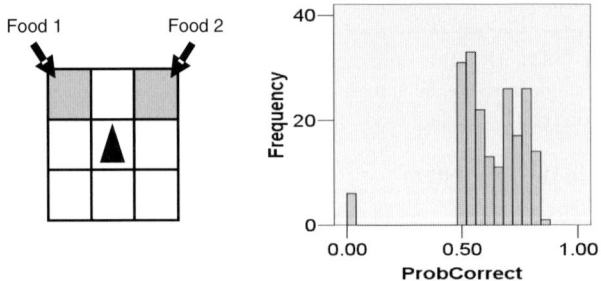

Fig. 3. (Left) The schematic structure of the comparison experiment. The agent is placed in the centre of the mini-world, facing "up". (Right) A histogram of the population performance in the quantity comparison experiment.

A few of the agents perform extremely badly, indicating that the evolved foraging solutions are brittle in the face of "evolutionary" change – perhaps emphasised as a consequence of a mutation bias against specialised structures [22]. The main bulk of the population distribution is also apparently bimodal; agents in the left-most cluster perform at roughly chance levels, whereas agents in the right-most cluster perform significantly above chance – only this latter group appear to discriminate quantity. The persistence of non-discriminating agents is unsurprising, since high rates of food collection can be achieved by sacrificing decision quality for decision speed. A visual inspection of the performance scores for agents in this cluster indicates strong asymmetry in their behaviour; many simply "choose" the right-hand square regardless of the food quantities presented.

2.4 Single-Subject Comparison Behaviour

Using the results of the previous section, we selected the best "discriminator" from the population and recorded its empirical performance in more detail. The results are displayed in fig. 4.

The agent's empirical behaviour displays certain characteristic phenomena that are also reliably found when both humans and animals compare quantities. As the minimum of the two quantities-to-be-compared increases (fig. 4a), there is an increase in discrimination error, this is an instance of the "Size Effect" [13]. As the numerical

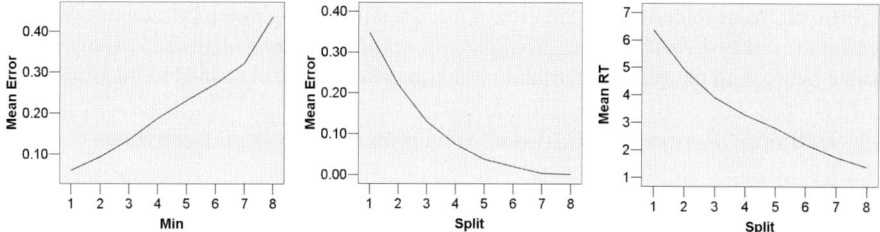

Fig. 4. Accuracy scores are rates of correct choices per 1,000 trials. (a) Mean accuracy vs. minimum quantity of food (Min) in a given trial. (b) Mean accuracy vs. numerical distance (Split) between food quantities. (c) Mean "reaction time" vs. numerical distance between quantities.

distance between the quantities increases (fig. 4b), there is a corresponding improvement in discrimination accuracy; this is an example of the "Distance Effect" [13].

Surprisingly, this agent also displays a Distance Effect for *reaction times*, defined as the number of time steps between the start of a comparison trial and the agent's selection of one of the two food quantities. Non-discriminating foragers can persist by sacrificing decision accuracy for decision speed, but this agent is capable of reversing that tradeoff, sacrificing decision speed in order to more reliably "go for more".

Though the agents are too simple to support a meaningful comparison with biological organisms, this behavioural correspondence is nevertheless encouraging. As mentioned prevously, most contemporary theories of quantity manipulation account for empirical phenomena (such as the Size and Distance effects) as a consequence of the way in which subjects represent quantities. A representational account of this agent's behaviour could therefore add a new dimension to this traditional debate, expanding the space of representational strategies that can account for the empirical phenomena.

3   Approximationist Representation

The key property of a representation is that it tracks some property of the environment. During the comparison experiment, the most salient properties are the values of the two food "options", which remain static throughout each trial. But a visual inspection of the agent's network dynamics reveals nothing remotely static – nothing that seems appropriate to represent these food values.

Our thesis is that although the behaviour of each unit is subject to chaotic variation, its *average* activity may still be interpreted as conveying representative information.

If a unit's average activity (relative to some food value) is a *functionally significant* representation, it should be possible to "fool" the agent by fixing the unit's activity to its average for a *different* food value. In deference to previous work (e.g. [23] and [24]) on lesion types in network analysis, we refer to this kind of interference as "Partial Informational Lesion" (PIL). Our Approximationist method relies on statistical analyses of the impact of PIL's on the agent's behaviour.

In the material that follows, we restrict the analysis to Localist "theories" of representation – to theories that *single* units can be interpreted as conveying functionally significant, representative information. This is a simplifying assumption, but not without biological justification; recent single-cell recording experiments [25] suggest that single neurons in the primate parietal cortex may be selectively tuned to quantity.

3.1 Defining "Representation-Like" Deviation from Normal Behaviour

Consider a simple example in which a hypothetical agent manages a quantity comparison using just two hidden units – unit A and unit B. Each unit is subject to extensive noise, but represents a corresponding quantity (Food A and Food B) by its average activity. As in the current work, our hypothetical system uses food values 1-9, signified by average unit activities of 0.1, 0.2…., 0.8, and 0.9. Suppose that we interfere with unit A so that its activity is *always* 0.5 – now, the agent will always perceive food A as taking the value '5'.

In some circumstances, this discrepancy between "actual" and "perceived" value of food A should reduce the agent's comparison performance. If the actual value of food A is '2', and the value of food B is '4', our agent will "think" (wrongly) that food A is larger and could make the wrong decision. There are also circumstances in which this intervention should improve the agent's performance. Suppose now that the true value of food A is in fact '2' and food B is '1'; the perceived and actual comparisons between the two quantities both have the same "answer" (i.e. food A is larger), but the perceived comparison is arguably easier because the numerical distance between '5' and '1' is greater than that between '2' and '1'.

This logic leads us to define two groups of comparison trial, relative to particular PIL's; *Consistent* trials are those for which PIL's should improve comparison performance, whereas *Inconsistent* trials are those for which PIL's should reduce comparison performance. The hypothesis that some unit's average activity does in fact "represent" can then be judged by reference to two "Representation Scores"; one for Consistent trials, and one for Inconsistent trials, calculated as in equation 3.

$$R_j^+ = \sum_{i=1}^{n} P_j(C,L) - \sum_{i=1}^{n} P_j(C,N)$$

$$R_j^- = \sum_{i=1}^{n} P_j(I,L) - \sum_{i=1}^{n} P_j(I,N)$$

$$(3)$$

where R_j is the representation score for the j-th unit, and $P_j()$ is a function that counts the number of correct comparison choices that the agent makes under the four possible conditions. The four conditions are:

1) (C, L): Consistent comparisons made while the unit was subject to a PIL (i.e. is Lesioned).
2) (C, N): Consitent comparisons, but where the unit is allowed to change freely (i.e. is Normal).
3) (I, L): Inconsistent comparisons made while the unit was subject to a PIL (i.e. is Lesioned).
4) (I, N): Inconsistent comparisons, but where the unit is allowed to change freely (i.e. is Normal).

A "positive" result is observed when R^+ (the "Consistent Score") is significantly positive, and R^- (the "Inconsistent Score") is significantly negative.

3.2 Identifying "Representation-Like" Deviation from Normal Behaviour

The current results are derived from the hypothesis that *single* units may represent food quantities. The average activities of these units – putatively localist representations – were collected during 100 repetitions of the comparison experiment described in section 2. This process associates each unit with one average activity for each food value (1-9) in each position (Food 1 or Food 2) – 18 values in total.

Fig. 5 displays scatter plots for the 2-dimensional representation scores of each of the agent's 25 hidden units. Each unit has a data point in both graphs. Each component of a unit's representation score is a mean average value. Scores that pass our "Filter test" will lie in the top-left quadrant of each graph; when applied to the units that correspond to these data points, PIL's improve the agent's performance during Consistent comparison trials, and reduce that performance during Inconsistent trials.

T-tests for paired samples (Lesioned vs. Normal in both Consistent and Inconsistent comparison conditions, N = 90 in both cases) confirm that the marked data points represent significant deviation from normal performance after the lesion ($p < 0.05$) in both Consistent and Inconsistent conditions.

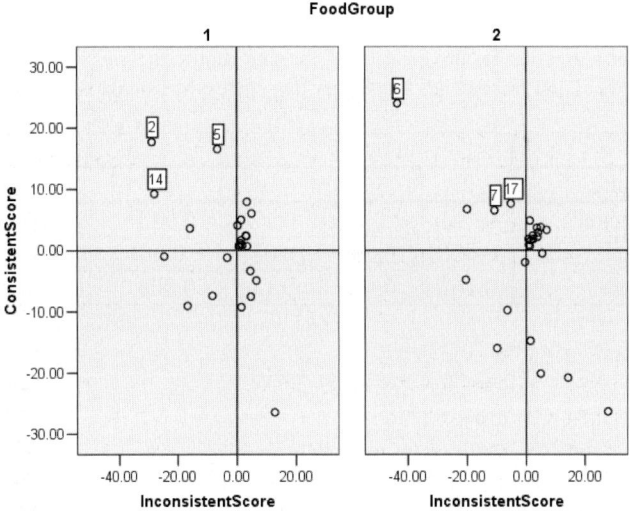

Fig. 5. Representation scores for the agent's 25 hidden units, relative to the hypotheses that each unit represents either food group 1 (left) or food group 2 (right). Scores in the top-left quadrant of the graph indicate a positive result. Data labels denote the unit numbers of associated data points.

Both of the food groups support multiple theories of unit-centric representation; relative to each food group there are three units that, when lesioned with PIL's, engender the behavioural deviations that we hoped to discover. A visual inspection of the associated representation scores is not sufficient to adjudicate between them. To make that step, we need to extend the method.

3.3 Comparing Candidate Theories

Returning once more to our hypothetical agent, consider what happends when we fix the value of unit A to '0.1'. In this case, the agent will "think" that Food A takes the value '1', regardless of the true value of Food A, and its comparison performance should reflect that perception. This logic is the foundation for our extension.

We can collect a pair of performance scores based on this hypothesis; in the "Lesioned" condition, the PIL is applied and Food A takes some value other than '1', whereas in the "Unlesioned" condition, no PIL is applied and Food A is always equal to '1'. After collecting analogous pairs for every other food value, relative to each of the two food groups, we have two sets of paired series of performance scores. To the extent that the PIL's have captured the agent's representational strategy, there should be a significant relationship between these paired series; we can capture that relationship using linear regression. If the relationship is significant, its "variance explained" (R^2) provides the metric that we need to compare competing theories.

Table 1 displays the results when this Comparison test is applied to the data-points highlighted in Fig. 5. The values are derived from series generated by 10 repetitions of each experiment; regressed series are 90 elements long.

Table 1. Linear regression results for each of the unit-centric theories identified by the Filter test. The shaded column corresponds to a theory that passes the Comparison test.

	Food 1			Food 2		
Unit	2	5	14	6	7	17
Significance	0.127	< 0.001	0.176	0.943	0.544	0.062
R^2	0.026	0.219	0.021	< 0.001	0.004	0.039
Beta	0.162	0.476	0.144	0.008	-0.065	-0.198

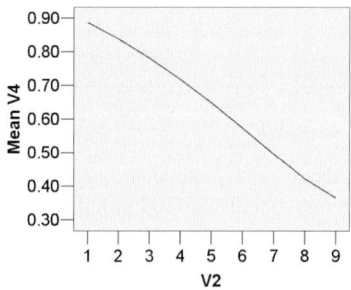

Fig. 6. The average activity of the agent's hidden unit '5', relative to the value of food group 1

Only one of the units passes this Comparison test – unit 5 does justify an interpretation of representing food group 1 by its average activity. The quality of that justification depends on the R^2 and Beta values associated with the regression; the quality increases as these variables approach the value '1'. This unit's activity (Fig. 6) is inversely proportional to the value of "Food 1", a pattern consistent with the "Accumulator" theory [26] of neural quantity representation.

4 Discussion

Conventional approaches to cognitive modeling tend to focus on the isolated, functional components of cognitive behaviour; methods that could facilitate more behaviourally integrated models have failed to attract great support among cognitive neuroscientists. One reason for this is the sense that the gulf between human / animal and artificial agent behaviour is too wide to permit useful comparisons. Another reason is that these techniques can challenge our fundamental notions of representation, which remains an important conceptual tool for understanding cognitive systems.

The current work lays the foundations of a methodological framework designed to address these apparent inconsistencies. We have evolved agents whose behavioural performance is reminiscent of that found in biological organisms, and offered an analytical framework that permits the recovery of classically "recognisable" representations from those agents. Critically, our method quantifies the extent to which a representational account of the agent's behaviour can be justified – the extent to which the theory captures the underlying causal process.

We chose to base the analysis on the thesis the average unit activities can be interpreted as conveying functionally significant representative information. Though potentially controversial, this thesis is at least consistent with the practice of single-cell recording experiments (e.g. [25]), which emphasise average neural behaviour at the expense of the apparently random [27] variation in specific spike trains. The real justification for the choice flows from the results of the analysis itself; despite its restricted (Localist) scope, we have identified a unit that appears to represent quantity by its average activity. We expect that the strength of this result can be improved by relaxing the Localist restriction, and work to implement that extension is currently underway.

Despite its limitations, the current version of this system yields an interesting implication concerning the symmetry assumption in conventional cognitive modeling. Contemporary models of quantity comparison are invariably "functionally symmetrical" in that both of the quantities-to-be-compared are treated in the same way; the current result exposes that assumption to unfavourable scrutiny. As mentioned in section 2.3, the evolved population reliably contains "non-discriminating" foragers. These agents display strongly asymmetrical behaviour, selecting food group 2 (initially on the agent's right) regardless of the food values. This strategy emerges rather earlier in evolutionary runs than does the more "discriminating" variant; the implication is that quantity comparison processes emerge within a behaviourally asymmetrical context.

Given the context of behavioural asymmetry, it seems natural to predict that an agent's representational strategy will also be asymmetrical – though preliminary, our results do support this prediction because there is no equivalent of unit '5' for food group 2. In other words, though we cannot say with confidence how the agent "represents", we can predict that its distributed representations will not be symmetrical.

References

1. Feyerabend, P.: Materialism and the mind-body problem. Review of Mataphysics, Vol. 17 (1963) 49-66
2. Feyerabend, P.: Mental events and the brain. Journal of Philosophy, Vol. 60 (1963) 295-296
3. Stich, S., Ravenscroft, I.: What is Folk Psychology? Cognition, Vol. 50 (1994) 447-68
4. Rorty, R.: Mind-Body Identity, Privacy, and Categories. Review of Metaphysics, Vol. 19 (1965) 24-54
5. Crowther-Heyk, H.: George Miller, language, and the computer metaphor. History of Psychology, Vol. 2 (1999) 37-64
6. Bechtel, W., Richardson, R. C. Discovering complexity: Decomposition and localization as strategies in scientific research. Chapter 2, Princeton: Princeton University Press (1993)
7. Stoianov, I., Zorzi, M., Umiltà, C.: The role of semantic and symbolic representations in arithmetic processing: Insights from simulated discalculia in a connectionist model. Cortex, Vol. 40 (2004) 194-196
8. Brooks, R. A.: Intelligence without representation. Artificial Intelligence, Vol. 47 (1991) 139-159
9. Beer, R. D.: Dynamical approaches to cognitive science. Trends in Cognitive Sciences, Vol. 4(3) (2000) 91-99
10. van Gelder, T. J.: Dynamic approaches to cognition. In Wilson, R., Keil, F. (editors), The MIT Encyclopedia of Cognitive Sciences. Cambridge MA: MIT Press (1999) 244-6
11. Harvey, I.: Untimed and misrepresented: Connectionism and the computer metaphor. Newsletter of the Society for the Study of Artificial Intelligence and Simulation of Behaviour (AISB Quarterly), Vol.96 (1996) 20-27
12. Cabeza, R., Nyberg, L.: Imaging cognition II: an empirical review of 275 PET and fMRI Studies. Journal of Cognitive Neuroscience, Vol. 12 (2000) 1-47
13. Zorzi, M., Stoianov, I., Umiltà, C.: Computational modeling of numerical cognition. In Campbell, J. editor, Handbook of mathematical cognition, chapter 5 London: Psychology Press (2005) 67-83
14. Dahaene, S., Cohen, L., Sigman, M., Vinckier, F.: The neural code for written words: a proposal. Trends in Cognitive Science, Vol. 9(7) (2005) 335-341
15. Naccache, L., Dahaene, S.: The priming method : imaging unconscious repetition priming reveals an abstract representation of number in the parietal lobes. Cerebral Cortex, Vol. 11 (2001) 966-974
16. Smolensky, P.: On Variable Binding and the Representation of Symbolic Structures in Connectionist Systems. Technical report CU-CS-355-87, Department of Computer Science, University of Colorado at Boulder, 1987
17. Beer, R. D.: Toward the evolution of dynamical neural networks for minimally cognitive behavior. In Maes, P., Mataric, M., Meyer, J., Pollack, J., Wilson, S., editors, Proceedings of the 4th International Conference on Simulation of Adaptive Behavior, MIT Press (1996) 421-429
18. Verguts, T., Fias, W.: Representation of number in animals and humans: a neural model. Journal of Cognitive Neuroscience, Vol. 16(9) (2004)
19. Gelman, R., Gallistel, C. R.: Language and the origin of numerical concepts. Science, Vol. 306(5695) (2004) 441-443
20. Butterworth, B.: What Counts, chapter 3, The Free Press: New York (1999) 144-147
21. Uller, C., Jaeger, R., Guidry, G., Martin, C.: Salamanders (Plethodon cinereus) go for more: rudiments of number in an amphibian. Animal Cognition, Vol. 6 (2003) 105-112

22. Watson, R.A., Pollack, J.B.: Coevolutionary Dynamics in a Minimal Substrate. Proceedings of the 2001 Genetic and Evolutionary Computation Conference, Spector, L., et al., editors, Morgan Kaufmann (2001)
23. Keinan, A., Sandbank, B, Hilgetag, C. C., Meilijson, I., Ruppin, E.: Fair attribution of functional contribution in artificial and biological networks. Neural Computation, Vol. 16(9) (2004) 1887-1915
24. Aharanov, R., Segev, L., Meilijson, I., Ruppin, E.: Localization of function via lesion analysis. Neural Computation, Vol. 14(4) (2003) 885-913
25. Nieder, A., Freedman, D. J., Miller, E. K.: Representation of the quantity of visual items in the primate prefrontal corext. Science 6 (2002) 1708-1711
26. Meck, W. H., Church, R. M.: A mode control model of counting and timing processes. Journal of Experimental Psychology: Animal Behaviour Processes 9 (1983) 320-324
27. Tomko G., Crapper, D.: Neuronal variability: non-stationary responses to identical visual stimuli. Brain Res, Vol. 79 (1974) 405-418

Modular Design of Irreducible Systems

Martin Hülse and Frank Pasemann

Fraunhofer Institute for Autonomous Intelligent Systems,
Schloss Birlinghoven,
53754 Sankt Augustin, Germany
{martin.huelse, frank.pasemann}@ais.fraunhofer.de
http://www.ais.fraunhofer.de

Abstract. Strategies of incremental evolution of artificial neural systems have been suggested over the last decade to overcome the scalability problem of evolutionary robotics. In this article two methods are introduced that support the evolution of neural couplings and extensions of recurrent neural networks of general type. These two methods are applied to combine and extend already evolved behavioral functionality of an autonomous robot in order to compare the structure-function relations of the resulting networks with those of the initial structures. The results of these investigations indicate that the emergent dynamics of the resulting networks turn these control structures into irreducible systems. We will argue that this leads to several consequences. One is, that the scalability problem of evolutionary robotics remains unsolved, no matter which type of incremental evolution is applied.

1 Motivation

In evolutionary robotics (ER) artificial neural networks (ANN) are frequently used as medium for the development of behavior control for autonomous systems by evolutionary algorithms (EA). As control structures, ANNs are applied with or without learning processes on feedforward as well as on recurrent connectivity structures [1,2,3].

The usage of ANNs in ER is mainly justified by their evolabilty and their ability to implement robust and generalized sensor-motor mappings. But the usage of ANNs is also suggested to support a modular organization and its incremental development of multi-functional behavior control [4,5].

One objective of this paper is to draw attention to a crucial aspect inherent in the mechanisms of incrementally evolved recurrent neural networks (RNN), which create multi-functional robot behavior. In the following we introduce experiments that are strictly focused on the combination and extension of structural and functional separated neural units. We call them basic modules. In order to combine or extend the behavior functionality of such basic modules we have evolved neural expansions or couplings between them. We call this approach a modular design.

Our experiments show that the behavioral functionality of the resulting networks is based on new dynamical properties. We call these new dynamics *emergent dynamics*, in order to emphasize that these dynamical properties can not

S. Nolfi et al. (Eds.): SAB 2006, LNAI 4095, pp. 534–545, 2006.

be generated by the single basic modules. Exactly these emergent dynamics turn a network into an irreducible system, although it was developed by a modular design. We call this principle *the modular design of irreducible systems.*

We believe that this principle is relevant for each nontrivial behavior control, developed in a self-organized process. Certainly, we are well aware that this principle contradicts approaches to complex systems research, that are based on the concept of "nearly decomposable" systems [6,7]. Although our incrementally evolved neural control structures are based on basic modules that are separated according to structure and function, they are far away from being "nearly decomposable." Therefore we see our results as a suggestion to think about concepts of modularity that go beyond structural organization of neural systems.

But before we come to discuss this issue we will introduce the properties of RNNs serving as control structures for autonomous robots, that act in the sensorimotor loop. We find the modular neurodynamics approach [8] the most fruitful to describe this issue within the theory of dynamical systems. This paper thus is organized as follows.

In the next section we introduce a description of RNNs as *parameterized sensor-driven dynamical systems.* It follows a section where a method (called schemas) is introduced that provide a formal description of multi-functional robot behavior according to the underlying neural dynamics. The fourth section gives a brief introduction of our evolutionary algorithm, called ENS^3 (evolution of neural systems by stochastic synthesis) [9]. We use this algorithm to develop the recurrent neural connectivity structures of RNNs. Further on, our techniques for the incremental evolution, called fusion and expansion, are based on this algorithm. Both techniques will be described in this section too. Section 5 goes on to examine experimental examples of incrementally (i.e., by expansion and fusion) evolved behavior control including the analysis of their underlying dynamical properties. The last chapter 6 concludes this paper with a discussion specifically focusing on the essential mechanisms of neural control structures underlying a concrete multi-functional behavior. This leads to conclusions relevant for each type of incremental evolution of nontrivial and multi-functional behavior control.

2   Neuromodules Acting in the Sensorimotor Loop

In the following we describe a recurrent neural network as a time discrete system. Further on, we apply the standard additive neuron with sigmoidal transfer function as neuron model. Now an arbitrary RNN can be defined by ordinary differential equations of the following type:

$$a_i(t+1) = \Theta_i + \sum_{j=1}^{n} w_{ij} \cdot \sigma(a_j(t)) \, , \; i = 1, \ldots, n \, . \tag{1}$$

The variable $a_i(t)$ describes the activation of neuron i at time step t, w_{ij} the strengths of the incoming synapse from neuron j, and Θ_i its bias term. As transfer function the standard sigmoid $\sigma(x) = \frac{1}{1+e^{-x}}$ is used.

In such a way we get a formal description of RNNs as nonlinear dynamical system with discrete time [10]. It is known from analytical investigations that already small RNNs (i.e., two neurons recurrently coupled) of this type can have asymptotically stable fixed points, periodic, quasi-periodic and chaotic attractors, even coexisting [11].

Taking a RNN with fixed bias and weight parameters then an initial state (i.e. neuron activations) is transformed by iterations into an asymptotic final state, called attractor. The choice of a different initialization might result in a different final state (as it is utilized in Hopfield-networks [12]). But if RNNs should be applied as control structures for autonomous robots, obviously only RNNs are of particular interest that get some inputs, derived from sensor data, and deliver signals for certain actuators.

A RNN that is extended in this way consists of three types of neurons: input, hidden and output neurons. Input neurons serve only as buffers for specific sensor signals. Therefore activation and output of an input neuron means the same. Neurons are called output neurons, if their output values are used as control signals for specific actuators. All the other neurons are called hidden neurons.

Notice, by simply adding one or more input neurons a RNN undergoes a qualitative change that can hardly be overemphasized.

We still assume constant bias and weight parameters. As formal description of the extended RNN we have now:

$$a_i(t+1) = \Theta_i + \left(\sum_{j=1}^{n} w_{ij} \cdot \sigma(a_j(t))\right) + \left(\sum_{j=1}^{m} w_{ij} \cdot I_j\right), \ i = 1, \ldots, n.$$

According to equation 1 we have only one new sum, which describes the influence of the input signals to the activation of neuron a_i. With: $\overline{\Theta}_i := \Theta_i + \left(\sum_{j=1}^{m} w_{ij} \cdot I_j\right)$, $i = 1, \ldots, n$ we get:

$$a_i(t+1) = \overline{\Theta}_i + \sum_{j=1}^{n} w_{ij} \cdot \sigma(a_j(t)), \qquad i = 1, \ldots, n. \tag{2}$$

In comparison with equation 1 we see that in general each hidden and output neuron of such an extended RNN has now a bias term $\overline{\Theta}_i$ that changes as the senor values undergo certain variations. But if the bias is changed, then the whole character of the RNN is altered. Because like the weights w_{ij}, the bias terms are parameters that define the transition rules of a RNN. Hence, the addition of input neurons turns a RNN into a parameterized dynamical systems. Its parameters are the input neurons, i.e. the sensors with their co-domains.

The change of the transition rules can lead to qualitative different attractors. With respect to our RNNs as time discrete systems we can distinguish four qualities of attractors: asymptotically stable fixed points, periodic, quasi-periodic and chaotic attractors [10]. Parameter values which cause a qualitative change of the attractor are called bifurcation points.

The quality of the system's attractor is very important, because it determines the system's dynamics. With other words, if we know the attractors of the pa-

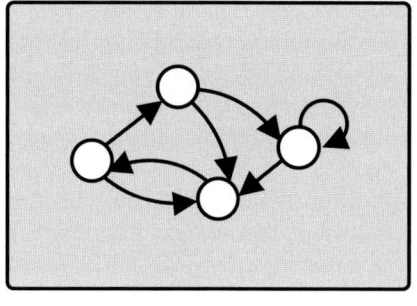 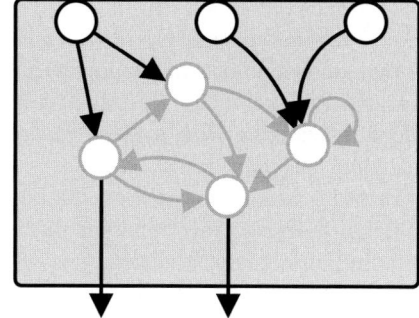

Fig. 1. A recurrent neural network or a concrete dynamical system (left) and an extension of this RNN to a neuromodule representing a whole class of dynamical systems. The extension to a NM is simply done by including three input neurons and the definition of output neurons (right).

rameter configurations and the bifurcation points, we understand much more of the signal processing that a RNN with continously given (sensor) inputs is performing.

The concept of *neuromodule* (NM) [8] was introduced to emphasize this qualitative difference between one RNN as a concrete dynamical system and a RNN with continuous inputs as parameterized dynamical system. This formalism is quite general and for the objectives of this paper it is not advisable to introduce it in its most general form. Here it is sufficient to utilize a more specific formalism that *defines a neuromodule as a RNN with fixed weight and bias terms and a specific input- and output-layer. The input neurons are the parameters of the neuromodule.*

As a NM is applied as control structure for an autonomous systems it acts in the sensorimotor loop. This basically means that the parameters of the NM underly permanent changes, due to the noise and significant changes of the sensor values, that represents behavior relevant changes of the external world. Therefore we call a neuromodule acting as control structure for autonomous robots a *sensor-driven parameterized dynamical system.*

Noise and significant changes of the sensor values are connected to two issues usually denoted as robustness and action selection. In our framework of NM robustness means that a NM has to generate the same dynamics although its parameters underly permanent disturbances. The action selection process of a NM has to be understood as an appropriately change of the system dynamics due to significant alterations of the parameter / sensor input resulting from the robot-environment interaction.

3   Schemas

The usage of a NM as control structure leads to the fact that not all parameter / sensor value configurations can be expected to emerge during the interaction of the robot within its environment. Obviously there are sub-spaces in the parameter space of a NM which are not relevant for the control of the robot.

Investigations on the structure-function relation of NM shall only take into account the behavior relevant parameter configurations. The dynamics that a NM is generating within these parameter domains is what we call the *behavior relevant dynamics* of a NM.

In order to stay focused on the behavior relevant dynamics we chose from an external point of view behavior patterns of the robot that we are interested in. But for further investigations we define these patterns according to the sensor value / parameter configurations of the NMs and their temporal regimes as they are typically appearing during the execution of this behavior pattern.

These configurations become the starting point for our analysis of the structure-function relation. We call such behavior patterns *schemas* in order to emphasize that they *are formally defined as parameter configurations and the temporal regime with respect to a given neuro-module.*

Notice, schemas ensure the grounding of our structure-function relation. Since this approach inhibits the selection of arbitrary parameter domains. Further on schemas avoid descriptions of behavior patterns from the point of view of an external observer [5].

The structure-function relation for a single schema can be described as (1) the clarification of the attractors of the NM in the parameter domain that is given by a specific schema and (2) the explanation how these attractors are provided by neural sub-structures of the NM. The structure-function relation of the overall robot behavior can be clarified with respect to the relations between a set of single schemas. Of primary interest are also the changes of attractors during the transitions between the schemas.

But before we apply this approach to an analysis of the structure-function relation we will introduce our methods of incremental evolution.

4   Fusion and Expansion

The following methods of incremental evolution of recurrent neural networks are based on the functionality of the structure evolution algorithm ENS^3 (*evolution of neural systems by stochastic synthesis*) [9]. This algorithm belongs to the class of evolutionary strategies [13] and was originally developed to evolve structure and optimize parameters of RNNs of general type simultaneously. It applies a direct coding of the structure and parameters of the evolved RNNs and is thus especially suited for the optimization of the resulting problem solving behavior of the RNNs.

Given problem-specific defined input-output-layers and a transfer function nothing else is determined. Hence, any kind of recurrences can be expected to emerge, i.e. self-connections and loops established by inhibitory and excitatory synapsis.

The variation and selection operators of the ENS^3-algorithm are stochastic. During the variation neurons and / or synapses are inserted or removed with a user-given probability. Likewise the real-valued parameters (e.g. weight and bias terms) of the individual networks are modified due to user-defined probabilities of change and intensities.

Due to the direct coding of neural network structure and its parameters, there is no need to start the ENS^3-algorithm with an empty network. Any arbitrary network can be used as initial network– as long as the input-output-structure and transfer function are compatible.

This gives us the opportunity to evolve arbitrary extensions of already given neural networks. Basically we can distinguish two types of such extensions. On the one hand, one existing RNN can be expanded to enrich its behavioral capabilities. On the other hand, two or more already existing networks can be coupled in order to get an effective coordination of the separate functionalities solving a global task [14]. The latter methods is called fusion and the first expansion method.

In the next section we introduce four RNNs resulting from these methods. As behavior control we chose a simple light seeking task as demonstration of method, because this task needs a combination of a negative (obstacle avoidance) and positive (photo) tropism.

The basic modules for these experiments are manually designed, inspired by our investigations of structure-function relations of former evolution tasks [14].

We run the expansion and fusion experiments in two different modes, called restrictive and semi-restrictive. In the restrictive mode the initial structures are not varied. Neither their structural elements (synapses and hidden neurons) nor their parameters (bias and weight terms) are varied. Hence, new behavioral functionality can only be achieved through new structural elements (hidden neurons and / or synapses). In the semi-restrictive mode the parameter values of the initial structures can be varied. Their structural elements stay untouched.

5   Structure-Function Relation of Incrementally Evolved Behavior Control

All of the following experiments were performed with the Khepera robot [15]. This wheel driven miniature robot has two DC-motors (control signals m_l, m_r). The motors are able to move the left and right wheel forward (positive signals) and backward (negative signals). The sensor data of the Khepera are delivered by eight infrared sensors measuring lights intensity and distances.

Based on these eight sensors we derive a minimal input structure for our NMs. For distance information we have only two input neurons I_1 and I_2. These two neurons deliver the distances values to obstacles on the robot's left (I_1) and right side (I_2). The light intensities are given by four additional input neurons $I_{3,...,6}$. The values of the sensors I_3 and I_5 indicate the light intensity detected on the left and the right side of the robot, I_4 the light intensity at the front, and I_6 at the rear.

All sensor values are mapped onto the interval $[0; 1.0]$. For the light sensors, a value 0.0 refers to darkness and 1.0 to the maximal measurable light intensity.

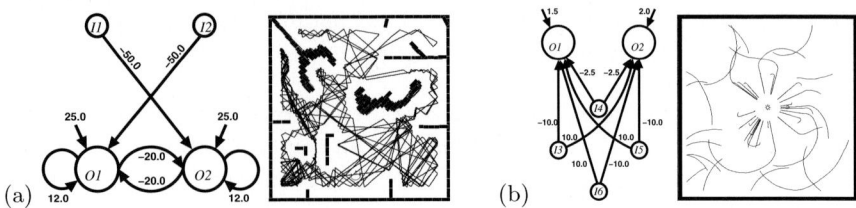

Fig. 2. The two basic modules $\mathcal{D}_\mathcal{O}$ (a) and $\mathcal{D}_\mathcal{L}$ (b) and the behavior in simulations. Notice, the right picture indicating the photo-tropism behavior shows several paths, including the cases when the robot was colliding with bounding walls.

While distance values are zero if no obstacle is detected. The value 1.0 represents a collision.

With respect to our transfer function $\sigma(x)$ (Eq. 1) all output values are in the open interval $(0; 1)$. As two output neurons O_1 and O_2 shall directly drive the motors, we apply the mapping: $m_{l,r} := 10 \cdot o_{1,2} - 5$, to derive positive and negative motor control signals. In such a way we derive integer values within the interval $[-5; 5]$ controlling the Khepera robot in a 2-dimensional simulation [16] and as real physical robot platform.

In the following upper symbols are used to refer to the neurons (e.g. I_n for input and O_n for output neurons) while the corresponding lower symbol refers to the neuron's output value.

5.1 Basic Modules

In order to compare the structure-function relation of the incrementally evolved neural control with respect to their underlying initial structure(s) we first briefly introduce the basic modules $\mathcal{D}_\mathcal{O}$ and $\mathcal{D}_\mathcal{L}$.

Module $\mathcal{D}_\mathcal{O}$ solves an obstacle avoidance task (Fig. 2(a)). Its dynamics is characterized by asymptotically stable fixed points. The only remarkably property is that in certain parameter domains two asymptotically stable fixed points are coexisting.

This coexistence becomes relevant during the transition between straight forward and turning movements as it is the case during the avoidance behavior of the robot. In such cases the coexisting fixed points generate hysteretic responses.

Also interesting the self-connections of the two output neurons and the recurrences between them create three different hysteresis elements. Their interplay leads to noise-filtering and to constant and context-sensitive turning angles [14].

The second basic module $\mathcal{D}_\mathcal{L}$ produces only a positive photo tropism. Its structure contains no recurrences. Hence, this module implements just a mapping generating a simple stimuli-response behavior.

5.2 Results Form the Expansion and Fusion Experiments

As demonstration of the expansion method we chose the obstacle avoidance module $\mathcal{D}_\mathcal{O}$ as initial structure and evolved a reactive light seeking behavior.

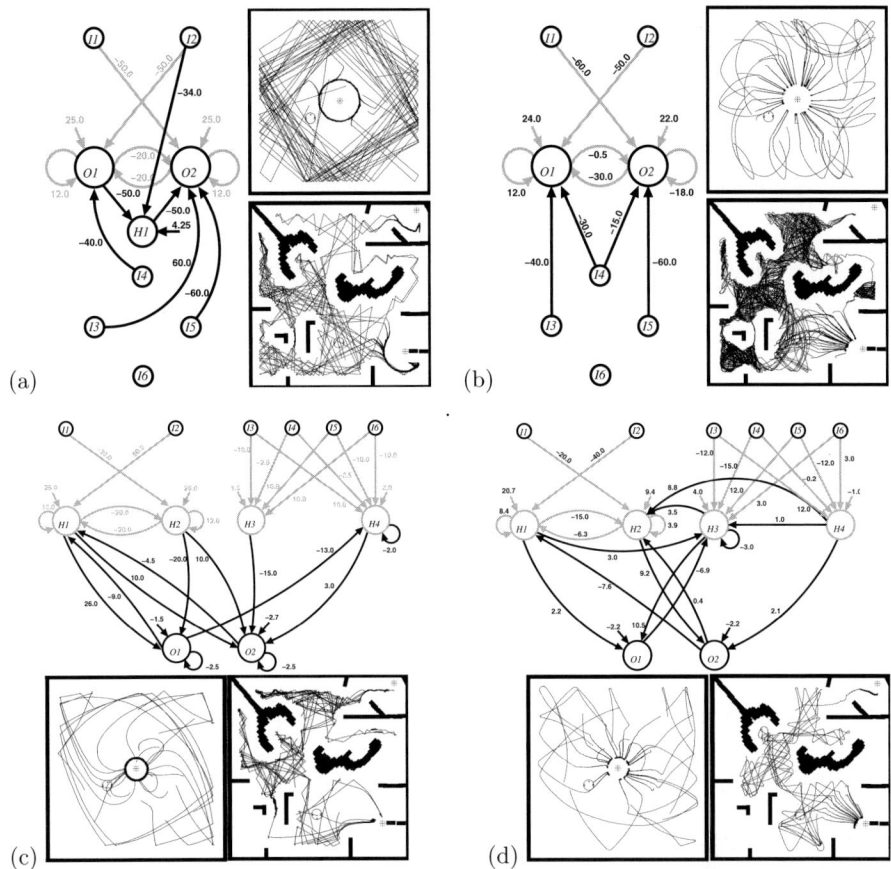

Fig. 3. Structure and their behavior in simulation of the incrementally evolved NMs $\mathcal{D}_{\mathcal{O} \Rightarrow \mathcal{L}}$ (a), $\mathcal{D}_{\mathcal{O} \rightarrow \mathcal{L}}$ (b), $\mathcal{D}_{\mathcal{O} \Leftrightarrow \mathcal{L}}$ (c) and $\mathcal{D}_{\mathcal{O} \leftrightarrow \mathcal{L}}$ (d). The grey colored elements of the NMs indicate those structure elements and their parameters which can not be modified by the variation operator during the evolutionary process.

Applying the fusion method to evolve the same behavior both basic modules were used to create one initial structure. In the resulting initial structure the former output neurons of the basic modules become hidden neurons. Further on, during the fusion process a development of synapses coming directly from the input neurons was inhibited. This forces the evolution of a neural coupling between the basic modules.

Examples of NMs resulting from the restrictive and semi-restrictive expansion and fusion experiments are shown in Figure 3. To refer to them the following symbols are used: $\mathcal{D}_{\mathcal{O} \Rightarrow \mathcal{L}}$ refers to the restrictive expanded module, the semi-restrictive expanded is represented by $\mathcal{D}_{\mathcal{O} \rightarrow \mathcal{L}}$. The modules resulting from the fusion experiments are symbolized with $\mathcal{D}_{\mathcal{O} \Leftrightarrow \mathcal{L}}$ for the restrictive and with $\mathcal{D}_{\mathcal{O} \leftrightarrow \mathcal{L}}$ for the semi-restrictive case.

As the pictures in Fig. 3 are indicating all the modules create the desired reactive light seeking behavior. But as we already mentioned, we are particularly interested in the attractors underlying the behavior relevant dynamics. In the next part we clarify this issue based on four schemas.

5.3   Dynamical Properties of Typical Schemas

The parameter configurations of the four schemas that have been investigated are the same for each NM. The schema *forward* is defined by minimal activations of all senor inputs / parameter values (including, like in all other cases too, the noise as it results from the sensors). This configuration corresponds to the straight forward movement of the robot, if no obstacle and light is detected.

A second schema *avoid* is defined by a high activation of either the left or the right distance sensor $I_{1,2}$ and minimal activations of all other input neurons. Such parameter configurations are present while the robot turns to the left or the right in order to avoid a collision with detected obstacles. (The case of dead ends, where both parameters I_1 and I_2 have high activation, needs an additional schema. This case is not considered in the following. See [14] for this issue.)

A third schema, called *approach*, is defined by parameter configurations which represent high activations of the left or right light sensor ($I_{3,5}$) while all other sensor values are minimal. This is related to an orientation of the robot to a light source.

The last schema that we have investigated is called *halt*. It is defined by a high activation of the frontal light sensor I_4, while all other sensor values are minimal.

Notice, although these qualitative definitions of the four schemas are the same for all modules, there are minor differences in the quantity. Consider the schema *halt*. Obviously, the activation of the frontal light sensor I_4 depends on the distance that a robot is standing in front of the light source. But this distance is a result of the interaction of the robot with the environment generated by the NM. Hence, the schemas quantitatively vary according to the concrete robot-environment interaction that a module is creating.

We simulated the NMs as dynamical systems in the domains that were given by these schemas. Simulation as dynamical systems basically means, that we decouple the NM from its sensorimotor-loop in order to compute the corresponding bifurcation diagrams and iso-periodic plots. The results of these simulations indicate the behavior relevant attractors.

Table 1 summarizes the results of our extensive simulations. Certainly, this summary can not explain how a specific attractor of a certain NM produces a behavior pattern. It is rather to point out the following general picture of our expansion and fusion experiments with respect to the behavior relevant dynamics.

The first thing that one can realize is that each NM utilizes qualitative different attractors. Furthermore, each module has at least one behavior relevant attractor that can not be generated by the basic module(s). On the other hand, the resulting behavior relevant attractors can undergo qualitative changes. And

Table 1. Attractors of the NMs for specific schemas, (FP = asymptotically stable fixed points, P-n = periodic attractor of period n, QP = quasi-periodic attractor). Notice, the mapping of $\mathcal{D}_{\mathcal{L}}$ is indicated with FP. Because its resulting sensor-motor mapping is "similar" to that of asymptotically stable fixed points.

module	schemas			
	forward	*avoid*	*approach*	*halt*
$\mathcal{D}_{\mathcal{O}}$	FP	FP	×	×
$\mathcal{D}_{\mathcal{L}}$	FP	×	FP	FP
$\mathcal{D}_{\mathcal{O}\Rightarrow\mathcal{L}}$	FP	FP	FP	P-6
$\mathcal{D}_{\mathcal{O}\rightarrow\mathcal{L}}$	FP & P-2	P-2	P-2	P-2
$\mathcal{D}_{\mathcal{O}\Leftrightarrow\mathcal{L}}$	FP	FP & P-n ($n > 20$)	FP	FP
$\mathcal{D}_{\mathcal{O}\leftrightarrow\mathcal{L}}$	FP	FP & QP	QP	FP

finally, qualitative different dynamics can be involved to generate one schema. This is mainly the case for module $\mathcal{D}_{\mathcal{O}\rightarrow\mathcal{L}}$ and its *forward* schema. The utilization of two qualities of attractors for one schema means, that the parameter values that define a schema are close to a bifurcation point. In module $\mathcal{D}_{\mathcal{O}\rightarrow\mathcal{L}}$ transitions between asymptotically stable fixed points and a period-2 attractor are already caused by the sensor noise. These noise induced phase transitions lead to slightly disturbed motor signals during the forward movement of the robot and might improve its exploration abilities. The indicated two attractors of schema *avoid* for $\mathcal{D}_{\mathcal{O}\Leftrightarrow\mathcal{L}}$ and $\mathcal{D}_{\mathcal{O}\leftrightarrow\mathcal{L}}$ refer to the fact that each one of them is responsible either for left turning or right turning.

Further investigations have given insights into which neural sub-structures cause the behavior relevant attractors. These results are in accordance with analytical investigations that are presented for instance in [11]. The period-6 attractor of $\mathcal{D}_{\mathcal{O}\Rightarrow\mathcal{L}}$ is an intrinsic property of its 3-ring, which is established by the two output neurons and hidden neuron H_1 (see Fig. 3 (a)). The period-2 attractor of the semi-restrictive expanded module $\mathcal{D}_{\mathcal{O}\rightarrow\mathcal{L}}$ results from the self-connection of output neuron O_2. This self-connection have become negative during the evolutionary process. It has been suggested in [17] that such single neurons with over-critical negative self-connections might be used as switchable oscillators. Here we see an application of such a switchable oscillator providing behavior control.

The quasi-periodic attractors of module $\mathcal{D}_{\mathcal{O}\leftrightarrow\mathcal{L}}$ is caused by the over-critical weights between neuron O_1 and H_3 (see Fig. 3 (d)). The dynamical properties of such *odd 2-loops* are well investigated [11]. They are called odd because the product of the two weights is negative. One can find two of such odd 2-loops also in module $\mathcal{D}_{\mathcal{O}\Leftrightarrow\mathcal{L}}$ (between H_1 and O_1, and between H_1 and O_2). They are the basis of the periodic attractors which becomes active during the avoidance behavior.

The behavior relevant asymptotically stable fixed points and the periodic and quasi-periodic attractors of the evolved networks are not intrinsic dynamical properties of their basic modules. They basically emerge from the recurrent neural couplings and extensions of the basic modules. Therefore we use the term *emergent dynamics* to refer to this essential property of the incrementally evolved NMs.

6   Multi-functionality as an Emergent Property

We have seen that multi-functional robot behavior can be implemented by small RNNs without learning processes. In contrast to others [3,5], we were able to show that the evolved multi-functionality is mainly provided by different qualities of dynamical properties, such as periodic and quasi-periodic attractors. Our investigations demonstrate that an understanding and a formal description of the nontrivial neural dynamical properties is well supported by the introduced concept of sensor-driven parameterized dynamical systems and the schema method, both grounded into the modular neurodynamics approach [8].

The presented structure-function relations of the evolved behavior control indicate, that nonlinearity and recurrences of our incrementally evolved couplings and extensions are the essential properties of the emerged multi-functionality. Therefore it follows, although we have used initial structures that were structural and functional separated, a relation between neural substructures and specific behavior patterns of the evolved modules does not exist. The incrementally evolved modules are irreducible.

Our results suggest, that an approach that is based on decomposition [6,7] in order to understand or to model neural behavior control can be misguiding and must fail at a certain point of complexity, if such an approach abstracts those properties that we have identified as the essentials of our neural systems, that are nonlinearity and recurrences.

Finally, incremental evolution is a basic mechanism for open-ended evolutionary processes to complex behavior control [18,19]. Hence, the introduced incrementally evolved multi-functionality based on nonlinear structural couplings or extensions of RNNs seems a promising approach. But as our results are indicating emergent dynamics has to be taken into account. There is still no theory that enables us to manage this type of emergence in general. As a consequence, it turns out that we can not know a priory if a given set of initial structures provide an incremental evolution of new behavior control. The scalability problem of ER remains unsolved in general.

References

1. Husbands, P. and Harvey, I. and Cliff, D. and Miller, G.: Artificial Evolution: A New Path for Artifical Intelligence? BRAIN AND COGNITION **34** (1997), 130 – 159.
2. Nolfi, S., and Floreano, D.: *Evolutionary Robotics: The Biology, Intelligence, and Technology of Self-Organizing Machines.* MIT Press, Cambridge, 2000.
3. Ziemke, T.: On 'Parts' and 'Wholes' of Adaptive Behavior: Functional Modularity Dichronic Structure in Recurrent Neural Robot Controllers. Proc. of the 6th Int. Conf. on Simulation of Adaptive Behavior (2000), 115–124.
4. Husbands, P. and Harvey, I. and Cliff, D.: Circle in the Round: State Space Attractors for Evolved Sighted Robots. Robotics and Autonomous Systems **20** (1995), 83–106.
5. Nolfi, S.: Using emergent modularity to develop control systems for mobile robots. Adaptive Behavior **5** (1997), 343–363.

6. Callebaut, W. and Rasskin-Gutman, D.: *Modularity*. Bradford Book, 2005.
7. Simon, H.: *The Science of the Artificial*. Cambridge University Press, 1969.
8. Pasemann, F.: Neuromodules: A dynamical systems approach to brain modellling. In: Herrmann, H. J. and Wolf, D. E. and Pöppel, E. *Supercomputing in brain research: From tomography to neural networks*, World Scientific, 1995.
9. Dieckmann, U.: Coevolution as an autonomous learning strategy for neuromodules. In: Herrmann, H. J. and Wolf, D. E. and Pöppel, E. *Supercomputing in brain research: From tomography to neural networks*, World Scientific, 1995.
10. Strogatz, S. H.: *Nonlinear Dynamics and Chaos*. Addison-Wesley, 1994.
11. Pasemann, F.: Complex dynamics and the structure of small neural networks. Network: Computation in Neural Systems **13** (2002), 195–216.
12. Hopfield, J. J. and Tank, D. W.: Computing with neural circuits: A model. Science **233** (1986), 625–633.
13. Bäck, T. and Schwefel, H.-P.: An overview on evolutionary algorithms for parameter optimization. Evolutionary Computation **1**, (1995), 1 – 23.
14. Hülse, M. and Wischmann, S. and Pasemann, F.: Structure and Function of Evolved Neuro-Controllers for Autonomous Robots. Connection Science **16** (2004), 249–266.
15. Mondada, F. and Franzi, E. and Ienne, P.: Mobile robots miniturization: a tool for investigation in control algorithms. Proc. of ISER' 93, Kyoto.
16. Michel, O.: Khepera Simulator, Package version 2.0. Freeware mobile robot simulator written at the University of Nice Sophia-Antipolis by Olivier Michel. Downloadable from the World Wide Web at `http://wwwi3s.unice.fr/~om/khep-sim.html`, 1995.
17. Pasemann, F.: Dynamics of a single model neuron. International Journal of Bifurcation and Chaos. **3** (1993), 271 – 278.
18. Beer, R. D.: An dynamical systems perspective on agent-environment interaction. Artificial Intelligence **72**, (1995), 173 – 215.
19. Bianco, R. and Nolfi, M.: Toward open-ended evolutionary robotics: evolving elementary robotic units able to self-assamble and self-reproduce. Connection Science **16** (2004), 227–248.

Spatially Constrained Networks and the Evolution of Modular Control Systems

Peter Fine, Ezequiel Di Paolo, and Andrew Philippides

Centre for Computational Neuroscience and Robotics (CCNR),
Department of Informatics, University of Sussex,
Brighton, BN1 9QG, UK
{p.a.fine, ezequiel, andrewop}@sussex.ac.uk

Abstract. This paper investigates the relationship between spatially embedded neural network models and modularity. It is hypothesised that spatial constraints lead to a greater chance of evolving modular structures. Firstly, this is tested in a minimally modular task/controller scenario. Spatial networks were shown to possess the ability to generate modular controllers which were not found in standard, non-spatial forms of network connectivity. We then apply this insight to examine the effect of varying degrees of spatial constraint on the modularity of a controller operating in a more complex, situated and embodied simulated environment. We conclude that a bias towards modularity is perhaps not always a desirable property for a control system paradigm to possess.

1 Introduction

Modular control systems can be used to enhance our ability to generate adaptive behaviour. Based on the knowledge that brains are often shown to contain highly distinct regions of processing activity, artificial mechanisms which promote modularity are often added to neural networks in order to build similarly decomposable structures. Yamauchi & Beer [18] use separate neural networks, each evolved to perform dedicated functions and later combined, to carry out a sequential learning task that could not otherwise be satisfied. Often, however, such an a priori decomposition of a task into modules is not available. In order to benefit from modularity in a more generic way, work such as Calabretta et al. [1] examines modular neurocontrollers which emerge from the evolutionary optimization of their networks, rather than from any hardwired structural constraint. They achieve this using a genetic operator which duplicates existing modules, thus allowing the topology to develop on its own towards the best performing modular architectures.

Often in evolutionary robotics, a neurocontroller is generated by fully connecting each of an agent's sensors to each of its interneurons, which are in turn fully linked to its motors. If recurrent connections are allowed, interneurons will be connected to every other interneuron. The functional structure of the network can then emerge by variations in the *strength* of those connections, some of which may drop to near 0 and impose an effectively topological constraint on the network's dynamics. Alternatively, it is possible to structure the connectivity of the network, perhaps using

S. Nolfi et al. (Eds.): SAB 2006, LNAI 4095, pp. 546–557, 2006.

a genetically mediated mechanism to specify which nodes are connected. Lindenmeyer systems, for example, can generate repeating, modular architectures using a compact genetic representation [10].

Another method of configuring neural network architectures involves embedding the neurons and connections within an abstract 'space', where they interact directly with only their nearby neighbouring components. It is apparent that the use of such a spatial constraint has important implications regarding the presence of modularity, without requiring any explicit modularity-generating mechanisms. For this reason, this paper concentrates on exploring modular control systems, within spatially embedded networks. We study the nature of the relationship between modularity and spatiality, and exploit the discovered inherent modularity found in spatially embedded systems to look at whether a bias towards modularity is necessarily a positive feature. Prior uses of spatially embedded networks are discussed in section 2. Since we will need to quantify the level of modularity related to a given form of spatial constraint, methods for measuring modularity are also discussed. This uncovers further distinctions regarding the nature and role of modularity in complex systems.

Section 3 then investigates the supposed relationship in a simple, necessarily modular task/controller scenario, using an evolutionary algorithm to compare the ability of spatial and non-spatial networks to generate modular solutions. Having found a correlation, we wished to examine its potential relevance to our understanding of a more realistic scenario. Section 4 uses data from a more complex, simulated robot task with no apparent necessity for a modular controller. The role of differing degrees of spatial embeddedness is investigated using these experiments, suggesting that in some cases, less modularity may be a positive attribute. Finally, Section 5 discusses the limitations of our approach whilst offering our conclusions.

2 Spatially Embedded Networks and Modularity

Cliff and Miller [2] embed the nodes in their neural networks within a 2-D planar surface. Neurons can then connect to other nodes which are nearby. They find that this provides a natural framework within which to map a control system onto the actuators and sensors of a simulated agent. This can also be seen as biologically inspired, since real neurons within the brain emit axons to connect to their neighbours (see [16] for a review). More recently, Philippides et al. [12] have studied more complex features of spatial neural networks by including a model of neurotransmitter diffusion across the plane, allowing neurons to interact both electrically and chemically with other local neurons.

This embedding feature has implications regarding the presence of modularity within network topologies. The notion of a neuron's local neighbourhood may encourage the division of network activity into distinct regions. One part of the spatial plane could implement a particular aspect of the control system, without interfering with (or being affected by) the other regions. Whilst a similar division of labour could potentially emerge from a fully connected network, it would be more difficult to generate, since without a spatial constraint a neuron is equally likely to interact with every other neuron in the network.

2.1 Measuring Modularity

The varied attempts to provide a metric for the quantification of modularity demonstrate different ways to conceive of its role in complex systems. One recent contribution by Newman & Girvan [11] searches for highly-interconnected communities within networks by first compiling them to abstract graphs of nodes and edges. This structural measure ignores the dynamics of the system under consideration, but nevertheless proved successful when applied to the modular architectures of a number of social interaction networks. Watson & Pollack [17] offer a measure which investigates the dynamical correlations between the node's behaviour. They claim that ignoring dynamics is liable to mislead, since some physical connections could be irrelevant to a network's overall behaviour. Conversely, even sparse connections between two topological clusters of nodes could play a dynamically highly significant role, undermining the distinctiveness of the supposed modules. Also, Polani et al. [14] present an information theoretic metric, which looks at the degree of 'mutual information' which is shared between subsystems.

Neurons often exhibit nonlinear interactions, and so in neural networks some connections will be likely ignored due to quiescent synapses. Others, however, may be exaggerated through phenomena such as bursting or hyper-excitability [8]. Measures which account for dynamical relationships will therefore no doubt be important in future studies of neural modularity. However we have reason to believe that a purely structural metric can still provide useful insights into the behaviour of the two cases studied in Sections 3 and 4 of this paper. In the first case, the network is made up of a simple spiking network which omits many of the more complex nonlinear features of real neurons. Combined with the fact that all weights were set to either 0 or 1, it is likely that neurons will influence each other in a quite homogenous manner across the network. A quantification of *structural* modularity may therefore tell us something about the modularity of the network's dynamics, as proved to be the case in Section 3.

The second case study involves networks comprising of two subsystems (gaseous and electrical, as described in Section 4). An analysis was undertaken to compare the structural relationships between these two systems with the level of overlap in their dynamical activity (presented in Philippides et al., [12]). They concluded that in this respect, the physical arrangement and dynamical behaviour was very similar despite some of the physical connections being inactive during the experimental trials. Whilst this is somewhat indirect and limited evidence towards a correspondence between the network's topology and its actual function, it suggests that if we restrict our analysis to structural organization of these neural networks, our conclusions may be informative to our understanding of a controller's behaviour. The validity of this approach forms an additional research question.

With this in mind, we used the structural measure of Newman & Girvan [11]. Since the spatial constraint operates by imposing restrictions on a network's physical connectivity, this measure provides the most direct way of studying how such constraints affect the development of the network. Additionally, it is clear that a network's physical topology will play a role in constraining its dynamics.

The measure proceeds by removing connections from a network, one by one. At each stage, the node with the highest degree of 'betweenness centrality' (a measure of the number of shortest paths between any two nodes which a given connection is a part of) is deleted. It is claimed that if a connection is frequently part of the shortest route between pairs of nodes, it is likely to be one of the sparse 'bridges' between different clusters of highly interconnected nodes. Eventually, as nodes are removed, the network breaks down into increasingly small and numerous separate graph components. Each discovered division into these component 'modules' is then examined and compared to see which provides the most appropriate division of nodes into clusters. The 'best' cluster division is determined in terms of the likelihood of a given node connecting to another member of its own module (rather than a different one). The proportion of intra-module connections versus inter-module links of the best rated division becomes a network's measured 'modularity rating', which should be used comparatively to determine whether a network is more or less modular than an alternative structure [6].

3   Minimally Modular Control Systems

As detailed above, we have reason to believe that spatially constrained networks offer the potential for clustered architectures, without requiring any mechanisms which 'create' or impose modularity. To investigate whether this is true in practice, a genetic algorithm was used to evolve controllers for a simple task. The emergence of modular topologies could thus be investigated without explicitly biasing the controller towards them, allowing the conditions under which they arise to be discussed.

For this, we required a minimal task/neurocontroller combination which necessitated a modular structure for successful completion, whilst being simple enough to analyse. In the course of previous investigations into spiking network dynamics, this task was found to require two distinct modules to be present within the controller. Note that the task only requires a modular controller when paired with the spiking neuron model presented; we do not claim that it represents a necessarily modular *task* by itself. The justification for the claim that success here requires modularity is presented after the description of the methods used.

3.1   Experimental Method

A simple simulated 'agent' was used to evolve a memory task, where it had to move to either the left or the right of its 'world', depending on its prior (since extinguished) sensory input. The agent was placed on a 1-dimensional line (200 units in length), where it could move to the left or right based on the output of two network nodes which were designated as outputs (the location x=0 corresponds to the far left, and x=200 the far right). Two different nodes were used as inputs, each receiving signals from one of two externally controlled sensors (labelled the left and the right sensor). For each evaluation of a genotype's fitness, two trials were run. In each, just one of the sensors was enabled (set to output a value of 1) for 2 'seconds', whilst the agent was held in its initial location at the middle of the line (x=100). The input was set to 0, and only then were the agent's motor neuron values used to move its location. High fitness was achieved if the agent moved to the left of the line when its left sensor had been previously enabled, and the right of the line in the right sensor trials.

$$F_l = \frac{\Delta t \sum_{t=0}^{T} \left(\frac{x_t}{\hat{X}} \right)}{T} \quad , \quad F_r = \frac{\Delta t \sum_{t=0}^{T} \left(\frac{\hat{X} - x_t}{\hat{X}} \right)}{T} \quad . \tag{1}$$

Equation (1) shows the fitness measures F_l and F_r for the left and right sensor trials respectively, which were averaged to give an agent's fitness. Δt corresponds to the timestep of integration, T is the length of the trial, x_t the agent's position at time t, and $\hat{X}$ the width of the line on which the agent resides.

A leaky integrate-and-fire neuron model was used (after [5], see [3] for use in evolutionary robotics setting). Sensor values were translated into input neuron spike trains using a Poisson process, with a probability of firing dependent on the sensor value plus noise. In addition to the 2 sensor and 2 motor neurons, 9 interneurons were included. The network activity is governed by:

$$\tau_m \frac{dV}{dt} = V_{rest} - V + g_{ex}(t)(E_{ex} - V) + g_{in}(t)(E_{in} - V) \quad . \tag{2}$$

$$g_{type}(t) \rightarrow g_{type}(t) + w_{ij} \quad . \tag{3}$$

$$\tau_{ex} \frac{dg_{ex}}{dt} = -g_{ex}, \quad \tau_{in} \frac{dg_{in}}{dt} = -g_{in} \quad . \tag{4}$$

$$\tau_{out} \frac{dM_x}{dt} = -M_x + M_G \left(\sum \delta(t_{now} - t_x) \right) \quad . \tag{5}$$

All parameters were evolved, unless otherwise specified. In Equation 2, τ_m is the membrane time constant [10ms, 40ms], V is the membrane potential, $V_{rest} = -70\text{mV}$ and the excitatory and inhibitory reversal potentials (E_{ex} and E_{in}) are 0mV and -80mV respectively. V 'spikes' when it reaches a normally distributed noisy threshold (V_{thresh}, mean [-60mV, -50mV] with deviation 1mV), and is followed by a random refractory period [2ms, 4ms] after which $V \rightarrow V_{rest}$. Equation 3 is applied when a neuron's inhibitory or exhibitory synapse receives a spike (type = in or ex respectively, and w_{ij} is the connection strength from the presynaptic neuron). Equation 4 describes the decay of the input conductances, with τ values [2ms, 4ms]. Finally, Equation 5 shows the motor output integrator (with x = left or right), capped at 1, where t_x is the time of the last spike in the corresponding motor neuron, M_G is the motor gain (scaled exponentially between [0.1, 50]) and $\tau_{out} = $ [40ms, 100ms].

The connectivity of the network was generated by embedding each node in a 2 dimensional plane using a genetically specified x and y coordinate. Each neuron maintained a number of connections n (evolved per neuron between 0 and 5), which were made to the n nearest neighbouring neurons (excluding self-connections, which were not permitted). To aid analysis, all connections were made with a weight of 1. A non-spatially embedded network was used as a control. Each neuron maintained an evolvable parameter in the range [0, 1] for every other neuron in the network. They were then connected to the n nodes for which this value was highest. Other control experiments were run using a more typical, fully connected architecture with evolvable weights, producing similar results to the first control network (results not shown). Note that the spatiality of the experimental condition was only exploited to generate the network architecture; once a trial had begun the network operated as in the control condition.

All controllers were optimized using a rank based genetic algorithm with elitism. The non-elite genotypes were mutated by adding a vector of random numbers of total length 2.5, taken from a Gaussian distribution around 0. Any mutations which took a gene beyond its 0 to 1 limits were discarded, and a different random mutation selected for that gene until the boundaries were satisfied.

The fact that this task requires a modular architecture stems from the way in which the neurons reset their internal state value after every spike with a fast timescale (order of milliseconds). In order to retain activity for longer (allowing the agent to move in relation to the extinguished sensory input at a timescale extending over seconds), more than one neuron would thus be required. In fact, an exhaustive search of each possible two and three neuron topology (with weights constrained to -1, 0 or 1) provided no architecture capable of showing reliable persistent activity after the input was removed. Only when four neuron circuits were considered was an adequate 'switch' behaviour found, which could be triggered by an input to one of the nodes, and maintained regular spiking activity through its recurrent dynamics after that input was removed.

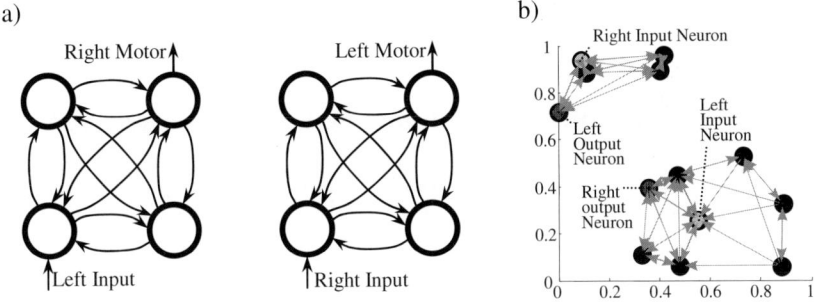

Fig. 1. a) Hand designed topology of spiking neurons which completes the task successfully. b) An example spatially embedded network, with a fitness of 0.825.

In order to complete the task, the agent would require two of these 'switches', one pairing the left sensor with the right motor (enabling the agent to move towards the left), and vice-versa. An example of this, using two four-node fully connected switches, is shown in Figure 1a. Whilst other mechanisms may exist, none were found throughout the extensive evolutionary simulations undertaken. The four neuron switches were somewhat robust to the removal of a limited number of connections. However the addition of a single connection linking any pair of nodes across different switches (thus breaking the perfect modularity) prevented the network from succeeding. Activity would 'leak' across that connection, triggering both switches (and therefore both motors) equally, preventing the agent from moving.

3.2 Results

A set of 24 evolutionary runs of each condition were undertaken, each with a population of 50 for 800 generations. Every run of the spatially constrained network reached a maximum fitness of 0.825, which was the same as the hand designed version (a score of 1 is not possible because the agent cannot move at the start of each trial). The control network, however, could only reach 0.67. These networks were only able to move to one side of the line when one particular sensor was enabled, and did not move at all in the other case. An analysis of the spiking dynamics of one of these control networks showed that one sensor and one motor neuron had been disconnected from the network, and all of the other neurons acted as a switch enabling movement in the direction mediated by the remaining connected motor neuron. Only when the spatial constraint was added were two distinct switches able to evolve, as shown in the example (Figure 1b), which was taken from generation 800 of one of the spatially embedded runs. This is corroborated by the results from Newman & Girvan's modularity measure, with the highest scoring spatial network from each run producing a mean score of 0.48, whilst the control network mean was 0.14 (significantly less modular, $p = 5.41 \times 10^{-15}$).

The above results demonstrate that even an apparently simple task, such as using a spiking network to 'remember' which direction an agent should move in, can necessitate a highly modular architecture. Crucially, this form of topology was not evolvable with a standard network generation technique. Only when embedded in space did the requisite modularity emerge, and it did so with apparent ease (in 100% of the runs carried out).

4 Controller Modularity in an Embodied, Active Vision Task

The previous results suggested a strong relationship between modularity and spatiality. However the use of an *a priori* modular task/controller, and the simplicity of the experiments leave open questions about the wider validity of this trend. The results of a complex scenario (which used a spatially embedded controller) were therefore analysed, looking at the modularity of the best performing controllers. Their use of a *partially* spatial controller (in addition to a fully embedded version) also allows for an interesting comparison to be made.

The experiments analysed in this section were carried out by Philippides et al. [12], and were an extension of the GasNet model (see reference for further details). GasNets are spatially embedded controllers, combining a standard neural network with a model of diffusing neurotransmitters. In addition to forming the topology of the network based on spatial factors (similarly to the spiking network experiments in Section 3), each node may act as a point source of gas. This diffuses across the plane, affecting nearby neurons by modulating their transfer function.

As in the minimal model presented in Section 3, each node in a GasNet maintains an evolvable x and y coordinate. Rather than connecting to its n nearest neighbours, it maintains two genetically specified 'cones' emanating from the neuron's position, for a limited distance across the plane. Positive connections are made to any neuron falling within the first of these, and negative connections to the second. The activity of these *electrical* connections are modelled as a discrete time, recurrent network which maintains a gain function k. Each neuron is also capable of producing one of two gases, either when its electrical output crosses a given threshold, or when gas in the vicinity reaches a certain level. Each of these parameters is genetically specified. The gases diffuse and disperse automatically. The summed concentration of the each gas is measured at each neuron's location. The level of the first gas is used to increase the gain of the transfer function applied to the neuron's electrical inputs, and the second gas decreases this value.

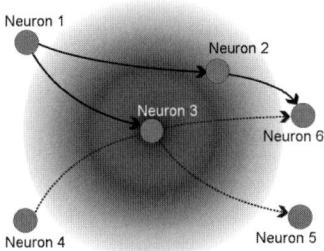

Fig. 2. An example GasNet electrical topology (not from experimental data), showing just the gas emitted by Neuron 3. The dashed lines denote negative connections.

The topologies were generated from the results of the experiment detailed in [12]. It involved a gantry robot starting from an arbitrary position and orientation in a black-walled arena. Equipped with a minimal vision system, taken from a forward-facing camera, it must navigate under extremely variable lighting conditions towards one shape (a white triangle), while ignoring the second shape (a white square).

Philippides et al. [12] also provide a modification of standard GasNets, taking inspiration from the way in which neurons emit Nitric Oxide (NO) in the mammalian cortex where it is generated by a network, or plexus, of fine fibres [13]. Rather than emanating from the neuron's location, the gas forms a uniform cloud, targeted over a potentially distant area, away from the source node. This was termed the *plexus* model, and used two extra evolvable parameters per neuron, determining the x and y coordinates of the centre of that neuron's gas cloud. This essentially means that whilst the electrical topology of the network is spatially constrained exactly as in the original GasNet, the gaseous connectivity (i.e. whether a pair of neurons influence each other through the production of gas) is not limited in this sense. Space still plays a role in this connectivity, because the gas from each neuron falls over a defined circular region of the plane. However this is a much more subtle constraint than in the original model, because that cloud can be located anywhere, instead of being restricted to being over nodes in the source neuron's locality.

The plexus GasNets were found to be significantly more evolvable for this task, both in terms of the chance of a given run producing a high fitness controller, and the number of generations required to do so. The authors characterize the reasons for the difference in terms of the lesser degree of coupling between the gaseous and electrical parts of the network's operation. They describe how during evolution changes in the gaseous structure of the network have a looser effect on the electrical systems (and vice versa). This is said to allow greater flexibility in 'tuning' each system against the other, with a corresponding greater level of evolvability.

Without wishing to question this factor, we propose a different, perhaps complementary hypothesis for the observed performance differences. In the plexus model, the spatiality of the electrical connections is preserved, but the gaseous links are no longer spatial (as described above). In light of the relationship between spatial constraints and modularity (see Section 3), it may be that the plexus networks are prone to a lower degree of modularity. With only a part of the structure restricted to neighbourhood-only connections, potentially parts of the overall network will be less modular and this may aid the development of a successful controller.

To determine whether this was the case, the best performing networks from 40 original GasNet runs and 40 plexus runs were compared. Firstly, the 7 original and 2 plexus runs which failed to reach the maximum fitness scores were removed, so as to only consider those networks which satisfied the task. Then, the excitatory and inhibitory electrical connectivity matrices were combined, so the resulting electrical matrix simply records whether a connection of either type was present or not (nodes connected by both positive *and* negative connections were assumed to be not connected, since the effect on the target neuron would be cancelled out). Also, the gaseous connectivity matrix was thresholded so that any non-zero level of influence between two nodes was regarded as a full connection, due to the Newman & Girvan metric's use of unweighted graphs. All connections were assumed to be undirected and self-connections were deleted, also due to further restrictions of the modularity measure.

4.1   Results

It is apparent from Figure 3 that the original GasNet produced overall topologies which were considerably more modular than the plexus model (confirmed by a T-test, $p = 6.24 \times 10^{-17}$). Visualisations of the combined topologies were used to ensure that

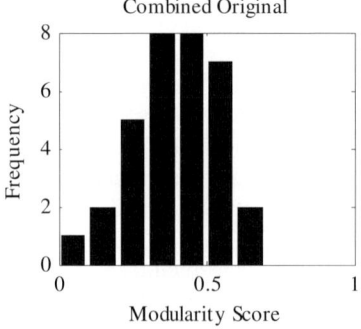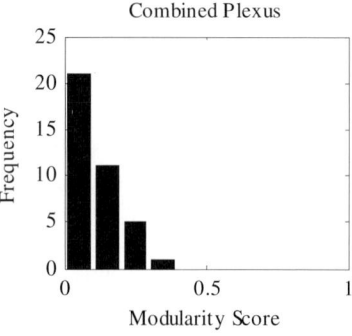

Fig. 3. Distribution of modularity scores for the combined electrical and gaseous topologies for the original and plexus models

the results of the modularity test did appeal to our own notions of modularity. For example, it was noticed that all of the plexus topologies consisted of one component (i.e., every node was reachable from every other node), whist 14 (out of 33) of the original GasNet runs produced a best performing controller with at least two network components. The examples shown in Figure 4 show a typical, best performing topology from each controller type.

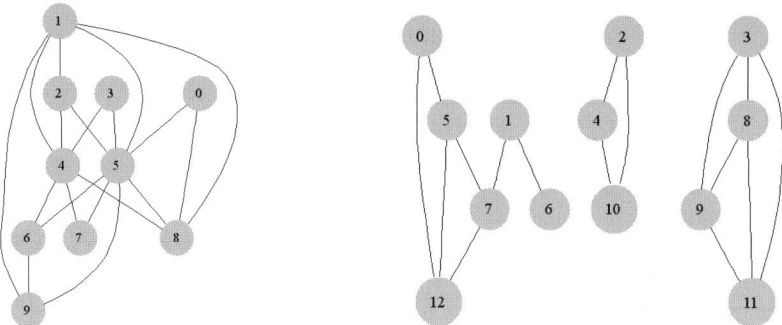

Fig. 4. Example combined topologies from the Plexus model (left), and the original GasNet (right). Generated by the GraphViz package [4].

Additionally, a statistically significant decrease in modularity for the plexus case was found when considering the gaseous interaction topologies alone, whereas the electrical connections were found to be statistically similar. This is to be expected, since the electrical systems were the same in both cases. The gaseous connectivity, however, was only spatially constrained in the original GasNet, which is likely to have lead to the increased level of gaseous modularity in that case. It would seem that networks which are *partially* spatially embedded have evolved topologies with a lower degree of modularity than fully embedded networks. These more interconnected networks were considerably more adept at evolving a suitable controller to complete the triangle/square detection task.

5 Discussion

Section 3 has demonstrated that spatially embedded neural networks are considerably more able to generate a modular solution to a minimally modular scenario than non-spatial neurocontrollers. Spatially constrained networks could thus provide a useful substrate within which to investigate the emergence of modularity, both in neural systems and potentially other forms of network. The form taken by the modules is, in some respects, free of assumptions imposed by the researcher. The use of pre-hardwired modules or specific, 'modularity-generating' mechanisms entails that the resultant modular properties of a network are liable to show a degree of bias in their configuration. Within a spatial network, however, the structure of any modules which ensue emerge from the same mechanisms which generate all aspects of the network's connectivity, and so could be used to study the emergence and form of modular architectures in general.

That said, the modules which form a spatially embedded network are manifestly structural constructions, rather than dynamical in nature. Whilst the physical topology of a network will undoubtedly influence its dynamical properties, structure does not tell the whole story. It is possible that a structurally modular connectivity could prove to be less flexible than a modularity which is determined by dynamics. One such example (Izhikevich, [9]) examines groups of neurons which are defined by stable, repeatable patterns of firing in large, randomly structured networks. Neurons can be a part of more than one dynamically determined 'module', and many more modules can occur than the number of neurons in a network. Structural modularity is nevertheless clearly an important feature to investigate, and biologically relevant, relating to cortical column work [7] amongst others. However future work should additionally investigate dynamical properties, possibly in relation to structural constraints. A measure of *dynamical* modularity would aid such a study.

Section 4 investigated the association between spatiality and modularity within a more complex scenario, including features such as a variable-length genotype and a requirement for embodied, situated perception and action in a simulated environment. In one experiment, this used a network with a similar level of constraint found in the spiking model. Another version used a looser kind of spatial constraint. Firstly, it was apparent that within this more realistic simulation, the fully embedded controllers produced highly modular networks. This was demonstrable using a measure of structural modularity, the results of which matched our intuitions gained from counting the number of components found in each network structure.

It turned out, however, that these highly modular solutions were considerably less evolvable than those of the other, partially embedded neurocontroller. Whilst modularity is clearly a desirable property in many cases, it seems that here too much inherent modularity could be a negative influence on the development and behaviour of the networks. The 'plexus' model may correspond to a compromise between spatially modular, and flexibly non-modular structural architectures. Of course, the reduction in modularity found in these networks could impact on the plexus model's capacity to evolve modular properties when highly modular controllers are required to satisfy a task. It would be interesting to determine whether they could, in fact, produce more modular topologies in a different experimental scenario which benefited from such an architecture.

This compromise could appeal to notions of near-decomposability (Simon, [15]), which discusses modularity in terms of subsystems which behave in distinct ways, but whose interdependencies are also of importance in describing a system's behaviour. Simon [15] describes these different levels of influence in terms of different timescales of interaction, which is not something that can be examined using the purely structural metric employed in this paper. More suitable tools would be required to make progress understanding in more general terms how different degrees of modularity arise from systems partially embedded in space.

Acknowledgements. This work has been supported by the Spatially Embedded Complex Systems Engineering (SECSE) project, EPSRC grant no EP/C51632X/1.

References

1. Calabretta R., Nolfi S., Parisi D. & Wagner G.P. (1998). Emergence of functional modularity in robots. From Animals to Animats V: Proceedings of the Fifth International Conference on Simulation of Adaptive Behavior. MIT Press, pp. 497-504.
2. Cliff, D., and Miller, G. F. (1995). Co-evolution of pursuit and evasion II: Simulation methods and results. From Animals to Animats IV: Proceedings of the Fourth International Conference on Simulation of Adaptive Behavior Cambridge, MA: MIT Press. pp. 506-515.
3. Di Paolo, E. (2003). Evolving spike-timing-dependent plasticity for single-trial learning in robots. Philosophical Transactions of the Royal Society of London, Series A: Mathematical, Physical and Engineering Sciences, 361(1811), pp. 2299-2319.
4. Gansner, E. R. & North, S. C. (1999). An Open Graph Visualization System and its Applications to Software Engineering. Software. Practice and Experience, 00(S1), 1–5.
5. Gerstner, W., Kreiter, A. K., Markram, H. & Herz, A. V. M. (1997). Neural codes: ring rates and beyond. Proc. Natl Acad. Sci. USA 94, pp. 12740-12741.
6. Guimera, R., Sales-Pardo, M. and Amaral, L. A. N. (2004). Modularity from Fluctuations in Random Graphs and Complex Networks. Phys. Rev. E 70, 025101.
7. Hubel, D. H., Wiesel, T. N. (1977). Functional architecture of macaque visual cortex. Proc. Roy. Soc. (London) 198B, pp. 1-59.
8. Izhikevich E. M. (2004). Which Model to Use for Cortical Spiking Neurons? IEEE Transactions on Neural Networks, 15. pp. 1063-1070.
9. Izhikevich, E. M. (2006). Polychronization: Computation With Spikes. Neural Computation 18, pp. 245-282.
10. Kitano, H. (1990), Designing neural networks using genetic algorithms with graph generation system. Complex Systems Vol. 4. pp. 461-476.
11. Newman, M. E. J. & Girvan, M. (2004). Finding and evaluating community structure in networks. Physical Review E 69, 026113.
12. Philippides, A., Husbands, P., Smith, T. and O'Shea, M. (2005). Flexible Couplings: Diffusing Neuromodulators and Adaptive Robotics. Artificial Life, 11(1&2):139-160.
13. Philippides, A., Ott, S., Husbands, P., Lovick, T. and O'Shea, M. (2005). Modeling co-operative volume signaling in a plexus of nitric oxide synthase-expressing neurons. Journal of Neuroscience 25(28). pp. 6520-6532.
14. Polani, D., Dauscher, P. and Uthmann, T. (2005). On a Quantitative Measure for Modularity Based on Information Theory. Advances in Artificial Life: 8th European Conference, ECAL 2005, Canterbury, UK. pp. 393. Springer Berlin / Heidelberg
15. Simon, H.A. (1969). The Sciences of the Artificial, Cambridge, MA. MIT Press.
16. Tessier-Lavigne, M. & Goodman, C. S. (1996). The molecular biology of axon guidance. Science. Nov 15;274(5290). pp. 1123-33.
17. Watson, R. A. and Pollack, J. B. (2005). Modular Interdependency in Complex Dynamical Systems. Artificial Life 11(4). pp. 445-457.
18. Yamauchi, B. and Beer, R.D. (1994). Integrating reactive, sequential and learning behavior using dynamical neural networks. From Animals to Animats 3: Proceedings of the Third International Conference on Simulation of Adaptive Behavior. pp. 382-391. MIT Press.

Evolving Spatiotemporal Coordination in a Modular Robotic System

Mikhail Prokopenko[1], Vadim Gerasimov[1], and Ivan Tanev[2]

[1] CSIRO Information and Communication Technology Centre,
Locked bag 17, North Ryde, NSW 1670, Australia
[2] Department of Information Systems Design, Doshisha University,
1-3 Miyakodani, Tatara, Kyotanabe, Kyoto 610-0321, Japan
mikhail.prokopenko@csiro.au

Abstract. In this paper we present a novel information-theoretic measure of spatiotemporal coordination in a modular robotic system, and use it as a fitness function in evolving the system. This approach exemplifies a new methodology formalizing co-evolution in multi-agent adaptive systems: information-driven evolutionary design. The methodology attempts to link together different aspects of information transfer involved in adaptive systems, and suggests to approximate direct task-specific fitness functions with intrinsic selection pressures. In particular, the information-theoretic measure of coordination employed in this work estimates the generalized correlation entropy K_2 and the generalized excess entropy E_2 computed over a multivariate time series of actuators' states. The simulated modular robotic system evolved according to the new measure exhibits regular locomotion and performs well in challenging terrains.

1 Introduction

Innovations in distributed sensor and actuator technologies, as well as advances in multi-agent control theory and studies of self-organization, support rapid growth in applications of complex adaptive multi-agent systems (MAS), such as modular robotics, multi-robot teams, self-assembly, etc. In particular, modular robots built of several similar building blocks (modules) become more and more attractive due to high versatility in their shapes, locomotion modes, tasks, and manipulation abilities [3,26,22,21,7]. This multi-faceted versatility increases robustness, adaptability, and scalability required in practical systems, ranging from search and rescue to space exploration. These requirements are achieved through a distribution of sensing, actuation and computational capabilities throughout the MAS such as a modular robotic system. This distribution forms a complex multi-agent network, enabling the desired responses to self-organize within the system, without central control. However, the main challenge with developing a self-organizing MAS is a design methodology for systematically inter-connecting a set of global system-level tasks, functions, etc. with localized sensors, behaviors, and actuators.

In this paper we further develop such a methodology originally sketched in [14], aiming at formalizing "taskless adaptation" of co-evolving multiple agents (robotic modules, network nodes, swarm elements, etc.). The co-evolution can be achieved in two

S. Nolfi et al. (Eds.): SAB 2006, LNAI 4095, pp. 558–569, 2006.

ways: via task-specific objectives or via generic intrinsic selection criteria. The generic information-theoretic criteria may vary in their emphasis: for example, we may focus on maximization of information transfer in perception-action loops [11,12]; minimization of heterogeneity in agent states, measured with the variance of the rule-space's entropy [25,17] or Boltzmann entropy in swarm-bots' states [1]; stability of multi-agent hierarchies [17]; efficiency of computation (computational complexity); efficiency of communication topologies [15,16]; efficiency of locomotion and distributed actuation [14,6,22,21], etc. The solutions obtained by information-driven evolution can be judged by their degree of approximation of direct evolutionary computation, where the latter uses task-specific objectives and depends on hand-crafting fitness functions by human designers. A good approximation will indicate that the chosen criteria capture the information-theoretic core of selection pressures. The main theme, however, is that different selection criteria incorporate information transfer within specific channels, and selecting some of these channels and not the others would guide information-driven evolutionary design.

Following [14] we apply here an information-theoretic measure of spatiotemporal coordination in a modular robotic system to an evolution of a sufficiently simple system: a modular limbless, wheelless snake-like robot (Snakebot) [22,21] without sensors. The only design goal of Snakebot's evolution, reported by Tanev and his colleagues, is fastest locomotion. Our immediate goal is information-theoretic approximation of this direct evolution. Specifically, we construct measures of spatiotemporal coordination of distributed actuators used by a Snakebot in locomotion. The measures are based on the generalized correlation entropy K_2 (a lower bound of Kolmogorov-Sinai entropy) and its excess entropy E_2 computed over a multivariate time series of actuators' states. The experiments reported by [14] confirmed that maximal coordination is achieved synchronously with fastest locomotion. In this paper we replace the direct measure with the information-theoretic measure of spatiotemporal coordination, and use the latter exclusively in evolving the Snakebot.

The following Section places this methodology in the context of previous studies, describes the proposed measures, and presents results, followed by conclusions.

2   Information Transfer as an Intrinsic Selection Pressure

An example of an intrinsic selection pressure is the acquisition of information from the environment: there is evidence that pushing the information flow to the information-theoretic limit (i.e., maximization of information transfer) can give rise to intricate behavior, induce a necessary structure in the system, and ultimately adaptively reshape the system [11,12]. The central hypothesis of Klyubin *et al.* is that there exists "a local and universal utility function which may help individuals survive and hence speed up evolution by making the fitness landscape smoother", while adapting to morphology and ecological niche. The proposed general utility function, *empowerment*, couples the agent's sensors and actuators via the environment. Empowerment is the perceived amount of influence or control the agent has over world, and can be seen as the agents potential to change the world. It can be measured via the amount of Shannon information that the agent can "inject into" its sensor through the environment, affecting future actions and future perceptions. Such a perception-action loop defines the agent's actuation channel, and, technically, empowerment is defined as the capacity of this actuation channel: the maximum mutual information for the channel over all possible distributions of the transmitted signal. "The more of the information can be made to appear in the sensor, the more control or influence the agent has over its sensor" — this is the main motivation for this local and universal utility function [12].

Heterogeneity in agent states is another generic pressure related to intrinsic coordination and self-organization. For example, it was measured with the variance of the rule-space's entropy [25] and applied to evolve the *spatiotemporal stability* of multi-cellular patterns in a sensor/communication network embedded within a self-monitoring impact sensing test-bed of an aerospace vehicle [9,17,24]. The study of spatiotemporal stability in evolving impact boundaries — continuously connected multi-cellular circuits, self-organizing in presence of cell failures and connectivity disruptions around damaged areas — employs both task-dependent graph-theoretic and generic information-theoretic measures in separating chaotic regimes from ordered dynamics. The task-dependent measure captured the impact boundary's connectivity in terms of the size of the average connected boundary fragment — an analogue of a largest connected sub-graph and its standard deviation over time. The intrinsic information-theoretic measure captured the diversity of transition rules invoked by the network cells during an impact boundary formation, using the Shannon entropy of the rules' frequency distribution:

$$H(X^t) = -\sum_{i=1}^{m} \frac{X_i^t}{n} \, log \frac{X_i^t}{n} \, ,$$

where n is the system size (the total number of cells), and X_i^t is the number of times the transition i was used at time t across the system. Both measures concurred in identifying complex dynamics, pointing to the same phase transition between chaos and order, for particular regions in a parameter-space. The entropy $H(X^t)$ can also be interpreted as the joint state transition entropy $H(S^t, S^{t+1})$, where S^t is the state of the cell at time t [17]. This opens a way to consider information transfer

$$I(S^t; S^{t+1}) = H(S^t) - H(S^t, S^{t+1}) \, ,$$

within the channel between a cell and itself at the next time-step.

An investigation of Baldassarre *et al.* [1], characterized coordinated motion in a swarm collective as a self-organized activity, and measured the increasing organization of the group on the basis of Boltzmann entropy. In particular, the emergent *common direction* of motion, with the chassis orientations of the robots spatially aligned, was observed to allow the group to achieve high coordination. Baldassarre *et al.* proposed a method to capture the spatial alignment via Boltzmann entropy by dividing the state space of the elements of the system into cells (e.g., cells of $45°$ each, corresponding to chassis orientations), measuring the number of elements in each cell for a given macrostate m, computing the number w_m of microstates that compose m, and calculating Boltzmann entropy of the macrostate as $E_m = k \, ln[w_m]$, where k is a scaling constant. This constant is set to the inverse of the maximum entropy which is equal to the entropy of the macrostate where all the elements are equally distributed over the cells. The results indicate that "independently of the size of the group, the disorganization of the group initially decreases with an increasing rate, then tends to decrease with a decreasing rate, and finally reaches a null value when all the robots have the same orientation" [1].

In this work, we advance from a purely spatial characterization (such as Boltzmann entropy of a macrostate distributing chassis orientations over the cells) to a spatiotemporal measure. The entropy measure proposed in our work is intended not only to capture spatial alignment of different modules, but also to account for temporal dependencies among them, such as travelling or standing waves in multi-segment chains observed by Ijspeert *et al.*. Importantly, we plan to focus on channels where information transfer contributes to a selection pressure.

We refer here to one more example of a selection pressure — efficiency of communication topologies — which can be interpreted as in terms of information transfer. One feasible average measure of a complex network's heterogeneity is given by the entropy of a network defined through the link distribution. The latter can be defined via the simple degree distribution — the probability P_k of having a node with k links. Similarly, one can capture the average uncertainty of the network as a whole, using the joint entropy based on the joint probability of connected pairs $P_{k,k'}$. Ultimately, the amount of correlation between nodes in the graph can be calculated via the mutual information measure, the information transfer [19], as

$$I(P; P') = H(P) - H(P|P') = \sum_{k=1}^{m} \sum_{k'=1}^{m} P_{k,k'} \log \frac{P_{k,k'}}{P_k P_{k'}}.$$

The reviewed examples highlight the possible role of information transfer in guiding selection of efficient perception-action loops, spatiotemporally stable multi-cellular patterns, and well-connected network topologies. We intend to demonstrate that spatiotemporal coordination in a modular robotic system can also be captured as information transfer, and apply such a measure to the system's evolution.

Before presenting our approach, we briefly review some studies of the relation between locomotion and rhythmic inter-modular coordination. Dorigo [7] describes an experiment in swarm robotics (SWARM-BOT) which also complements standard self-reconfigurability with task-dependent cooperation. Small autonomous mobile robots (s-bots) aggregate into specific shapes enabling the collective structure (a swarm-bot) to perform functions beyond capabilities of a single module. The swarm-bot forms as a result of self-organization "rather than via a global template and is expected to move as a whole and reconfigure along the way when needed" [7]. One basic ability of a swarm-bot, immediately relevant to our research, is *coordinated motion* emerging when the constituent independently-controlled modules coordinate their actions in choosing a common direction of motion. Our focus is on how much locomotion can be "patterned" in an aggregated structure. Regardless of an environment (aquatic, terrestrial or aerial), locomotion is achieved by applying forces generated by the rhythmic contraction of muscles attached to limbs, wings, fins, etc. Typically, a locomotory gait is efficient when all the involved muscles contract and extend with the same frequency in different phases. For example, Yim *et al.* [26] investigated a snake-like (serpentine) sinusoid gait, where forward motion is essentially achieved by propagating a waveform travelling down the length of the chain. Tanev and his colleagues [22,21] demonstrated emergence of side-winding locomotion with superior speed characteristics for the given morphology as well as adaptability to challenging terrain and partial damage.

3   Spatiotemporal Coordination of Actuators

Snakebot is simulated as a set of identical spherical morphological segments, linked together via universal joints. All joints feature identical angle limits, and each joint has two attached actuators. In the initial standstill position of Snakebot, the rotation axes of the actuators are oriented vertically (vertical actuator) and horizontally (horizontal actuator). These actuators perform rotation of the joint in the horizontal and vertical planes respectively. No anisotropic friction between the morphological segments and the surface is considered. Open Dynamics Engine (ODE) was chosen to provide a realistic simulation of the mechanics of Snakebot. Given this representation, the task of designing the fastest Snakebot locomotion can be rephrased as developing temporal patterns of desired turning angles of horizontal and vertical actuators for each joint, maximizing the overall speed. Previous experiments of evolvable locomotion gaits with fitness measured as either velocity in any direction or velocity in forward direction [22] indicated that side-winding locomotion — locomotion predominantly perpendicular to the long axis of Snakebot (Figures 1 and 2) — provides superior speed characteristics for the considered morphology. The actuators states (horizontal and vertical turning angles) are constrained by the interactions between segments and the terrain. The *actual turning angles* provide an underlying time series for our information-theoretic analysis: horizontal turning angles $\{x_t^i\}$ and vertical turning angles $\{y_t^i\}$ at time t, where i is the actuator index, S is the number of joints, $1 \leq i \leq S$, and T is the considered time interval, $1 \leq t \leq T$. Since we deal with actual rather than ideal turning angles, the underlying dynamics in the phase-space may include both periodic and chaotic orbits.

   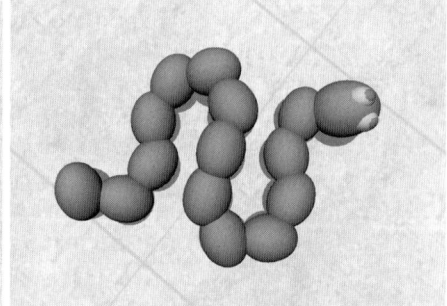

Fig. 1. Side view of the Snakebot   **Fig. 2.** Top view of the Snakebot

We intend to estimate "irregularity" for each of the multivariate time series $\{x_t^i\}$ and $\{y_t^i\}$. Each of these time series, henceforth denoted for generality $\{v_t^i\}$, contains both spatial and temporal patterns, and minimizing the irregularity over both space and time dimensions should ideally uncover the extent of spatiotemporal coordination among actuator states.

For any given actuator i, a simple characterisation of the "regularity" of the time series $\{v_t\}$ is provided by the auto-correlation function. However, the auto-correlation is limited to measuring only linear dependencies. We consider instead a more general approach. One classical measure is the Kolmogorov-Sinai (KS) entropy, also known as metric entropy [13]: it is a measure for the rate at which information about the state of the system is lost in the course of time. In other words, it is an entropy per unit time, an entropy rate or entropy density. Suppose that the $d-$dimensional phase space is partitioned into boxes of size r^d. Let $P_{i_0 \dots i_{d-1}}$ be the joint probability that a trajectory is in box i_0 at time 0, in box i_1 at time Δt, ..., and in box i_{d-1} at time $(d-1)\Delta t$, where Δt is the time interval between measurements on the state of the system (in our case, we may assume $\Delta t = 1$, and omit the limit $\Delta t \to 0$ in the following definitions). The KS entropy is defined by

$$K = - \lim_{r \to 0} \lim_{d \to \infty} \frac{1}{d \Delta t} \sum_{i_0 \dots i_{d-1}} P_{i_0 \dots i_{d-1}} \ln P_{i_0 \dots i_{d-1}} , \qquad (1)$$

and more precisely, as a supremum of K on all possible partitions. This definition has been generalized to the order-q Rényi entropies K_q [18]:

$$K_q = - \lim_{\Delta t \to 0} \lim_{r \to 0} \lim_{d \to \infty} \frac{1}{d \Delta t (q-1)} \ln \sum_{i_0 \dots i_{d-1}} P^q_{i_0 \dots i_{d-1}} . \qquad (2)$$

It is well-known that $K = 0$ in an ordered system, K is infinite in a random system, and K is a positive constant in a deterministic chaotic system. Grassberger and Procaccia [10] considered the correlation entropy K_2 in particular, and capitalized on the fact $K \geq K_2$ in establishing a sufficient condition for chaos $K_2 > 0$. Their algorithm estimates the entropy rate K_2 for a univariate time series. For our analysis we need to introduce a spatial dimension across multiple Snakebot's actuators. An estimate of the spatiotemporal entropy density can be obtained as

$$K = - \lim_{d_s \to \infty} \lim_{d_t \to \infty} \frac{1}{d_s} \frac{1}{d_t} \sum_{V(d_s, d_t)} p(V(d_s, d_t)) \ln p(V(d_s, d_t)) , \qquad (3)$$

where $V(d_s, d_t)$ are "patterns" of spatial size d_s and time length d_t [2]. Our objective, an estimate of spatiotemporal generalized correlation entropy, can be obtained as

$$K_2 = - \lim_{d_s \to \infty} \lim_{d_t \to \infty} \frac{1}{d_s} \frac{1}{d_t} \ln \sum_{V(d_s, d_t)} p^2(V(d_s, d_t)) . \qquad (4)$$

In achieving this objective, we follow Grassberger-Procaccia method [10] of computing correlation integrals, but use the multivariate time series with S actuators (joints) and T time steps in the following approximation:

$$K_2^{d_s d_t}(S, T, r) = \ln \frac{C_{d_s d_t}(S, T, r)}{C_{d_s (d_t+1)}(S, T, r)} + \ln \frac{C_{d_s d_t}(S, T, r)}{C_{(d_s+1) d_t}(S, T, r)} , \qquad (5)$$

where correlation integrals are generalized as

$$C_{d_s d_t}(S, T, r) = \frac{1}{(T-1)T(S-1)S} \sum_{l=1}^{T} \sum_{j=1}^{T} \sum_{g=1}^{S} \sum_{h=1}^{S} \Theta(r - \|V_l^g - V_j^h\|) . \qquad (6)$$

Here Θ is the Heaviside function (equal to 0 for negative argument and 1 otherwise), and the vectors V_l^g and V_j^h contain elements of the observed time series $\{v_t^i\}$ for each actuator (the spatial dimension), "converting" or "reconstructing" the dynamical information in two-dimensional data to information in the $d_s d_t$-dimensional embedding space [20]. More precisely, we use spatiotemporal delay vectors $V_k^i = (v_k^i, v_k^{i+1}, v_k^{i+2}, \ldots, v_k^{i+d_s-1})$, whose elements are time-delay vectors $v_k^i = (v_k^i, v_{k+1}^i, v_{k+2}^i, \ldots, v_{k+d_t-1}^i)$, and the spatial index i is fixed [14]. The norm $\|V_l^g - V_j^h\|$ is the distance between the vectors in the $d_s d_t$-dimensional space, e.g., the maximum norm:

$$\|V_l^g - V_j^h\| = \max_{\sigma=0}^{d_s-1} \max_{\tau=0}^{d_t-1} (v_{l+\tau}^{g+\sigma} - v_{j+\tau}^{h+\sigma})$$

Put simply, correlation integral $C_{d_s d_t}(S, T, r)$ computes the fraction of pairs of vectors in the $d_s d_t$-dimensional embedding space that are separated by a distance less than or equal to r. In order to eliminate auto-correlation effects, the vectors in equation (6) should be chosen to satisfy $|l - j| > L$, for an integer L, and $|g - h| > M$, for an integer M, in order to exclude auto-correlation effects among temporally close delays or closely coupled segments [23]. The standard temporal delay reconstruction [20] is recovered by setting $d_s = 1$ [4].

The correlation entropy K_2 (the generalized entropy rate) measures the irregularity or unpredictability of the system. A complementary quantity is the *excess entropy E* [8,5] — it may be viewed as a measure of the apparent memory or structure in the system. The generalized excess entropy E_2 is defined by considering how the finite-template (finite-delay and finite-extent) entropy rate estimates $K_2^{d_s d_t}(S, T, r)$ (equation (5)), converge to their asymptotic values K_2 (equation (4)). It is estimated for a fixed spatial extent D_s and a given time range D_t as:

$$E_2(D_s, D_t, S, T, r) = \sum_{d_s=1}^{D_s} \sum_{d_t=1}^{D_t} (K_2^{d_s d_t}(S, T, r) - K_2) . \tag{7}$$

For regular locomotion the asymptotic values should be zero (while non-zero entropies would indicate non-periodicity, i.e. deterministic chaos). It was shown that the excess entropy also measures the amount of historical information stored in the present that is communicated to the future [5,8]. In other words, it can be represented as asymptotic mutual information between two adjacent $d_s d_t$-dimensional half-planes

$$\lim_{d_s, d_t \to \infty} I(V_{-d_t}^g; V_0^h) =$$

$$\lim_{d_s, d_t \to \infty} I((v_{-d_t}^g, v_{-d_t}^{g+1}, v_{-d_t}^{g+2}, \ldots, v_{-d_t}^{g+d_s-1}); (v_0^h, v_0^{h+1}, v_0^{h+2}, \ldots, v_0^{h+d_s-1}))$$

where $v_k^i = (v_k^i, v_{k+1}^i, v_{k+2}^i, \ldots, v_{k+d_t-1}^i)$. This alternative representation establishes that the proposed measure may estimate information transfer within the space of actuators: the more information between the spatiotemporal past and the spatiotemporal future is transferred, the more coordination is achieved. If $g = h$ in the last expression, the transfer is purely between the temporal past and the temporal future. Otherwise, if $g \neq h$, we are concerned with how much information contained in the past of one group of actuators is injected into the future of another group of actuators.

When dealing with non-zero entropy rates K_2, one may consider *relative excess entropy*:

$$e_2(D_s, D_t, S, T, r) = \sum_{d_s=1}^{D_s} \sum_{d_t=1}^{D_t} \frac{K_2^{d_s d_t}(S, T, r) - K_2}{K_2 + \epsilon} . \qquad (8)$$

where ϵ is a small constant (e.g., $\epsilon = 0.03$), balancing the relative excess entropy e_2 for very small entropy rates K_2. The relative excess entropy e_2 attempts to "reward" the structure (coupling) in the locomotion and "penalise" its non-regularity.

4   Results

In this section we present experimental results of Snakebot's evolution based on estimates of the excess entropy E_2 (equation (7)) and the relative excess entropy e_2 (equation (8)). The Genetic Programming (GP) techniques employed in the evolution are described elsewhere [22,21]. In particular, the genotype is associated with two algebraic expressions, which represent the temporal patterns of desired turning angles of both the horizontal and vertical actuators of each morphological segment. Because locomotion gaits, by definition, are periodical, we include the periodic functions sin and cos in the function set of GP in addition to the basic algebraic functions. The selection is based on a binary tournament with selection ratio of 0.1 and reproduction ratio of 0.9. The mutation operator is the random subtree mutation with ratio of 0.01. Snakebots evolve within a population of 200 individuals, and the best performers are selected according to the excess entropy values, over a number of generations.

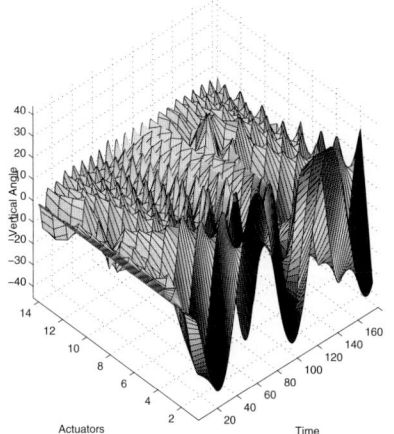   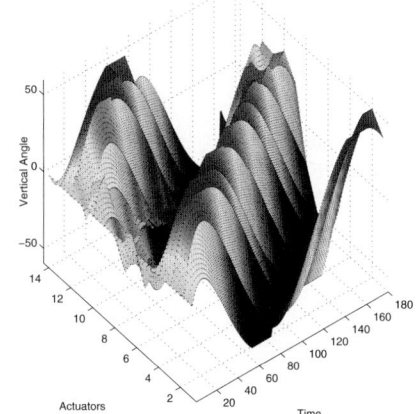

Fig. 3. First offspring: actuator angles   **Fig. 4.** Evolved solution: actuator angles

Figures 3 and 4 contrast (for vertical actuators) actual angles used by the first offspring and the final generation. Similarly, Figures 5 and 6 contrast the spatiotemporal correlation entropies produced by the first offspring and the evolved solution. It can be easily observed that more regular angle dynamics of the evolved solution manifests

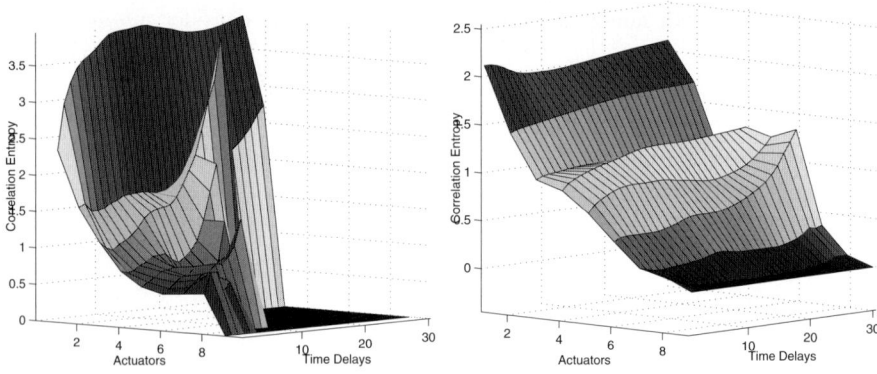

Fig. 5. First offspring: correlation entropy   **Fig. 6.** Evolved solution: correlation entropy

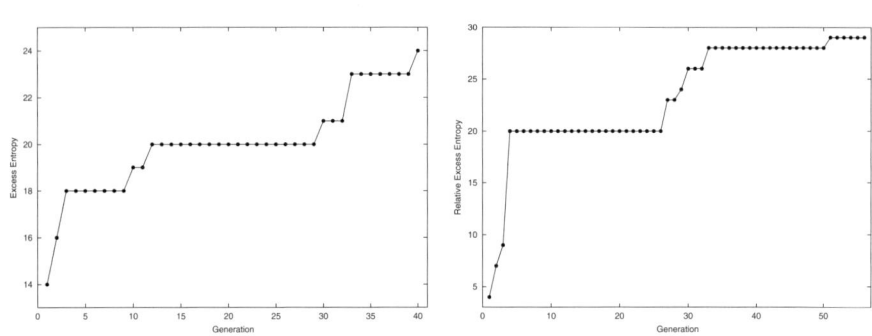

Fig. 7. Snakebot fitness over time: the best per-**Fig. 8.** Snakebot fitness over time: the best performer in each generation, using excess entropy former in each generation, using relative excess entropy

itself as more significant excess entropy. Figures 7 and 8 show typical fitness growth towards higher excess entropies estimated as E_2 (equation (7)) and the relative excess entropies e_2 (equation (8)), for two different experiments. It should be noted that there are well-coordinated Snakebots which are moving not as quickly as the Snakebots evolved according to the direct velocity-based measure, i.e. the set of fast solutions is contained within the set of well-coordinated solutions. This means that the obtained approximation of the direct fitness function by the information-theoretic selection pressure towards regularity is sound but not complete.

In certain circumstances, a fitness function rewarding coordination may be more suitable than a direct velocity-based measure: a Snakebot trapped by obstacles may need to employ a locomotion gait with highly coordinated actuators but near-zero absolute velocity. In fact, the obtained solutions exhibit reasonable robustness to challenging terrains, trading-off some velocity for resilience to obstacles. In particular, the evolved Snakebot shown in Figure 9 is able to traverse ragged terrains with obstacles three times as high as the segment diameter, move through a narrow corridor (only twice as wide as the segment diameter), and overcome various extended barriers. In addition, the

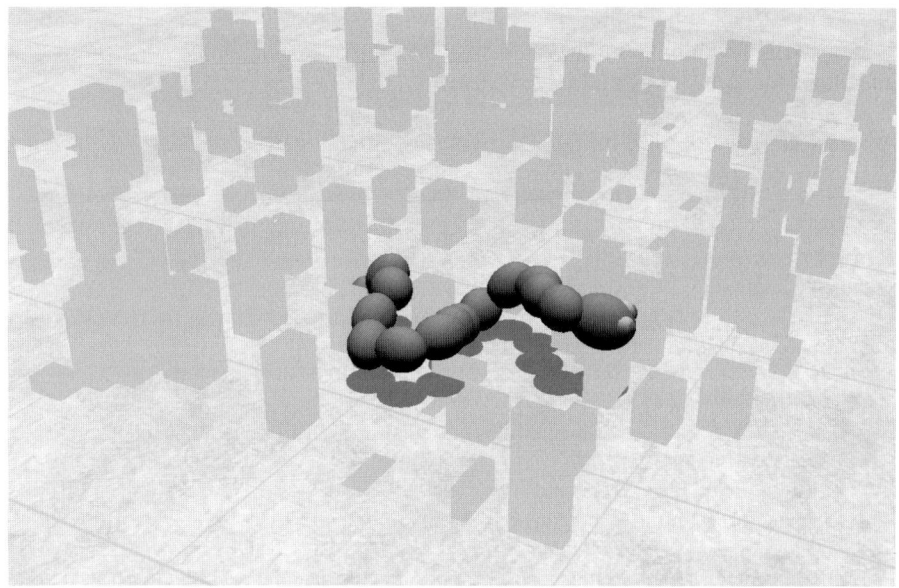

Fig. 9. Snakebot negotiating a terrain with obstacles

Snakebot is robust to failures of individual segments: e.g., it is able to move even when every third segment is completely incapacitated, albeit with only a half of the normal speed. Interestingly enough, the relative excess entropy is increased in partially damaged Snakebots, as the amount of transferred information in the coupled locomotion has to increase. Moreover, there appears to be a strong correlation between the number of damaged (evenly spread) segments s and the resulting relative excess entropy $e_2^s \approx \beta\, s$, where the coefficient β of the linear fit is approximately equal to the relative excess entropy of a non-damaged Snakebot e_2^0. This observation opens a way for Snakebot's self-diagnostics and adaptation: the run-time value of e_2 may identify the number of damaged segments, enabling a more appropriate response.

5   Conclusions

We modelled a specific step towards a theory of information-driven evolutionary design, using information-theoretic measures of spatiotemporal coordination in a modular robotic system (Snakebot). These measures estimate the generalized correlation entropies K_2 computed over a time series of actuators' states and the spatiotemporal excess entropies E_2. As expected, increased coordination of actuators is achieved by agents with faster locomotion. However, the set of fast solutions is a subset of the set of well-coordinated solutions. A more precise approximation of fast locomotion is a subject of future work. In parallel, we are investigating other tasks adaptation to which may require a high degree of actuators' coordination : e.g., rugged terrain traversal, energy-efficient locomotion, etc. Both directions essentially require identification of channels through which the information transfer among system's components is optimized. We believe that development of adequate information-theoretic criteria, such as the measure of spatiotemporal coordination of distributed actuators, will contribute to design guidelines for co-evolving multi-agent systems.

Acknowledgements. The third author was supported in part by the National Institute of Information and Communications Technology of Japan.

References

1. G. Baldassarre, D. Parisi, and S. Nolfi. Measuring coordination as entropy decrease in groups of linked simulated robots. Preprint, 2005.
2. G. Boffetta, M. Cencini, M. Falcioni, and A. Vulpiani. Predictability: a way to characterize complexity. *Physics Reports*, 356:367–474, 2002.
3. H. Bojinov, A. Casal, and T. Hogg. Multiagent control of self-reconfigurable robots. *Artificial Intelligence*, 142:99–120, 2002.
4. R. Carretero-González, S. Ørstavik, and J. Stark. Quasidiagonal approach to the estimation of lyapunov spectra for spatiotemporal systems from multivariate time series. *Phys Rev E Stat Phys Plasmas Fluids Relat Interdiscip Topics*, 62(5) Pt A:6429–6439, 2000.
5. J.P. Crutchfield and D.P. Feldman. Regularities unseen, randomness observed: The entropy convergence hierarchy. *Chaos*, 15:25–54, 2003.
6. R. Der, U. Steinmetz, and F. Pasemann. Homeokinesis - a new principle to back up evolution with learning. *Concurrent Systems Engineering Series*, 55: Computat. Intelligence for Modelling, Control, and Automation:43–47, 1999.
7. M. Dorigo. Swarm-bot: An experiment in swarm robotics. In P. Arabshahi and A. Martinoli, editors, *Proceedings of SIS 2005 – 2005 IEEE Swarm Intelligence Symposium*, pages 192–200. IEEE Press, 2005.
8. D.P. Feldman and J.P. Crutchfield. Structural information in two-dimensional patterns: Entropy convergence and excess entropy. *Physical Review E*, 67, 051104, 2003.
9. M. Foreman, M. Prokopenko, and P. Wang. Phase transitions in self-organising sensor networks. In W. Banzhaf, T. Christaller, P. Dittrich, J.T. KIm, and J. Ziegler, editors, *Advances in Artificial Life - Proceedings of the 7th European Conference on Artificial Life (ECAL)*, volume 2801 of *Lecture Notes in Artificial Intelligence*. Springer Verlag, 2003.
10. P. Grassberger and I. Procaccia. Estimation of the kolmogorov entropy from a chaotic signal. *Phys. Review A*, 28(4):2591, 1983.
11. A. S. Klyubin, D. Polani, and C. L. Nehaniv. Organization of the information flow in the perception-action loop of evolved agents. In *Proceedings of 2004 NASA/DoD Conference on Evolvable Hardware*, pages 177–180. IEEE Computer Society, 2004.
12. A. S. Klyubin, D. Polani, and C. L. Nehaniv. All else being equal be empowered. In M.S. Capcarrère, A.A. Freitas, P.J. Bentley, C.G. Johnson, and J. Timmis, editors, *Advances in Artificial Life, 8th European Conference, ECAL 2005, Canterbury, UK, September 5-9, 2005, Proceedings*, volume 3630 of *LNCS*, pages 744–753. Springer, 2005.
13. A.N. Kolmogorov. Entropy per unit time as a metric invariant of automorphisms. *Doklady Akademii Nauk SSSR*, 124:754–755, 1959.
14. M. Prokopenko, G. Gerasimov, and I. Tanev. Measuring spatiotemporal coordination in a modular robotic system. In L.M. Rocha, M. Bedau, D. Floreano, R. Goldstone, A. Vespignani, and L. Yaeger, editors, *Proceedings of Artificial Life X*. in press, 2006.
15. M. Prokopenko, P. Wang, M. Foreman, P. Valencia, D. Price, and G. Poulton. On connectivity of reconfigurable impact networks in ageless aerospace vehicles. *Robotics and Autonomous Systems*, 53:36–58, 2005.

16. M. Prokopenko, P. Wang, and D. Price. Complexity metrics for self-monitoring impact sensing networks. In J. Lohn, D. Gwaltney, G. Hornby, R. Zebulum, D. Keymeulen, and A. Stoica, editors, *Proceedings of 2005 NASA/DoD Conference on Evolvable Hardware (EH-05)*, pages 239–246. IEEE Computer Society, 2005.

17. M. Prokopenko, P. Wang, P. Valencia, D. Price, M. Foreman, and A. Farmer. Self-organizing hierarchies in sensor and communication networks. *Artificial Life*, 11:407–426, 2005.

18. A. Rényi. *Probability theory*. North-Holland, 1970.

19. R.V. Solé and S. Valverde. Information theory of complex networks: On evolution and architectural constraints. In E. Ben-Naim, H. Frauenfelder, and Z. Toroczkai, editors, *Complex Networks*, volume 650 of *Lecture Notes in Physics*, pages 189–210. Springer-Verlag, 2004.

20. F. Takens. Detecting strange attractors in turbulence. *Dynamical systems and turbulence*, 898:366, 1981.

21. I. Tanev. Learned mutation strategies in genetic programming for evolution and adaptation of simulated snakebot. *Genetic Evolutionary Computation - Gecco 2005, Proc.*, pages 687–694, 2005.

22. I. Tanev, T. Ray, and A. Buller. Automated evolutionary design, robustness, and adaptation of sidewinding locomotion of a simulated snake-like robot. *IEEE Transactions On Robotics*, 21:632–645, 2005.

23. J. Theiler. Spurious dimension from correlation algorithms applied to limited time-series data. *Physical Review A*, 34(3):2427–2432, 1986.

24. P. Wang and M. Prokopenko. Evolvable recovery membranes in self-monitoring aerospace vehicles. In S. Schaal, A. Ijspeert, A. Billard, S. Vijayakumar, J. Hallam, and J.-A. Meyer, editors, *From Animals to Animats VIII - Proceedings of the 8th International Conference on the Simulation of Adaptive Behaviour, Los Angeles*, pages 509–518. Cambridge, MA: A Bradford Book/MIT Press, 2004.

25. A. Wuensche. Classifying cellular automata automatically: Finding gliders, filtering, and relating space-time patterns, attractor basins, and the z parameter. *Complexity*, 4(3):47–66, 1999.

26. M. Yim, K. Roufas, D. Duff, Y. Zhang, C. Eldershaw, and S. Homans. Modular reconfigurable robots in space applications. *Autonomous Robots*, 14:225–237, 2003.

Spiking Neural Controllers
for Pushing Objects Around

Răzvan V. Florian[1,2]

[1] Center for Cognitive and Neural Studies (Coneural),
Str. Saturn nr. 24, 400504 Cluj-Napoca, Romania
florian@coneural.org
http://www.coneural.org/florian
[2] Babeş-Bolyai University, Institute for Interdisciplinary Experimental Research,
Str. T. Laurian nr. 42, 400271 Cluj-Napoca, Romania

Abstract. We evolve spiking neural networks that implement a seek-push-release drive for a simple simulated agent interacting with objects. The evolved agents display minimally-cognitive behavior, by switching as a function of context between the three sub-behaviors and by being able to discriminate relative object size. The neural controllers have either static synapses or synapses featuring spike-timing-dependent plasticity (STDP). Both types of networks are able to solve the task with similar efficacy, but networks with plastic synapses evolved faster. In the evolved networks, plasticity plays a minor role during the interaction with the environment and is used mostly to tune synapses when networks start to function.

1   Introduction

Genuine, creative artificial intelligence can emerge only in embodied agents, capable of cognitive development and learning by interacting with their environment [1]. Before the start of the learning process, the agents need to have some innate (predefined) drives or reflexes that can induce the exploration of the environment. Otherwise, the agents might not do anything once emerged in their environment, and learning would not be possible. In the experiments presented in this paper, we evolve a basic drive for a simple simulated agent that is able to interact with the objects in its environment. This drive could be used in future research to bootstrap the ontogenetic cognitive development of the agent.

The agent is controlled by a spiking neural network [2,3]. Among classes of neural network models amenable to large scale computer simulation, recurrent spiking neural networks are an attractive choice for implementing control systems for embodied artificial intelligent agents [4]. Spiking neural networks have more computational power per neuron than other types of neural networks [5,6,7]. Several studies [8,9,10,11] have shown that spiking neural networks achieve better performance for the control of embodied agents than continuous time recurrent neural networks or McCulloch-Pitts networks. More importantly, spiking neurons have a closer resemblance to real neurons than other neural models, which allows a bidirectional transfer of concepts and methodologies between neuroscience

S. Nolfi et al. (Eds.): SAB 2006, LNAI 4095, pp. 570–581, 2006.

and artificial neural systems. Biological examples may suggest architectures and learning mechanisms for artificial models that would improve their performance. In the reverse direction, theories developed during the study of embodied artificial neural networks may lead to new concepts and explanations regarding the activity of real neural networks [12].

Evolved spiking neural networks have been used in the last few years for the control of simulated or real robots, but more rarely than other types of neural networks [8,13,14,15,16,17,18,9,19,20,21,22,23,24,10,11,25,26,27]. Among previous evolutionary studies, only one explored the properties of a plastic spiking neural network [18,9,19]. Very few studies used spiking neural controllers for embodied agents that were not evolved, but were taught using other learning methods [28,29,30,31,32].

This paper presents experiments where we evolved spiking neural networks with static as well as with plastic synapses. These networks are one of the largest spiking neural networks evolved to date. The evolved controllers display interesting minimally-cognitive capabilities, being able to discriminate relative object size.

2   The Agent, Its Environment and Its Task

2.1   The Simulator

The agent and its environment were simulated using Thyrix, an open source simulator specifically designed for evolutionary and developmental experiments for embodied artificial intelligence research [33]. The simulator provides a two-dimensional environment with simplified, quasi-static (Aristotelian) mechanics, and supports collision detection and resolution between the objects in the environment.

2.2   The Agent's Morphology

The agent's morphology was chosen as the simplest one that allows the agent to push the circular objects in its environment without slipping of objects on the surface of the agent. Slipping may appear, for example, if a circle pushes another circle, and the pushing force is not positioned exactly on the line connecting the centers of the two circles.

We wanted maximum simplicity both for economy (in order to need less computing time for evolution) and for having few degrees of freedom, which allows a simpler analysis of the behavior of the agent. However, we have tried to respect the principle of ecological balance [34] in the design of the agent's morphology and sensorimotor capabilities.

Thus, the agent is composed of two circles, connected by a variable length link. The link is "virtual", in the sense that it provides a force that keeps the two circles together, but it does not interact with other objects in the environment, i.e. external objects can pass through it without contact. With this morphology, the agent can easily push other circles in its environment, by keeping them

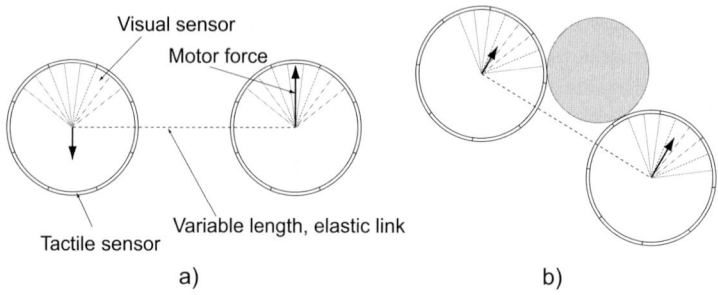

Fig. 1. a) The agent's morphology. b) The agent pushing a ball.

between its two body-circles, without the need of balancing them to prevent slipping. The agent was named Spherus and the code that defines it is available in the open source Thyrix simulator (http://www.thyrix.com).

2.3   The Agent's Effectors and Sensors

The agent can apply forces to each of its two body-circles. The forces originate from the center of the circles and are perpendicular to the link connecting them. Two effectors correspond to each of the two body-circles, one commanding a forward-pushing force, and one commanding a backward-pushing force. These effectors allow thus the agent to move backward or forward, to rotate in place, and, in general, to move within its environment. A fifth effector commands the length of the virtual link connecting the two body-circles, between zero and a maximum length. If the actual length of the link is different from the commanded length, an elastic force (proportional with the difference between the desired and actual length) acts on the link, driving it to the desired length.

The agent has contact sensors equally distributed on the surface of its two body-circles (8 contact sensors per circle, spanning a 45° angle each). The activation of the sensors is proportional to the sum of the magnitudes of the contact forces acting on the corresponding surface segment, up to a saturation value. Each circle also has 7 visual sensors, centered around the "forward" direction. Each sensor has a 15° view angle, originating from the center of the circle. The activation of the sensors is proportional to the fraction of the view angle covered by external objects. The range of the visual sensors is infinite.

The agent also has proprioceptive sensors corresponding to the effectors. Each body-circle has two velocity sensors, measuring the velocity in the forward and backward directions, respectively. The sensors saturate at a value corresponding to the effect of the maximum motor force that can be commanded by the effectors. The agent also has a proprioceptive sensor that measures the actual length of the link connecting the two body-circles, that saturates at the maximum length that can be commanded by the link effector.

Thus, the agent has a total of 5 effectors and 35 sensors (16 contact sensors, 14 visual sensors, and 5 proprioceptive ones). Each sensor or effector can have an activation between 0 and 1.

2.4   The Environment

In the experiments presented in this paper, the environment consisted of one agent and 6 circles ("balls") that the agent can move around. The spatial extension of the environment was not limited. The balls have variable radiuses (varying linearly between $r_1 = 0.06$ m and $r_2 = 0.26$ m), comparable in size to the radius of the agent's body-circles (0.1 m).

During each trial, the agent and the balls were positioned randomly in the environment, without contact, in a rectangular perimeter of 6 m by 4 m.

2.5   The Task

The task of the agent was to move alternatively each of the balls in its environment, on a distance as long as possible, in limited time (100 s of simulated time). More specifically, the fitness of each agent was computed as the sum of the distances on which each ball was moved, but with a threshold of $d_t = 2$ m for each ball. Thus, the agent had to move all balls, instead of just detecting one ball and pushing it indefinitely. The sum of distances may thus range between 0 and $6\,d_t = 12$ m.

This task was considered to implement a seek-push-release drive, that might be used in future experiments to bootstrap more complex behaviors, such as arranging the balls in a particular pattern, sorting the balls by size, or categorizing different kinds of objects.

If the agent moves in straight line at the maximum speed corresponding to the maximum forces it can produce, pushing the six balls for equal time, and if we neglect the time needed for taking curves, seeking the balls, switching between balls, the distance that it may cover in the limited time is 55.945 m. Given the existence of distances between balls, the fact that the speed is lower when taking curves, that the agent has to release the balls when switching them, we can see that the task is relatively difficult. From the perspective of an external observer, it may require the coordination of the motor effectors for attaining high speeds, the evaluation of the distance or time spent pushing a certain ball, and eventually the memorization of either objects' sizes or positions, that prevents the repeated pushing of the same balls.

To determine the fitness of a particular individual, we have averaged its performance on three trials, with random initial configurations of the balls.

3   The Controller

3.1   The Spiking Neural Network

The controller of the agent consisted of a recurrent spiking neural network. The controller had as input the activations of the agent's sensors, and as output the activations of the agent's effectors. The network was implemented by a fast, event-driven spiking neural simulator, inspired by Neocortex [35].

The network consisted of leaky integrate-and-fire neurons [3] with a resting and reset potential of -65 mV, a threshold potential of -40 mV, a resistance of

10 MΩ and a decay time constant of 10 ms. The network was fully connected: all neurons were connected to all neurons in the network, except input neurons, which had only efferent connections; there were no self-connections. There were 50 hidden neurons, in addition to the 70 input neurons and the 5 output neurons. The network was thus composed of 125 neurons and 6875 synapses.

The simulator used discrete time with a resolution of 1 ms. At each time step, only the neurons that received spikes were updated (hence the event driven nature of the updating of the network). If the updated neurons fired, their spikes were stored in a list. This spike list was used during the next time step to update the affected postsynaptic neurons. Thus, the spikes propagated within an axonal delay of one time step.

3.2   Spike-Timing-Dependent Plasticity

During some of the experiments, the neural network featured spike-timing-dependent plasticity (STDP). STDP is a phenomenon that was experimentally observed in biological neural systems [36,37,38,39]. The changes of the synapse efficacies depend on the relative timing between the postsynaptic and presynaptic spikes. The synapse is strengthened if the postsynaptic spike occurs shortly after the presynaptic neuron fires, and is weakened if the sequence of spikes is reversed, thus enforcing causality. Notably, the direction of the change depends critically on the relative timing.

We have modeled STDP following the method of [40]. The values of the parameters used were $A_+ = 0.005$, $A_- = 1.05\,A_+$, and $\tau_+ = \tau_- = 50$ ms. Following [9], we implemented directional damping for the synapse efficacies. The synapse efficacies w, which were variable due to STDP, were limited to the interval $[0, w_{max}]$, where w_{max} could be either positive or negative, and was a genetically determined maximum (in absolute value) efficacy.

4   The Agent-Controller Interface

In interfacing a spiking neural controller with an embodied agent, a conversion of the analog input to binary spikes and then of spikes to an analog output has to be performed. Following [9], the analog values of the sensor activations were converted to a spike train using a Poisson process with a firing rate proportional to the activation. The maximum firing rate of the input neurons was set to 100 Hz.

The spikes of the motor neurons were converted to an analog value by a leaky integrator of time constant $\tau = 10$ ms. The maximum value of the effector activation, 1, corresponded to a firing rate of the motor neuron of 100 Hz.

Each sensor of the agent, of activation s, $0 \leq s \leq 1$, drove two input spiking neurons, one being fed with activation s and the other with activation $1-s$. Thus, both the activation of the sensor and its reciprocal was fed to the network. The reason of this duplication of the sensory signal in the spiking neural network is twofold. First, this allows the network to be active even in the absence of

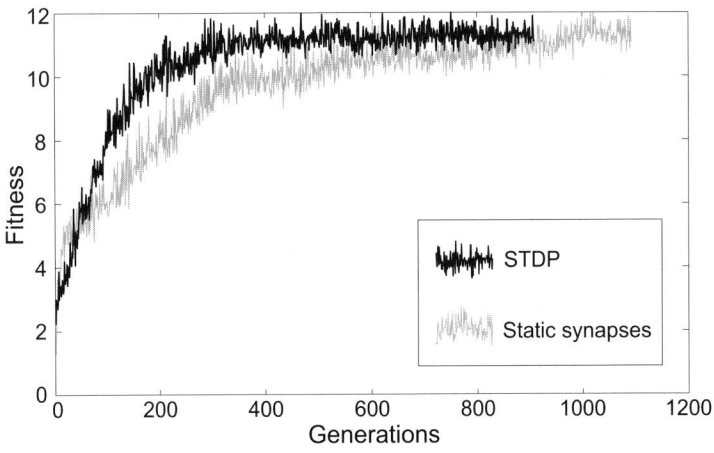

Fig. 2. Best fitness of the networks from a population over generations

sensory input. For example, if the agent is in a position where nothing activates its sensors (there is no object in its visual range, no tactile contact etc.), there must be however some activity in the neural network, in order for the effectors to be activated and the agent to orientate to stimuli. Second, this mechanism implies that the total input of the network is approximately constant in time (the number of spikes that are fed to the network by the input). This simplifies the selection of the network's parameters and the analysis of the network's behavior.

5   The Evolutionary Algorithm

The parameters determined by evolution were the values of the synaptic efficacies w (in the non-plastic case), or the values of the maximum (in absolute value) synaptic efficacies w_{max} (in the STDP case). The genome directly encoded these values for the 6875 synapses. We used a standard evolutionary algorithm, with a population of 80 individuals, truncation selection (the top 25 % individuals reproduced) and elitism. 10 % of the offspring resulted from mating with single cut crossover. Mutation was applied uniformly to all genes.

6   Results

Networks with both static and plastic synapses evolved to solve the required task, with the fitness of the best individuals reaching a plateau at about 11.3, very close to the maximum possible of 12 (see Fig. 2). Plastic networks evolved faster, in terms of generations, than networks with static synapses. However, the simulation of plastic networks required a higher computational effort. Only one evolution has been performed for each case (STDP, and respectively static synapses), because of the required computing time (a few weeks on a standard PC).

7   Behavioral Analysis

The agents controlled by the evolved networks seek the closest ball, push it for a while, then release it and seek another ball. They use visual information for seeking the balls: when no ball is pushed, and a ball enters the visual field, the agents go towards it. If there is no ball in the visual field, the agents rotate in circles in order to visually scan the environment. When they push a ball, they keep the link that unite their two body-circles extended, in order to have most of their visual field not occupied by the pushed ball. However, the link is not extended to maximum, for not letting the ball pass trough it. The agents also move circularly when pushing a ball, in order to seek other balls in the environment. When another ball enters the visual field, they go towards it while still pushing the first ball, and release it only when they are close to the new ball. Release is performed by extending the link. Again, they use visual information in order to seek new balls while pushing one. For example, if a single ball is placed in the environment, they keep pushing it in circles indefinitely, without releasing it. The behavior of the evolved agents is thus relatively complex, from an external observer perspective, requiring the composition of three sub-behaviors: seek, push, release. Although the agents may come more than once to push a particular object, their strategy leads them to alternatively push, in most cases, all objects in their environment. A movie displaying the behavior of an evolved agent is available online at `http://coneural.org/reports/object_pushing/object_pushing.avi`.

The balls that the agents push have the same density, and thus larger balls are heavier. Since the environment obeys a quasistatic physics, velocity is inversely proportional to mass, for a given force. In order to optimize their behavior for solving the task, the agents have to push the balls as hard as possible, and, for constant (maximum) forces, they have to push larger (heavier) balls longer periods of time than the smaller ones, to move them on similar distances during a limited time interval. It is interesting that this behavior — pushing for longer periods the larger balls — actually emerges during evolution. On average, the balls from the set encountered during evolution are pushed on the same distances. This is illustrated in Fig. 3 a). We tried to uncover the mechanisms that determine this behavior by subjecting the evolved agents to several "psychological" experiments, where the reality they were accustomed to (through evolution) was modified. More precisely, we modified systematically the radiuses and/or the densities of the balls. It can be seen (Fig. 3) that the average distance on which a ball is pushed does not depend exclusively on its characteristics, such as radius or mass, but on the characteristics of the whole set of balls. The average distances do not depend exclusively on the geometry of the environment, but also on the interaction of the agent with the balls. There is no particular parameter on the basis of which the agent estimates for how long it should push a ball, but the constance of the average distances emerges as a property of the complex dynamical system constituted by the agent, its environment, and the neural network, a property which is found and selected by the evolution. However, an external observer could argue that the evolved

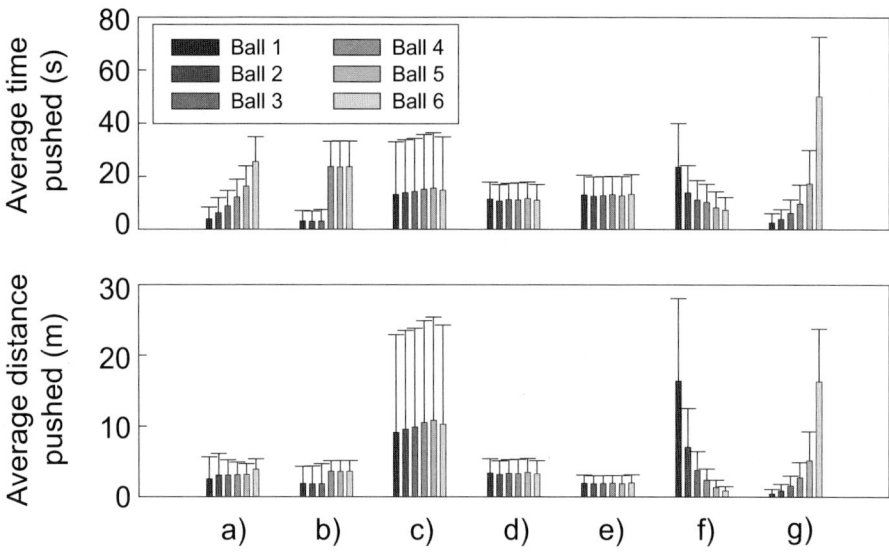

Fig. 3. Average (over 500 trials) and standard deviation of the average time and of the average distance a ball is pushed during a trial. a) The balls are like during the evolution, having radii varying linearly between r_1 and r_2. b) First three balls have radius r_1 and the other three have radius r_2. c)-e) The balls have equal radii: c) r_1; d) $r_a = (r_1 + r_2)/2$; e) r_2. f) The balls have equal radii r_a but variable masses (corresponding to radii varying linearly between r_1 and r_2, at default density). g) The balls have equal masses (corresponding to a radius r_a at default density) but radii varying linearly between r_1 and r_2. The relatively high variability of the displayed quantities is due by the variability of initial positions of the balls in the environment.

agents discriminate relative object size, by pushing larger objects for longer periods.

8 The Role of Plasticity

The evolved networks featuring STDP had a performance similar to the one of networks with static synapses, but evolved slightly faster. The improvement in evolution speed observed in plastic networks could be explained by a number of factors, including random exploration (smoothing) of the fitness landscape in the surroundings of individuals [41]. It is interesting to investigate whether plasticity also has an active role in determining the networks' performance. We have thus frozen the plasticity in networks evolved with STDP, either completely or after 1 s of activity (i.e., the first 1% of the duration of a trial). The results are presented in Fig. 4: freezing completely the plasticity leads to an important (80%) loss of performance, while freezing it after a short time that allows plasticity to act reduces performance only with about 33%. This means that most of the role of STDP in our evolved networks consists in tuning synapses to quasi-

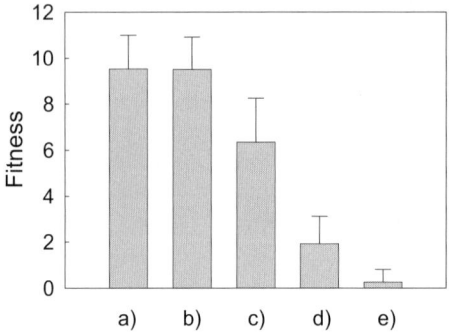

Fig. 4. Average fitness (over 500 trials) and standard deviation: a) Best network evolved with static synapses. b) Best network evolved with STDP. c) Best network evolved with STDP; plasticity is freezed after 1 s from the beginning of each trial. d) Best network evolved with STDP; plasticity is completely freezed. e) Random networks, not subjected to evolution.

stable, adaptive strengths, rather than contributing actively to the network's dynamics.

9 Conclusion

We have successfully evolved fully connected spiking neural networks consisting of 125 neurons and 6875 synapses, that allow a simple agent to alternatively seek, push and then release the 6 balls in its environment. This is one of the largest spiking neural network successfully evolved for the control of an embodied agent reported in the literature. This was possible because we used a fast agent-environment simulator especially designed for evolutionary and developmental experiments, and a fast event-driven neural network simulator. The evolved agents display interesting, minimally-cognitive behavior [42,43], by switching as a function of context between the three sub-behaviors (seek, push, release) and by being able to discriminate relative object size. We evolved networks with either static synapses or synapses featuring STDP. Plasticity proved to play a minor role in the dynamics of the evolved networks, contributing mostly to speeding up the evolutionary process and to tuning the synapses at the beginning of the networks' activity.

The evolved drive could be used in future experiments to bootstrap the development of more complex behaviors, by using, for example, reinforcement learning [32] to shape further the agent's behavior.

Acknowledgements

This work was partly supported by Arxia SRL and by a grant of the Romanian Government (MEdC-ANCS). We thank to Raul Mureşan for providing the code of the Neocortex spiking neural network simulator and for useful feedback.

References

1. Florian, R.V.: Autonomous artificial intelligent agents. Technical Report Coneural-03-01, Center for Cognitive and Neural Studies, Cluj, Romania (2003)
2. Maas, W., Bishop, C.M., eds.: Pulsed neural networks. MIT Press, Cambridge, MA (1999)
3. Gerstner, W., Kistler, W.M.: Spiking neuron models. Cambridge University Press, Cambridge, UK (2002)
4. Florian, R.V.: Biologically inspired neural networks for the control of embodied agents. Technical Report Coneural-03-03, Center for Cognitive and Neural Studies, Cluj, Romania (2003)
5. DasGupta, B., Schnitger, G.: Analog versus discrete neural networks. Neural Computation **8** (1996) 805–818
6. Maass, W., Schnitger, G., Sontag, E.D.: A comparison of the computational power of sigmoid and boolean threshold circuits. In Roychowdhury, V.P., Siu, K., Orlitsky, A., eds.: Theoretical Advances in Neural Computation and Learning. Kluwer Academic Publishers (1994) 127–151
7. Maas, W.: Networks of spiking neurons: the third generation of neural network models. Neural Networks **10** (1997) 1659–1671
8. Floreano, D., Mattiussi, C.: Evolution of spiking neural controllers for autonomous vision-based robots. In Gomi, T., ed.: Evolutionary Robotics IV. Springer-Verlag, Berlin (2001)
9. Di Paolo, E.A.: Spike timing dependent plasticity for evolved robots. Adaptive Behavior **10** (2002) 243–263
10. Saggie, K., Keinan, A., Ruppin, E.: Solving a delayed response task with spiking and McCulloch-Pitts agents. In: Advances in Artificial Life: 7th European Conference, ECAL 2003 Dortmund, Germany, September 14-17, 2003. Volume 2801 of Lecture Notes in Computer Science. Springer, Berlin / Heidelberg (2003)
11. Saggie-Wexler, K., Keinan, A., Ruppin, E.: Neural processing of counting in evolved spiking and mcculloch-pitts agents. Artificial Life **12**(1) (2005) 1–16
12. Ruppin, E.: Evolutionary embodied agents: A neuroscience perspective. Nature Reviews Neuroscience **3** (2002) 132–142
13. Floreano, D., Schoeni, N., Caprari, G., Blynel, J.: Evolutionary bits'n'spikes. In Standish, R.K., Bedau, M.A., Abbass, H.A., eds.: Artificial Life VIII: Proceedings of the Eight International Conference on Artificial Life. MIT Press, Boston, MA (2002)
14. Floreano, D., Zufferey, J.C., Mattiussi, C.: Evolving spiking neurons from wheels to wings. Proceedings of the 3rd International Symposium on Human and Artificial Intelligence Systems, Fukui, Japan (2002)
15. French, R.L.B., Damper, R.I.: Evolving a nervous system of spiking neurons for a behaving robot. In: Proceedings of Genetic and Evolutionary Computation Conference (GECCO 2001), San Francisco, CA (2001) 1099–1106
16. French, R.L.B., Damper, R.I.: Evolution of a circuit of spiking neurons for photo-taxis in a Braitenberg vehicle. In Hallam, B., Floreano, D., Hallam, J., Hayes, G., Meyer, J.A., eds.: From animals to animats 7: Proceedings of the Seventh International Conference on Simulation of Adaptive Behavior. MIT Press, Cambridge, MA (2002) 335–344
17. Damper, R.I., French, R.L.B.: Evolving spiking neuron controllers for phototaxis and phonotaxis. In Raidl, G., ed.: Applications of Evolutionary Computation, EvoWorkshops 2003. Volume 2611 of Lecture Notes in Computer Science. Springer, Berlin (2003) 616–625

18. Di Paolo, E.A.: Evolving spike-timing dependent plasticity for robot control. EP-SRC/BBSRC International Workshop: Biologically-inspired Robotics, The Legacy of W. Grey Walter, WGW'2002. HP Labs, Bristol, 14 - 16 August 2002 (2002)

19. Di Paolo, E.A.: Evolving spike-timing dependent plasticity for single-trial learning in robots. Philosophical Transactions of the Royal Society A **361** (2003) 2299–2319

20. Roggen, D., Hofmann, S., Thoma, Y., Floreano, D.: Hardware spiking neural network with run-time reconfigurable connectivity in an autonomous robot. In: 2003 NASA/DoD Conference on Evolvable Hardware (EH'03). (2003) 199

21. Van Leeuwen, M., Vreeken, J., Koopman, A.: Evolving vision-based navigation on wheeled robots. Institute for Information and Computing Sciences, Utrecht University (2003)

22. Katada, Y., Ohkura, K., Ueda, K.: Artificial evolution of pulsed neural networks on the motion pattern classification system. Proceedings of the 2003 IEEE International Symposium on Computational Intelligence in Robotics and Automation (CIRA), July 16 - 20, 2003, Kobe, Japan (2003) 318–323

23. Katada, Y., Ohkura, K., Ueda, K.: An approach to evolutionary robotics using a genetic algorithm with a variable mutation rate strategy. In: Proceedings of The 8th International Conference on Parallel Problem Solving from Nature (PPSN VIII). (2004) 952–961

24. Soula, H., Beslon, G., J.Favrel: Evolving spiking neural nets to control an animat. In: Proceedings of International Conference of Artificial Neural Networks and Genetic Algorithm 2003 - Roanne, France. (2003)

25. Hagras, H., Pounds-Cornish, A., Colley, M., Callaghan, V., Clarke, G.: Evolving spiking neural network controllers for autonomous robots. In: Proceedings of the 2004 IEEE International Conference on Robotics and Automation, New Orleans, USA. (2004)

26. Federici, D.: A regenerating spiking neural network. Neural Networks **18**(5–6) (2005) 746–754

27. Federici, D.: Evolving developing spiking neural networks. In: Proceedings of CEC 2005 - IEEE Congress on Evolutionary Computation. (2005)

28. Damper, R., Scutt, T.: Biologically-motivated neural learning in situated systems. In: Proceedings of the 1998 IEEE International Symposium on Circuits and Systems (ISCAS '98). (1998)

29. Damper, R., French, R.L.B., Scutt, T.: Arbib: An autonomous robot based on inspirations from biology. Robotics and Autonomous Systems **31**(4) (2000) 247–274

30. Soula, H., Alwan, A., Beslon, G.: Obstacle avoidance learning in a spiking neural network. In: Last Minute Results of Simulation of Adaptive Behavior, Los Angeles, CA (2004)

31. Soula, H., Alwan, A., Beslon, G.: Learning at the edge of chaos: Temporal coupling of spiking neuron controller of autonomous robotic. In: Proceedings of AAAI Spring Symposia on Developmental Robotics, Stanford, CA (2005)

32. Florian, R.V.: A reinforcement learning algorithm for spiking neural networks. In Zaharie, D., Petcu, D., Negru, V., Jebelean, T., Ciobanu, G., Cicortaş, A., Abraham, A., Paprzycki, M., eds.: Proceedings of the Seventh International Symposium on Symbolic and Numeric Algorithms for Scientific Computing (SYNASC 2005), IEEE Computer Society (2005) 299–306

33. Florian, R.V.: Thyrix: A simulator for articulated agents capable of manipulating objects. Technical Report Coneural-03-02, Center for Cognitive and Neural Studies, Cluj, Romania (2003)

34. Pfeifer, R., Scheier, C.: Understanding intelligence. MIT Press, Cambridge, MA (1999)
35. Mureşan, R.C., Ignat, I.: The "Neocortex" neural simulator: A modern design. International Conference on Intelligent Engineering Systems, September 19-21, 2004, Cluj-Napoca, Romania (2004)
36. Markram, H., Lübke, J., Frotscher, M., Sakmann, B.: Regulation of synaptic efficacy by coincidence of postsynaptic APs and EPSPs. Science **275**(5297) (1997) 213–215
37. Bi, G.Q., Poo, M.M.: Synaptic modifications in cultured hippocampal neurons: Dependence on spike timing, synaptic strength, and postsynaptic cell type. Journal of Neuroscience **18**(24) (1998) 10464–10472
38. Bi, G.Q.: Spatiotemporal specificity of synaptic plasticity: cellular rules and mechanisms. Biological Cybernetics **87** (2002) 319–332
39. Dan, Y., Poo, M.M.: Spike timing-dependent plasticity of neural circuits. Neuron **44** (2004) 23–30
40. Song, S., Miller, K.D., Abbott, L.F.: Competitive hebbian learning through spike-timing-dependent synaptic plasticity. Nature Neuroscience **3** (2000) 919–926
41. Turney, P.: Myths and legends of the Baldwin effect. Proceedings of the Workshop on Evolutionary Computing and Machine Learning at the 13th International Conference on Machine Learning (ICML-96), Bari, Italy (1996) 135–142
42. Beer, R.: Toward the evolution of dynamical neural networks for minimally cognitive behavior. In Maes, P., Mataric, M., Meyer, J., Pollack, J., Wilson, S., eds.: From animals to animats 4: Proceedings of the Fourth International Conference on Simulation of Adaptive Behavior. Volume 421–429. MIT Press, Cambridge, MA (1996)
43. Slocum, A.C., Downey, D.C., Beer, R.D.: Further experiments in the evolution of minimally cognitive behavior: From perceiving affordances to selective attention. In Meyer, J.A., Berthoz, A., Floreano, D., Roitblat, H.L., Wilson, S.W., eds.: From animals to animats 6: Proceedings of the Sixth International Conference on Simulation of Adaptive Behavior. MIT Press, Cambridge, MA (2000) 430–439

Hierarchical Cooperative CoEvolution Facilitates the Redesign of Agent-Based Systems

Michail Maniadakis[1,2] and Panos Trahanias[1,2]

[1] Inst. of Comp. Science, Foundation for Research and Technology-Hellas (FORTH),
71110 Heraklion, Crete, Greece
[2] Department of Computer Science, University of Crete,
71409 Heraklion, Crete, Greece
{mmaniada, trahania}@ics.forth.gr

Abstract. The current work addresses the problem of redesigning brain-inspired artificial cognitive systems in order to gradually enrich them with advanced cognitive skills. In the proposed approach, properly formulated neural agents are employed to represent brain areas. A cooperative coevolutionary method, with the inherent ability to co-adapt substructures, supports the design of agents. Interestingly enough, the same method provides a consistent mechanism to reconfigure (if necessary) the structure of agents, facilitating follow-up modelling efforts. In the present work we demonstrate partial redesign of a brain-inspired cognitive system, in order to furnish it with learning abilities. The implemented model is successfully embedded in a simulated robotic platform which supports environmental interaction, exhibiting the ability of the improved cognitive system to adopt, in real-time, two different operating strategies.

1 Introduction

Brain-inspired computational systems are recently employed to facilitate cognitive abilities of artificial organisms. The brain of mammals consists of interconnected modules with different functionalities, implying that models with distributed architecture should be designed. In this context, a modular design approach is followed by [1,2], to develop distributed brain-like computational models.

The construction of large scale models is difficult to be accomplished by developing from scratch complicated structures. An alternative approach could be based on implementing partial models of brain areas which are gradually refined to more efficient ones. Along this line, existing approaches suffer in terms of scalability, because they lack a systematic procedure to support the progressively more complex design procedure. In contrast, they follow a manual design approach and thus they can not be used as a long-term modelling framework.

We have recently proposed a new computational framework to design distributed brain-inspired structures [3]. Specifically, the model consists of a collection of self-organized neural agents, each one representing a brain area. The performance of agents is specified in real-time according to the interaction of the composite model with the external world, simulating epigenetic learning. The

S. Nolfi et al. (Eds.): SAB 2006, LNAI 4095, pp. 582–593, 2006.

self-organization dynamics of epigenetic learning are designed by an evolutionary process which simulates phylogenesis. Following the phylogenetic/epigenetic approach, the objective adopted during the evolution of agents, is to enforce the development of brain area like performance, after a certain amount of environmental interaction. Instead of using a unimodal evolutionary process, we employ a Hierarchical Cooperative CoEvolutionary (HCCE) approach which is able to highlight the specialties of brain areas, represented by distinct agents. The agent-based coevolutionary framework has been utilized to develop models that reproduce computationally biological findings [4], and additionally to integrate partial models formulating gradually more complex ones [5].

The present study investigates the ability of the agent-based coevolutionary framework to facilitate redesign steps, enriching existing models with gradually more advanced features. The ability of partial redesign is an important characteristic for an effective and successful computational framework that aims to support long-term design processes. This is because initial design steps impose constraints to the computational structure that may harm forthcoming modelling efforts. Hence, it is necessary to have a consistent design method that reformulates systematically partial structures, and additionally guarantees the cooperation of the refined components (and potentially some completely new) with the unchanged preexisting ones.

The proposed computational framework is particularly appropriate to support redesign steps because of the distributed architecture it follows. Specifically, due to the combination of agent-based modelling with the distributed HCCE design methodology, we are able to address and specify explicitly the special features of each component in the model. As a result, when partial redesign steps are necessary, we are provided with a systematic mechanism to reconfigure subcomponents according to an enhanced set of design objectives.

The rest of the paper is organized as follows. In the next section, we present the neural agent structures used to represent brain areas, and the hierarchical cooperative coevolutionary scheme which supports the design of agents. Then, we present the results of the proposed approach on redesigning partly an artificial cognitive system in order to furnish it with reinforcement learning abilities. Finally, conclusions and suggestions for future work are drawn in the last section.

2   Method

The design of brain-inspired structures is based on the argument that the behavior of animals is a result of phylogenetic evolution, and epigenetic environmental experience [6]. Phylogenetic evolution is facilitated by the HCCE design approach, while epigenetic learning is facilitated by the self-organization dynamics of the computational model. Both of them are described below.

2.1   Computational Model

Two different neural agents provide a computational framework which supports modelling: (i) a cortical agent to represent brain areas, and (ii) a link agent to support information flow across cortical modules.

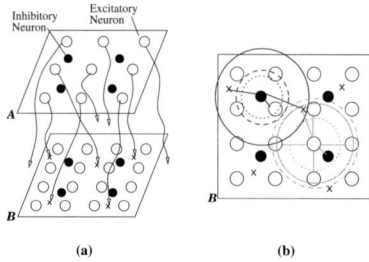

(a)   (b)

Fig. 1. Schematic representation of the computational model. Part (a) illustrates a link agent which supports information flow from cortical agent A to B. Part (b) illustrates synapse definition in cortical agent B.

Link Agent. The structure of the link agent is properly designed to support connectivity among cortical modules. Using link agents, any two cortical modules can be connected, simulating the connectivity of brain areas.

Each link agent is specified by the projecting axons between two cortical agents (Fig 1(a)). Its formation is based on the representation of cortical modules by planes with excitatory and inhibitory neurons (see below). Only excitatory neurons are used as outputs of the efferent cortical agent. The axons of projecting neurons are defined by their (x, y) coordinates on the receiving plane. Cortical planes have a predefined dimension and thus projecting axons are deactivated if they exceed the borders of the plane. This is illustrated graphically in Fig 1(a), where only the active projections are represented with an × on their termination. As a result, the proposed link structure facilitates the connectivity of sending and receiving cortical agents supporting their combined performance.

Cortical Agent. Each cortical agent is represented by a rectangular plane. A cortical agent consists of a predefined population of excitatory and inhibitory neurons, which all follow the Wilson-Cowan model with sigmoid activation. Both sets of neurons, are uniformly distributed, defining an excitatory and an inhibitory grid on the cortical plane. On the same plane there are also located the axon terminals from the projected cortical agents.

All neurons receive input information either from i) projecting axons, or ii) excitatory neighbouring neurons, or iii) inhibitory neighbouring neurons. The connectivity of neurons follows the general rule of locality. Synapse formation is based on circular neighbourhood measures. A separate radius for each of the three synapse types, defines the connectivity of neurons. This is illustrated graphically in Fig 1(b), which further explains the example of Fig 1(a). Neighbourhood radius for i) axons is illustrated by a solid line, for ii) excitatory neurons by a dashed line, and for iii) inhibitory neurons by a dotted line. Sample neighbourhoods for excitatory neurons are illustrated with grey, while neighbourhoods for inhibitory neurons are illustrated with black.

The performance of cortical agents is adjusted by environmental interaction, similar to epigenetic[1] learning ([7]). To enforce experience-based subjective learn-

---
[1] Epigenesis here, includes all learning processes during lifetime.

ing, each set of synapses is assigned a Hebbian-like learning rule defining the self-organization dynamics of the agent. This is in contrast to the most common alternative of genetically-encoded synaptic strengths which prevents experience based learning. We have implemented a pool of 10 Hebbian-like rules that can be appropriately combined to produce a wide range of functionalities [3].

Reinforcement Learning. Reinforcement learning models are very popular in robotic applications in recent years. Computational models similar to the one described above have been demonstrated to exhibit reinforcement learning abilities (e.g. [8]). The idea is based on treating the reward as an ordinary signal which can be properly given as input in pre- and post- synaptic neurons to coordinate their activations. In other words, the external reinforcement signal takes advantage of the internal plasticity dynamics of the agent, in order to modulate its performance accordingly.

2.2   Hierarchical Cooperative CoEvolution

Similar to a phylogenetic process, the structure of agents can be specified by means of an evolutionary method. However, using a unimodal evolutionary approach, it is not possible to explore effectively partial components, which represent brain substructures. To alleviate that, coevolutionary algorithms have been recently proposed that facilitate exploration, in problems consisting of many decomposable components [9]. Specifically, coevolutionary approaches involve many interactive populations to design separately each component of the solution. These populations are evolved simultaneously, but in isolation to one another. Partial populations are usually referred as *species* in the coevolutionary literature, and thus this term will be employed henceforth.

The design of brain-inspired structures fits adequately to coevolutionary approaches, because separate coevolved species can be used to perform design decisions for each substructure representing a brain area. As a result, coevolution is able to highlight the special features of each brain area, and additionally the cooperation within computational modules.

We have presented a new evolutionary scheme to improve the performance of cooperative coevolutionary algorithms, employed in the context of designing brain-inspired structures [3,4]. We employ two different kinds of species to support the coevolutionary process encoding the configurations of either a Primitive agent Structure (PS) or a Coevolved agent Group (CG). PS species specify partial elements of the model, encoding the exact structure of either cortical or link agents. A CG consists of groups of PSs with common objectives. Thus, CGs specify configurations of partial solutions by encoding individual assemblies of cortical and link agents. The evolution of CG modulates partly the evolutionary process of its lower level PS species to enforce their cooperative performance. A CG can also be a member of another CG. Consequently several CGs can be organized hierarchically, with the higher levels enforcing the cooperation of the lower ones.

The HCCE-based design method for brain modelling is demonstrated by means of an example (Fig 2). We assume the existence of two cortical agents con-

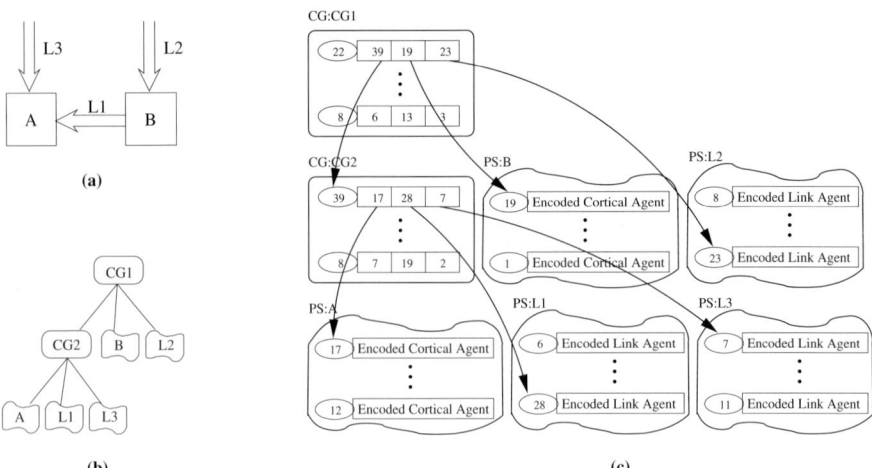

Fig. 2. An overview of the hierarchical coevolutionary scheme, with CG species tuning the evolutionary processes of PS species

nected by three link agents representing their afferent and efferent projections (Fig 2(a)). One hypothetical HCCE process employed to specify agent structure is illustrated in (Fig 2(b)). CGs are illustrated with oval boxes, while PSs are represented by ovals.

All individuals in all species are assigned an identification number which is preserved during the coevolutionary process. The identification number is employed to form individual assemblies within different species. Each variable in the genome of a CG is joined with one lower level CG or PS species. The value of that variable can be any identification number of the individuals from the species it is joined with. PSs encode the structure of either cortical or link agents. The details of the encoding have been presented in [3], and thus they are omitted here due to space limitations. A snapshot of the exemplar HCCE process described above is illustrated in (Fig 2(c)). Identification numbers are represented with an oval. CGs enforce cooperation of PS structures by selecting the appropriate cooperable individuals among species.

In order to test the performance of a complete problem solution, populations are sequentially accessed starting by the higher level. The values of CG individuals at various levels are used as guides to select cooperators among PS species. Then, PS individuals are decoded to specify the structure of cortical and link agents, and the performance of the proposed overall solution is tested on the desired task.

Furthermore, the proposed HCCE scheme allows the employment of separate fitness measures for different species. This matches adequately to the distributed agent-based modelling of brain areas, because different objectives can be defined for different components of the system, preserving their autonomy. As a result, the hierarchical coevolutionary scheme addresses explicitly the special roles of agents, facilitating any potential redesign of their structure.

For each species s, a fitness function f_s is designed to drive its evolution. All PS species under a CG share a common f_s. Specifically a partial fitness function $f_{s,t}$ evaluates the ability of an individual to serve task t, while the overall fitness function is estimated by:

$$f_s = \prod_t f_{s,t} \qquad (1)$$

Furthermore, the cooperator selection process at the higher levels of hierarchical coevolution will probably select an individual to participate in many assemblies. (e.g. the case of individual 28 of PS species L1, of Fig 2(c)). Let us assume that an individual participates in K assemblies which means that it will get K fitness values $f_{s,t}$. Then, the ability of the individual to support the accomplishment of the t-th task is estimated by:

$$f_{s,t} = max_k\{f_{s,t}^k\} \qquad (2)$$

where $f_{s,t}^k$ is the fitness value of the k-th solution formed with the membership of the individual under discussion.

The above equations describe fitness assignment in each species of the hierarchical coevolutionary process. Just after testing the assemblies of cooperators and the assignment of their fitness values, an evolutionary step is performed independently on each species, to formulate the new generation of its individuals. This process is repeated for a predefined number of evolutionary epochs, driving each species to the accomplishment of each own objectives and additionally enforcing their composite cooperative performance.

3   Results

The present experiment demonstrates the effectiveness of the agent-based coevolutionary framework to redesign computational structures, furnishing them with gradually more advanced cognitive skills. In order to prove the validity of the result, a mobile robot is utilized to support environmental interaction. Specifically, we employ a two wheeled robotic platform equipped with 8 uniformly distributed distance, light and positive reward sensors.

In our previous work [4], we have described the utilization of the HCCE scheme to model working memory (WM) development and how it is employed to accomplish delayed response tasks. In short, a light cue is presented to the robot and the latter has to memorize the side of light cue appearance in order to make a future choice, related to 90° turning, left or right (similar tasks have been also discussed in other studies e.g. [10]). Two different response strategies can be defined. According to the Same-Side (SS) response strategy, the robot should turn left if the light cue appeared at its left side, and it should turn right if the light source appeared at its right side. Evidently, the complementary response strategy can be also defined, named Opposite-Side (OS), which implies that the robot should turn left if the light cue appeared at its right side, and it should turn right if the light source appeared at its left side.

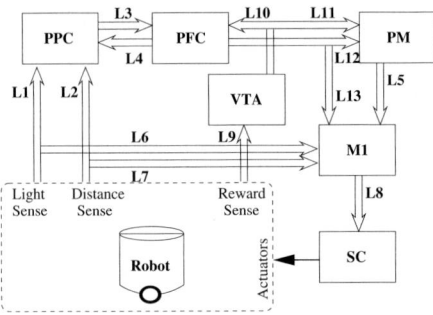

Fig. 3. A schematic demonstration of the computational model

The HCCE design mechanism has been employed to implement computational models exhibiting either the SS or the OS response strategy [4]. In both cases, the models are developed with the inborn ability to develop the correct response strategy. The question that now arises, is if we can design a single computational system that is able to adopt either the SS or the OS response strategy during life-time. The current study investigates the redesign of existing models that exhibit predefined behaviors, in order to enrich them with the ability to adapt their response strategy, as it is indicated by properly located reward signals.

3.1   Experimental Setup

The present experiment aims at extending the computational structure described in [4], thus developing an improved system with learning abilities. The composite model is illustrated in Fig 3. In order to facilitate the design procedure, we avoid designing the composite model from scratch. Particularly, the current experimental process keeps in their original formulation the components which are less involved in the learning procedure (namely, Posterior Parietal cortex (PPC), Primary Motor cortex (M1), and Spinal Cord (SC)). The biological structures mostly involved in the learning process are Prefrontal and Premotor cortices (PFC, PM) [11]. The module representing PFC was also present in our previous experiment, and it needs to be redesigned in order to be furnished with learning facilities. PM is a new module that needs to be designed from scratch. Both PFC and PM modules receive information related to the reward stimuli, adapting accordingly the motion orders to the lower levels of the motor hierarchy. Furthermore, an additional module to strengthen reward information is added, modulating effectively PFC, PM operation. This module could represent Ventral Tegmental Area (VTA) that guides learning in neocortex [12].

Learning the Opposite-Side Strategy. The training process of the robot is separated to several trials. Each trial includes two sample-response pairs, testing the memorization of two different sample cues by the robot (left or right side of light source appearance), and the selection of the appropriate delayed response in each case. Particularly, during the response phase, the light source disappears, and the robot drives freely to the end of the corridor where it has to make a turn choice. In the OS training process, the response is considered correct, if the robot turns to opposite side of light cue appearance. In the case that the robot makes the correct choice, it drives to the reward area receiving a positive reinforcement that modulates its belief regarding the correct response strategy.

The learning of the OS response strategy is tested for T trials, each one consisting of M simulation steps. The success of the training process is evaluated by:

$$E_{tr} = \left(\sum_{T,left} \sum_{M} r \right) \left(\sum_{T,right} \sum_{M} r \right) \left(1 - \sqrt{\frac{B}{2 \cdot T \cdot M}} \right)^3 \tag{3}$$

The first term seeks for maximum reward stimuli when the correct response of the robot is considered the left side, while the second seeks for maximum reward when the correct response is the right side. The higher the reward the robot has received, the more successful was the reinforcement training process. The last term minimizes the number of robot bumps on the walls.

Additionally, HCCE facilitates the employment of partial criteria highlighting the special roles of cortical agents in the composite model. Specifically, we use a partial criterion that addresses the development of WM-like activation patterns on PFC. Two different states a, b are defined, associated with the two possible sides of light source appearance. For each state, separate activation averages, p_l, are computed, with l identifying PFC excitatory neurons. The formation of WM patterns is evaluated by:

$$E_{wm} = \left(\frac{v_a}{m_a} + \frac{v_b}{m_b} \right) \cdot \min \left\{ \sum_{\substack{l \\ p_l^a > p_l^b}} (p_l^a - p_l^b), \sum_{\substack{l \\ p_l^b > p_l^a}} (p_l^b - p_l^a) \right\} \tag{4}$$

m_a, v_a, m_b, v_b are the means and variances of average activation at states a, b. The first term enforces consistent activation, while the second supports the development of separate activation patterns for each state a, b.

Another criterion addresses the development of different activation patterns in PM structure. They are related to the different higher level motion commands that should be passed to M1. Two different states r, l are defined, associated with the commands of right or left turning. For each state, separate activation averages, p_k, are computed, with k identifying PM excitatory neurons. The successful development of distinct activation patterns for the right and left turning is measured by:

$$E_c = \left(\frac{v_r}{m_r} + \frac{v_l}{m_l} \right) \cdot \min \left\{ \sum_{\substack{k \\ p_k^r > p_k^l}} (p_k^r - p_k^l), \sum_{\substack{k \\ p_k^l > p_k^r}} (p_k^l - p_k^r) \right\} \tag{5}$$

The explanation of the measure is similar to eq (4).

Finally, an additional criterion highlights the development of different patterns on the VTA structure, related to the two possible locations of the reward signal. Two different states x, y are defined, associated with the right or left reward

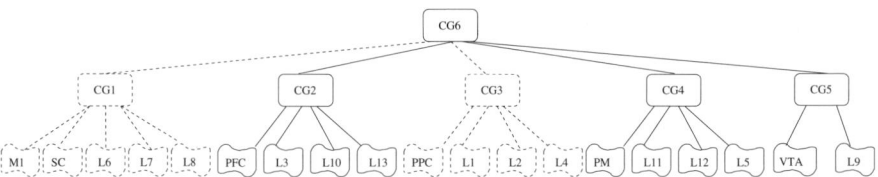

Fig. 4. An overview of the extended Hierarchical Cooperative CoEvolutionary process employed to design the composite computational model

location. For each state, separate activation averages, p_t, are computed, with t identifying VTA neurons. This is described by:

$$E_r = \left(\frac{v_x}{m_x} + \frac{v_y}{m_y} \right) \cdot \min \left\{ \sum_{\substack{t \\ p_t^x > p_t^y}} (p_t^x - p_t^y), \sum_{\substack{t \\ p_t^y > p_t^x}} (p_t^y - p_t^x) \right\} \tag{6}$$

The explanation of the measure is similar to eq (4).

Learning the Same-Side Strategy. Just after testing the performance of the robot on learning the OS strategy, the computational structure is re-initialized, and we test if it is able to adopt the SS response strategy, by means of a different set of reward stimuli. The process is again separated to T trials, and it is very similar to the one described for OS training. Specifically, each trial includes two sample-response pairs, but this time, due to the SS strategy the reward stimulus is located to the same side that the light cue appeared.

The measure evaluating the adoption of the SS strategy by the robot is the same with the one described in eq. (3). Furthermore, additional evaluation measures similar to those described in eqs (4), (5), (6) highlight the roles of PFC, PM, AmpR structures in the composite model. Overall, we employ two different sets of measures, namely $E_{wm,os}$, $E_{c,os}$, $E_{r,os}$, $E_{tr,os}$ and $E_{wm,ss}$, $E_{c,ss}$, $E_{r,ss}$, $E_{tr,ss}$ evaluating the ability of the robot to adopt either the OS or the SS strategy during the reward-based training process, and the distinct role of substructures in the composite model.

3.2   Computational Modelling

We turn now to the design of the model by means of the HCCE scheme. The hierarchical coevolutionary process that re-designs and extends the computational model is illustrated in Fig 4. The species below CG1 and CG3 are depicted with a dotted line, in order to demonstrate that we keep their original structure (formulated in our previous experiment [4]) and they are not evolved in the current coevolutionary design procedure. According to the experimental scenario followed in the present study, two behavioral tasks $t1$, $t2$, are employed to validate respectively the adoption of either the OS or SS response strategies.

Specifically, the fitness function employed for the evolution of $CG2$ and its lower level species, evaluates the success of OS and SS learning procedures, and the development of WM activity in PFC. Following the formulation introduced in eqs. (1), (2), this is described mathematically by:

$$f_{CG2}=f_{CG2,t1}\cdot f_{CG2,t2} \text{ with } f_{CG2,t1}^{k}=E_{wm,os}\cdot E_{tr,os}, \quad f_{CG2,t2}^{k}=E_{wm,ss}\cdot E_{tr,ss} \quad (7)$$

where k represents each membership of an individual in a proposed solution.

The agent structures grouped under $CG4$ serve the success on OS, SS learning, and the development of the appropriate higher level motion commands on PM. Thus, the fitness function employed for the evolution of $CG4$ is:

$$f_{CG4}=f_{CG4,t1}\cdot f_{CG4,t2} \text{ with } f_{CG4,t1}^{k}=E_{c,os}\cdot E_{tr,os}, \quad f_{CG4,t2}^{k}=E_{c,ss}\cdot E_{tr,ss} \quad (8)$$

where k is as above.

The agent structures grouped under $CG5$ support OS, SS learning and the development of different reward patterns on VTA. Thus, the fitness function employed for the evolution of $CG5$ is:

$$f_{CG5}=f_{CG5,t1}\cdot f_{CG5,t2} \text{ with } f_{CG5,t1}^{k}=E_{r,os}\cdot E_{tr,os}, \quad f_{CG5,t2}^{k}=E_{r,ss}\cdot E_{tr,ss} \quad (9)$$

where k is as above.

Finally, the top level CG enforces the integration of partial configurations in a composite model, aiming at the cooperation of substructures in order to facilitate the accomplishment of both learning processes, and additionally highlighting the role of each cortical agent in the composite model. The fitness function employed for the evolution of $CG6$ is defined accordingly, by:

$$f_{CG6}=f_{CG6,t1}\cdot f_{CG6,t2} \text{ with } \begin{array}{l} f_{CG6,t1}^{k}=E_{tr,os}\cdot\sqrt{E_{wm,os}\cdot E_{c,os}\cdot E_{r,os}}, \\ f_{CG6,t2}^{k}=E_{tr,ss}\cdot\sqrt{E_{wm,ss}\cdot E_{c,ss}\cdot E_{r,ss}} \end{array} \quad (10)$$

where k is as above.

The hierarchical coevolutionary process described above, employed populations of 200 individuals for all PS species, 300 individuals for $CG2$, $CG4$, $CG5$, and 400 individuals for $CG6$. After 70 evolutionary epochs the process converged successfully. Sample results of robot learning to adopt the OS and SS strategies are illustrated in Figs 5, 6. In both cases, the response of the robot in the first two trials (columns 2,3) are incorrect. However, in the third trial (column 4), the robot tries another strategy which is successful, and it is continued for all the remaining trials. As a result, HCCE successfully redesigns the computational structure, formulating an improved model with reinforcement learning abilities.

Overall, the present experimental procedure demonstrates the power of the HCCE-based design mechanism to refine an existing computational structure in order to enhance its functionality. The same results demonstrate also that the distributed design mechanism is particularly appropriate to enforce the cooperation among new and preexisting components. As a result, HCCE can be consistently employed to facilitate the success of complex, long-term design procedures.

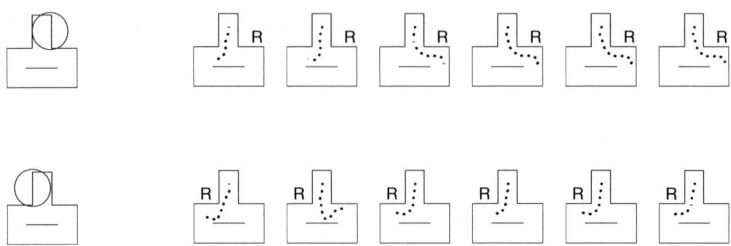

Fig. 5. A sample result of robot performance in the Same-Side response task. The first column illustrates sample cues. The rest columns (2-7) demonstrate the response of the robot in consecutive trials. The "R" depicts the side of the reward. Snapshots in the first line illustrate robot responses when light sample cue appears to the right, while the second line illustrates robot responses when light sample appears to the left.

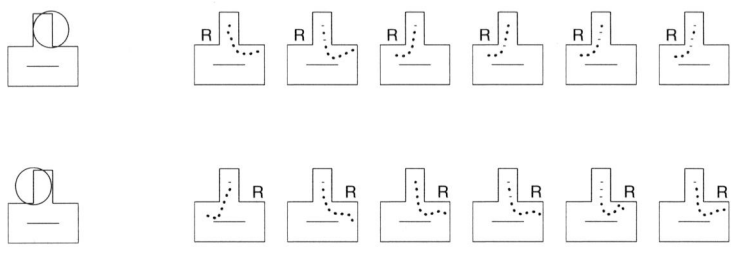

Fig. 6. A sample result of robot performance in the Opposite-Side response task. The first column illustrates sample cues. The rest columns (2-7) demonstrate the response of the robot in consecutive trials. The "R" depicts the side of the reward. Snapshots in the first line illustrate robot responses when light sample cue appears to the right, while the second line illustrates robot responses when light sample appears to the left.

4   Conclusions

The work described in this paper, addresses the development of cognitive abilities in artificial organisms by means of implementing brain-inspired models. Specifically, we introduce a systematic computational framework for the design and implementation of brain-like structures. This is based on the employment of neural agent modules to represent brain areas, and an HCCE-based design methodology to facilitate both the design of partial models and their further advancement in gradually more complex ones.

Due to the distributed architecture followed by both the agent-based model and the HCCE design methodology, the proposed computational framework is able to address explicitly the structure of system components. Hence it is able to add new components in the model, and re-design some of the pre-existing ones in order to advance gradually the capabilities of the model.

We believe that by exploiting the proposed approach, a powerful method to design brain-inspired structures can emerge. Further work is currently underway to investigate the suitability of our approach in large scale modelling tasks.

References

1. Krichmar, J., Edelman, G.: Brain-based devices: Intelligent systems based on principles of the nervous system. In: Proc. 2003 IEEE/RSJ Int. Conference on Intelligent Robots and Systems. (2003) 940–945
2. Kozma, R., Wong, D., Demirer, M., W.J., F.: Learning intentional behavior in the k-model of the amygdala and ethorinal cortex with the cortico-hyppocamal formation. Neurocomputing **65-66** (2005) 23–30
3. Maniadakis, M., Trahanias, P.: Modelling brain emergent behaviors through co-evolution of neural agents. accepted for publication, Neural Networks Journal (2006)
4. Maniadakis, M., Trahanias, P.: Distributed brain modelling by means of hierarchical collaborative coevolution. In: Proc. IEEE Congress on Evolutionary Computation, (CEC). (2005) 2699–2706
5. Maniadakis, M., Trahanias, P.: Design and integration of partial brain models using hierarchical cooperative coevolution. In: Proc. International Conference on Cognitive Modelling, ICCM. (2006)
6. Geary, D., Huffman, K.: Brain and cognitive evolution: Forms of modularity and functions of mind. Psych. Bulletin **128** (2002) 667–698
7. Cotterill, R.: Cooperation of the basal ganglia, cerebellum, sensory cerebrum and hippocampus: possible implications for cognition, consciousness, intelligence and creativity. Progress in Neurobiology **64**(1) (2001) 1 – 33
8. Blynel, J., Floreano, D.: Levels of dynamics and adaptive behaviour in evolutionary neural controllers. In: Proc. of the Seventh International Conference on Simulation of Adaptive Behavior (SAB). (2002) 272–281
9. Potter, M., De Jong, K.: Cooperative coevolution: An architecture for evolving coadapted subcomponents. Evol. Computation **8** (2000) 1–29
10. Ziemke, T., Thieme, M.: Neuromodulation of reactive sensorimotor mappings as a short-term mechanism in delayed response tasks. Adaptive Behavior **10**(3-4) (2002) 185–199
11. Murray, E., Bussey, T., Wise, S.: Role of prefrontal cortex in a network for arbitrary visuomotor mapping. Experimental Brain Research **113** (2000) 114–129
12. Kandel, E.R., Schwartz, J., Jessell, T.M.: Principles of Neural Science. Mc Graw Hill (2000)

Bubbleworld.Evo: Artificial Evolution of Behavioral Decisions in a Simulated Predator-Prey Ecosystem

Thomas Schmickl and Karl Crailsheim

Department for Zoology, Karl-Franzens-University Graz,
Universitaetsplatz 2, A-8010 Graz, Austria
schmickl@nextra.at ,
karl.crailsheim@uni-graz.at
http://zool33.uni-graz.at/schmickl/

Abstract. This article presents a multi-agent simulation of an abstract ecosystem which is inhabited by two species: a predator species and a prey species. Both species show the typical behaviors found in such an ecological relationship that are: hunting behavior and escaping behavior. In the simulation, the actors make behavioral decisions according to "genetically fixed" weighting parameters. These parameters determine which prey item is selected by the predator and which predators are avoided the most by prey. Thus these parameters shape the decisions performed by both species. We incorporated artificial evolution by allowing successful animals to pass their features to their offspring, a process that includes mutation and recombination of these "genes". The simulation shows that different kinds of optimal behavioral choices emerge out of artificial evolution, when the simulation is run with different physiological and morphological parameters of the actors.

1 Introduction

In nature, the behaviors of animals are shaped by two processes. On the one hand, behaviors can be optimized during life-time of an individual by learning. In some cases, these learned patterns can be transferred among animals, even without reproduction of an organism (memes, see [1]). On the other hand, another very important process of adaptation is biological evolution [2] that works via natural selection, inheritance and mutation. Most behaviors of animals follow intrinsic patterns, which are passed and shaped as their corresponding genes pass from generation to generation [1][3]. Such evolutionary adaptations of behaviors are difficult to study in real life because - except some rare cases - they work over extremely long periods. Computer simulation allows a fascinating approach to study these processes, because simulations can be performed for hundreds or thousands of generations within hours of run time, thus they allow comparative studies of artificial evolution under controlled and reproducible conditions.

Since the beginning of the research field called "artificial life" [4][5], the emergence of behaviors and the shaping of behaviors have been studied in a variety of computational experiments. In some approaches behaviors were studied in specific environments that do not resemble real ecosystems very much, like the 'core-war'

S. Nolfi et al. (Eds.): SAB 2006, LNAI 4095, pp. 594–605, 2006.

system, the 'coreworld' system, the 'tierra' system, or the 'avida' system. For an overview about these systems, see [4]. Other studies tried to evolve behaviors produced by artificial neural networks which control the behavior of the agents. Some approaches used tournaments-like contests to evolve the morphology and the behaviors of their agents [6], [7]. None of theses approaches showed sophisticated behavior of prey or of predators, least of all in a way that produces a predator-prey system that is comparable to well known ecological models [8][9][10][11].

The approach of *bubbleworld* is significantly different from the approaches mentioned above. We started with a fully developed simulation of a predator-prey ecosystem that involves intraspecific competition among both simulated species. Our simulation (called "*bubbleworld*") incorporates developmental processes of the prey species ("*bubbles*") and a simulated metabolism of the predator species ("*sorgs*"). This metabolism forces the predator to act "optimal" otherwise the predator dies because it runs out of energy. Figure 1 shows a screenshot of the "world" of bubbles and sorgs at runtime. Based on this ecological-focused multi-agent simulation, we developed *bubbleworld.evo* which differs from *bubbleworld* by allowing inheritance, mutation, recombination of some of the individuals' features. All factors implemented in *bubbleworld* do also exist in *bubbleworld.evo*. We implemented the basic behavioral rules of our two modeled species in a way that is comparable to hunting behaviors found in nature: A predator selects one focal prey animal and starts to chase it. If it catches the focal prey, it swallows the prey, digests it for some time (while resting) and earns energy this way. If the predator does not catch the focal prey, it can make a new decision after some time steps by selecting another prey item.

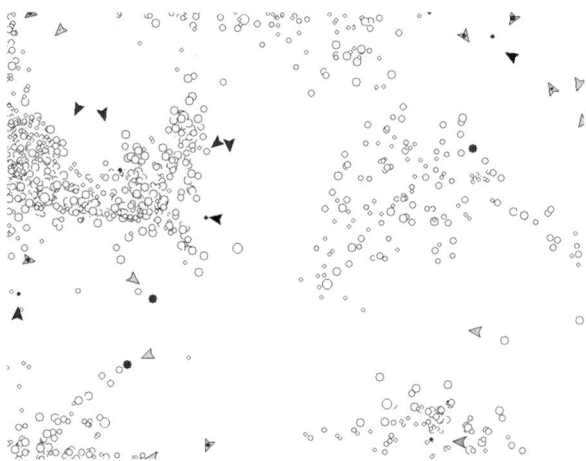

Fig. 1. A screenshot of the multi-agent simulation *bubbleworld* at runtime. Hundreds of prey items (bubbles, circle-shaped agents) try to avoid the predators (sorgs, triangular-shaped agents). The bubbles drawn with filled color are currently targeted due to the decision process described in equation 2. The brighter the predators are colored, the higher their energy level is and the faster they are. The bigger the prey is, the faster it can escape, but the more energy gain it offers to its predators.

The prey animals show even simpler behavior: They flee away from the predators. What makes our study interesting is the process how a predator chooses the focal prey out of all the prey animals that are within its sensory radius. Several aspects have to be taken in account for the predator, because wrong decisions will lead to unsuccessful hunts and will cost energy without gain. A frequently unsuccessful predator dies, the genes that led to a sub-optimal decision can disappear from the "gene-pool" this way. Successful predators grow faster and thus reproduce more frequently; ultimately "good" features will accumulate in the population.

The situation gets more complex as there are 4 criterions that have to be assessed and weighted correctly with the prey-selection decision performed by the predators: The distance to the prey, the angle to the prey, the size of the prey and the current speed of the prey. The overall decision is the result of these factors and of the inherited weights that the predator associates with them. Also the prey has to make decisions in the case that more than one predator is in sight: This decision is done by choosing the optimal escape direction. An optimal decision has to account for the threats associated to the different predators. These threats depend on the distances and on the current orientations of the predators. Obviously the decisions made by predators and prey are somehow inter-linked. If prey animals "know" the way predators choose their focal prey, the decision of the optimal escape direction can be optimized, and vice versa. So we even expect to find some aspects of co-evolution.

2 The Model

Our simulation represents a habitat that incorporates two species that live together in a predator-prey relationship. Both species show intra-specific competition to allow density-regulated growth of the populations (as they are described in [8][9][10][11]). The simulation is a multi-agent simulation [12], what means that behaviors are programmed based on local sensory input of the agents (simulated animals).

In our simulation, there is an autotrophic species called "*bubbles*". These agents grow slowly over time. The faster they move the less energy they can invest into growing, thus the slower they grow. Maximum growth rate of resting bubbles is one state unit (= size unit) per time step. Minimum growth rate of a bubble (moving at full speed) is 0.66 state units per step. A bubble that has reached 1000 state units is able to reproduce. The "mother" bubble dies and produces 1-10 offspring, depending on the number of bubbles within its sensory radius, which is 150 pixels. The offspring start at a size of one unit. The bigger a bubble is, the faster it can move. The minimum speed for a small (young) bubble is 3 pixels per step; the maximum speed is 5 pixels per step. A bubble can accelerate with 1 pixels/step^2 and can slow down 0.5 pixels/step^2. A bubble can turn at a maximum extent of 0.1 degrees per time step. Fig. 2a depicts the "normal" life cycle of a bubble in our simulation.

The "*sorgs*" are the predatory species in our simulated predator-prey system, they hunt for bubbles and feed on them. A sorg is born with an energy state level of 300 energy units. Per time step it consumes some energy depending on its speed, according to equation 1. A sorg that reaches a level of zero energy units dies. A sorg that catches a bubble, swallows it and digests it (Fig. 2b), by retrieving energy from this process. The bigger the caught bubble is, the longer this digestion process takes and the more energy is gained by the sorg. Bigger bubbles are harder to catch, as they are faster, but they represent more energy gain. In addition, the longer digestion time leads to longer hunting pauses, which in turn allows the other bubbles to grow and reproduce. A sorg's maximum speed depends on its energy state. In the default settings, the maximum speed of a very fit sorg is set to 5.3 pixels per step and the maximum speed of a sorg that is almost out of energy is set to 3.3 pixels per time step. Maximum acceleration of a sorg is 1 pixels/step^2, maximum deceleration is 0.5 pixels/step^2. A sorg can turn its direction at a maximum extent of 0.1 degrees per time step. A sorg that reaches a level of 1000 energy unit dies after it has reproduced into 1-3 offspring, which start with 300 energy units in turn. The number of offspring is regulated by the local sorg density: the more sorgs are within the sensory radius (180 pixels), the lower is the number of offspring.

$$cost = 1.2 \cdot \frac{speed_{current}}{speed_{max}} + 0.2 \qquad (1)$$

Please note that all of the parameters mentioned above are adjustable in *bubbleworld* and that the description above represents the "default settings". In some simulation experiments described in this article, we deviated from these defaults. Table 1 shows all values of default settings and parameter names.

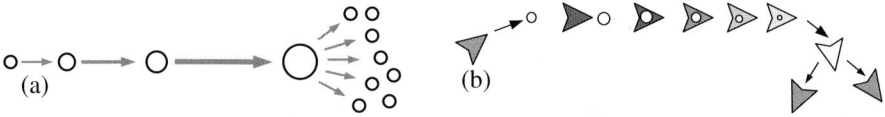

Fig. 2. (a) The life cycle of a successful bubble: The bubble starts with a size of 1 unit and moves slowly. It grows over time and increases its maximum speed. As soon as the bubble reached a size of 1000 units, it reproduces into 1-10 offspring, which start their lives again small and slow.(b) The life cycle of a successful sorg: The sorg is born with a low energy level (dark color) and is therefore quite slow. It selects a small (and therefore also slow) bubble, chases it and catches it. The bubble gets digested and the sorg gains energy (color of sorg brightens as the bubble shrinks). As soon as the sorg reaches a level of 1000 energy units, it reproduces and 1-3 sorgs with low energy level are born. This picture is simplified, because in *bubbleworld*, a sorg has to hunt successfully several times before it reaches the reproductive state.

2.1 Modeling the Behaviors of Predators and Prey

Both species show a typical behavior that can be frequently found in nature. The predators select one specific prey animal and test it (hunt it). The focus of a predator can change, but this is not performed in every time step: the probability to select a new focal prey is 1/20 per time step. Thus, on average, the prey is tested for 20 time steps. As it is depicted in figure 3a, there are 4 criterions (variables) evaluated during this decision making: The distance of the prey to the predator, the angle of the turning curve that has to be performed to focus on the prey, the size of the prey and the actual surplus, that is the maximum speed of the sorg compared to the maximum speed of the bubble. As shown in equation 2, these variables are used to calculate a "quality" index of the prey: the prey with the highest quality index is chosen as focal prey. The most important point in the decision making process is that all four factors are weighted by the four weighting-factors called *w(distance)*, *w(angle)*, *w(size)* and *w(speedsurplus)*. The higher one of these weights is (in relation to the other weights!), the more dominant the associated factor is represented in the decision making process. Equation 2 shows the quality that is associated to the bubble *i* by a sorg.

$$quality_i = -angle_i \cdot w_{angle} - distance_i \cdot w_{distance}$$
$$+ size_i \cdot w_{size} + speedsurplus_i \cdot w_{speed} \tag{2}$$

A similar decision is made by the bubbles when they decide on their optimal escaping direction. The escape vector is the sum of the weighted escape vectors away from all sorgs that are within the sensory range of the bubble. These individual escape vectors are weighted by two factors which account for the distance of each sorg and for the angle this sorg has to turn to focus on the bubble. Figure 3b depicts the decisions performed by a bubble attacked by two sorgs:

$$vector_{result} = \sum_{i=1}^{sorgcount} \left(vector_i \cdot \left(\frac{w_{distance} \cdot (r_{sensory} - distance_i)}{r_{sensory} + 1} + \frac{w_{angle} \cdot (\pi - angle_i)}{\pi} \right) \right) \tag{3}$$

As can be seen in equation 3, two weighting-factors called *w(distance)* and *w(angle)* are associated with the variables "distance" and "angle". The variable $angle_i$ describes the angle the sorg *i* has to turn to target the bubble. The variable $distance_i$ represents the distance from sorg *i* to the bubble that makes its decision. $r_{sensory}$ represents the radius of the bubbles sensory range.

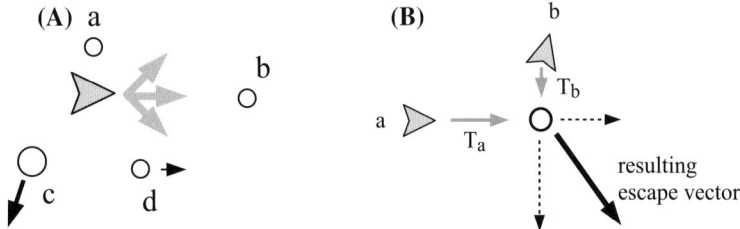

Fig. 3. (A) The decisions made by sorgs: Bubble (a) is the closest one, bubble (b) requires the smallest turning angle. Bubble (c) is the biggest one and offers the highest energy gain. Bubble (d) has the slowest speed. (B) The decisions made by bubbles: The vectors Ta and Tb point from sorgs (a) and (b) to the bubble. The length of the two resulting escape-vectors depend on the values of the two weights *w(distance)* and *w(angle)*. The resulting overall escape-vector of the bubble is calculated by the sum of the individual escape-vectors. The bubble depicted in this figure bases its decision mostly on the factor "distance", thus *w(distance)* >> *w(angle)*.

2.2 Artificial Evolution

In the runs with artificial evolution, the weights are not global parameters for the whole population, they are stored within an artificial chromosome that resides inside of each bubble and inside of each sorg. These chromosomes are binary strings containing 32 bits per "gene", each gene represents one weighting factor used in equations 2 and 3. Upon reproduction, these chromosomes are passed on to the offspring. They are affected by a random mutation of one bit per reproduction. To avoid too big changes within one mutation step, we used grey coding for the binary information, as it is described in [13]. In addition to that, with a probability of 1/10 per reproductive event, another (randomly chosen) species-mate is selected for recombination of "genetic material" and a cross-over event is generated. This cross-over is done by transferring a random-sized fragment from a randomised position of the "donor" animal to a randomised position in the offspring's' chromosome.

Table 1. Settings of the fixed parameters in the two setups used for artificial evolution. The basic unit for distances is "pixels" the basic unit for energy status and maturation status is called "state units". The basic unit for time is called "step". Maximum maturation is one "state unit" per time step. A bubble needs 1000 state units to reach the reproductive phase.

Parameter name	Units	Setup1 (default abilities of both species)		Setup2 (agile bubbles and fast sorgs)	
		Bubbles	Sorgs	Bubbles	Sorgs
$P_{reproduce}$	1/step	0.01	0.01	0.01	0.01
Maturation step	steps	1000	1000	1000	1000
Max. offspring	animals	10	3	10	3
Max speed (young)	pixels/step	5	5.3	5	6
Max speed (old)	pixels/step	3	3.3	3	4
Turn angle	degrees	0.1	0.1	0.2	0.1
$r_{sensory\text{-}range}$	pixels	150	180	150	180
$P_{focus\text{-}change}$	1/step	---	0.05	---	0.05
Start state	state units	1	200	1	200
Max. gain	state units	---	400	---	400
Max digestion time	steps	---	150	---	150

2.3 Experimental Setup

Two evolutionary setups were performed in the experiments described in this article. These simulation runs were performed with different (morphological) parameters of bubbles an sorgs which result in different abilities of the simulated animals: the changed parameters affect their maximum speed and their maximum turning angle.

Setup1 (default settings): In this experiment, the basic settings of morphological and physiological parameters of bubbles and sorgs were used. The most important point is that in this setup, the bubbles and sorgs were of equal agility, that means they showed identical maximum turning angles per time step. In this setup, the sorgs were slightly (0.3 pixel per step) faster than bubbles to allow the sorgs to hunt successfully.

Setup2 (agile bubbles, fast sorgs): In this experiment, the maximum turn angle per time step of the bubbles was doubled, what led to more agile bubbles which were able to out-manoeuvre sorgs by simply moving in a curve with a small radius. To compensate for that disadvantage, the maximum speed of sorgs was increased significantly (one pixel per time step).

In total 46 successful repetitions of artificial evolution were performed, a successful repetition has to reach time step 200,000 without having one of the populations died out. Initially, the evolutionary runs started with totally randomised values (uniform random values between 0 and 1) of the weighting parameters *w(distance)*, *w(angle)*, *w(speedsurplus)*, *w(size)*, and *w(angle)*. After the end of each run, the values of the weighting parameters of the survived bubbles and sorgs were investigated. For interpreting these values, the reader has to keep in mind that the relative differences between two weights modulate the decision made by a sorg or by a bubble. A bubble having *w(distance)*=1 and *w(angle)*=1 decides in exactly the same way than a bubble with the values *w(distance)*=0.1 and *w(angle)*=0.1.

3 Results

In total, we performed 26 evolutionary runs according to setup 1 (default settings) and 33 evolutionary runs according to setup 2 (agile bubbles / fast sorgs). In two cases the simulation runs of setup 1 were unsuccessful, what means that one or both species died out through the evolutionary process. In setup 2, this happened more frequently (in 11 out of 33 cases), showing that the predator-prey relationship has a less stable equilibrium point in this setup. All runs lasted for 200,000 steps.

3.1 Comparison of the Evolved Weights in the Two Setups

In a first analysis, we plotted the values of the two weights *w(distance)* and *w(angle)* of bubbles and of sorgs in a x/y phase-plot (Figs. 4,5, and 6). We found 26620 bubbles at time step 200,000, so we reduced the number of plotted bubbles in figure 4 by randomly removing 19 out of 20 bubbles, thus we displayed every 20th bubble that survived. The dashed diagonal line in the figures 4,5, and 6 indicate those combinations of weighting factors that lead to an equal weighting of the associated variables. E.g., a dot in the upper left triangle of figure 4 corresponds to a bubble that gives the distance of a sorg a higher priority than the current orientation of the sorg. The further the dot is away from the dashed diagonal line, the higher the priority of the sorg's distance in the decision making of such a bubble is. As can be seen in figure 4 the bubbles evolve a decision strategy in setup 1 that favors the factor "distance" significantly over the factor "angle". This means, that the bubbles always flee away from the nearest sorg, regardless of the direction the sorg is facing to. In contrast to that, in setup 2 the artificial evolution led to a significantly different decision strategy: The bubbles still favor "distance" a little bit more over "angle" in their decision making, but in this setup, the orientation of the sorg plays an almost similar important role.

A similar picture is found in the sorgs decision making after 200,000 time steps of evolution (Fig. 5). In setup 1, the sorgs choose mainly the closest bubble. But in contrast to the bubbles (Fig. 4), the angle the sorg has to turn towards the bubble (variable "angle") has a significant weight. In setup 2, the weight associated with the variable "distance" decreased, while the weight of the variable "angle" increased. This is a plausible result, because the sorgs are faster in this setup and the bubbles have a higher turning ability (agility) in this setup. The investigation of the two other factors

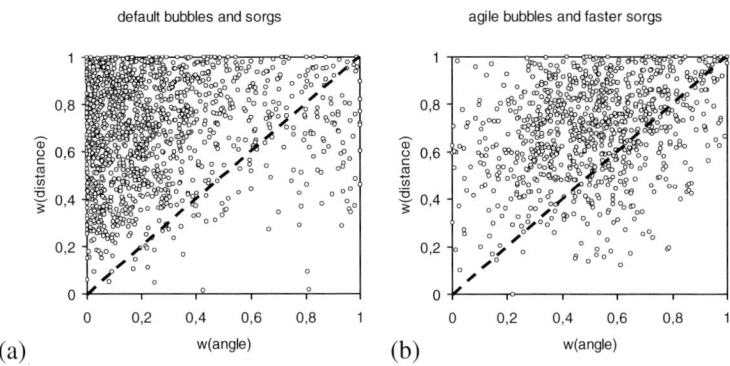

Fig. 4. Weighting factors of bubbles after the artificial evolution. (a) In setup 1, the evolution led to bubbles that use primarily the distance of sorgs as an important factor in their decision making. (b) In setup 2, the orientation of the sorg plays a significantly higher role.

evaluated during the decision making of the sorgs revealed more interesting details (Fig. 6). In setup 1, the variable "speed_surplus" gets almost ignored (weight close to zero), and the variable "size" is weighted also on a very low extent. In contrast to that, these weights increased significantly in setup 2. The faster sorgs now select bigger bubbles and currently slow-moving bubbles with a significantly higher frequency. This is a sound result, because this decision strategy allows the sorgs to exploit their increased speed abilities.

Table 2. Median and IQR (=interquartile range) values of the evolved values of the weighting factors in the decision making process of bubbles and sorgs

weight	Setup1 (default abilities of both species)		Setup2 (agile bubbles and fast sorgs)	
	Bubbles	Sorgs	Bubbles	Sorgs
w(distance)	0.74±0.33	0.83±0.17	0.72±0.29	0.80±0.24
w(angle)	0.19±0.32	0.24±0.24	0.50±0.32	0.47±0.38
w(size)	---	0.32±0.33	---	0.56±0.33
w(speed_surplus)	---	0.11±0.14	---	0.38±0.12

As a summary of the results of the evolutionary runs, we compiled the final relative values that evolved for w(distance)/w(angle) and for w(size)/w(speed_surplus) in figure 7. It shows that both species evolved a higher dominance of the factor "distance" over the factor "angle" in the default setup (setup 1). Also the variable "size" had a higher dominance over the variable "speed_surplus" in setup1 compared to setup 2. All of these differences were found to be statistically significant (t-test, $p<0.01$). Table 2 shows a summary of the median value of weights that evolved in both setups. In a final simulation run we investigated whether or not the evolved values of the weighing factors in our two setups are able to create plausible population dynamics in *bubbleworld*. As can be seen in figure 8, well known population

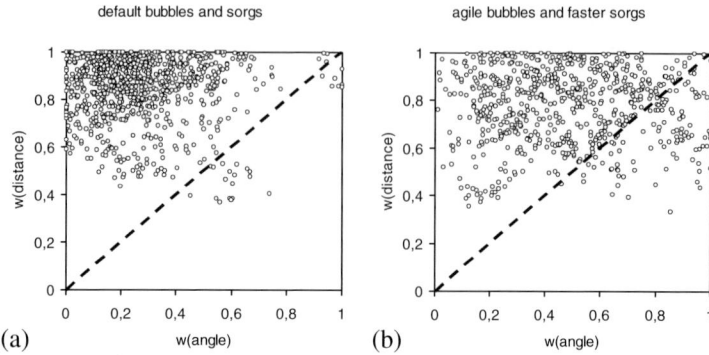

Fig. 5. Weighting factors of sorgs after the artificial evolution – part 1. (a) In setup 1, the evolution led to sorgs that use primarily the distance of bubbles in their decision making. On average the weight of the variable "distance" was approx. 4 times higher than the weight of the factor angle. (b) In setup 2, the weight of the variable "angle" increased significantly.

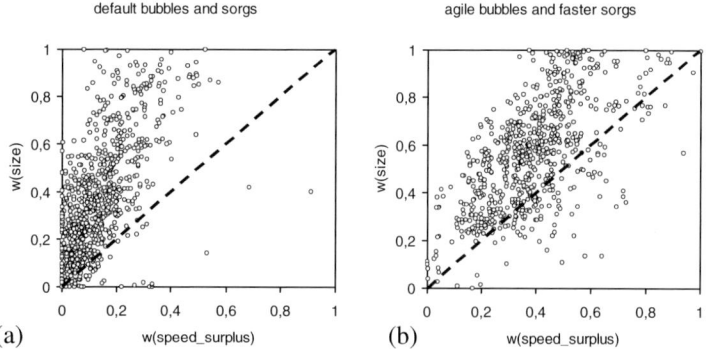

Fig. 6. Weighting factors of sorgs after the artificial evolution – part 2. (a) In setup 1, the evolution led to sorgs that mainly ignore the speed of the bubbles. At a small extend, the variable "size" is used in decision making. (b) In setup 2, the weight of the variables "size" and "speed-surplus" increased significantly, with the result that bigger bubbles are attacked more frequently, allowing the sorgs to draw advantage from their increased maximum speed.

dynamics for predator-prey ecosystems are found. Figure 8 shows also a simulation of the classical Lotka-Volterra model (with intraspecific competition and with a stochastic component). As can be seen by the subfigures 8b, 8d, 8f, the proportions of predators to prey in *bubbleworld* resemble the well known trajectories for the classical differential-equation model.

Finally, we investigated whether or not the evolved values of the weighing factors in our two setups are able to create plausible population dynamics in *bubbleworld*. As can be seen in figure 8, well known population dynamics for predator-prey systems are found. Figures 8e and 8f show asimulation of the classical Lotka-Volterra model (with intraspecific competition and a stochastic faktor, taken from [9]):

$$\frac{dN_{Prey}}{dt} = r_{Prey} \cdot \left(1 - \frac{N_{Prey}}{K_{Prey}}\right) \cdot \delta - s_{Prey} \cdot N_{Prey} \cdot N_{Pred} \tag{4}$$

$$\frac{dN_{Pred}}{dt} = r_{Pred} \cdot N_{Pred} \cdot N_{Prey} \cdot \delta - s_{Pred} \cdot N_{Pred} \tag{5}$$

In equations 4 and 5, N_{Prey} and N_{Pred} represent the population sizes of prey and of predators. r_{Pey} and r_{Pred} represent their growth rates, s_{Prey} and s_{Pred} represent their death rates. K_{Prey} represent the "carrying capacity" of the habitat, which is determined by intraspecific competition. δ represents a stochastic component. These equations are well known to describe a predator-prey system that tends towards an equilibrium point, but that is forced to oscillate due to stochastic factors. In *bubbleworld*, the growth rates mentioned above are represented by the reproduction process of bubbles and sorgs (see figure 2). As in equation 4, the bubbles in *bubbleworld* can only die if they are eaten by sorgs. The carrying capacity is implemented by lowering the reproductive rate of bubbles and sorgs according to the number of species mates that are within their sensory range. And the stochastic component is mainly represented by the "random walk" that is the basic behavioural program of bubbles and sorgs in times when they are not involved in a hunting episode. As can be seen by the subfigures 8b, 8d, 8f, the proportions of predators to prey in *bubbleworld* resemble the well known trajectories for the classical differential-equation model.

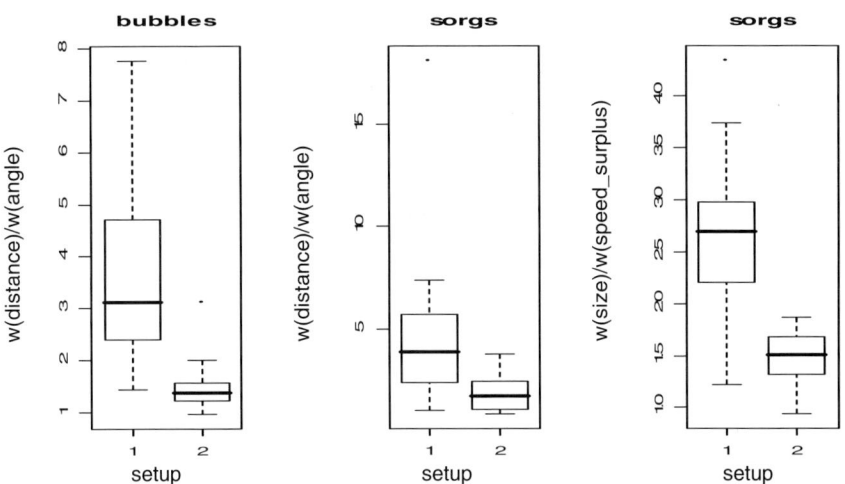

Fig. 7. Results of both setups in the evolutionary runs. $N_1=24$, $N_2=22$. In each single run, we calculated the median ratio of the weights that are depicted in the figures 4,5, and 6. The evolved relative values of w(*distance*) vs. w(*angle*) were significantly higher than in setup 2. Also the evolved relative values of w(*size*) vs. w(*speed_surplus*) were significantly higher in setup 1 than in setup 2 (t-test, p<0.01).

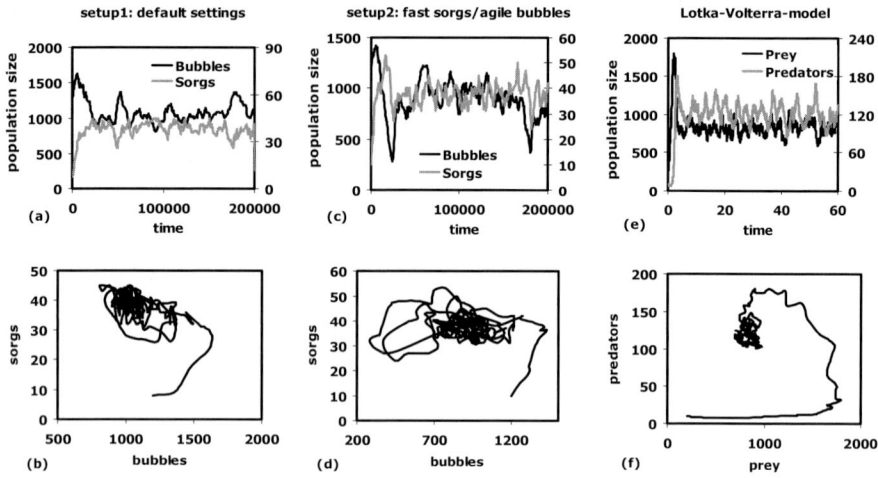

Fig. 8. Population dynamics of sorgs and bubbles. (a,b,c,d): In two exemplary simulations using the median evolved parameters listed in table 2 (without any evolution), well-known population dynamics [8][9][10][11] for a predator-prey relationships can be observed. For comparison, one additional simulation of the classical Lotka-Volterra differential-equation model is depicted in the subfigures (e) and (f).

4 Discussion

In conclusion, we showed that successful evolution can be simulated with *bubble-world.evo*. The evolution accounts for different abilities of the predators and prey in the two simulated setups. Finally, the evolution produced animals that are able to generate a predator-prey system that shows systems dynamics that are comparable to the examples found in literature for real-world ecosystems [8] [9] [10][11]. This is important, because recent studies in "artificial life" were able to evolve creatures that show interesting behaviors, but emergence of classical predator-prey dynamics was never found in these approaches. We achieved this result mainly by giving our agents a nature-like behavior and by evolving not the behavioral patterns itself but by evolving the decision mechanism that triggers the behaviors in a certain manner. Although the results of *bubbleworld.evo* are quite promising, it was not possible to find two interesting processes so far: significant levels of co-evolution over time and the emergence of distinct species that show different successful behaviors. The main reason for that is that the spectrum of possible behaviors in *bubbleworld.evo* is limited and therefore it does not represent an "open-ended" evolution (as is discussed in [7]).

The prey-selection mechanism of predators was already investigated by [14], but without artificial evolution and the analysis focused on the decision-making of the predator only. In [15], the evolution of a forager that moves in a "braitenberg"-vehicle-like way was demonstrated. In [16] artificial neural networks were evolved which produced low-level behaviors, but these studies did not lead to classical predator-prey dynamics. The approach closest to *bubbleworld.evo* so far is described in

[17], but in this approach the autotrophes (prey) reproduced at a certain fixed rate and were immobile, so that principally no co-evolution of behaviors could emerge in this system. Future versions of *bubbleworld.evo* will increase the potential behavioral repertoire of the agents by describing the movement patterns by repulsive and attracting forces. Such a behavioral representation can also describe herding and flocking behaviors. It is also planned to allow age-dependent changes in the parameters that shape behavior and decisions. This way, younger (and slower) sorgs can follow different hunting strategies than older (faster) sorgs.

References

1. Dawkins, R.: Das egoistische Gen. Springer-Verlag, Berlin Heidelberg New York (1978).
2. Darwin, C. R.: The Origin of Species. John Murray, London. (1859).
3. Alcock, J.: Animal Behavior: An Evolutionary Approach. Sinauer Associates, Inc. (1998).
4. Adami, C.: Introduction to Artificial Life. Springer Verlag, New York (1998).
5. Levy, S.: Artificial Life. Pantheon Books, Random House, Inc. (1992).
6. Sims, K.: Evolving 3D Morphology and Behavior by Competition. In: Brooks, R., Maes, P., (eds.): Articial Life IV Proceedings. Cambridge, MA: MIT Press (1994).
7. Komosinski, M., Rotaru-Vaga, A. : From Directed to Open-Ended Evolution in a Complex Simulation Model. In: Bedau, M.A., McCaskill, J.S., Packard, N.H., Rasmussen, S. (eds.): Artificial Life VII. MIT Press, Cambridge, MS., London, U.K. (2000) 293 – 299.
8. Wilson, W.: Simulating Ecological and Evolutionary Systems in C. Cambridge University Press (2000).
9. Bernstein, R.: Population Ecology. An Introduction to Computer Simulations. Wiley & Sons Ltd. (2003)
10. Edelstein-Keshet, L.: Mathematical Models in Biology. McGraw Hill (1988).
11. Wissel, C.: Theoretische Ökologie. Springer Verlag (1989).
12. Woolridger, M.: An Introduction to Multiagent Systems. John Wiley & Sons, Chichester, England (2002).
13. Kennedy, J., Eberhart, R.C.: Swarm Intelligence. Academic Press, San Francisco, San Diego, New York, Boston, London, Sydney, Tokyo, (2001).
14. Nishimura, S.I.: Studying Attention Dynamics of a Predator in a Prey-Predator System. In: Bedau, M.A., McCaskill, J.S., Packard, N.H., Rasmussen, S. (eds.): Artificial Life VII. MIT Press, Cambridge, MS., London, U.K. (2000) 337 – 342.
15. Hutt, B., Keating, D.: Artificial Evolution of Visually Guided Foraging Behaviour. In: Adami, C., Belew, R.K., Kitano, H., Taylor C.E. (eds.): Artificial Life VI. MIT Press, Cambridge, MS., London, U.K. (1998) 393 – 397.
16. Channon, A.D., Damper R.I.: Evolving Novel Behaviors via Natural Selection. In: Adami, C., Belew, R.K., Kitano, H., Taylor C.E. (eds.): Artificial Life VI. MIT Press, Cambridge, MS., London, U.K. (1998) 384-388.
17. Gracia, N., Pereira, H., Lima, J.A., Rosa, A.: Gaia: An Artificial Life Environment for Ecological Systems Simulation. In: Langton, C.G., Shimohara, K. (eds.): Artificial Life V. MIT Press, Cambridge, MS., London, U.K. (1997) 124-131.

Incremental Evolution of Target-Following Neuro-controllers for Flapping-Wing Animats

Jean-Baptiste Mouret, Stéphane Doncieux, and Jean-Arcady Meyer

Université Pierre et Marie Curie-Paris 6, UMR7606, AnimatLab/LIP6, Paris,
F-75015 France;CNRS, UMR7606, Paris, F-75015, France
{jean.baptiste.mouret, stephane.doncieux, jean-arcady.meyer}@lip6.fr

Abstract. Using an incremental multi-objective evolutionary algorithm and the ModNet encoding, we generated working neuro-controllers for target-following behavior in a simulated flapping-wing animat. To this end, we evolved tail controllers that were combined with two closed-loop wing-beat controllers previously generated, and able to secure straight flight at constant altitude and speed. The corresponding results demonstrate that a wing-beat strategy that consists in continuously adapting the twist of the external wing panel leads to better manoeuvring capabilities than another strategy that adapts the beating amplitude. Such differences suggest that further improvements in flying control should better rely on some sort of automatic incremental evolution procedure than on any hand-designed decomposition of the problem.

1 Introduction

Birds continuously demonstrate capabilities which would be of great interest for most Unmanned Aerial Vehicles (UAVs). They are highly agile, able to take off without any runway, to settle on a branch, and to exploit thermals like a sailplane. Consequently, bio-inspired flapping-wing platforms could represent useful trade-offs able to benefit from both the manoeuvrability of helicopters and the energy efficiency of standard airplanes.

However, taking inspiration from birds to design a flapping-wing UAV requires a deep understanding of complex aerodynamic principles that nature learned to exploit in about 150 millions years of evolution. In particular, the numerous degrees of freedom of such UAV must be carefully synchronized to produce adequate thrust and lift forces. Additionally, the corresponding rhythmic movements must be continuously adapted to unpredictable changes in the surrounding air mass. To tackle such issues, that are currently not solved by traditional engineering approaches, the ROBUR project of the AnimatLab [7] aims at designing the morphology and control of a flapping-wing animat through artificial evolution.

In a previous work [15], we used a multi-objective evolutionary algorithm to generate wing-beat neuro-controllers able to secure a straight and horizontal flight at constant speed, even in cases where the flying animat was artificially slowed down. Two efficient control strategies emerged, but with no indication about which one should be used in more challenging conditions.

S. Nolfi et al. (Eds.): SAB 2006, LNAI 4095, pp. 606–618, 2006.

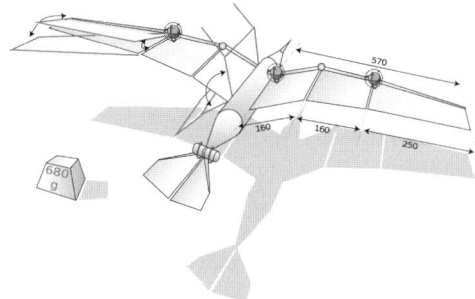

Fig. 1. The simulated bird is modelled using cones, cylinders and rigid panels. A wings internal panel can be moved along the dihedral and the twist axis, while its external panel can be moved along the twist and the sweep axis.

The goal of the research effort described in this article was to extend this work to the control of target-following behavior. To this end, we used an incremental approach that already proved to be efficient at designing wheeled [19,2], legged [12] and flying [1] robots. We thus capitalized on the previously evolved wing-beat controllers, and let evolution combine their effect with that of newly generated tail controllers that would force the animat to orient itself towards a targeted direction. We also assumed that the corresponding results would help better assessing the relative advantages of the two wing-beat strategies just mentioned.

The corresponding experiments called upon a realistic aerodynamic simulator that computes lift and drag forces whatever the local airflow direction. The underlying aerodynamic model has been validated in a wind-tunnel for a fixed-wing UAV.

The simulated bird was made of cones and cylinders which made up its body, and of rigid panels that composed its wings and tail (figure 1). The total wingspan was 124 cm. Additional relevant details are to be found in [15].

This article starts with a summary of the previous results we obtained with wing-beat controllers. The next section describes the evolution of tail neuro-controllers for target-following. Sample trajectories and typical neural networks are then exposed, and the corresponding results are discussed.

2   Wing-Beat Controllers

Contrary to airplanes which use their wings to sustain themselves, and a propeller to create thrust, birds use their wings to create both an upward and a thrust forces. A wing-beat is made of two distinct phases, the down-stroke and the up-stroke. During the down-stroke, the wings are fully extended and powered downward. The twist is tilted down during this phase, particularly towards the tip. As a consequence, the lift force, created by the pressure difference around the airfoil, is oriented forward and upward. During the up-stroke, the wing is partially folded, to reduce the drag. Additional information on these bird-flight kinematics can be found in [16,10].

We previously evolved wing-beat neuro-controllers able to exploit such forces and to maintain a flying animat at a constant speed and altitude despite external disturbances. The design of these controllers drew inspiration from the work of biologists on Central Pattern Generators (CPGs) [5,3,13] and called upon both non-linear oscillators [3,13] and standard McCulloch and Pitt's neurons. These controllers could use a speed sensor as input, and four actuators as outputs: the dihedral and the twist of a wing's internal panel, the twist and the sweep of its external panel. The symmetry between wing-beats was forced.

Evolutionary algorithms are the only means that allow to optimize both the topology and the parameters of this kind of neural-networks. The problem to be solved requiring multiple trade-offs – from the energy consumption mini-mization to the maximization of accelerations – the multi-objective evolutionary algorithm MOGA [8] was used, together with ModNet[6], a modular encoding scheme adapted to the task of evolving neural networks. The corresponding fit-ness function depended upon six objectives [15], which were evaluated in two stages.

Two different classes of optimal strategies emerged that both relied on the same kinematic principles. According to the first strategy, the twist of a wing's external panel is increased when an acceleration is required, hence increasing the thrust component of the lift force. The wing being folded during the up-stroke, the twist of this panel is not changed. The analysis of the corresponding controllers showed that they implement a simple proportional control of the external twist as a function of the difference between the current speed and the targeted speed.

According to the second wing-beat strategy, all degrees of freedom of a wing exhibit a sinusoidal movement. When an acceleration is required, the amplitude of the oscillations is increased and, consequently, the magnitude of the upward and forward components of the lift force are increased.

Videos of some animats exhibiting these strategies can be downloaded from our website: `http://animatlab.lip6.fr/RoburEvolvingEn`. To the best of our knowledge, this is the first time that closed-loops controllers are obtained for flapping-wing flight.

3   Target Following

The detailed kinematics used by birds to change the direction of their flight largely remain an open question, the answer to which probably varies across bird species. It has been suggested that the tail might be used for such use, in a manner similar to the use of elevators in an aircraft [18]. However, some birds succeed to fly without their tail, and mostly rely on wing movements for that. For instance, it has been shown that pigeons use down-stroke velocity asymmetries and rapid alternating wing movements to turn [21,20].

Likewise, although the control of standard airplanes and UAVs is a widely studied topic, described in many textbooks like [14] for instance, it remains to be proved that classical control methods can be directly used to control an artificial flapping-wing bird because the dynamics of such an engine are complex to model, and because wing-beats generate a lot of parasite movements, especially when orienting.

Additionally, it turns out that radio-controlled ornithopters built by hobby-ists exploit their tail to execute simple manoeuvres, as many radio-controlled airplanes do.

For all these reasons, and because the control of a tail can easily be decoupled from the wing-beat control, in a first approach towards implementing useful flying capacities, we chose to evolve tail controllers that could be combined with the wing-beat controllers previously evolved (section 2). More precisely, the objective of these additional controllers was to orient the artificial bird towards a target point, for instance a GPS way-point or a visual landmark, while keeping its altitude constant. To this end, we used exactly the same methodology, calling upon MOGA and ModNet softwares, that the one evocated above and that led to efficient flapping-wing controllers.

To the best of our knowledge, only two papers previously dealt with the gener-ation of kinematics for a flapping-wing artificial bird able to turn [22,17]. But the goal of these research efforts was to create visually convincing movements and not to design closed-loop controllers. Consequently, the optimization of the kinematics parameters characterizing each trajectory were computationally too greedy to be tested as competitive approaches to the one that has been chosen here.

3.1   Sensors and Actuators

In the absence of any a priori knowledge concerning the sensors that birds use to control their flying manoeuvres, we allowed evolution to incorporate, or not, four sensors in the neural controllers it would generate: an altitude sensor - assessing the difference between the current and targeted altitude - a direction sensor - reorient-ing the animats relative direction to the target - a roll sensor and a pitch sensor.

Furthermore, in the absence of precise specifications of real sensors to be embedded on a real platform, in this preliminary stage we chose to use ideal sensors that would prevent evolution from exploiting specific characteristics of specific devices.

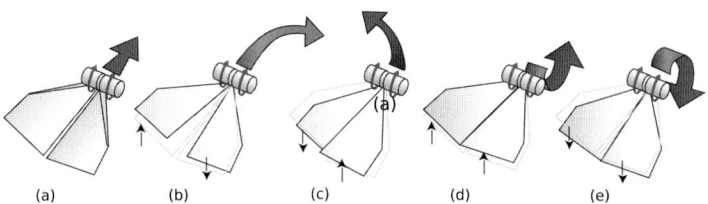

Fig. 2. A v-tail is used to control the artificial bird. (a) When the the panels are in the neutral position, the animat move along a straight line. (b) When the left panel is raised and the right one lowered, the animat turns right. (c) Symetrically, when the left panel is lowered and the right one rised, the animat turns left. (d) When both panels are rised, the animat goes up. (e) Symetricallly, when both panels are lowered, the animat goes down.

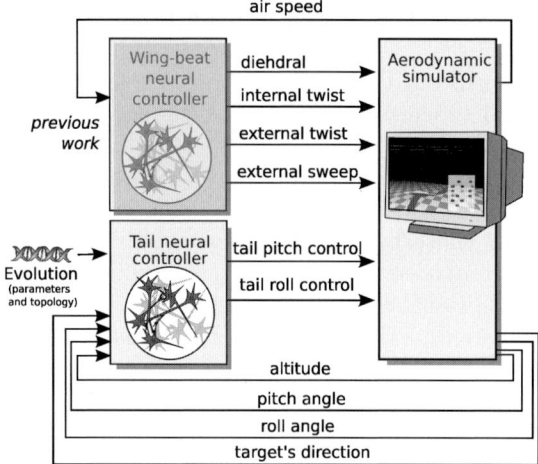

Fig. 3. Overview of the control loop. The wing-beat controller has been evolved in a previous work.

Besides the wing characteristics evocated above, an ideal servo-mechanism was used to move each of the two panels that constituted the animats V-shaped tail. To provide an intuitive control of the bird, these two actuators were mixed together in a way similar to that used in radio-controlled sailplanes. Thus, two virtual actuators are provided, one to control the pitch angle and one to act on the yaw and roll angles. Figure 2 displays some typical reactions of the simulated bird to various tail configurations.

3.2 Fitness

A multi-objective evolutionary algorithm was used to seek controllers able to secure a constant-altitude flight while pointing towards a given target. Figure 3 shows an overview of the corresponding control loop.

Let us denote by N the total length of the evaluation. The bird started its flight at $15m.s^{-1}$ and the position of its center of gravity $g(n)$ was measured at each time-step n. The altitude objective was written as:

$$O_{alt} = -\frac{1}{N} \sum_{n=0}^{n=N} |\mathbf{g}_z(n) - \mathbf{Z}| \tag{1}$$

where $\mathbf{g}_z(n)$ denotes the altitude of the artificial bird at time-step n and Z the desired altitude.

Let first define the vector $\mathbf{v}(n)$ which links the current position $\mathbf{g}(n)$ to the target position $\mathbf{T}$:

$$\mathbf{v}(n) = \mathbf{T} - \mathbf{g}(n). \tag{2}$$

We then defined the second objective, which rewarded how close the simulated bird was to the desired direction:

$$O_{tar} = -\frac{1}{N} \sum_{n=0}^{n=N} |atan2(\mathbf{v}_y(n), \mathbf{v}_x(n)) - \theta(n)| \qquad (3)$$

The $atan2(y, x)$ function calculates the arc tangent of $\frac{y}{x}$ except that the signs of both arguments are used the determine the quadrant of the result.

These two objectives had to be maximized, with an optimal value of 0. They might be evaluated for k different targets. In this case, the results of successive evaluations were summed together.

$$O_{alt} = \frac{1}{k} \sum_{i=0}^{i=k} O_{alt,i} \qquad (4)$$

$$O_{tar} = \frac{1}{k} \sum_{i=0}^{i=k} O_{tar,i} \qquad (5)$$

3.3 Experimental Setup

We used the objectives just defined in conjunction with the MOGA algorithm and the ModNet encoding. Population's size was 350 with 60 % of elites. Six different targets were used, towards which the animat was expected to fly. The total evaluation length was 18000 time-steps, simulating 54 seconds of flight (9 seconds for each target).

Thanks to the use of a so-called model-module pool[6,15], the neural controllers that were evolved could incorporate and modify across successive generations any number of three different modules:

- a "derivative" module, which computed an approximation of the derivative of its input signal;
- an "integral" module, which computed an approximation of the integral of its input signal;
- a generic module, made of standard McCulloch and Pitt's neurons, with an evolvable structure.

Using two previously evolved controllers, each exhibiting a different wing-beat strategy (section 2), three evolutionary runs of 500 generations were performed to evolve tail controllers likely to complement them. About 12 hours on 20 Pentium at $2Ghz$ were required for each run.

4 Results

4.1 First Wing-Beat Controller

Starting with a wing-beat controller that adapted the twist of the external panel to maintain a targeted flying speed, the individuals that constituted the Pareto front of the last generation segregated in two populations: those that obtained good results on the altitude objective, and those that were efficient with respect to the target objective. The behavior of two randomly-selected individuals representative of each of these populations is shown on Figure 4.

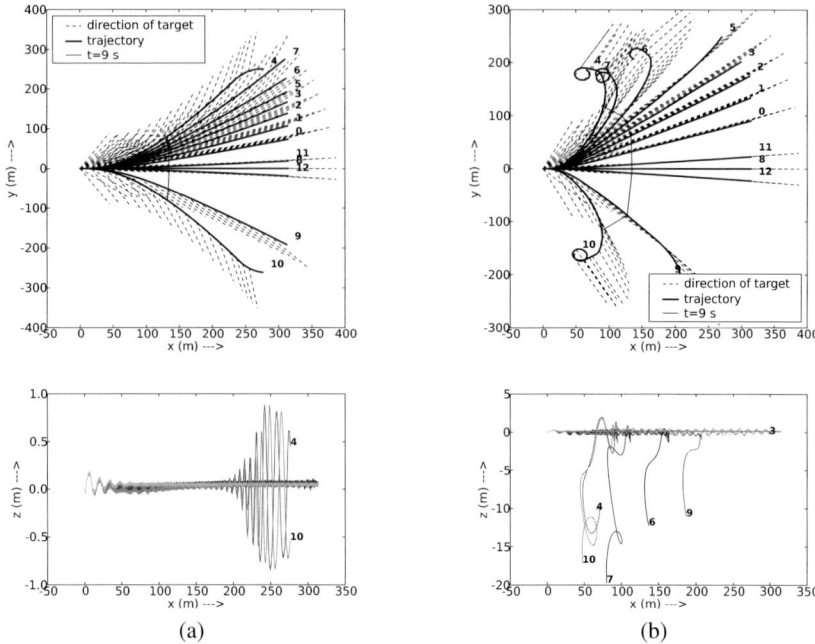

Fig. 4. (a) Top: Examples of target-following trajectories for an individual using the first type of wing-beat controller. Bottom: This individual obtained good results on the altitude objective. (b) Trajectories of another individual (Top) with good results on the target objective (Bottom).The targets that were used here were different from those that served to evaluate the individuals. For flights not exceeding the corresponding evaluation period (9 sec), the corresponding behaviors were satisfactory. Beyond this period, some trajectories led to stalling, particularly when the targeted orientation angle was high.

While both individuals succeeded to execute light turns, especially during the evaluation period (9 s), they didn't deal successfully with larger changes of directions beyond the evaluation period. They used a different approach to handle the latter case. The first individual, which had the best fitness for the altitude objective, stopped orienting when the required change in direction was too large. As a consequence, it did not loose altitude, but at the price of not aligning correctly with the target. The second individual, which had a better fitness on the target objective, often started orienting in the right direction, but ended stalling along a spiral trajectory. However, because such events occurred after the evaluation period, this individual was not much penalized with respect to the altitude objective.

We performed a Multiperturbation Shapley value Analysis (MSA) [11] to understand the inner workings of the first individual's tail controller (figure 5). The most useful neurons were those numbered 0, 1, 2, 3, 4 and 5, while the other neurons didn't seem to contribute a lot to the animat's orienting behavior. In particular, neurons 1, 3 and 4 had a large contribution to the first objective,

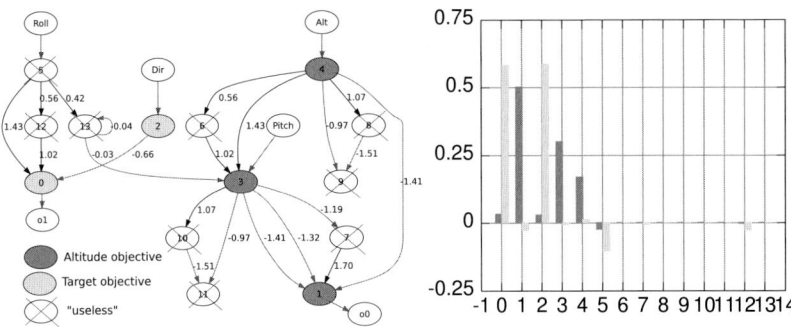

Fig. 5. Left: the neural network that produced the trajectories shown in figure 4 (a). Right: the results of the corresponding MSA. The y-axis gives the numbers identifying the neurons. The x-axis gives the Shapley values. High values indicate important contributions to each of the two considered objectives (altitude and target). Grey levels characterize which neuron is important to the realization of which objective.

while neurons 0 and 2 mostly contributed to the second objective only. This result indicates that the two objectives were decoupled by the evolutionary process, a conclusion that is confirmed by the close observation of the organization of this controller (Figure 5), which appears as almost split in two separate networks. The first one, on the left side of the figure, controls the flight direction using the target's direction sensor. The second one, on the right side, controls the elevator. These two independent controllers are linked by a connexion with a very low weight, which explains the null Shapley value of neuron 11.

By assigning a null or negative contribution to neurons 5, 12 and 11, the MSA also indicates that the roll sensor is not useful to this controller. Such could be also the case with the pitch sensor, as it will be shown later for an other controller, but the MSA is not conclusive on this point because one cannot decide if the utility of neuron 3 must be attributed to information brought by the altitude sensor, by the pitch sensor, or by both sensors. Be that as it may, we believe that the role of the roll sensor would be much greater if the animat had to fly in an unstable air mass, for instance to secure a constant roll angle despite external perturbations. This hypothesis could be tested by adding perturbations during the evaluation procedure, as we did in [15].

4.2   Second Wing-Beat Controller

Figure 6 shows the behavior of two individuals populating the Pareto front of the last generation, starting with a wing-beat controller that adapted the amplitude of the oscillations to maintain a targeted flying speed. These individuals were randomly-selected among those that respectively obtained good results on the altitude objective, and good results on the target objective.

The first individual exhibits a behavior similar to the one displayed on figure 4 because it stopped orienting when the required changes of direction were large.

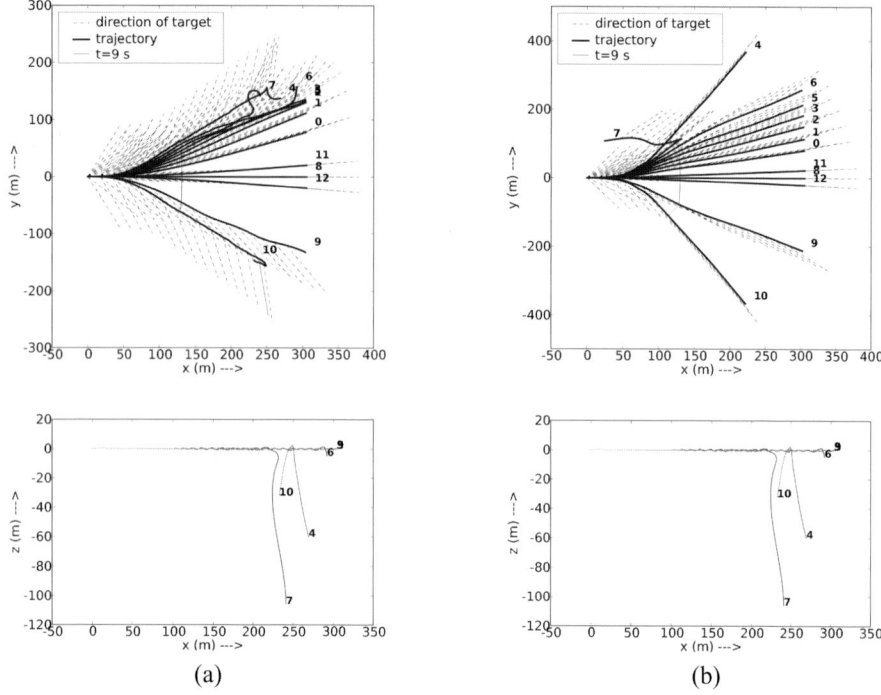

Fig. 6. (a) Top: Examples of target-following trajectories for an individual using the second type of wing-beat controller. Bottom: This individual obtained good results on the altitude objective. (b) Trajectories of another individual (Top) with good results on the target objective (Bottom). Both individuals performed correctly during the evaluation period (9s).

However, it stalled after having flied for about $200m$. Moreover, the error between the animat's direction and the targeted one was larger on figure 6 (a) than on figure 4 (a).

The second individual got the best fitness evaluation according to the angle objective. Surprisingly, although the animat succeeded to perform the largest turns (labeled 9 and 10), it couldnt avoid climbing up. This was due to the fact that the wing-beat controller reached a saturated state, according to which the wings were steadily flapped with a maximum strength, thus producing a maximum thrust. By flying faster, the bird generated more lift and, accordingly, went up. This saturated state was not reached for every target, as demonstrated by the stalling trajectory 7.

The analysis of the corresponding neural controller (figure 7) indicates that it is also split in two sub-networks, one controlling the direction and the other controlling the altitude. For this controller, both the roll and the pitch sensors seem useless. Again, these two sensors would probably be required for flights in an unstable air mass.

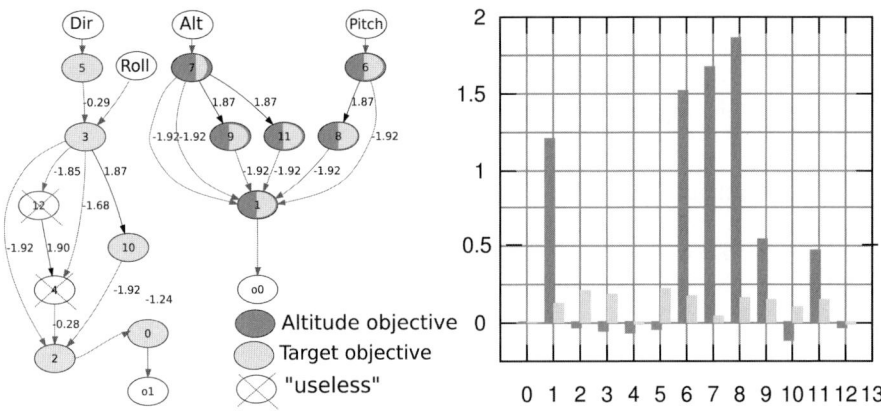

Fig. 7. Left: The neural network that produced the trajectories shown in figure 6 (a).
Right: Its MSA results.

4.3   Objective Space

Figure 8 displays the two objective scores attained by each animat generated
during the three evolutionary runs, for the two wing-beat strategies.

It thus turns out that individuals exploiting the first strategy are distributed in
most of the objective space, with a higher concentration close to the Pareto front.
The shape of this front expresses the necessity of a trade-off, since no individual
gets an optimal score on both objectives. The evolutionary process generated a lot
of such trade-off solutions situated near the optimum, i.e., at coordinates 0,0.

The exploration of the objective space by individuals exploiting the second
strategy was quite different. First, the best scores thus attained are substantially
lower than in the previous case, especially on the target objective. Second, the
explored solutions cover a much limited range of possible scores.

The horizontal line with a target objective value of -0.6 corresponds to individuals that did not turn at all. These individuals quickly disappeared when we

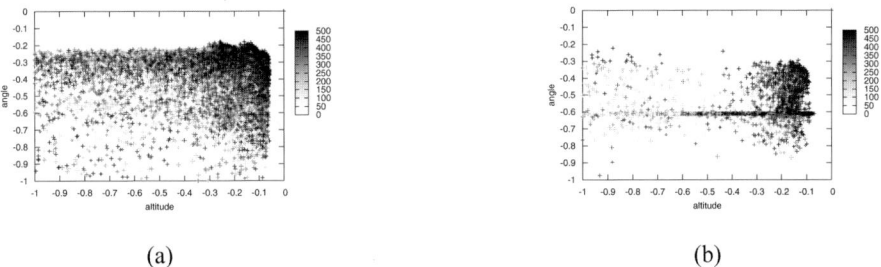

(a)   (b)

Fig. 8. Exploration of the objective space by all individuals exploiting the first (a) or
the second (b) wing-beat strategy. Each dot corresponds to an individual. Grey levels
denote the last generation for which this individual was present in the population.

used the first wing-beat controller, but they were still present after 500 generations when the second controller was used.

All these results suggest that optimizing a target-following behavior is more difficult with the second controller than with the first.

5   Discussion

The above results demonstrate that, once neural controllers for straight-line flight have been evolved, it is possible to capitalize on the corresponding networks to evolve additional tail controllers that exhibit minimal target-following capacities. However, as mentioned above, better results would probably be obtained by relaxing the symmetry constraint that was imposed here on wing movements, and by exploiting the manoeuvrability capacities thus offered. Such approach might entail the joint evolution of wing-beat and tail control, a task that seems highly challenging according to current technology. To raise its chances of success, the recourse to some sort of automatic incremental methodology seems mandatory. Indeed, it has been shown here that one cannot rely on fundamental principles or empirical knowledge to decide which, among two available wing-beat controllers, would be better suited to pave the way to additional flying manoeuvrability. Nevertheless, their aptitudes for this endeavour ultimately turned out to be quite different.

Two main reasons motive the use of incremental evolutionary approaches to evolutionary robotics:

- The bootstrap problem. In many real-life situations, an intermediate action is required – e.g. pushing a button to switch-on a light – before receiving any reward – e.g. going to the light to get food. By decomposing the problem into sub-tasks, one may guide the evolutionary process towards satisfying solutions that would otherwise be hard or impossible to discover.
- The search space problem. By imposing intermediary stages, one reduces the size of the search space, hence speeding-up the evolutionary process.

Some promising work has been carried out to automatize such a procedure using cooperative co-evolution [9,4]. We plan to assess the applicability of similar approaches to flying behavior in the near future.

6   Conclusion

Although evolving flying robots seems a greater challenge than evolving crawling, walking, or swimming artefacts, we have shown that a suitable evolutionary algorithm, combined with an efficient coding and a two-stage approach, made it possible to generate close-loop controllers for a target-following flying behavior. However our results suggest that future improvements should better rely on some sort of automatic incremental evolution procedure than on any hand-designed decomposition of the considered problem.

Acknowledgements. This work was supported by a UPMC BQR grant.

References

1. G. J. Barlow, C. K. Oh, and E. Grant. Incremental evolution of autonomous controllers for unmanned aerial vehicles using multi-objective genetic programming. In *Proceedings of the 2004 IEEE Conference on Cybernetics and Intelligent Systems (CIS)*, pages 688–693, Singapore, 1-3 December 2004. IEEE.
2. J. Chavas, C. Corne, P. Horvai, J. Kodjabachian, and J.-A. Meyer. Incremental evolution of neural controllers for robust obstacle-avoidance in Khepera. In P. Husbands and J.-A. Meyer, editors, *Proceedings of The First European Workshop on Evolutionary Robotics - EvoRobot98*, pages 227–247. Springer-Verlag, 1998.
3. A. H. Cohen, P. J. Holmes, and R. H. Rand. The nature of the coupling between segmental oscillators of the lamprey spinal generator for locomotion : A mathematicaal model. *Journal of Mathematical Biology*, 13:345–369, 1982.
4. Edwin D. de Jong and Jordan B. Pollack. Ideal evaluation from coevolution. *Evolutionary Computation*, 12(2), 2004.
5. F. Delcomyn. Neural basis for rhythmic behaviour in animals. *Science*, 210:492–498, 1980.
6. S. Doncieux and J.-A. Meyer. Evolving neural networks for the control of a lenticular blimp. In G. R. Raidl, S. Cagnoni, J. J. Romero Cardalda, D. W. Corne, J. Gottlieb, A. Guillot, E. Hart, C. G. Johnson, E. Marchiori, J.-A. Meyer, and M. Middendorf, editors, *Applications of Evolutionary Computing, EvoWorkshops2003: EvoBIO, EvoCOP, EvoIASP, EvoMUSART, EvoROB, EvoSTIM*. Springer Verlag, 2003.
7. S. Doncieux, J.-B. Mouret, L. Muratet, and J.-A. Meyer. The ROBUR project: towards an autonomous flapping-wing animat. In *Proceedings of the Journées MicroDrones*, Toulouse, 2004.
8. C. M. Fonseca and P. J. Fleming. Genetic algorithms for multiobjective optimization : formulation, discussion and generalization. In *Proceedings of the Fourth International Conference on Evolutionary Programming*, pages 416–423, 1993.
9. G. L. Haith, S. P. Colombano, J. D. Lohn, and D. Stassinopoulos. Coevolution for problem simplification. *Proceedings of the Genetic and Evolutionary Computation Conference*, 1999.
10. T.L. Hedrick, B.W. Tobalske, and A.A. Biewener. Estimates of circulation and gait change based on a three-dimensional kinematic analysis of flight in cockatiels (nymphicus hollandicus) and ringed turtle doves (streptopelia risoria). *Journal of Experimental Biology*, 205:1389–1409, 2002.
11. A. Keinan, B. Sandbank, C. C. Hilgetag, I. Meilijson, and E. Ruppin. Fair attribution of functional contribution in artificial and biological networks. *Neural Computuation*, 16(9):1887–1915, 2004.
12. J. Kodjabachian and J.-A. Meyer. Evolution and development of neural networks controlling locomotion, gradient-following, and obstacle-avoidance in artificial insects. *IEEE Transactions on Neural Networks*, 9:796–812, 1997.
13. K. Matsuoka. Mechanisms of frequency and pattern control in the neural rhythm generators. *Biological Cybernetics*, 56:345–353, 1987.
14. D. McLean. *Automatic Flight Control Systems*. Prentice Hall, New York, 1990.
15. J.-B. Mouret, S. Doncieux, T. Druot, and J.-A. Meyer. Evolution of closed-loop neuro-controllers for flapping-wing animats. 2006. submitted for publication.
16. U. M. Norberg. *Vertebrate Flight*. Springer-Verlag, 1990.
17. Y. S. Shim, S. J. Kim, , and C. H. Kim. Evolving flying creatures with path following behaviors. In *The 9th International Conference on the Simulation and Synthesis of Living Systems (ALIFE IX)*, pages 125–132, 2004.

18. A. L. R. Thomas. On the tails of birds. *BioScience*, 47(4):215–225, April 1997.
19. J. Urzelai, D. Floreano, M. Dorigo, and M. Colombetti. Incremental robot shaping. *Connection Science Journal*, 10(384):341–360, 1998.
20. D. R. Warrick, M. W. Bundle, and K. P. Dial. Bird maneuvering flight: Blurred bodies, clear heads1. *INTEG. AND COMP. BIOL.*, 2002.
21. D. R. Warrick and K. P. Dial. Kinematic, aerodynamic and anatomical mechanisms in the slow, maneuvering flight of pigeons. *The Journal of Experimental Biology*, 201:6552013672, 1998.
22. J. Wu and Z. Popoviè. Realistic modeling of bird flight animations. *ACM Trans. Graph.*, 22(3):888–895, 2003.

Evolution and Adaptation of an Agent Driving a Scale Model of a Car with Obstacle Avoidance Capabilities

Ivan Tanev[1], Michal Joachimczak[2], and Katsunori Shimohara[1]

[1] Department of Information Systems Design, Doshisha University,
1-3 Miyakodani, Tatara, Kyotanabe 610-0321, Japan
{itanev, kshimoha}@mail.doshisha.ac.jp
[2] Institute of Oceanology, Polish Academy of Sciences, Department of Genetics and Marine Biotechnology, Powstancow Warszawy 55, 81-712 Sopot, Poland
mjoach@iopan.gda.pl

Abstract. We present an approach for evolutionary design of the driving style of an agent, remotely operating a scale model of a car running in a fastest possible way. The agent perceives the environment from a video camera and conveys its actions to the car via standard radio control transmitter. In order to cope with the video feed latency we propose an anticipatory modeling in which the agent considers its current actions based on the anticipated intrinsic (rather than currently available, outdated) state of the car and its surrounding. The driving style is first evolved offline on a software model of the car and then adapted online to the real world. An online evolutionary adaptation of the offline-obtained best styles to the needs to avoid a small obstacle results in lap times that are virtually the same as the best lap times achieved on the same track without obstacles. Presented work is a step towards the automated design of the control software of remotely operated vehicles capable to find an optimal solution to various tasks in different environmental situations. The results, also, can be seen as an attempt to explore the feasibility of developing a framework of adaptive racing games in which the human competes against a computer with matching capabilities, both operating physical, scale models of cars.

1 Introduction

The success of the computer playing games (like chess [5]) has long served as touchstone of the progress in the field of artificial intelligence (AI). The expanding scope of applicability of AI, when the latter is employed to control the individual characters (agents) which are able to "learn" the environment and to adopt an adaptive optimal (rather than a priori preprogrammed) playing tactics and strategy include soccer [10], F1 racing [14], etc. [3]. Focusing in the domain of car racing, in this work we consider the problem of designing a driving agent, able to remotely control a scale model of a racing car, which runs in a fastest possible way. Our work is motivated by the opportunity to develop an agent, able to address some of the challenges, which a human driver of racing car faces. In order to provide a fastest laps times around the circuit, the driver needs to define the best driving (racing) line, or the way the car enters, crosses the apex, and exits the turns of the circuit. Moreover, realizing the once defined optimal line, the driver has to make a precise judgment about the current

S. Nolfi et al. (Eds.): SAB 2006, LNAI 4095, pp. 619–630, 2006.

state (i.e., position, orientation and velocity) of the car and the environment, and to react quickly and precisely.

The *objective* of our work is an automated evolutionary design of the functionality of driving agent, able to remotely operate a scale model of racing car (hereafter referred to as "car") running in a fastest way around. The agent should be able to control the car in (i) a consistent way and (ii) to avoid small, static obstacles. An agent with such capabilities would open up an opportunity to build a framework of adaptive racing games in which the human competes against a computer with a matching capabilities, with both of them remotely operating scale models, rather than simulated cars. The proposed evolutionary approach could be also applied for automated design of the control software of remotely operated vehicles capable to find an optimal solution to various tasks in different environmental situations.

Achieving the objective implies that the following tasks should be addressed: (i) developing an approach that allows the agent to adequately control the scale model of the car addressing the challenge of controlling a fast moving artifact via closed control loop with a finite feedback latency; (ii) formalizing the driving style and defining the key parameters that describe it, and (iii) developing an algorithm paradigm for automated definition of the fastest driving style by setting its key parameters to their optimal values.

The related work done by Wloch and Bentley [14] demonstrates the feasibility of applying genetic algorithms for automated optimization of the setup of the simulated racing car. However, neither the adaptation of the driving style to the setup of the car (i.e., a co-evolution of the driving style and the setup of the car) nor the use of a physical (scale) model of a car was considered in their work. Conversely, Togelius and Lucas [12] used scale models of cars in their research to demonstrate the ability of the artificial evolution to develop optimal neuro-controllers with various architectures. However, the effects of the inherent latencies in the video feedback on either the precision or the speed of the car was beyond the scope of their work. In our previous work [13], we used an evolutionary approach to optimize the controller of a scale model of a car. Although we did consider the feedback latency and proposed a way to alleviate its detrimental effect on the drivability of the car, we have considered neither the driving consistency nor the avoidance of obstacles or "guardrails", which are relevant for the real-world applications.

The remaining of the article is organized as follows. Section 2 introduces the hardware configuration used in our work. Section 3 elaborates on the anticipatory model employed by the driving agent to alleviate the detrimental effect of the feedback latency on the performance of the agent. Section 4 discusses the key attributes of driving style with obstacle avoidance capabilities and the proposed approach of genetic algorithms employed to automatically evolve them. Section 5 draws a conclusion.

2   System Configuration

2.1   The Car

We choose the 1:24 scaled model of an F1 racing car, with the bodywork stripped from decals and repainted for more reliable image tracking (Figure 1).

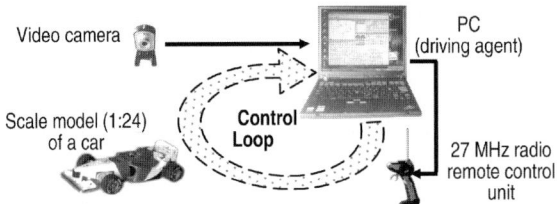

Fig. 1. System configuration

This off-the-shelf car features a simple two-channels radio remote control (RC) with functionality including "forward", "reverse", and "neutral" throttle control commands and "left", "right" and "straight" steering controls. The car has the following three favorable features: (i) a wide steering angularity, (ii) a spring suspension system in both front and rear wheels, and (iii) a differential drive. The former feature implies a high maneuverability of the car. The torsion spring of the rear suspension of the car functions as an elastic buffer, which absorbs the shocks, caused by the sharp and often violent alterations in the torque generated by the car's motor. These torque alterations occur during the pulse-width modulation (PWM) of the throttle, by means of which the driving agent regulates the speed of the car within the range from zero to the maximum possible value. In addition, torque alterations occur when the "reverse" throttle command is applied for braking of the car that still runs forward. The absorption of these shocks is relevant for the smooth transfer of the torque from the motor to the driving wheels of the car without an excessive wheelspin, achieving a good traction under braking and acceleration. Moreover, the absorptions of the shocks caused by the frequent torque alterations are important for the longevity of the transmission of the car. The last mentioned feature - differential rear wheels drive implies that that the torque of the motor is split and delivered to the rear wheels in a way that allows them to rotate at different angular speeds when necessary, e.g., under cornering. Therefore, the car turns without a rear wheels spin, which results in a smooth entrance into the turns and a good traction at their exits.

The main mechanical characteristics of the car are shown in Table 1. Characteristics #6-#9 are experimentally obtained from the car running on the considered track surface - a polyvinyl chloride carpet with coefficients of static and kinematic friction between the rubber tires and the surface $\mu_S=0.8$ and $\mu_K=0.7$, respectively. The tires of the turning car, operated at, and beyond the limits of the friction (grip, adhesion) forces, slide to some degree across the intended direction of traveling. The dynamic weight redistribution causes the grip levels at the front and rear wheels to vary as the turning car accelerates on "forward" or decelerates on either "neutral" or "reverse" throttle commands. This, in turn, yields different sliding angles for the front and rear wheels, causing the fast turning car to feature either a neutral steering (the slide angles of both axles assume the same values), an oversteer (slide angle of the front wheels are narrower than that of the rear ones) or understeer (slide angle of the front wheels are greater than that of the rears). In addition to the degradation of the maneuverability of the car, the sliding of the wheels results in a significant braking momentum that, in turn, reduces the velocity of the turning car if the latter enters the turn too fast. Moreover, the increased actual turning radius due to sliding of either a neutral or understeering car means that the car might enter the run-off areas or even hit the guardrails on tight corners of the track, which, in turn, might result either in a damage of the car, or lost of momentum (or both). Therefore, sliding limits the average velocity of the cornering car (due to the lower, than intended speeds along longer, than intended arcs), which may have a detrimental effect on the overall lap times. On the other hand, the sum of the driving and centrifugal forces applied to the rear wheels of a turning car under braking (i.e, when "reverse" throttle command is applied on cornering) may exceed the reduced (due to the weight redistribution) grip limits of the rear wheels, causing the car to oversteer. Depending on the severity and duration of the oversteer, the car might either turn into the corner smoothly with a nominal or slightly lower turning radius, or spin out of control. The complexity of the effects of the various handling attitudes of the car on the lap time renders the task of optimizing the driving style of the agent quite challenging, which, in turn, additionally motivated us to consider an automated heuristic approach to address it.

Table 1. Mechanical characteristics of the car

Parameter	Value
1) Car: model and scale	Auldey F1, 1:24
2) Dimensions (l x w x h), mm	200 x 86 x 54
3) Wheelbase, mm	130
4) Steering angle, degrees	30
5) Mass, g	310
6) Max straight line velocity, mm/s (scaled, km/h)	2000 (172)
7) Acceleration on full throttle, mm/s^2	800
8) Deceleration on reverse, mm/s^2	-2000
9) Deceleration due to the mechanical drag on throttle lift-off, mm/s^2	-600

2.2   Perceptions and Actions of the Agent

The perceptions of the agent are obtained from a video camera mounted overhead. The camera features a CCD sensor and lenses with wide field of view (66 degrees), which allows to cover a sufficiently wide area of about 2800mm x 2100mm from an altitude of about 2200mm. The camera operates at 320x240 pixels mode, with a video sampling interval of 30ms. The camera is connected to the personal computer (PC) through a PCMCIA-type video capture board.

The agent's actions (a series of steering and throttle commands) are conveyed to the car via standard two-channels radio control transmitter operating in 27MHz band. The four mechanical buttons (two buttons per channel) of the transmitter are electronically bypassed by transistor switches activated by the controlling software. Transistors are mounted on a small board, connected to the parallel port of the PC.

3   Anticipatory Modeling

3.1   Outdated Perceptions

The delays introduced in the control loop (Figure 1) by the latency of the video feed imply that the current actions of the driving agents are based on outdated perceptions, and consequently, outdated knowledge about its own state and the surrounding environment. For the hardware used in our system, the average estimate of the aggregated latency is about 90ms, which results in a maximum error of perceiving the position of the car of about 180mm when the later runs at its maximum speed of 2000mm/s. The latency also causes an error in perceiving the orientation (bearing) and the speed of the car. As demonstrated in [13], the cumulative effect of these errors renders the tasks of precisely following even simple O-, 8-, and S-shaped routes hardly solvable.

3.2 Software Simulator. Anticipating the State of the Car and the Environment

In order to investigate the detrimental effect of latency on the performance of the driving agent, and to verify the effectiveness of the proposed approach for its alleviation, we developed a software simulation of the car and tracks. The additional rationales behind the development of the software simulation include (i) the possibility to verify the feasibility of certain circuit configurations without the need to consider the risks of possible damage to the environment or the car (or both), and (ii) the opportunity to "compress" the runtime of the fitness evaluation in the eventual implementation of agent's evolution [6][8]. Furthermore, while operating the real car, the driving agent, as elaborated below, continuously applies the kernel of the developed simulator - the *internal model* of the car and the environment in order to anticipate the car's intrinsic state from currently available (outdated) perceptions. The software simulator takes into consideration the Newtonian physics of the (potentially sliding) car and the random uniform noise of +/-1 pixel (equal to the experimentally obtained value) incorporated in the "tracking" of the modeled car. No deviation of the estimated average latency of 90ms of the video feedback is modeled in the current version of the system.

In the proposed approach of incorporating an anticipatory modeling [11], the driving agent considers its current actions based on *anticipated* intrinsic (rather than currently available, outdated) state of the car and surrounding environment. The agent *anticipates* the intrinsic state of the car (position, orientation, and speed) from the currently available outdated (by 90ms) state by means of iteratively applying the history of its own most recent actions (i.e., the throttle and steering commands) to the internal model of the car. It further anticipates the perception information related to the surrounding environment, (e.g., the distance and the bearing to the apex of the next turn) from the viewpoint of the anticipated intrinsic position and orientation of the car. The approach is somehow related to the dead reckoning in GPS-based vehicle navigation [1]. In our previous work [13] we demonstrated that compared to the system without anticipatory modeling, the implementation of the latter contributes to the significant improvement of the drivability of the car on simple O-, 8-, and S-shaped routes.

4 Experimental Results

4.1 Attributes of the Driving Style

We consider the driving style as the driving line, which the car follows before, around, and after the turns in the circuits combined with the speed, at which the car travels along this line. Our choice of driving styles' parameters is based on the view, shared among the high-performance drivers from various racing teams in different formulas, that (i) the track can be seen as a set of consequent turns they need to optimize divided by simple straights (ii) the turns and both the preceding and following straights should be treated as a single whole [2]. Therefore, we introduce the following key attributes of the driving style: (i) straight-line gear - the gear at which the car approaches the turn, (ii) turning gear, (iii) throttle lift-off zone – the distance from the apex at which the car begins slowing down from the velocity corresponding to the straight line gear to the velocity of the turning gear, (iv) braking velocity - the threshold, above which the car being in the throttle lift-off zone, applies brakes (i.e., reverse throttle command) for slowing down, and (v) approach (homing) angle – the bearing of the apex of the turn. Higher values of the latter parameter yield wider driving lines featuring higher turning radiuses. Viewing the desired values of these attributes as values that the agent has to maintain, the functionality of the agent can seen as issuing such a control sequence that result in the perceived state of the car and environment to match the desired values of the corresponding parameters.

4.2 Evolving Driving Styles

Assuming that the key parameters of optimal driving style around different turns of a circuit will feature different values, our objective of automatic design of optimal driving styles can be rephrased as an automatic discovery of the optimal values of these parameters for each of the turns in the circuit. This section elaborates (i) on the proposed evolutionary approach for automatic discovery of these optimal values on the software simulator of the car and (ii) on the method used to adapt the evolved solution to the concrete physical characteristics of the real scaled model of the car on the real track.

GA. Genetic algorithm (GA) [4] is a naturally inspired domain-independent problem-solving approach in which a population of individuals representing the parameters of the candidate solution to the problem (individuals' genotypes) is evolved applying the Darwinian principle of reproduction and survival of the fittest. The fitness of each individual is based on the quality with which the candidate solution is performing in a given environment.

Genetic Representation. The genotype in the proposed GA encodes for the evolving optimal values of the key parameters of the driving style for each of the turns of a given circuit. In order to allow for the crossover operation to swap not only the values of a particular parameter, but also the *complete set* of driving style parameters associated with particular turn with the *complete set* of parameters of another turn, and consequently, to protect the higher granularity building blocks from the destructive effects of crossover, we implement a hierarchical, tree-based representation of the genotype as a parsing tree, as usually employed in genetic programming. A sample genotype, represented as XML/DOM formatted text, is shown in Figure 2. The main parameters of GA are shown in Table 2.

   The sample circuit considered in our experiments of evolving driving styles of the agent operating both the software model and the real scale model of the car is shown in Figure 3a. The circuit features a combination of one high-speed (3), one medium-speed (1) and two low-speed hairpin turns (2 and 4), represented in the figure with

```
<?xml version="1.0" ?>
 - <GP xmlns:xs="http://www.w3.org/2001/XMLSchema-instance" xs:noNamespaceSchemaLocation="GPSchema.xsd">
    - <DStyle ind="3">
       + <Turn ind="4">
       - <Turn ind="21">
            <StraightLineGear>4</StraightLineGear>
            <ApproachingMode>0</ApproachingMode>
            <ApproachingAngle>8</ApproachingAngle>
            <ApproachingAngleThreshold>7</ApproachingAngleThreshold>
            <ThrottleLiftOffZone_x10>29</ThrottleLiftOffZone_x10>
            <BrakingVelocity_x10>186</BrakingVelocity_x10>
            <TurningGear>3</TurningGear>
            <DistToCurrSwitchToNext_x10>12</DistToCurrSwitchToNext_x10>
         </Turn>
       +<Turn ind="38">
       +<Turn ind="55">
    </DStyle>
 </GP>
```

Fig. 2. Sample genotype represented as XML/DOM-formatted text. The sub-tree with the values of attributes of the second turn of the circuit is shown expanded.

their respective apexes. The series of turns 4-1-2 form a challenging, technical S-shaped sector of right, left, and right turn. The length of the track, measured between the apexes of the turns is about 3800mm. The walls ("guardrails") are virtual in that they are not physically constructed on the track. Consequently, in both cases (simulated and real car), "hitting" the walls in no way effects the dynamics of the car. However, each "crash" is penalized with 0.4s (about 10% of the expected lap time), added to the actual lap time. This reflects our intention to evolve driving styles that avoid the potentially dangerous crashes into the eventual real walls rather than trying to exploit the occasional benefits of bouncing from them.

Table 2. Main parameters of STGA

Category	Value
Population size	100 individuals
Selection	Binary tournament, selection ratio 0.1, reproduction ratio 0.9
Elitism	Best 4 individuals
Mutation	Random sub-tree mutation, ratio 0.01
Trial interval	Single flying lap for the software model and two flying laps for the real car
Fitness	Average lap time in milliseconds
Termination criteria	Number of generations = 40

Offline Evolution of Driving Styles. The fitness convergence results of the offline evolution on the software anticipatory model of the car, aggregated over 50 independent runs of GA, are shown in Figure 3b. As figure illustrates, the best lap time average over all runs of GA improved from 4770ms to about 4200ms (i.e., about 14%) within 40 generations, which for a single run of STGA consumes about 24 minutes of runtime on PC with 3GHz CPU, 512MB RAM and Windows XP OS. For the measured average speed of about 1100mm/s the achieved average reduction of lap time by 570ms corresponds to an advantage of about 63cm (more than 3 lengths of the car) per lap.

Porting the Evolved Solution to the Real Car. For the considered task of evolving driving styles, the process of porting the solution evolved offline can be viewed as a process of adaptation to the changed fitness landscape of the task. In our approach we employ the same GA framework, as used for offline evolution, for a phylogenetic adaptation of the already obtained set of good solutions to the changes in the fitness landscape caused by switching from the simulated world into the reality. At the beginning of the adaptation the GA is initialized with a population comprising 20 best-of-run driving styles obtained from the offline evolution. In order to address the challenges of (i) guaranteeing an equal initial conditions for the time trials of all candidate solutions and (ii) automatic positioning of the real car before each time trial, we employ a time trial comprising an out-lap followed by a series of flying timed laps, and finally, an in-lap in a way similar to the current qualifying format in the car racing formulas. After crossing the start-finish line (shown immediately after the turn 4 in Figure 3a) completing the final timed lap governed by the current driving style, the car enters the in-lap and slows down under "Neutral" throttle command. Depending on the speed at the start-finish line, the car comes to a rest at a point somewhere between turns 1 and 2. At this point, which can be seen as an improvised pit, the next driving style which has to be evaluated is loaded into the agent's controller, and the car starts its out-lap. Controlled by the new driving style, the car negotiates turns 2, 3 and 4. During the out lap the car covers a distance from the pit stop to the start-finish line, which is sufficient enough to cancel out any effect of the previously evaluated driving style on the performance of the current driving style. In order to compensate for the eventual small inconsistence of the lap time, a total amount of 2 timed laps are conducted during the time trial of each driving style, and the average lap time is considered as a corresponding fitness value.

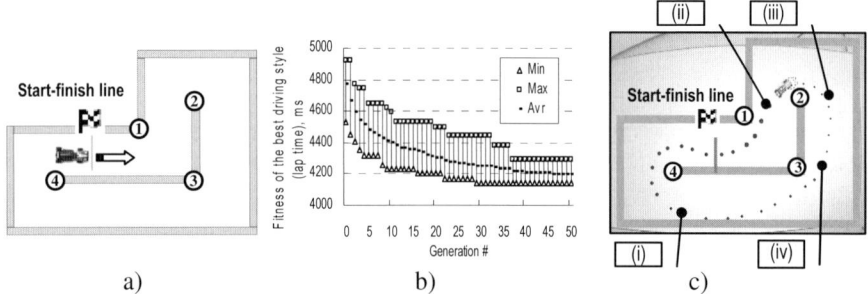

Fig. 3. Sample circuit used for evolution of driving style of agent (a), the fitness convergence characteristics (b) of offline evolution of driving styles on this circuit and (c) the emergent features of an online evolved best driving style.

The online evolution of the initial population of 20 best-of-run solutions obtained offline was allowed to run until no improvement in fitness value of the best driving style have been registered for 4 consecutive generations. A single run has been completed, and improvement of the aggregated fitness value of the best solution from the initial value 4930ms (due to the initial hitting of the "walls" by the fast agents evolved offline and currently operating the scale model of the car) to 4140ms (average speed of 1120mm/s, scaled to about 97km/h) has been observed within 10 generations. The emergent features of the evolved best driving style of the anticipatory agent are shown in Figure 3c. As illustrated in the figure, (i) the car starts its flying lap entering the turn 4 relatively wide and exiting it close to the apex, which allows (ii) to negotiate the turn 1 and the following turn 2 using the shortest possible driving line. The car exits the turn 2 in a way (iii) that allows for a favorable orientation at the entrance of the following turn 3 and (iv) an early acceleration well before its apex, contributing to the achievement of the faster speed down the back straight between turns 3 and 4. The car uses the full width of the track and enters the turn 4 wide and exits it close to its apex preparing for the first turn of the next flying lap.

The experimental results of the lap times of a sample best driving agent over 300 consecutive laps (30 runs of 10 laps each) on the circuit shown in Figure 3a indicate the adequate consistence of the control. The relative standard deviation of the lap time is low (0.038, i.e., a standard deviation of 160ms of an average lap time of 4140ms), which is important for the efficiency of the evolutionary optimization [9].

In order to estimate the human-competitiveness of the evolved solution, we conducted the experiments with the best (in terms of both speed and consistence) of four human operators. The operators were given enough time to learn the optimal driving of the car on the sample track. The average lap time over 30 runs (of 10 laps each), achieved by human was 4150ms, which is virtually the same as the time achieved by the evolved solution. However, the standard deviation of the lap time of the human-controlled car was 270ms, indicating poorer consistence than that of the driving agent.

Evolution of Optimal Obstacle Avoidance. Obstacle avoidance is a key capability of any mobile robot. However, depending on *what* the characteristics of the obstacle are (large or small, static or moving), *whether* the artifact is a priori aware of it or not, and *when* it is introduced to the scene (before the trial or at runtime), the implementation of obstacle avoidance requires an addressing of numerous algorithmic and technological challenges. In this very preliminary work we consider the simplest case of a *static* obstacle with *known* properties (position and size), introduced to the scene *before* the time trial. For the considered car-racing domain, the problem of optimal obstacle avoidance can be viewed as discovering the driving line of circumnavigating an obstacle and the speed along this line that result in a minimal lap time around a predefined circuit.

Based on the repulsive potential field approach of obstacle avoidance, we view the steering the car away from the obstacle as correction of the desired angle of approach (Desired_A_A) of the apex of the following turn. The parameterization of the maneuver is shown in Figure 4. As figure illustrates, the steering correction is initiated when the car enters the obstacle zone. The degree of this correction depends on the angular distance between the car and the obstacle (Figure 4, parameter A_O) as follows: the correction A_C of Desired_A_A is set to its maximal (initial) value (Figure 4, parameter A_{CI}) when the bearing of the obstacle is minimal (i.e., $A_O=0$, when the car travels head on into the obstacle), and decreases inversely proportionally to zero with the increase of the bearing A_O to its maximal value (i.e., $A_O=90$ degrees when the car is lined-up with the obstacle). In addition to the steering correction, a throttle control is also applied to maintain the desired velocity V_O while negotiating the obstacle. The evolutionary optimization of the obstacle avoiding implies, in addition to the values of general parameters of the driving style, an automated discovery of the values of the key parameters of the obstacle avoidance that result in a fastest lap around the circuit. The obstacle avoidance parameters are the direction of avoidance (left or right), radius of the obstacle zone (Z_O), initial correction of the apex approach angle (A_{CI}), and speed inside the obstacle zone (V_O).

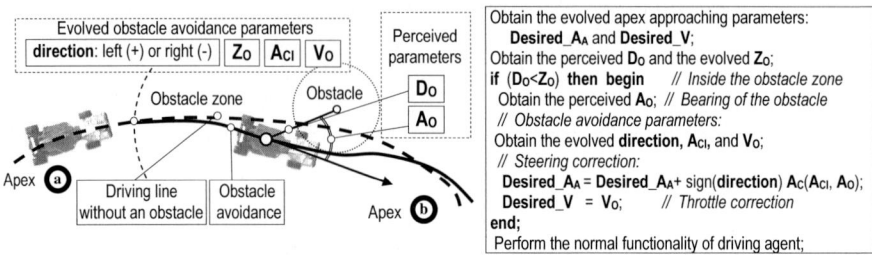

Fig. 4. Parameterization (left) and the algorithm (right) of the obstacle avoidance maneuver

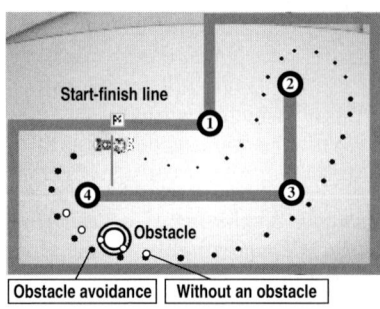

Fig. 5. The traces of the sample best evolved obstacle avoiding driving style adapted to the real world. The dark trailing circles in (b) depict the trajectory of the center of the car. The time-stamp interval between each of these circles is 120ms (4 sampling intervals).

In our experiment, a round obstacle with a diameter of 180mm (about two car widths) is placed 400mm before the apex of the turn 4 (Figure 3a) on the driving lines of the sample 10 best-of-run driving styles. Analogous to the collisions with the walls, "hitting" the obstacle has no effect on the dynamics of the car. Rather, the agent is penalized with 0.4s added to the actual lap time. GA, employed to adapt the best evolved driving styles to the modified fitness landscape caused by the need to avoid an obstacle, is initialized with a population comprising 90 randomly created individuals, plus 10 best-of-run driving styles (with randomly initialized part of chromosome that encodes for the obstacle avoidance parameters) obtained from the experiments as elaborated earlier in this section. However, conversely to the canonical GA, we implemented the genetic operations that alter only the parameters of driving styles, that are relevant to the obstacle avoidance, i.e., parameters, associated with the turns 2, 3, and the obstacle itself. The experimental results of 20 independent runs indicate that due to the reduced search space of such biologically plausible "facilitated variations" [7] the evolutionary adaptation of the driving style is quick (within one to three generations) delivering lap times of 4200ms (from the initial 4900ms), i.e. the same lap times as for the circuit with no obstacles.

The online adaptation of the initial population of 20 best-of-run solutions obtained offline was performed in a way analogous to the case of circuit with no obstacles. An improvement of lap time of the best solution from the initial value of about 5230ms to

4220ms was observed within 4 generations. The snapshot of the traces of the best-evolved driving style is shown in Figure 5.

5 Conclusions

The objective of this work is an automatic evolutionary design of driving agent, able to remotely operate a scale model of racing car running in a fastest possible way. The agent's actions are conveyed to the car via simple remote control unit. The agent perceives the environment from live video feed of an overhead camera. In order to cope with the inherent video feed latency we implemented an approach of anticipatory modeling in which the agent considers its current actions based on anticipated intrinsic (rather than currently available, outdated) state of the car and surrounding environment. We formalized the notion of driving style with obstacle avoidance capabilities and defined the key parameters, which describe it. We demonstrated the feasibility of applying genetic algorithms to evolve the optimal values (i.e., yielding fastest lap times) of these parameters first on a software simulator of the car and then in the real world. An evolutionary adaptation to the changes in the fitness landscape caused by the need to avoid a small static obstacle with a priori known properties result quickly in lap times that are virtually the same as the best lap times achieved on the same track without obstacles. Presented work can be viewed as a step towards developing a framework of a racing game in which the human competes against a computer, both of them remotely operating scale models of racing cars. In our future work we are planning to incorporate more challenging obstacles that should be negotiated in a robust way regardless of their number, current locations, moving directions or velocities.

References

[1] Abbott, E. and Powell D.: Land-vehicle Navigation Using GPS, The Proceedings of the IEEE, January (1999), vol 87, no 1, 145-162
[2] Frere, P.: Sports Cars and Competition Driving, Bentley Publishing (1992)
[3] Funge, J. D.: Artificial Intelligence for Computer Games, Peters Corp. (2004)
[4] Goldberg, D.: Genetic Algorithms in Search, Optimization, and Machine Learning, Addision-Wesley (1989)
[5] IBM Corporation: Deep Blue, URL: http://www.research.ibm.com/deepblue/ (1997)
[6] Jacobi, N.: Minimal Simulations for Evolutionary Robotics, Ph.D. thesis, School of Cognitive and Computing Sciences, Sussex University (1998)
[7] Kirschner, M. W. and Gerhart J.: The Plausibility of Life. Resolving Darwin's Dilemma, Yale University Press (2005) 336 pages
[8] Meeden, L. and Kumar, D.: Trends in Evolutionary Robotics, Soft Computing for Intelligent Robotic Systems, edited by L.C. Jain and T. Fukuda, Physica-Verlag, New York, NY (1998) 215-233
[9] Miller, B. L., and Goldberg, D. E.: Genetic Algorithms, Tournament Selection, And The Effects Of Noise. Complex System, 9(3), (1995) 193-212
[10] Robocup: URL: http://www.robocup.org/02.html (2005)
[11] Rosen, R.: Anticipatory Systems, Pergamon Press (1985)

[12] Togelius J., and Lucas S. M.: Evolving Controllers for Simulated Car Racing, Proceedings of IEEE Congress on Evolutionary Computations (CEC-2005), Edinburgh, UK, September 2-5 (2005) 1906-1913

[13] Tanev, I., Joachimczak, M., Hemmi H. and Shimohara, K.: Evolution of the Driving Styles of Anticipatory Agent Remotely Operating a Scaled Model of Racing Car, Proceedings of the 2005 IEEE Congress on Evolutionary Computation (CEC-2005), Edinburgh, UK, September 2-5 (2005) 1891-1898

[14] Wloch, K. and Bentley, P.: Optimizing the Performance of a Formula One Car Using a Genetic Algorithm, Proceedings of the 8th International Conference on Parallel Problem Solving from Nature, Birmingham, UK, September 18-22, (2004) 702-711

Evolving Robot's Behavior by Using CNNs

Eleonora Bilotta[1], Giuseppe Cutrí[2], and Pietro Panano[3]

[1] University of Calabria, 87036 Arcavacata di Rende, Italy
bilotta@unical.it
[2] University of Turin, 10124 Turin, Italy
giuseppe.cutri@unito.it
[3] University of Calabria, 87036 Arcavacata di Rende, Italy
piepa@unical.it

Abstract. This paper deals with a new kind of robotic control, based on Chua's nonlinear circuit called Cellular Neural Network (CNN). A CNN is a net of coupled circuits, connected in a grid structure, which inherits its features and properties from the well known Artificial Neural Network and Cellular Automata. It has been demonstrated that CNNs are able of universal computation, many cognitive processes such as pattern recognition, features extraction, image processing, and mathematical simulations of nonlinear equations such as Navier-Stokes equations, reaction-diffusion equations, and so on. Using an approach like Evolutionary Robotics, we evolved, instead of Neural Networks, CNNs by using Genetic Algorithms (GAs), for controlling the behavior of an hexapod robot in a simulated environment. We developed a Java3D software in which physical simulations are carried on by using different kind of robots. In this program, a module for evolving the robot's behavior by GAs has been implemented. Furthermore, many advanced sensors and actuators complete the evolution of the robot's behavior. The evolved behavior of our robots is very similar to that of real insects, and we analyzed the pathways these agents perform in the simulated environment.

1 Introduction

Evolutionary Robotics [1], a new branch of Robotics, models robotic behavior using a biological approach, in which the control system is based on the well known Artificial Neural Networks (ANNs), evolved through Genetic Algorithms (GAs).

We replaced the ANN based control system with CNNs. We used this method since CNNs have a powerful hardware implementation (response time is about few nanoseconds), so it will be possible to build a real robot with an integrated powerful analog brain based on CNNs.

CNNs have been already used to reproduce brain activities and a lot of works in this way can be found in [2], [3] and [4], where the authors simulate the Central Pattern Generator (CPG), a brain area whose task is to control locomotion, by resolving the Turing's reaction-diffusion equations (that happen in the CPG), modeled on top of the CNN status equation. An other method for CNN parameterization can be found in [5], which is the first example of evolution of CNNs through Genetic Algorithm.

S. Nolfi et al. (Eds.): SAB 2006, LNAI 4095, pp. 631–639, 2006.
© Springer-Verlag Berlin Heidelberg 2006

So using the approach of Evolutionary Robotics and CNNs, we developed a robotic control based on CNNs, whose parameters are evolved by GAs.

In order to obtain this aim, we developed a simulated virtual environment to perform simulations into a 3D space. We called this application RoVEn (acronym of Robot Virtual Environment).

In the next Sections, a basic introduction on CNNs and how they work is presented, followed by a very brief overview of Genetic Algorithms and, in particular, how this technique has been used for our aim. Then we will present RoVEn and how an example robot (an hexapod) was realized in order to verify the efficacy of the proposed method. At the end some results and conclusion are presented.

2   Cellular Neural Networks

Cellular Neural Networks [6] are dynamical systems composed by a lattice of analog non linear circuits called cells. The structure of a CNN is a grid of one, two or three dimensions, in which each cell is connected to the nearest cells of the grid. In this paper we refer only to two-dimensional CNNs.

Given a CNN of LxK cells, it is possible to define a parameter r (ray) that specifies the length of the *neighborhood* of cells that are connected to each other. In a more formal way, we define $N_r(i,j)$ as the set of cells that belong to the neighborhood of ray r of the cell $C(i,j)$ or

$$N_r(i,j) = \begin{cases} C(l,k) : \max(|l-i|,|k-j|) \leq r \\ 1 \leq l \leq L; 1 \leq k \leq K \end{cases}$$

The contour so defined exhibits a symmetric property, in the sense that if $C(i,j) \in N(l,k)$ then $C(l,k) \in N(i,j)$ for every $C(i,j)$ and $C(l,k)$ of the net.

A typical example of i,j cell is shown in Fig. 1. The notation u, x and y means respectively the input, the state and the output of the cell. In particular, the voltage on the node v_{xij} of the cell $C(i,j)$ is called state of the cell and its initial value must be minus or equals to 1; the voltage on the node v_{uij}, called input of the cell, must be constant, with amplitude minus or equals to 1; finally, the voltage on the node v_{yij} is called output.

Looking at Fig. 1, one can see that each cell has:

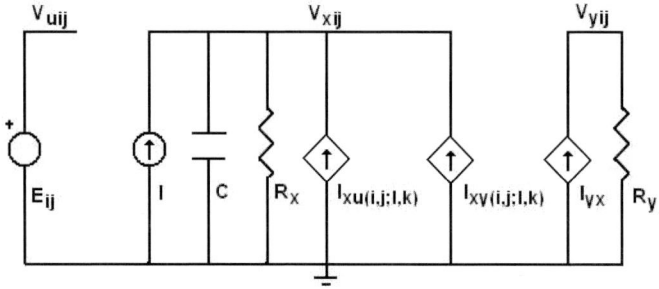

Fig. 1. The circuital model of a CNN cell

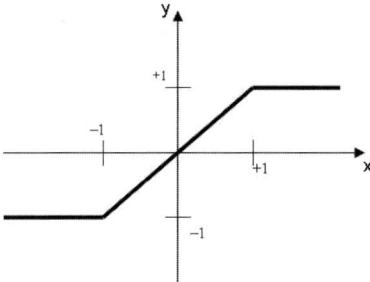

Fig. 2. The f function

- an independent voltage generator E_{ij};
- an independent current generator I;
- a linear capacitor C;
- two linear resistors R_x and R_y;
- $(2r+1)^2$ linear current generators voltage controlled, that are coupled with the other cells of the neighborhood through the v_{ulk}, v_{ylk} and v_{xlk} voltages.

In particular, $I_{xy}(i,j;l,k)$ and $I_{xu}(i,j;l,k)$ are the generators with the characteristics:

$$I_{xy}(i,j;l,k) = A(i,j;l,k)V_{ylk}$$
$$I_{xu}(i,j;l,k) = B(i,j;l,k)V_{ulk}$$

Moreover, the only non-linear element in each cell is the current generator voltage controlled:

$$I_{yx} = \frac{1}{R_y}f(V_{xij})$$

Where f, represented in Fig. 2, is defined as:

$$f(x_{ij}) = \frac{1}{2}(|x_{ij}+1| - |x_{ij}-1|)$$

2.1 CNN Model

Using the Kirchkoff's laws for the circuit on Fig. 1, it is possible to obtain the mathematical model of a CNN cell:

$$C\frac{dv_{xij}(t)}{dt} = -\frac{1}{R_x}v_{xij}(t) + A(i,j;l,k)\,v_{yij}(t) + B(i,j;l,k)\,v_{uij} + I \quad (1)$$

Let's notice that:

- a generic cell is *central* if its neighborhood has $(2r+1)^2$ cells; *of frontier* otherwise;
- cells have local interconnections, but cells not directly connected can influence each other through the propagative effects of the continuum-time dynamics of the CNN;

– the dynamics of a CNN has three *feedback* mechanisms: on the output, on the input and on the state of the system. These feedback effects depend by the parameters $A(i, j; l, k)$ and $B(i, j; l, k)$, called respectively *feedback operator* or *feedback template* and *feed-forward operator* or *feed-forward template*. Templates parameters stand for the weights of the state equations and, with the parameter I called bias or threshold, they complete the CNN specification.

2.2   CNN Matrices

Equation 1 can be rewritten in the form:

$$\dot{x}_{ij}(t) = -\frac{1}{R_x C} x_{ij}(t) + A(i, j; l, k)\, y_{ij}(t) + B(i, j; l, k)\, u_{ij} + z \qquad (2)$$

where z is the threshold. Templates can be expressed in a matrix form; for example, for $r = 1$ we have:

$$A = \begin{pmatrix} A(i, j; i-1, j-1) & A(i, j; i-1, j) & A(i, j; i-1, j+1) \\ A(i, j; i, j-1) & A(i, j; i, j) & A(i, j; i, j+1) \\ A(i, j; i+1, j-1) & A(i, j; i+1, j) & A(i, j; i+1, j+1) \end{pmatrix}$$

$$B = \begin{pmatrix} B(i, j; i-1, j-1) & B(i, j; i-1, j) & B(i, j; i-1, j+1) \\ B(i, j; i, j-1) & B(i, j; i, j) & B(i, j; i, j+1) \\ B(i, j; i+1, j-1) & B(i, j; i+1, j) & B(i, j; i+1, j+1) \end{pmatrix}$$

The parameters of the circuit can be scaled and it is often convenient to rewrite the equation 2 in the following dimensionless normalized form, that is the cell state equation:

$$\dot{x}_{ij}(t) = -x_{ij}(t) + A * y_{ij}(t) + B * u_{ij} + z$$

As we will show, it is not so easy to find the coefficients of these matrices. This is the main aim of our work, in order to create a robotic control based on these dynamical systems.

3   Genetic Algorithms

Genetic Algorithms, invented by John Holland [7], are optimization methods used to solve a large class of problems where the classic techniques do not supplies a valid support. GAs have been used for evolving robot's control by a large number of researchers [8], [9], [10].

CNN can be totally specified by a set of parameters as well. These parameters, put in a string, become the DNA of the system, where each of these features stands for a gene. In particular, we used as genotype of the CNN the feedback and feed-forward matrices. Instead, phenotype is a robot whose CNN parameters are specified.

An important phase of the GA method is the evaluation of the fitness of a genotype. In our problem, this can be done by simulating the robot behavior in a physical simulator and then extracting some information from this simulation (e.g. Lagrangian parameters values and theirs derivate, distance covered, etc.). Such information are the variables of a function called *fitness function* that assigns a value to a genotype.

So, in our case, the scheme of a Genetic Algorithm is the following:

1. Generation of initial population of robots, whose parameters of the CNN are randomly generated;
2. Fitness evaluation on robot's behavior calculating the fitness values for each robot;
3. Selection of a percentage of robots with the best fitness values;
4. Reproduction of new borns whose genotype is obtained by *cross-over* and random *mutation*;
5. Go to step 2: allows for the repetition of the cycle.

The fitness function will be given after introducing the system we have developed.

4   RoVEn

RoVEn is a software allowing for the creation of complex robotic structures, starting from simple rigid bodies, that can be generated as regular 3D geometries or imported from VRML files. Subsequently, it is possible to connect these bodies each other trough spherical joints that contain servo-motors.

We used Java3D™ framework [11], for the graphical part, and the Open Dynamic Engine (ODE) [12] for the physical simulation.

Spherical joints can be appropriately designed with some properties which allow for binding each Degrees Of Freedom (DOF) of the joint.

Other parameters, called Elevation Random Range (ERR), Azimuth Random Range (ARR) and Tilt Random Range (TRR) permit to choose the range of angles according to which the body rotates. These parameter values are chosen arbitrarily by the program when the simulations start. This procedure is activated for changing initial conditions at any simulation.

It is possible to connect to the robot various kind of sensors, among which *position sensor* and *camera sensor* play an important role in the simulation. The first is a sensor returning its position in the absolute coordinates system. The second is an image acquisition camera that returns a picture of $n x n$ pixels (where n is a user selected value) of the world, taken from its point of view.

Once a robotic structure is created, it is possible to control each actuator by a *control system* that can be obtained by programming it in the Java™ programming language. The control system can be put into RoVEn as a plug-in, and then sensors and actuators can be connected by a *control panel*. As will be explained below, the main aim of our work is the development of such control system in order to model robot's behavior using Cellular Neural Networks. The reason that brought us to develop a new simulation software is that in RoVEn it is easy to obtain control systems *genetically*. In fact, we developed a module that permits to run a sequence of simulations, generated by the GA procedure, directly in the RoVEn environment. In this way we are able to evolve these control systems in order to perform assigned tasks.

5   Hexapod Robot

The robot we have used as prototype (Fig. 3) has been built by a body to which are connected six legs, each of them is made by two pieces. Each of these piece is connected by spherical joints. For each leg, a first spherical joint connects the leg to the body (internal joint), while the second one connects the second piece of the leg to the first one (external joint) (Fig. 4).

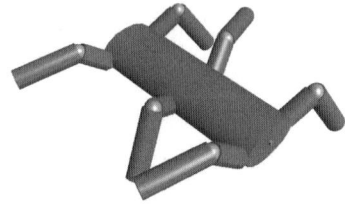

Fig. 3. The hexapod robot

Even if each joint is spherical, in order to simplify the problem, we set up the properties EB and TB of the internal joints to the couple of values -1, 1 that hinder two DOF. So the joints become a cylindrical hinge. With this configuration (that leaves free only the azimuth), the internal joints can only move the legs ahead and behind. Furthermore, the properties AB and TB of the external joints are set up to the couple of values -1, 1 in order to hinder two DOF except for the elevation rotation (that raises and lows the second part of the leg).

Moreover, the elevation is bounded in a range of values that goes from -45° to 0°, and the azimuth is bounded in a range of values that goes from -20° to +20°. This is done to prevent that one leg can touch another one and create problems in the simulator.

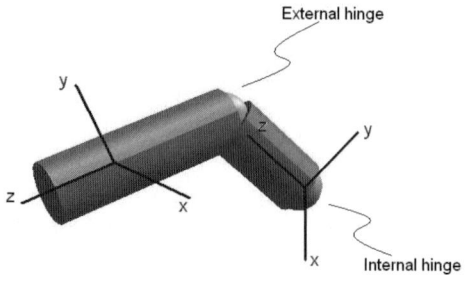

Fig. 4. Robot's leg

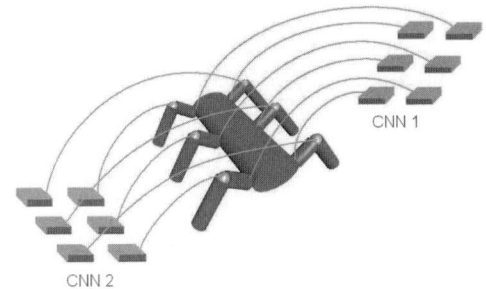

Fig. 5. Connection scheme of CNNs and motors

Table 1. Connection scheme of CNN cells and motors

CNN 1	CNN 2
IFLJ IFRJ	EFLJ EFRJ
IMLJ IMRJ	EMLJ EMRJ
IBLJ IBRJ	EBLJ EBRJ

A position sensor is attached in the front side of the robot in order to know its position when required.

5.1 CNN Control System

Together with RoVEn, we developed simple control systems that are supplied directly by the program. For example, one of this permits to control actuators and to read sensors status by a graphical interface; another one permits to assign as input, for each parameter of a motor, a time-series, taken from a text file.

The CNN Control System (CCS) we developed is a software version of the CNN circuit, in which one can connect to each cell any sensor in the virtual environment as input; while the output of the cells can be the input for actuators. This connections are realized by using functions of the RoVEn SDK.

In our experiments we used two CCS 3x2 (six cells); one CNN is connected to the internal joints, and one is connected to the external joints (Fig. 5). Since the robot has six legs, there are 6 + 6 hinges, each one has a motor connected to a cell of a CNN. The output of each cell is the input for the motors as shown in Table 1.

In this table, the following convention is used: the first letter indicates an internal joint (I letter) or an external one (E letter); the second letter indicates a frontal, middle or back joint (F, M, B letters); the third letter indicates a left or right joint (L or R letters), finally the last letter (J) represents a joint.

Moreover, the first CNN takes as input the value of the angular sensors inside the hinges – the association cell-hinge is the same as the output – while the second CNN takes as input the output of the first (first cell of CNN 1 is connected to the first cell of CNN 2, second cell of CNN 1 is connected to the second cell of CNN 2, and so on).

6   Evolution

The GA scheme above presented can be improved to obtain more efficient robots. Here we present the features we used in our implementation:

- Single point cross-over: the cross-over of the step 4 of the GA is done by dividing in the same point the genotype strings of the two parents, and then obtaining a new child, choosing the first piece from parent 1 and the second piece from parent 2;
- Mutation: a percentage of genes (usually not more than 3%) randomly chosen is changed into a new random value;
- Elite strategy: a percentage of the population with best fitness is copied without been modified in the next generation;
- Fitness overestimation: fitness value for the genotype i is given by:

$$\max(val_{i,j-1}, val_{i,j})$$

where $val_{i,j-1}, val_{i,j}$ are respectively the fitness value of the genotype i at the previous $j-1$ and the current j generation.

Note that with the fitness overestimation, the graph of the fitness values for each generation, represents a monotonic function.

Since our aim is to evolve robots able to walk straight forward, the fitness function is given by the distance covered by the robot. Actually, it is done by checking the state of the position sensor in the front side of the robot's body. As the simulation time is the same for each robot, the distance divided by time gives the mean velocity, so robots have higher fitness as they walk faster.

The feedback matrix is set up with all coefficients equals to 0 except for the central one, which is equals to 1. The feed-forward matrix is evolved and the threshold is given by the mean of the thresholds of the parents.

In our experiments, we ran evolution for about 50 iterations. Evolving both CNN simultaneously can bring some undesirable instability phenomena (i.e. the CNN evolution cannot have time to synchronize with the other one because

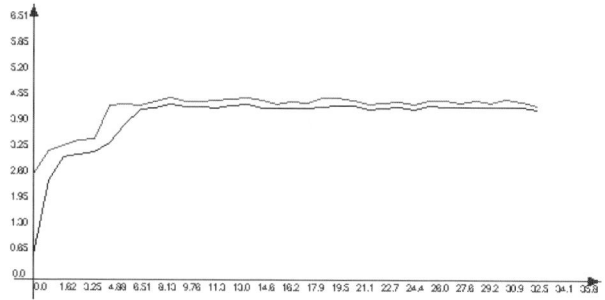

Fig. 6. Fitness diagram for the best evolution obtained. The abscissa report the number of the generation and the ordinate report the fitness value.

it is mutating too fast). For this reason the evolution is done considering one CNN at time. At the beginning, each robot has one CNN whose parameters are generated randomly, according to some conditions specified in the user interface (in our experiments values of each coefficients must be in a range of values equals to -1, 1). The other CNN is static, and the feed-forward value is chosen with a 1 for some coefficients and 0 elsewhere. At the end of the evolution the two CNN are switched and the GA works on the second CNN. This procedure is done for 3 or 4 times.

7   Results and Conclusions

As result we got very interesting robots whose behavior is similar to real insects. In fact, the robot's legs alternate the movements of the legs in a coordinated way. In this way, the robot always maintains both static and dynamic stability.

Many experiments have been run. The best evolved robot has got a fitness of 4.1 cm in 20 seconds and its fitness diagram is shown in Fig. 6.

The presented method has given satisfactory results, demonstrating that the evolutionary process, used for evolving Artificial Neural Networks can be applied effectively to Cellular Neural Networks as well.

With this method, many classical robotic problems – such as locomotion on different terrain, obstacle avoidance, etc. – can be studied, while the robotic structure can be varied as well.

References

1. Nolfi, S., Floreano, D.: Evolutionary Robotics - The Biology, Intelligence, and Technology of Self-Organizing Machines. The MIT Press (1997)
2. Arena, P., Fortuna, L., Branciforte, M.: Realization of a Reaction-Diffusion CNN Algorithm for Locomotion Control in an Hexapode Robot. Journal of VLSI Signal Processing (1999)
3. Arena, P., Basile, A., Fortuna, L., Frasca, M., Patané L.: A CNN Approach for Controlling a Roving Robot. CLAWAR (2003)
4. Manganaro, G., Arena, P., Fortuna, L.: Cellular Neural Networks – Chaos, Complexity and VLSI Processing. Springer (1998)
5. Kozek, T., Roska, T., Chua, L.O.: Genetic Algorithm for CNN Template Learning. IEEE Transactions on Circuits and Systems (1993)
6. Chua, L.O.: CNN: A Paradigm for Complexity. World Scientific Publishing Co. Pte. Ltd (1998)
7. Holland, J.: Adaptation in natural and artificial systems. Penguin Books (1993)
8. Bianco, R., Nolfi, S.: Evolving the neural controller for a robotic arm able to grasp objects on the basis of tactile sensors. Adaptive Behavior (2004)
9. Nolfi, S., Marocco, D.: Evolving robots able to visually discriminate between objects with different sizes. International Journal of Robotics and Automation (2002)
10. Miglino, O., Lund, H.H., Nolfi, S.: Evolving mobile robots in simulated and real environments. Artificial Life (1995)
11. http://java.sun.com/products/java-media/3D/
12. http://www.ode.org

Collective and Social Behaviours

Experimental Study on Task Teaching to Real Rats Through Interaction with a Robotic Rat

Hiroyuki Ishii[1], Motonori Ogura[1], Shunji Kurisu[1], Atsushi Komura[1],
Atsuo Takanishi[2], Naritoshi Iida[3], and Hiroshi Kimura[3]

[1] Graduate School of Science and Engineering, Waseda University
3-6-1, Okubo, Shinjuku-ku, Tokyo, Japan
hiroyuki@ruri.waseda.jp
http://www.takanishi.mech.waseda.ac.jp/
[2] Department of Mechanical Engineering / H.R.I., Waseda University
[3] Graduate School of Letter, Arts and Science, Waseda University

Abstract. "Learning" has been well studied in several research areas such as psychology, brain science, computer science and robotics. In these studies, many experiments using animals have been performed. On the other hand, several researchers have been studying adaptive interactions and task learning through interactions between humans and robots. We then focus on adaptive interactions between animals and robots. The purpose of our research is to develop a framework of adaptive interactions between animals and robots through interaction experiments between rats and a robotic rat. We propose a novel behavior generation algorithm for the robot to enable it to autonomously teach a simple behavior task to rats as an example of adaptive interaction. This algorithm was implemented in the robot and the experimental setup, and then verified through the experiment.

1 Introduction

Many studies on "learning" are performed in various research areas such as psychology, brain science, computer science and robotics. Before the 19th century, "learning" was studied by philosophers. Early studies on "learning" from a positive scientific viewpoint started in psychology. In the earlier part of the 20th century, "learning" was studied in animal psychology.

Since the theory of evolution published by Darwin (1809-1882), animal psychology has been playing a very important part and contributing to clarifying the human mind [1]. Many studies on animal behavior focusing on their learning ability or mechanisms have been performed, and many effective results are reported. For instance, Thorndike (1874-1949) established the concept of "trial and error," a basic principle of learning psychology, through experiments using the "puzzle box" ("problem box") [2]. Skinner (1904-1998) developed the "Skinner Box" and conducted experiments using rats. He then advanced Thorndike's concept and established the concept of "Operant Conditioning" [3]. From the latter half of the 20th century, brain

S. Nolfi et al. (Eds.): SAB 2006, LNAI 4095, pp. 643–654, 2006.
© Springer-Verlag Berlin Heidelberg 2006

scientists and neuroscientists started to study the learning mechanisms of animals. For instance, Rudy and Sutherland reported the relationship between brain activity and behavior learning [4] [5].

Referring these studies, some computer scientist then started to study recreating the learning ability of animals as artificial intelligence. For instance, the model of "reinforcement learning" was developed based on "operant conditioning" [6]. Doya then expanded it and implemented it in small mobile robots named "cyber rodent" [7] [8]. Using them, he has been trying to represent adaptive behavior of rodents. Yamada and Yamaguchi also implemented the reinforcement learning in an AIBO for natural and adaptive interaction with humans [11]. Furthermore, several researchers have been studying adaptive interactions between humans and robots [9] [10]. These robots also have some kind of adaptive models inspired from animals. Robots that learn and obtain new skills and behavior through interactions with humans and the environment are useful for people without special knowledge of robotics. These robots are expected as nursing-care robots and personal robots in the recent aging society.

As stated above, the studies on animals have given many useful ideas to the researchers in intelligent robotics and artificial intelligence. We then considered that experimental studies on adaptive interaction between robots and animals from animal psychological viewpoint could contribute to developing framework of that. Thus, the aim of our study is to develop frameworks in adaptive interactions between animals and robots through the interaction experiment between them. We believe these frameworks would give several useful ideas to design adaptive interactions between humans and robots. We selected rats as the first target because of their abundance in previous works in animal psychology. Some researchers have also been studying interactions between animals and robots. In United States, Sanjay developed a behavior model of rats and then integrated this model into a small mobile robot [12]. In Switzerland, the studies on social interaction between cockroaches and a robot have been performed [13].

Since 1995, we have been developing experimental setups and small mobile robots as robotic agents. Using them, we have been conducting several interaction experiments between rats and the robots. In 2003, we developed WM-6, a rat-robot that had two levers as shown in Figure 2 and an experimental setup for long-time experiments (over 24 hours or more) as shown in Figure 1. We then conducted the interaction experiment and succeeded in conditioning (training) the rats to perform a simple task, which was pushing the levers on WM-6 to obtain food as shown in Figure 3. In this experiment, the experimenter pushed the levers on WM-6 in front of the rats in order to demonstrate the lever-pushing task. The rats then modified their behavior and finally learned the rule between lever pushing and food feeding [14].

We then considered that the robot could teach the lever-pushing task to the rats by itself instead of being demonstrated by the experimenter. Therefore, a novel behavior generation algorithm for a robot that enables it to autonomously condition rats to perform a simple task was developed introducing the idea of "shaping" [15]. In this paper, we describe the behavior generation algorithm for a robot to teach a behavior task as an example of learning through interaction. We then report the experimental evaluation of this algorithm. Finally, we propose some techniques to design robot's behavior in the situation that it teaches behavior tasks to animals. We also discuss how to expand this algorithm into several kinds of adaptive interactions.

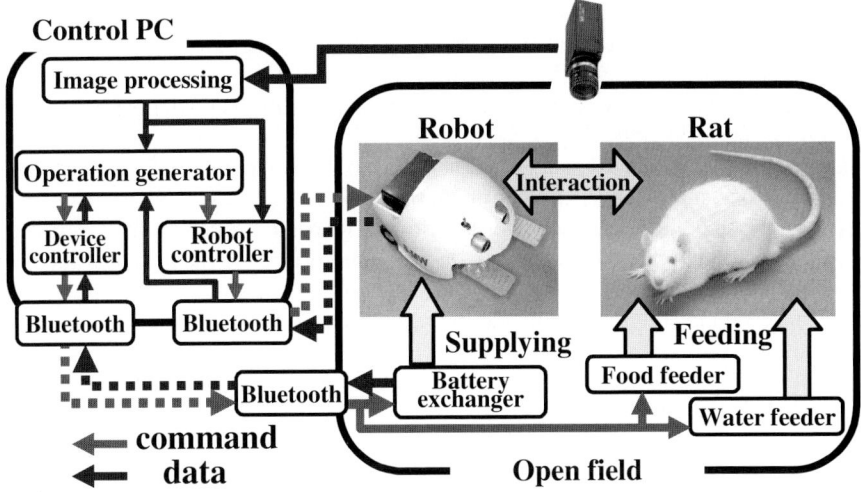

Fig. 1. Overview of the experimental setup

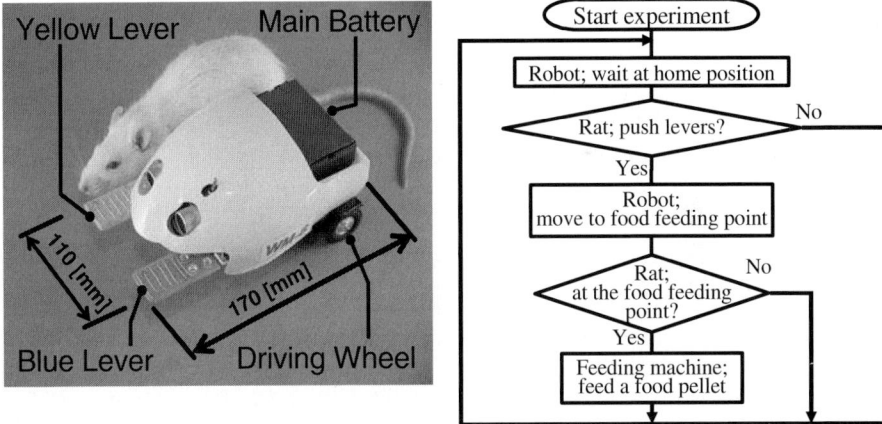

Fig. 2. Real rat and a robotic agent WM-6

Fig. 3. Task to obtain food; the rat obtains food when it pushes the lever on the robot

2 Shaping

"Shaping" is a method of operant conditioning proposed by Skinner [16]. This is one of the most effective methods of conditioning subjects, such as animals or humans, to perform difficult or complex behavior tasks that would rarely occur spontaneously. Skinner said it is possible to condition the performance of such difficult behavior tasks by dividing the conditioning process into small steps. In the first step, the subjects are conditioned to perform a simple behavior task that is associated with the required complex task by selective reinforcement. In the second step, the subjects are conditioned to perform a slightly difficult behavior task that is more strongly associated with the required task. In this way, by increasing the difficulty of the conditioning in a step-by-step manner, it is possible to condition the subjects to perform a complex behavior task. "Shaping" is also used as a technique in behavior therapy to treat mental disorder and psychological problems in human beings [17].

Using this method, Skinner conditioned a rat to push a lever on the wall in the Skinner box to obtain food. At the first step, he conditioned the rat to look at the lever to obtain food. He then increased the difficulty of conditioning step by step and finally succeeded in conditioning the rat to push the lever to obtain food.

3   Robotic Agent; WM-6

We developed a small mobile robot as a robotic agent, WM-6 (Waseda Mouse No. 6) as shown in Figure 2. The mobility and dimensions are almost equal to those of mature rats as shown in Table 1. WM-6 is wirelessly controlled by a PC and consists of two levers, two driving wheels, a microcontroller, a Bluetooth wireless communication module, and a Li-ion battery.

3.1   Mobile Mechanism

WM-6 has 2 drive wheels and 1 passive omni-directional ball caster. Each drive wheel is separately actuated by a DC motor with planetary gear reduction and is mounted on the left and right sides of the rear portion while the ball caster is mounted on the center of the front end. Due to this mobile mechanism, WM-6 is non-holonomically constrained.

3.2   Power Supply

A Li-ion battery is selected as the power supply unit for WM-6, since it is simple to measure the remaining battery level using the battery voltage. WM-6 has a Li-ion battery pack (7.2 [v], 1500 [mAh]). WM-6 operates constantly for a minimum of 120 [min] with one fully charged battery. In addition, the battery exchanger (described in Chapter IV) automatically exchanges the battery on the robot without human handling. Therefore, it is possible to perform the interaction experiments for over 120 [min] without human interruption to exchange the battery.

Table 1. Specifications of WM-6

Weight	g	540
Length	mm	170
Width	mm	85
Height	mm	100

Max speed	m/s	1.0
Max rotational speed	deg/s	270

3.3 Electronic System

WM-6 has a microcontroller PIC (16F877, 20 MHz, Microchip ltd.) and a Bluetooth communication unit. The microcontroller controls the directions and velocities of the left and right wheels separately (via DC motors) according to instructions sent from the PC. In addition, the microcontroller measures the battery voltage and states of each lever (described later) before sending these data to the PC every 100 [ms].

Bluetooth communication supports two-way communication and its energy consumption is low; it is thus suitable for small mobile robots. WM-6 has a standby battery for the Bluetooth communication unit to maintain the power supply while the Li-ion battery is being exchanged.

3.4 Interaction Module; Levers

WM-6 has two levers to interact with rats. Since Skinner's experiment, levers have been used in many experiments using rats. Pushing levers is not an innate behavior of rats, meaning that it is highly likely that pushing the levers observed in experiments is an intentional behavior.

The dimensions of the levers are 20 x 30 [mm] identical to the Skinner box, and they are colored blue and yellow respectively for image processing. These levers consist of touch sensors that are electrically connected to the microcontroller and the logic level of each touch sensor is also sent to the PC through the microcontroller. It is possible to use these data as variables on the operation generator module in the PC. For example, in Figure 3, WM-6 moves to the front of the food-feeding machine when the lever is pushed. We consider these two to be levers are input devices of WM-6.

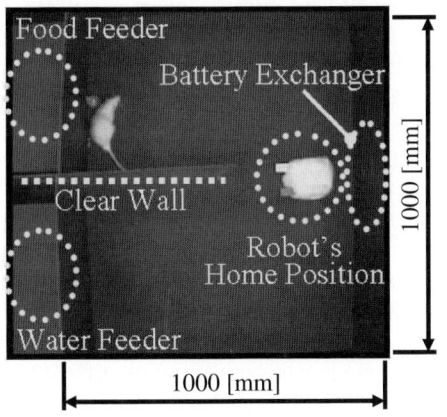
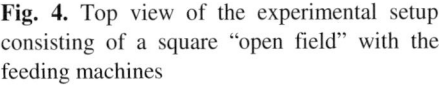
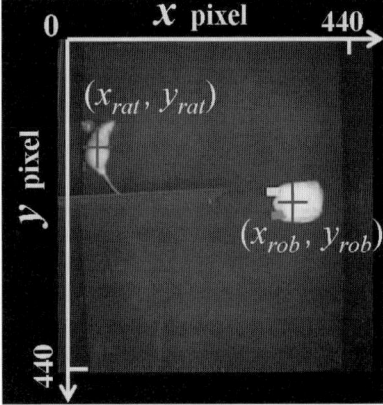

Fig. 4. Top view of the experimental setup consisting of a square "open field" with the feeding machines

Fig. 5. Positions of rat and WM-6 computed by the image processing

4 Experimental Setup

The experimental setup as shown in Figure 1 was developed. The interaction experiments between the rats and WM-6 are conducted in an "open field," 1000 [mm] square flat area surrounded by a wall of 450 [mm] high (Figure 4). The food-feeding machine, water-feeding machine and the battery exchanger are mounted on the wall. A CCD camera is positioned above the open field and sends images of the experiment to the PC every 30 [ms]. The PC automatically controls WM-6, the food-feeding machine, the water-feeding machine and the battery exchanger.

4.1 Food-Feeding Machine and Water-Feeding Machine

The food-feeding machine consists of a microcontroller PIC (16F877, 20 [MHz]) and a stepping motor. This machine releases a food pellet of 45 [mg] into a plastic bowl on the field when it receives an instruction sent from the PC.

4.2 Battery Exchanger

The battery exchanger consists of a microcontroller PIC (16F877, 20 [MHz]), electromagnets for attracting the batteries, 2-DOFs arm and a charger. WM-6 moves to the front of the battery exchanger when the battery on the robot is running low. The PC then sends an instruction to the battery exchanger. After that, the arm attracts the dead battery on the robot via electromagnets and exchanges it for a fully charged battery on the charger.

4.3 Control PC

Software that automatically controls WM-6 and all the machines is installed in a PC (CPU; Pentium IV 3.4 GHz, OS; Windows XP) that has an image-processing board and two Bluetooth communication units. This software consists of some software modules involving an image-processing module, operation generator module, robot controller module and device controller module. Therefore, it automatically conducts the interaction experiments and records the data without human intervention.

The image-processing module receives images from the CCD camera via the image-processing board. This module then computes the gravity points of the rat and the robot respectively every 100 [ms] (Figure 5). This module also saves the positions of the rat and the robot, and their movement distances respectively every 1 [sec] using CSV format.

The operation generator module generates the motion of WM-6 and the operation of the experimental setup based on pre-programmed patterns (e. g. Figure 3). For these patterns, experimenters can use variables such as the robot's position, the rat's position, the state of each lever on the robot, and the battery voltage of the robot. The behavior generation algorithm for autonomous teaching is included in this module. It is described in the next chapter.

The robot controller module determines the robot's movements according to the motions generated by the operation generator module. This module then controls WM-6 to move to the target point by controlling the directions and the velocities of each motor according to the distance and angle relative to the target point from the current point.

5   Behavior Generation Algorithm for Teaching

A novel behavior generation algorithm that enables WM-6 to teach lever-pushing task to rats is being developed introducing the idea of "shaping." The lever-pushing task is shown in Figure 3. In this task, rats have to push levers on the robot to obtain food, while the robot usually remains at its home position. When the rat pushes the levers on the robot, the robot moves to the front of the food-feeding machine and then stays there for three seconds. During the time that the robot stays there, the food-feeding machine releases a food pellet if the rat moves there. After these three seconds, the robot returns to the home position.

This behavior task looks simple and easy to learn. However, it is much harder for rats than the lever-pushing task in the Skinner box due to the movement of the robot. In fact, the rats never learned this behavior task without any teaching in our previous experiments [14]. Thus, the robot has to autonomously behave to show the rules between the lever-pushing task and food feeding.

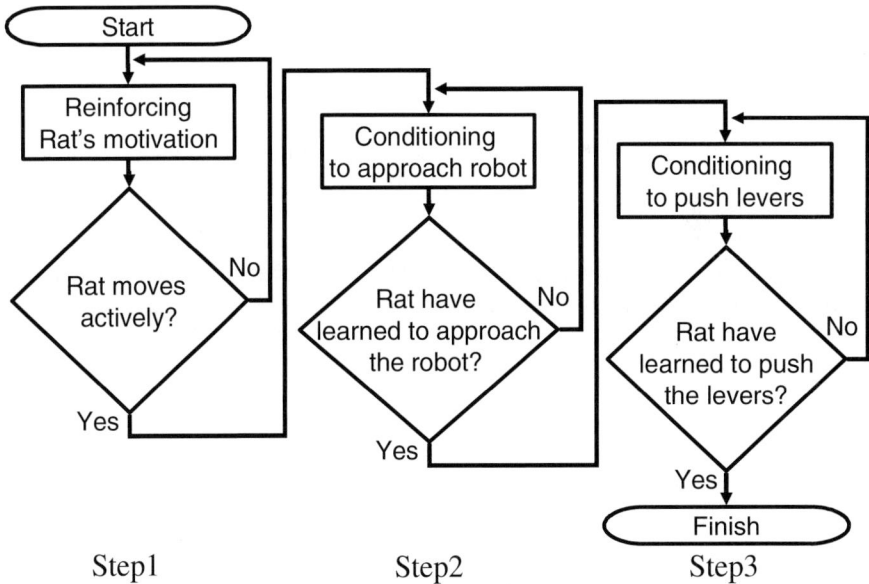

Fig. 6. Behavior generation algorithm for teaching lever-pushing

5.1   Design of Behavior Generation Algorithm

In psychology, the method of "shaping" is used for this kind of complex task learning. We designed the behavior generation algorithm for the robot introducing the idea of "shaping." Therefore, the learning process until the rats learn the lever-pushing task is divided into three steps (Step 1, Step 2, and Step 3). We then determined the target behavior or task in each step. We also constructed operational patterns of the setup and robot in each step to increase the chances of the target behavior appearing in the rats as shown in Figure 6.

Step 1: Reinforcement of the rat's motivation

The target behavior of this step is active movement, the simplest kind of behavior. Rats rarely move in an environment that they have never previously experienced due to their natural sense of caution. Therefore, in this step, the food-feeding machine routinely releases ten food pellets every 1 hour. In our previous experiment, the rats that had obtained food in the open-field moved actively compared to those that had not. Therefore, we believe that these routine feedings are effective in reinforcing the rat's motivation to move. When the total movement distance of the rat exceeds 50 [m], this step is finished and the next step is started.

Step 2: Conditioning the rat to approach to WM-6

The target behavior of this step is the approach to the robot. To attract the rat's interest in WM-6, the robot routinely moves to the front of the food-feeding machine and this machine then releases a food pellet at the moment at which the robot arrives. We believe that the rat learns the relationship between the robot and the feedings through these routine movements and feedings. It is then expected that the rat would be interested in WM-6 and hence approach it.

The rat's approach to the robot is detected by image processing. When d_{rr}, the distance between the rat and the robot, is less than D_{ap}, the threshold of approach detection, the approach is detected. When the rat's approach to WM-6 is detected, the robot moves to the front of the food-feeding machine. To reinforce the approach of the rat to the robot, the food-feeding machine then releases a pellet. After that, the robot returns to the home position. When the number of detections and reinforcements of approaches exceeds 200, this step is finished and the next step is started.

Step 3: Conditioning the rat to push the levers

The target behavior of this step is pushing the levers on WM-6. At the beginning of this step, when the rat approaches the robot, the robot moves to the front of the food-feeding machine and this machine then releases a pellet. After 200 reinforcements, D_{ap}, the threshold of approach detection, is reduced every time the rat approaches the robot. We believe that the rat would then approach the robot more closely, and it is subsequently expected that the rat would occasionally push the levers on the robot.

When the rat pushes the levers on WM-6, the robot moves to the front of the food-feeding machine and this machine releases a pallet. In this way, the rat would be conditioned to push the levers on the robot. When the number of times that the lever is pushed exceeds 200, this step is finished. In this way, we believe that rat can be conditioned to push levers to obtain food.

5.2 Implementation

This behavior generation algorithm is implemented in the operation generator module in the PC. The behavior of WM-6 and the operation of the setup are automatically generated according to this algorithm. The PC then automatically controls the experimental setup and WM-6 without any human operation or intervention. Thus, the experimenter has to just release the rat that has no experience of experiment into the experimental setup. The setup and the robot then autonomously condition the rat to perform lever-pushing task in a couple of days.

6 Experimental Evaluation

An experimental evaluation for the behavior generation algorithm for teaching was performed with the experimental setup.

6.1 Procedure

The experiment is performed autonomously using the experimental setup implementing the behavior generation algorithm for teaching. Three rats were used in this experiment. They were male albino rats without any experimental experience and bred singly in breeding cages. Before the experiment, they were made hungry by food restriction. Table 2 shows the conditions of the rats at the start of this experiment.

Each trial was conducted using a single rat autonomously without any human intervention. The rat was released into the experimental setup at the start of each trial. Until the number of times that the lever was pushed exceeded 200, the rat was left there.

6.2 Results

In this experiment, all three rats learned to push the levers on WM-6 to obtain food. In the case of Rat 1, the time required to learn the lever-pushing task was 9360 min (156 hour), in the case of Rat 2, it was 3260 (55 hour), and in the case of Rat 3, it was 5320 min (89 hour). Table 3 shows the time that these rats required to finish each step. The cumulative number of movements, approaches to the robot, and lever pushings of each rat are shown in Figure 7.

Table 2. Condition of the rats used in the evaluation experiment

	Experimental Group		
	Rat 1	Rat 2	Rat 3
Age [weeks]	75	15	15
Weight [g]	320	270	260

Table 3. Experimental result; time required for finishing each step [min]

	Rat 1	Rat 2	Rat 3
Step 1	430	50	240
Step 2	6080	1850	1950
Step 3	9360	3260	5320

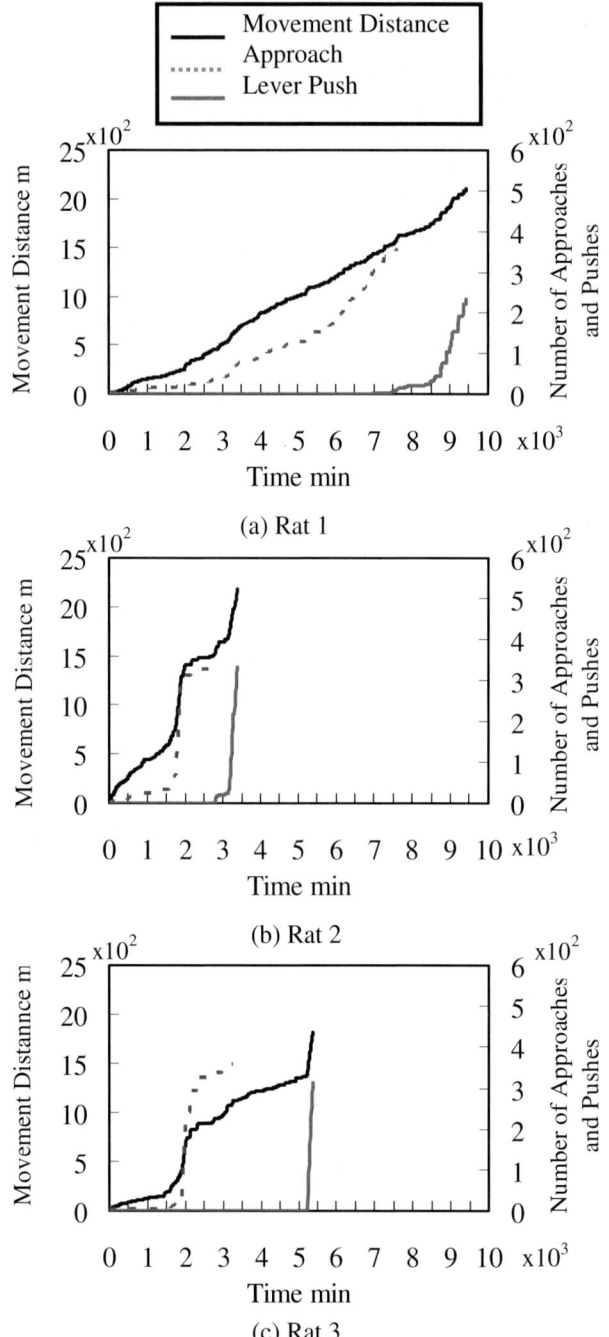

Fig. 7. Experimental Result; cumulative number of movement, approach to the robot, and lever pushing of each rat

6.3 Consideration

All three rats learned the lever-pushing task to obtain food, whilst the rats in the previous experiments in which the robot did not behave autonomously never learned [14]. Therefore, the behavior generation algorithm for teaching was verified.

The times required to learn the lever-pushing task are different for each rat. Rat 1 required the longest time to learn the task of all three. The learning curves of Rat 1, as shown in Figure 7 (a), slowly increase from the beginning to the end, while rapid points of increasing are found in those of Rat 2 and Rat 3. As shown in Table 2, the age of Rat 1 and those of Rat 2 and 3 are different. Generally, a rats' lifespan is two or three years. Therefore, we consider that Rat 1 is over middle age, and that Rats 2 and 3 are young. Thus, we believe that the differences in their learning curves show differences in their learning ability depending on age.

7 Summary and Discussion

The purpose of our study is to develop frameworks of adaptive interactions between animals and robotic agents. As an example of this, we developed a behavior generation algorithm for a robot to enable it to teach lever-pushing task to rats introducing the idea of "shaping." We then implemented it in the experimental setup and the robotic agent WM-6, and experimentally verified it. In the experiment, the robot understood rat's learning level and changed the "step" depending it. At the moment, the rat learned the behavior task through the interactions with the robot. Thus, we believe this experiment can be said a simple example of adaptive interaction.

Through this experiment, we proposed three ideas to design robot's behavior in the situations that it teaches behavior task to animals. The first is that robot should change its behavior depending learning level of the animals. The second is that the learning process should be divided into several small steps based on the concept of "shaping." The third is that design of robot's behavior is important to let the rat obtain new behavior.

The behavior generation algorithm for teaching that we developed is just for the lever-pushing task. However, the design method of behavior generation algorithms for the robotic agents based on the idea of "shaping" is useful for many kinds of robotic agents. The next step of our study is to develop the design method much more clearly. Therefore, we should consider methods of dividing behavior tasks into small steps and constructing operation patterns in each step.

We also have another idea of multi-branching shaping. In this paper, we used one-way and one-line without-branching "shaping." We consider the idea of multi-branching shaping, which has some options in each step, is useful and effective for task teaching. A major problem in implementation is choosing the most suitable branch. Therefore, we have to develop an algorithm that enables robots to choose the most suitable branch in their options interacting with rat.

Acknowledgements

A part of this study was conducted at the Humanoid Research Institute (HRI), Waseda University. The authors would like to express their thanks to Okino Industries LTD, OSADA ELECTRIC CO.LTD, SHARP CORPORATION, Sony Corporation, Tomy

Company LTD and ZMP IMC. We also would like to express our thanks to Solid Works Corp., Advanced Research Institute for Science and Engineering of Waseda University.

References

[1]   R. A. Books, "From Darwin to Behaviourism," Cambridge University Press, 1984
[2]   E. L. Thorndike, "Animal Intelligence," 1911.
[3]   B. F. Skinner, "The Behavior of Organisms," An American Experimental Analysis, 1938.
[4]   Rudy JW, Sutherland RJ. "The hippocampal formation is necessary for rats to learn and remember configural discriminations," Behavioral Brain Research 34; 97-109, 1989.
[5]   McDonald RJ, Ko C H, Hong NS. "Attenuation of context-specific inhibition on reversal learning of a stimulus response task in rats with neurotoxic hippocampal damage," Behavioral Brain Research 136, pp. 113-126, 2002.
[6]   Sutton R.S. Barto A.G., "Reinforcement Learning: An Introduction (Adaptive Computation and Machine Learning)," Bradford Books, 1998.
[7]   Uchibe E., and Doya K., "Competitive-Cooperative-Concurrent Reinforcement Learning with Importance Sampling," Proc. of International Conference on Simulation of Adaptive Behavior: From Animals and Animats, pages 287-296, 2004.
[8]   Doya K. and Uchibe E., "The Cyber Rodent Project: Exploration of Adaptive Mechanisms for Self-Preservation and Self-Reproduction," Adaptive Behavior, vol. 13, no. 2, pages 149-160, 2005.
[9]   Suga Y., Ikuma Y., Nagao D., Ogata T., and Sugano S., "Interactive Evolution of Human-Robot Communication in Real World," Proceeding of IEEE/RSJ Interanational Conference on Intelligent Robots and Systems (IROS2005), August, 2005
[10] Ogata T., Sugano S., and Tani J., "Open-end Human-Robot Interaction from the Dynamical Systems Perspective -Mutual Adaptation and Incremental Learning," Advanced Robotics, VSP and Robotics Society of Japan, Vol.19, No. 6, pp. 651-670, July, 2005.
[11] Yamada S. and Yamaguchi T., "Training AIBO like a Dog," Proc. of the 13th International Workshop on Robot and Human Interactive Communication (ROMAN-2004), pp.431-436, 2004
[12] Sanjay S. Joshi, Jeffery Schank, Nicolas Giannini, Lisa Hargreaves, and Randall Bish, "Development of Autonomous Robotics Technology for the Study of Rat Pups. Proc. of the 2004 IEEE Int'l Conference on Robotics and Automation, pp.2860-2864, 2004.
[13] Colot A., Caprari G. and Siegwart R., "InsBot: Design of an Autonomous Mini Mobile Robot Able to Interact with Cockroaches," Proc. of the 2004 IEEE Int'l Conference on Robotics and Automation, pp. 2418-2423, 2004.
[14] Ishii H., Aoki T., Moribe K., Nakasuji M., Miwa N., Takanishi A., "Interactive Experiments between Creature and Robot as a Basic Research for Coexistence between Human and Robot," Proc. of the 12th Int'l IEEE Workshop on Robot and Human Interactive Communication, CD-ROM, 2003.
[15] Ishii H., Nakasuji N., Ogura M., Miwa H., Takanishi A., "Experimental Study on Automatic Learning Speed Acceleration for a Rat using a Robot" Proc. of 2005 IEEE International Conference on Robotics and Automation, 2005.
[16] Colman A.M., "Dictionary of Psychology," Oxford University Press,
[17] B. F. Skinner, "Science of Human Behavior," 1953.

Believability Testing and Bayesian Imitation in Interactive Computer Games

Bernard Gorman[1], Christian Thurau[2],
Christian Bauckhage[3], and Mark Humphrys[1]

[1] Dublin City University, Glasnevin, Dublin 9, Ireland
{bgorman, humphrys}@computing.dcu.ie
[2] Bielefeld University, D-33501 Bielefeld, Germany
cthurau@techfak.uni-bielefeld.de
[3] Deutsche Telekom AG, 10587 Berlin, Germany
christian.bauckhage@telekom.de

Abstract. In imitation learning, agents are trained to carry out certain actions by examining a demonstration of the task at hand. Though common in robotics, little work has been done in translating these concepts to computer games. Given that present-day games generally use antiquated AI techniques which can often lead to stilted, mechanical and conspicuously artificial behaviour, it seems likely that approaches based on the imitation of human players may produce agents which convey a more humanlike impression than their traditional counterparts. At the same time, there exists no formal method of quantifying what *constitutes* a 'humanlike' impression; an equivalent of the Turing test is needed, with the requirement that an agent's appearance and behaviour be capable of deceiving an observer into misidentifying it as human. The aims of this paper are thus threefold; we describe an approach to the imitation of strategic behaviour and motion, propose a formal method of quantifying the degree to which different agents are perceived as 'humanlike', and present the results of a series of experiments using these two systems.

1 Introduction

Imitation learning, as the name suggests, refers to the acquisition of skills or behaviors through examination of a demonstrator's execution of a given task. Imitative techniques have been adopted by many researchers in robotics as a means of 'bootstrapping' their machines' intelligence, providing them with a high level of competence after a comparatively short training period [1]. Demiris and Hayes [2], for instance, train an apprentice robot to navigate a maze by imitating the actions of a demonstrator agent. Schaal [3] proposes a control-based approach to imitating a tennis swing from demonstration. Fod, Mataric and Jenkins [4] outline various statistical approaches to deriving movement primitives from observed human motion.

Despite the interest exhibited by the robotics community, however, very few attempts have been made to apply these principles to interactive computer games. Indeed, even the most modern games still predominantly rely on symbolic

S. Nolfi et al. (Eds.): SAB 2006, LNAI 4095, pp. 655–666, 2006.

artificial intelligence techniques that were developed several decades ago [5,6]. Given that many modern games allow the recording of entire sessions, and that – rather than limb movement data, as is common in robotic imitation – these recordings encode the frame-by-frame behaviour of the player under complex, rapidly-changing conditions and in competition with opponents of comparative skill, it becomes clear that computer games are an ideal platform for research in imitation learning. In this paper, we detail part of our work in this area; a Bayesian-based approach to the derivation and imitation of human strategic behaviour and motion patterns in commercial computer games. In conjunction with the believability-testing system described below, we then demonstrate its effectiveness in producing convincingly humanlike game agents.

When evaluating imitation agents, three distinct metrics are applicable: i) **statistical** analysis of the accuracy with which the observed behaviours are reproduced; ii) **believability testing** to verify whether the cloned agent effectively conveys the impression of being human; iii) **performance-based** assessment of the imitation agent in direct competition against other agents and human players. This paper concerns itself with believability testing. A significant impediment to work in this field is the lack of a formal, rigorous standard for determining how 'humanlike' an artificial agent is, or any strict means of comparing the believability of different agents. While some contributions compare observers' reactions to artificial and human players [7], these have invariably been of a very limited, informal and often inconclusive nature. Imitation learning holds obvious potential as a method of producing more credible agents, but there has thus far been no means of empirically assessing this credibility; the need for a perception- and behavior-based analogue to the Turing test is clear [8]. To address this need, we introduce a formal method of quantifying the degree to which cognitive agents are perceived as 'humanlike', and of facilitating the objective comparison of different agents. This method has been designed to minimize the subjectivity associated with such surveys, and to produce a *believability index* weighted according to both the observer's experience and the certainty with which the agents are identified.

The *first-person shooter* (FPS) genre – wherein players explore a 3D environment littered with weapons, bonus items, traps and pitfalls, with the objective of defeating as many opponents as possible – was chosen for our work on the basis that it provides a relatively *direct* mapping of human decisions onto agent actions. Due to its prominence within the literature [9], we opted to use iD Software's QUAKE II® as our testbed. In order to extract the required data from its recorded *DM2* or *demo* file format – consisting of the network traffic received during the game - and to realise the in-game agents (or *bots*, in game vernacular), we employ our own QASE API and its MatLab-integration facilities [10].

2   Imitation Learning - Methodology

In this section, we outline our current approach to imitating human movement and strategic behavior in QUAKE II® . The individual components of this ap-

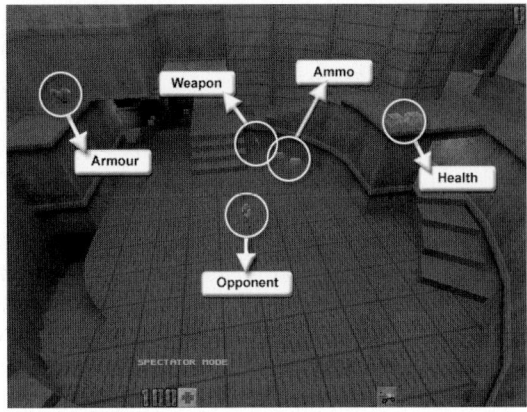

Fig. 1. Typical QUAKE II® environment

proach were introduced in previous publications [11,12], while a forthcoming contribution describes their integration. Here, we briefly review the system; readers are referred to the earlier publications for additional details.

2.1   Behaviour Model

The current model focusses on two core aspects of human behaviour; *strategic planning* and *motion modelling*. A number of investigations [13,7] have found that the ability of an agent to exhibit long-term strategic planning faculties is a crucial factor in determining how humanlike it appears. The importance of *motion modelling* is equally evident - human players frequently exhibit actions other than simply moving along the environment surface, including jumps, weapon changes and discharges, crouches, etc. In many cases, the player can only attain certain goals by performing one or more such actions; they therefore have an important *functional* element. From the perspective of creating a believable agent, it is also vital to reproduce the *aesthetic* qualities they encode.

2.2   Learning Goal-Oriented Strategic Behaviours

In QUAKE II® , experienced players traverse the environment methodically, controlling important areas of the map and collecting *items* to strengthen their character. Thus, we define the player's long-term goals to be the items scattered at fixed points around each level. By learning the mappings between the player's status and his subsequent item pickups, the agent can adopt observed strategies when appropriate, and *adapt* to situations which the player did not face.

We first read the set of all player locations $l = [x, y, z]$ from the recording, and cluster the points using a *fast k-means* to produce a *goal-oriented* discrimination of the level's topology. We also construct an $n \times n$ matrix of edges E, where n is the number of clusters, and $E_{i,j} = 1$ if the player was observed to move from node i to node j and 0 otherwise. The player's *inventory* – the list of what quantities of which items he currently possesses – is also read from the demo at each timestep, and unique state vectors are obtained; these *inventory prototypes* represent the varying situations faced by the player during a game. We can now construct a set of *paths* which the player followed while in each such situation.

Having obtained the different paths pursued by the player in each inventory state, we turn to reinforcement learning to learn his behaviour. The topological map of the game environment may now be viewed as a *Markov Decision Process*, with the clusters corresponding to states and the edges to transitions.In this scenario, the MDP's actions are considered to be the choice to move to a given node from the current position. Thus, the transition probabilities are $P(c' = j | c = i, a = j) = E_{ij}$ where c is the current node, c' is the next node, a is the executed action, and E is the edge matrix. We assign an increasing reward to consecutive nodes in every path taken under each prototype, such that the agent will be guided along similar paths to the human when facing similar situations. With the transition probabilities and rewards in place, we now run a modified version of the *value iteration algorithm* in order to compute the utility values for each node in the topological map under each inventory state prototype.

A number of other features of human planning behaviour must also be taken into account. Principal among these are the human player's intuitive *weighing* of strategic objectives, and his understanding of *object transience* – that is, a collected item will be unavailable until the game regenerates it after a fixed interval. To model these, we introduce a weighted *fuzzy clustering* approach and an *item activation* variable:

$$m_p(s) = \frac{a(o_p)d^{-1}(s, p)}{\sum a(o_i)d^{-1}(s, i)} \tag{1}$$

where m is the membership, s is the current inventory state, p is a prototype inventory state, P is the number of prototypes, a is 1 if the object o at the terminal node of the path associated with prototype p is present and 0 otherwise, and d^{-1} is an inverse-distance or proximity function. The membership distribution implicitly specifies the agent's current goals, which will later facilitate integration with the Bayesian motion-modeling system. The final utilities are:

$$U(c) = \gamma^{e(c)} \sum V_p(c)m_p(s), \quad c_{t+1} = \max_y U(y), \quad y \in \{x | E_{c,x} = 1\} \tag{2}$$

where $U(c)$ is the final utility of node c, γ is the discount, $e(c)$ is the number of times the player has entered cluster c, $V_p(c)$ is the original value of node c in state prototype p, and E is the edge matrix.

2.3   Bayesian Motion Modelling

It is not sufficient to simply identify the player's goals and the paths along which (s)he moved to reach them; it is also necessary to capture the actions executed by the player in pursuit of these goals. In a previous contribution [12], Thurau et al describe an approach based on Rao, Shon & Meltzoff's Bayesian inverse-model for action selection in infants and robots [14]. The choice of action at each timestep is expressed as a probability function of the subject's current position c_t, next position c_{t+1} and goal c_g:

$$P(a_t|c_t, c_{t+1}, c_g) = \frac{P(c_{t+1}|c_t, a_t)P(a_t|c_t, c_g)}{\sum_u P(c_{t+1}|c_t, a_u)P(a_u|c_t, c_g)} \quad (3)$$

This model fits into the strategic navigation system almost perfectly; the clusters c_t and c_{t+1} are chosen by examining the utility values, while the current goal state is implicitly defined by the membership distribution. In order to derive the probabilities, we read the sequence of actions taken by the player as a set of vectors v such that $v = [\varDelta\text{yaw}, \varDelta\text{pitch}, \text{jump}, \text{weapon}, \text{firing}]$. We then cluster these action vectors to obtain a set of *action primitives*, each of which amalgamates a number of similar actions performed at different times into a single unit of behavior.

Several important adaptations must be made in order to use this model in the game environment. Firstly, Rao's model assumes that transitions between states are instantaneous, whereas multiple actions may be performed in Quake II® while moving between successive clusters; we therefore express $P(c_{t+1}|c_t, a_t)$ as a soft-distribution of all observed actions on edge $E_{ct,ct+1}$ in the topological map. Secondly, Rao assumes a single unambiguous goal, whereas we deal with multiple weighted goals in parallel. We thus perform a similar weighting of the probabilities across all active goal clusters. Finally, Rao's model assumes that each action is independent of the previous action. In Quake II®, however, each action is constrained by that performed on the preceding timestep; we therefore introduce an additional dependency in our calculations. The final probabilities are computed as follows:

$$\sum_g m_g P(a_t|c_t, c_{t+1}, c_g) \frac{P(a_t|a_{t-1})}{\sum_u P(a_u|a_{t-1})} \quad (4)$$

3   Believability Testing

As discussed earlier, there exists no standard method of gauging the 'believability' of game bots, nor of objectively comparing this quality in different agents; given that one of the central aims of our work lies in improving the believability of such agents, this is clearly a shortcoming which needs to be addressed. The most obvious means of determining the degree to which agents are perceived as human is to conduct a survey. This, of course, immediately raises questions of subjectivity, experimenter influence, and so on. In order to produce a credible assessment of agent believability, any proposed system must be designed with these concerns in mind. Our aims, then, are as follows: i) to construct a framework which facilitates rigorous, objective testing of the degree to which game agents are perceived as human; ii) to formulate a *believability index* expressing this 'humanness', and allowing comparisons between different agents.

The system developed to fulfil these criteria is described below. We outline the structure of the survey and its applicability to the testing of agents in general, using our own experiments to illustrate key concepts; we then describe these experiments and their results in greater detail. The test itself can be taken at http://reynard.computing.dcu.ie/sab_tests/

3.1   Structure of the Believability Test

To counteract any potential observer bias, the test takes the form of an anonymous online survey. Respondents are first presented with detailed instructions covering all aspects of the test. Before starting, they are further required to estimate their experience in first-person shooter games, at one of five different levels. Subjective judgements are avoided by explicitly qualifying each experience level:

1. Never played, rarely or never seen
2. Some passing familiarity (played / seen infrequently)
3. Played occasionally (monthly / every few months)
4. Played regularly (weekly)
5. Played frequently (daily)

Upon proceeding to the test itself, the respondent is present with a series of pages, each of which contains a group of video clips. Each group shows similar, but not identical, sequences of gameplay from the perspective of the in-game character. This approach was adopted due to concerns that asking respondents to view individual clips in isolation, with no basis for comparison against similar samples, would lead to a significant amount of subjectivity and guesswork. Within each group, the clips may depict any combination of human and artificial players; the respondent is required to examine the behaviour of the character in each clip, and indicate whether (s)he believes it is a human or artificial player. The clips are marked on a gradient, as follows:

1: Human, 2: Probably Human, 3: Don't Know, 4: Probably Artificial, 5: Artificial

This rating is the central conceit of the survey, and will later be used to compute the believability index. The respondent is also asked to specify how many times (s)he viewed the clip (to a maximum of 3 times), and to provide an optional comment explaining his/her choice. In cases where (s)he indicates that (s)he believes the agent to be artificial, (s)he will be further asked to rate how "humanlike" (s)he perceives its behaviour to be, on a scale of 1 to 10. This more subjective rating is not involved in the computation of the believability index, but may be used to provide additional insight into users' opinions of different agents. Having completed all required sections on each page, the user submits his/her answers and moves on to the next.

3.2   Subjectivity, Bias and Other Concerns

Aside from the observer effect, there are several areas in which the potential for subjectivity and the introduction of bias exist. Since our aim is to provide

Fig. 2. Extract from the main believability test screen

an objective measure of believability, these must be eliminated or minimized. A number of these issues are discussed below.

The first obvious pitfall lies in the selection of video clips. The selector may deliberately choose certain clips in an effort to influence the respondents. To guard against this, we first ensure that the number of samples is sufficient to embody a wide variety of behaviours, and secondly, we cede control of the selection of the specific behaviours to an unbiased arbiter. In our case, we wished to compare the believability of our imitation agents against both human players and traditional rule-based bots; thus, we first ran numerous simulations with the traditional agent – over whose behavior we had no control – to generate a corpus of gameplay samples, and then proceeded to use human clips embodying similar behavior both in the believability test and to train our imitation agents.

Similarly, the order in which the videos are presented could conceivably be used to guide the respondents' answers. To prevent this, we randomize the order in which the groups of clips are displayed to each user, as well as the sequence of clips within each page; the test designer thus has no control over the order of the samples seen by the user. Additionally, the filenames under which the clips are stored are randomized, such that the respondent cannot determine the nature of each clip based on examining the webpage source (e.g. clip 1 always human, clip 2 always artificial, etc).

Another issue concerns the possibility that users will choose the 'Probably' options in a deliberate effort to artificially minimize their error and 'beat' the test, or that they will attempt to average out their answers over the course of the survey – that is, they may rate a clip as 'human' for little reason other than that they rated several previous clips as 'artificial', or vice-versa. To discourage this, we include notes on the introduction page to the effect that the test does not adhere to any averages, that the user's ratings should be based exclusively upon their perception of the character's behavior in each clip, and that the user should be as definitive as possible in their answers. A related problem is that of user fatigue; as the test progresses, the user may begin to lose interest, and will consequently invest less effort in each successive clip. We address this by including a feature enabling users to save their progress at any point, allowing them to complete the survey at their convenience.

It is also imperative to ensure that the test is focused upon the variable under investigation – namely, the believability of the agent's movement and behavior.

As such, the survey must be structured so as not to present 'clues' which might influence the respondents. For instance, the tester should ensure that all clips conform to a standard presentation format, so that the respondent cannot discern between different agents based on extraneous visual cues - different players may have used different in-game character models, individual player names, etc. To this end, we run a script over the demo files to remove all such indicators and homogenize them to a common format.

In the specific case of our imitation agents, this requirement that all extraneous indicators be removed raises a conflict between two of our goals in conducting the survey. If the players in two of the three clips we use on each page begin from the same location and exhibit near-identical behavior, the respondent may conclude through pure logical deduction that (s)he is probably viewing a human and imitation agent, and consequently that the remaining clip is more likely to be a traditional artificial agent. Note that this might not necessarily be true, but even an incorrect answer based on factors other than believability will adversely affect the accuracy of the results. We circumvent this problem by training imitation agents with different (but similar) samples of human gameplay to those actually used in the test. The resulting clips are therefore comparable, but do not 'leak' any additional information; respondents must judge whether or not they are human based solely on their appearance. At the same time, however, we obviously wish to test how accurately our agents can capture the aesthetic appearances of their human exemplars. To satisfy both requirements, a small minority of imitation agents *are* trained using the same human data as presented in the survey; in the experiments described below, 2 of the imitation agents were direct clones, while the remainder were trained on different data.

3.3   Evaluation of Results

Before evaluating the results of the survey, one should ensure that there have been a substantial number of responses with a decent distribution across all experience levels; a good 'stopping criterion' is to run the test until the average experience level is at least 3 (i.e. a typical gamesplayer). Standard analyses (precision, recall, etc) can be carried out on the results; however, as mentioned earlier, we also wish to formulate a believability index which is specifically designed to express the agent's believability as a function of user experience and the certainty with which the clips were identified.

Recall that each clip is rated on a scale of 1 (definitely human) to 5 (definitely artificial). Obviously, the true value of each clip is always either 1 or 5. Thus, we can express the degree to which a clip persuaded an individual that the visualised character was human as the normalised difference between that person's rating and the value corresponding to 'artificial':

$$h_p(c_i) = \frac{|r_p(c_i) - A|}{max(h)} \qquad (5)$$

where $h_p(c_i)$ is the degree to which person p regarded the clip as depicting a human, $r_p(c_i)$ is person p's rating of clip i, A is the value on the rating scale which corresponds to 'artificial', and $max(h)$ is the maximum possible difference between a clip's rating and the value of 'artificial'. In other words, $h_p(c_i)$ will be 0 if the individual identified a clip as artificial, 1 if he identified it as human, and somewhere in between if he chose one of the 'Probably' or 'Don't Know' options. We now weight this according to the individual's experience level:

$$w_p(c_i) = \frac{e_p h_p(c_i)}{avg(e)} \tag{6}$$

where e_p is the experience level of person p and $avg(e)$ is the mean experience level. Finally, we sum the weighted ratings across all clips and respondents, and take the average:

$$b = \frac{\sum_p^n \sum_i^m w_p(c_i)}{nm} \tag{7}$$

where b is the believability index, n is the number of individual respondents, and m is the number of clips. The believability index is, in essence, a weighted representation of the degree to which a given type of clip was regarded as human, in the range $(0, 1)$. In order to express the *strength* of the result and to facilitate comparison between agents evaluated in different surveys, we also compute a *confidence index* as follows:

$$c = \frac{avg(e)}{max(e)} \tag{8}$$

where $avg(e)$ is the average experience of the respondents, and $max(e)$ is the maximum experience level; the confidence index is thus conditioned upon a sufficient level of expertise among respondents. In the context of the survey, then, a 'good' result for an AI agent would involve a high value of b for both the agent and human clips, together with a confidence index of 0.6 or greater (indicating that respondents were, on average, significantly experienced).

4   Experiments

In this section, we detail an experiment carried out using the believability test in conjunction with our imitation agents. The purpose of this experiment was twofold; first, to evaluate the believability-test framework itself, and second, to examine how believable our imitation agents were in comparison with human players and traditional rule-based artificial agents.

The experiment consisted of 15 groups, with 3 clips in each; these clips were, on average, approximately 20 seconds in length. We first ran numerous simulations involving the rule-based artificial agent to derive a set of gameplay samples, and then used similar samples of human players both in the test itself and to train our imitation agents. The rule-based agent used was the QUAKE II® Gladiator bot, which was chosen due to its reputation as one of the best bots available.

It should be noted that, since our imitation mechanism is designed to imitate strategic navigation and human motion, combat was omitted from consideration in this study. As one of our respondents commented, this filters out one variety

Table 1. Believability/Confidence indices, Recall and Precision values. Recall values consider classification as 'human' to be the desired results. Precision is estimated over [human or imitation] identified as human, and rule-based agent identified as artificial.

Clip Type	Believability	Confidence	Recall (%)	Precision (%)
Human	0.69		68.08	78.39
Imitation	0.69	0.64	68.81	
Artificial	0.36		36.69	50.87

of behavior from the agent's repertoire, and has the effect – as with the original Turing test – of reducing the opportunities for an observer to detect artificialities. While the test can be used to accurately gauge how well our system captures human strategy and movement, a further study involving combat behaviours is essential. See Future Work for further discussion.

With the video clips in place, the URL of the survey site was distributed to the mailing lists of several colleges in Ireland and Germany. After a one-week test period, we had amassed a considerable number of responses. After discarding incomplete responses, we were left with 20 completed surveys, totalling 900 individual clip ratings; the average experience level of respondents was 3.2.

As can be seen from Tab. 1, the survey produced a very favourable impression of our imitation agents compared to the artificial agent. The believability indices for human, imitation and traditional artificial clips were 0.69, 0.69 and 0.36, respectively. In other words, the imitation agents were misidentified as human 69% of the time, while the rule-based agents were mistaken as human in only 36% of cases (weighted according to experience). Clips which actually *did* depict human players were also identified 69% the time. Essentially, it seems that respondents were generally unable to discern between the human players and our imitation agents. These results are corroborated by the recall values, which indicate that both the human and imitation clips were classified as human in approximately 68% of cases, while the rule-based agent was classified as human only 36.69% of the time. Since the human sources used to train the imitation agents were different than those human clips presented as part of the test, this implies that the results are based on the general abilities of the imitation mechanism, rather than any factors unique to the clips in question.

Further indication of the imitation agents' effectiveness is evident in the graph of believability against experience level shown in Fig. 3; as experience level rises, respondents correctly identify human clips as human more frequently, and misidentify the traditional agent as human less frequently. The identification of imitation agents as human, by contrast, closely parallels that of genuine human clips. These trends may be explained by the fact that more experienced players have a greater knowledge of characteristically human behaviours – smooth strafing, unnecessary jumping, pausing to examine the environment, and similar idiosyncrasies – which the traditional agent would not exhibit, but which would be captured and reproduced by the imitation bots. This interpretation is supported by many of the comments submitted by respondents, including those shown in Table 2.

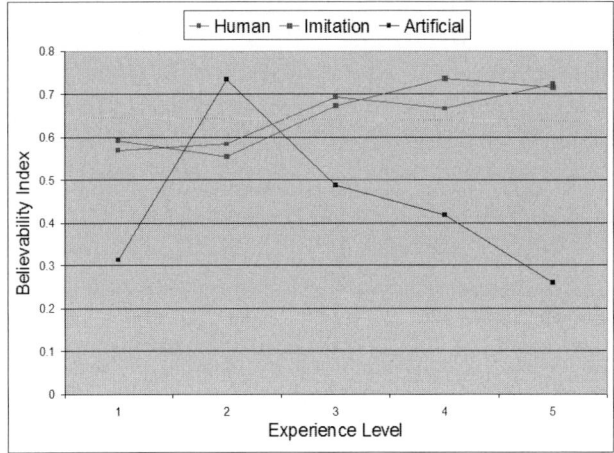

Fig. 3. Variation of believability with experience level

Table 2. Sample comments from imitation clips misidentified as human

Experience	Comment
5	Bunny hop for no reason, also seems to be scanning for enemies
5	Fires gun for no reason , so must be human
5	Unnecessary jumping
5	Stand and wait. Ai wouldn't do this (?)
5	Human as they knew how to Rocket jump
5	The rocket jump and the short sequence of backward running at the end suggest this was human

In conclusion: while further testing (mainly of combat behaviours) is required, the results of the believability study suggest that our imitation agents exhibit far greater 'humanness' than even a well-regarded rule-based agent, and indeed are comparable to genuine human players. We consider this to be strong evidence in support of our original premise; namely, that imitation learning has the potential to produce more believable game agents than traditional AI techniques.

5   Summary and Future Work

In this paper, we proposed a formal method of quantifying the degree to which different agents are perceived as 'humanlike', in the form of a web-based survey and an objective metric based on both the respondents' level of experience and the accuracy with which the players/agents were identified. Through our experiments, we verified the effectiveness of the believability-testing system; we further showed that our imitation-learning approach produces game bots which are capable of conveying a significantly more humanlike impression than traditional agents, and are often almost indistinguishable from genuine human players.

Clearly, the next stage in our work must concentrate on imitating combat behaviours, and integrating them into the existing imitation mechanism. Beyond this, tests based on the third metric described in the introduction will also be conducted – that is, in-game performance-based evaluation of the imitation bots, in direct competition with human players and other artificial agents.

References

1. Schaal, S.: Is Imitation Learning the Route to Humanoid Robots? Trends in Cognitive Sciences **3**(6) (1999) 233–242
2. Hayes, G., Demiris, J.: A Robot Controller Using Learning by Imitation. In: Proc. of the 2nd Int. Symposium on Intelligent Robotic Systems. (1994) 198–204
3. Schaal, S.: Movement Planning and Imitation by Shaping Nonlinear Attractors. In: 12th Yale Workshop On Adaptive And Learning Systems. (2003)
4. Fod, A., Matarić, M., Jenkins, O.: Automated Derivation of Primitives for Movement Classification. Autonomous Robots **12**(1) (2002) 39–54
5. Laird, J.E., v. Lent, M.: Interactice Computer Games: Human-Level AI's Killer Application. In: Proc. AAAI. (2000) 1171–1178
6. Fairclough, C., Fagan, M., MacNamee, B., Cunningham, P.: Research Directions for AI in Computer Games. Technical report, Trinity College Dublin (2001)
7. Livingstone, D., McGlinchey, S.: What Believability Testing Can Tell Us. In: In Proc. Int. Conf. on Computer Games: AI, Design and Education. (2004)
8. Livingstone, D.: Turing's Test and Believable AI in Games. CiE **4**(1) (2006) 6
9. J.E. Laird, J.D.: Creating Human-like Synthetic Characters with Multiple Skill-Levels: A Case Study Using the Soar Quakebot. In: Proc AAAI. (2000)
10. Gorman, B., Fredriksson, M., Humphrys, M.: QASE – An Integrated API for Imitation and General AI Research in Commercial Computer Games. In: Proc. CGAMES Int. Conf. Computer Games. (2005) 207–214
11. Gorman, B., Humphrys, M.: Towards Integration of Strategic Planning and Motion Modelling in Interactive Computer Games. In: Proc. Int. Conf. Computer Game Design & Technology. (2005) 92–99
12. Thurau, C., Paczian, T., Bauckhage, C.: Is Bayesian Imitation Learning the Route to Believable Gamebots? In: Proc. GAME-ON North America. (2005) 3–9
13. Laird, J.E.: Using a Game to Develop Advanced AI. IEEE Computer (2001) 70–75
14. Rao, R., Shon, A., Meltzoff, A.: A Bayesian Model of Imitation in Infants and Robots. In Dautenhahn, K., Nehaniv, C., eds.: Imitation and Social Learning in Robots, Humans, and Animals: Behavioural, Social and Communicative Dimensions. Cambridge University Press, (2004)

Asynchronous Cyclic Pursuit[*]

Andaç T. Şamiloğlu[1,2], Veysel Gazi[1], and Buğra Koku[2]

[1] TOBB University of Economics and Technology,
Department of Electrical and Electronics Engineering,
Söğütözü Cad., No: 43, Söğütözü, 06560 Ankara, Turkey
[2] Middle East Technical University, Mechanical Engineering Department,
İnönü Bulvarı, Çankaya, Ankara, Turkey

Abstract. In this article we study the convergence of the positions of a multi-agent system in a cyclic pursuit under asynchronism and time delays. Each agent is assumed to operate on an infinite sequence of behaviors modeled by a finite state machine, which is represented by a discrete asynchronous mathematical model on a higher-level. The results on the convergence of the synchronous model are used in the proof of convergence of the asynchronous system. Numerical simulations are also performed to verify the theoretical results.

1 Introduction

Recent robotics research has been focusing on multi-agent systems or basically groups of autonomous mobile agents. Such systems are of interest for several reasons: (i) Tasks may be too complex or sometimes impossible for a single agent to achieve; (ii) Performance of the system may be improved by using multiple agents; (iii) The agents of a multi-agent system may be easier to build, cheaper, more flexible, and more fault tolerant than a single agent designed for each separate complex task; (iv) The constructive, synthetic logic developed for cooperative mobile robotics can also be beneficial in the problems of other sciences; especially for social sciences including organization theory, economics, cognitive psychology or life sciences like theoretical biology and animal ethology [1]. The references in [1] and [2] provide comprehensive reviews of multi-agent systems.

The output of multi-agent systems research has implications on many fields of (engineering) applications such as terrestrial, space and oceanic exploration, military surveillance and rescue missions, and other automated collaborative operations. The desired approach in solving such engineering problems is achieving the global objective or emergent behavior by simple local rules/interactions. However, determination of agent level simple interaction rules that yields the desired global behavior is a challenging problem that has not been solved yet. On this subject one of the earliest famous study was performed by Reynolds [3].

[*] This work was supported by the Scientific and Technological Research Council of Turkey (TÜBİTAK) under grant 104E170.

S. Nolfi et al. (Eds.): SAB 2006, LNAI 4095, pp. 667–678, 2006.
© Springer-Verlag Berlin Heidelberg 2006

He introduced a model and wrote a program called *boids* (or bird-oids) that simulates a flock of birds in flight. The behavior-based techniques used by Reynolds were also studied by Balch and Arkin [4]. They designed reactive behaviors to implement multivehicle formations in combination with rules for collision avoidance and other navigational goals.

Bruckstein worked on the behaviors of ants in [5]. He investigated how the path connecting the anthill and a food location becomes a straight line after a pioneer ant shows the way to the food. A study on the efficiency in chemotaxis due to schooling behavior was performed by Grünbaum in [6].

The very first scientist worked on the mathematics of *pursuit curves* was the French scientist Bouguer (c. 1732) [7]. In 1877, Lucas asked what trajectories would be generated if three dogs, initially placed at the vertices of an equilateral triangle, were to run one after the other? Brocard showed that each dog's pursuit curve would be that of a logarithmic spiral and that the dogs would meet at a common point (*Brocard point*) [7]. Klamkin and Newman [8] showed that, three bugs in cyclic pursuit which are not initially collinear, will meet at a point and this meeting will be mutual. Behroozi and Gagnon [9] later on proved that if all the bugs have the same speed and a nonmutual capture occurs, then this capture should be a head on collision. Richardson [10] showed that for the n-bugs problem, the head on collision is possible even for non-collinear initial positions but the probability of this collision is zero if the initial positions of the bugs are determined due to a smooth probability distribution. Similarly, Bruckstein, Cohen, and Efrat [11] considered a deterministic continuous pursuit in cyclic order and with preassigned varying speeds.

A study on the aggregation problem is performed in [12] with agents that are anonymous, homogeneous, memoryless, and lack communication capabilities. In a similar study in [13] the authors showed that asynchronous autonomous agents which have limited visibility and no memory, would gather at the same location in finite time provided that they have a compass. The problem of aggregation or gathering to a point is studied also by several other researchers within different contexts and under different names such as synchronization, consensus seeking, rendezvous, and others [14,15,16,17,18].

A systematic analysis of probabilistic aggregation strategies in swarm robotic systems is presented in [19], which considers four basic behaviors of the agents -*obstacle avoidance*, *approach*, *repel*, and *wait*- for aggregation. Similarly, in [20], the effects of different evolutionary parameters on the performance and scalability of system are studied.

A particular version of pursuit problem is studied in [21] for a system of n wheeled vehicles which are subject to a single nonholonomic constraint. The study provides a full stability analysis for the special case when $n = 2$ and how the global behavior of the system can be shaped through appropriate controller gain assignments. The same authors showed in [22] that the equilibrium formations of the system are generalized regular polygons and studied the local stability of these equilibrium polygons. The authors extend their work by studying the stability of equilibrium formations for multiple unicycle systems in cyclic pursuit in [23] and provide a complete local stability analysis for the general case $n \geq 2$. The study of Lin, Broucke, and Francis [24] is similar to these in means of the convergence of agents under certain conditions.

In this paper we focus on the problem of a multi-agent system performing cyclic pursuit with asynchrony in motion and sensing. We use a finite state machine to describe the sequence of behaviors of each agent and a discrete asynchronous mathematical model on a higher-level. After presenting the proof of convergence for synchronous cyclic pursuit model, we analyze the properties of the asynchronous model. Finally, we provide some numerical simulations to illustrate the results of the study.

2   Asynchronous High-Level Model

Consider the architecture shown in Figure 1 which consists of three behaviors: *wait, sense and compute,* and *move*. During the *sense and compute* behavior the ith agent gets (measures or receives by other means) the relative position of the $i+1^{th}$ agent and computes its own next desired position or way-point. During the *move* behavior the ith agent moves towards the computed way-point. During the *wait* behavior, the agent doesn't move or basically stays in place. These behaviors are arbitrated by using a finite state machine (FSM) in an infinite loop and in the sequence shown in Figure 1.

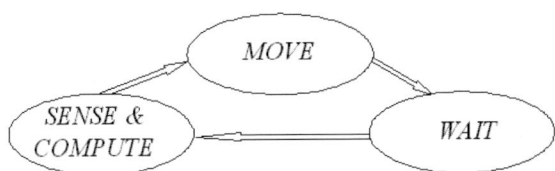

Fig. 1. Finite State Machine Model

Since here we are concerned with cyclic pursuit the computed next positon of the ith agent is always towards the sensed position of the $i + 1^{th}$ agent and during the move state the agent moves towards this way-point. We assume that each agent has a low level control which guarantees that the agent reaches the computed way-point in a finite time. We are not concerned with the low level dynamics and how the low-level control is implemented. Therefore, the analysis below is applicable for many systems with variety of different low-level vehicle dynamics including heterogenous swarms/systems (i.e. swarms consisting of more than one type of agents). Moreover, we ignore the issue of collisions between the agents. The resulting sequence of behaviors can be summarized as: Move towards the pursued agent. Wait for a predetermined time interval. Then sense the location of the next agent and move again towards that agent.

In this system we assume that the agents are ordered from 1 to n. Agent i pursues agent $i + 1$ modulo n. In other words, the last (n^{th}) agent pursues the first one. The agents are assumed to move in 2-D space and the position of the agents is given by

$$z_i(t) = [x_i(t), y_i(t)]^T \in R^2, i = 1, 2, ..., n \qquad (1)$$

Note, however, that this is not a critical assumption and the results developed will be valid also for $z_i(t) \in R^m$ $(m = 1, 2, ...)$ for a finite positive integer m.

Recall that during the *sense and compute* behavior the ith agent gets the position of the $i + 1^{th}$ agent and then computes its own next desired position or way point. However, during these sensing and computing processes of the ith agent the $i + 1^{th}$ agent may be in its *move* state and therefore the measured position of the next agent may be outdated. Moreover, the measurement of the position of the next agent may itself incure some delay. Whether ultrasonic, infrared or other type of sensors are used the propagation delay of the signals may lead to measurement of old (outdated) positions. Similarly delay will be also present even if the positions are obtained by inter-agent communication or by other means such as global positioning system. Therefore, the modeling of the dynamics of agents in cyclic pursuit should be designed including the position sensing delays. Referring to this phenomena we introduce the variables $\tau_{i+1}(t)$ which satisfy $0 \le \tau_{i+1}(t) \le t$ in order to represent the delay in the position measurements. In other words, we assume that at time t agent i knows $z_{i+1}(\tau_{i+1}(t))$ instead of the actual $z_{i+1}(t)$ about the position of agent $i + 1$. In other words, $z_{i+1}(\tau_{i+1}(t))$ is the *perceived position* of agent $i + 1$ by agent i at time t. Also since each agent operates on its own local clock following the state machine cycle on Figure 1 without a need for synchronization with the other agents, we introduce a set of time indices T^i, $i = 1, 2, ..., n$, at which the agent i updates its way-point z_i. It is assumed that at the other instances the agent i does not perform way-point calculation (it might be in one of other states/behaviors at these time instants). With these in mind the "high-level" dynamics of each agent can be represented as

$$z_i(t+1) = (1-p)z_i(t) + p\,z_{i+1}(\tau_{i+1}(t)), \qquad t \in T^i \qquad (2)$$
$$z_i(t+1) = z_i(t), \qquad t \notin T^i$$

where p is the gain satisfying $0 \le p \le 1$ and as mentioned above the variables $\tau_{i+1}(t)$, $i = 1, \ldots, n$, are used to represent the time index of the position information of the $i + 1^{th}$ agent. These variables satisfy $0 \le \tau_{i+1}(t) \le t$ for all $t \in T^i$ and for all i. If agent i has not yet obtained any information about the $i + 1^{th}$ agent's position and still has the initial position information, then $\tau_{i+1}(t) = 0$ whereas $\tau_{i+1}(t) = t$ means that agent i has the current position information of the $i + 1^{th}$ agent. The difference between the current time t and the value of the variable $\tau_{i+1}(t)$ is the delay occurring due to the sensory, computing and/or communication processes or other reasons.

In equation (2), the elements of the set $T^i \subset \{0, 1, 2, ...\}$ are the indices of the sequence of ordered physical times $\mathcal{T} = \{t_0, t_1, t_2, ...\}$ similar to the times of events in discrete-event systems where $t_i < t_{i+1}$ are the time instants at which the events in the system occur. The times t_i do not need to be equally spaced, i.e., the intervals $t_{i+1} - t_i$ do not have to be equal. Referring to the FSM model in Figure 1 during the time interval between two subsequent indices of T^i the agent performs its *move, wait,* and *sense and compute* behaviors. As expected the completion of the sequence of the behaviors may take different time intervals for different agents and for the same agent at different steps. For instance the distance of the way-point of the agents may change at each step and so the *move* states may last for different amounts of times. Since behaviors of agents last for different time intervals, each agent has its own time set, T^i and these time sets are independent. However, it is possible to have $T^i \cap T^j \neq \emptyset$ for $i \neq j$ which means that sometimes two or more agents may update their state simultaneously. Note that the set T^i is needed only for analysis purposes and in order to implement the iteration in (2) it is not required for the agents to know it. Similarly, the agents do not need to know neither the sets T^i nor the set of physical times T. Therefore, there is no need for a global clock or means for synchronization for implementing equation (2) and each agent can operate based on its internal logic and using only its local clock without a need for synchronization. Before analyzing the convergence performance of this proposed asynchronous model, we will focus on the synchronous case in the following section, after which the asynchronous case will be analyzed in detail.

3   Convergence Under Total Synchronism

In this section we assume that the agents are synchronized and analyze the systems behavior based on this assumption. From practical point of view synchronism is hard to implement in swarm of individual agents with decentralized control since each agent has different duration of states. Still we analyze the convergence of the synchronous case because later in the following section we will use the results from this section to establish the stability of the asynchronous case. We start with the assumption of no delay in the position information. In particular we assume that $\tau_{i+1}(t) = t$ for all i and that $T^i = T = \{0, 1, ...\}$ for all i. In other words, all of the agents will move at the same time instants and each one knows the current position information of the agent it pursues. With respect to this assumption the dynamics of the model become

$$z_i(t+1) = (1-p)z_i(t) + p\,z_{i+1}(t) \tag{3}$$

Writing these equations in matrix form we obtain.

$$z(t+1) = Az(t) \tag{4}$$

where $z(t) = [z_1(t)\ z_2(t)\ \cdots\ z_n(t)]^T \in \mathbb{R}^{n \times 2}$ and

$$A = \begin{bmatrix} 1-p & p & 0 & \cdots & 0 \\ 0 & \ddots & \ddots & \ddots & \vdots \\ \vdots & \ddots & \ddots & \ddots & 0 \\ 0 & & \ddots & \ddots & p \\ p & 0 & \cdots & 0 & 1-p \end{bmatrix}$$

The stability of equation (4) depends on the eigenvalues of A. All eigenvalues of the state matrix A should lie within the unit circle. To show that this is the case we will use a result from matrix theory. In particular, we will use Gerchgorin's Theorem [25] which we present below for the convenience of the reader.

Gershgorin's Theorem. Let $A_{n \times n} = [a_{ij}]$, and let

$$R_i(A) \equiv \sum_{j=1, j \neq i}^{n} |a_{ij}|, \quad 1 \leq i \leq n \tag{5}$$

denote the *deleted absolute row sums* of A. Then, all the eigenvalues of A are located in the union of n discs

$$\bigcup_{i=1}^{n} \{z \in \mathbb{C} : |z - a_{ii}| \leq R_i(A)\} \equiv G(A)$$

where $\mathbb{C}$ denotes the complex plane.

Therefore, for an $n \times n$ square matrix A, n circles can be drawn with centers at the diagonal elements of A, i.e., a_{ii}, $i = 1; 2; ...; n$ and with radius of each of the circles equal to the sum of the absolute values of the other elements in the same row, that is, $\sum_{j \neq i} |a_{ij}|$. Such circles are called Gershgorin's discs. Then all the eigenvalues of A lie in the region formed by the union of all the n discs. From Gershgorin's Theorem we know that all the eigenvalues of the matrix A in (4) are located within discs centered at $(1 - p)$ and having radius p. Then as seen in Figure 2a the vector of points, s in the smaller circle can be formed as $s = (1-p) + \alpha e^{j\theta}$ where $\alpha \leq p$ is the distance of the eigenvalue to the Gershgorin's disc center. Then it can be shown that $|s| = (1 - p)^2 + \alpha^2 + 2(1 - p)\alpha \cos(\theta)$. Moreover, since $\alpha \leq p$ we have

$$(1 - p)^2 + \alpha^2 + 2(1 - p)\alpha \cos(\theta) \leq (1 - p)^2 + p^2 + 2(1 - p)p \cos(\theta)$$

and if $p \leq 1$ then $(1 - p)^2 + p^2 + 2(1 - p)p \cos(\theta) \leq 1$ and $(1 - p)^2 + \alpha^2 + 2(1 - p)\alpha \cos(\theta) \leq 1$ or basically $|s| \leq 1$ is satisfied. Therefore, the eigenvalues of matrix A are within the unit circle if $p \leq 1$. The circles that enclose the location of eigenvalues for the values of $p = 0.25$, 0.50, 0.75, and 1.00 are plotted in Figure 2b. Note that the circle for $p = 1$ is indeed the unit circle.

Another issue to note here is that one of the eigenvalues of the matrix A is always on the unit circle at $\lambda = 1$ and the convergence point of the system depends on that eigenvalue. We can simply show that for the $n \times n$ state matrix A in (4) the characteristic polynomial is $P = (1 - p - \lambda)^n + p^n(-1)^{n-1}$. Note that one of the roots of this characteristic polynomial is always $\lambda = (1-p)+p = 1$ (as stated above) while all the other eigenvalues are within the unit circle. The eigenvector corresponding to this eigenvalue is $\alpha = [1 \ 1 \ ... \ 1]^T$. Now, the solution of (4) can be written as

$$z(t) = (\lambda_1)^t \alpha_1 c_1 + (\lambda_2)^t \alpha_2 c_2 + ... + (\lambda_n)^t \alpha_n c_n \tag{6}$$

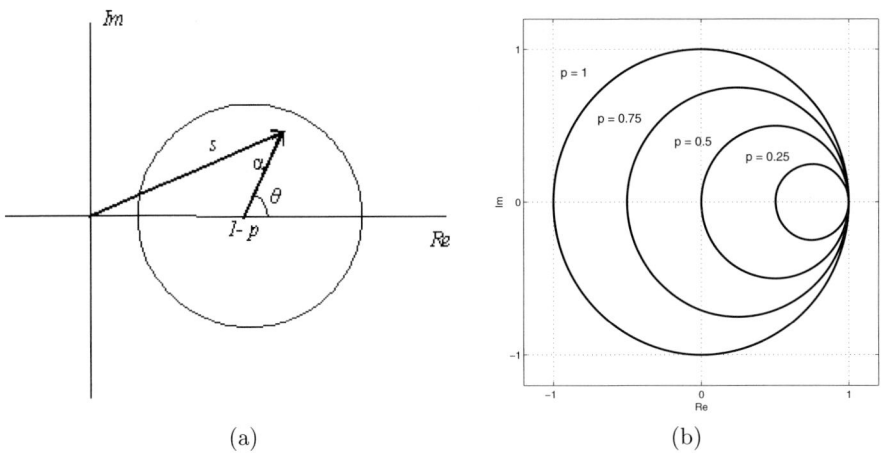

Fig. 2. (a) Gershgorin disc with center at $1 - p$ and radius p. (b) Gershgorin discs for $p = 0.25$, 0.50, 0.75, and 1.00.

where λ_i are the eigenvalues of A and α_i are the corresponding eigenvectors and c_i are arbitrary constants which depend on the initial conditions. (Actually, since $z_i(t) \in \mathbb{R}^2$, $c_i = [c_{1i}, c_{2i}] \in \mathbb{R}^2$ are constant row vectors). Let $\lambda_1 = 1$ be the eigenvalue on the unit circle while $|\lambda_i| < 1$, $\forall i = 2, \ldots, n$ are the other eigenvalues. Then the solution in (6) will converge to:

$$\lim_{t \to \infty} z(t) = \alpha_1 c_1 = [c_1^T \; c_1^T \; \cdots \; c_1^T]^T$$

which means that all agents will reach to the same point, $c_1 \in \mathbb{R}^2$ as $t \to \infty$.

Now based on the above convergence result of the synchronous system we will define a sequence of (contracting) sets which will be useful later on in the proof of the asynchronous case. Let us define

$$Y(t) = \{y \in \mathbb{R}^2 | m(t) \le y \le M(t)\} \subset \mathbb{R}^2 \tag{7}$$

where $m(t) = \min_{i=1,\ldots,n}\{z_i(t)\}$ and $M(t) = \max_{i=1,\ldots,n}\{z_i(t)\}$ where the inequality sign and the *minimum* and *maximum* operators are operated elementwise. Note that the sequence $m(t)$ is non-decreasing and the sequence $M(t)$ is non-increasing. In other words, we have $m(t+1) \ge m(t)$ and $M(t+1) \le M(t)$ for all t. Moreover, one can show that there exists a finite $\mu > 0$ such that $m(t + \mu) > m(t)$ and $M(t + \mu) < M(t)$ for all t. In fact it is guaranteed that a decrease in $M(t)$ and an increase in $m(t)$ occurs in a few time steps. We will do the below analysis as if $m(t) \in \mathbb{R}$ and $M(t) \in \mathbb{R}$ but similar analysis will hold also for the $m(t) \in \mathbb{R}^2$ and $M(t) \in \mathbb{R}^2$ case.

If $M(t)$ and $m(t)$ are not equal, (in other words, agents have not converged yet to a common point) then $M(t) > m(t)$. Let $I_M(t) = \{i | z_i(t) = M(t)\}$ and $I_m(t) = \{i | z_i(t) = m(t)\}$ denote the sets of agents located at time t at the maximum and the minimum, respectively. Also, denote with $\#(M) = |I_M(t)|$

and $\#(m) = |I_m(t)|$ the number of agents in these sets. Note that at least one of the agents in $I_M(t)$ and $I_m(t)$ is pursuing an agent outside of its corresponding set. Therefore, $\#(M)$ and $\#(m)$ both decrease at each step. This guarantees that $M(t)$ will decrease in at most $\#(M) - 1$ steps and $m(t)$ will increase in at most $\#(m) - 1$ steps from time t. The worst case occurs when half of the agents are at the maximum and the remaining half are at the minimum. Then both $M(t)$ and $m(t)$ do not change for $n/2 - 1$ steps. This implies that the interval between $M(t)$ and $m(t)$ contracts in at most $\mu = n/2$ time units. Then it is clear that $Y(t+\mu) \subset Y(t)$. Defining $Z(k) = Y(k\mu)$ as the sequence of contracting sets and in the light of the preceding convergence analysis, we may write

$$c_1 = Z \subset \ldots Z(k+1) \subset Z(k) \subset \ldots \subset Z(0)$$

Now for every k let us define $\bar{Z}(k)$ such that

$$\bar{Z}(k) = \underbrace{Z(k) \times Z(k) \times \ldots \times Z(k)}_{n} \subset \mathbb{R}^{n\times 2} \tag{8}$$

and note that $\alpha_1 c_1 = \bar{Z} \subset \ldots \bar{Z}(k+1) \subset \bar{Z}(k) \subset \ldots \subset \bar{Z}(0)$ is satisfied. We will use these definitions in the next section.

4   Convergence Analysis of the Asynchronous Model

In this section, we analyze the convergence properties of the asynchronous system. As mentioned before here each agent performs the behaviors at totally different time instants. Formally z_i's are updated at $t \in T^i$ where T^i for each agent are independent. Moreover, the sensing/measurement process may incur delays. We start with an assumption which establishes a bound on the maximum possible time delay as well as guarantees uniformity in the updates of the agents. The analysis here is based the results on parallel and distributed computation in [26].

Assumption 1. *There exists a positive integer B such that*
(a) For every i and every $t \geq 0$, at least one of the elements of the set $\{t, t+1, \ldots, t+B-1\}$ belongs to T^i.
(b) There holds $t - B < \tau_{i+1}(t) \leq t \quad \forall i \quad t \geq 0, t \in T^i$

Assumption 1 is a fairly realistic assumption since in any practical system the measurement/communication delays must be bounded. If an agent is unable to receive information for an unbounded amount of time from its neighbor which it tries to pursue, then it may not be able to follow/pursue it and the pursuit behavior looses its meaning. Similarly, in order for the system to work properly every agent should be able to move to its next way-point and complete the cycle in Figure 1 in a finite amount of time. Note, however, that the agents do not need to know the value of B.

Theorem 1. *For the multi-agent system in cyclic pursuit described by the equation in (2) under Assumption 1 as $t \to \infty$ the positions of all the agents will converge to a common point or basically*

$$\lim_{t \to \infty} z_i(t) = c \quad \forall i = 1 \dots n \tag{9}$$

where c is some constant.

Proof. Given time $t_k \in T$ such that $z_i(t) \in Z(k)$ for all $i = 1, 2, ..., n$ and $t \geq t_k$, we will show that there exists a time t_{k+1} such that $z_i(t) \in Z(k+1)$ for all $i = 1, 2, ..., n$ and $t \geq t_{k+1}$. Therefore, let us assume that there exists a time $t_k \in T$ such that $z_i(t) \in Z(k)$ for all $i = 1, 2, ..., n$ and $t \geq t_k$. Consider agent i; from the asynchrony we know that there may be time delay in sensing the position of agent $i+1$ by agent i. Therefore, even though $z_{i+1}(t_k) \in Z(k)$, it might be the case that, $z_{i+1}(\tau_{i+1}(t_k)) \notin Z(k)$. However, by Assumption 1, the delay in the position information update is bounded by B steps. Therefore, at time t_k we have

$$t_k - B < \tau_{i+1}(t_k) \leq t_k \quad \forall i \quad t_k \geq 0, t_k \in T$$

Furthermore, for all $t \geq t_1 = t_k + B$ and for each agent $i = 1, 2, ..., n$ it is guaranteed that

$$t_k < \tau_{i+1}(t) \leq t \quad \forall \, t \geq t_1,$$

implying that

$$z_{i+1}(\tau_{i+1}(t)) \in Z(k) \quad \forall \, t \geq t_1$$

Recall also from Assumption 1 that the update of the positions of each agent is subject to the delay which is at most B steps. Then at time $t_2 = t_1 + B = t_k + 2B$, all the agents will have updated their position information. If the agent is at *maximum* and not pursuing an agent at *maximum*, then $z_i(t_2)$ will decrease and if the agent is at *minimum* and not pursuing an agent at *minimum*, then $z_i(t_2)$ will increase. Therefore, if there are at most one agent at each *maximum* and *minimum*, then from the result for the synchronous case in the preceding section, the position set will contract, implying $Z(t_2) \subset Z(k)$. However, recall the worst condition of agent topology in the synchronous convergence problem; the position sets were to converge in at most $\mu = n/2$ amount of steps. Applying this worst condition for the synchronous case together with the discussion above, we find that the position sets are guaranteed to contract in at most $2\mu B$ steps. Let us define $t_{k+1} = t_k + 2\mu B = t_k + nB$. Then it is guaranteed that $z_i(t) \in Z(k+1) \subset Z(k)$ for all $t \geq t_{k+1} = t_k + 2\mu B$. Since at the initial state we have $z_i(0) \in Z(0) \; \forall i = 1, 2, ..., n$ the induction is complete. Then using the result above we have

$$c = Z \subset ... \subset Z(k+1) \subset Z(k) \subset ... \subset Z(0)$$

which implies the convergence of agents to a common point, $c \in \mathbb{R}^2$.

5   Simulation Examples

We simulated the cyclic pursuit for 5 agents. We performed simulation for both the synchronous and asynchronous cases in order to see the differences between the two cases and in particular the effects of asynchronism. The initial positions of agents are $S = \{(7, 2), (-4, 6), (-9, -4), (-2, -7), (4, -6)\}$. The gain p for the updates is selected to be $p = 0.05$. In the synchronous case the agents converge to $Z_f = [-0.8750, -1.8365]$ after sufficiently long simulation interval. The trajectories of the agents are shown on Figure 3a. For the asynchronous case we used the same initial positions of agents and gain (p) value. In order to achieve asynchronism in simulation and also to simulate the delays in sensing and processing we integrated a probability mechanism that decides whether to update the position information of the $i + 1^{th}$ agent in the system. In the following simulation sample the probability of update is chosen to be % 20. The result of the simulation for this case is on Figure 3b. The agents converge to $Z_f = [-0.2565, -1.9317]$. The convergence point in this case is different from the synchronous case, since the asynchronism in the actions of agents leads to pursuing of next agent with old position information and take the *move* action with some delay. This results in, a different sequence of contradicting sets and therefore different final position.

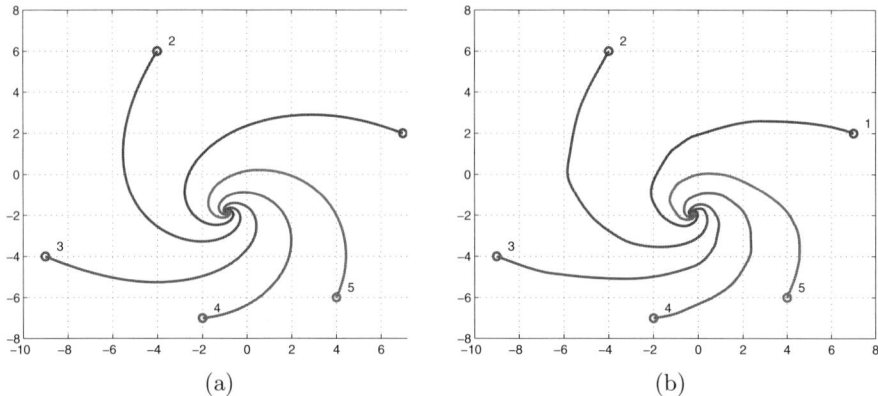

(a)   (b)

Fig. 3. (a) Simulation results for synchronous convergence. (b) Simulation results for asynchronous convergence.

Moreover, in order to measure or compare the performance of these two systems we plotted the sum of the distances between the agent positions

$$e(t) = \sum_{i=1}^{n} \sum_{j=1, j \neq i}^{n} ||z_i(t) - z_j(t)||^2$$

in Figure 4 for both synchronous and asynchronous cyclic pursuits. It is seen that the synchronous cyclic pursuit converges faster than the asynchronous one.

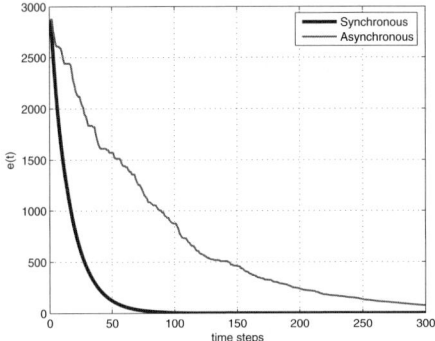

Fig. 4. Convergence performance of synchronous and asynchronous pursuits

This is an expected result when we consider the delays in actions and position sensing of agents during asynchronous pursuit. Although not shown in Figure 4 $e(t)$ converges to zero in the asynchronous case as well.

Note that in the implementation here the value of the probability of update and sensing dictates the value of B. As this probability decreases the value of B increases and this decreases the speed of convergence.

6   Conclusions

In this paper we showed the convergence of the positions of n agents in cyclic pursuit with asynchronous dynamics to a common point. We assumed that the agents perform fundamental behaviors modeled by a finite state machine consisting of *wait*, *sense and compute*, and *move* states. To reach the proof of convergence of asynchronous pursuit we started with the convergence of synchronous cyclic pursuit. Then using the result for the synchronous case we showed that the asynchronous system will converge as well, despite the asynchronism and the time delays.

It is claimed that if convergence to a point is feasible, then more general formations are achievable as well [24]. However, it is not clear whether its possible or not to achieve convergence to any geometric formation using cyclic pursuit under asynchronism and time delays and this needs to be investigated further.

References

1. Cao, Y.U., Fukunaga, A.S., Kahng, A.B.: Cooperative mobile robotics: Antecedents and directions. Autonomous Robots **4**(1) (1997) 7–23
2. Mataric, M.: Issues and approaches in the design of collective autonomous agents. Robotics and Autonomous Systems **16** (1995) 321–331
3. Reynolds, C.W.: Flocks, herds, and schools: A distributed behavioral model. Comput. Graph. **21**(4) (1987) 25–34
4. Balch, T., Arkin, R.C.: Behavior-based formation control for multi-robot teams. IEEE Trans. Robot. Automat. **14** (1998) 926–939

5. Bruckstein, A.M.: Why the ant trails look so straight and nice. Mathematical Intelligencer **15**(2) (1993) 59–62
6. Grübaum, D.: Schooling as a strategy for taxis in a noisy environment. Evolutionary Ecology **12** (1998) 503–522
7. Bernhart, A.: Polygons of Pursuit. Scripta Mathematica (1959)
8. Klamkin, M.S., Newman, D.J.: Cyclic pursuit or "the three bugs problem". The American Mathematical Monthly **78**(6) (1971) 631–639
9. Behroozi, F., Gagnon, R.: Cyclic pursuit in a plane. Journal of Mathematical Physics **20**(11) (1979) 2212–2216
10. Richardson, T.J.: Non-mutual captures in cyclic pursuit. Annals of Mathematics and Artificial Intelligence **31** (2001) 127–146
11. Bruckstein, A.M., Cohen, N., Efrat, A.: Ants, crickets and frogs in cyclic pursuit. Center Intell. Syst., Technion-Israel Inst. Technol. (1991)
12. Gordon, N., Wagner, I.A., Bruckstein, A.M.: Gathering multiple robotic a(ge)nts with limited sensing capabilities. Lecture Notes in Computer Science **3172** (2004) 142–153
13. Flocchini, P., Prencipe, G., Santoro, N., Widmayer, P.: Gathering of asynchronous oblivious robots with limited visibility. Lecture Notes in Computer Science **2010** (2001) 247–258
14. Moreau, L.: Stability of multiagent systems with time-dependent communication links. IEEE Trans. on Automatic Control **50**(2) (2005) 169–182
15. Jadbabaie, A., Lin, J., Morse, A.S.: Coordination of groups of mobile autonomous agents using nearest neighbor rules. IEEE Trans. on Automatic Control **48**(6) (2003) 988–1001
16. Ren, W., Beard, R.W.: Consensus seeking in multi-agent systems under dynamically changing interaction topologies. IEEE Trans. on Automatic Control **50**(5) (2005) 655–661
17. Lin, J., Morse, A.S., Anderson, B.D.O.: The multi-agent rendezvous problem - the asynchronous case. In: Proc. of Conf. Decision and Control, Atlantis, Paradise Island, Bahamas (2004) 1926–1931
18. Sepulchre, R., Palay, D., Leonard, N.E.: Collective motion and oscillator synchronization. In V.J. Kumar, N.E. Leonard, A.M., ed.: in Proc. of the 2003 Block Island Workshop on Cooperative Control, Springer-Verlag (2003)
19. Soysal, O., Sahin, E.: Probabilistic aggregation strategies in swarm robotic systems. In: Proc. of the IEEE Swarm Intelligence Symposium, Pasadena, California (2005)
20. Bahceci, E., Sahin, E.: Evolving aggregation behaviors for swarm robotic systems: A systematic case study. In: Proc. of the IEEE Swarm Intelligence Symposium, Pasadena, California (2005)
21. Marshall, J.A., Broucke, M.E., Francis, B.A.: A pursuit strategy for wheeled-vehicle formations. Proceedings of the 42nd IEEE Conference on Decision and Control (2003) 2555–2560
22. Marshall, J.A., Broucke, M.E., Francis, B.A.: Formations of vehicles in cyclic pursuit. IEEE Transactions on automatic control **49**(11) (2004) 1963–1974
23. Marshall, J.A., Broucke, M.E., Francis, B.A.: Pursuit formations of unicycles. Automatica **42**(1) (2006) 3–12
24. Lin, Z., Broucke, M., Francis, B.: Local control strategies for groups of mobile autonomous agents. IEEE Transactions on Automatic Control **49**(4) (2004) 622–629
25. HORN, R.A., R.JOHNSON, C.: Matrix Analysis. Cambridge University Press (1992)
26. Bertsekas, D.P., Tsitsiklis, J.N.: Parallel and Distributed Computation: Numerical Methods. Athena Scientific, Belmont, MA (1997)

Evolved Homogeneous Neuro-controllers for Robots with Different Sensory Capabilities: Coordinated Motion and Cooperation

Elio Tuci[1], Christos Ampatzis[1], Federico Vicentini[2], and Marco Dorigo[1]

[1] IRIDIA, CoDE, Université Libre de Bruxelles,
Avenue F. Roosevelt 50, CP 194/6, 1050 Bruxelles, Belgium
{etuci, campatzi, mdorigo}@ulb.ac.be
http://iridia.ulb.ac.be/
[2] Robotics Lab, Mechanics Dept.,
Politecnico di Milano, Milano, Italy
federico.vicentini@polimi.it

Abstract. This paper tackles the issue of designing homogeneous neuro-controllers with artificial evolution in order to control groups of robots that differ in terms of sensory capabilites. In order to accomplish a common goal, the agents have to complement the partial "view" they have of the environment. The results obtained prove that the agents are capable of cooperating and coordinating their actions in order to carry out a navigation task. A preliminary analysis of the mechanisms underlying the group behaviour is provided.

1 Introduction

Embodied autonomous systems are relatively recent methodological tools which can be used to investigate various aspects of social interactions and behavioural coordinations in artificial and natural organisms (see [9,2]). In this type of systems, social behaviour is investigated by firstly determining the characteristics of the agents' embodiment (e.g., sensory and motor capabilities of the agent) and the world that they inhabit, and by subsequently looking at how the these features influence social skills.

This approach is particularly prominent in a subset of embodied autonomous systems, generally referred to as Evolutionary Robotics models (ER, see [7]). Roughly speaking, ER is a methodological tool to automate the design of robots' controllers. ER is based on the use of artificial evolution to find sets of parameters for artifical neural networks that guide the robots to the accomplishment of their objective, avoiding dangers. Owing to its properties, ER can be employed to look at the effects that the physical interactions among embodied agents and their world have on the *evolution* of individual behaviour and social skills. In the recent past, ER has been used in the context of social behaviour to investigate issues concerning the evolution of communication in groups of agents required to solve tasks that demanded coordination and cooperation (see [8,10,1,11,5]). Following this line of investigation, we are interested in further exploring the evolution

S. Nolfi et al. (Eds.): SAB 2006, LNAI 4095, pp. 679–690, 2006.

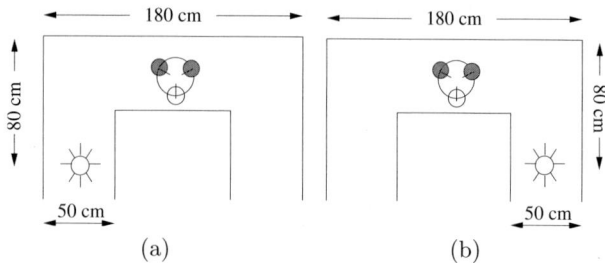

Fig. 1. (a) *Env. L*; (b) *Env. R*. See text in Sec. 1 for details

of social skills. In particular, we focus on a context in which a group of agents with different sensory capabilities are required to share their "knowledge" of the world to accomplish a common task. We consider the following experiment: three robots are placed in an arena, as shown in Fig. 1. The arena is composed of walls and a light that is always turned on. The light can be situated at the bottom left corridor (*Env. L*) or at the bottom right corridor (*Env. R*). The robots are initialised with their centre anywhere on an imaginary circle of radius 12 cm centred in the middle of the top corridor, at a minimum distance of 3 cm from each other. Their initial orientation is always pointing towards the centroid of the group. The goal of the robots is (i) to navigate towards the light whose position changes according to the type of environment they are situated in, (ii) to avoid collisions.

The peculiarity of the task lies in the fact that the robots are equipped with different sets of sensors. In particular, two robots are equipped with infrared and sound sensors but they have no ambient light sensors. These robots are referred to as R_{IR} (see Fig. 2a). The other robot is equipped with ambient light and sound sensors but it has no infrared sensors. We refer to this robot as R_{AL} (see Fig. 2b). Robots R_{IR} can perceive the walls and other agents through infrared sensors, while the robot R_{AL} can perceive the light. Therefore, given the nature of the task, the robots are forced to cooperate in order to accomplish their goal. In principle, it would be very hard for each of them to solve the task solely based on their own perception of the world. R_{AL} can hardly avoid collisions; R_{IR} can hardly find the light source. Thus, the task requires cooperation and coordination of actions between the different types of robots. Notice that the reason why we chose the group to be composed of two R_{IR} and one R_{AL} robot is that this intuitively seems to be the smallest group capable of spatially arranging itself adaptively in order to successfully navigate the world. Although the robots differ with respect to their sensory capabilities, they are homogeneous with respect to their controllers. That is, the same controller, synthesised by artificial evolution, is cloned in each member of the group. Both types of robots are equipped with a sound signalling system (more details in Sec. 2). However, contrary to other studies (see [5,1]), we do not assume that the agents are capable of distinguishing their own sound from that of the other agents. The sound broadcasted into the environment is perceived by the agent through omnidirectional microphones.

Therefore, acoustic signalling is subject to problems such as the distinction between own sound from those of others and the mutual interference due to lack of turn-taking (see [8]).

The results of our study show that a quite robust and effective phototactic strategy evolves in spite of each of the agents being deprived of essential elements to accomplish the task. The successful strategies are based on cooperation and coordination of actions among the agents. The mutual coordination results particularly striking so that, as already emphasised in a similar model [8], it turns out to be very hard to speak in terms of causality. For example, (a) phototaxis is induced in the group by robot R_{AL}, but this behaviour seems to be effectively displayed by the robot R_{AL} only if it is situated in a social context—i.e., surrounded by robots R_{IR}; (b) angular movement introduces rhythm in acoustic perception, which *per se*, is not sufficient to coordinate the movements of the group. However, coordinated actions come about by the fusion of perception of sound and patterns in infrared proximity sensors. In conclusion, from these simulations, we learn something about the relationship between individual and social skills, and the potentiality of the system which can be further exploited to study the evolution of more complex forms of social interactions in similar circumstances (e.g., groups of morphologically heterogenous robots).

2   The Simulated Agents

The controllers are evolved in a simulation environment which models some of the hardware characteristics of the real *s-bots*. The *s-bots* are small wheeled cylindrical robots, 12 cm of diameter, equipped with a variety of sensors, and whose mobility is ensured by a differential drive system (see [6] for details). Robot R_{IR} makes use of 12 out of 15 infrared sensors (Ir_i) of an *s-bot*, while

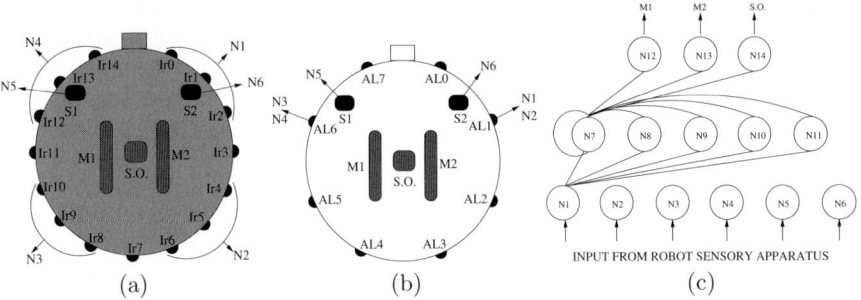

(a)   (b)   (c)

Fig. 2. (a) The simulated robots R_{IR}; (b) The simulated robots R_{AL}; (c) the network architecture. Only the connections for one neuron of each layer are drawn. The input layer of R_{IR} takes readings as follows: neuron N_1 takes input from $\frac{Ir_0+Ir_1+Ir_2}{3}$, N_2 from $\frac{Ir_4+Ir_5+Ir_6}{3}$, N_3 from $\frac{Ir_8+Ir_9+Ir_{10}}{3}$, N_4 from $\frac{Ir_{12}+Ir_{13}+Ir_{14}}{3}$, N_5 from S_1, and N_6 from S_2. The input layer of R_{AL} takes readings as follows: N_1 and N_2 take input from AL_1, N_3 and N_4 take input from AL_6, N_5 from S_1, and N_6 from S_2. M_1 and M_2 are respectively the left and right motor.

robot R_{AL} uses the ambient light sensors (AL_1) and (AL_6) positioned at $\pm 67.5°$ with respect to the orientation of the robot (see Fig. 2). The signal of the infrared sensor is a function of the distance between the robot and the obstacle. Light sensor values are simulated through a sampling technique.

All robots are equipped with a sound output (S.O.) that is situated in the centre of the body of the robot, and with two omnidirectional microphones (S_1 and S_2), placed at $\pm 45°$ with respect to the robot's heading. Sound is modelled as an instantaneous, additive field of single frequency with time-varying intensity ($\eta \in [0.0, 1.0]$) which decreases with the square of the distance from the source, as previously modelled in [8]. Sound intensity is regulated by the firing rate of neuron $N14$ (see Sec. 3 for details). Robots can perceive signals emitted by themselves and by other agents. The modelling of the perception of sound is inspired by what described in [8]. There is no attenuation of intensity for self-produced signals which can in principle be loud enough ($\eta = 1.0$) to make it impossible for a robot to perceive sound signals emitted by others. The perception of sound emitted by others is affected by a "self-shadowing" mechanism which is modelled as a linear attenuation without refraction, proportional to the distance travelled by the signal within the body of the receiver (see [8] for details).

Concerning the function that updates the position of the robots within the environment, we employed the Differential Drive Kinematics equations, as presented in [3]. 10% uniform noise was added to all sensor readings, the motor outputs and the position of the robot. The characteristics of the agent-environment model are explained in detail in [12].

3   The Controller and the Evolutionary Algorithm

The agent controller is composed of a network of five inter-neurons and an arrangement of six sensory neurons and three output neurons (see Fig. 2c). The sensory neurons receive input from the agent sensory apparatus. Thus, for robots R_{IR}, the network receives the readings from the infrared and sound sensors. For robots R_{AL}, the network receives the readings from the ambient-light and sound sensors. The inter-neuron network (from N_7 to N_{11}) is fully connected. Additionally, each inter-neuron receives one incoming synapse from each sensory neuron. Each output neuron (from N_{12} to N_{14}) receives one incoming synapse from each inter-neuron. There are no direct connections between sensory and output neurons. The network neurons are governed by the following state equation:

$$\frac{dy_i}{dt} = \begin{cases} \frac{1}{\tau_i}(-y_i + gI_i) & i \in [1,6] \\ \frac{1}{\tau_i}\left(-y_i + \sum_{j=1}^{k} \omega_{ji}\sigma(y_j + \beta_j) + gI_i\right) & i \in [7,14]; \ \sigma(x) = \frac{1}{1+e^{-x}} \end{cases} \tag{1}$$

where, using terms derived from an analogy with real neurons, y_i represents the cell potential, τ_i the decay constant, g is a gain factor, I_i the intensity of the sensory perturbation on sensory neuron i, ω_{ji} the strength of the synaptic connection from neuron j to neuron i, β_j the bias term, $\sigma(y_j + \beta_j)$ the firing rate. The cell potentials y_i of the 12th and the 13th neuron, mapped into $[0,1]$ by a sigmoid function σ and then linearly scaled into $[-6.5, 6.5]$, set the robot motors output. The cell potential y_i of the 14th neuron, mapped into $[0, 1]$ by a sigmoid function σ, is used by the robot to control the intensity of the sound emitted η. The following parameters are genetically encoded: (i) the strength of synaptic connections ω_{ji}; (ii) the decay constant τ_i of the inter-neurons and of neuron N_{14}; (iii) the bias term β_j of the sensory neurons, of the inter-neurons, and of the neuron N_{14}. The decay constant τ_i of the sensory neurons and of the output neurons N_{12} and N_{13} are set to 0.1. Cell potentials are set to 0 any time the network is initialised or reset, and circuits are integrated using the forward Euler method with an integration step-size of $dt = 0.1$.

A simple generational genetic algorithm is employed to set the parameters of the networks [4]. The population contains 80 genotypes. Generations following the first one are produced by a combination of selection with elitism, recombination and mutation. For each new generation, the three highest scoring individuals ("the elite") from the previous generation are retained unchanged. The remainder of the new population is generated by fitness-proportional selection from the individuals of the old population. Each genotype is a vector comprising 84 real values (i.e., 70 connection weights, 6 decay constants, 7 bias terms, and a gain factor). Initially, a random population of vectors is generated by initialising each component of each genotype to values chosen uniformly random from the range $[0,1]$. New genotypes, except "the elite", are produced by applying recombination with a probability of 0.3 and mutation. Mutation entails that a random Gaussian offset is applied to each real-valued vector component encoded in the genotype, with a probability of 0.15. The mean of the Gaussian is 0, and its standard deviation is 0.1. During evolution, all vector component values are constrained to remain within the range $[0,1]$. Genotype parameters are linearly mapped to produce network parameters with the following ranges: biases $\beta_i \in [-4, -2]$ with $i \in [1,6]$, biases $\beta_i \in [-5,5]$ with $i \in [7,14]$; weights $\omega_{ij} \in [-6,6]$ with $i \in [1,6]$ and $j \in [7,11]$, weights $\omega_{ij} \in [-10,10]$ with $i \in [7,11]$ and $j \in [7,14]$; gain factor $g \in [1,13]$. Decay constants are firstly linearly mapped into the range $[-1.0, 1.3]$ and then exponentially mapped into $\tau_i \in [10^{-1.0}, 10^{1.3}]$. The lower bound of τ_i corresponds to the integration step-size used to update the controller; the upper bound, arbitrarily chosen, corresponds to about 1/20 of the maximum length of a trial (i.e., 400 s).

4   The Fitness Function

During evolution, each genotype is translated into a robot controller, and cloned in each agent. Then, the group is evaluated six times, three trials in *Env. L*, and three trials in *Env. R*. The sequence order of environments within the six trials has no bearing on the overall performance of the group since each robot controller is reset at the beginning of each trial. Each trial (e) differs from the others in the initialisation of the random number generator, which influences the robots' starting position and orientation, and the noise added to motors and sensors. Within a trial, the robot life-span is 400 simulated seconds (4000 simulation cycles). In each trial, the group is rewarded by an evaluation function f_e which seeks to assess the ability of the team to approach the light bulb, while avoiding collisions and staying within the range of the robots' infrared sensors. By taking inspiration from the work of Quinn et al. [11], the fitness score is computed as follows:

$$f_e = KP\left(\sum_{t=i}^{T}[(d_t - D_{t-1})(tanh(S_t/R))]\right);$$

As in [11], the simulation time steps are indexed by t and T is the index of the final time step of the trial. d_t is the Euclidean distance between the group location at time step t and its location at time step $t = 0$, and D_{t-1} is the largest value that d_t has attained prior to time step t. S_t is a measure of the team's dispersal beyond the infrared sensor range R ($R = 24.6$ cm) at time step t. Recall that robot R_{AL} has no infrared sensors. Therefore, it does not have a direct feedback at each time-step of its distance from its group-mates. Nevertheless, the sound can be indirectly used by this robot to adjust its position within the group. If each robot is within R range of at least another, then $S_t = 0$. Otherwise, the two shortest lines that can connect all three robots are found and S_t is the distance by which the longest of these exceeds R. $tanh()$ assures that, as the robots begin to disperse, the team's score increment falls sharply.

$P = 1 - (\sum_{i=1}^{3} c_i/c_{max})$ if $\sum_{i=1}^{3} c_i \leq c_{max}$ reduces the score in proportion to the number of collisions which have occurred during the trial. c_i is the number of collisions of the robot i and $c_{max} = 4$ is the maximum number of collisions allowed. $P = 0$ if $\sum_{i=1}^{3} c_i > c_{max}$. The team's accumulated score is multiplied by $K = 3.0$ if the group moved towards the light bulb, otherwise $K = 1.0$. Note that a trial was terminated early if (a) the team reached the light bulb (b) the team distance from the light bulb exceeded an arbitrary limit set to 150 cm, or (c) the team exceeded the maximum number of allowed collisions c_{max}.

5   Results

Ten evolutionary simulations, each using a different random initialisation, were run for between 1000 and 1500 generations of the evolutionary algorithm. The termination criterion for each run was set to a time equal to 86400 seconds of CPU time. Experiments were performed on a cluster of 32 nodes, each with 2 AMD Opteron244^{TM} CPU running GNU/Linux Debian 3.0 OS. In order to have a better estimate of the behavioural capabilities of the evolved controllers, we post-evaluate, for each run, the genotype with the highest fitness. The entire set of post-evaluations should establish whether a group of robots is capable of reaching the light in *Env. L* and *Env. R*. In particular, the robots of a successful group should be capable of coordinating their movement and of cooperating, in order to approach the light bulb without colliding with each other or with the walls. A trial is successfully terminated when the centroid of the group is south of the light bulb. During post-evaluation, each of the best ten evolved controllers is subject to a set of 1200 trials in both environments. The number of post-evaluation trials per type of environment (i.e., 1200) is given by systematically varying the initial positions of the three robots according to the following criteria: (i) we defined four different types of spatial arrangements in which the robots are placed at the vertices of an imaginary equilateral triangle inscribed in a circle of radius 12 cm and centred in the middle of the top corridor (see Fig. 3b); (ii) for each spatial arrangement, we identified three possible relative positions of the robot R_{AL} with respect to the walls' corridor (see white circle in Fig. 3b); (iii) for each of these (four times three) initial positions, the post-evaluation is repeated one hundred times. The initial orientation of each robot is determined by applying an angular displacement randomly chosen in the interval $[-30°, 30°]$ with respect to a vector originating from the centre of the robot and pointing towards the centroid of the group. The four times three different arrangements take into account a set of relative positions among the robots and between the robots and the walls so that the success rate of the group is not biased by these elements. During post-evaluation, the robot life-span is more than twice longer than during evolution (i.e., 1000 s, 10000 simulation cycles). This should give the robots enough time to compensate for possible disruptive effects induced by initial positions never or very rarely experienced during evolution. At the beginning of each post-evaluation trial, the controllers are reset (see Sec. 3 for details).

The results of the post-evaluation phase are shown in Fig. 3a. We notice that the best controller is the one produced by run n. 2, achieving a performance over 90% in *Env. L* and *Env. R*. Runs n. 4, 9, and 10 display a performance over 80%, run n. 1, and 7 displays a performance around 75% in both environments.

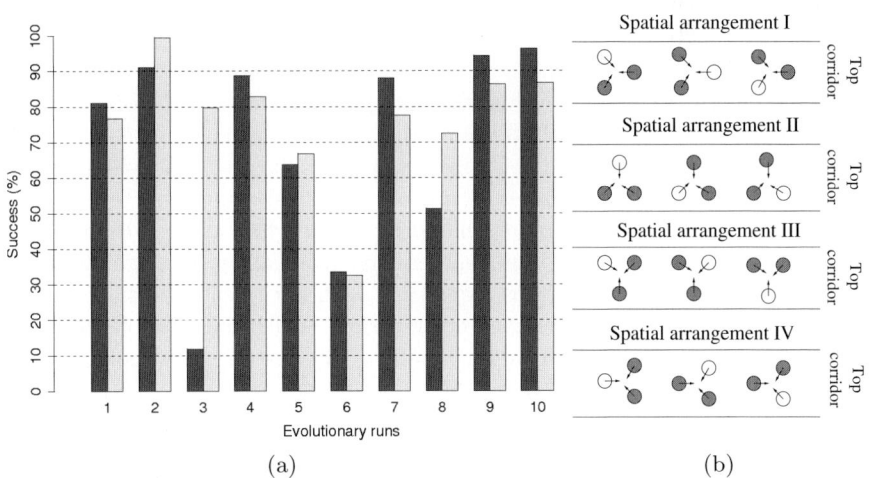

(a)   (b)

Fig. 3. (a) Results of post-evaluation showing the percentage of success of the best evolved controllers of each run over 1200 trials per type of environment. White bars refer to *Env. L*, and black bars to *Env. R*. (b) The robots' initial positions during the post-evaluation phase. White circles refer to R_{AL}, grey circles refer to R_{IR}.

Table 1. Further results of the post-evaluation test, showing for the best evolved controllers of each run: (i) the percentage of unsuccessful trials due to exceeded time limit without the group having reached the target (columns 2, and 3); (ii) the percentage of unsuccessful trials which terminated due to collisions (columns 4, and 5); (iii) the average and standard deviation of the final distance of the centroid of the group to the light during the unsuccessful trials (respectively columns 6, 8 for *Env. L*, and columns 7, 9 for *Env. R*). Note that in all trials the initial distance between the centroid of the group and the light is equal to 85.14 cm.

| run | (%) of failure due to time limit | | (%) of failure due to collisions | | Distance to the light | | | |
| | | | | | avg | | std | |
	Env. L	*Env. R*	*Env. L*	*Env. R*	*Env. L*	*Env. R*	*Env. L*	*Env. R*
n. 1	15.75	20.08	3.17	3.17	22.22	24.29	14.35	22.20
n. 2	1.42	0.00	7.50	0.42	71.63	87.96	20.47	0.46
n. 3	19.17	4.67	69.00	15.50	45.11	36.67	23.06	13.73
n. 4	0.00	4.92	11.25	12.25	62.72	52.80	22.24	23.98
n. 5	20.75	11.33	15.58	21.83	48.83	47.31	38.70	25.72
n. 6	43.83	61.67	22.58	5.75	35.11	30.76	16.12	12.80
n. 7	0.00	10.17	12.00	12.33	67.60	42.93	15.98	28.05
n. 8	36.33	3.58	12.33	23.92	31.22	58.94	22.05	23.46
n. 9	0.67	7.50	5.00	6.25	55.49	29.10	22.73	17.29
n. 10	0.00	6.42	3.67	6.83	54.72	50.59	18.08	32.26

Note that when looking at the performances of the best evolved controllers, as shown in Fig. 3a, one has to take into account the arbitrary criteria we chose to determine whether or not a group of robots is successful in any given trial. We should recall that, in order to be successful, no robot has to collide with the walls or with the other robots. This is a very strict condition, which, given the nature of the task, demands each agent to be very accurate in coordinating its movement. Further post-evaluation tests proved that, if we allow the group to make a certain number of collisions (i.e., four collisions) before defining a trial as a failure, then several controllers would result almost always successful in both types of environment—the data of these post-evaluation tests are not shown. Whether or not the robots should be allowed to collide or the extent to which a single collision invalidates the performance of the group, are issues that extend beyond the interest of this paper, and shall not be discussed any further. Instead, we focus on other performance measures which tell us more about the characteristics of the best evolved controllers. For instance, by looking at the data shown in Table 1, we notice that, for all the runs, the majority of the failures are due to collisions. Exceptions are run n. 1 and 6 which seem to be only minimally affected by this factor (see columns 4 and 5, Table 1). If we look at the average distances to the light (see columns 6 and 7, Table 1) and the relative standard deviations (see columns 8 and 9, Table 1), we can see that the robots guided by these controllers seem to be capable of covering much of their initial distance to the light. Therefore, the small percentage of collisions is indeed the result of an effective coordination of actions among the agents invalidated by the lack of time to complete the task due to a slow phototactic movement, rather than, for instance, a consequence of the lack of movement of the group towards the light.

In the rest of this section, we concentrate on the analysis of the controller of run n. 2, which proved to be the most effective at the first post-evaluation test. In particular, we try to understand more about the mechanisms used by the robots to coordinate their actions and to complement the partial view that each of them has of the world.

5.1 Further Analysis of the Best Evolved Navigation Strategy

In an effort to understand how the robots manage to cooperate and coordinate their actions in order to solve the task, we repeated the post-evaluation test described at the beginning of Sec. 5 for groups of robots controlled by controller run n. 2. However, in these series of tests, the robots are deprived in various ways of sensory information which may or may not turn out to be crucial for the achievement of the task.[1] Recall that, only by "paying attention" to the sound signals emitted by robots R_{IR}, robot R_{AL} can avoid collisions. Sound signalling *and/or* coordination through the infrared sensors might play a significant role in guiding the group towards the target.

First, we run two tests, referred to as *Test A* and *Test B*, which should reveal to us whether the robots employ effective navigational strategies based on cooperation and coordination of actions or rather fixed phototactic movement which may work as well given that the dimensions of the corridors and the positions of the lights in the two worlds do not vary. In *Test A*, the best controller of run n. 2 is cloned on three robots R_{IR}. Consequently, the robots have no means to know where the light is placed. As shown in Table 2, the group was 100% unsuccessful due to time limit exceeded without having reached the target (see columns 2 and 3, Table 2). Moreover, in both environments, the average distance between the centroid of the group and the light does not differ much from the initial distance (see columns 6 and 7, Table 2). The rather small standard deviation confirms that this group of robots seems not to make any significant movement away from its initial position (see columns 8 and 9, Table 2). Indeed, it seems to be the presence of a robot R_{AL}—missing in the group in this test—that triggers the movement and guides the group towards the target. Not surprisingly, the robots are very effective in avoiding collisions (see columns 4 and 5, Table 2). In *Test B* a single robot R_{AL} is controlled by the best controller of run n. 2. The results tell us that R_{AL}, if left without robots R_{IR}, systematically collides with walls (see columns 4 and 5, Table 2). *Test A* and *B* suggest that the successful

---

[1] In the post-evaluation tests in which alterations concern the agents' received sound signal, or the nature of the group (i.e., what types of robot are part of the group), and/or the characteristics of the environment, the changes are applied after 10 s (i.e., 100 simulation cycles) from the beginning of each trial. This give time to the controllers to reach a functional state different from the initial one, arbitrarily chosen by the experimenter, in which the cell potential of the neurons is set to 0 (see Sec. 3).

Table 2. Results of different post-evaluation tests for the best evolved controller of run n. 2. See text in Sec. 5.1 for details. Note that in all trials the initial distance between the centroid of the group and the light is equal to 85.14 cm.

Test	(%) of failure due to time limit		(%) of failure due to collisions		Distance to the light			
					avg		std	
	Env. L	Env. R	Env. L	Env. R	Env. L	Env. R	Env. L	Env. R
A	100	100	0.00	0.00	85.32	85.50	8.24	8.31
B	0.00	0.00	100.00	100.00	90.83	104.90	3.64	5.36
C	100	100	0.00	0.00	122.57	127.94	5.19	4.73
D	2.42	0.00	46.17	0.58	75.06	84.83	25.32	4.90
E	0.25	0.00	12.08	6.75	70.03	37.03	22.33	23.06

strategies of run n. 2 are based on effective coordination of actions and cooperation among the different types of agents of the group. In brief, there is neither phototaxis nor any other movement along the corridors if robot R_{AL} is missing in the group. There is neither obstacle avoidance nor successful phototaxis if a single robot R_{AL} is left alone in this simple world. Surprisingly, while robots R_{IR} retain their capability to avoid obstacles if situated in an "odd" group of all R_{IR} robots, a single robot R_{AL} can hardly perform phototaxis if left alone. This is shown by the results of *Test C* in which a single robot R_{AL} is placed in a boundless arena (no walls) with only a light at around 85 cm away from it. As proved by the final distance to the light (see column 6 and 7, Table 2), a single robot R_{AL} is not capable of approaching the light source. Oddly enough, it displays an anti-phototactic movement. In summary, the different types of robot complement each other not only to accomplish the task, but also to carry out those functions for which they are more apt (e.g., phototaxis in R_{AL}).

In *tests D and E*, the best controller run n. 2 is cloned on a group of three robots, in which, as during evolution, two are R_{IR} and one R_{AL}. Contrary to the evolution, in *test D*, the robots R_{IR} only hear their own sound; in *test E*, the robots R_{IR} can potentially perceive the sound emitted by the robot R_{AL} but they can not hear each other's sound. These tests should help us to understand more about the significance of sound signalling. Data in Table 2 show that, in *Test D*, robots do not systematically fail to reach the target. Although the performance in *Env. L* is severely disrupted with almost 50% of unsuccess rate, in *Env. R* the group performance is not touched by the alterations we applied to the system (see columns 2, 3, 4 and 5, Table 2). The failure in *Env. L*, is mostly due to collisions, which seem to occur rather far away from the lights (see columns 6, and 8, Table 2). In summary, the sound received by the robots R_{IR} from robot R_{AL} seems to play a significant role in carrying out obstacle avoidance in *Env. L*.

In *Test E*, we immediately notice that the rate of failure is rather low (see columns 2, 3, 4 and 5, Table 2). The success rate turns out to be quite similar to that achieved in the evolutionary conditions in which all the robots can hear the sound emitted by all the others. It seems fair to conclude that (i) communication through sound signalling among the members of the group is required in order to successfully approach the target; (ii) successful strategies of controller run n. 2

are only marginally based on communication through sound signalling between the robots R_{IR}. Initially, we thought that this latter phenomenon was a side effect of the spatial arrangement of the group during navigation. For instance, if the robots form a chain in which the robot R_{AL} is in the middle position and the other two robots R_{IR} are at the two ends of the chain, then the latter robots may not hear each other because of the distance between them. Consequently, preventing the robots R_{IR} from hearing each other can not affect in any way a navigational strategy of a group that does not rely on this element. However, by looking at the spatial arrangement of the robots during navigation, we saw that, within a trial, they tend to dispose themselves in various spatial configurations in which the two robots R_{IR} do perceive each other's signals. This implies that robots R_{IR} might be capable of discriminating among agents of different type (i.e., R_{IR}, R_{AL}). However, this and other issues related to management of the coordination and cooperation of the group can not be inferred from this preliminary analysis, and they need to be further investigated.

6   Conclusions

In a context in which robots differ in their sensory capabilities, cooperation and coordination of actions evolved for the group to achieve a common goal. Behavioural capabilities of the single agents become effective in a social context in which mutual dependencies at various operational levels characterise the system more than causal explanations. The agents (i) emit sound signals that are not too loud to hinder the perception of the sound emitted by the others, but loud enough to be captured by the other robots if relatively close to the emitter; (ii) negotiate a common direction of movement; and (iii) navigate safely (i.e, without collisions) towards the target. The "dynamic speciation" of the homogenous controller, whose mechanisms underpin sensory-motor coordination and social interactions in structurally different agents, is particularly significant. From an engineering point of view, these results suggest that homogeneous controllers can be efficiently exploited to control morphologically identical as well as morphologically different groups of robots. This element can be also exploited in case of hardware failure, in which an on-line re-assignment of association between agent's sensors and network's input neurons might provide a robust mechanism to preserve the functionality of multi-robot systems. Moreover, a better coordination of actions might be achieved by varying the characteristics of the sound and/or morphological features of the sound signalling systems—e.g., the number and/or the position of the loudspeakers and microphone. Finally, further investigations need to be carried out to provide a deeper operational explanation of the properties of the system. Does the variability in the emission of sound reflect a simple "vocabulary" grounded on sensor-motor activity of the agent? This issue is an interesting subject for future investigations.

Acknowledgements

E. Tuci and M. Dorigo acknowledge European Commission support via the*ECAgents* project, funded by the Future and Emerging Technologies programme (grant IST-1940), and by COMP2SYS, a Marie Curie Early Stage Training Site (grant MEST-CT-2004505079). The authors thank their colleagues at IRIDIA for stimulating discussions and feedback during the preparation of this paper, and the two anonymous reviewers for their helpful comments. M. Dorigo acknowledges support from the Belgian FNRS, of which he is a Research Director, and from the "ANTS" project, an "Action de Recherche Concertée" funded by the Scientific Research Directorate of the French Community of Belgium. The information provided is the sole responsibility of the authors and does not reflect the Community's opinion. The Community is not responsible for any use that might be made of data appearing in this publication.

References

1. G. Baldassarre, S. Nolfi, and D. Parisi. Evolving mobile robots able to display collective behaviour. *Artificial Life*, 9:255–267, 2003.
2. A. Cangelosi and D. Parisi, editors. *Simulating the Evolution of Language*. Springer Verlag, London, UK, 2002.
3. G. Dudek and M. Jenkin. *Computational Principles of Mobile Robotics*. Cambridge University Press, Cambridge, UK, 2000.
4. D. E. Goldberg. *Genetic Algorithms in Search, Optimization and Machine Learning*. Addison-Wesley, Reading, MA, 1989.
5. D. Marocco and S. Nolfi. Emergence of communication in embodied agents: Co-adapting communicative and non-communicative behaviours. In A. Cangelosi, G. Bugmann, and R. Borisyuk, editors, *Modeling language, cognition and action: 9th Neural Computation and Psychology Workshop*. World Scientific, London, UK, 2005.
6. F. Mondada, G. C. Pettinaro, A. Guignard, I. V. Kwee, D. Floreano, J.-L. Deneubourg, S. Nolfi, L. M. Gambardella, and M. Dorigo. SWARM-BOT: A new distributed robotic concept. *Autonomous Robots*, 17(2–3):193–221, 2004.
7. S. Nolfi and D. Floreano. *Evolutionary Robotics: The Biology, Intelligence, and Technology of Self-Organizing Machines*. MIT Press, Cambridge, MA, 2000.
8. E. Di Paolo. Behavioral coordination, structural congruence and entrainment in a simulation of acoustically coupled agents. *Adaptive Behavior*, 8(1):27–48, 2000.
9. R. Pfeifer and C. Scheier. *Understanding Intelligence*. MIT Press, Cambridge, MA, 2001.
10. M. Quinn. Evolving communication without dedicated communication channels. In J. Kelemen and P. Sosik, editors, *Advances in Artificial Life: 6th European Conf. on Artificial Life*, volume 2159 of *LNCS*, pages 357–366. Springer Verlag, Berlin, Germany, 2001.
11. M Quinn, L. Smith, G. Mayley, and P. Husbands. Evolving controllers for a homogeneous system of physical robots: Structured cooperation with minimal sensors. *Phil. Trans. of the Royal Soc. of London, Series A*, 361:2321–2344, 2003.
12. F. Vicentini and E. Tuci. Swarmod: a 2d s-bot's simulator. Technical Report TR/IRIDIA/2006-005, IRIDIA, Université Libre de Bruxelles, 2006. This paper is available at http://iridia.ulb.ac.be/IridiaTrSeries.

Robot Learning in a Social Robot*

Salvador Dominguez[1], Eduardo Zalama[2],
Jaime Gómez García-Bermejo[2], and Jaime Pulido[1]

[1] Cartif Foundation, Computer Vision and Robotics Division,
Parque Tencológico de Boecillo parcela 2005, 47151 Boecillo, Valladolid, Spain
{saldom, jaipul}@cartif.es
http:/www.cartif.es
[2] University of Valladolid, Department of Systems Engineering and Control,
Paseo del Cauce s/n 47011 Valladolid, Spain
{ezalama, jaigom}@eis.uva.es

Abstract. In this paper, research work on Arisco is described. Arisco is a social robot built around a robotic head with gesture ability, visual and auditive perception and learning. It is intended for interacting with people. The general architecture is first described in the paper. Then, the learning capacity of Arisco is addressed. It learns and performs associations between different stimulus responses through several dynamic neural networks, guided by motivational drives. Main contribution of this paper is the integration in a real robot of conditioning learning models based on a neural competitive network. A number of experiments are discussed, covering stimulus competition, habituation and first and second order conditioning.

1 Introduction

Social robots have been developed during the last few years as robots with human-like abilities, which make human-robot interaction easier [1,2,3]. Robots with large perception capability, equipped with advanced communication interfaces based on natural language, expressivity and gesture recognition, makes communication easier not only for people with low technological skills, but also for people with decreased attention capability. For example, it has been shown that these agents help with problems like autism, thus, social robots can be used as a therapeutic tool [6].

Learning the consequences of its own actions is especially important when an animal or an intelligent machine has to operate in an unknown (or partly unknown) environment.

Models of *classical conditioning* and *operant conditioning* have been conceived from the field of psychology to explain how an organism can achieve autonomous behaviour in a changing environment [9,10]. In the classical conditioning paradigm, learning happens by repeatedly associating a Conditioning Stimulus (CS), which

---

* This work has been partly supported by the Spanish Ministry of Science and Technology (project numbers DPI2002-04377-C02-01, DPI2005-06911), and by the Castilla y León local Government.

S. Nolfi et al. (Eds.): SAB 2006, LNAI 4095, pp. 691–702, 2006.

usually has no particular significance for the animal, with an Unconditioned Stimulus (UCS) which has significance and leads to an Unconditioned Response (UCR). For example, a dog that repeatedly hears a bell before being fed will eventually begin salivating when just hearing that bell. The response that is elicited by the CS after classical conditioning is known as the Conditioned Response (CR). In the operant conditioning paradigm, the animal learns the consequences of its actions. More specifically, it learns to exhibit, more frequently, behaviour that has led to a reward, and less frequently, behaviour that has led to a punishment. For example, a hungry cat in a cage from which some food can be seen, will learn to press a lever that allows it to escape the cage and reach the food. In this situation, the animal cannot simply wait for things to happen, but must generate different behaviours and learn which ones are effective. The main problem is how to learn which behaviour has produced the reward. Furthermore, associations should be learned through time, since stimuli and rewards arrive continuously.

In the present work, a model of classical conditioning and operant conditioning based on a competitive neural network, which learns new stimulus-response associations, is presented. Research has been performed on Arisco, a social robot who interacts with people [4]. A general robot description is given in section 2. Behaviours and the learning architecture are presented in section 3. Experiments are presented and discussed in section 4. Finally, conclusions are summarized in section 5.

2 Robot Description

Arisco (see figure 1) is a robotic head with a capacity for gestures inspired physically in Kismet [2] but with different internal functioning. Kismet behavior system is organized into loosely layered, heterogeneous hierarchies of behavior groups without learning. Arisco behaviors are activated depending on the competitive networks after stimulus-response association.

Fig. 1. Arisco robot

Arisco has 17 degrees of freedom, for neck movement and gesture production, and is equipped with 2 sphere-shaped webcams (at eyes), two microphones (at ears), one presence sensor and one short range, proximity sensor. Three microcontroller based boards to control the whole head, interface the sensors and provide the incoming sound direction. These boards and the webcams are connected to a laptop computer under Knoppix 3.6, through a single usb port. This computer performs high-level planning, behaviour control, visual processing (*see* section 3.1), and voice command recognition (through an enhanced spectrogram matching procedure [7]).

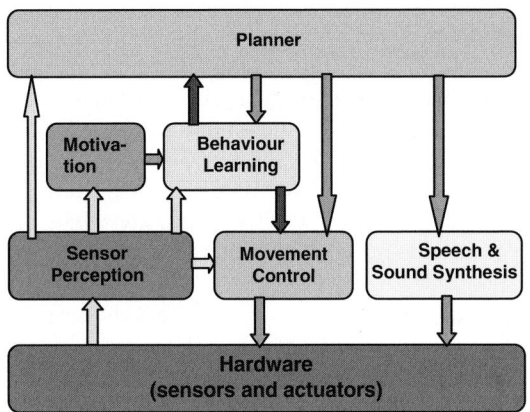

Fig. 2. Overall organization of the robot

The overall architecture of the robot is shown in figure 2. There are four main levels:

- Hardware level. This is the lower level, and includes all sensors and actuators in the head.
- Reactive level. The information coming from the head sensors is processed at this level. Also, the information coming from higher levels is here interfaced to the hardware level, through the *movement control module* and the *sound and speech generation module*. The former provides all actuator synchronization towards visual tracking [8], sound tracking [5] and gesture production (angry, sad, and happy); the latter provides speech from text and other sound effects (music etc.).
- Deliberative level. At this level, external stimuli (from sensors) are processed, and motivational stimuli are generated upon the *internal needs* of the robot. All these stimuli are processed through learning neural networks in the *behaviour and learning module*, so that the currently active behaviours of the robot are chosen. These behaviours are fed to both the planning level, and the *movement control module* within the reactive level, for gesture and tracking movement production.
- Planning level. High-level commands are here interfaced to the lower levels according to the active behaviour, the sensorial information and internal variables such as the date, time and even information retrieved from Internet (such as weather information). The robot profile is defined at this level (a receptionist, in our case) for dialogue planning. Dialogue is then executed according to the user requests, and the external and motivational stimuli of the robot. Thus, an interactive, bidirectional dialogue with the surrounding humans is maintained in defined context environments.

3   Behaviours

The *behaviour and learning* module has two main subsystems: (i) the attentional behaviour subsystem, which selects the attention focus within the incoming sensor information; and (ii) the emotional behaviour subsystem, which selects the *state of mood* corresponding to a given situation. Both subsystems are controlled through competitive neural network maps, two separate computational maps [12]. The sensor information is filtered at the input (stimulus) layer through neuron competition, and the dominant behaviour is selected at the output layer depending on the weights between both layers. These weights are self-tuned at each cycle, so that the learning process takes place through time. Furthermore, some behaviours are directly wired to a certain stimulus (without adaptive weights). They are *innate* reflex behaviours, and correspond to self-protection instincts in animals (e.g. taking a hand away from pain).

3.1   Attentional Behaviour

This behaviour consists of selecting and tracking an attention point within the environment around the robot [4]. This is done by selecting the most salient stimulus among the following:

- Sudden sound (and its incoming direction)
- Highly-saturated colour region.
- Human face (along with its estimated distance).
- Light source (bright saturation point).
- A random point (useful when no noticeable feature is detected).
- An image region whose HS histogram is close to that of a recently detected feature (useful for a continuous tracking of a momentarily disappeared feature).

3.2   Emotional Behaviour

The emotional behaviour subsystem selects one emotional behaviour from the incoming-stimulus amplitude. Only a few representative, easily recognized behaviours are required [11] (e.g. neutral, happy, angry, sad, astonished). Furthermore, an associated activity degree determines how happy, angry etc. the robot is. The obtained behaviour is fed to the hardware level through the *movement control module*, so that different face gestures are produced. It should be noted that the attention point tracking and the gesture production are controlled in a parallel way: the face expression depends on the emotional behaviour (but not on the tracked feature), and the tracked feature depends on the active attentional behaviour.

Proper cinematic and dynamic parameters, randomness ranges and motor positions corresponding to the different gestures are set up in the said movement control module. Furthermore, reference values for neck, eyes and vision motors, suitable for the attention point tracking are also generated by this module.

3.3   Neural Network Architecture

The attentional behaviour and the emotional behaviour are implemented separately through the neural network architecture presented in Fig. 3. It consists of a habituation network coupled to a computational map. The network maintains the stimulus variation through time and the competitive interactions in short term memory, STM, and the stimulus to behaviours association in long term memory, LTM. Both STM and LTM dynamics have been modelled through differential equations [Eqs. 1-8].

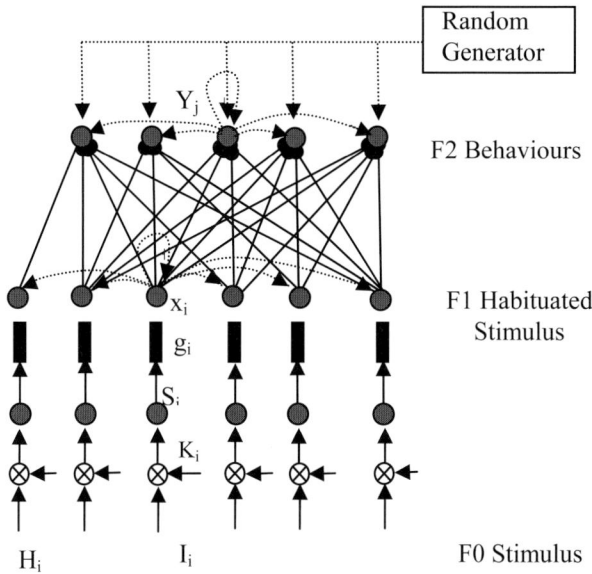

Fig. 3. Neural network architecture

Stimuli from sensors are placed at F0 level, e.g. a certain color region, a human face, or a resemblance figure between spectrograms (of the current word and a voice-command stored in an internal database). Some stimuli, I_i, lead to conditioned behaviours, so they are connected through adaptive weights. Other Stimuli, H_i, lead to reflex behaviour, so they are directly connected (e.g. getting scared at a sudden sound, such as a clap, or receiving positive/negative reinforcement for intensifying/attenuating a given behaviour). All stimuli are [0,1] normalized, and relative salience is modulated through K_i gains. Constant gain values have been selected in this work but variable values coming from a motivational module are also possible (e,g, face detection stimulus gain could increase when the robot is motivated for social interaction). Furthermore, stimuli may be either persistent through time (such as a face or colour region detection) or sporadic (such as a clap or word recognition). In the present model, sensor information is processed through a filtering neuron so that the stimulus decays slowly through time, because a certain time in the network is required for learning (STM memory). This is particularly convenient for sporadic stimuli. The filter equation is

$$\frac{dS_i}{dt} = -aS_i + aK_iI_i \tag{1}$$

where S_i is the neuron activity (i.e. the attenuated sensorial information) and a is the rate of decay (so that the neuron activity decays at rate a when there is no excitation).

The described neural network has habituation capability: the interest in a continuous stimulus decays through time. For example, given that the robot is following a certain colour, the robot's interest in this stimulus eventually decreases through time, under the interest of a new stimulus; say a human face for example. The habituation equations used in this work are based on the *Slow Transmitter Habituation and Recovery model* [12]. In concrete, the habituation from F0 to F1 layer follows

$$\frac{dg_i}{dt} = E(1-g_i) - F S_i g_i \tag{2}$$

where S_i is the ith stimulus input and g_i is the transmitter amount at the ith channel. The transmitter amount decays from maximum value (unity) at the beginning towards $E/(E+F S_i)$ value, proportionally to the S_i input activity level. Once the input stimulus stops ($S_i = 0$), the transmitter recharges towards unity. E and F determine the transmitter charging and discharging rate respectively. Thus, habituation happens according to $S_i g_i$.

Sensorial stimuli feed an "On-Centre Off-Surround" competitive network [11]: "On centre" refers to positive feedback at the current neuron, and "Off Surround" refers to lateral inhibition of neighbouring neurons (see F1 level in figure 3). Grossberg [12] developed the multiplicative model taking the Hodgking-Huxley [14] equation that describes the neuron membrane behaviour. The equation that describes the multiplicative network is

$$\frac{dx_i}{dt} = -Ax_i + (B-x_i)[S_i g_i + f(x_i)] - x_i \sum_{i \neq j} f(x_j) \tag{3}$$

$$f(x_i) = Dx^2 \tag{4}$$

where x_i is the activity of neuron i. Ax_i in (3) is the decaying rate. The second term is the "on centre" term, which pushes the neuron activity towards a maximum saturation value, B. Last term represents the lateral inhibition of neighbouring neurons. Different dynamic behaviour is obtained with different $f(x_i)$ choices (linear, faster than linear, slower than linear or sigmoid). In this work, a parabolic function has been selected (eq. **4**), so that the winning neuron is reinforced over the others ("winner takes all"). Input to the behaviour level follows

$$Y_j = \sum_{j=1}^{N} x_j z_{ij} \tag{5}$$

where z_{ij} are the adaptive connections, learned through an outstar learning law [12]:

$$\frac{dz_{ij}}{dt} = -\gamma z_{ij} + \beta(y_i - z_{ij})x_j . \tag{6}$$

γ es the forgetting factor, β is the learning rate, and y_j is the jth neuron activity in the behaviour layer. Finally, the behaviour layer is modelled by a competitive neural layer similar to the stimulus layer:

$$\frac{dy_j}{dt} = -Ay_j + (B - y_j)[T_j + f(y_j)] - y_j \sum_{i \neq j} f(x_j) \tag{7}$$

$$f(y_i) = Dy_j^2 \tag{8}$$

The behaviour layer is activated by a random generator in the first learning stages (see figure 3), so that the robot can develop an exploratory behaviour. This kind of behaviour has been identified as an intrinsic learning mechanism in the early months for children, for basic visu-motor and speech-aural coordination learning.

4   Experimental Results

In this section, some experimental results of Arisco's learning are addressed. Arisco's face expressions corresponding to its different states of mood are shown in Fig. 4. Conditioning of these states upon visual and audio stimuli are specifically discussed.

| Sad | Happy | Bored | Angry | Scared |

Fig. 4. Arisco emotional behaviours

Differential equations have been solved through a trapezoidal approximation: given

$$\frac{dx}{dt} = g(x) \ , \tag{9}$$

the equivalent discrete equation is

$$x(kh) = x((k-1)h) + \frac{[g(kh) + g((k-1)h)]h}{2} \tag{10}$$

where h is the time integration step and k is a positive integer ($0 < kh < t$).

An experiment on second order conditioning is hereby presented: a UCS which reflexly leads to a UCR is associated to a CS. In concrete, a clap UCS has been selected, which leads to a scared UCR. A red card shown to the robot is the CS. The experiment objective consists in making Arisco scared by just showing it the red card. This is achieved by clapping repeatedly while the red card is in sight. The input stimulus corresponding to a red card is presented in Fig. 5.a (in red). Stimulus intensity decays through time, thanks to the habituation, so that new stimulus arrival is favoured. The activity corresponding to four successive claps from 23 to 51 seconds is shown in blue.

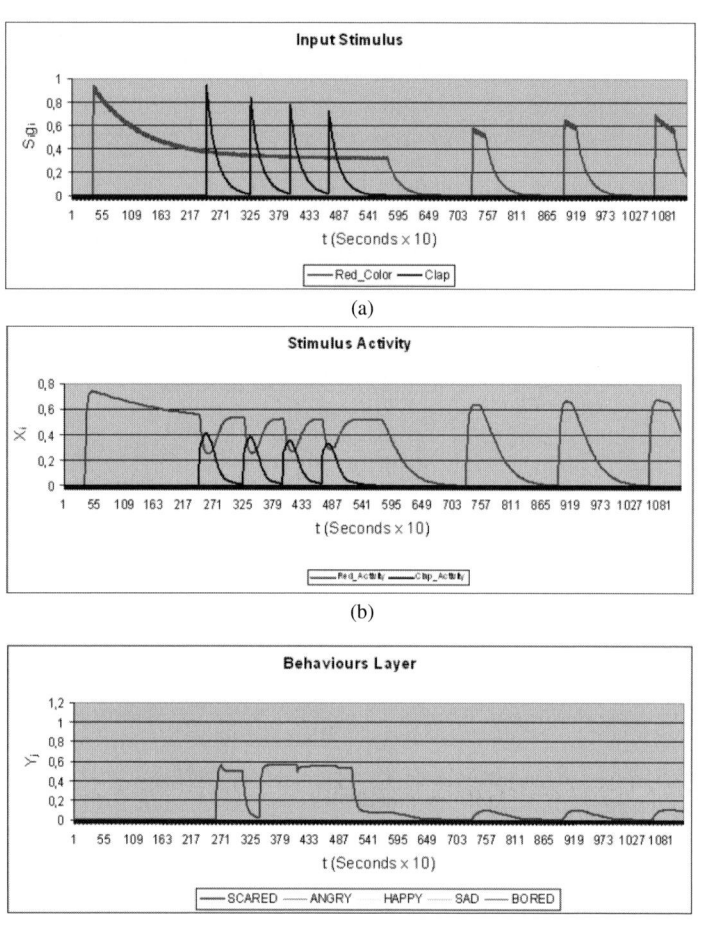

Fig. 5. First order conditioning experiment corresponding to a UCS stimulus (a clap) and a CS stimulus (a red card). (a) Input stimulus to competitive network after filtering and habituation. (b) Stimulus activity in the competitive layer. (c) Response behaviour. Equation parameters are $h=0.1$, $A=0.3$, $B=1.0$, $C=1.0$, $D=5.0$, $E=0.04$, $F=0.08$, $\beta=0.01$, $\gamma=0.0001$.

Claps get Arisco scared, as a consequence of a reflex behaviour (fig. 5.c). Both red card and clap stimulus in the competitive layer are shown in fig. 5.b. When claps happen, the activity corresponding to the red colour detection decreases while the activity corresponding to claps increases. Weight learning (according to equation **6**) occurs within the time period during which both the scared behaviour and the activity of the red colour are active. From 71 seconds on, only the red card is shown (without any clap). The habituated input stimulus and the stimulus activity in the competitive layer are shown in fig. 5.b and 5.c: the scared behaviour is activated every time the red card is presented to the robot (the behaviour intensity being below that of the reflex behaviour; this intensity is proportional to the elapsed learning time, i.e. the stimulus association time).

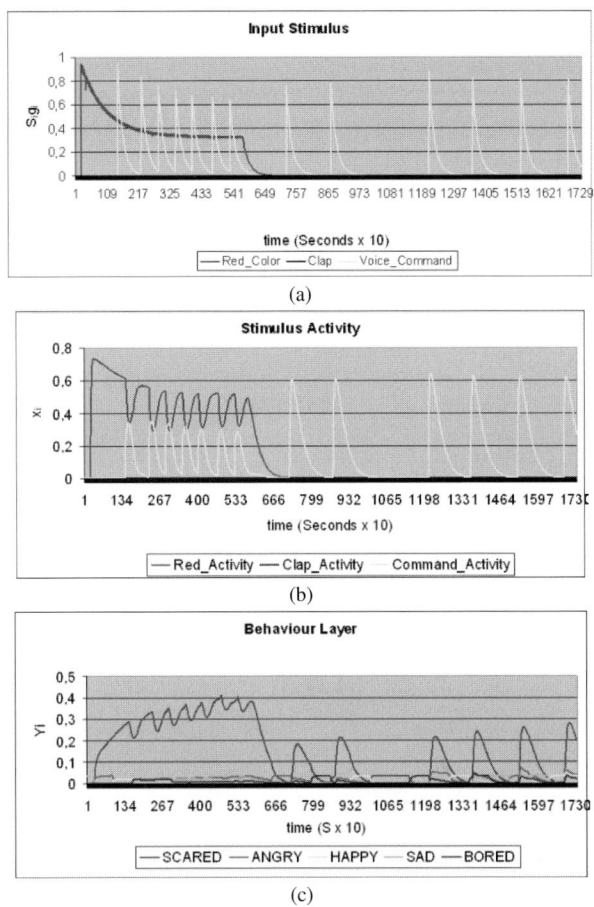

(a)

(b)

(c)

Fig. 6. Second order conditioning experiment corresponding to a CS1 stimulus (red card) previously conditioned to scared behaviour and CS2 stimulus ("scared" pronunciation). (a) Input stimulus to competitive network after filtering and habituation. (b) Stimulus activity in the competitive layer. (c) Response behaviour.   Equation parameters: $h=0.1$, $A=0.3$, $B=1.0$, $C=1.0$, $D=5.0$, $E=0.04$, $F=0.08$, $\beta=0.1$, $\gamma=0.0001$.

An experiment on second order conditioning is now discussed. The experiment target consists of checking the association to an *already associated* stimulus. After the experiment above, Arisco gets scared every time it sees a red card. Now, the robot is told "scared" each time it sees the red card, so that in the future it will get scared by just hearing "scared".

The stimulus corresponding to the red card detection while the robot is told "scared" is shown in fig. 6.a. Stimulus activity in the competitive layer is shown in Fig. 6.b. Red card stimulus is already conditioned, so Arisco gets scared (Fig. 6.c). Now "scared" is associated with the behaviour previously learned through weight adjustment. Then, the scared behaviour begins to stand out after adaptation, when the robot is just told "scared", in spite of the short learning time (as seen from 65 seconds on in Fig. 6.c).

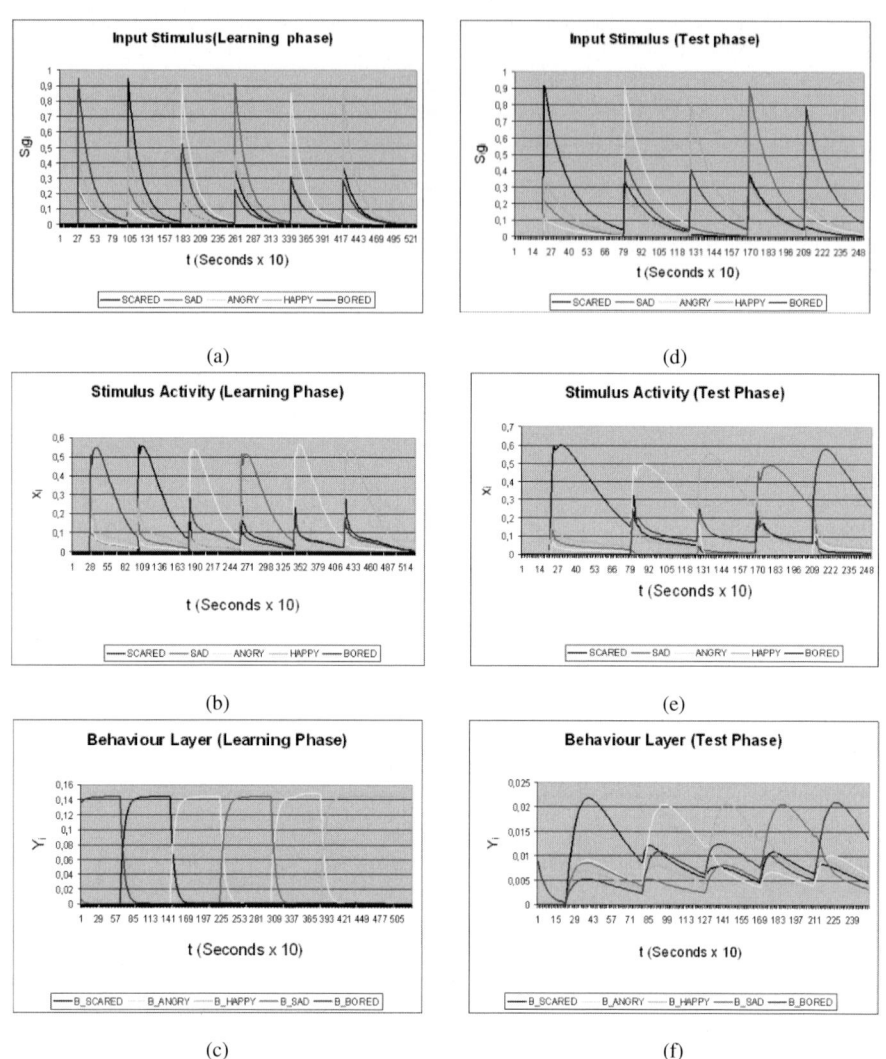

Fig. 7. Operant conditioning experiment. (a) Voice utterances corresponding to the behaviours. (b) Stimulus activity in competitive layer. (c) Random behaviour generation. (d) Voice utterances (test phase). (e) Stimulus activity in competitive layer (test phase) (f) Behaviour obtained from F0-F1-F2 pathways. Equation parameters: $h=0.1$, $A=0.3$, $B=1.0$, $C=1.0$, $D=5.0$, $E=0.04$, $F=0.08$, $\beta=0.01$, $\gamma=0.0001$.

An experiment on operant conditioning is finally discussed. The robot randomly generates different behaviours during a given time. The aim of the experiment is to test how the robot associates the generated behaviours to the incoming stimuli. The behaviours activated by the random generator are shown in Fig. 7.c (bored, scared, angry, sad and happy). For example, the robot is told "bored" when its facial expression corresponds to the bored behaviour.

The network is input the voice-command recognition (see Fig. 7.a). The input competition and the simultaneous behaviour-to-stimulus association learning is shown in Fig. 7.b. A test is performed after the initial learning phase: the robot is told the previously learned words, in a random order (Fig. 7.d). Stimulus competition is shown in Fig. 7.e, and the behaviour activity obtained from equations (**5,7,8**) is shown in Fig. 7.f. The winning behaviour is always that corresponding to the correct word.

Some videos of the experiments can be downloaded from the site http://www. eis.uva.es/~eduzal/arisco/sab06.html .

5 Conclusions

In this paper, current research on a robotic head with basic social and learning abilities have been presented. The head has a number of facial features and can show basic emotional behaviours, recognize voice commands, locate the incoming sound direction, detect human faces and colour regions, and track movement. A learning architecture model based on competitive neural networks and outstar learning law, has been described. Though classical conditioning and operant conditioning learning methods have been studied exhaustively, they have not been used very much in real robots. The proposed architecture can do associations between any kind of stimuli and behaviors. Suitable performance of the model has been verified by means of different experiments of first order, second order and operant conditioning. Future work will address low level behaviour learning (face expressions, eye-head coordination, etc) and the development of a motivational module for priming stimuli according to Arisco's internal state and homeostatic signals. Motivation can also guide learning through reinforcement signals.

References

1. Fong, T., Nourbakhsh, I., Dautenhahn, K.: A survey of socially interactive robots: concepts, design, and applications, Technical Report No. CMU-RI-TR-02-29, Robotics Institute, Carnegie Mellon University, (2002)
2. Breazeal, C.: Designing Sociable Robots, MIT Press, Cambridge, MA, (2002)
3. Kozima, H., Yano, H.: A robot that learns to communicate with human caregivers. In: First International Workshop on Epigenetic Robotics. Lund Sweeden. (2001)
4. Dominguez, S. , Zalama, E., García-Bermejo, J. G.: Arisco un Robot Social con Vocación de Recepcionista. VII Workshop on Phiscal Agents.Las Palmas de Gran Canaria. 27-28 April. Spain (2006)
5. Tapias, S., Dominguez, S., Zalama, E., García-Bermejo, J. G.: Desarrollo de un Sistema de Localizacion de Fuentes Sonoras Basado en Microcontroladores. Proceedings de XXVI Jornadas de Automática. (2005)
6. Werry, I., Dautenhahn, K., Ogden, B., Harwin, W.: Can social interaction skills be taught by a social agent? The role of a robotic mediator in autism therapy, in: Proceedings of the International Conference on Cognitive Technology, (2001)
7. Orabona, F., Metta, G., Sandini, G.: Object-based Visual Attention: a Model for a Behaving Robot. 3rd International Workshop on Attention and Performance in Computational Vision, San Diego, CA, USA. (2005)

8. Dominguez, S., Zalama, E., García-Bermejo, J. G.: Desarrollo de una Cabeza Robótica con Capacidad de Seguimiento Visual e Identificacion de Personas. Proceedings de XXVI Jornadas de Automática. Alicante, Spain. (2005)

9. Rescorla, R. A., Wagner, A. R.: A theory of pavlovian conditioning: variations in the effectiveness of reinforcement and nonreinforcement. In Black, A. H., & Prokasy,W. F. (Eds.), Classical Conditioning II, chap. 3, pp 64–99. Appleton, New York. (1972)

10. Sutton, R. S., Barto, A. G.: Toward a modern theory of adaptive networks: Expectation and prediction. Psychological Review, vol 88, pp 135–170. (1981)

11. Grossberg, S.: Some Nonlinear Networks capable of Learning a Spatial Pattern of Arbitrary Complexity. Proceedings of the National Academy of Sciences, USA, Vol 59 pp 368-372. (1968)

12. Grossberg, S.: Studies of Mind and Brain: Neural Principles of Learning, Perception, Development, Cognition and Motor Control. Reidel Press, Boston, MA, (1982)

13. Hodgkin, A.L., Huxley, A.F. A quantative description of membrane current and its application to conduction and excitation in nerve. Journal of Physiology, 117:500-544, (1952)

Integration of an Autonomous Artificial Agent in an Insect Society: Experimental Validation

Grégory Sempo[1], Stéphanie Depickère[2], Jean-Marc Amé[1],
Claire Detrain[1], José Halloy[1], and Jean-Louis Deneubourg[1]

[1] Unit of Social Ecology, CP231, Université Libre de Bruxelles
Av. F. Roosevelt, 50, 1050 Bruxelles, Belgium
gsempo@ulb.ac.be
[2] INLASA, Entomologia médica
Rafael Zubieta N°1889 (Lado del Estado Mayor General)
Miraflores Casilla M - 10019, La Paz – Bolivia

Abstract. In mixed societies of robots and cockroaches, several insect-like-robot (Insbot) and animals interact in order to perform collective decision-making. Many gregarious species are able to collectively select a resting site without any leadership. The key process is based on the modulation of the probability of leaving the shelter according to the total population under this shelter and its light intensity. It is important that cockroaches perceive the robot as a "congener". This recognition is mainly based on a chemical blend. The aim of this study is to validate experimentally (1) the behavioral patterns expressed by the cockroaches in presence of shelters and of an Insbot, and (2) the important role played by the chemical blend on collective decision-makings.

1 Introduction

With the rapid development of biology and biotechnology comes the growing need for systems where intelligent artificial and living agents cooperate. Controlling these interactions is therefore becoming a key challenge. Designing such synergetic societies requires studying new forms of information processing, problem-solving as well as synergetic behaviors between living beings and machines. Animal societies will be one of the first biological systems where living agents and autonomous artifacts will cooperate to solve problems. The machine does not replace the animal but collaborates and bring new capabilities to the mixed society. On one hand, the artificial systems bring new types of sensors, actuators and communication possibilities to the living systems; on the other hand the animals bring their cognitive and biological capabilities to the artificial systems.

This study is a part of the Leurre project [1] [2] from which the main objective is to prove that it is possible to develop and control mixed-societies composed of insects (cockroaches) and small insect-like robots (the Insbots). Previous related works have studied interactions between robots and different animal species such as rats [3], or dogs [4]. As regards gregarious species, one may mention the use of smart collars to

S. Nolfi et al. (Eds.): SAB 2006, LNAI 4095, pp. 703–712, 2006.

hopefully control the herding behavior of cattle [5] or the Robot Sheepdog [6] to control a flock of ducks by moving them safely to a pre-determined position.

The challenge of the Leurre project [2] consists in the development of a mixed society where several robots and several animals are able to interact in order to perform collective decision-making. One example of such collective decision is shelter selection by groups of cockroaches [7] [8].
To reach this objective, it means:

1. to build of robots able to interact with animals;
2. that the artificial agents will be able to respond to the signals emitted by the animals and to produce signals able to induce behavioral responses of the animals;
3. that the interactions designed for the artificial agents or used by natural agents, are of the type mutual inhibition, mutual activation and competition for limited resources [9].

Then, our first goal was to obtain social interaction between an artificial and an animal agent. Indeed, we will lure animals in such a way that robots are socially integrated in the animal society, so that they are considered by animals "as one of them". This means that robot and animals influence each other, that they both contribute to the collective decision of the mixed society. Groups of cockroaches exhibit such collective decision-making: e.g. when several identical shelters are present, the group chooses only one of them. Many experiments have been performed with different species [7] [8]. The large American species, *Periplaneta americana* is our reference species to test mixed societies. The classical example of collective decision-making is when all group members choose the same solution among identical alternatives. This decision is collegial without any leadership or anthropomorphic procedure such as voting.

Contrary to the traditional lures that are elaborated following qualitative observations of the animal, the development of as well the hardware and the behaviour/software of our robot results from quantitative studies of the individual and collective behaviors of the cockroach. At the individual level, behavioral sciences have shown that animals' interactions could be rather simple signals and that it is possible to interact with animals not only by mimicking their whole behaviors but also by making specifically designed artifacts. In this respect, the challenge was to determine the pertinent communication channels needed to integrate the robot within the animal group and to validate the repeatability of results in real experiments with animals.

Concerning cockroaches, several studies have shown that recognition between individuals is based on the chemical compounds present on their body. Consequently, the robot was wearing a paper dress containing the cuticular extract of cockroaches [10]. At a collective level, the cockroach *P. americana* is a gregarious insect that forms large cluster of individuals in dark resting site. This cluster formation results from self-organized amplification processes based on the modulation of the probability to move according to two factors: the number of surrounding individuals and the local light intensity. It means that this probability decreases with the number of congeners present and with the level of darkness [7] [8]. Then like cockroaches, Insbot will behave according to this simple rule.

The aim of this study is to illustrate the building blocks needed to validate experimentally the behavioral patterns expressed by the cockroaches in presence of shelter and in presence of an Insbot. We will highlight the crucial role played by cuticular extracts of cockroach on the formation of mixed group.

Finally, we will test whether the Insbot is well integrated in the animal society and whether it does not disturb the collective decision-making and aggregation patterns. Therefore among, all the interactions, we will study the interactions between cockroaches and the influence of shelters and Insbots on cockroaches' behavior (Fig. 1).

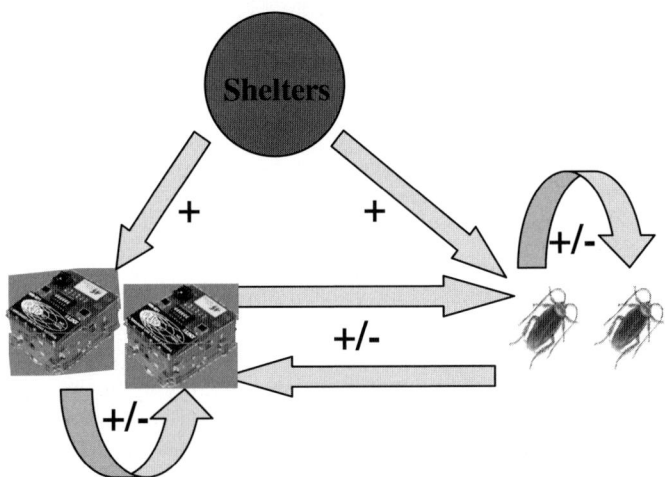

Fig. 1. All possible interactions between Insbots, cockroaches and shelters. A distinction was made either the reaction induce by the interaction was an attraction (+) or a repulsion (-).

2 Methods

Cockroaches (*Periplaneta americana*) were raised in transparent boxes (length: 80 cm; width: 40 cm; height: 100 cm). Water and food pellets were provided *ad libitum*. Cockroaches were kept under laboratory conditions at 25°C ± 1 in a 12h:12h light:dark cycle.

2.1 Experimental Setup and Procedure

Experiments were performed with adult male cockroaches. Following on the experiment conditions, 1, 10 or 30 adult cockroaches were picked up from the rearing box and isolated for 48 hours in the dark. They has access to water and food pellets. Animals with any external damage (e.g. missing antennal segments or legs) were discarded. After this isolation period, we introduced cockroaches in a circular arena (diameter: 100 cm) delimited by a black polyethylene ring (height: 20 cm, thickness: 1 cm) (Fig. 2a,b). To confine cockroaches in this experimental arena, its inner surface was covered by an electric fence (alternation of positively and negatively charged black aluminum layers (19 V, 0.2 A)).

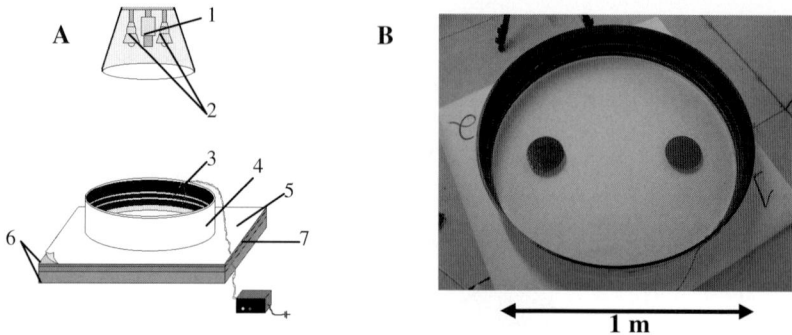

Fig. 2. A: Experimental setup composed of top camera (1), lamp (2), electrical fence (3), polyethylene ring (4), paper sheet (5), vibration absorbent layers (6) and wooden layer (7). Shelters are not represented on the illustration. The distance between lights and ground floor is 1.5 m. **B:** Experimental arena with two identical shelters.

The environmental conditions of the experimental setup were finely tuned in order to fit with cockroach and Insbot sensing capabilities. Firstly, temperature around the experimental setup was maintained at 20°C ± 1 since the walking speed of insects, like that of any insect, is highly sensitive to this factor. Secondly, since cockroaches mark the floor with a blend of different molecules (mainly hydrocarbons [10][11]), the white paper sheet (120g/m²) covering the ground of the arena (Fig. 2a) was renewed before each experiment. This prevents bias in cockroach behavior induced by the chemical marking deposited passively by congeners. In addition, the paper sheet was placed on two phonic layers which reduce the vibrations that could frighten the cockroaches. Finally, to prevent any disturbance of the cockroaches and the robots, lighting was exclusively artificial and produced by four neon lamp bulbs poor in IR (Fig. 2a, Philips ambiance Pro, 20 Watts, 355 lux ± 5 at ground level).

Two Plexiglas discs (diameter: 15 cm) were suspended 3 cm above the ground and positioned at 23 cm from the edge, symmetrically to the centre of the arena (Fig. 2b). Before each experiment, discs were cleaned with denatured ethanol (ethanol + ether). To obtain different luminosities under the shelters, we covered it with one (the light shelter with underneath 100 lux ± 5) or two layers (a.k.a. dark shelter with underneath 75 lux ± 5) of a red filter (Rosco color filter, E-Colour #019: Fire). The size of one shelter is large enough to contain up to 30 individuals, we did not observed any overcrowding. Due to the absence of red light-sensitive cells in their compound eye [12], cockroaches perceive an area illuminated by red light as shadow. When placed in an enlightened arena, these cockroaches have a higher probability to stop as soon as they enter the shadow area [13].

Concerning Insbot, they are able to distinguish the four main features of the setup: the light intensity under of shelters, the arena walls, the cockroaches and the other robots [14]. Behaviours of the Insbot are divided into two categories: reactive behaviours and higher level behaviours. At the exception of reactive behaviours (e.g. obstacle avoidance and wall following), every 500 ms, the robot had a probability to perform each of the three actions (turning, moving, or stopping) according to its position (centre or periphery; under a dark or bright) and the number of surrounding individuals (cockroaches or robots). These probabilities were determined from the observation of cockroach's behaviours (for detail of Insbot's behaviour see [14]).

Fig. 3. Two cockroaches and a chemically marked Insbot under a dark shelter (diameter: 15cm)

In addition the Insbot was wrapped into a paper dress (Whatman, grade 1 filter paper, 8.5cm) which covered the whole surface of the Insbot excepting sensors (Fig. 3). Onto this paper dress, we laid extracts of chemical compounds necessary to induce the aggregation process and to maintain the cohesion of aggregates. These compounds are mainly cuticular hydrocarbons which were extracted by immersing adult cockroaches in dichloromethane [10]. Sixty microliters of these extracts were laid on the tested Insbot, which is equivalent to the amount of compounds present on the cuticle of one cockroach (marked Insbot). A control test was carried out by using a paper-dressed Insbot covered impregnated with sixty microliters of dichloromethane only (unmarked Insbot).

The experiment began when cockroaches and/or an Insbot were introduced at the centre of the arena. All individual displacements were recorded by a camera for three hours (Fig. 1a, Fire-I Digital camera, Unibrain).

3   Results

3.1   Validation of the Experimental Setup Homogeneity

We had to test in our experimental setup whether two identical shelters have the same probability to be occupied by individuals. Indeed, the presence of external landmarks could lead to a higher selection rate by agents (animal or artificial) of one of the two identical shelters. In this latter case, we could not make a distinction between the respective contribution of the position of the shelter in the arena and of tested factors (luminosity, number of individuals it contains) on the observed aggregation pattern. Then, the analysis of the cockroach distribution under the shelters after 180 minutes of experiments, we can exclude the existence of any bias related to the position of the shelter in the arena which may favor its selection by the group at the expenses of the other shelter. Indeed, cockroaches have an equal probability (number of cockroaches under a shelter / total number of cockroaches under the two shelters) of 0.48 and 0.52 (305 cockroaches tested; Chi-square goodness of fit: $\chi^2_{0.05,1} = 0.40$, p > 0.05) to be found respectively under the left or the right light shelter.

Insbot can be influenced in its choices by any landmark external to the arena (for detail of sensor sensitivity see [14]). Tests with isolated Insbot (11 experiments for a total of around 8 hours of experiments) show that Insbot have an equal probability of 0.48 and 0.52 (number of entries = 147; Chi-square goodness of fit: $\chi^2_{0.05,1} = 0.15$, p > 0.05) to enter respectively in the left or in the right light shelter.

3.2 Perception of Shelters as Resting Site

The fraction of the total population (for one individual and groups of 10 or 30 cockroaches) under two identical dark shelters is significantly greater than the probability expected in case of homogeneous distribution of individuals in the arena (Table 1). This expected probability, that also assumes that the cockroaches do not interact together, is equal to the ratio between the area of the shelters (353.4 cm²) and the area of the arena (7853.8 cm²). Then, due to the low light intensity under shelters in comparison with the uncovered part of the arena, the two shelters constitute the only heterogeneities susceptible to focus the cockroach aggregation.

In table 1, we observe a high proportion of cockroaches aggregated under the shelters confirming that these heterogeneities are well perceived as resting sites. Indeed, one cockroach has a four times higher probability to be found out under shelters than expected from random. Furthermore, there is an enhancing group effect: when tested in groups of 10 or 30 individuals, more than a half of cockroaches are staying under the shelters. Hence the probability of a cockroach to leave a shelter decreases as the number of aggregated conspecifics increases. This confirms that the spatial distribution of the animals is not the simple summation of individuals' resting preferences but also outcomes from interattraction effects.

Table 1. Comparison between probability of being under shelters resulting from a random homogeneous distribution of individuals in the arena and the observed probability of presence under shelters for 1, 10 or 30 cockroaches

	Probability of presence	
	Random	Observed
1 cockroach (37 replicates)	0.045	0.20
10 cockroaches (30 replicates)	0.045	0.56
30 cockroaches (25 replicates)	0.045	0.62

This interattraction which plays a key-role in collective decision, is not only based on physical contacts but also on chemical attraction, due to the perception of cuticular blend of group members. Therefore, we have investigated whether the addition of these chemical compounds on Insbot may improve its integration within a cockroach's group.

3.3   Chemical Marking and Insbot Integration in Mixed Society

To test to which extent the marking of Insbot with cuticular extracts of cockroaches is needed to allow the integration of the Insbot in a cockroach group, we compared individual and collective behavior of robot and insects in experimental tests with 10 cockroaches without Insbot, 10 cockroaches with an unmarked Insbot and 10 cockroaches with an Insbot chemically marked.

3.3.1   Influence on the Individual Behavior of Cockroach
The level of acceptance of an Insbot by one cockroach is assessed by the mean time during which they stay close together under the same shelter.

The cockroaches are able to discriminate between an unmarked and a marked Insbot. Indeed, the mean time of contact between a cockroach and a marked Insbot is similar to the mean time between two cockroaches but lower than that between a cockroach and an unmarked Insbot (Fig. 4. Kruskal-Wallis test: KW = 14.3, 235 replicates, $p < 0.001$. Dunn's multiple comparison test: $p < 0.05$ only for the comparison cockroach-cockroach vs cockroach-unmarked Insbot). The lack of agonistic behavior by the cockroach as well as its prolonged association with the marked Insbot confirms the successful acceptance of this marked artificial agent.

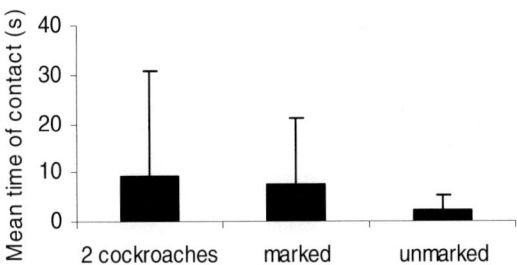

Fig. 4. Mean time of contacts (± S.D.) under shelters between a cockroach and either another cockroach, a chemically marked Insbot or an unmarked Insbot. The duration of interactions between cockroach and robot are significantly different for the marked and unmarked robot.

3.3.2   Influence on the Collective Behavior of Cockroaches
The integration of an Insbot in a group of cockroaches is assessed by its influence on the clustering behavior of cockroaches and its presence within animal clusters.

The unmarked Insbots seem to disturb the cockroach aggregation pattern. With an unmarked or a marked robot, respectively 40% or 70% of mixed society experiments ended by at least 66% of cockroaches under shelters (Fig. 5). This result shows that unmarked robot prevents the cockroaches from being under shelter.

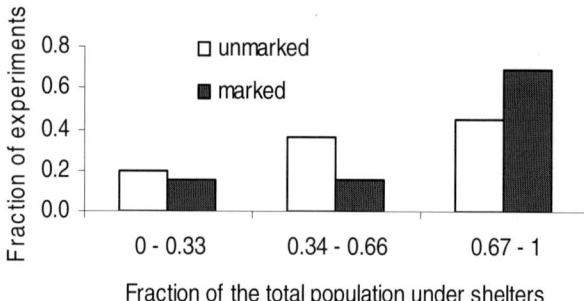

Fig. 5. Frequency distribution of experiments according to the proportion of aggregated cockroaches under shelter for experiments with marked or unmarked Insbot. The unmarked robots tend to lower the occurrence of large aggregates.

If we assumed that Insbot marked or not, has no influence on the aggregative choice of cockroaches, one would expect that the percentage of aggregated cockroaches would be similar in clusters including a robot or not. However, our observations show that the unmarked Insbot has a repulsive effect while the marked one attracts animals. Indeed, in only 30% of experiments, the majority of cockroaches (more than 2/3 of the total cockroach population) are with an unmarked Insbot under the same shelter (Fig. 6). On the other hand, around 60% of experiments with marked Insbot ended with the majority of sheltered cockroaches under the same shelter (Fig. 6).

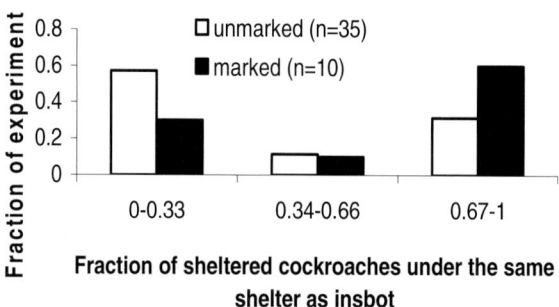

Fig. 6. Frequency distribution of experiments as a function of the proportion of cockroaches under the same shelter as either a marked or an unmarked Insbot. The insects prefer to aggregate by segregating the unmarked Insbots. If the robots are marked, the insects prefer to aggregate in the same shelter as the Insbots i.e. without segregation.

4   Conclusion

Many gregarious species are able to perform collective decision such as the selection of one resting site without being guided by the leader, simply due to local -often amplifying- interactions between group members. These collective decisions are not the simple summation of individual behavior (e.g. movement pattern, resting preference) but are deeply governed by a network of positive feedback loops. Hence, it is of the utmost importance to identify the nature and the specificities of amplifying interactions between group members in order to implement them in artificial agents and to build functional mixed societies. In accordance with the goals of the Leurre project [2], we are able to manage the behavioral and chemical parameters of the Insbot in order to integrate it in insect groups.

The good integration of robot in animal group through the formation of mixed aggregates namely depends on chemical recognition of robots by the cockroaches as a full member of the group.

We have validated the ability of the chemical lure put on Insbot to mimic interattraction between cockroaches. We highlight the crucial role played by cuticular extracts of cockroach on the formation of mixed group. Without chemical marking, the robot disturbs cluster formation and lead to spatial segregation from the insects. By contrast, with chemical marking, cockroaches share a shelter with the robot that could probably be perceived by the insects as a "congener". Since the presence of one marked Insbot increases the resting time of a nearby cockroach, this artificial agent can nucleate the aggregation of animals and manipulate their spatial distribution. Besides, aggregation is a well known auto-catalytic process in which clustering increases with the number of already aggregated individuals. Therefore, in the future we will investigate how collective response of a mixed society will change with increasing number of artificial agents.

Acknowledgment

The LEURRE project [2] is funded by the Future and Emerging Technologies programme (IST-FET) of the European Community, under grant IST-2001-35506. G. Sempo is funded by "La Fondation des Treilles", France and is research associate from the ECagents project of the European Community (IST-1940). C. Detrain and J.-L. Deneubourg are research associate from the Belgian National Fund for Scientific Research.

References

1. Colot, A., Caprari, G., Siegwart, R.: InsBot : Design of an Autonomous Mini Mobile Robot Able to Interact with Cockroaches. Proceedings of the 2004 IEEE International Conference on Robotics & Automation. New Orleans, LA (2004) 2418-2423
2. LEURRE Project official website, *http://leurre.ulb.ac.be*
3. Ishii, H., Nakasuji, M., Ogura, M., Miwa, H., Takanishi, A.:.Accelerating Rat's Learning Speed Using a Robot - The robot autonomously shows rats its functions, Proceedings of the 2004 International Workshop on Robot and Human Interactive Communicaiton, Kurashiki, Okayama Japan (2004)
4. Kubinyi, E., Miklósi, A.., Kaplan, F., Gacsi, M., Topal, J., Cs anyi, V.: Can a dog tell the difference? dogs encounter AIBO, an animal-like robot in two social situations. Proceedings of the seventh international conference on simulation of adaptive behavior on From animals to animats. MIT Press, Cambridge (2002)

5. Butler, Z., Corke, P., Peterson, R., Rus, D.: Virtual fences for controlling cows. Proceedings of the 2004 IEEE international Conference on Robotics & Automation New Orleans, LA April (2004) 4429- 4436

6. Vaughan, R., Sumpter, N., Henderson, J., Frost, A., Cameron, S.: Experiments in automatic flock control. Robotics and Autonomous Systems, 31 (2000) 109-117

7. Ame, J.M., Rivault, C., Deneubourg, J.L.: Cockroach aggregation based on strain odour recognition. Anim. Behav., 68 (2004) 793-801

8. Jeanson, R., Rivault, C., Deneubourg, J.L., Blancos, S., Fournier, R., Jost, C., Theraulaz, G.: Self-organized aggregation in cockroaches. Anim. Behav., 69 (2004) 169-180

9. Saffre, F., Halloy, J., Deneubourg. J.-L.: The Ecology of the Grid. Second International Conference on Autonomic Computing (ICAC'05) (2005) 378-379

10. Saïd, I., Costagliola, G., Leoncini, I., Rivault, C.:Cuticular hydrocarbon profiles and aggregation in four *Periplaneta* species (Insecta: Dictyoptera). J. Insect. Physiol., 51 (2005) 995-1003

11. Leoncini, I., Rivault, C.:Could species segregation be a consequence of aggregation processes. Example of *Periplaneta americana* (L.) and *P. fuliginosa*. Ethology, 111 (2005) 527—540

12. Mote, M.I., Goldsmith, T.H.: Spectral sensitivities of color receptors in the compound eye of the cockroach *Periplaneta. Journal of Experimental Zoology*, 173 (1970) 137-145

13. Meyer, D.J., Margiotta, J.F., Walcott, B.: The shadow response of the cockroach *Periplaneta americana. Journal of Neurobiology*, 12 (1981) 93-96

14. Tâche, F., Asadpour, M., Caprari, G., Karlen, W. and Siegwart, R.: Perception and Behavior of InsBot: Robot-Animal Interaction Issues. Proceedings of the IEEE International Conference on Robotics and Biomimetics, Hong Kong SAR and Macau SAR (2005)

Collective Decision-Making Based on Individual Discrimination Capability in Pre-social Insects

Jean-Marc Amé, Jesus Millor, José Halloy,
Grégory Sempo, and Jean-Louis Deneubourg

Service d'Ecologie sociale, CP 231, Université Libre de Bruxelles,
Bd Fr. Roosevelt, B-1050 Bruxelles, Belgium
jldeneub@ulb.ac.be

Abstract. Gregarious insects, like cockroaches, aggregate in shelters during their resting period. How do individuals reach a collegial decision? What is the relation between the distributions of the individuals and the parameter values characterizing the population and the environment? With a model based on experimental data, we demonstrated that the collegial decision is based on the relation between the individual resting time in a shelter and the population in this shelter. We extended this model to the case where different sub-groups may interact and where the crowding effect under the shelters influences the aggregation. This second model shows that depending on the interaction between the sub-groups and the crowding effect, different patterns are observed such as segregation of the different sub-group or the aggregation of the whole population.

1 Introduction

Grouping is the most common collective behaviour among living organisms. This phenomenon extends over the entire diversity of taxon and spans many biological characteristics like life-history strategy, degree of mobility... [1][2]. Many species from bacteria to higher vertebrates form groups more or less stable in time and space in response to environmental heterogeneities and environmental constraints or to attraction between individuals [3]. The level of interactions among individuals in a population relies on the spatial distribution of individuals that influences the structure and the organization of populations [4]. Aggregation can be defined as a higher temporal and spatial density of individuals than in the surrounding area [5][2]. The origin and the stability of social aggregates result from mutual inter-individual interactions which are mediated by information transfer between individuals. This can induce emergent group behaviours, patterns or functions that are not merely the sum of the individual behaviours [6]. Self-organized systems allow understanding how non-linear interactions can lead to complex and non-intuitive behaviours even with basic rules or information transfer at the individual level [2].

In cockroach species, during the diurnal phases of their rhythm of activity the most widespread collective behaviour is gregariousness [7-9]. Studies on cockroaches and

S. Nolfi et al. (Eds.): SAB 2006, LNAI 4095, pp. 713–724, 2006.

especially on *Blattella germanica* have shown that clustering results from inter-attraction between individuals in response to a signal mediated by chemical cues [10][11] and that this aggregation can be consider as a social phenomenon [12]. Different benefits result from aggregation such as reduction of stresses, increase in efficiency of alarm responses and antipredator behaviour, faster development and more efficient reproduction. Cockroaches tend to gather in shelters during their resting period. In *Eublaberus distanti* individuals are able to discriminate individuals belonging from the same strain or from another strain [13]. Ishii & Kuwahara [14] have shown that groups of cockroach larvae were able to select an aggregation shelter according to its odour conditioning in binary choices tests. Different strains of cockroaches imply different individual chemical signals per strains. Recent studies based on recognition of cuticular hydrocarbons profiles in *Blattella germanica* show that strains discriminate the signals and this leads to resting shelter selection. In a recent experimental binary choices study based on odour discrimination between species or strains of cockroaches, Leoncini and Rivault [15] have shown that segregation can occur when the carrying capacity of shelters is a limiting factor. In relation to the crowding effect sub-groups aggregate under the same shelter or segregate between the two shelters.

The bases to model this kind of collective phenomena have been introduced in previous studies [16-18]. We present here collective decision making linked to aggregation problems between different sub-groups of individuals by taking into account the crowding effect. Our aim is to show that inter-attraction can lead, in relation to the limited carrying capacity of shelters, to different patterns of aggregation from a homogeneous distribution of individuals to segregation between sub-groups. Modulating the level of inter-attraction between strains and the crowding effect, we show in this study that either segregation or aggregation can be emergent patterns due to local interaction between individuals without global knowledge of the system.

We describe the differential equations model that derives from the previous ones [16-18] and the stochastic description of the model by using master equations to take into account the fluctuations characterizing such systems and determine the probability distribution of individuals from each sub-group under shelters.

This model study will be useful to determine how the properties observed at the individual-level can explain the patterns that emerge at the collective-level without an active signal.

2   Formalizing the Model

2.1   Meanfield Formulation of the Model

Previous experiments provided data to build a dynamical model of aggregation based on the individual behaviour [18][19]. In previous studies on mechanisms that induce collective choices in binary choice tests, we have analyzed on one hand similar models based on strain recognition without a crowding effect [16][17]. On the other hand, we have analyzed the effect of crowding on one strain on binary choices and multiple choices tests and describe the patterns that appear [18]. The present model mixes these studies and takes into account the crowding effect of shelters in binary choices tests for two strains and thus the effect of limited space due to the carrying capacity of shelter on strain repartition.

We present first the general model for p strains ($i=1...p$) and two shelters ($j=1,2$) with a limited carrying capacity (S_j). We assume that the number of individuals from each strain is equal to N and that the maximum number of individuals that each shelter can harbor is equal ($S_j=S$). $X_{i,j}$ is the number of individuals of the strain i under the shelter j.

$$\sum_{j=1}^{2} X_{i,j} = N \qquad \forall i = 1...p \tag{1}$$

At each time step, each individual in the shelter j has a probability $Q_{i,j}$ to leave this shelter and to explore the arena. It has the same probability at each time step to encounter and to join the shelter h (R_h).

Neglecting the time outside shelters, we can write the evolution of the number of individuals of strain i under shelter j as follow:

$$\frac{dX_{i,j}}{dt} = -R_h Q_{i,j} X_{i,j} + R_j Q_{i,h} X_{i,h} \qquad j = 1,2 \quad h = 1,2 \quad h \neq j \qquad \forall i = 1...p \tag{2}$$

The probability to join the shelter j is given by:

$$R_j = \mu \left(1 - \sum_{i=1}^{p} \frac{X_{i,j}}{S} \right) \qquad j = 1,2 \quad \forall i = 1...p \tag{3}$$

Where μ is the maximal kinetic constant for entering in shelter j. R_j is equal to 1 as we have neglected the time outside the shelter thus individuals dynamic only depend to the probability to leave a shelter to reach directly another one.

The probability $Q_{i,j}$ for one individual belonging to one strain to leave a shelter is in relation to the number of individuals present under this shelter. Experimental tests showed that larvae prefer their own strain odour to that of other strains [19]. In this case, the influence of individuals belonging to the same strain can be more important than that of individuals belonging to other strains. Thus the basic model must be completed with parameters of inter-attraction between strains i and l: β_{il}.

We suppose that the interaction of strain i on strain l is the same that l on i and that each strain has the same interaction with others, therefore $\beta_{il}= \beta_{li}= \beta$.

The parameters of inter-attraction inside a strain already present in the single strain model are always considered equals to 1 ($\beta_{ii}= 1$). To express that an individual of one strain tends to stay more with an individual of the same strain than with individual of another strain, $0 \leq \beta \leq 1$. If $\beta = 0$, we have p independent strains with no inter-attraction between them. If $\beta=1$, we have p strains that interact in the same way with others.

The experimental results show that the probability $Q_{i,j}$ of leaving shelter j decreases with the density of individuals $\dfrac{X_{i,j}}{S}$ under this shelter [16]:

$$Q_{i,j} = \frac{\theta}{\left(1 + \rho\left[\dfrac{X_{i,j}}{S} + \sum_{\substack{l=1 \\ l \neq i}}^{p} \dfrac{\beta X_{l,j}}{S}\right]\right)^{n}} \qquad j = 1,2 \quad \forall i = 1...p \tag{4}$$

where θ is the maximal probability of leaving the shelter j per unit of time, and ρ is a reference surface ratio for estimating the carrying capacities. From personal measures, we assume that $n \approx 2$ [16].

We can resume the evolution of the number of individuals of each strain under the two shelters after normalization ($x_i = X_i/N$) as follow:

$$\sum_{j=1}^{2} x_{i,j} = 1 \quad \forall i = 1...p$$

$$\frac{dx_{i,j}}{dt} = -\frac{\theta x_{i,j}}{1 + \rho\left(\dfrac{x_{i,j}}{\sigma} + \beta\sum_{\substack{l=1 \\ l \neq i}}^{p} \dfrac{x_{l,j}}{\sigma}\right)^2}\left(1 - \frac{\sum_{l=1}^{p} x_{l,h}}{\sigma}\right)$$

$$+ \frac{\theta x_{i,h}}{1 + \rho\left(\dfrac{x_{i,h}}{\sigma} + \beta\sum_{\substack{l=1 \\ l \neq i}}^{p} \dfrac{x_{l,h}}{\sigma}\right)^2}\left(1 - \frac{\sum_{l=1}^{p} x_{l,j}}{\sigma}\right) \tag{5}$$

$$where \begin{cases} j = 1 \quad h = 2 \\ j = 2 \quad h = 1 \end{cases} \quad with \quad \sigma = \frac{S}{N}$$

Shelters can't hold more than the total number of individuals per strains (N). These conditions imply for p strains that $2S \geq pN$ or $\sigma \geq \dfrac{p}{2}$ (e.g. $p=1$ $\sigma \geq 0.5$; $p=2$ $\sigma \geq 1$).

2.2 Stochastic Formulation of the Model

A stochastic description of the model can be done by using master equations to take into account the fluctuations characterizing such systems and determine the probability distribution of individuals from each strain under shelters.

$X_{i,j}$ is the number of individuals of strain i under shelter j. The systems is characterized by $(N+1)^p$ states Ω per strain i ($1...p$) and per shelter j:

$$\Omega(X_{i,j}) = X_{1,j}, ..., X_{i,j}, ...X_{p,j}$$

Thus, we associate to each state $\Omega(n)$ a probability (P) to be in this state at time t:

$$P(\Omega(n),t) \qquad \forall i = 1...p$$

We can define the transition probability between states W: $W(\Omega(n)|\Omega(n'))$.

The following dynamical equation counts the processes leading to the state $\Omega(n)$ and the processes removing it from this state:

$$\frac{dP(\Omega(n),t)}{dt} = \text{contribution of transition} \quad \Omega(n') \to \Omega(n) \quad n \neq n' \tag{6}$$
$$- \text{contribution of transition} \quad \Omega(n) \to \Omega(n')$$

$$\frac{dP(\Omega(n),t)}{dt} = \sum_{\substack{n' \\ n' \neq n}} (P(\Omega(n'),t)W(\Omega(n')|\Omega(n))) - \sum_{\substack{n' \\ n' \neq n}} (P(\Omega(n),t)W(\Omega(n)|\Omega(n'))) \tag{7}$$

with $P(\Omega(n),t=0) = P_0(\Omega(n))$

In term of probability, the incoming individuals on site j per unit of time of state $\Omega(n)$ come from transitions of all states $\Omega(n')$ where the probability of occupation at time t is $P(\Omega(n'),t)$ to state $\Omega(n)$ with a transition probability $W(\Omega(n')|\Omega(n))$. Else the outgoing of individuals from site j per unit of time of state $\Omega(n)$ is proportional to its probability of occupation at time t $P(\Omega(n),t)$ with a transition probability $W(\Omega(n)|\Omega(n'))$.

At each time step, $X_{i,j}$ can either unchanged or vary by only 1 or –1, corresponding to the individual movements between the shelters:

$$\Omega(X_{i,j} - 1) = X_{1,j}, ..., X_{i,j} - 1, ..., X_{p,j}$$
$$\Omega(X_{i,j} + 1) = X_{1,j}, ..., X_{i,j} + 1, ..., X_{p,j}$$

For two sites, assuming $X_{i,2}=N-X_{i,1}$ we can define the contributions of transition for state $\Omega(X_{i,1})$ per unit time under shelter 1. For example $W_i(\Omega(X_{i,1}-1)|\Omega(X_{i,1}))$ is the transition probability per unit time of going from state $\Omega(X_{i,1}-1)$ to state $\Omega(X_{i,1})$. It corresponds to the movement of an individual of strain i between the shelter 2 and the shelter 1. $P(\Omega(X_{i,1}-1),t)$ is the probability of being at state $\Omega(X_{i,1}-1)$ at time t.

$$W_i(\Omega(X_{i,1}-1)|\Omega(X_{i,1})) = \frac{\theta(X_{i,2}+1)}{1 + \rho(\frac{X_{i,2}+1}{S}) + \beta\sum_{\substack{l=1 \\ l \neq i}}^{p}\frac{X_{l,2}}{S})^2} \left[1 - \frac{\left(\sum_{i=1}^{p}X_{i,1}\right)-1}{S}\right] \tag{8}$$

To obtain the contribution of transition to the state $\Omega(X_{i,1})$, we sum over all states that can lead to this state in a single step, corresponding to the movement of individuals of the p strains. Similarly the contribution from the state $\Omega(X_{i,1})$ per unit time is the product of the probability of being in state $\Omega(X_{i,1})$ at time t, times the sum of the transition probabilities per unit time from $\Omega(X_{i,1})$ to all other states accessible from $\Omega(X_{i,1})$.

Thus the stochastic evolution per unit of time of the number of individuals of strain i under the shelter 1 is given by the equation (9):

$$\frac{dP(\Omega(X_{i,1}),t)}{dt} =$$

$$\sum_{i=1}^{p} (W_i(\Omega(X_{i,1}-1)|\Omega(X_{i,2}))P(\Omega(X_{i,1}-1),t) + W_i(\Omega(X_{i,1}+1)|\Omega(X_{i,1}))P(\Omega(X_{i,1}+1),t))$$

$$-\sum_{i=1}^{p} (W_i(\Omega(X_{i,1})|\Omega(X_{i,2}-1))P(\Omega(X_{i,1}),t) + W_i(\Omega(X_{i,1})|\Omega(X_{i,1}+1))P(\Omega(X_{i,1}),t))$$

(9)

3 Results

3.1 Meanfield Model

From the general model (Eq. 5), for two strains ($i=2$) we can resume the evolution of the number of individuals of these strains under the two shelters as follow:

$$\frac{dx_{i,j}}{dt} = -\frac{\theta x_{i,j}}{1+\rho(\frac{x_{i,j}}{\sigma}+\beta\frac{x_{l,j}}{\sigma})^2}(1-\frac{x_{i,h}+x_{l,h}}{\sigma})$$

$$+\frac{\theta x_{i,h}}{1+\rho(\frac{x_{i,h}}{\sigma}+\beta\frac{x_{l,h}}{\sigma})^2}(1-\frac{x_{i,j}+x_{l,j}}{\sigma})$$

(10)

$$where \begin{cases} i=1 \ l=2 \\ i=2 \ l=1 \end{cases} et \begin{cases} j=1 \ h=2 \\ j=2 \ h=1 \end{cases}$$

With two strains and two identical shelters, the model has nine stationary solutions that correspond to different distributions of individuals under shelters. We resume below these analytical solutions and their analytical stability.

(a) The symmetrical state

These first solutions correspond to the dispersal of individuals between shelters (dispersal) with an equal number of individuals on each strain on both shelters:

$x_{1,1}=x_{1,2}=x_{2,1}=x_{2,2}=0.5$.

The symmetrical state exists for all values of σ and β and whatever the value of the parameter ρ. This state is stable when no other states exist (Fig. 1a & 2); see below for the conditions of existence of the aggregative and the segregative states. Thus for small values of σ ($S\approx N$) and $\beta > 0.3$, and for huge values of σ, the shelters collect half the number of individuals of each strain (equipartition).

(b) The two aggregative states

Another group of two solutions is asymmetrical (heterogeneous), with an unequal number of individuals whatever the strain under each shelter: one shelter is selected, the two strains aggregate under the same shelter:

$x_{1,1}=x_{2,1}; \ x_{1,2}=x_{2,2}$.

These two aggregative states are always stable when they exist (Fig. 1a & 2).

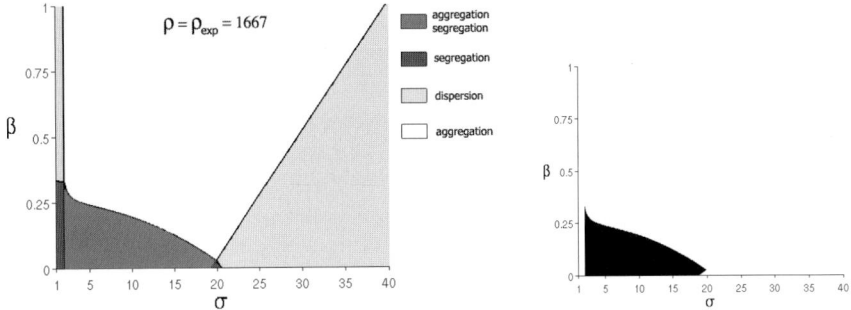

Fig. 1. Existence and stability of the different states for two strains and two shelters in relation to (σ, β). Assuming equation (5), $\sigma \geq 1$ and $\rho = \rho_{exp} = 1667$. a- the homogeneous state, the aggregative and segregative states. b- the four mixed states (always unstable).

(c) The two segregative states

One group of solutions, are symmetrical (but heterogeneous), with an equal number of individuals whatever the strain under each shelter. This means that one shelter was selected by one strain, the other shelter by the other strain. The two strains segregate:

$$x_{1,1} = x_{2,2}; \; x_{1,2} = x_{2,1}.$$

The existence of the two segregative states depends on the level of interaction between strains and on the carrying capacity of shelters (Fig. 1a & 2). The following inequality gives the domain of existence of those states:

For low values of σ and β, these states are stable but when β increases and when these states exist, they become unstable.

(d) The four mixed states

This group of 4 solutions correspond to an unequal number of individuals of each strain under each shelters: $x_{1,1} \neq x_{2,2} \neq x_{1,2} \neq x_{2,1}$.

The four mixed states exist if the aggregative and the segregative states exist in the same time and are always unstable (Fig. 1b).

The segregative solutions are always stable when they exist in the same range of values that the symmetrical state only i.e. for small values of σ ($\sigma < 2$, Fig. 1), and in a range of huge value of σ ($\sigma \approx 20$ for $\rho = 1667$) and small values of β (Fig. 1a & 2). Despite the aggregative states, even if σ is small, the segregative states exist while the coefficient of inter-attraction between strains is small ($\beta < 0.3$) but their stability is limited with the existence of the four mixed states that is always unstable (Fig. 1b & 2, $\sigma = 5$). The segregative states disappear when $\sigma > 21$ for $\rho = 1667$. The aggregative states exist from $\sigma = 2$ but disappear in relation to the inter-attraction and the carrying capacity of shelters due to the non limited space of shelters (Fig. 1a). For $S >> N$, the symmetrical state become the stable state.

3.2 Stochastic Study

The numerical resolution of master equations aims to follow the time evolution of the distribution of individuals under shelters and gives the probability of each states

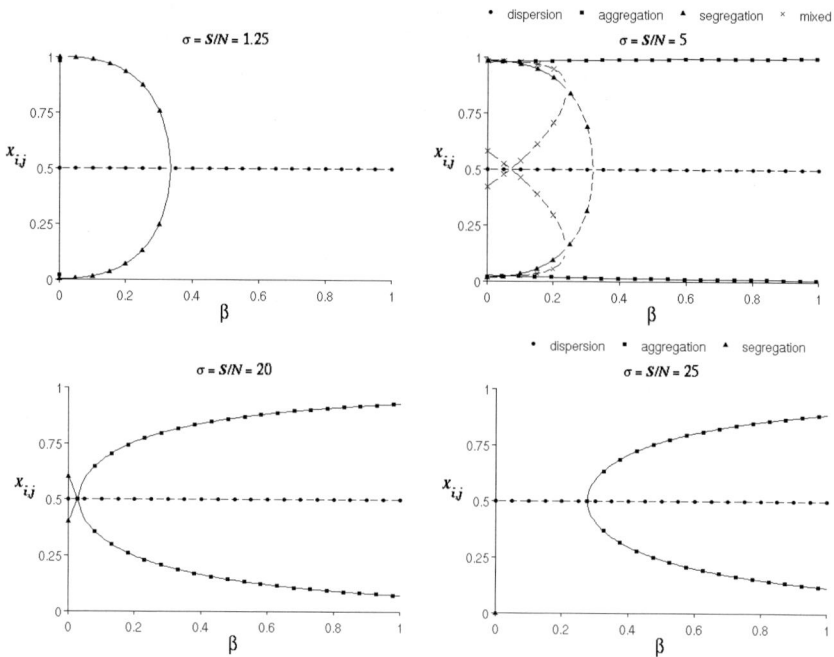

Fig. 2. Bifurcation diagrams as a function of β for four values of σ and $\rho = 1667$. Dashed lines represent unstable branches and solid lines, stable branches. (●) represent the symmetrical state (dispersion), (▲),the segregative states, (■), the aggregative ones and (×), the four mixed states.

$\Omega(X_{i,j})$ at the stationary regime. The time to reach the stationary regime increases in relation to N.

For small values of σ ($\sigma < 2$) and β lower than 0.3 (Fig. 3a & 3b), the segregative states are selected; for β greater than 0.3, only the symmetrical state exist and is selected (Fig. 3c). So for a fixed carrying capacity, the greater is N, the more the segregation is selected if the strains few interact.

For medium values of σ and small values of β we have coexistence of the aggregative states and the segregative ones (Fig. 3a & 3b) but for very small values of β the segregative states is more often selected and for upper values of β the aggregative states is more selected than the other ones (Fig. 3 & Fig. 4).

In a range of values of σ between 2 and 19, the aggregative states and/or the symmetrical state are selected in relation to β (Fig. 3). The greater is σ in this range and/or β the smaller is the probability to select the segretative states instead of the aggregative ones (Fig. 3).

However for huge values of σ ($\sigma > 20$), in a first hand the segregative states disappear and in a second hand the aggregative ones in relation to β. For $\sigma \geq 40$, the symmetrical state is the only existing state as predicted in the meanfield model (Fig. 3 and see Fig. 1a).

This stochastic model shows similar results with experiments [15]. In those experiments, we can assume that $S = 25$ and by approximation with the results of the model that $\beta \approx 0.15$. For two strains, with populations of 5 ($\sigma = 5$) and 10 ($\sigma = 2.5$) individuals and for this small value of β, the segregation decreases face to the aggregation in relation to the increase of σ (Fig. 3b).

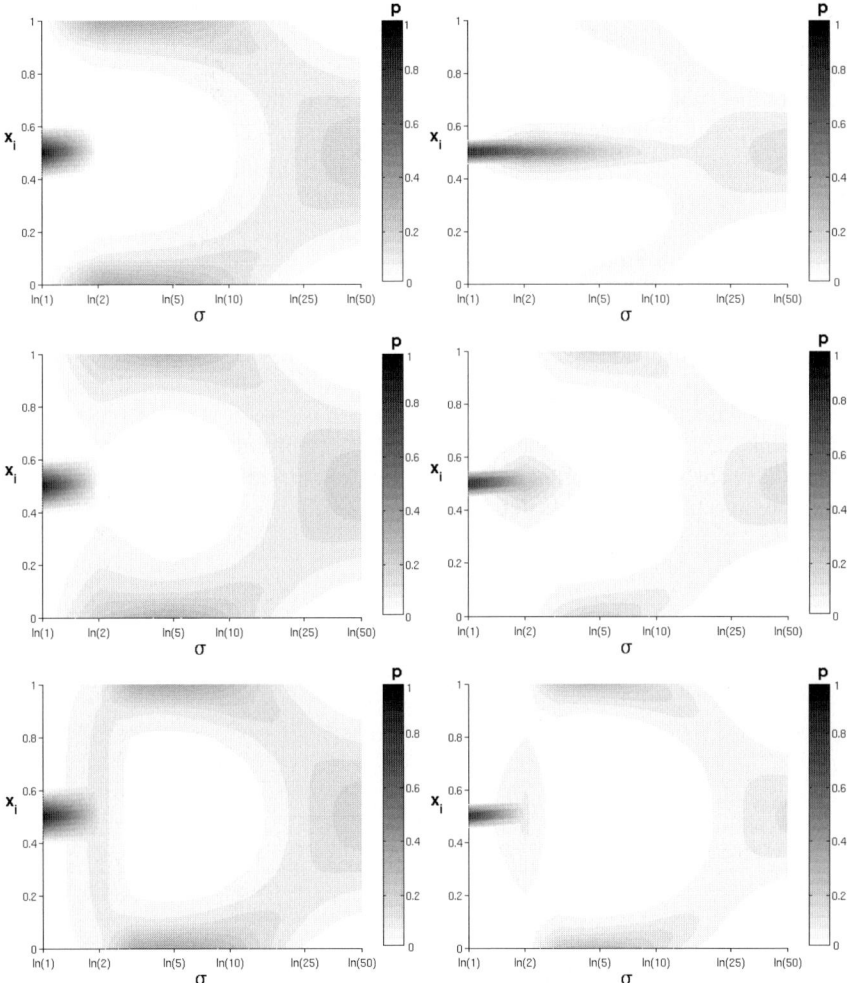

Fig. 3. Stochastic resolution of the two-strain model for $\rho = 1667$. The figures give the probabilities to be in state $\Omega(X_i)$ associated to the proportions of individual in relation to σ (logarithmic scale on this figure). In the left column, x_i is the proportion of individuals under one shelter for one of the two strain; in the right column, x_i is the proportion of individuals under one shelter for both strains. a- probabilities for $\beta = 0.05$; b- probabilities for $\beta = 0.15$; c- probabilities for $\beta = 0.35$.

4 Discussion

Results show that solutions are not qualitatively different from the model without crowding effect [16][17]: collective patterns that emerge at the collective level without and with the crowding effect are identical. Whereas the crowding effect which is represented by the carrying capacity of shelters (S) limits the range of existence and of stability of the aggregative and segregative states for huge values of S. At the stationary regime, only 3 parameters characterize the model: σ (S/N), β (inter-attraction parameter) and ρ (reference surface ratio) that are linked to group properties but not to the individual behaviour. β is the parameter of interaction between strains (or between species of cockroaches). When $\beta > 0.3$, there is no difference between a two-strains group or a one-strain group. For $\beta < 0.3$ and small value of σ, the segregative states are stable. For larger values of σ, both aggregative and segregative states are stable, but the greater is the values of σ, the lower is the probability of observing the segregative states.

Modulating the inter-attraction β, the total number of individuals (N) and the carrying capacity of the shelter (S), a diversity of solutions is generated (emergence of segregation, co-existence of segregation and aggregation, …) without any modulation of the individual behaviour and individual knowledge of the global system: the greater the number of individuals under a shelter is, the lower the probability of leaving this shelter. However, the crowding effect (environmental constraint) plays a role on the probability of joining a shelter by limiting to a critical value the number of individuals accepted under a shelter.

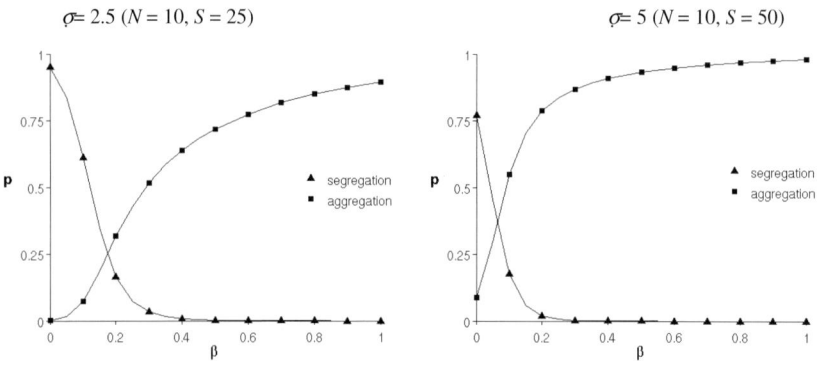

Fig. 4. Probability to reach the aggregative states or the segregative states at the stationary regime with 80% of individuals of each strains under shelters for $N = 10$ and $S = 25$ ($\sigma = 2.5$) and 50 ($\sigma = 5$). (▲) represents the segregative states and (■), the aggregative ones.

This study shows that this kind model based on minimal rules of inter-attraction between individuals can explain the distribution of groups in a population (clustering and gregariousness between animal species or strains). By taking account positive feedbacks due to individual behaviours and negative feedbacks due to environmental constraints, collective choices can lead to the segregation or the aggregation of groups of animals [2][17][18]. This should describe a relevant and generic process for understanding the dynamics of aggregation and segregation between sub-groups or species without the need of sophisticated behaviours modulated by the density [15][17]. Self-organised mechanisms govern many cases of aggregation and collective choice. Behavioural positive feedbacks, based on different type of signals (pheromone, silk, mechanical…), are the keystones of the dynamics of aggregation [20-24]. Our hypothesis is that for all these cases, the coupling between crowding and these positive feedbacks leads to a diversity of patterns similar to those of the cockroaches.

References

1. Parrish, J.K. & Edelstein-Keshet, L. 1999. Complexity, pattern, and evolutionary trade-offs in animal aggregation. *Science* **284**, 99-101.
2. Camazine, S., Deneubourg, J.L., Franks, N., Sneyd, J., Theraulaz, G. & Bonabeau, E. 2001. *Self-Organization in Biological Systems*. Princeton Univ. Press, Princeton.
3. Parrish, J.K., Viscido, S.V. & Grünbaum, D. 2002. Self-organized fish schools: An examination of emergent properties. *Biol. Bull.* **202**, 296-305.
4. Okubo, A. 1980. *Diffusion and Ecological Problems: Mathematical Models*. Springer-Verlag, Berlin.
5. Southwood, T.R.E. 1966. *Ecological Methods*. London, Methuen.
6. Parrish, J.K., Hamner, W.M. & Prewitt, C.T. 1997. Introduction - from individuals to aggregations: unifying properties, global framework, and the holy grails of congregation. *In*: Animal Groups in three Dimensions (Parrish, J. K. & Hamner, W. M., eds). Cambridge Univ. Press, Cambridge, pp. 1-14.
7. Bell, W.J., Parsons, C. & Martinko, E.A. 1972. Cockroach aggregation pheromones: analysis of aggregation tendency and species specificity (Orthoptera: Blattidae). *J. Kans. Entomol. Soc.* **45**, 414-421.
8. Rivault, C. 1990. Distribution dynamics of *Blattella germanica* in a closed urban environment. *Entomol. Exp. Appl.* **57**, 85-91.
9. Appel, A.G. 1995. *Blattella* and related species. *In*: Understanding and Controlling the German Cockroach (Rust, M.K., Owens, J.M. & Reierson, D.A., eds). Oxford Univ. Press, Oxford, pp. 1-20.
10. Ishii, S. & Kuwahara, Y. 1968. Aggregation of German cockroach *Blattella germanica* nymphs. *Experientia* **24**, 88-89.
11. MacFarlane, J.E. & Alli, I. 1987. The effect of lactic acid and volatile fatty acids of the aggregation behaviour of *Periplaneta americana*. *Comp. Biochem. Physiol.* **86**, 45-47.
12. Ledoux, A. 1945. Etude expérimentale du grégarisme et de l'iinter-attraction sociale chez les Blattidés. *Annales des Sciences Naturelles Zoologie et Biologie Animale* **7**, 76-103.
13. Brossut, R. 1979. Gregarism in cockroaches and in *Eublaberus* in particular. *In*: Chemical Ecology: Odour Communication in Animals (Ritter, F.J., ed.). Elsevier, Amsterdam, pp. 237-246.
14. Ishii, S. & Kuwahara, Y. 1967. An aggregation pheromone of the German cockroach *Blattella germanica* (Orthoptera: Blattellidae). 1. Site of the pheromone production. *Appl. Entomol. Zool.* **2**, 203-217.
15. Leoncini, I. & Rivault, C. 2005. Could species segregation be a consequence of aggregation processes? Example of *Periplaneta americana* (L.) and *P. fuliginosa* (Serville). *Ethology* **11**(5), 527.

16. Amé, J.M., Rivault, C. & Deneubourg, J.L. 2004. Cockroach aggregation based on strain odour recognition. *Anim. Behav.* **68**, 793-801.
17. Millor, J., Amé, J.M., Halloy, J. & Deneubourg, J.L. 2006. Individual discrimination capability and collective decision-making. *J. Theor. Biol.* **239**, 313-323.
18. Amé, J.M., Halloy, J., Rivault, C., Detrain, C., Deneubourg, J.L. Collegial decision making based on social amplification leads to optimal group formation. *PNAS* **103** (15), 5835-5840.
19. Rivault, C. & Cloarec, A. 1998. Cockroach aggregation: discrimination between strain odours in *Blattella germanica*. *Anim. Behav.* **55**, 177-184.
20. Saffre, S., Furey, R., Krafft, B. & Deneubourg, J.L. 1999. Collective decision-making in social spiders: dragline-mediated amplification process acts as a recruitment mechanism. *J. Theor. Biol.* **198**, 507-517.
21. Detrain, C. & Deneubourg, J.L. 2002. Complexity of environment and parsimony of decision rules in insect societies. *Biol. Bull.* **202**, 268-274.
22. Depickère, S., Fresneau, D. & Deneubourg, J.L. 2004. A Basis for Spatial and Social Patterns in Ant Species: Dynamics and Mechanisms of Aggregation. *J. Insect Behav.* **17**, 81-97.
23. Jeanson, R., Deneubourg, J.L. & Theraulaz, G. 2004. Discrete dragline attachment induces aggregation in spiderlings of a solitary species. *Anim. Behav.* **67**, 531-537.
24. Sumpter, D.J.T. & Pratt, S.C. 2003. A modelling framework for understanding social insect foraging. *Behav. Ecol. Sociobiol.* **30**, 109-123.

Economic Optimisation in Honeybees: Adaptive Behaviour of a Superorganism

Ronald Thenius, Thomas Schmickl, and Karl Crailsheim

Department for Zoology, Karl-Franzens University
Graz,Universitätsplatz 2, A-8010 Graz, Austria
theniusr@stud.uni-graz.at, schmickl@nextra.at,
karl.crailsheim@uni-graz.at

Abstract. A honeybee colony has to work highly efficient to survive. Most of a honeybee's energetic demands are satisfied by consuming carbohydrates which are collected by forager bees in form of nectar from flowering plants. The storage of this nectar is performed by another specialised group of bees. The size of the two workgroups (foragers and receivers) are precisely regulated by dances performed by forager bees, a process that represents adaptive behaviour of a superorganism. We implemented these mechanisms in a simulation of a honeybee colony to investigate the possible advantages of bigger colonies in nectar foraging.

1 Introduction

1.1 Biological Background

Honeybees (*Apis mellifera L.*) are eusocial insects that live in large colonies of up to tens of thousands of individuals. A honeybee colony is a self-organizing system without a central regulatory unit. The individuals show age-polyethism, division of labour and task partitioning [1]. A cohort of bees which are specialized on the same task is called a "temporal caste".

Forager bees fly out of the hive to collect nectar, pollen, water or resins in the surrounding environment. The forager caste is subdivided into two groups: scouts, which explore the environment for new food sources, and recruits, which exploit "already known" food sources. Nectar receiver bees wait near the hive's entrance to accept nectar from the returning forager bees, which transport the nectar in their crops. This nectar is then processed and stored in the upper parts of the colony.

1.1.1 Selection of Food Sources
Usually, a colony has to choose among several food sources of varying qualities. The quality of a source can be described by the following equation:

$$qi = \frac{g - c}{c} \tag{1}$$

whereby qi is the quality-index of the found source; g is the energetic gain of the collected nectar, measured in Joules [2]. The gain depends on the sugar concentration

S. Nolfi et al. (Eds.): SAB 2006, LNAI 4095, pp. 725–737, 2006.
© Springer-Verlag Berlin Heidelberg 2006

of the found source (mol/l)and on the collected volume (µl); c represents the costs of the foraging trip, measured in Joules. The costs depend on the weight of the bee [mg], on the distance to fly (m) and on the speed of flight.

Foragers returning from a successful foraging trip perform "waggle-dances" near the hive entrance, an area that is called "dance-floor". These dances communicate the direction, the distance and the type of the source to several dance-following bees (recruits). The duration of a "waggle-dance", as well as the probability to fly to this source again is modulated by the quality of the source [1].The longer a dance lasts, the higher is the probability, that one or more recruits follow the dance, and are recruited to the source themselves. In this way a self-organizing system for optimal source exploitation is accomplished.

The relation between dance duration and source quality [2] can be shown in a "dance-response-curve" (Fig. 1). These dance-response-curves vary from individual to individual. Foragers with a big slope of the dance-response-curve dance very intense for a high quality source, and very little or not for a bad quality source, versus foragers with a small slope of the dance-response-curve, which dance little for sources of all qualities.

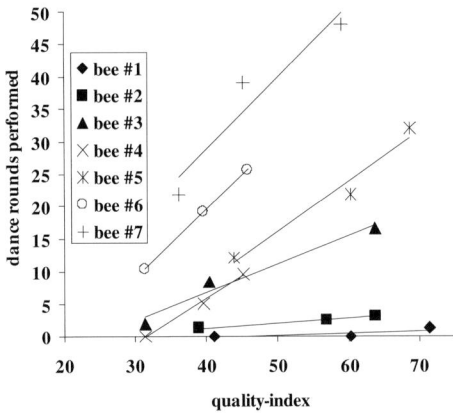

Fig. 1. Correlation between the quality of a nectar source and the dance rounds performed by honeybees. Redrawn from [2].

1.1.2   Regulation of the Size of the Worker Cohorts

To allow the colony to work efficiently, it is crucial to regulate the size of the worker cohorts of foragers and receivers. Honeybees have evolved an astonishingly simple and robust way to perform this regulation: if the search for a receiver bee takes too long after a successful foraging flight, the forager bee performs a "tremble-dance", which recruits additional receiver bees (Fig. 2). If the forager bee was able to transfer its nectar load quickly, it performs a "waggle dance" to recruit additional forager bees. If the size of the two working cohorts are balanced, moderate waiting times for foragers are most likely; after a waiting time of about 50 seconds, the forager performs no dance at all, thus the sizes of the two cohorts stay unchanged. If returning forager bee has to search for a receiver bee more than about 50 seconds, it performs a "tremble dance" to recruit additional receiver bees.

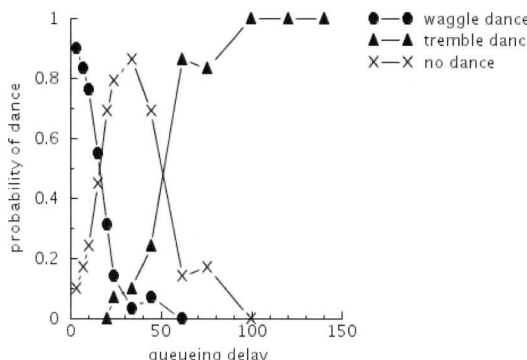

Fig. 2. Correlation between the queuing delay experienced by a forager while waiting for a receiver bee and the probabilities to perform waggle or tremble dances. Redrawn from [2].

In this way an auto-balanced, self-organizing system for optimal source exploittation is accomplished:

1   The nectar influx is restricted to the current receiving workforce.
2   The receiving workforce is regulated according to the current influx.
3   The nectar sources of highest profitability are exploited to the greatest extent.

1.2 Motivation

The aim of this study is to investigate the influence of colony size on the adaptive regulation of the foraging-process and workload-balancing mechanisms, as they can be found in honeybee colonies. Why do Honeybees under natural conditions live only in colonies bigger than about 1000 individuals? Is there a correlation between honeybee behaviour, decision making and colony size?

2 Material and Methods

2.1 The Simulation Platform

We extended our discrete-time multi-agent simulation platform of a honeybee colony "honeybee foraging simulation" (HoFoSim,[3][4][5]). In HoFoSim, only foragers were implemented as individual agents, receivers were implemented as a global property of the system. The new model "honeybee forager and receiver simulation" (HoFoReSim) adds receiver bees to the system, that are implemented as agents, with their own metabolism.

The advantages of the HoFoReSim simulation platform are:

•   It allows the high adaptivity to a variety of questions.
•   It enables us to study the interactions between the forager and receiver bees
•   It includes the nature-like individuality of honeybees regarding e.g., crop size, weight of bees, and dance-response curves.
•   The detailed simulation of the forager bee and receiver bee metabolism are implemented.

We chose Netlogo 3.0 as programming environment [6].

2.1.1 Implementation of Honeybees

In our model, each honeybee in the colony is simulated by a single agent, which is driven by a finite-state machine consisting of a variety of behavioural states[1] and transitions of these states. The whole collective of all agents is further on called "colony". Each agent is implemented with a metabolism. The metabolic rate changes corresponding to the behavioural state of the agent. Furthermore the metabolic rate is influenced by the actual weight of the agent, consisting of its body weight and the nectar load. An agents' metabolism consumes nectar as its energy-source. The simulation of metabolic activity together with the nectar income allows us to measure the economic success of a colony. To test the reliability of the simulation, we successfully compared simulated experiments to published empirical data (e.g., [7]).

In our model, we turned special attention on the following behavioural aspects of honeybees:

- the duration of different activities in and outside the hive,
- the probability of changing from one behavioural state to another,
- the dancing behaviour of forager-bees,
- the exact simulation of biological relevant attributes, such as weight, stomach-size etc.,
- the metabolic rates of the different behaviours.

2.1.2 The Simulation Environment

As shown in figure 3 the simulation environment is divided into two areas: the hive, where forager-receiver-interaction, dances and other colony internal activities take place, and the outside environment with nectar feeders placed in it. Each patch of the hive area represents a comb area of 0.75cm x 0.75cm, a patch outside the hive

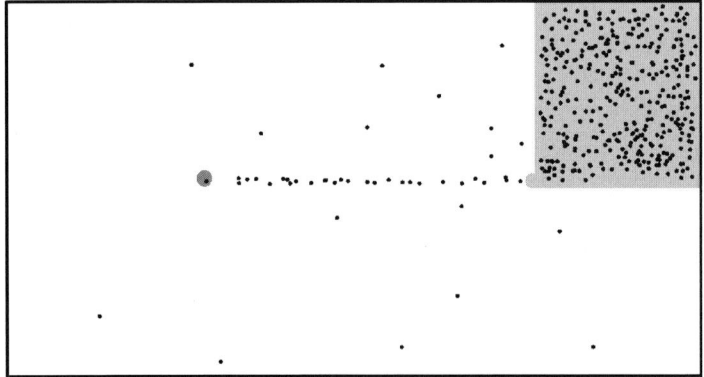

Fig. 3. Screen-shot of the simulation. **Gray area on the right**: Hive, where interactions between foragers and receivers, and the dances take place **Small dots outside the hive**: Bees flying in the environment. **Big dot outside the hive:** Nectar feeder, on which bees upload nectar.

---

[1] e.g., "foraging", "dancing", "storing", "receiving", etc.

represents an area of 2.7m x 2.7m . Outside the hive several nectar feeders can be placed with up to 600 meters distance form the hive and have variable nectar concentrations. For the current work we used only one nectar feeder of constant nectar concentration.

2.1.3 The Behavioural States and Transitions of Forager Bees

The finite state automaton simulating the forager bees includes several behavioural states, differing in the corresponding agent behaviour and metabolic rate. Table 1 shows the most important behavioural states of forager bees together with the associated activity level. We implemented two activity levels:

- "low" for all running and walking activities of bees (inside of the hive), and
- "high" for flying activities of the bees (foraging flights to and from the food source and random scouting flights).

Table 1. States of a forager bee and the corresponding metabolic rate

behavioural state	behaviour	activity level
in hive w/o info	random walk in hive	low
dance following	random walk in hive, sensitive for waggle dances	low
in hive with info	random walk in hive	low
foraging	direct flight to the source	high
returning	direct flight home	high
searching for receiver	random walk in the hive, sensitive for receiver bees	low
unloading to receiver	passing nectar to the receiver bee	low
dancing	random walk in the hive, offering information to nearby bees in state "dance following"	low
scouting	random flight	high

Transitions from one behavioural state of forager bees to another (Table 2) are triggered by one of the following mechanisms:

- by fixed time delays (T), meaning a fixed number of time steps after with a agent switches from one behavioural state into another[2],
- by a fixed probability (P), meaning a fixed probability per time step, with which an agent switches from one behavioural state into another[3],
- by an internal trigger (I), meaning an internal sensor changing the behavioural state of an agent[4], or
- by an external triggers (E), meaning an agent-agent or an agent-environment interaction changing the behavioural state of an agent[5].

---

[2] e.g., leaving the hive after a dance.
[3] e.g., become a dance follower.
[4] e.g., return to colony when stomach is full or mostly empty.
[5] e.g., entering the colony.

Table 2. Behavioural transitions of forager bees (for abbreviation see text above)

to \ from	in hive w/o info	dance following	in hive with info	foraging	return with info	return w/o info	search for receiver	unload to receiver	dancing	scouting
in hive w/o info	-	-	P	-	-	E	-	I	-	-
dance following	P	-	-	-	-	-	-	-	-	-
in hive with info	-	-	-	-	-	-	-	-	-	-
foraging	-	E	T	-	-	-	-	I	T	-
return with info	-	-	-	T	-	-	-	-	-	E
return w/o info	-	-	-	I	-	-	-	-	-	I
search for receiver	-	-	-	-	E	-	-	-	-	-
unload to receiver	-	-	-	-	-	-	E	-	-	-
dancing	-	-	T	-	-	-	-	I	-	-
scouting	P	-	-	-	-	-	-	-	-	-

2.1.4 The Behavioural States and Transitions of Receiver Bees

The finite state automaton of the receiver bees (table 3) is very much similar to the state automaton of the forager bees (see 2.1.2). All behavioural states of receiver bees correspond to the "low" activity level, because these bees never leave the hive in our model.

Table 3. The most important behavioural states of receiver bees and the associated activity level

behavioural state	behaviour	activity level
waiting	random walk in hive, sensitive for nearby foragers in state "searching for receiver"	low
load from forager	getting nectar from a forager in state "unload to receiver"	low
storing	random walk and search deposition place	low
idle	random walk, insensitive for any foragers	low

Similar to the behavioural transitions of forager bees (2.1.2) the behavioural transitions of receiver (Table 4) bees can be triggered by fixed time delays (T), by a fixed probability (P), by an internal trigger (I),or by an external trigger (E).

Table 4. Behavioural transitions of receiver bees (for abbreviation see text above)

to \ from	wait for a forager	load from a forager	storing	idle
wait for a forager	-	E	T	E
load from a forager	I	-	-	-
storing	-	I	-	-
idle	P	-	-	-

Receiver bees accept nectar from returning forager bees. When a receiver bee´s crop is full, the receiver bee starts to store the nectar in a honey-store, what is represented by the state "storing", while which the receiver bee is not available for any other bee-bee interaction. Depending on the colony's nectar need the time a receiver bee spends with storing nectar varies from 10 minutes to 28 minutes [2]. The bigger the colony's honey reserves are, the smaller is the available space to store nectar, and the longer a storer bee has to search for appropriate space.

2.1.5 Activity Levels and Metabolic Rates

The metabolic rate of a honeybee depends on its actual weight and the activity level she is in. In our model, the empty weight WE of a honeybee is 83.32 ± 6.89 [8]. The maximum crop volume is 45.9 ± 8.7 [9]. The weight WL of the actual nectar load L of the sugar-concentration LC is calculated by the equation 2 (based on own measurements, data not shown):

$$WL = L * (0.065 * LC + 1.0026). \tag{2}$$

Due to this, the actual weight of a bee WA is calculated by the equation 3:

$$WA = WE + WL. \tag{3}$$

Based on [2] the metabolic rate of a bee MR is calculated by the equation 4 and 5, depending which activity level the bee is in:

High activity level:

$$MR = (0.00287 * (WA^{0.629})(2.827 * LC) \tag{4}$$

Low activity level:

$$MR = (0.00248 * (WA^{0.492})) / (2.827 * LC) \tag{5}$$

2.2 The Simulation Settings

We performed three kinds of experiments: The first experiment addresses the question how a honeybee colony reacts to different sets of start configurations regarding the relation of foragers to receivers in the colony. The second experiment addresses the question how effective colonies work with these different initial sizes of worker cohorts. Finally the third experiment addresses the question how colonies of different size can react to changes in the environment.

In all experiments one single feeder was placed in the simulation environment. This feeder was placed in a distance of 340 m (=127 patches) to the colony and offered a sugar solution with a concentration of 2.5 mol/l. This distance of the feeder to the colony leads to a foraging cycle length of about 2.8 minutes

2.2.1 Experiment 1: Reactions of a Colony to Different Start Configurations

To test the reactions of a colony to different start configurations, we performed several simulation runs with colonies of different forager to receiver ratios (F:R). This ratio we varied between 1:4 and 4:1. The total number of foragers and receivers in the colony was kept constant with 1000 bees. The number of recruited foragers and receivers after 3 hour of simulation time was recorded.

2.2.2 Experiment 2: Influence of Different Colony Size on the Mean Queuing Delay Inside the Colony

To test the economics of colonies of different sizes, we performed several simulation runs with colonies from 40 bees in the colony up to 1600 bees. The forager to receiver starting ratio was kept constant at 1:1. The mean queuing delays of foragers and receivers, and the net nectar gain of the colony was measured.

2.2.3 Experiment 3: Reactions to Changes in the Environment

To investigate the ability of a colony to adapt to sudden changes in the environment, we spontaneously doubled the length of the foraging flight cycle as proposed by [10], and switched it back to initial level after 2 hours. We measured the number of forager bees recruited to the source. For this experiments the colonies started with an optimal workload balance between foragers and receivers. If a colony has a suboptimal workload balance, effects of changes in flight cycle length may be masked by regulation mechanisms of the colony. Due to this, changes in the flight cycle length could take maximal effect on the number of recruited forager bees under conditions of optimal workload balance.

3 Results

3.1 Experiment 1: Reactions of a Colony to Different Start Configurations

We found, that the starting condition of a colony has small influence on the ratios of recruited foragers to receivers after 3 hours. Colonies with a forager to receiver ratio

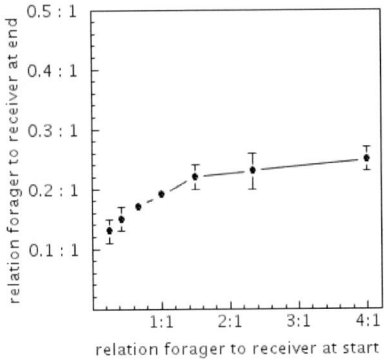

Fig. 4. Relation of recruited foragers to recruited receivers after 3 hours in colonies with different start configurations, respective the relation of foragers to receivers in the colony. Dots show mean values; Error bars indicate standard deviations.

of F:R = 1:4 ended up at a recruited forager to recruited receiver ration of 1:0.13 ± 0.02. Colonies with F:R = 4:1 ended up at a recruited forager to recruited receiver ration of 1:0.25 ± 0.02 (Fig.4).

3.2 Experiment 2: Influence of Different Colony Size on the Mean Queuing Delay of Foragers and Receivers Before Unloading

The mean queuing delay of foragers waiting for a receiver to unload the nectar is negatively correlated with the number of bees in the colony (Fig. 5A). Also the mean queuing delay of receivers waiting for a forager decreases with an increase of bees in the colony (Fig. 6A). Because the forager bees draw information about the workload balance of the colony from the experienced queuing delays while waiting for a

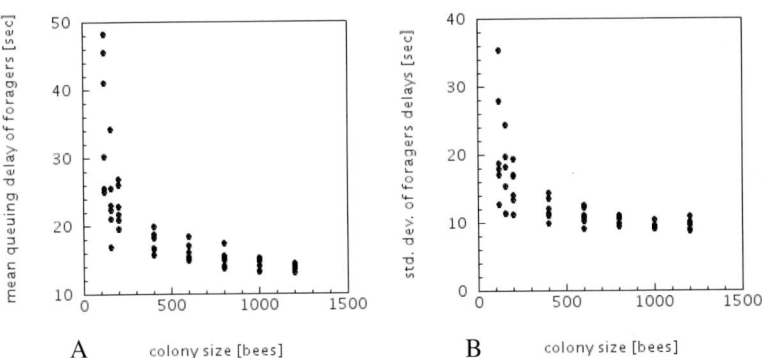

A   B

Fig. 5. Mean values and standard deviations of queuing delays experienced by forager bees. Mean values and standard deviations decrease with an increase of the colony size. Only results of colonies of 120 or more bees are displayed for depicting reasons.

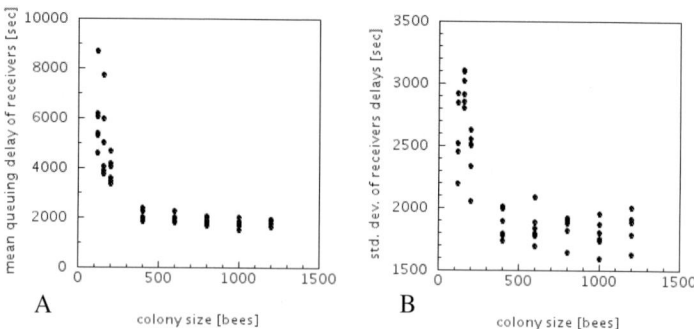

A   B

Fig. 6. Mean values and standard deviations of queuing delays experienced by receiver bees. Mean values and standard deviations decrease with an increase of the colony size. Only results of colonies of 120 or more bees are displayed for depicting reasons. Queuing delays of receivers are much longer than the queuing delays of forager bees (Fig. 5), for explanation see discussion.

receiver (Fig. 2), we also measured the standard deviation of queuing delays, and found that these standard deviations are also decreasing with an increase of bees in the colony, for both, foragers (Fig. 5B) and receivers (Fig. 6B).

3.2.1 Economy of Honeybee Colonies of Different Size

We measured the net nectar gain of the colony by the volume of nectar gained per bee per hour. In our simulation small colonies (< 400 bees)[6] worked very inefficient (Fig. 7). The efficiency of the colonies increased with their size, reaching a plateau at a colony size of about 400 bees. Bigger colonies did not gain more nectar per bee than about 7μl / bee / hour.

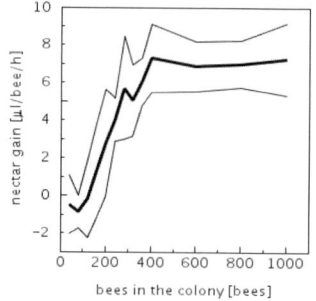

Fig. 7. Net nectar gain of simulated honeybee colonies of different size. The net nectar nectar gain increases with the size of the colony, reaching a plateau at a colony size of about 400 bees (Bold Line: Mean values; Thin lines indicate standard deviations; n = 6).

---

[6] Real world honeybee colony consist of much more than the two modelled temporal casts, for explanation see discussion.

3.3   Reactions to Changes in the Environment

We found, that bigger colonies (Fig. 8B) react to a doubling of the foraging flight cycle length with a stronger and more precise recruitment of forager bees than smaller colonies (Fig. 8A). Colonies with 200 bees (Fig. 8A) show very little reaction, during the phase of elongated foraging flight cycle, which lasted from hour 2 to hour 4 of the experiment. Colonies with higher numbers of bees (1000 bees, Fig. 8B) show a strong recruitment of forager bees within the first hour of the phase of elongated foraging flight cycle.

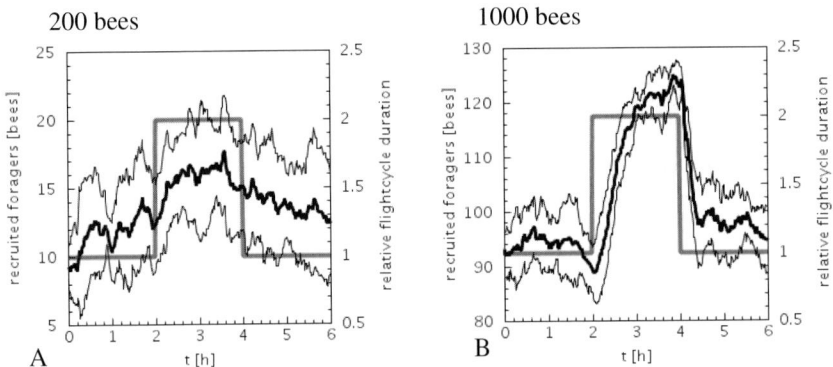

Fig. 8. A: Number of recruited forager bees in a colony of 200 bees. During the phase of elongated foraging flight cycle (hour 2-4) there is hardly any reaction visible in the number of recruited forager bees. **B:** Number of recruited forager bees in a colony of 1000 bees. During the phase of elongated foraging flight cycle the colonies react with massive forager recruitment. When the foraging cycle length drops back to initial level, the number of recruited foragers drops also back to normal level. Bold lines: Mean values (left y-axis); Thin lines indicate standard deviations; Gray line: relative flight cycle duration (right y-axis); n = 6.

4   Discussion

Our studies show, that our model of honeybee colony is very stable concerning the workload balancing between foragers and receivers, mostly independent of the configuration of the colony (Fig. 4). A mismatch between the forager and the receiver cohort is compensated by the waggle-tremble-dance recruitment system, mentioned in chapter 1.1. The results of our studies differ significantly from the work of [10], which show a much clearer correlation between the forager-to-receiver ratio and the queuing delays, with minimum queuing delays at a ratio of foragers to receivers of 1:1 (under conditions of equal length of foraging cycle and storing cycle). This differences are based on the much more detailed simulation used for our studies. Our simulation includes the majority of the known recruitment and workload-balancing mechanisms of a honeybee colony, like dances, or the possibility of a honeybee to be "idle", what leads to a reservoir of workforce. Our results, concerning the stability of a colony against internal mismatches, brought up the question, how different colony sizes affect the colony's efficiency. We found, that the waiting times experienced by the foragers reflect the colony situation less precisely in small colonies, than in bigger ones (Fig.5, 6). Due to the importance of waiting times of forager bees in the regulatory system of the colony (chapter 1.1), we found that bigger colonies have much higher quality of the information (less noise in the communication channel) about the status of the workload balance of the colony available. This also affects the efficiency of the colony, as is expressed by the foraging success (Fig. 7) and the ability of the colony to react to changes in the environment (Fig. 8). The much longer queuing delays of receiver bees are based on the low probability of the receiver bees to switch from state "wait for a forager" to "idle", which is necessary to keep receivers waiting while long periods with no returning foragers.

In our model colonies of less than 400 bees (foragers and receivers) are not able to work at maximum efficiency. Compared to the situations in a real world honeybee colony, we estimate, that a colony with less than about 1000 bees can not work optimal. Please keep in mind, that a real world honeybee colony consist of much more than the two modelled temporal casts e.g., nurses, wax builders, cleaners, guards. Honeybees in temperate climatic zones also have to spend big amounts of energy on temperature regulation of the colony, thus we maybe still underestimate the necessary minimum colony size for optimal foraging in real world honeybees. Nevertheless the simulation gives us a deeper insight into the adaptiveness of bigger amounts of individuals in a honeybee colony, due to single honeybee behaviour.

Our simulation study showed the importance of incorporating all relevant feedback of the focal system into the model. Our model is the first published model that incorporates all important feedback involved in the honeybee foraging system in a bottom-up individual-based approach. The fact that the model (HoFoReSim) models foragers and receivers, the metabolisms of these bees and the relevant behaviours of these bees (dances, flights,...) significantly increases the reliability of our results. Based on our model we can successfully investigate the implications of individual behaviours on the colony's efficiency. This way, the model allows us to investigate the adaptive behaviour of a superorganism, as was demonstrated in this article.

Acknowledgements

The writing of this article was supported by the "Fonds zur Förderung der Wissenschaftlichen Forschung (FWF)", project no. P15961-B06 and by the EU IST-FET-open project (IP) 'I-Swarm', no. 507006. Simulation experiments presented here were supported by the "University Graz / BMBWK Infrastructure Investment Program, Projekt-Nummer TUGP1" (Hard- and Software)".

References

1. Frisch, K. v. Tanzsprache und Orientierung der Bienen. Springer Verlag, Berlin, Heidelberg, New York, (1965)
2. Seeley, T. A. The Wisdom of the Hive. Havard University Press, Cambridge, Massachusetts, London, England (1995), ISBN 0674953762.
3. Schmickl, T. and Crailsheim, K. Cost of environmental fluctuations an benefits of dynamic foraging decisions in honey bees. Adapitve Behavior 12, (2004), 263-277

4. Thenius, R., Schmickl, T. and Crailsheim K. The "Dance or Work" problem: Why do not all honeybees dance with maximum intensity. Lecture Notes in Artificial Intelligence 3690, (2005), 246–255.
5. Schmickl, T., Thenius, R., and Crailsheim, K. Simulating swarm intelligence in honey bees: foraging in differently fluctuating environments. Proceedings of the 2005 conference on Genetic and evolutionary computation, (2005), 273 - 274, ISBN 1-59593-010-8
6. Wilensky, U. 1999. NetLogo. http://ccl.northwestern.edu/netlogo/. Center for Connected Learning and Computer-Based Modeling, Northwestern University. Evanston, IL.
7. Seeley, T. A., Camazine, S., and Sneyd, J. Collective decision-making in honey bees: how colonies choose among nectar sources. Behavioral Ecology and Sociobiology 28, (1991), 277–290.
8. Hrassnigg, N. and Crailsheim, K. Differences in drone and worker physiology in honeybees (Apis mellifera). Apidologie 36, (2005), 255-277.
9. Huang, M. and Seeley, T. D. Multiple unloadings by nectar foragers in honey bees: a matter of information improvement or crop fullness? Insectes Sociaux 50, (2003), 1–10.
10. Anderson, C. and Ratnieks F. Task Partitioning in Insect Societies.: I. Effect of colony size on queueing delay and colony ergonomic efficiency. The American Naturalist, 154, (1999), 521-535.

Cumulative Cultural Evolution: Can We Ever Learn More?[*]

Paul Vogt[1,2]

[1] Induction of Linguistic Knowledge / Computational Linguistics section
Tilburg University, P.O. Box 90153, 5000 LE Tilburg, The Netherlands
[2] Language Evolution and Computation Research Unit, School of Philosophy,
Psychology and Language Sciences, University of Edinburgh, U.K
paulv@ling.ed.ac.uk

Abstract. This paper investigates the dynamics of cumulative cultural evolution in a simulation concerning the evolution of language. This simulation integrates the iterated learning model with the Talking Heads experiment in which a population of agents evolves a language to communicate geometrical coloured objects by playing guessing games and transmitting the language from one generation to the next. The results show that cumulative cultural evolution is possible if the language becomes highly regular, which only happens if the language is transmitted from generation to generation.

1 Introduction

Our knowledge seems to be ever more increasing, our social networks are getting more complex, technology is advancing all the time, etc. In short: many aspects of our culture seem to be getting ever more complex. This paper investigates some dynamics of this *cumulative cultural evolution* by exploring a recent model that simulates the cultural evolution of language.

Cultural evolution is often characterised in terms similar to Darwin's [1] evolution theory, see, e.g., [2,3]. Simply put, this means that culture evolves based on the principles of variation, competition and selection. Darwinian explanations have been applied to various aspects of cultural evolution, such as linguistics [4].

Boyd and Richerson [2] have argued that social learning is favoured if it is less costly or more accurate than individual learning of such skills or knowledge. Furthermore, if the cost decreases or accuracy increases for learning skills socially over generations, such skills may improve accumulatively over time. Translating this to language evolution, this view supports a gradual evolution of complexity in languages, see, e.g., [5]. Recent computational studies on the cultural evolution of language have shown how languages may become increasingly complex in terms of, e.g., grammatical structures when it is transmitted from one generation to the next [6,7]. One typical change that seems to have occurred in these models

---

[*] This study has been supported by an EC Marie Curie fellowship and a VENI grant funded by the Netherlands Organisation for Scientific Research (NWO).

S. Nolfi et al. (Eds.): SAB 2006, LNAI 4095, pp. 738–749, 2006.

is that languages change in order to become more learnable, as hypothesised in, e.g., [8]. If languages are learnt easier (or faster), this both decreases the learning costs and increases learning accuracy.

Using the model introduced in [7], this paper illustrates how the iterated learning of compositional structures from initially holistic languages can lead to a cumulative cultural evolution until an optimal (cognitive) platform is reached. Although this model has been used previously to study various aspects regarding the evolution of compositionality (e.g., [7,9,10]), this study is unique in the way the dynamics of the evolution is analysed. This is done using different parameter settings and measures highlighting how this model can contribute to our understanding of how cumulative cultural evolution can work. The next section will present the model. Section 3 will present and discuss experimental results, which show that indeed learning cost decreases and learning accuracy increases. Conclusions are provided in Section 4.

2   Iterated Learning and Language Games

As the model of this paper simulates a transition from holistic languages to compositional languages, it is good to start with a definition of such languages:

Holistic languages are languages in which parts of expressions have no functional relation to any parts of their meanings. For instance, there is no part of the expression "bought the farm" that relates to any part of its meaning has died.

Compositional languages are languages in which parts of expressions do have a functional relation to parts of their meanings and the way they are combined. For instance, the part "John" in "John loves Mary" refers to a guy named John, likewise "loves" and "Mary" have their own distinctive meanings. In addition, this sentence has a different meaning in English when the word-order changes, as in "Mary loves John".

Following Wray's [5] hypothesis that modern languages were preceded by holistic protolanguages, Kirby and colleagues have shown how an initially holistic protolanguage can transform into a compositional language if the language is iteratively transmitted from one generation to the next through a *bottleneck* (which means that the next generation only observes a small subset of the language from the previous generation) [6]. This *bottleneck effect* is understood by realising that holistic languages are unstable over time when the next generation only observes a small subset, while compositional languages are stable (see Fig. 1). This, of course, is only possible if the learners in this model can extract compositional structures from their (possibly holistic) input.

The basic principle of this *iterated learning model* (ILM) [6] is that the population consists of two groups: adults and children. The children acquire the language by observing the adults' speech directed to them. At the end of an iteration, during which the children have observed a part of the language, all adults are removed, all children become adults and new children are added to

Type	$G(n)$	Utter.	$G(n+1)$
Hol.	ab-00	ab-00	ab-00
	cd-01	cd-01	cd-01
	fg-10	fg-10	fg-10
	fb-11		??-11
Comp.	ab-00	ab-00	ab-00
	ac-01	ac-01	ac-01
	db-10	db-10	db-10
	dc-11		dc-11

Fig. 1. This figure illustrates why holistic languages (upper part) are unstable when a population of generation $G(n+1)$ only observes three of the four utterances (Utt.) from generation $G(n)$'s language (i.e. word-meaning mappings). In this case, if generation $G(n+1)$ wishes to communicate about meaning 11 (meanings in this example are to be read as bit-strings), then this generation will have to create a new word. If the language is compositionally structured, as in the bottom part of this figure, observing only three out of four instances would allow the next generation to reconstruct the entire previous language. Hence transmitting a compositional language through a bottleneck is more stable than transmitting holistic languages.

the population. This process is then repeated, such that language is transmitted over subsequent generations. In earlier ILMs (e.g., [6]), this transmission was entirely *vertical*, because adults only directed their speech to children, who only listened until they were adults themselves. With such a condition, the only way to achieve a bottleneck is by the experimenter setting a parameter regulating that children only see a fragment of the language of a given size.

It has been shown that when communication within one iteration is *isotropic*, which means that communication goes in all directions from adult to adult, adult to child, child to adult and child to child, then compositionality can evolve as a stable system without the need for the experimenter to impose a bottleneck [9].[1] This is because when children need to speak, they may face the consequences of the bottleneck if they need to speak about previously unseen meanings (cf. meaning 11 in Fig. 1). This *implicit bottleneck* is then a more natural bottleneck and may in part explain why in horizontal models of language evolution compositionality evolves, see, e.g., [11]. In this model, which is a simulation of the Talking Heads experiment [12], the population plays a large number of *guessing games* to develop a language that allows the population to communicate about their world, which contains a number of coloured geometrical shapes.

It is impossible to present all details of the model in this paper; the interested reader is referred to [7,9]. The guessing games are played by two agents: a speaker and a hearer. Both agents are presented a small number of objects randomly sampled from the world. These objects constitute the *context* of the game; the world contains a total of 120 objects (12 colours combined with 10 shapes). Each agent individually categorises the perceptual features of each object using

[1] In [9] this type of transmission was called *horizontal transmission*, but since in [3] this refers to transmission only within one generation, the term isotropic is preferred.

1	S → greensquare/(0,1,0,1)	0.2
2	S → A/rgb B/s	0.8
3	A → red/(1,0,0,?)	0.6
4	B → triangle/(?,?,?,0)	0.7

Fig. 2. This example grammar contains rules that rewrite a non-terminal into an expression-meaning pair (1, 3 and 4) or into a compositional rule that combines different non-terminals (2). The meanings are 4-dimensional vectors, where the first 3 dimensions relate to the RGB colour space (rgb) and the 4th relate to the shape feature (s). The question marks are wild-cards. Each rule has a rule score that indicates its effectiveness in past guessing games. Only sentences of 2 constituents are allowed in this grammar.

a method based on the discrimination game, whose details are irrelevant to the scope of this paper. Suffices to say that each object is categorised such that it is distinctive from all other objects in the context. If distinctive categorisation fails, a new category is constructed for which the object's perceptual features serve as exemplars. (Note that initially, each agent has no categories at all; these are all constructed by these discrimination games.) Categories are represented as prototypical points in a 4-dimensional space, each dimension relating to a perceptual feature, which are the red, green and blue components of the RGB colour space and a shape feature.

Once the agents have categorised the objects in the context, the speaker selects one object at random as the *topic* of the communication. This agent then searches its grammar for ways to encode an expression that conveys the topic's meaning. The grammar (Fig. 2) consists of simple rewrite rules that associate forms with meanings either holistically (e.g., rule 1) or compositionally (e.g., rule 2 combined with rules 3 and 4). The grammar may be redundant in that there may be rules that compete to encode or decode an expression (cf. [11,13]). The speaker searches for those (compositions of) rules that match the topic's meaning and if more than one are found, he selects the rule that has the highest rule score. If the speaker fails to encode an expression this way, a new form is invented as an arbitrary string and is associated with the topic's meaning or – if a part of the meaning matches some non-terminal rule – with the rest of this meaning.

In turn, the hearer tries to decode the expression by searching her own grammar for (compositions of) rules that match both the expression and a category relating to an object in the current context. If there are more such rules, the hearer selects the one with the highest score, thus guessing the object intended by the speaker. The hearer then points to this object, and if this is the object intended by the speaker, the speaker acknowledges success; otherwise, the speaker points to the topic allowing the hearer to acquire the correct meaning.

If the guessing game was successful, both the speaker and hearer increase the scores of the rules they used and lower the scores of those rules that compete with the used rules. If the game has failed, the scores of used rules are lowered and the hearer acquires the proper association between the heard expression and the topic's meaning. To this aim, the hearer tries the following three steps until one step has succeeded:

1. If a part of the expression can be decoded with a part of the topic's meaning, the rest of the expression is associated with the rest of the meaning. For instance, if the hearer of the grammar shown in Fig. 2 hears the expression "redcircle" meaning (1,0,0,.5), the part "red"-(1,0,0,?) can be decoded, so the hearer adds rule B→circle/(?,?,?,.5) to its grammar.

2. If the above failed, the hearer searches its memory, where she stores all heard or produced expression-meaning pairs, to see if there are instances that are partly similar to the expression-meaning pair just heard. If some similarity can be found, the hearer will break-up the expression-meaning pairs containing the similarities – following certain heuristics, thus forming new compositional rules. Suppose, for instance, the hearer had previously heard the expression-meaning pair "greensquare"-(0,1,0,1), and now hears the expression-meaning pair "yellow square"-(1,1,0,1). The hearer can then break up these pairs based on the similarity "square"-(?,1,0,1), thus forming rules S→C/r D/gbs, C→green/(0,?,?,?), C→yellow/(1,?,?,?) and D→square-(?,1,0,1). Note that this is not the ideal break up, since it breaks apart the red component of the RGB colour space from the blue and green components and the shape feature. The next section shows that over time such mistakes diminish as a result of competition and selection.

3. If the above adaptations both fail, the heard expression-meaning pair is incorporated holistically, leading to a new rule such as S→yellowcircle/(1,1,0,.5).

At the end of these steps, the hearer performs a few post-processes to remove any multiple occurrences of rules and to update the grammar such that other parts of the internal language relates more consistently to the new knowledge. Full details of the model are found in [7,9].

3   Cultural Evolution of Language

In order to see how improvements accumulate culturally, a series of simulations were carried out in which the above model was run for 20 iterations (or generations) of 100,000 guessing games each. In these simulations the population size was set to 50 (25 adults and 25 children). Ten simulations were run with different random seeds. Although most different runs were quite similar, a few revealed some noticeable differences (see [10] for a discussion on these differences). Therefore, only the results of a few single runs are presented here.

Before presenting the results, a few measures are defined:

Communicative success measures the number of successful guessing games within a time window of 50 games.

Compositionality measures the number expressions decoded or encoded using a compositional rule over a time window of 50 games.

Similarity measures the number of games in which both agents used the same syntactic structure within a time window of 100 games. A syntactic structure is considered similar if the words and the linguistic categories used are the same and in the same order. (A linguistic category is characterised by the dimensions that make up the conceptual space of a non-terminal node.)

Learning period measures the number of guessing games it takes within each iteration for communicative success to exceed 0.8 for the first time.

All these measures (except learning period) are normalised to a value between 0 and 1. In addition, the relative frequencies of rule types used during successive periods of 10,000 guessing games are analysed. As the agents can break up the 4-dimensional conceptual space in two conceptual spaces (or linguistic categories) of lower dimension without knowing how, 15 different rule types (including the holistic type) can develop. Only 5 rule types are inspected in this paper (all other had insignificant frequencies). These are:

I: S→rgbs   holistic rule
II: S→A/r B/gbs   red v. green, blue & shape
III: S→B/gbs A/r   green, blue & shape v. red
IV: S→C/rgb D/s   colour v. shape
V: S→D/s C/rgb   shape v. colour

Figure 3 shows the results for the first 10 iterations of one simulation (the results did not change much during the final 10 iterations and are therefore not shown). The top left graph shows the evolution of communicative success, which rises to a more or less stable level of about 85-90% during the first 5-6 iterations. Each iteration is marked by a sharp decrease due to the drastic change of the population when all adults are replaced by the children and new language-less children are introduced. Compositionality (top right graph) increased more rapidly in initial stages, but also kept on rising during the first 5-6 iterations until a stable level around 93% was reached. Similarity (bottom left) evolved more similar to communicative success and reached a stable level of around 80%.

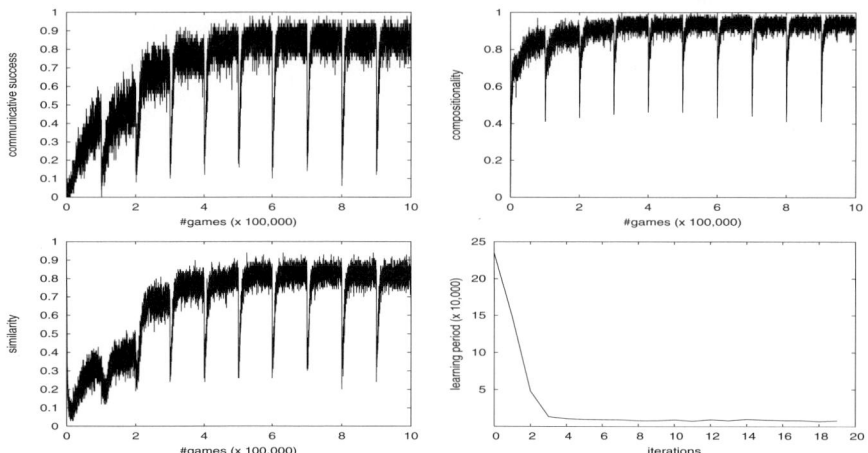

Fig. 3. The results of the first simulation. The graphs show communicative success (top left), compositionality (top right), similarity (bottom left) and learning period (bottom right). See the text for details.

All these three graphs clearly show that levels reached at the end of one generation are rapidly reached in the next generation, and – while there is still room for improvement – these levels further increase at a slower rate. It was expected that after a plateau was reached, learning speeds will continue to improve. This expectation was based on earlier visual inspection of similar graphs. However, closer analysis of the learning period reveals that this was not the case (Fig. 3 bottom right). Clearly, learning took quite a while in the first iteration, but rapidly decreased during the next 5 iterations and then remained more or less at the same level.

The graphs concerning communicative success and similarity clearly show some stagnation in development during the second iteration. Figure 4 (left) shows why this is the case: During the first two iterations there is a lot of competition between different rule types, as there are still a substantial number of holistic rules and rules of type II and III abound, before rules that differentiate colours from shapes start to dominate in iteration 3. Recall that agents break apart holistic rules when they find a similarity in the expression and the meaning. Previous analysis of the model has shown that the probability of finding in two different games a similarity in any conceptual space, other than the colour and shape spaces, is 1.5 to 4 times larger than the probability of finding a similarity in the colour or shape space [7]. Most likely, compositions of type II and III are found. However, although such combinations are more frequently encountered, they are less efficient than rules of type IV and V (combining colour and shape), because rules of type II and III require almost 4 times as many different rules than those of type IV and V. Moreover, a meaning that only takes the red component of the RGB colour space cannot be applied with all possible meanings that can be constructed in the other 3 dimensions. (Note that this is mainly due to skew distribution of features in the RGB space.)

Once rules of type IV and V start to dominate, the other rules' relative frequencies start to diminish rapidly. In this simulation, there is first an almost complete dominance of rule type V over rule type IV, but near the end of the third iteration, the relative frequency of type IV starts to become substantial too. Once these two rule types completely dominate (usage is then almost 100%), communicative success and similarity can further improve. Drops in the usage

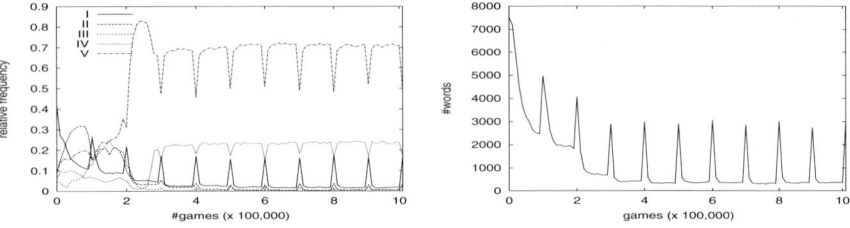

Fig. 4. The evolution of the relative frequency with which different rule types were used (left) and the number of different words used (right). Both were measured for successive periods of 10,000 guessing games.

of these rules coincide with the replacement of adults with new children, after which the new children first invent quite some new words holistically when they lack the linguistic knowledge to encode an expression for some meaning.

This tendency is also reflected in figure 4 (right) which shows the evolution of the number of different words used in the language during successive periods of 10,000 games. The evolution first shows an immense increase of different words: up to around 7,500 words are used in the first 10,000 games. Ignoring the rise in words after the replacements of half the population, the number of different words then decreases sharply until it stabilises around 200 words. As has been observed in [14] for simulations evolving non-compositional vocabularies, the decrease of words coincides with an increase in communicative success. Interestingly, it has been suggested that the large number of words created early in development (which to a lesser degree is also true shortly after the start of a new iteration) increases the chances of finding similarities and thus give a boost toward the development of compositional structures [10]. This, in addition to the fact that the number of words created is proportional to the population size, explained the observation that the level of compositionality was substantially larger for populations of sizes 40-100 than for populations of less than 40 agents [10].

Comparing the 10 different runs revealed that 7 out of 10 yielded results similar to the run discussed; the other 3 evolved similar to the one shown in Figure 5. This figure shows the results of another run with the same conditions, though plotted for the entire 20 iterations. Here the first 5 iterations are similar to the previous run, though communicative success reaches a lower level. Moreover, communicative success then slowly decreases, also slowing down the development

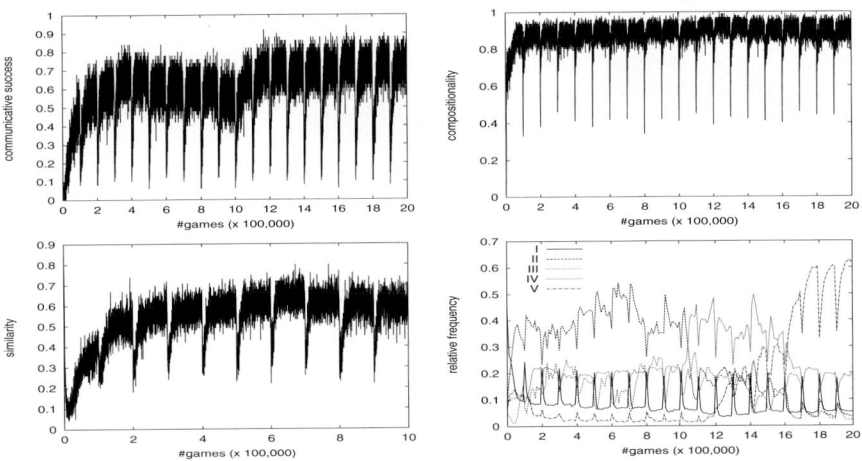

Fig. 5. The results of the second simulation. The figures show communicative success (top left), compositionality (top right), similarity (bottom left) and the relative frequency with which certain rule types were used during successive periods of 10,000 games (bottom right).

of similarity. From the 11th iteration, communicative success rises again slightly to a seemingly stable level around 0.7 reached in iteration 13. Similarity, however, suddenly decreases from the 11th iteration and only starts to rise again from iteration 15.

The difficulties arising during this evolution are apparently caused by the strong competition between different rule types. Type II evolves as the most dominant rule during the first 10 iterations, followed by a period of dominance of rule type IV, which is finally taken over by rule type V. During this final period, the competition with other rules seems to weaken. Interestingly, the transition from rule type II to IV coincides with the decrease in similarity. This may be explained by realising that for some time type II has established itself in the language, though in competition with other types. When another type takes over, this type is in strong competition with the previous type, so that different agents are likely to favour one type over another, meaning that in a given situation one agent may use type II while another uses type IV. This, thus, lowers similarity, even though communicative success has improved during this period. This is related to the fact that well structured rules of type IV are taking over. Given the decrease in competition with other rules near the end and the higher frequency of type V rules, it is expected that the results would have improved further if the simulation was run longer.

So far, the simulations were carried out for 20 iterations of 100,000 games each. However, what will happen if we look at one iteration that is run for 1 million games? Figure 6 shows the results of a simulation with that condition. In order to make an interesting comparison with the case where language is transmitted from one generation to the next, this figure also shows results of a simulation run for 2 iterations of 500,000 games each. Both simulations show a

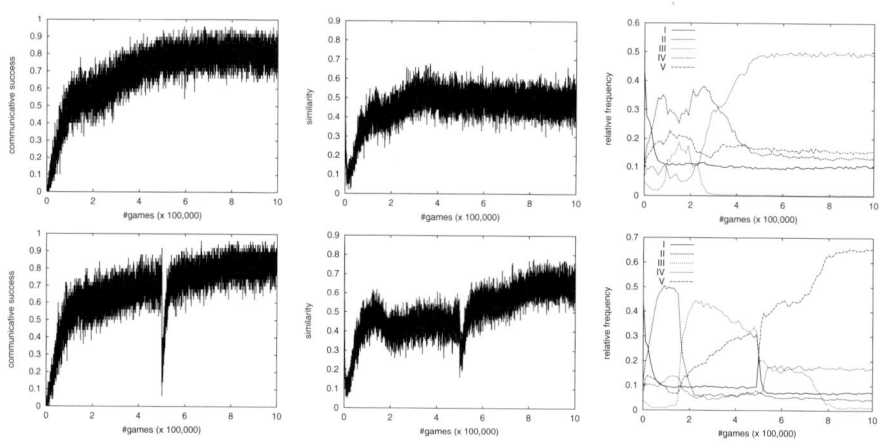

Fig. 6. The results of an experiment to compare two conditions of less iterations, each lasting for more guessing games. The top graphs show the communicative success (left), similarity (middle) and relative frequency of rules (right) for 1 iteration of 1 million games. The bottom graphs show the same measures for 2 iterations of 500,000 games.

very similar evolution of communicative success (apart from the discontinuity in the second case). Similarity, however, tells a very different story. In the 1 iteration case, similarity does not further increase, but instead decreases slowly after approximately 300,000 games. When there are 2 iterations, similarity tends to increase further in the second iteration.

The rightmost graphs in Figure 6 show that in the 1 iteration case, there is first a lot of competition leading to many fluctuations, but after about 500,000 games, the system tends to become stable and the population never seems to get out of this system that has 4 rule types being used quite frequently (note the different scale on the y-axis compared to the one shown in Fig. 4). Although the communication system is successful to some degree, different agents have acquired different grammars, thus reducing similarity. In the second condition, there is also a lot of competition and fluctuation during the first iteration. Now before the system stabilises, the population is changed and the new agents quickly learn the well structured parts of the language from their ancestors and further improve this language.

This comparison thus shows that if there would be no population turnover and all agents are equally old (which is the case for the 1st iteration of all simulations), there comes a point in which there is hardly any improvement to gain. This is most likely because at some point, each agent has acquired a system that works sufficiently well, which *is* the case, because communicative success is fairly high. Only when a new generation starting from afresh enters the population, there is room for improvement. The 'good' structure – the one that combines colour with shape – is relatively to learn. Each colour can be combined with any shape. Hence, each time it is used in previously unseen situations, this structure allows successful interaction. Successful interaction increases the rule score, thus allowing this type of rule to win the competition with other rule types. A rule type that combines the red component with the other dimensions (i.e. rule type II or III) does not have that property, because if, for example, you have a category whose red component has value 1 (i.e. redness), only $\frac{5}{12}$ of all categories in the complementary dimensions can be combined with this one.[2] So, the likelihood of applying this rule type successfully in unseen cases is lower than the odds for success when rule type IV or V is used.

4   Discussion and Conclusions

This paper investigates the dynamics of cumulative cultural evolution in language using a computer model that simulates the evolution of compositional structures in language. This model, first introduced in [7], integrates the iterated learning model [6] with the Talking Heads experiment [12]. The results show that initially holistic languages can evolve into compositional languages by transmitting the language from one generation to the next. Unlike results obtained with the ILM in which language is transmitted vertically [6], no transmission

---

[2] In the model 5 out of 12 colours have a red component of value 1.

bottleneck needs to be imposed when the language is transmitted isotropically, because then language learners face an implicit bottleneck [9].

The model is based on the idea that the evolution of language is a cultural analogy of Darwinian evolution [4]. In addition, the ontogenetic development of language is also similar to Darwinian evolution (cf. *Neural Darwinism* [15]) in that individuals acquire many variants of the language by communicating with different individuals and making errors in acquisition. These different variants then compete and are selected based on their effectiveness in communication. The same principles work at the cultural level, where the main elements subject to evolution are the word forms. So, there are two different levels of evolution interacting with each other (see [16] for an interesting discussion of such systems).

At the cultural level a set of conventions are formed that map expressions to objects in the agents' world. These expressions are segmented internally to form compositions with parts of the objects' meanings. These meanings are constrained by the visual input to the agents and the way these are processed in terms of features (in the model these features are similar for all agents). Initially, there are different (possibly ill-structured) rule types competing with each other. The simulations show that when a next generation learns from such a mixed language, they rapidly acquire the well-structured rule types and improve the structure of the rest of the language. The improvement primarily concerns the way meanings are segmented (i.e. the improvement from using rule type II or III to types IV or V). These improvements are triggered by a similarity in form, in addition to a similarity in meaning. Hence, there is a co-evolution of form (or syntax) and meaning (or semantics) in that on the one hand evolution of syntax is based on exploring similarities in form, though constrained by semantic structures and on the other hand evolution of semantics is based on exploring similarities in meaning, though constrained by syntactic structures [7].

As predicted by Boyd and Richerson [2], these simulations show a cumulative cultural evolution when the costs of social learning are low and accuracy is high. This can be concluded affirmatively because only when well structured compositional systems (i.e. those systems that have rules combining colours with shapes) develop, communicative success can further improve. Learning these well structured systems is less costly, because they are acquired more rapidly than unstructured ones, and they are also learnt more accurately, given the increase in similarity in language use of different individuals.

Prior to this study, it was anticipated – based on visual inspection of the data – that once a plateau of optimal communicative success was reached, learning speed will improve over subsequent generations. This, however, was not confirmed by this study. Once an optimal plateau was reached, learning speeds remained more or less constant over different generations. Explanations for this are that the agents have either reached a cognitive plateau on which no further improvements are possible, or because the evolved language reflects the structure of the environment so well that no pressures on further improvements are present.

Concluding, this study suggests that cumulative cultural evolution of language is possible if 1) the language is transmitted from one generation to the next, 2) the language evolves to well structured systems so it can be easily learnt by members of the population and 3) there is still room for improvement in the cognitive capacity of individuals or perhaps that there is a need for improvement because the structure does not reflect the environmental structure. These findings may well extend to other domains of cultural evolution.

References

1. Darwin, C.: The Origin of Species. John Murray, London (1959)
2. Boyd, R., Richerson, P.: The origin and evolution of cultures. Oxford University Press, Oxford (2005)
3. Cavalli-Sforza, L.L., Feldman, M.W.: Cultural Transmission and Evolution: A quantitative approach. Princeton University Press, Princeton, NJ (1981)
4. Croft, W.: Explaining Language Change: An evolutionary approach. Longman, Harlow (2000)
5. Wray, A.: Protolanguage as a holistic system for social interaction. Language and Communication **18** (1998) 47–67
6. Kirby, S., Smith, K., Brighton, H.: From UG to universals: linguistic adaptation through iterated learning. Studies in Language **28**(3) (2004) 587–607
7. Vogt, P.: The emergence of compositional structures in perceptually grounded language games. Artificial Intelligence **167(1–2)** (2005) 206–242
8. Deacon, T.: The Symbolic Species. W. Norton and Co., New York, NY. (1997)
9. Vogt, P.: On the acquisition and evolution of compositional languages: Sparse input and the productive creativity of children. Adaptive Behavior **13(4)** (2005) 325–346
10. Vogt, P.: Stability conditions in the evolution of compositional languages: issues in scaling population sizes. In Bourgine, P., Képès, F., Schoenauer, M., eds.: Proceedings of the European Conference on Complex Systems, ECCS'05. (2005)
11. Batali, J.: The negotiation and acquisition of recursive grammars as a result of competition among exemplars. In Briscoe, E., ed.: Linguistic Evolution through Language Acquisition: Formal and Computational Models. Cambridge University Press, Cambridge (2002) 111–172
12. Steels, L., Kaplan, F., McIntyre, A., Van Looveren, J.: Crucial factors in the origins of word-meaning. In Wray, A., ed.: The Transition to Language, Oxford, UK, Oxford University Press (2002)
13. De Beule, J., Bergen, B.K.: On the emergence of compositionality. In Cangelosi, A., Smith, A., Smith, K., eds.: The Evolution of Language: Proceedings of the 6th International Conference on the Evolution of Language. (2006)
14. Baronchelli, A., Loreto, V., Dall'Asta, L., Barrat, A.: Bootstrapping communication in language games: Strategy, topology and all that. In Cangelosi, A., Smith, A., Smith, K., eds.: Proceedings of Evolang 6, World Scientific Publishing (2006)
15. Edelman, G.M.: Neural Darwinism. Basic Books Inc., New York (1987)
16. Tamariz, M.: Evolutionary dynamics in language form and language meaning. In Cangelosi, A., Smith, A.D.M., Smith, K., eds.: The Evolution of Language: Proceedings of the 6th International Conference on the Evolution of Language, World Scientific Press (2006)

Agents Adopting Agriculture:Modeling the Agricultural Transition

Elske van der Vaart, Bart de Boer, Albert Hankel, and Bart Verheij

Artificial Intelligence, Rijksuniversiteit Groningen, Grote Kruisstraat 2/1
9712 TS Groningen, The Netherlands
elskevdv@ai.rug.nl

Abstract. The question "What drove foragers to farm?" has drawn answers from many different disciplines, often in the form of verbal models. Here, we take one such model, that of the ideal free distribution, and implement it as an agent-based computer simulation. Populations distribute themselves according to the marginal quality of different habitats, predicting settlement patterns and subsistence methods over both time and space. Our experiments and our analyses thereof show that central conclusions of the ideal free distribution model are reproduced by our agent-based simulation, while at the same time offering new insights into the theory's underlying assumptions. Generally, we demonstrate how agent-based models can make use of empirical data to reconstruct realistic environmental and cultural contexts, enabling concrete tests of the explanatory power of anthropological models put forward to explain historical developments, such as agricultural transitions, in specific times and places.

1 Introduction

To us modern agriculturalists, "Why farm?" seems like a non-question. Intensive food production is what supports our large, complex societies. It frees many of us to become specialists, enriching life in ways beyond mere provisioning: as doctors, entertainers, scientists. Without crop cultivation, our current population densities and growth rates would be impossible to sustain.

From that perspective, the advantages of agriculture over hunting-gathering, our earlier subsistence method, appear obvious. The daily toil of foraging for wild foods can only result in an existence best characterized as "nasty, brutish and short", as Thomas Hobbes once put it. Our ancestors' eventual switch from foraging to farming can then be explained simply by people discovering how to accomplish it.

The problem with this reasoning is that it rests on false assumptions. Our food crops today have characteristics carefully selected for by humans. On the whole, they are annuals, easy to sow and easy to harvest. The first farmers had much less to work with. If we take maize as an example, its likely precursor, *teosinte*, a wild cereal, produces a harvest only every other year, with tiny, brittle cobs and seeds nearly impossible to extract from their rock-hard casings [8].

In fact, it has become increasingly clear that the first agriculturalists probably worked harder [3], enjoyed less diverse diets [8] and experienced more disease [1]

S. Nolfi et al. (Eds.): SAB 2006, LNAI 4095, pp. 750–761, 2006.

than their immediate hunter-gatherer predecessors. So what drove those first farmers, ten thousand years ago in the Fertile Crescent [5], to take to cultivating crops and raising livestock, considering the hardships it imposed? Why there? Why then? And what about the other independent centers of agriculture, like the Andes, or the Far East [5]? Were the causes the same, or different?

Kennett and Winterhalder [10] present a collection of papers on behavioral ecology approaches to these questions. In this context, behavioral ecologists consider the ecological and evolutionary roots of subsistence change and how it helps individuals adapt to their environments. The mathematical and graphical theories of behavioral ecology are meant to capture the decisions and tradeoffs of individuals, and yet often include none, as they model only population-level behavior. As such, Winterhalder and Kennett [14, page 19] conclude, "…although there are at present no agent-based models of domestication or agricultural origins, behavioral ecology adaptations of the agent-based approach appear an especially promising avenue for research".

This paper is the result of a first attempt to realize that promise. To this effect, we have taken one of the models from [14] and implemented it as an agent-based computer simulation. Our research questions were simple. If we follow the principles of the model in question, will our agents behave like the theory predicts? And if so, what can the simulation teach us about its underlying assumptions? The model we selected for this treatment, that of the *ideal free distribution*, is a theory of habitat choice. It assumes that any area can be divided into a number of discrete habitats, differentiated by their *suitability*, and that populations will distribute themselves according to the *marginal quality* of those habitats [14].

Figure 1A provided an illustration of use of suitability curves in the ideal free distribution. Imagine human colonists, arriving on a pristine island, consisting of two habitats: the coastline and a mountain. The mountain is covered with dense forests; the coastline is sunny and rife with fish. If the assumptions of the ideal free distribution hold, all colonists will initially settle on the coast, because of its greater 'suitability' for human residence (*d0*; numbers in brackets refer to specific densities in Figure 1A). As population pressure rises, the coastal habitat becomes less and less attractive due to crowding and resource depletion, reducing its 'marginal quality'. Eventually, the mountain's suitability will rival that of the coast (*d1*). From this density onwards, people should start settling in the mountains, with further population growth spread equally over both habitats [inspired by 11, 12].

Let us further hypothesize that our island is home to a small stand of wild cereals, initially ignored by the colonists in favor of other, more easily procured food sources. As population grows, however, and these other food sources are exploited to carrying capacity, suitability of both habitats for human *foragers* drops considerably. At some point, artificially increasing the amount of wild cereal by planting and tending it may become worthwhile. At this density (*d2*), any additional residents should take to the mountains, where this type of food production is possible. In fact, these first farmers might actually *increase* the suitability of that habitat for other agriculturalists by clearing large sections of the forest, which should result in more people switching, until some maximum optimal density is reached (*d3*) [11, 12].

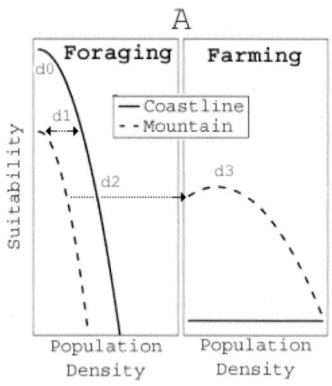
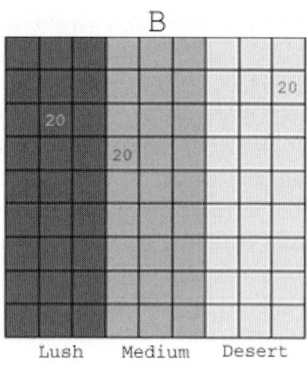

Fig. 1. A: Suitability of example habitats relative to population density and subsistence method, numbered densities refer to descriptions in text [after 11, 12] B: Simplified 9x9 model scape, 3 bands of 20 agents shown

Essentially, by comparing the 'suitability curves' (see Figure 1A) of different habitats with respect to changes in population density, the ideal free distribution model can make predictions about human settlement patterns over both time and space. By adding suitability curves for farming, the model can also predict people's subsistence strategies, as well as where they will adopt them and when they will switch [11, 12].

Using archeological evidence to construct appropriate suitability curves, the ideal free distribution theory has recently been applied to questions of agricultural origins in Eastern Spain [11] and Oceania [12]. In theory, realistic agent-based simulations, designed to mimic actual environmental and cultural conditions, offer possibilities for testing the explanatory power of such verbal models.

In this paper, we report on a first step in that direction. We have built a simple three habitat system and populated it with digital foragers and farmers, incorporating as much empirical data as possible. Our two goals are to show that 1) this simulation is capable of reproducing the predictions of the ideal free distribution model and 2) that it can at the same time provide insights into the validity and applicability of its underlying assumptions.

Our simulation, loosely based on Epstein and Axtell's sugarscape model [6], consists of a square grid of patches, divided into three discrete habitats. Each patch is inhabited by both *prey* and *cereal*, simulated by reasonably realistic growth rates, population sizes and energy values. The three habitats are 7x21 patches each, and differ only in the maximum densities of prey and cereal that can be supported by each patch. A reduced version of this model is Figure 1B.

Every time step, representing a year, the agents that inhabit these habitats must obtain enough food to survive, which they can accomplish by hunting prey, gathering cereal, or starting farms. Every subsequent time step, prey and cereal grow back, and the agents select new patches, preferring the most suitable ones. Reproduction grows the agent population until prey and cereal become seriously depleted, prompting adaptation in the form of habitat migration or subsistence change.

2   The Simulation Model

Considering that our explicit aim is to generate insight into an anthropological theory, it is of prime importance to ensure that any results will make sense to anthropologists. To this end, we have attempted to keep our model as realistic as possible, given the constraints of the data available to us. Our agent-based examination of the ideal free distribution involves a *scape* consisting of *habitats*, *food resources* to populate those habitats, *agents* that utilize these food resources and a set of *behavioral rules* that specify agents' subsistence behavior.

To capture the essential aspects of the ideal free distribution theory, the quality of our habitats must change with population density, and the agents must be able to make informed choices about which habitat is currently most suitable. Winterhalder *et al.*'s [13] mathematical model of optimal foraging theory, which considers 'the interaction of human population, diet selection, and resource depletion', provides a realistic quantification of both these aspects.

In this model [13], different food sources become increasingly hard to obtain as they become scarcer, increasing *search times*, which naturally occurs as population pressure rises. This reduces foraging efficiency and thus negatively impacts habitat suitability, which is measured as *net acquisition rates* [11]. In [13], the model is applied to questions of diet choice in a fixed area; in our simulation, we expand it to answer questions of habitat choice in a larger region, by explicitly including the dimension of space.

2.1   Characteristics and Properties of the Habitats, Food Resources and Agents

Scape and Habitats. To simulate hunter-gatherer behavior, one must first simulate a world of food to hunt and gather. Our world consists of 21x21 patches, each representing 300 square kilometers, with time steps (*epochs*) corresponding to a year. The edges of the world are true edges. According to [3] most foraging groups build base camps, which they may move once or twice a year. A forager can cover about ten kilometers and still return to such a base camp the same day; using this as a radius, we can calculate a home range of approximately 300 square kilometers [13]. This is what determines our patch size. Our three habitats are 7x21 patches each, colloquially termed the '*lush*', '*medium*' and '*desert*' habitats, ordered from left to right in the scape and differentiated only by their carrying capacities for different food resources (see Figure 1B).

Food Resources. Our hunter-gatherers have two dietary options, *prey* and *cereal*, which populate each patch. The prey has characteristics inspired by large ungulates; the cereal is loosely based on wild barley. Here, energy value (e_i) is the calories provided by a single prey or kilo of cereal; *handle time* (h_i) is the total time required to catch, clean and cook a prey once spotted, or harvest, thresh and prepare a kilo of cereal gathered once located; *carrying capacity* (C_i) is the maximum prey or cereal population that can be supported by a square kilometer of patch and growth rate (r_i) specifies the intrinsic rate of increase of each, where the index i specifies prey p or cereal c. The prey values were calculated from anthropological data in [13]; the cereal characteristics represent ballpark figures, chosen as conscientiously as possible using data from [7, 8, 9].

Table 1. Characteristics of Prey & Cereal

Type	Energy value (e_i) (Cals)	Handle time (h_i) (minutes)	Carrying capacity (C_i) (no or kilos/km²)			Growth rate (r_i) (no or kilos/year)
			Lush	Medium	Desert	
Prey (p)	13800	235	8.0	5.4	2.6	0.7
Cereal (c)	3390	120	600	400	200	1

Both prey and cereal grow according to equation (1), taken from [13] were $p_i(t)$ is the population size of food type i at time step t, C_i is the maximum carrying capacity for food type i and r_i is its intrinsic rate of increase. The equation is applied per patch. This results in cereal and prey slowly growing towards their carrying capacities, then stabilizing. In principle, there is no influence of cereal and prey densities between patches, unless either cereal or prey completely disappears from a patch. In that case, each of its eight neighboring patches that *does* still have a viable population of the resource in question has a 10% chance of repopulating the depleted patch each year. A 'viable population' is either 100 prey or 400 kilos of cereal; these are also the initial population sizes of each for a repopulated patch.

$$p_i(t+1) = p_i(t)\frac{C_i e^{r_i}}{C_i(1 - s_i t(1 - e^{r_i}))} \ . \tag{1}$$

Agents. The agents in our simulation represent small bands of hunter-gatherers. Every band starts as a group of 20 people. Every times tep, each of those 20 people has an 0.02 chance of reproducing, which is supposedly roughly characteristic of actual hunter-gatherers [13]. Once a band is made up of 40 individuals, it splits in two, each again representing 20 people. Every time step, the bands must forage 2000 kilocalories (Cal) for each of their group members, for each day of the year. Maximum foraging time is fixed at 14 hours per day. If there is shortfall, group size is scaled down to the number of people adequately fed. Foraging, however, also has costs; 4 Cal per minute (c_s) spent searching for prey or cereal, and 6 Cal per minute (c_h) spent catching or harvesting them [13].

Table 2. Agent Characteristics

Characteristic	Value
Group size	20 – 40 (people)
Min. energy	2000 (Cals)
Max. forage time	14 (hours)
Growth rate	0.02 (per year)

Characteristic	Value
Search cost (c_s)	4 (Cals/min)
Catch cost (c_c)	6 (Cals/min)
Search speed (s_s)	0.5 (km/hour)
Search radius (s_r)	0.0175 (km)

Search Times. The catch times of prey and cereal are fixed; the search times for each depend on population density [13] within a patch. The rarer a food source has become, the longer it takes to locate. It is also dependent on the speed at which agents search (s_s), and their search radius (s_r) as they do so (i.e. 'How far can they see?') (see Table 2). Search times are then calculated using equation (2), where s_i is the search time for food type i and d_i its current population density.

$$s_i = \frac{1}{s_s \cdot s_r \cdot 2 \cdot d_i} \text{ (in minutes/prey) .} \qquad (2)$$

Net Acquisition Rates. Now that we have specified the time it takes to catch (h_i) and find (s_i) each type of food source, their respective energy values (v_p and v_c) and the energy costs incurred in obtaining them (c_h and c_s), we can calculate the 'net acquisition rates' (NAR) for each; that is, how much energy an agent gains for each hour spent hunting for prey or gathering cereal. The higher the net acquisition rate, the more efficient foraging for that food type is. The net acquisition rate of food source i at time t is then equal to the values in Table 3, as calculated per habitat.

Table 3. Net Acquisition Rates

	NAR 'Lush' Habitat (Cals/hour)	NAR 'Medium' Habitat (Cals/hour)	NAR 'Desert' Habitat (Cals/hour)
Prey (p)	965	674	281
Cereal (c)	1298	1256	792
Farming (f)	677	605	481

$$NAR_i = \frac{e_i}{s_i + h_i} - \frac{s_i}{s_i + h_i} c_s - \frac{h_i}{s_i + h_i} c_h . \qquad (3)$$

2.2 Behavioral Rules

Our agents now have properties, a scape of habitats to move about in, and two types of food resources to forage for. In this section, we define the rules that guide their behavior. The ideal free distribution is a model of habitat choice, which assumes that individuals populate habitats according to their marginal quality. Our agents thus need to have a sense of what makes a patch *suitable*, which depends on their *dietary preferences*. Also relevant is the *range* in which agents can evaluate patches, and what the costs and benefits are of *food production*.

Suitability & Dietary Preferences. A patch's suitability may be measured by 'the production of young or rate of food intake' [14] of the initial occupant. Our agents rank patches by prey density first, cereal density second, reflecting the large percentage of meat found in most foragers' diets [4], and the greater prestige associated with hunting over gathering as it is observed in most hunter-gatherer cultures. Once a patch has been selected, agents hunt prey and gather cereal in proportion to their net acquisition rates; as prey becomes scarcer relative to cereal, it is consumed less (see equation (4)). Consumption of a food source stops if its net acquisition rate drops below zero.

$$\text{percentage of food source } i \text{ in diet} = NAR_i \Big/ \sum_{\text{All food sources } j} NAR_j \qquad (4)$$

Range. Every time step, each band starts in one patch, and may choose to move to another, which is always the most suitable patch it has knowledge of. This represents a small group of hunter-gatherers moving its base camp once a year. But which patches are considered to be 'in range'? Realistically, one might assume that these hunter-gatherers only have some sense of the area just outside their home range, and hence can only evaluate their own patch and the eight surrounding ones.

However, one of the ideal free distribution's explicit assumptions is that '…all individuals have the information to select and the ability to settle in the most suitable habitat available.' [12]. This would be best modeled by each agent having perfect knowledge of the suitability of each other patch in the scape. In our experiments, we try both options. Costs of moving are not considered, as ideal free distribution model assumes that those costs are '…negligible, when compared to the benefits of optimizing long-term habitat choice.' [10].

Food Production. Given that our simulation is intended to provide insights into behavioral ecology approaches to agricultural transitions, we must model some form of food production. Using [7, 8] as sources, some educated guesswork allows us to derive the additional kilos a square hectare of cultivated cereal might yield (c_k^+, where c_k is the wild harvest), as well as the time it takes to produce a kilo of cereal by farming (t_f) (equations (5)&(6)), where A is the number of hectares of cereal that can be tended by working an hour daily. This is *excluding* harvest time, which is considered to be identical to that of wild cereal, as given by Table 1.

$$c_k^+ = 0.25(1000 - c_k) \quad \text{(in kilos/hectare)} . \tag{5}$$

$$t_f = \frac{1}{h \cdot (c_k + c_k^+) \cdot 365} \quad \text{(in hours/kilo)} . \tag{6}$$

If we assume that it takes approximately half an hour a day to tend a hectare of cereal, and that both tending and harvesting cereal are strenuous activities [8], costing 6 Cals of energy/minute (c_h), we can then derive net acquisition rates for farming cereal in the three different habitats, as demonstrated by table 3. This means that the efficiency of farming is *independent* of population density. The area which is suitable for agriculture is bounded, however, at 10% of each patch [7]. Food production is only practiced if its net acquisition rate becomes higher than that of foraging for cereal; it then enters the diet in accordance with equation (4). Agents can thus forage exclusively, farm exclusively, or practice some mixture of both.

3   Experiments and Results

Our first goal is to ascertain to what degree our simulation reproduces the predictions of the ideal free distribution. To this end, we run the model in three different configurations. First, as a single food source environment, where agents can only forage for cereal (Experiment I). Second, as a hunter-gatherer society, where both cereal and prey are available, but switching to food production is impossible (Experiment II).

Third, as the full simulation, where agents can forage or farm as desired (Experiment IIIa). In these three experiments, agents have access to the suitability of every patch in the scape, to mimic the ideal free distribution model's assumption of 'perfect information' [12]. As a test of the consequences of this assumption, we will also run the full simulation with bands that can only evaluate the suitability of their own patch and its eight neighboring patches (Experiment IIIb).

In each of these experiments, we initialize the model by seeding five bands at random locations throughout the scape. Bands are awarded the opportunity to select patches in fixed order, with older bands first and younger bands last. This represents the process of 'daughter populations' splitting off and seeking new habitats. Unless otherwise stated, an experiment consists of 10 runs of 3000 epochs each.

3.1 Experiment I: Gathering

Setup & Predictions. In this first experiment, agents must make their livelihoods exclusively by gathering, which means they rank patches by cereal density only. Our agents have perfect information and free access to every patch on the scape; the ideal free distribution model straightforwardly predicts that they should colonize the 'lush' region first, followed by the 'medium' area and tailed by the 'desert' habitat, at a speed that maximizes rate of food intake throughout the scape.

Results. All bands are immediately drawn to the 'lush' area, where population grows until about epoch 270 ($\mu = 267$, $\sigma = 5$), when the first bands migrate to the 'medium area'. Expansion into the 'desert area' follows around epoch 310 ($\mu = 310$, $\sigma = 5$), with population still growing, reaching a maximum of approximately 0.39 ($\mu = 0.39$, $\sigma \approx 0$) agents/km^2. Finally, around epoch 330 ($\mu = 326$, $\sigma = 8$), carrying capacity of all three habitats is simultaneously exhausted, resulting in a massive population crash. The survivors pull back towards the 'lush' area, and the process restarts, with cycles of approximately 270 epochs ($\mu = 267$, $\sigma = 31$). Comparing average gather times between habitats reveals no significant differences in any of the runs (two-tailed t-test, $p > 0.05$ for all pairs of average gather times). Figure 2A shows one run's first cycle of agent densities and gather times over all three habitats.

Discussion. The agents are as ideally free distributed as possible. All agents, all over the scape, work nearly equally hard during each time step, having distributed themselves unevenly over the habitats to do so.

3.2 Experiment II: Hunting and Gathering

Setup & Predictions. Now, agents have both prey to hunt and cereal to gather, which agents do relative to both food sources' net acquisition rates (equation (4)). In essence, this means they have a preference for a mixed diet, but are willing to be flexible as resource densities change. Patches are selected by prey density first, cereal density second. The order in which habitats are settled is easily predicted as 'lush', then 'medium', then 'desert', but what an "ideal" distribution is in this situation, is unclear.

Results. All agents immediately relocate to the 'lush habitat', expanding into the 'medium' habitat around epoch 160 ($\mu = 162$, $\sigma = 5$) and the 'desert' habitat about 40 epochs later ($\mu = 200$, $\sigma = 6$), with population steadily increasing until approximately epoch 300 ($\mu = 310$, $\sigma = 40$), when carrying capacity is once again simultaneously exhausted all over the scape, causing massive drops in agent totals. As an example. the first cycle of one run is shown in the bottom panel of figure 2B.

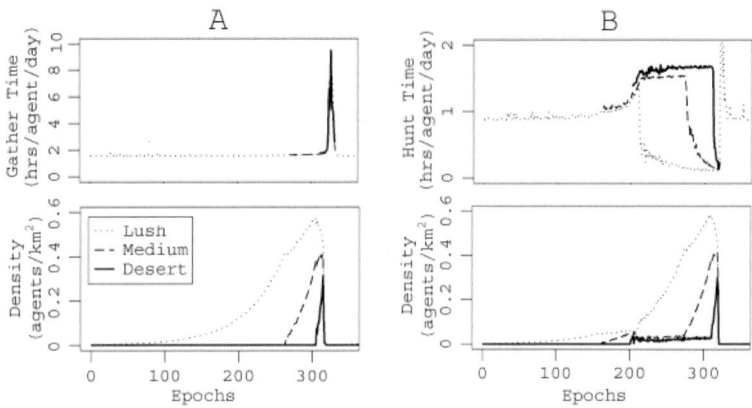

Fig. 2. A: Top, average time spent gathering in the first cycle of a sample run of Experiment I; note how the lines for the different habitats actually cover each other; Bottom, average agent densities of the same run and cyle. B: Top, average time spent hunting in the first cycle of a sample run of Experiment II; Bottom, average agent densities of same cycle and run.

The first migration to the second and third habitats is the result of agents following prey; the quick population increase in the 'lush' habitat after epoch 200 is the result of prey having become so depleted that many agents are having to live without meat; with no reason to follow prey to the lesser habitats, they crowd the 'lush area' until its cereal is so depleted that it becomes worthwhile to gather in the 'medium' and 'desert' regions as well. Figure 1B shows an example, but a similar process occurs in all cycles and runs.

Average foraging times, however, vary significantly between habitats. If we compare the time spans where all three regions were populated, agents in the 'lush area' worked an average of almost two hours ($\mu = 1.88$, $\sigma \approx 0$) per day, agents in the 'medium' area foraged for over two and half hours ($\mu = 2.67$, $\sigma \approx 0$) while agents in the 'desert' area spent almost three-and-a-quarter hours foraging faily ($\mu = 3.21$, $\sigma \approx 0$).

Discussion. Judged by the differences in the agents' average workloads, the obtained distribution can hardly be considered ideal for 'rate of food intake'. On closer inspection, however, it appears that the distribution may be slightly fairer than it appears. The reason the agents in the 'lush' area forage so little, is that there is no prey left to catch. If we consider the average energy from hunting in epochs where all three habitats are occupied, agents in the 'lush' area eat the least meat ($\mu = 49$ Cals, $\sigma = 1$) and agents in the 'medium' area the most ($\mu = 93$ Cals, $\sigma = 1$), while agents in the 'desert' area are in the middle ($\mu = 81$ Cals, $\sigma = 1$).

3.3 Experiment III: Hunting, Gathering and Farming

Setup & Predictions. Agents can now hunt, gather or farm, with gathering initially twice as efficient as farming. Agents still prefer the mixed diets of Experiment II, with food production only considered in case its net acquisition rate outranks that of gathering cereal. Initially, agents should distribute themselves over the three habitats as they do in Experiment II, but rather than population crashing, agents should switch to farming when prey and cereal first run out, starting in the 'lush' region, then the 'medium area', then the 'desert' habitat.

This system is essentially deterministic, as it is independent of initial conditions and involves so many agents that random fluctuations cancel each other out (Pearson's correlation coefficient of agent totals per epoch: 0.99 per any two runs). As such, it seems safe to report on only one run per configuration. We run this model both with perfect (IIIa) and local information (IIIb).

Results, IIIa. In the perfect information condition, agents first settle in the 'lush' habitat, colonizing the 'medium' habitat in epoch 143 and the 'desert' habitat in epoch 181. Food production starts in epoch 215, in the 'lush' area, where the number of farm hectares steadily climbs until epoch 459, when every single patch of 'lush' arable land is in use, prompting new farmers to migrate towards the 'medium' habitat, and later to the 'desert' habitat (epoch 492).

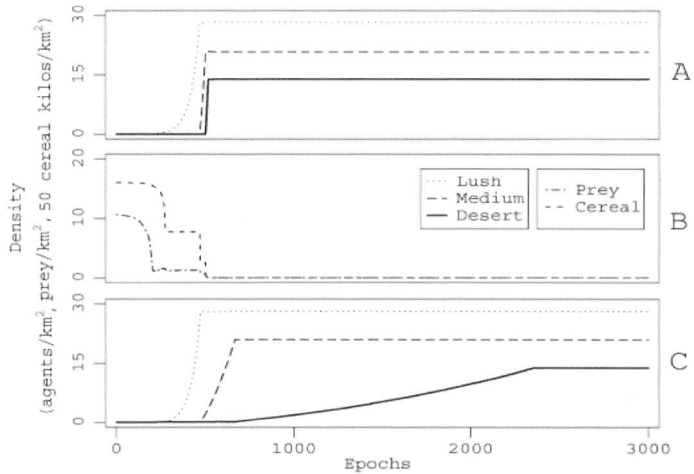

Fig. 3. A: Agent population densities of Experiment IIIa, foragers & farmers with perfect information; B: Prey and cereal densities of Experiment IIIa, C: Agent population densities of Experiment IIIb, foragers & farmers with local information

By epoch 507, all possibilities for agriculture have been seized, and agent densities stabilize at an average of 20 agents per square kilometer (Figure 3A). In the meantime, both wild cereal (epoch 499) and prey (epoch 508) have become extinct (see Figure 3B). Results of the 'local information' (IIIb) configuration runs are qualitatively similar, with the glaring exception of the timing. It takes up to epoch 2685 for the 'desert' habitat to fully fill up with farmers (Figure 3C).

Discussion. The extinction of prey and wild cereal is the result of equation (4), which allows farmers to continue "hunting and gathering on the side" once food production enters the diet. The large time difference in completion of the agricultural transition between the 'perfect information' and 'local information' condition is caused by lack of distribution possibilities. The high population densities of the farmed habitats cause large numbers of new bands to be formed, but they have nowhere to go if they cannot see the relatively unpopulated 'desert' area.

4 Conclusions

The question of why our ancestors first switched from foraging to farming has fascinated scientists for decades. In this paper, we hoped to offer only the suggestion of a new method, not a new model, to the study of this fascinating issue. We took up Winterhalder & Kennett's [14] suggestion to implement a behavioral ecology model as an agent-based simulation, consisting of three discrete habitats, and evaluated if our simulation could reproduce the model's predictions. We have shown that:

In the simplest possible implementation of the model, with only one food source and complete information, an ideal free distribution of the agents does indeed emerge. Agents in all three habitats spend an equal amount of time gathering food. Hence, under these conditions, the ideal free distribution model is retained.

If there are two food sources: highly desirable but hard to catch prey and less desirable, but easy to collect cereal, agents in different habitats end up spending different amounts of time. At first sight this contradicts the ideal free distribution model. However, agents that spend more time have more meat in their diets. It thus seems that the distribution remains ideal if a more complete definition of 'ideal' is used.

Agriculture can and does emerge in our simulations, and it emerges in the way that is predicted by the ideal free distribution. Furthermore, wild prey and cereal go extinct, meaning that agents cannot go back to their original hunting-gathering life style. This conforms to the ratchet effect that is observed in human populations. When the assumption of global information is lifted, populations no longer distribute themselves ideally. This causes the transition to agriculture to take longer.

It seems that, generally, our simulation has fulfilled our two initial goals. We have reproduced the predictions of the ideal free distribution theory in two settings – a single prey system and the transition to agriculture, assuming perfect information – and generated insight into its underlying assumptions in two others. Namely, the paucity of 'rate of food intake' as a general measure of habitat suitability, and the fact that assuming global information rather changes the model's predictions, at least in our simulation. For the future, we think that agent-based simulation can do more than just confirm predictions and point out potential shortcomings in existing theories – they can actually help solve them.

Realistically representing a preference for meat is practically impossible in a verbal model, but easy to implement in an agent-based simulation. The same goes for specifying the amount of information available about other habitats. Anthropological findings may often offer some idea about what is plausible in any given historical or environmental setting, but quantifying the effects of these local variables in a verbal model is difficult, and usually very hard to verify empirically. Agent-based models offer an easy way of simulating specific conditions and cultural practices (for a recent example, see [2]), and thus seem to offer much explanatory power when it comes to considering agricultural transitions in specific times and places.

Acknowledgements

We would like to thank Gert Kootstra for his valuable input in the early phases of this project, and Tim Dorscheidt for many spirited discussions.

References

1. Armelagos, G.J., Goodman, A.H., Jacobs, K.H.: The Origins of Agriculture: Population Growth During a Period of Declining Health. Population and Environment 13(1) (1991) 9-22
2. Axtell, R.L., Epstein, J.M., Dean, J.S., Gumerman, G.J., Swedlund, A.C., Harburger, J., Chakravarty, S., Hammond, R., Parker, J., Parker, M.: Population Growth and Collapse in a Multi-Agent Model of the Kayenta Anasazi in Long House Valley. Proceedings of the National Academy of Sciences of the United States of America 99 (2002) 7275-7279
3. Cohen, M.N.: The Food Crisis in Prehistory. Yale University Press, New Haven (1977)
4. Cordain, L., Miller, J.B., Eaton, S.B., Mann, N., Holt, S.H.A., Speth, J.D.: Plant-Animal Subsistence Ratios and Macronutrient Energy Estimations in Worldwide Hunter-Gatherer Diets. American Journal of Clinical Nutrition 71 (200) 682-692
5. Diamond, J.: Guns, Germs, and Steel. W.W. Norton & Company, New York London (1997)
6. Epstein, J.M., Axtell, R.: Growing Artificial Societies: Social Science from the Bottom Up. MIT Press, Cambridge (1996)
7. Flannery, K.V.: Origins and Ecological Effects of Early Domestication in Iran and the Near East. In: Ucko, P.J., Dimbleby, G.W.: The Domestication and Exploitation of Plants and Animals. Duckworth, London (1969) 73-100
8. Flannery, K.V.: The Origins of Agriculture. Annual Review of Anthropology 2 (1973) 271-310
9. Harlan, J.R., Zohary, D.: Distribution of Wild Wheats and Barley. Science 153 (1966) 1075-1080
10. Kennett, D.J., Winterhalder, B. (eds.): Behavorial Ecology and the Transition to Agriculture. University of California Press, Berkeley Los Angeles London (2006)
11. Kennett, D.J., Anderson, A., Winterhalder, B.: The Ideal Free Distribution, Food Production, and the Colonization of Oceania. In: Kennett, D.J., Winterhalder, B. (eds): Behavioral Ecology and the Transition to Agriculture. University of California Press, Berkeley Los Angeles London (2006) 1-21
12. McClure, S.B., Jochim, M.A., Barton, C.M.: Human Behavioral Ecology, Domestic Animals, and Land Use during the Transition to Agriculture in Valencia, Easter Spain. In: Kennett, D.J., Winterhalder, B. (eds): Behavioral Ecology and the Transition to Agriculture. University of California Press, Berkeley Los Angeles London (2006) 197-216
13. Winterhalder, B., Baillargeon, W., Cappelletto, F., Daniel, I.R., Prescott, C.: The Population Ecology of Hunter-Gatherers and Their Prey. Journal of Anthropological Archaeology 7 (1988) 289-328
14. Winterhalder, B., Kennett, D.J.: Behavorial Ecology and the Transition from Hunting and Gathering to Agriculture. In: Kennett, D.J., Winterhalder, B. (eds.): Behavioral Ecology and the Transition to Agriculture. University of California Press, Berkeley Los Angeles London (2006) 265-288

Adaptive Behavior in Language
and Communication

Noisy Preferential Attachment and Language Evolution

Samarth Swarup[1] and Les Gasser[1,2]

[1] Dept. of Computer Science
[2] Graduate School of Library and Information Science,
University of Illinois at Urbana-Champaign,
Urbana, IL 61801, USA
{swarup, gasser}@uiuc.edu

Abstract. We study the role of the agent interaction topology in distributed language learning. In particular, we utilize the replicator-mutator framework of language evolution for the creation of an emergent agent interaction topology that leads to quick convergence. In our system, it is the links between agents that are treated as the units of selection and replication, rather than the languages themselves. We use the Noisy Preferential Attachment algorithm, which is a special case of the replicator-mutator process, for generating the topology. The advantage of the NPA algorithm is that, in the short-term, it produces a scale-free interaction network, which is helpful for rapid exploration of the space of languages present in the population. A change of parameter settings then ensures convergence because it guarantees the emergence of a single dominant node which is chosen as teacher almost always.

1 Introduction

The study of communication and language is an important aspect of the study of adaptive behavior. Predefined languages for multiagent systems may not be appropriate as they reflect the designer's viewpoint rather than the agents', and are unable to adapt to changing environmental conditions and task definitions. It is much more desirable for the agents to be able to create and maintain their own language. This is not an easy task, however, as the mechanisms of language evolution are far from being well understood.

The last decade or so has seen increasing application of computational methods to the study of language evolution [1], [2], [3]. The main mathematical approach, meanwhile, is to apply models of biological evolution to the evolution of language(s) [4]. In this case, the languages themselves are considered the units undergoing selection and mutation. These models have been used to address questions about convergence [5], and the emergence of syntax [6], for example.

One of the main problems in language evolution, which has received little attention so far, is how to get a population of agents to converge to a common language, without globally imposing some kind of hierarchy on the population. In other words, how does the topology of agent interactions affect the convergence to a common language?

S. Nolfi et al. (Eds.): SAB 2006, LNAI 4095, pp. 765–776, 2006.

The topology clearly has an important role to play in convergence. For example, if the population is split into two disjoint subgroups, then they cannot converge onto a single language except by chance. Even if the topology consists of a single component, multiple languages might co-exist in the population, especially if the rate of change of the language (in response to environmental changes, for example) is high in relation to diameter of the network. In other words, languages might be changing faster than they can propagate across the network.

In this work we show that we can take advantage of evolutionary dynamics to actually construct the agent interaction topology on the fly. This is done by a subtle change of focus. Instead of the languages being treated as the units of selection and replication, we treat the interaction links between agents as the units undergoing selection and replication.

The rest of this paper is organized as follows. We first describe some recent work investigating the role of the interaction topology in the convergence of language and emergence of social conventions in multi-agent systems. This is followed by a discussion of our model for generating agent interaction topologies, which is based on the evolutionary framework described by the replicator-mutator equation. We show that this mathematical model is valid through some simple simulations. Then we go on to do a language learning experiment using simple recurrent neural networks. Finally we discuss the possibilities for expanding on this work to include situatedness and further numerical exploration of the theoretical model.

2   Related Work

There has been some significant work on the convergence of a population of agents to a particular language. Komarova et. al [7], and Lee et al. [8] have studied the problem from the point of view of population dynamics, while Dall'Asta et al. [9] have studied the dynamics of the naming game [10] on small-world networks, and Lieberman et al. [11] have introduced evolutionary graph theory, which is the study of evolutionary processes on graphs.

The model of Dall'Asta et al., while very interesting, is not really an evolutionary model since there is no notion of selection or variation in it. Therefore we will not discuss it further here.

Lee et. al studied the role of the interaction topology on the convergence of a population to a single language. This study looked at a set of specific interaction topologies, including fully connected, linear, von Neumannn lattice, and a bridge topology. Using the model of Komarova et. al, which assumes random pairwise interactions between all agents, they empirically studied the critical learning fidelity threshold for language convergence in the various topologies. Although several different interaction topologies were used, the topologies were not emergent and were specified beforehand by the creator of the experiments. In addition, the agents did not learn a language from interactions with other agents, but rather neighbors of high fitness agents were transformed into copies of the high fitness agent with some fixed probability. Lieberman et al. put their work on a firmer theoretical basis by studying the probability of fixation (i.e. the probability that a fitter language, if it appears by mutation, will be adopted by the entire population) on a graph. They showed that some graphs can be selection amplifiers, in that the probability of fixation can be made as high as possible, and also that some graphs are selection suppressors.

Here we ask the question, can the agents generate the topology on the fly, while still ensuring that the emergent topology leads to rapid convergence?

3 Agent Interaction Topologies and Convergence

The agent interaction topology is a weighted directed *influence* graph which describes the influence of an agent on the language of another agent. Such a graph captures constraints such as spatial locality, agents' knowledge of each others' existence, interaction choice preferences, etc.

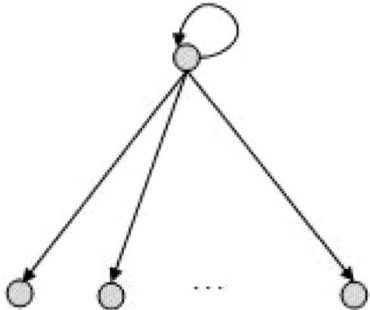

Fig. 1. The interaction graph for quickest convergence. One agent teaches the language to all the other agents. The circles represent agents, and the directed arrows represent the influence of one agent on the language of the other agent.

A simple strategy for rapid convergence would be to designate a special agent from which each of the other agents learn their language, as in figure 1. This corresponds essentially to a pre-imposed or designed language, which may or may not be of the highest objective quality. Such a centralized system is brittle in practice because a) the teacher agent has to be responsible for adapting the language to keep up with changing tasks, environments and needs of all the agents, b) communicative load on the teacher increases at least linearly with population size, reducing scalability, and c) the centralized teacher is a single failure point. Multi-agent systems are generally distributed and open, which means that there is no central control point, and agents may enter and leave the population at any time. This means that although desirable for its speed, uniformity, and certainty, the interaction topology shown in figure 1 is both undesirable and unrealistic for a general multi-agent system.

It would be much better if agents could develop their interaction topologies on the fly, by selecting interaction partners autonomously. We would still like, however, to have some guarantee of convergence, and of rapid convergence. In this regard, we next discuss the Noisy Preferential Attachment algorithm which we can use initially to generate scale-free topologies, and also to guarantee convergence. The final emergent topology, as we will see, looks a lot like fig. 1, but the crucial distinction is that it is emergent. Thus, e.g., if the central agent left the population, another would emerge to take its place.

4   Noisy Preferential Attachment

We first derive a variant of the replicator-mutator equation (RME), the RME without Death (RME-WD). Then we show that the preferential attachment model of small-world network generation is a special case of the RME-WD. We then use this equivalence to give a version of the preferential attachment algorithm, called Noisy Preferential Attachment, which we will later use to generate the agent interaction topology.

The Replicator-Mutator Equation (RME) describes the rate of change of the proportion of *types* (genomes, languages) in a population undergoing replication and mutation.

Suppose there are N types in a population of n individuals. Let f_i be the fitness of an individual of type i. Since fitness includes both frequency-independent and frequency-dependent components, it is written as,

$$f_i = w_i + \sum_{j=1}^{N} a_{ij} x_j, \qquad (1)$$

where x_j is the proportion of individuals of type j in the population, and w_i is a measure of intrinsic fitness of the language which might be related to learnability, expressiveness, etc.. The matrix $A = [a_{ij}]$ is known as the payoff matrix, and can be thought of as the payoff or reward achieved by an individual of type i in an interaction with an individual of type j. In the case of languages, A can be thought of as a measure of intelligibility, i.e. the degree to which a speaker of language i understands a speaker of language j.

The total number of individuals added to the population in a time step is $\sum_{j=1}^{N} f_j x_j n$. Further, replication is imperfect. With a small probability, replicating an individual of type i results in an individual of type j. This is quantified by a matrix $Q = [q_{ij}]$, where q_{ij} is the probability that replication of an individual of type i results in an individual of type j. In the case of languages, this corresponds to learning fidelity. In a limited interaction between individuals, the learner may not learn exactly the teacher's language.

Suppose also that the size of the population is held constant at n, by removing an equal number of individuals uniformly randomly, as are added to the population. This is because, in the case of language learning, when an agent learns a new language, it necessarily replaces the old language of that agent. Then the number of individuals of type i that are removed in one time step is $x_i \sum_{j=1}^{N} f_j x_j n$.

Putting all these terms together, we get the rate of change of the proportion of individuals of type i in the population, as

$$\dot{x}_i = \sum_{j=1}^{N} f_j x_j q_{ij} - x_i \phi, \tag{2}$$

where $\phi = \sum_{j=1}^{N} f_j x_j$. This is the Replicator-Mutator Equation (RME).

4.1   Replication-Mutation Without Death

If there is no death, the number of individuals of type i at the next time step is,

$$x_i'(n + \sum_{j=1}^{N} f_j x_j n) = x_i n + \sum_{j=1}^{N} f_j x_j q_{ij} n.$$

Here $n + \sum_{j=1}^{N} f_j x_j n$ is the total size of the population at the next time step, and $\sum_{j=1}^{N} f_j x_j q_{ij} n$ is the number of new individuals of type i. Rearranging, and letting $\phi = \sum_{j=1}^{N} f_j x_j$, we get

$$x_i'(1 + \phi) = x_i + \sum_{j=1}^{N} f_j x_j q_{ij}$$

$$x_i'(1 + \phi) - x_i(1 + \phi) = \sum_{j=1}^{N} f_j x_j q_{ij} - x_i \phi$$

Thus the rate of change of the proportion of type i is,

$$\dot{x}_i = \frac{\sum_{j=1}^{N} f_j x_j q_{ij} - x_i \phi}{1 + \phi} \tag{3}$$

This is the Replicator-Mutator Equation without Death. Note that, since $f_j \geq 0 \; \forall \; j$, the denominator on the right-hand side is always positive. Therefore the critical points of the RME-WD are the same as those of the RME.

The general form of the RME is very difficult to study because of the large number of parameters (all the entries of the A and the Q matrices). Often a special symmetrical case is studied, where the A matrix is set to have diagonal values equal to 1, and off-diagonal values $a << 1$, and the Q matrix is similarly set to have diagonal values p (close to 1), and off-diagonal values $(1 - p)/(N - 1)$. A complete analysis of the critical points is possible in this *fully symmetric* case [12].

In particular, when p is less than a critical threshold, the system has only one attractor, where all types are present in equal proportion in the population. For large values of N and small values of a, this threshold is approximately 0.5. Above this value, the attractor turns into a repeller, and the only stable attractors that emerge correspond to the situation where a single type dominates the population. Note that in the absence of death, and presence of mutation, there will always be all types present in the population, but the *proportion* of one of the types goes towards one. Since the system is fully symmetric, it could be any one of the N types that eventually dominates the population, and which attractor the system falls into depends on the initial conditions, and statistical fluctuations.

We now shed some light on the transient behavior of the RME-WD, under certain special conditions, by showing its equivalence with the preferential attachment algorithm of small-world network generation. This means that, under the right initial conditions, if we create a network (as we will later describe) using the RME-WD, the network will be a scale-free network (in the short term).

4.2   Preferential Attachment and the Underlying Probabilistic Model

The Preferential Attachment algorithm is the most commonly cited model of small-world network generation [13]. Small-world networks are graphs which have three properties: a small diameter, a high clustering coefficient, and a power-law degree distribution. The clustering coefficient is defined as the average fraction of neighbors of a node that are also neighbors of each other. Barabasi and Albert showed that a small-world network can be generated by preferential attachment, as follows.

We start the network with a small number of nodes and links, say two of each, randomly connected. At each step, we add a node to the network and add a link from the new node to one of the pre-existing nodes with probability proportional to the number of in-links that node already has. Thus, the probability of node i acquiring a new link is,

$$P(i) = \alpha x_i^\gamma, \tag{4}$$

where x_i is the proportion of in-links that go to node i, γ is a constant, and α is a normalizing term. γ is generally set to 1, in which case α is also 1.

This process results in a small-world network. There are a couple of things worth noting here. First, since new nodes don't have any in-links, the probability of acquiring any in-links is zero for these nodes. To get around this problem, every node is assumed to have one pseudo-link, i.e. the number of in-links for each node for the purposes of preferential attachment, begins at 1. Second, since new nodes are added at every time step, the number of links remains approximately equal to the number of nodes in the network. In later work, Albert and Barabasi modified the preferential attachment algorithm to allow rewiring of links with some small probability, and also to allow adding links without adding nodes with some small probability [14], but the essential algorithm remains the same as that described above.

The underlying probabilistic model is an instance of a Polya's urn model, as described below (and also in [15]).

Imagine a set of N urns which are all empty except for one, which has one ball in it. We now add balls one by one. A ball is put into urn i with probability proportional to the number of balls already in that urn (plus one "pseudo-ball").

This process is clearly equivalent to the preferential attachment algorithm with the caveat that we have fixed the number of urns to be N. An urn represents a node and a ball represents an in-link. In the short-term, i.e. while the number of balls is of the same order as the number of urns, this probabilistic model represents the small-world network generation process.

We now add a further step to it to make it equivalent to the RME-WD. We introduce a *transfer matrix*, Q, which is the same as the mutation matrix in the RME. Suppose a ball is added to urn i at time step t. Then a ball is taken out of urn i and moved to any of the urns with probability q_{ij}. This is similar to later versions of the preferential attachment model which include rewiring.

This probabilistic model captures the RME-WD dynamics if we consider urns to correspond to types and balls to individuals in the population. Since it is the balls that correspond to the individuals undergoing replication and mutation, we have to set the payoff matrix, A, equal to the null matrix in this case to get the linear dependence of $P(i)$ on x_i. Note that in this case, the RME-WD loses its frequency-dependent aspect. If we set $A = I$, the identity matrix, $P(i)$ varies as the square of the proportion of individuals of type i. If the off-diagonal elements of A are set to be non-zero, then $P(i)$ acquires additional second-degree terms.

We call this extended (but still finite) version of preferential attachment, Noisy Preferential Attachment (NPA) [16], because of the introduction of the mutation matrix into the probabilistic model. A caveat is in order here too: there is no notion of pseudo-links (or pseudo-balls) in this model. New nodes (types) are introduced into the graph (population) by the mutation process. This means that the number of nodes increases much more slowly that it does in the preferential attachment case. Therefore to generate a large network, the initial state needs to include a fairly large number of nodes with non-zero number of in-links. Alternatively, the mutation rate needs to have a high value.

5   Using NPA to Generate Agent Interaction Topologies

We use the NPA algorithm to generate the agent interaction topology on the fly in two stages as follows. Initially the agents have no knowledge of (the quality of) each other's languages. Therefore the first stage is an exploration phase, which sets up the second convergence phase. In the exploration phase, an agent Alice chooses another agent, Bob, as a teacher with probability proportional to Bob's fitness. The fitness of an agent is equal to the number of times that agent has been chosen as a teacher. The fitness can also include a term that is independent of the frequency of selection as teacher, but for now, we ignore this term since our current simulations are ungrounded. With probability $(1 - p)$, Alice switches to a uniformly randomly chosen teacher. This is similar to the notion of exploration-exploitation in reinforcement learning. The intuition is that if a lot of agents are choosing a particular agent as teacher, then choose that agent as a teacher because a lot of agents consider its language to be good. However, the proportions might be misleading near the beginning of the process because the actual counts will be low. Therefore it makes more sense to explore rather than exploit at the beginning of the distributed language learning process, i.e. it makes sense to start out with a high value of the mutation rate, $(1 - p)$, and switch to a low value when the process has been going on for a while.

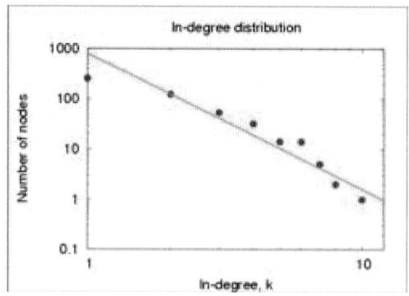

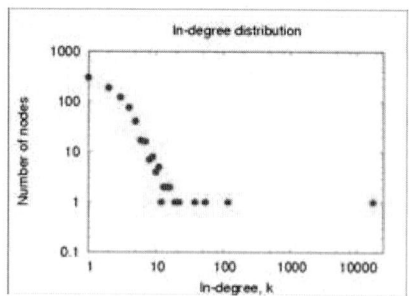

Fig. 2. The degree distribution for $p =$ 0.3 and $\gamma = 1$, after 1000 links have been added. The number of nodes in the graph is 1000 as well. The distribution is clearly a power law.

Fig. 3. The degree distribution after 20,000 links have been added to the graph. One node is clearly dominant.

Figure 2 shows a simulation in which we have a population of 1000 agents, i.e. a graph with 1000 nodes. Initially, one link is randomly added to start the process off. The initial value of p is 0.3, and γ, which is the exponent of the proportion in the preferential selection equation, is set to 1. We add one link at each time step, and figure 2 shows the in-degree distribution after 1000 links have been added. The graph is plotted on a log-linear scale, and the distribution is clearly a power-law. Therefore, at this stage, the graph is a scale free network.

At this point we start the second stage, by changing the value of p to 0.95, and the value of γ to 2. The intuition is that once the space of languages has been sufficiently explored, we can switch to the "convergence mode", where we trust the statistics of interactions that have been established in the first stage to guide us to a good overall language.

As we continue adding links, the node with highest degree becomes the dominant node. Figure 3 shows the degree distribution after 20,000 links (total) have been added. We can see that a single node has acquired a far larger proportion than the rest, and because of the frequency-dependent effect, the proportion of links acquired by this nodes will continue to increase towards 1 as we continue adding links to the graph. This means that the population will converge to the language of this agent. If this agent later gets removed from the population, the next most "fit" agent will become the dominant agent. It may possibly have a different language, though.

6   A Language Learning Experiment

We now do a simulation where we have a population of agents trying to converge onto a common language by learning from each other. The agent interaction topology is generated as described above. The agents use simple recurrent neural networks to generate, parse, and learn sentences. Each simple recurrent network has 5 inputs, 3 hidden layer nodes, and 5 outputs. There are 5 symbols in the "languages", {a,b,c,d,e}, and we use a 1-of-n encoding, i.e. the symbol a is encoded as the vector $[1, 0, 0, 0, 0]$ at the input of the neural network. A sentence is generated from a simple recurrent network by setting its internal state to 0.5 and giving it a random initial input vector. The output of the neural net is then fed back to its input and this process is repeated until we have generated as many symbols as we want. The weights of the neural networks are initialized randomly in the range $[-0.5, 0.5]$.

The population size was set to 100, and the experiment was run for 1000 time steps. At each time step, an agent is selected in sequential order, and it chooses a teacher according to the NPA algorithm. It receives a sample of 100 sentences of length 10 from the teacher and trains on this sample to convergence or for 100 epochs, whichever comes first. Every 50th time step, we collect 5 randomly generated sentences from each agent to form a testing set and the one-step symbol prediction error is calculated for each agent on this testing set. These are summed up to indicate the error (the inverse of convergence) of the entire population. This value is plotted in figure 4.

The parameters for the NPA algorithm were set in a manner similar to the previous section. Since there are 100 agents, i.e. 100 nodes in the graph of the agent interaction topology, we set $p = 0.3$ and $\gamma = 1$ for the first 100 steps, and then changed these values to $p = 0.95$ and $\gamma = 2$ for the remainder of the simulation. As we see in figure 4, the error only starts dropping after time step 100. However, after that the error drops quite rapidly and reaches almost zero by time step 1000. As a comparison, we also plot the error with uniformly random teacher selection. We see that convergence is attained much faster with the NPA algorithm.

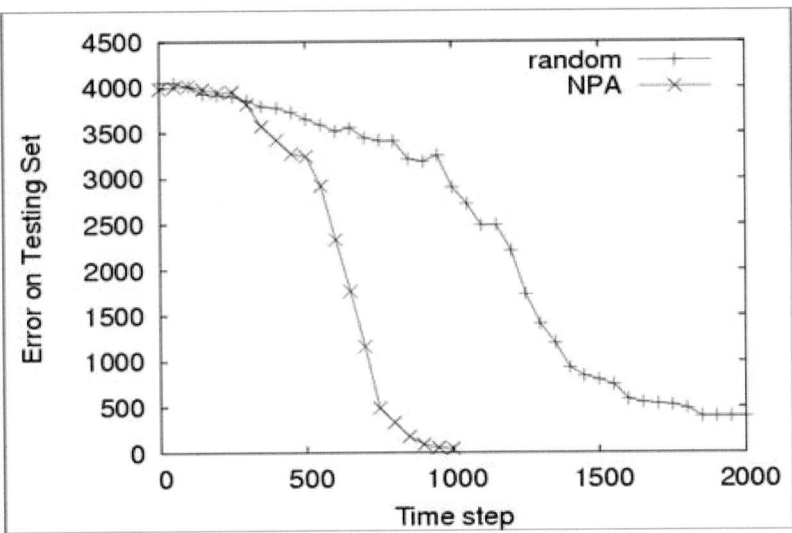

Fig. 4. The one step symbol prediction error summed over all the agents on the testing set. The testing set is generated by sampling 5 sentences from each of the agents at that particular time step. The low prediction error at the end ($\sim 1\%$) indicates almost perfect convergence of the language.

7   Discussion and Future Work

The two stages of the distributed convergence mechanism described in this work combine the ideas of convergence to a common social convention and convergence to a common language, through the mechanism of frequency-dependent (or preferential) selection.

The goal of the first, or exploratory, stage is to evaluate the languages that are present in the population and collectively decide on a single language, embodied by a single teacher, as the language to converge upon. The second stage then focuses on all the agents learning this one language, again in a distributed way, by simply changing two global parameters of the system: the probability of switching to a random teacher rather than the preferentially selected one, and the exponent which determines how strongly the preferential selection mechanism works. The underlying theoretical model guarantees that the appropriate parameter values will result in both a power law initial distribution of links, and the dominance of one teacher in the second stage.

Another important point, which was not mentioned in the language learning experiment, is that learning is non-trivial. With all learning architectures and algorithms, including the simple recurrent networks trained with the delta rule and backpropagation that we have used, some languages are easier to learn than others. In the work of Lee et al., e.g., this is captured by the learning fidelity parameter [8]. However, the point is that in practice, this parameter is non-uniform across the language space. "Simple" languages can be learned with greater fidelity than more complex ones. For example, it is much easier to converge upon point attractors, i.e. languages consisting of a single symbol, with simple recurrent networks, than to converge upon other kinds of attractors: e.g. languages consisting of alternating symbols or other more complex languages. In the language learning experiment shown, it is much harder to attain convergence when the selected teacher has a complex language, and convergence depends on having appropriate parameters settings for the neural networks (such as number of hidden layer nodes, learning rate, momentum, etc.).

There are many details that remain to be fleshed out in the model. The algorithm given is not truly a distributed algorithm since it hasn't really been specified from the point of view of a single agent. We need to explicate how the exploration phase works at the beginning when the agents have no information about each other. We can imagine that each agent maintains a list of its own estimate of all the other agents' fitness values. They can all be initialized to zero, and then a few of these get filled in by a "recommendation" mechanism corresponding to the preferential selection. In other words, an agent chooses a teacher randomly (or chooses a neighbor), and then gets referred to a better teacher by this one based on the teacher's knowledge about the population. This is where a scale-free (or small-world if possible) nature of the agent interaction topology helps. The small diameter of such a network means that it is easy for an agent to find a good teacher using such a referral mechanism.

In future work, we intend to explore a more grounded language learning case. As a first step, we need to investigate the effects of making the frequency-independent part of fitness non-zero. A second concern is the quality of language that is converged upon. As pointed out above, there is an inherent bias in the population towards learning simple languages, as learning fidelity is higher in this case. There has to be a corresponding pressure towards language *complexification*, perhaps from a task based reward, otherwise the learned language will almost surely be the simplest possible.

8   Conclusions

We have outlined a system capable of converging onto a common language in a distributed manner. It relies on the framework of language evolution, with a change of focus: instead of treating the languages themselves as the individuals undergoing replication, selection and mutation, we treat the links in the agent interaction topology as the evolutionary units.

We further showed that under certain special conditions, we can recover the preferential attachment algorithm of small-world network generation from the replicator-mutator system of evolution. This allowed us to give a two stage model of distributed language learning, based on our Noisy Preferential Attachment algorithm. In the first stage, the agents explore the languages present in the population and generate a scale-free network of interaction. In the second stage, the parameters are changed to allow rapid convergence by letting a dominant "teacher" node emerge through the same evolutionary dynamics.

We demonstrated through a simple language learning experiment using simple recurrent networks, that such a system can converge to a common language autonomously and rapidly.

In the near future we intend to explore the parameter space of the system more thoroughly, both numerically and theoretically, through more grounded simulations.

Acknowledgements

We thank Kiran Lakkaraju for help with the programming of the language learning experiment, and for very helpful feedback. We thank the UIUC Language Evolution Group for very helpful and stimulating discussions. This work was supported in part by NSF grant IIS-0340996.

References

1. Steels, L.: The evolution of communication systems by adaptive agents. In Alonso, E., D.K., Kazakov, D., eds.: Adaptive Agents and Multi-Agent Systems: Adaptation and Multi-Agent Learning. LNAI 2636, Springer Verlag, Berlin (2003) 125–140
2. Smith, K., Kirby, S., Brighton, H.: Iterated learning: A framework for the emergence of language. Artificial Life 9(4) (2003) 371–386
3. Batali, J.: Computational Simulations of the Emergence of Grammar. In: Approaches to the Evolution of Language: Social and Cognitive Bases. Cambridge University Press, Cambridge (1998)

4. Nowak, M.A., Komarova, N.L.: Towards an evolutionary theory of language. Trends in Cognitive Sciences **5**(11) (2001) 288–295
5. Komarova, N.L.: Replicator-mutator equation, universality property and population dynamics of learning. Journal of Theoretical Biology **230** (2004) 227–239
6. Nowak, M.A., Plotkin, J.B., Jansen, V.A.A.: The evolution of syntactic commmunication. Nature **404** (2000) 495–498
7. Komarova, N., Niyogi, P., Nowak, M.: Evolutionary dynamics of grammar acquisition. Journal of Theoretical Biology **209**(1) (2002) 43–59
8. Lee, Y., Collier, T.C., Taylor, C.E., Stabler, E.E.: The role of population structure in language evolution. In: Proceedings of the 10th International Symposium on Artificial Life and Robotics. (2005)
9. Dall'Asta, L., Baronchelli, A., Barrat, A., Loreto, V.: Agreement dynamics on small-world networks. Europhysics Letters (2006)
10. Steels, L.: A self-organizing spatial vocabulary. Artificial Life **2**(3) (1996) 319–332
11. Lieberman, E., Hauert, C., Nowak, M.A.: Evolutionary dynamics on graphs. Nature **433** (2005) 312–316
12. Mitchener, W.G.: Bifurcation analysis of the fully symmetric language dynamical equation. Journal of Mathematical Biology **46**(3) (2003) 265–285
13. Barabási, A.L., Albert, R.: Emergence of scaling in random networks. Science **286** (1999) 509–512
14. Albert, R., Barabási, A.L.: Topology of evolving networks: Local events and universality. Physical Review Letters **85**(24) (2000) 5234–5237
15. Chung, F., Handjani, S., Jungreis, D.: Generalizations of Polya's urn problem. Annals of Combinatorics **7** (2003) 141–153
16. Swarup, S., Gasser, L.: Unifying network and evolutionary dynamics. In Preparation (2006)

The Emergence of Communication by Evolving Dynamical Systems

Steffen Wischmann[1] and Frank Pasemann[2]

[1] Bernstein Center for Computational Neuroscience,
Bunsenstr. 10, D-37073 Göttingen, Germany
steffen_wischmann@web.de
http://www.chaos.gwdg.de/~steffen
[2] Fraunhofer Institute for Autonomous Intelligent Systems,
Schloss Birlinghoven, D-53754 Sankt Augustin, Germany
frank.pasemann@ais.fraunhofer.de

Abstract. In the context of minimally cognitive behavior, we used multi-robotic systems to investigate the emergence of communication and cooperation during the evolution of recurrent neural networks. The networks are systematically analyzed to identify their relevant dynamical properties. Evolution efficiently adapts these properties through small structural changes within the networks when specific environmental conditions are altered, such as the number of interacting robots. The findings signify the importance of reducing the predefined knowledge about resulting behaviors, dynamical properties of control, and the topology of neural networks in order to utilize the strength of the Evolutionary Robotics approach to Artificial Life.

1 Introduction

The dynamical systems approach to cognition [1,2,3] aims at the study of natural cognitive systems as dynamical systems. While concrete dynamical models of cognitive phenomena are still under construction, "one powerful way to improve our intuitions, clarify the key issues and sharpen the debate is through a careful study of simpler idealized models of *minimally cognitive behavior*, the simplest behavior that raises issues of genuine cognitive interest" [4]. We consider minimal cognition as metabolism-independent sensorimotor behavior [5] and presuppose that cognitive behavior generally results from perception-action couplings [6].

The Evolutionary Robotics [7,8] approach to Artificial Life [9] aims at the emergence of such perception-action couplings during the evolution of *complete brain-body-environment systems* [4,10].

As artificial brain structures we utilize recurrent neural networks (RNNs) which can be described as parameterized dynamical systems [11]. We distinguish two types of parameters. The first type concerns parameters of single neurons (bias terms), the structural coupling (the topology), and the strength of these couplings (the synaptic weights). These parameters are shaped by an evolutionary algorithm. The second type of parameters are characterized by the sensor

S. Nolfi et al. (Eds.): SAB 2006, LNAI 4095, pp. 777–788, 2006.

states of a robot, which are represented by the activation of input neurons. They dynamically change during the interaction of a robot with the environment.

Here, we will investigate the dynamical properties of evolved RNNs and their relation to observable collective behavior in groups of robots, especially by systematically exploring the sensor input activations provoked by robot-environment interactions. The detailed analysis of communication underlying dynamical properties of recurrent neural networks and their relation to structural changes during evolutionary processes distinguishes our work from pioneer studies on the evolution of emergent communication among artificial agents [12,13] as well as from more recent studies [14,15]. These studies describe significant results but they mainly focus on a detailed analysis at the behavioral level of communicating artificial agents.

In earlier studies we utilized RNNs to coordinate conflicting behaviors in very large robot groups [16]. There, we *manually designed* a local communication system between several robots in order to synchronize individual internal neural rhythms which determine the behavior of each robot. For the following experiments, we used the same robot platform, but implemented a much simpler task in order to investigate how communication can *emerge* as the basis of cooperation by reducing the predefined knowledge assigned to the evolutionary process. Furthermore, we investigate how evolution shapes certain parameters of behavior underlying dynamical systems and how it adapts these parameters to specific changes of the environment.

2   The Ingredients for the Emergence of Communication

To keep the analysis of evolved RNNs, concerning the dynamical properties and their relation to behavior, still tractable, we use a neuron model with only two parameters, a bias term and a synaptic self-weight [17]. A network consisting of n units is then defined as a parameterized discrete-time dynamical system:

$$a_i(t+1) = \theta_i + \sum_{j=1}^{n} w_{ij}\, f(a_j(t)) \,, \quad i = 1, \ldots, n \,, \tag{1}$$

where $a_i \in \mathbf{R}$ denotes the activity of neuron i, w_{ij} the synaptic strength of the connection from neuron j to neuron i, and θ_i its fixed bias term. The output $o_i = f(a_i)$ of a unit i is given by a sigmoid transfer function, here by $f :=$ $tanh$ (i.e., $o_i \in (-1, 1)$). Although this neuron model is rather simple, already small recurrent networks of this type can generate complex dynamics, such as periodic, quasi-periodic, or even chaotic attractors [11]. For the evolution of these dynamical systems we used an implementation (see [18] for details) of the evolutionary algorithm ENS^3 [19]. This algorithm optimizes the parameters intrinsic to the RNN, such as synaptic weights, bias terms, *and the topology of the network*. For evolution and analysis a physical simulation environment was created. There, we implemented important properties, such as noise of sensors and motors, in accordance with results of measurements done with the physical robot.

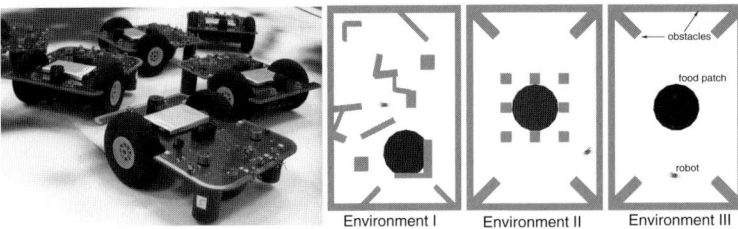

Fig. 1. The physical robot *Do:Little* (left) and three simulation environments of decreasing complexity (I → III)

We utilized the *Do:Little* robot (illustrated in Fig. 1) as an artificial creature. The advantages of this robot are its simple but very reliable sensor and motor capabilities. Besides infrared sensors for obstacle detection and floor sensors for measuring the gray scale of the ground, we especially made use of its robust communication system. It consists of a stereo microphone which can detect the direction of sound signals emitted by nearby robots. In every interval of the robot's update cycle (100 ms) the robot can produce several sound signals of the same frequency. Signals are differentiated by a unique sequence of pulses within one update cycle. The advantage of not coding different signals with different frequencies is that the robots are able to detect their sound signals very reliable even in rather noisy environments. Thus only acoustic signal peaks can be detected but no continuous sound signals. We will see later how this constraint will influence the evolution of communication behavior if we change the population size.

Neurons $I1$, $I2$, and $I3$ represent the left, right, and back infrared sensors, respectively. The sensor inputs are linearly mapped onto $[-1, 1]$, where -1 means no obstacle detection and $+1$ indicates very close obstacles. Neuron $I4$ represents the floor sensor. The inputs are also linearly mapped onto $[-1, 1]$, where -1 indicates white colored and $+1$ black colored ground. For communication we only used one acoustic signal. The angle α of a perceived sound signal to the heading direction of the robot is represented by $o(I5) = 0.5 \cdot (1 + sin(\alpha))$ and $o(I6) = 0.5 \cdot (1 - sin(\alpha))$. The speed of the left and right wheel are calculated by $c \cdot (o(O1) - o(O2))$ and $c \cdot (o(O3) - o(O4))$, respectively, where c is a speed factor. Important for the understanding of the described communication systems is that the robot emits a single sound signal when $o(O5)$ switches from a negative to a positive value.

In the following experiments we wanted to know more about the minimal requirements necessary to provoke the emergence of communication within a population of robots during the evolution of their control architectures. Hence, we defined the following simple task: A single robot can increase its fitness by exploring the environment and finding patches of food while avoiding collisions with obstacles and other robots. Thereby an individual can benefit from the behavior of other robots if they cooperate. Such robots, sharing a common environment, are conspecific because they are identical with respect to their morphology and control, and the selection process during evolution is group based because the mean fitness of all robots in a group is taken[1]. However, this does not necessarily mean that only cooperative behavior can be successful or communication will inevitably emerge. For instance, even solitary behavior can be efficient if each individual is able to locate food patches reliably without running into obstacles or other robots.

To overcome the well known bootstrap problem of the evolutionary approach to the development of behavior [7], we applied a so called semi-restrictive incremental method [18]. Therefor, in the first evolutionary step the task of a single robot was to explore its environment as good as possible without running into obstacles. For this task robots were equipped with infrared sensors for avoiding obstacles. The topology of the neural network was not determined, only input and output neurons were defined. Structural elements, such as synapses and hidden neurons, could freely emerge in between. In the second evolutionary step, robots could additionally access a floor sensor for detecting black food patches on the ground. We selected several different RNNs which were successful in solving the exploration task as a basis for evolving RNNs which are now supposed to force the robots to stay on a food patch as soon as they find one. During this second evolutionary process, already existing structural elements were not allowed to be removed (whereas their parameters could change), but new structural elements could emerge within the *whole* network which now also has new sensor inputs. The same technique was applied for the last step, where robots, in addition, could access a speaker and a stereo microphone for emitting and sensing sound signals. RNNs resulting from the preceding step provided the basis for this evolutionary run.

Consequently, after the first evolutionary step we always put a certain predefined knowledge in each subsequent step. However, this was only done to provide basic behaviors for the evolution of more complex behaviors, for which we never defined how a network eventually should look like. Therefore, we argue that the emergence of communication during evolution was neither explicitly forced by a given network structure nor by the fitness function. Hence, the remaining constraint was the design of the environment.

At first, we thought that a complex environment, such as Environment I, shown in Fig. 1, would enforce the emergence of cooperation. In this environment it is rather complicated for a solitary individual to quickly find the food patch. Once an individual find it perchance, it could use its communication system to guide the others. Surprisingly, even after many repetitions of the evolutionary process no cooperation emerged. In our opinion this is because of the bootstrap problem [7]. It may take too much time until an individual finds the food patch. And consequently, even when it then starts to call other robots, this would not significantly increase the performance compared to robust solitary behavior. Therefore, a stepwise refinement of the communication system (note, the robots had to learn signal-

---

[1] The fitness of a single robot i is $F_i = 600\frac{k_i}{T}$, where T is the number of evaluation time steps and k_i is defined by how often the robot is able to find a food patch in T (whenever the robot finds a food patch, it recharges its virtual battery and is than replaced randomly within the environment).

ing *and* the appropriate responses to other signals) may become very improbable during evolution (this first assumption has to be verified in future work). Thus, we decreased the environmental complexity by removing obstacles and placing the food patch in the center (Environment II), but even there no cooperation emerged during evolution. Only further removal of obstacles (Environment III) enabled the emergence of cooperation, as it will be discussed in the following section. Note, that although all RNNs were evolved in the rather simple Environment III, the resulting cooperative behavior was robust enough that, in the end, we could also observe better performance in the more difficult Environment I and II compared to solitary behavior without any additional optimization.

3   Dynamics of Evolved Communication Systems

3.1   Communication in Small Groups of Robots

One small sized network resulting from the evolution of robot groups containing 10 individuals is drawn in Fig. 2A (we call individuals with this RNN as control architecture individuals of type $A1$). For completeness, the whole network and its parameters are given, but in the following we will concentrate only on the communication system intrinsic to the RNN. By means of an odd loop with over-critical synaptic weights[2], the sound generating output neuron $O5$ is connected with a hidden neuron ($H1$). This loop acts as a switchable oscillator [11] depending on the value of $I4$, the floor sensor input. $I4$ is equal to -1.0 as long as the robot is moving on white ground. As we can see in the bifurcation diagram (Fig. 2C) the oscillation, caused by a period-4 attractor, is switched on by an increased activation of $I4$. The bifurcation point is very close to $I4 = -1.0$, and therefore, it can already be crossed by noise of the floor sensor. However, in order to emit a sound signal at least two points of the periodic orbit have to be in the negative and in the positive domain. This is only the case for $I4 > -0.7$ (never reached by sensor noise only). As a constraint of the environment, food patches always provoke sensor signals of $I4$ within $[0.8, 1.0]$. For these values the output of $O5$ oscillates as shown in Fig. 2B. Thus, communication is context-sensitive: whenever a robot detects a food patch, it emits a sound signal every 4 time steps. This signal triggers a positive taxis in nearby perceiving robots through the input neurons $I5$ and $I6$. Consequently, these robots will approach the signaling robots until they reach the food patch where they then also immediately start signaling. Therefore, communication is unidirectional, because the signaling of one robot alters the behavior of another robot, but this behavioral change does not influence the behavior of the signaler.

Another RNN with a completely different solution for context-sensitive communication is shown in Fig. 3A. We call individuals with this control to be of type $A2$. There, communication is realized by utilizing sensor noise. If no obstacle is close to the left side of the robot ($I1 = -1.0$), the robot will stay on

---

[2] Here, due to the use of *tanh* as activation function, over-critical means a synaptic weight $|w_{ij}| > 1.0$. See [3] for details.

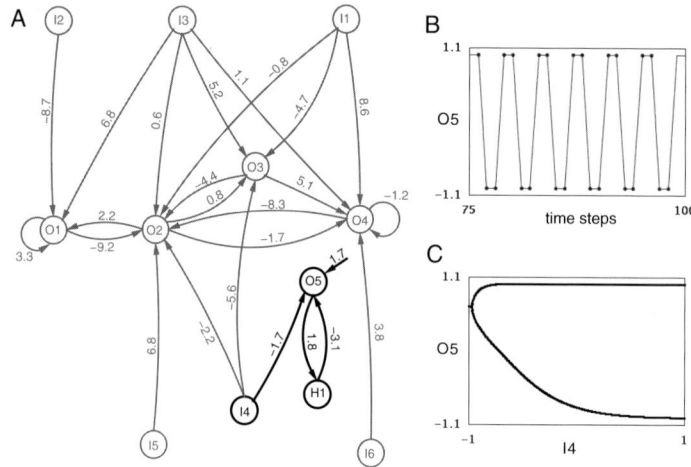

Fig. 2. A: RNN of *A1* individuals (group size: $n = 10$ individuals during evolution) with inputs from left ($I1$), right($I2$), and back($I3$) distance sensors, left ($I5$) and right ($I6$) sound sensors, and the floor sensor ($I4$). Output neurons $O1$ & $O2$ and $O3$ & $O4$ steer the left and right wheel, respectively. $O5$ controls the signaling. $H1$ is a hidden neuron. B: Signals of $O5$ when a robot stays on a food patch ($I1 = I2 = I3 = -1.0; I4 = 1.0; I5 = I6 = 0.0$). C: Bifurcation diagram for $O5$ by varying $I4$ ($I1 = I2 = I3 = -1.0; I5 = I6 = 0.0$).

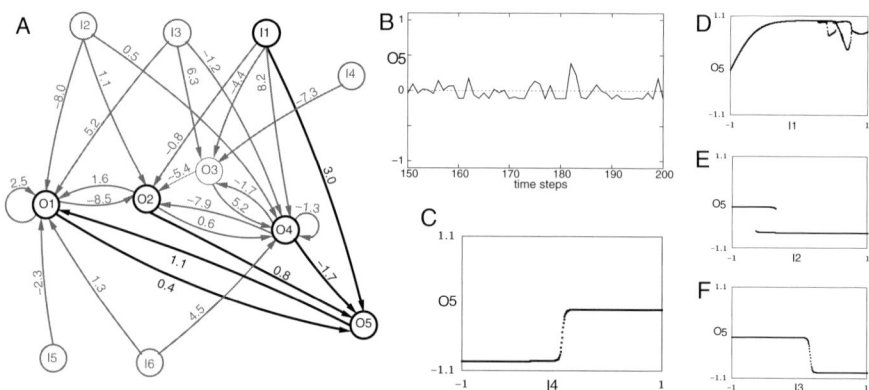

Fig. 3. A: RNN of *A2* individuals ($n = 10$ during evolution, cf. Fig. 2A). B: Signals of $O5$ when the robots stays on a food patch (cf. Fig. 2B). C: Bifurcation diagram for $O5$ by varying $I4$ (cf. Fig. 2C). D-F: Bifurcation diagram for $O5$ by varying each distance sensor ($I1, I2, I3$) input, where $I4 = 1.0; I5 = I6 = 0.0$ (not varied distance sensor inputs are set to -1.0).

a detected food patch because the outputs of $O1$ and $O2$ are equal (within the upper saturation domain of the transfer function, i.e., 1.0) as well as the outputs of $O3$ and $O4$ (within the lower saturation domain, i.e., -1.0). Correlating these values with the connections projecting to $O5$ (emphasized in Fig. 3A) leads to an activation of $O5 = 0$. In this situation the asymptotically stable fixed point (SFP in the following) is in the *linear domain* of the transfer function. Hence, the neural activation is highly sensitive to the sensor noise of $I1$. Therefore the output crosses the zero-line randomly from the negative to the positive domain (see Fig. 3B). Diagram C shows how the SFP of $O5$ is shifted from the *lower saturation domain* to the *linear domain* when the robot detects a food patch via $I4$. Once the robot stays on a food patch, this SFP can be shifted away depending on the activation of $I1, I2$, and $I3$ (Fig. 3D-F). That means, if another robot approaches a signaling robot, and consequently activates its infrared sensors accordingly, the signaling will cease.

Using noise for behavior control of autonomous robots is usually not wanted and engineers try to eliminate it from their systems as often as possible. In our example, it is a quite efficient solution for signaling. Infrared sensors are always noisy, and we tested different noise levels in the simulation environment[3] with the result that the behavior does not qualitatively change when we vary the noise level between 2% and 10%. In contrast to the prevention of noise in most technical applications, for biological systems it is well known that noise can significantly enhance sensorimotor patterns by means of a mechanism called Stochastic Resonance [20].

When we compare $A1$ and $A2$ with respect to their performance depending on the group size (see Fig. 5), we see that the more individuals are interacting in the same environment, the better $A2$ performs compared to $A1$ (if $n > 7$). The reason is the described constraint of the physical communication system, namely the ability for perceiving only sound signal peaks. The more individuals of $A1$ are signaling at the same time, the higher the probability that their signals will sum up to a continuous signal which cannot be perceived anymore by other robots still searching for food. Already four individuals of $A1$ can produce a continuous signal when they are all signaling with different phases. Note, this is not simply an artifact of the simulation. Experiments with physical robots have also shown that the maximal frequency, where two subsequent signals can be distinguished, is $5Hz$.

In contrast to the constant period-4 signals in $A1$, the individuals of $A2$ signal rather randomly. Consequently, the probability of producing a continuous signal for a longer time period is rather low in larger groups. Additionally, whenever a food patch becomes crowded, signaling robots will perceive nearby robots by their infrared sensors which in turn will stop their signaling, as we have discussed above (Fig. 3D-F). Thus, communication can no longer be described as unidirectional because signaling of food patch locations will attract other robots which in turn influence this signaling behavior as soon as they come close to the signaler.

Nevertheless, both control architectures resulted from the evolution with a group size of 10 individuals. For this size the performance difference between $A1$ and $A2$ is not as significant as it becomes with increased group size (Fig. 5).

---

[3] The noise level of the physical infrared sensors is between 4% and 6%.

3.2   Evolutionary Changes of Communication in Larger Groups

The results of the previous section suggest that performance may improve when we repeat the evolution with a larger group size. More interesting than a simple performance improvement would be to see how the communication system, as a part of the RNN, will change when we increase the number of interacting individuals. Therefore, we started a new evolutionary run with the control architecture of *A1* as initial structure (which was more appealing because of its independence of sensor noise and its lower fitness at larger population size). We increased the group size to 25 individuals and allowed again parameter changes of the initial RNN as well as the emergence of new structural elements.

One resulting RNN is shown in Fig. 4A. When we compare the structural elements responsible for the communication system with the initial RNN of *A1* (Fig. 2A), we notice the same odd loop between *H*1 and *O*5 with over-critical synaptic weights. In addition, we found an over-critical self-connection at *H*1. With the given weight configuration this module exhibits quasi-periodic oscillations (Fig. 4B) which are switched on by an increased activation of the floor sensor *I*4 (see Fig. 4C). We applied a power spectrum analysis to the time series in Fig. 4B and found a mean period length of about 8.7 time steps. The period of time between emitting two subsequent sound signals is now almost twice as long as in *A1*. Although this is presumably a coincidence, the correlation is interesting because the group size used in evolution of *B1* is also almost twice as large as it was used for the evolution of *A1*.

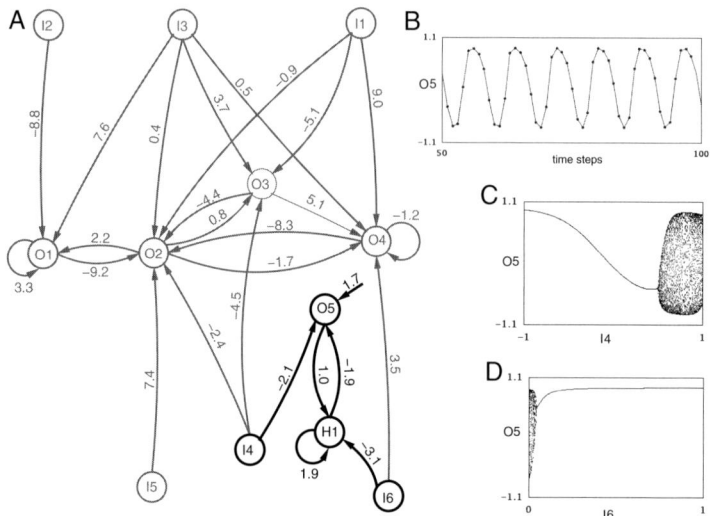

Fig. 4. A: RNN of *B1* individuals ($n = 25$ during evolution, cf. Fig. 2A). B: Signals of *O*5 when the robot stays on a food patch (cf. Fig. 2B). C: Bifurcation diagram for *O*5 by varying *I*4 (cf. Fig. 2C). D: Bifurcation diagram for *O*5 by varying *I*6 with $I1 = I2 = I3 = -1.0; I4 = 1.0; I5 = 0.0$.

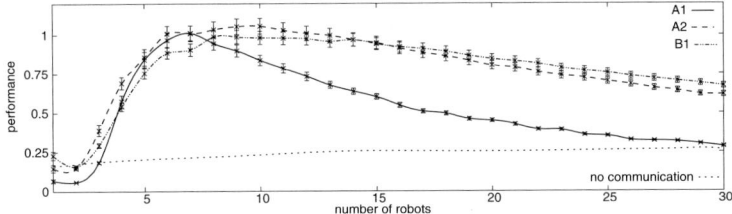

Fig. 5. Performance of *A1*, *A2*, and *B1* individuals. For each group size several simulation runs ($N = 25$) were performed, each lasting 18,000 time steps. Performance is the mean number of finding a food patch in 600 time steps for a single robot (mean of all robots in a group is taken). The mean performance of *A1*, *A2*, and *B1* individuals where the communication system was deactivated is drawn as a reference (dotted line).

Another new structural element is the connection between *I6*, the left microphone input, and *H1*. Whenever a robot is staying on a food patch, and therefore *I4* > 0.8, the described quasi-periodic oscillation (see Fig. 4B-C) is responsible for sound emission. As soon as another nearby robot also starts signaling, *I6* will become activated (even when the other robot is to the most right side, which is due to the high noise, approx. 30%, of the sound direction detection). Then, as we can see in the bifurcation diagram of Fig. 4D, the quasi-periodic attractor switches to a SFP, thus the oscillation will cease. Because the sound signal of signaling robots lasts only one time step, these oscillations immediately start again in the next time step (*I6* = 0). This reset mechanism will lead to a synchronization of the signaling among robots which stay together on the food patch (a mechanism very similar to the synchronization of internal neural rhythms described in [16]). That means, if there are many robots on a food patch, they will not produce a continuous sound signal as it is the case for robots of type *A1*. One can see the improvement of the performance with respect to *A1* in Fig. 5. However, the performance of *B1* is not significantly higher compared to *A2* (although *A2* was only evolved with a population size of 10). The next section will discuss this surprising robustness of *A2* against more complicated environmental conditions which did not occur during its course of evolution.

3.3   Discussion

The results presented in section 3.1 demonstrate an example of the phenomena called natural drift, which is well known from evolution of biological systems [6]. We started several evolutionary runs for the rather simple task of exploration and obstacle avoidance. The initial conditions were always the same (empty RNN structure, fitness function, environment). Evolution came up with solutions being very different concerning the network's topology, but very similar concerning the observed behavior. Secondly we started again several evolutionary runs, with different RNNs as initial structure which resulted from the preceding step. The task was slightly more complex (i.e., individuals had to find food patches and should stay there). Again, several RNNs resulting from this evolution were selected as initial structures for the final evolutionary step. This time individuals could develop a communication system in order to cooperate. During this runs we still allowed the emergence of new structural changes within the initial RNNs. And we presented two completely different solutions of emergent communication (*A1* and *A2* individuals) to the same task as a result of different structural changes. Both perform well with respect to the given fitness function and group size with which evolution took place. However, individuals utilizing noise for communication (*A2*) do also perform well in conditions they were not confronted with during their course of evolution. They posses the intrinsic property to be robust against increased population size not only because of the lowered probability of producing unrecognizable continuous sound signals, but also because of the described indirect bidirectional communication behavior. We argue, such solutions can hardly be found when too much predefined knowledge about the fitness function and topology of RNNs is assigned to the evolutionary process.

In section 3.2 we demonstrated another striking result of the described experiments and analysis: the evolutionary adaptivity of RNNs, as dynamical systems for behavior control, to varying environmental conditions, such as the number of interacting robots. We observed how small structural changes within such networks lead to an adaptation of the communication mechanism. During the evolution with small group sizes a context-sensitive communication system developed which is based on a simple two neuron loop that provided period-4 oscillations (*A1* individuals). In this case robots directly communicate the discovery of a food patch unidirectionally to other robots. The behavior was sufficient to improve the performance of the robot group, as we have defined it there. However, by changing the environmental conditions, that is, by increasing the number of robots, it turned out that this strategy was not sufficient enough anymore. In our experiments this was especially due to the physical implementation of the sound perception system. Confronted with this constraint and the larger group size, small structural changes refined this solution. These changes lead to quasi-periodic oscillations of longer periods which then are also synchronized among interacting agents (*B1* individuals). Communication is not longer just a simple stimulus-response action. It is direct bidirectional: the act of signaling also directly influences the signaling behavior of other robots.

4   Conclusions

In this paper we showed how communication among interacting autonomous robots emerges by evolving dynamical systems, like recurrent neural networks, in the context of *complete brain-body-environment systems* [4,10]. We have seen that only small structural changes are necessary to alter previously solitary behavior to cooperative behavior among communicating robots.

The presented communication system and collective behavior are indeed rather simple. And often the answer to what is necessary in the Artificial Life approach to place autonomous robots into the same category as animals is to keep up climbing the complexity ladder. However, we agree with [21] in that this is "not the most practical answer .. [because].. seeking such complexity blindly, by typically restricting the search to achieving more complex behaviors, does not accomplish much". Therefore, our approach is to build simple, and therefore still tractable, models of *minimally cognitive behavior* [22] and to increase their complexity as soon as our understanding improves [23].

There are also some relations to biological systems, we can draw from our observed processes of *evolutionary adaptation*. We setup a communication system in the robot where the perception is limited to signal peaks. This was thought to be a disadvantage for the development of behavior, although it is of great advantage for the interaction with highly noisy real-world environments. In the end, by evolving dynamical systems, we have to argue, that such physical constraints are not necessarily a disadvantage for the development of behavior. Evolution finds solutions which integrate the properties of such physical system very well. On a more abstract phenomenological level, we can compare our artificial system with biological systems, for instance with the synchronized flashing of male fireflies during mating [24]. This astonishing collective process is in general also based on pulse coupled oscillators [25]. Although the process which leads to this synchronization is now well understood, the evolutionary reason why thousand of fireflies synchronize their flashing can only be assumed. One possible explanation is that females are stronger attracted by sudden bright pulses than by a clutter of single flashes. It is also well known that humans, or animals in general, react to sudden changes in their environment stronger than to sustained sensory inputs (it is also known that persistent stimuli can attenuate sensation [26]). Here, we unintentionally put this property into our system. There was no other choice than utilizing only acoustic changes instead of continuous signals. And evolution found solutions able to handle this handicap and adapt to changes in the environment in a very efficient way considering the size of the resulting networks. We argue that such solutions are hardly found when too much predefined knowledge about the topology and the dynamics are assigned to such systems, however compelling this may seem in order to speed up the evolutionary process.

References

1. Port, R., van Gelder, T.: Mind as Motion. MIT Press (1995)
2. Thelen, E., Smith, L.: A Dynamic Systems Approach to the Development of Cognition and Action. MIT Press (1994)
3. Pasemann, F.: Neuromodules: A Dynamical Systems Approach to Brain Modelling. In: Supercomputing in Brain Research: From Tomography to Neural Networks. World Scientific (1995) 331–347
4. Beer, R.D.: The dynamics of active categorical perception in an evolved model agent. Adaptive Behavior **11**(4) (2003) 209–243
5. van Duijn, M., Keijzer, F., Franken, D.: Principles of minimal cognition. Adaptive Behavior **14**(2) (2006) in press
6. Maturana, H.R., Varela, F.J.: The Tree of Knowledge: The Biological Roots of Human Understanding. rev. edn. Shambhala (1992)
7. Nolfi, S., Floreano, D.: Evolutionary Robotics: The Biology, Intelligence, and Technology of Self-Organizing Machines. MIT Press (2000)

8. Harvey, I., Di Paolo, E.A., Wood, R., Quinn, M., Tuci, E.: Evolutionary robotics: A new scientific tool for studying cognition. Artificial Life **11**(1–2) (2005) 79–98
9. Langton, C.: Artificial Life: An Overview. MIT Press (1995)
10. Pfeifer, R., Scheier, C.: Understanding Intelligence. MIT Press (1999)
11. Pasemann, F.: Complex dynamics and the structure of small neural networks. Network : Computation in Neural Systems **13**(2) (2002) 195–216
12. MacLennan: Synthetic ecology: an approach to the study of communication. In Langton, C., Taylor, C., Farmer, J., Rasmussen, S., eds.: Proc. Artificial Life II, Addision-Wesley (1991)
13. Werner, G., Dyer, D.: Evolution of communication in artificial organisms. In Langton, C., Taylor, C., Farmer, J., Rasmussen, S., eds.: Proc. Artificial Life II, Addision-Wesley (1991)
14. Quinn, M., Smith, L., Mayley, G., Husbands, P.: Evolving controllers for a homogeneous system of physical robots: Structured cooperation with minimal sensors. Philosophical Transactions of the Royal Society of London, Series A: Mathematical, Physical and Engineering Sciences **361**(1811) (2003) 2321–2344
15. Marocco, D., Nolfi, S.: Emergence of communication in teams of embodied and situated agents. In: Proc. of the 6th Int. Conference on the Evolution of Language. (2006)
16. Wischmann, S., Hülse, M., Knabe, J., Pasemann, F.: Synchronization of internal neural rhythms in multi-robotic systems. Adaptive Behavior **14**(2) (2006) in press
17. Pasemann, F.: Dynamics of a single model neuron. International Journal of Bifurcation and Chaos **2** (1993) 271–278
18. Hülse, M., Wischmann, S., Pasemann, F.: Structure and function of evolved neurocontrollers for autonomous robots. Connection Science **16**(4) (2004) 249–266
19. Dieckmann, U.: Coevolution as an autonomous learning strategy for neuromodules. In: Supercomputing in Brain Research: From Tomography to Neural Networks. World Scientific (1995) 427–432
20. Gammaitoni, L., Hänggi, P., Jung, P., Marchesoni, F.: Stochastic resonance. Reviews of Modern Physics **70**(1) (1998) 223–287
21. Di Paolo, E.A.: Organismically-Inspired Robotics: Homeostatic Adaptation and Teleology Beyond the Closed Sensorimotor Loop. In: Dynamical Systems Approach to Embodiment and Sociality. Advanced Knowledge International (2003) 19–42
22. Beer, R.D.: Toward the evolution of dynamical neural networks for minimally cognitive behavior. In: From Animals to Animats 4: Proc. of the 4th Int. Conference on Simulation of Adaptive Behavior. (1996) 421–429
23. Beer, R.D.: Arches and stones in cognitive architecture. Adaptive Behavior **11**(4) (2003) 299–305
24. Camazine, S., Deneubourg, J.L., Franks, N.R., Sneyd, J., Theraulaz, G., Bonabeau, E.: Self-Organization in Biological Systems. Princeton University Press (2001)
25. Strogatz, S.H., Stewart, I.: Coupled oscillators and biological synchronization. Scientific American **269**(6) (1993) 102–109
26. Bays, P.M., Flanagan, J.R., Wolpert, D.M.: Attenuation of self-generated tactile sensations is predictive, not postdictive. PLoS Biology **4**(2) (2006)

Origins of Communication in Evolving Robots

Davide Marocco and Stefano Nolfi

Institute of Cognitive Science and Technologies, CNR,
Via San Martino della Battaglia 44, Rome, 00185, Italy
{davide.marocco, stefano.nolfi}@istc.cnr.it

Abstract. In this paper we describe how a population of simulated robots evolved for the ability to solve a collective navigation problem develop individual and social/communication skills. In particular, we analyze the evolutionary origins of motor and signaling behaviors. Obtained results indicate that signals and the meaning of the signals produced by evolved robots are grounded not only on the robots sensory-motor system but also on robots' behavioral capabilities previously acquired. Moreover, the analysis of the co-evolution of robots individual and communicative abilities indicate how innovation in the former might create the adaptive basis for further innovations in the latter and vice versa.

1 Introduction

The development of embodied agents able to interact autonomously with the physical world and to communicate on the basis of a self-organizing communication system is a new exciting field of research ([13], [1], [10], [3], [9], for a review see [2], [11], [14] and [7]). The objective is that to identify methods of how a population of agents equipped with a sensory-motor system and a cognitive apparatus can develop a grounded communication system and use their communication abilities to solve a given problem. These self-organizing communication systems may have characteristics similar to that observed in animal communication [5] or human language.

In this paper we describe how a population of simulated robots evolved for the ability to solve a collective navigation problem develop individual and social/communication skills. In particular, we analyze the evolutionary origins of motor and signaling behaviors. Obtained results indicate that the signals and the meaning of signals produced by evolved robots are grounded not only on robots sensory-motor system but also on robots' behavioral capabilities previously acquired. Moreover, the analysis of the co-adaptation of robots individual and communicative abilities indicate how innovations in the former might create the adaptive basis for further innovations in the latter and vice versa.

In the next section we describe the experimental setup (for more details on the experiments and on the characteristic of the communication system at the end of the evolutionary process, see [4]). In section 3, we describe the evolutionary origin of the communication system used by evolved robots. Finally, in section 4, we summarize the main results and we briefly discuss the implications of these experiments.

S. Nolfi et al. (Eds.): SAB 2006, LNAI 4095, pp. 789–803, 2006.

2 The Experimental Set-Up

A team of four simulated robots placed in an arena of 270x270cm (Fig.1, Left) are evolved for the ability to find and remain in the two target areas by equally dividing between the two targets. Robots communicate by producing and detecting signals up to a distance of 100cm. A signal is a real number with a value ranging between [0.0, 1.0].

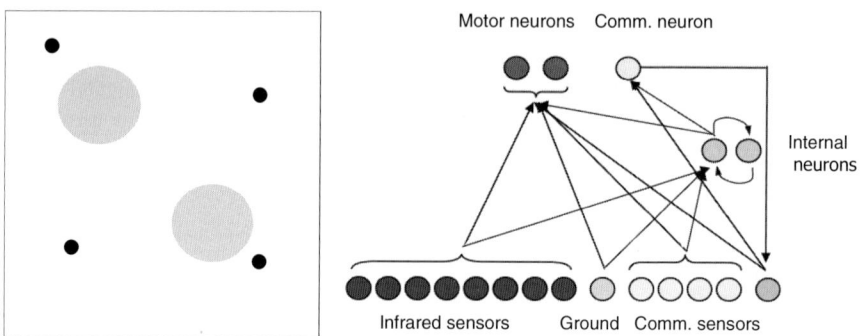

Fig. 1. Left: The environment and the robots. The square represents the arena surrounded by walls. The two gray circles represent two target areas. The four black circles represent four robots. Right: The neural controller of evolving robots.

Robots' neural controllers (Fig. 1, Right) consist of neural networks with 14 sensory neurons that encode the activation states of 8 infrared sensors, 1 ground sensor (that binarily encodes the color of the ground), 4 communicative sensors (that encode the value of the signals produced by other robots from four corresponding orthogonal directions (i.e. frontal [315°-44°], rear [135°-224°], left [225°-314°], right [45°-134°]), and the activation state of the communication neuron at times t-1 (i.e. each robot can hear its own emitted signal at the previous time step). These sensory neurons are directly connected to the three motor neurons that control the desired speed of the two wheels and the value of the communication signal produced by the robot. The neural controllers also include two internal neurons that receive connections from the sensory neurons and from themselves and send connections to the motor and communicating neurons [8]. The three motor neurons encode the desired speed of the two wheels of the robot and the value of the signal emitted by the robot.

The output of motor neurons is computed according to the logistic function (2), the output of sensory and internal neurons is computed according to function (3) and (4), respectively (for more details on these activation functions and on the relation with other related neural models see [6]).

$$A_j = t_j + \sum_i w_{ij} O_i \tag{1}$$

$$O_j = \frac{1}{1+e^{-Aj}} \qquad (2)$$

$$O_j = O_j^{(t-1)} \tau_j + I_j \left(1 - \tau_j\right) \qquad (3)$$

$$O_j = O_j^{(t-1)} \tau_j + \left(1 + e^{-A_j}\right)^{-1} \left(1 - \tau_j\right) \qquad (4)$$

With A_j being the activity of the j_{th} neuron, t_j being the bias of the j_{th} neuron, w_{ij} the weight of the incoming connections from the i_{th} to the j_{th} neuron, O_i the output of the i_{th} neuron, $O_{j(t-1)}$ being the output of the j_{th} neuron at the previous time step, τ_j the time constant of the j_{th} neuron, and I_j the activity of the j_{th} sensors.

The free parameters of the robots' neural controllers have been evolved through a genetic algorithm. Each team of four robots was allowed to "live" for 20 trials (each trial lasting 100 seconds, i.e. 1000 lifecycles of 100 ms each). At the beginning of each trial the position and the orientation of the robots was randomly assigned outside the target areas. The fitness of the team of robots consists of the sum of 0.25 scores for each robot located in a target area and a score of -1.00 for each extra robot (i.e. each robot exceeding the maximum number of two) located in a target area. The total fitness of a team is computed by summing the fitness gathered by the four robots in each time step.

The initial population consisted of 100 randomly generated genotypes that encoded the connection weights, the biases, and the time constants of 100 corresponding neural controllers. Each parameter was encoded with 8 bits and normalized in the range [-5.0, +5.0], in the case of connection weights and biases, and in the range [0.0, 1.0], in the case of time constants. Each genotype was translated into 4 identical neural controllers that were embodied in the four corresponding robots, i.e. teams were homogeneous and consisted of four identical robots. For a discussion about this point and alternative selection schemas see [7]. The 20 best genotypes of each generation were allowed to reproduce by generating five copies each, with 2% of their bits replaced with a new randomly selected value. The evolutionary process lasted 2000 generations (i.e. the process of testing, selecting and reproducing robots is iterated 2000 times). The experiment was replicated 10 times starting by 10 different initial populations.

By analyzing the fitness thorough out generations we observed that evolving robots are able to accomplish their task to a good extent in all replications from generation 500 on (evolving robots are able to find and remain in the two target areas by equally dividing between the two areas in 58.3% of the trials). Further increases of performance observed from generation 500 on, are due to slight improvements with respect to the ability to solve the task faster (the average time required by the four robots of all replications to reach the two target areas goes from 74s to 67s in generation 500 and 2000, respectively) and better (the percentage of trials in which the task is solved correctly increase from 58.3% to 67.5%, in generation 500 and 2000 respectively).

By comparing these results with the results obtained in a control condition in which robots were not allowed to detect signals (i.e. in which the state of the communication sensors was always set to 0.0) we observed that, in all replications, the fitness reach a stable state after 150 generations, which is significantly lower than the case in which robots are allowed to communicate (i.e. robots are able to solve the problem only in 36.7% of the trials after 2000 generations).

The comparison between the results obtained in the normal and in the control condition in which robots are not allowed to detect other robots' signals indicates how the possibility to produce and detect other robots' signals is necessary to achieve optimal or close to optimal performance.

In the next subsection we will analyze the evolutionary origins of robots ability to solve their task and of the communication system displayed by evolved individuals.

3   Origins and Evolution of a Self-organized Communication System

To understand the evolutionary origins of robots' communication system we analyzed the motor and signaling behavior of evolving robots through out generations. To reconstruct the chain of variations that led to the final evolved behavior we analyzed, for each replication, the lineage of the best individual of the last generation (i.e. the 1999 individuals, one for each generation, that constitute the ancestors of the best individual of generation 2000). Below we report the results of this analysis by focusing in particular on the best replication of the experiment. The analysis of the other replications of the experiment (not shown) produced qualitatively similar results (although the values of the signals serving a given function and the length of different evolutionary phases vary significantly).

As shown in Fig. 2 and Fig. 3, in the case of the best replication of the experiment, the fitness quickly increases by reaching high level performance during the first 50 generations (the team of robots of generation 50 is able to solve the problem in 64%

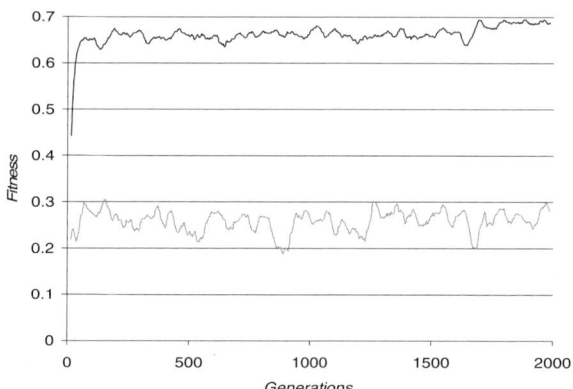

Fig. 2. Fitness of the lineage of the best individual of generation 2000 through out generations in the case of the most successful replication of the experiment. The black and gray lines represent the performance in a normal and no-signal condition (in which robots are not allowed to detect other robots' signals). Lines indicate the moving average over 30 generations. Each individual have been tested for 100 trials.

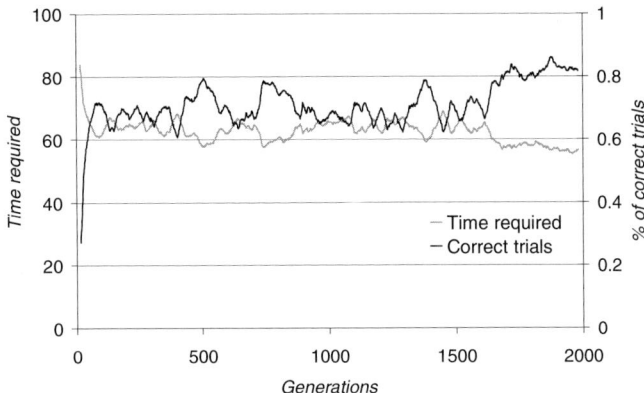

Fig. 3. Percentage of trials in which robots accomplish the task successfully (within 100s) and average time required by the robots to reach the target areas by equally dividing between the two areas throughout generations. Average performance obtained by testing each team for 100 trials.

of the trials and the average time required to reach the two target areas is 65.4s). From generation 50 to generation 1700, the fitness remains rather stable, beside small and unstable increases in performance. From about generation 1700 on, performance stabilizes again on a slightly higher value with respect to previous generations (the team of robots of generation 2000 are able to solve the problem in 79% of the trials and the average time required to reach the two target areas is 57.6s).

By analyzing the motor and signaling behavior through out generations in the case of the best replication of the experiment (the same replication shown in Fig 2 and 3) we observed the following phases:

Generation 1. At this stage robots move in the environment by producing curvilinear trajectories and by avoiding obstacles (in most of the cases). Robots produce two stable signals with a value of 0.53 and 0.33 when they are located inside or outside a target area, respectively, and far from other robots. Moreover, robots produce highly variable signals when they interact with other robots located nearby.

In particular, when a robot located outside a target area starts to detect the signal emitted by another robot, it modifies the signal produced by a stable signal with a value of about 0.33 (a signal that we will call **A** that is produced by robots that do not detect signals produced by other robots) to an highly variable signal with an average value of 0.28 (a signal **B** that is produced by robots detecting the signal **A** or **B** produced by another robot). Signal **B** increases robots' exploratory abilities (i.e. the probability to reach target areas). Indeed, by testing the robots in a normal condition and in a control condition in which they are not allowed to produce the signal **B**, we observed that the average time spent by the robots to reach a target area for the first time is 58.8s and 70.4s, in the normal and control condition respectively. Therefore, the functionality of signal **A** is that to trigger the production of signal **B**. The functionality of signal **B** is that to increase robots navigation ability, as described above.

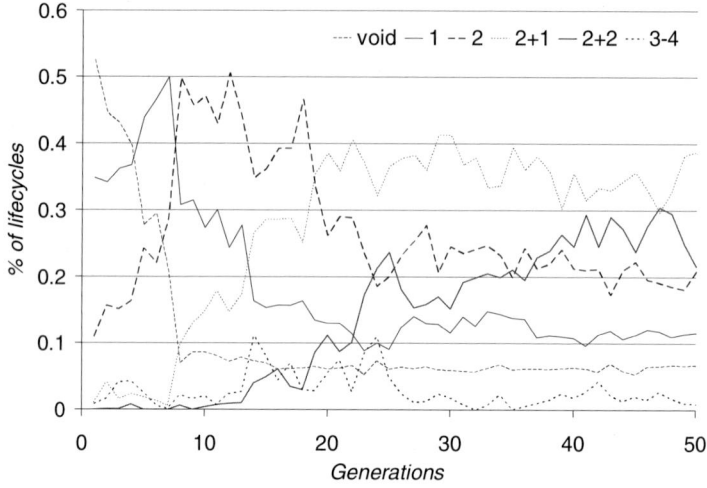

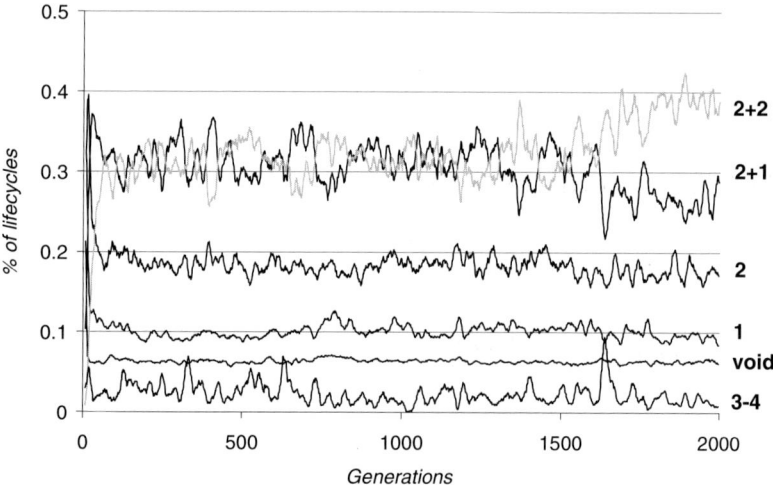

Fig. 4. Percentage of lifecycles spent by a team of robots in 6 possible states: void = all robots are outside target areas, 1 = only one robot is located in a target area, 2 = two robots are located in target areas (either in the same or in two different areas), 2+1 = 3 robots are located inside two different target areas, 2+2 = all robots are located inside the target area equally divided between the two areas, 3-4 = 3 or 4 robots are located in the same target area. The data refer to the lineage of the best individual of the last generation for the best replication of the experiment. Each robot have been tested for 100 trials lasting 1000 lifecycles. Top graph: data up to generation 50. Bottom graph: data from generation 0 to 2000.

We do not assign an identification letter to the other signals produced by robots at this stage since these signals does not seem to have any clear adaptive function. Some of these signals however, for example an highly varying signal (average value 0.45) and a stable signal (with a value of about 0.53) produced by robots located in a target area interacting or not interacting with other robots located nearby will acquire a functional role in successive generations.

On the basis of these individual and social behaviors (i.e. an individual obstacle avoidance behavior, individual exploration behavior, and a social behavior based on signals that alters other robots' trajectories in a way that enhances their chance to reach target areas) robots are able to spend almost half of their lifetime on target areas (Fig. 4). A typical behavior observed at this stage is shown in Fig.5 (Gen. 1).

Generations 2-7. During this phase robots progressively evolve an individual ability to remain in target areas. Indeed, at generation 7, robots located on target areas rotate on the spot so to remain there for the rest of the trial. Moreover, robots produce several differentiated signals. However, as in the previous phase, only two of these signals have an adaptive function.

As in the case of generation 1, robots located outside target areas produce a signal **A** (a signal with a value of about 0.34 produced by robots located far from other robots), and a signal **B** (a varying signal with an average value of 0.24 produced by robots interacting with other robots located outside target area). We keep the same labels introduced above since, although the value and the effect of the signals slightly varied, the functionality of the signals is very similar to that of the signals described in the previous section. As for generation 1, the functionality of signal **A** is that to trigger the production of signal **B** and the functionality of signal **B** is that to increase robots' ability to reach target areas. Moreover, as in the case of generation 1, robots located on target areas produce two non-adaptive signals: (1) an highly varying signal with an average value of about 0.73 (produced by robots that interact with other robots located nearby outside target areas), and (2) a stable signal with a value of about 0.82 (produced by robots that do not detect signals produced by other robots). These two signals do not have any adaptive function, but rather produce a decrease in robots' performance. Indeed, the production of these two signals reduce the chances that robots located outside target area join target areas that already contain a single robot.

As a result of the newly developed individual behavior that allows robots to remain on target areas, however, the percentage of lifecycles in which one or two robots are located on a target area increases from 35% to 45% and from 10% to 22%, respectively (see Fig. 4). A typical behavior observed at this stage is shown in Fig.5 (Gen. 7).

Generations 8-14. The development of an individual ability to remain on target areas developed in previous generations posed the adaptive basis for the development of a cooperative behavior that allows robots located on a target area alone to attract other robots toward the same target area. As we said in the previous section, the highly varying signal produced at generation 7 by robots located inside a target area interacting with other robots located outside the area reduced the chances that the latter robots join the area. At generation 14, however, this highly varying signal is not produced anymore. This innovation results from the fact that robots located outside target

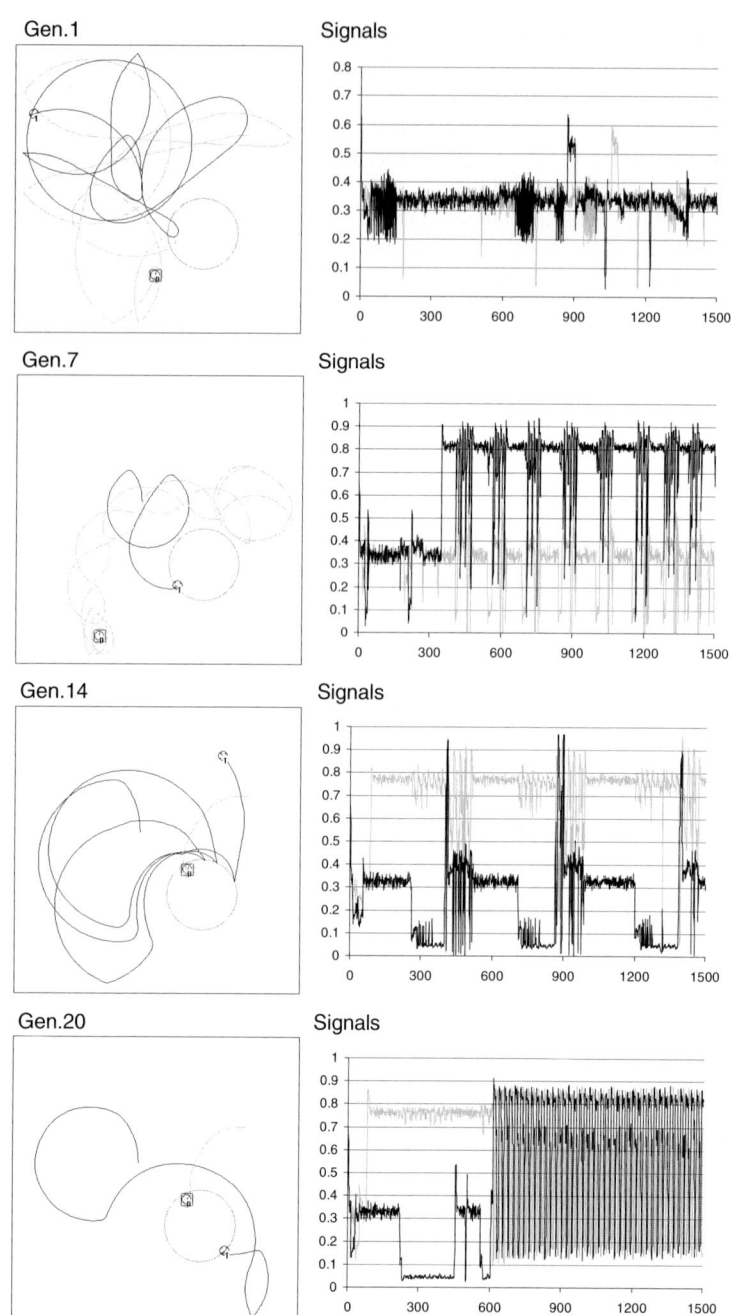

Fig. 5. Motor and signaling behavior observed at different generations. Left: the trajectory produced by two robots tested in an environment including a single target area. Right: the signals produced by the two robots during the test shown in the left part of the figure. The motor trajectory and the signal of the first and of the second robot are shown with black and gray lines, respectively.

areas interacting with robots located inside target areas now produce signal **D**, (i.e. a signal with a value of 0.04). Since producing a signal with an almost null value is equivalent to stop signaling, the production of signal **D** implies that robots located inside a target area alone now produce a signal **C** (a stable signal with a value of about 0.78) independently of whether they interact or not with another robot located nearby outside the target area. Since the signal **C,** produced by a robot located inside a target area, increases the chances that other robots will enter in the same target area, the innovation that allows robots located outside the target area to switch their signaling behavior off (as soon as they detect signal **C**) produces a significant adaptive advantage.

To summarize, during this phase robots develop an ability to produce a new signal (signal **D**) whose functionality is that to allow robot located inside target area to keep producing signal **C** even when other robots are located nearby. This in turn allows robots to exploit the effect of signal **C,** that consists in attracting other robots toward the source of the signal (i.e. toward the corresponding target area). This effect of signal **C** on other robots motor behavior already existed in previous generations. However it could not be exploited since robots located in target area were able to produce signal **C** only when no other robots were located in the communicative range.

The acquisition of an ability to switch signaling behavior off leads to a specialization of the role of the two interacting robots since, in these situations, the robot located in the target area and producing the signal **C** acts as a speaker and the robot located outside the target area producing signal **D,** acts as a hearer. The social interaction between the two robots in this circumstance, therefore, can be described as a form of information exchange (in which a speaker robot located inside a target area informs the hearer robot on the location of the target area and in which the hearer robot reacts to the signal by moving toward the direction of the area) or as a form of manipulation (in which the speaker robot drives the hearer robot toward the target area by exploiting the tendency of the hearer robot to alter its motor trajectory as a result of a detected signal).

At this stage, robots are not still able to remain in a target area in couple (see Fig. 5, Gen. 14). In fact, as soon as a second robot reaches a target area, the two robots start to produce two different signals (i.e. two highly varying signals with an average value of 0.63 for the former and 0.38 for the latter robot) that are maladaptive since they increase the chances that one of the two robots abandons the area.

As a result of the innovations occurring during this phase (that mainly consist in the variations that leads to the production of signal **D**) the percentage of lifecycles in which two and three robots are located on a target area increases from 22% to 50% and from 0% to 18% (Fig. 4).

Generations 15-20. The development of an ability to attract nearby robots toward target areas that contain a single robot described in the previous section leads to an increase in performance but also poses new adaptive opportunities, namely the need to develop an ability to remain into target areas that contain a single robot and the need to produce a signal that keep other robots away from a target area that contains two robots. These two problems are solved in this phase through variations that allow robots to not exit from target areas when they detect the signal produced by another robot located in the same target area. This is achieved through the development of a new signal **E** (an highly varying signal with an average value of 0.61 produced by robots located in a target area that contains two robots). Signal **E** plays two adaptive functions: (1) it does not push the other robot located in the target area out (unlike the signals previously produced in this circumstance), and (2) it reduces the chances that other robots located outside the target area will join the area itself. Interestingly, signal **E** (i.e. the signal produced by two interacting robots located in the same target area) allows the two robots to generate an information (i.e. that encode the fact that the area contains two robots) that is not directly available to none of the two robots.

Generations 21-1700. During this long evolutionary phase the performances of the robots, the number of signals, and the functionalities of signals remain rather stable. Evolving robots display close to optimal performance, few simple but crucial individual behaviors (that allow the robots to explore the environment, avoid obstacles, and remain into target areas) and an effective communication system that now includes 5 signals (i.e. signals **A**, **B**, **C**, **D**, and **E** described in previous sections) that modulate the robots' behavior by producing an enhanced exploratory behavior, a target approaching behavior, and a target avoidance behavior. Since each of these individual and communication abilities provides a clear adaptive advantage, all of them are preserved during the rest of the evolutionary process.

Despite of that, some characteristics of the individual behavior exhibited by the robots, the value of the signals serving a given function, and the impact of signals on other robots' behavior vary significantly.

Variations of individual behaviors mainly concern how robots explore the environment while they do not detect signals produced by other robots. This fact can be explained by considering that robots' ability to find target areas on their own plays a limited adaptive value at this stage in which individuals posses a reliable ability to find target areas by exploiting the signals produced by other robots. Variation on individual behavioral abilities, however, can be tolerated only within limits. To illustrate this point let us consider how robots' individual exploratory behavior varies during this phase. As we reported above, robots located outside target areas tend to produce a curvilinear trajectory and to avoid obstacles. The combination of these two behaviors allow the robots to explore different parts of the environment and to encounter target areas relatively quickly. The turning angle with which robots move forward, however, should be sufficiently large so to avoid turning on the same position indefinitely. The turning angle of the robots in this circumstance is indeed a character that is subjected to significant variations until a certain threshold is reached. Variations that overcome the threshold tend to be maladaptive since they lead to robots that are unable to explore the environment without the help of other robots (as shown in Fig. 6, Gen. 225). However, their negative effects only manifest in robots that do not receive the necessary social help during their lifetime. As a consequence, these variations might be retained and might cause a drop in performance in successive generations until characters similar to those previously lost are restored (for an example, see Fig. 6, Gen. 226-230). This analysis illustrates how individual behavior, such as individual abilities to explore the environment, does not only poses the evolutionary basis for the emergence of the communication system, but still plays a fundamental role when the communication system is established. This individual behavior, in fact, also constitutes a pre-requisites for the ability of the robots to collect information to be communicated or to create the conditions for receiving useful signals.

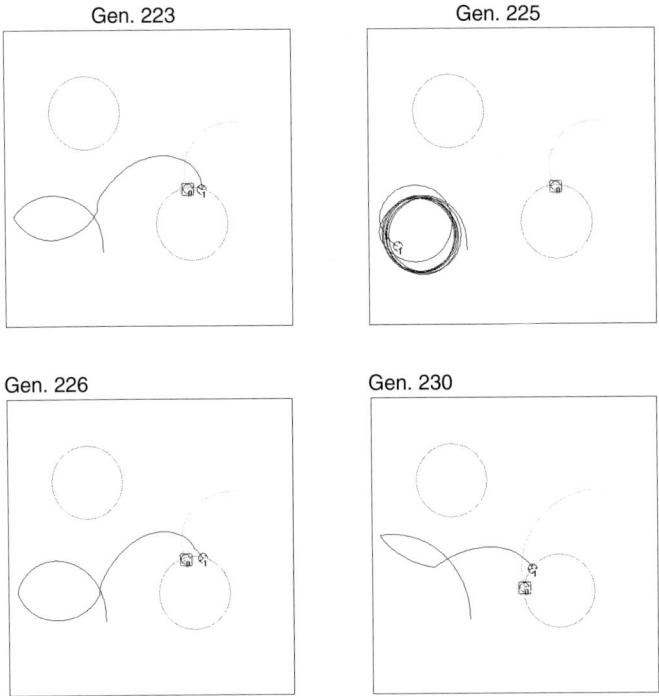

Fig. 6. Behavior exhibited by robots of different generations. For space reasons only the behavior produced by two robots of different generation is displayed. As can be seen, at generation 225 robots lose their ability to explore the environment and keep circling in the same area. The exploration ability is recovered in successive generations.

Other characteristics that significantly vary during this phase are the value of the signals and the way in which signals affect robots behavior. Although the functionality of the five signals described above remains rather constant during this phase, the value associated to each signal significantly vary (Fig. 7). This fact can be explained by considering that the functionality of a signal depend both on the value of the signal and the effect that the signal produces on robots. The possibility to co-adapt the value of signals and the impact of a signal on robots' motor and signaling behavior, ensures that the functionality can be preserved while the signals and their effects co-vary.

In principle, these neutral variations could lead to new organizations of the communication system, that might represent a pre-requisite for further innovations of individual and communicative abilities. Some preliminary evidences suggest that this is indeed one of the reasons that explain the evolutionary transition that leads to better behaviors in the next phase (see below). This evidences however are only preliminary and should be integrated with further analysis that we plan to conduct in the future.

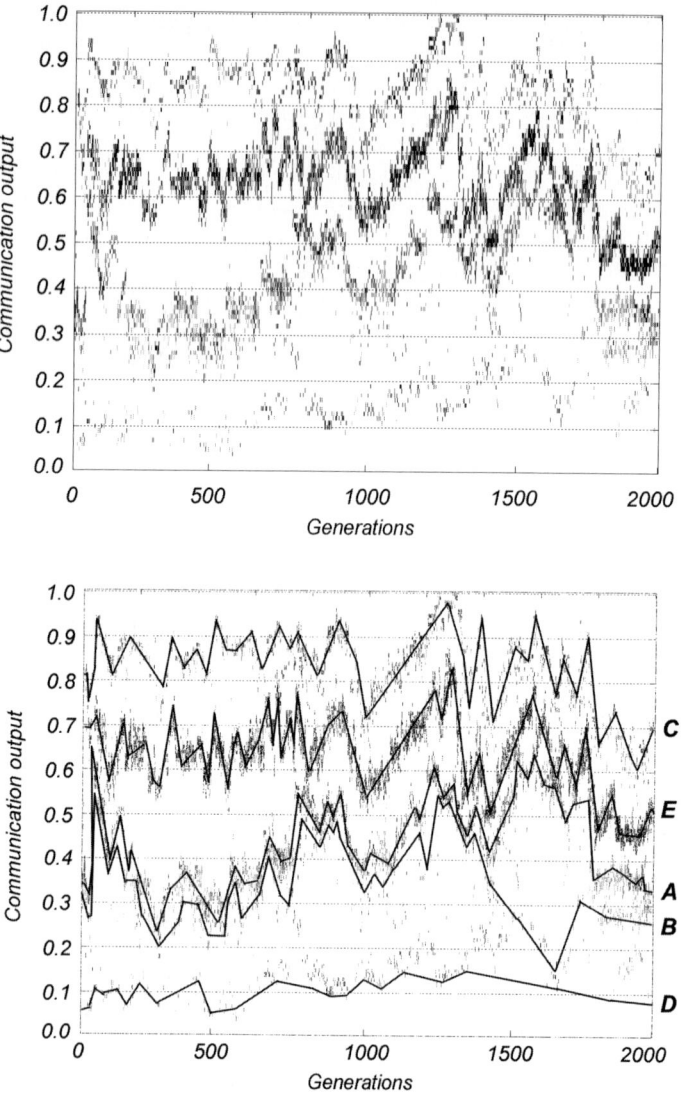

Fig. 7. The value of the signals produced by robots throughout generations. Each point represents the mean value of each signal. The gray scale of each point indicates the variance of the signal with respect of the mean value (i.e. the darkness of each point is proportional to the variability of the corresponding signal). The data displayed on the graph have been obtained by filtering out signals that are produced only occasionally. Oscillatory signals have been identified through a wavelet analysis. The bottom figure displays the same data of the top figure with a superimposed schematization of signals average value through out generations.

Generations 1700-2000. After a long phase in which performances remain rather stable, a small but stable increase in performance is observed from generation 1700 on (Fig. 2). Indeed, as shown in Fig. 3, the percentage of times in which the robots are able to accomplish their task correctly increase from 62% to 79% and the average time required for solving the problem decreases from 67s to 57s during this evolutionary phase.

The evolutionary transition that leads to this improvement involves a significant reorganization of the values of the five signals (see Fig. 7). Although the number and the general functionality of the signals remains the same, from generation 1700 on, the values of the five different signals are distributed on a wider range and the value of each signal is distributed on a smaller range with respect to previous generations.

Other variations occurring in this phase might affect the way in which signals are exploited. In particular, from generation 1700 on robots often display: (a) an enhanced ability to avoid target areas that already contains two robots without remaining plugged into unfruitful conditions, (b) an ability to reach the target areas faster by taking the risk to end up in a target area that already contains two robots but by also being able to exit from these areas, (c) an enhanced ability to negotiate situations in which robots concurrently receive signals from several robots. However, further analysis should be conducted to clarify the nature and the adaptive role of the innovations occurring during this phase.

4 Conclusion

In this paper we described how a population of simulated robots, evolved for the ability to solve a collective navigation problem, develop an effective communication system. By analyzing the evolutionary origins of motor and signaling behaviors we observed that the co-adaptation of robots' motor and communicative abilities plays a crucial role on the evolutionary dynamic.

In some cases the development of new motor skills poses the basis for the successive development of new social abilities. For instance, the development of an ability to remain in target areas constitutes a pre-requisite for the development of an ability to communicate the location of the area to other robots so to increase the chances that other robots will join the same target area. In other cases, the development of social/communication abilities pose the basis for the development of new motor skills. For instance the development of an ability to detect the number of robots located in a target area through bi-directional signaling interactions creates the basis for the development of an effective avoidance behavior that allow robots to avoid entering in crowded target areas and to look for another target areas.

Interestingly the co-adaptation process of motor and social/communicative abilities may potentially lead to open-ended evolutionary dynamics in which innovations create the adaptive basis for further innovations thus leading to a progressive increase in performance and to a progressive complexification of agents abilities. Indeed, while during the first phase of the evolutionary experiment robots can only rely on few environmental cues (that provide information on whether they are located on a target area or not and whether they are close to obstacles), in later generation they can exploit a much larger number of cues (that, for example, provide information also on the location of target areas and on the number of robots located in target areas).

Finally, we observed how the complexification of robots' motor and social skills involve different aspects, and can be characterized along several dimensions: (a) an increase in the number of elementary behaviors exhibited by the robots, (b) an increase in the number of signals produced by robots, (c) an increase in the number of ways in which the same signal affect robots' behaviors in different contexts, (d) a differentiation of the modalities with which communication is regulated (e.g. the transformation of symmetrical interaction forms in which communicating robots act concurrently as speakers and hearers to specialized asymmetrical interaction forms in which one robot acts as a speaker and one robot acts as an hearer).

From a scientific point of view, these types of experiments and results can allow us to understand better how 'meanings' originate and how signals are grounded in agents sensory-motor and behavioral abilities. From an application point of view, these methods can allow us to develop a new generation of artifacts able to solve practical problems by cooperating and communicating on the basis of a self-organized communication system.

Acknowledgments

The research has been supported by the ECAGENTS project funded by the Future and Emerging Technologies program (IST-FET) of the European Community under EU R&D contract IST-1940.

References

1. Cangelosi A., Parisi D.: The emergence of a 'language' in an evolving population of neural networks. Connection Science. Taylor & Francis (1998) 10: 83-97.
2. Kirby S.: Natural Language from Artificial Life. Artificial Life. The MIT Press, Cambridge (2002) 8(2):185-215.
3. Marocco D., Cangelosi A., Nolfi S.: The emergence of communication in evolutionary robots. Philosophical Transactions of the Royal Society A, London (2003) 361: 2397-2421.
4. Marocco D., Nolfi S.: Emergence of Communication in Embodied Agents Evolved for the Ability to Solve a Collective Navigation Problem. Connection Science. Taylor & Francis (in press).
5. Marocco D., Nolfi S.: Communication in Natural and Artificial Organisms. Experiments in evolutionary robotics. In: Lyon C., Nehaniv C., Cangelosi A. (eds.): Emergence of Communication and Language, Springer-Verlag (in press).
6. Nolfi S.: Evolving robots able to self-localize in the environment: The importance of viewing cognition as the result of processes occurring at different time scales. Connection Science. Taylor & Francis (2002) (14) 3:231-244.
7. Nolfi S.: Emergence of Communication in Embodied Agents: Co-Adapting Communicative and Non-Communicative Behaviours. Connection Science. Taylor & Francis (2005) 17 (3-4): 231-248.
8. Nolfi S., Marocco D.: Evolving Robots Able To Integrate Sensory-Motor Information Over Time. Theory in Biosciences. Urban & Fischer Verlag, Berlin, (2002) 120: 287-310.
9. Quinn M., Smith L., Mayley G., Husbands P.: Evolving controllers for a homogeneous system of physical robots: Structured cooperation with minimal sensors. Philosophical Transactions of the Royal Society A, London (2003) 361: 2321-2344.
10. Steels L. The Talking Heads Experiment. Antwerpen, Laboratorium (1999). Limited Pre-edition.

11. Steels L. Evolving grounded communication for robots. Trends in Cognitive Science, Elsevier (2003) 7(7): 308-312.
12. Steels L., Kaplan F.: AIBO's first words: The social learning of language and meaning. Evolution of Communication. John Benjamins (2001) 4: 3-32.
13. Steels L., Vogt P.: Grounding adaptive language games in robotic agents. In: Husband P., Harvey I. (eds.): Proceedings of the 4th European Conference on Artificial Life. The MIT Press, Cambridge (1997).
14. Wagner K., Reggia J.A., Uriagereka J., Wilkinson G.S.: Progress in the simulation of emergent communication and language. Adaptive Behavior Sage Publications (2003), 11(1): 37-69.

The Complexity of Finding an Optimal Policy for Language Convergence

Kiran Lakkaraju[1] and Les Gasser[2]

[1] Department of Computer Science
University of Illinois, Urbana-Champaign
klakkara@uiuc.edu
[2] Graduate School of Library and Information Science
University of Illinois, Urbana-Champaign
gasser@uiuc.edu

Abstract. An important problem for societies of natural and artificial animals is to converge upon a similar language in order to communicate. We call this the language convergence problem. In this paper we study the complexity of finding the optimal (in terms of time to convergence) algorithm for language convergence. We map the language convergence problem to instances of a Decentralized Partially Observable Markov Decision Process to show that the complexity can vary from P-complete to NEXP-complete based on the scenario being studied.

1 Introduction

Language is a collective property of the society. A language is inherently a communicative system (although it has some non-communicative interactions with agents, Clark ([1]) suggests that in addition to a communicative function, language can serve as a tool to reshape the computational space that our brains must handle), that allows agents to interchange information.

In this work we study how a set of initially diverse (in terms of languages) agents can come to an agreement upon a single language. We refer to this as the *language convergence* problem.

Previous work in this area has focused on the convergence rate of a particular algorithm. Each agent has a learning algorithm which will learn a language based on examples of sentences from other agents. The algorithm for convergence usually specifies a set of agents that each agent can interact with, and the parameters of the learning algorithm.

In this paper we want to explore the question, how hard is it for an agent to learn how to converge? We do not want to know how to converge in a specific setting, but rather how to converge in a whole set of situations. For instance, we want a *policy* for the agent that will tell it whom to interact with in order for the agent to be able to communicate with the entire society after a period of time.

Other work has focused on evaluating single algorithms to determine if, when an agent follows a specific policy, will the entire society converge. For instance,

S. Nolfi et al. (Eds.): SAB 2006, LNAI 4095, pp. 804–815, 2006.

Cucker, Smale and Zhou give bounds on how many other agents each agent must interact with in order for the entire population to converge, given a policy where each agent is to learn from the sentences it gathers of other agents languages ([2]). Steels creates a simulation and empirically shows the convergence of a population of agents in [3]

Our work differs from the above because we want to study the higher order problem of how hard it is to learn an algorithm for convergence, not how long it takes to converge using a particular algorithm. We want to find the optimal algorithm for convergence. The optimal algorithm, when implemented by an agent, will result in the quickest convergence. To study the complexity of finding an optimal algorithm for convergence we show how to map instances of language convergence problems to instances of a Decentralized Partially Observable Markov Decision Process (*Dec-POMDP*) [4]. The optimal algorithms for convergence correspond to the optimal joint policies of a *Dec-POMDP* . We make use of previous complexity results for finding the optimal joint policy for a *Dec-POMDP* ([4], [5])

By mapping language convergence scenarios to the *Dec-POMDP* model we can gain insight on the computational complexity of finding an optimal solution. This provides us with insight on the worst-case complexity of solving these language convergence scenarios.

In this paper, four language convergence scenarios are examined, single goal oriented, multiple goal oriented, teacher-student, and teacher-student with population observation. Each scenario can be modeled as a type of *Dec-POMDP* .

In Section 2 we go over the *Dec-POMDP* model and the mapping to the language convergence situation. Next, we examine four different language convergence scenarios mapped to the *Dec-POMDP* . Section 3 examines the complexity of the four different language convergence scenarios. Finally we talk about some related work, future work, and conclusions.

2   Language Convergence as a *Dec-POMDP*

In this section we describe the language convergence problem as a *Dec-POMDP* .

A *Dec-POMDP* is very similar to a POMDP except that in a *Dec-POMDP* the state changes based on the actions of multiple agents. In a *Dec-POMDP* , we have a set of agents embedded in an environment, modeled as a global state. The agents can execute actions that produce a change in the environment and possibly a reward. Each agent makes its own observation about the environment at each time step. In a POMDP there is only a single entity controlling the system. While the process is controlled by multiple agents, there is only one reward which is based on the single global state.

The structure of the behavior of the agents is:

1. Each agent, in parallel, observes the environment. This generates an observation for each agent.
2. Each agent, in parallel, chooses an action by using their policies and the observation they have just perceived.

3. The global state changes, based on the current global state, and the actions of every agent.
4. One reward is generated for all the agents based on the previous state, the actions executed, and the resulting state.

Formally, a *Dec-POMDP* is a tuple:

$$M =< S, A, P, R, \Omega, O, T >$$

where (assuming the number of agents is n):

- S is a finite set of states, with initial state s^0. The state at the current time will be called the *global state*.
- $A = \{A_i | A_i$ is a finite set of actions for agent i.$\}$
- P is a transition function, giving the probability $P(s'|s, a_1, \ldots, a_n)$ of moving from state s to state s', given actions $a_1, \ldots, a_n$, where a_i is the action executed by agent i.
- R is a global reward function, giving the system-wide reward $R(s, a_1, \ldots, a_n, s')$ when actions $a_1, \ldots, a_n$ cause the state-transition from s to s'.
- $\Omega = \{\Omega_i | \Omega_i$ is a finite set of observations for agent i $\}$
- O is an observation function, giving the probability $O(o_1, \ldots, o_n | s, a_1, \ldots, a_n, s')$ that each agent i observes o_i when actions $a_1, \ldots, a_n$ cause the state transition from s to s'. Where o_i is the observation of agent i.
- T is the time-horizon (finite or infinite) of the problem.

A joint policy $< \delta_1, \ldots, \delta_n >$, is a set of local policies, δ_i where

$$\delta_i : \Omega_i^* \rightarrow A_i \tag{1}$$

The joint policy specifies a policy for each agent that will determine the action an agent should take at each time step based on the sequence of observations it has made. Figure 1 is an illustration of the *Dec-POMDP* model.

See [5] for a full description of various classes of the *Dec-POMDP* . Roughly, we can characterize the various sub-classes of *Dec-POMDP* by how much of the global state each agent can observe (from each agent fully observing the global state, to each agent only observing its own "local" state) and the accuracy to which they can view the states (from viewing the state itself to viewing an observation of it). Different combinations of these properties induce different complexities when solving the *Dec-POMDP* .

A *Dec-POMDP* has independent transitions if the global state can be factored into n components such that the actions of an agent affects only its component. An independent transition *Dec-POMDP* will be referred to as an IT, *Dec-POMDP*

A *Dec-POMDP* has independent observations if the state can be factored into n components such that the observations of an agent depend only upon its component and the actions it has executed. An independent observation *Dec-POMDP* will be referred to as an IO,*Dec-POMDP*

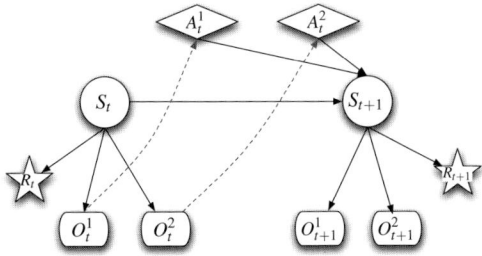

Fig. 1. Illustration of a *Dec-POMDP* . S_t represents the state of the environment at time t. A_t^1 and A_t^2 represent the actions that agents 1 and 2 executed at time t. R_t is the reward at time t, and O_t^1 and O_t^2 are the observations at time t. The dotted arrows between the observations and actions indicate that these observations were known when making the decision on which action to execute at the next time step.

Following [5], we assume that the same decomposition of the state holds for the independent transition and independent observations. We will refer to an agents component of the state as its *partial view* or its *local state*. The partial view of an agent will be denoted by S_i.

An example of an IO, IT *Dec-POMDP* is a simple gridworld situation. Suppose multiple agents are wandering around a 2-d gridworld. The state of the system would be the aggregate locations of each agent. The partial view of each agent would be its location. It is easy to see that any action an agent does (for instance "Move North") will only affect its own state, thus satisfying the independent transition property. We can further have each agent observe only its own location, thus satisfying the independent observation property.

If each agent can determine the global state based only on its sequence of observations, we say the *Dec-POMDP* is *Fully-Observable*. In our gridworld example, this would be like each agent knowing the locations of every other agent based only on its own observations.

If there is a mapping from the aggregate observations of every agent to the current state, then we say the *Dec-POMDP* is *Jointly Fully-observable* A jointly fully-observable *Dec-POMDP* is called a *Dec-MDP* . The gridworld example above is actually a *Dec-MDP* . The aggregation of each agents observations is, by definition, the state of the system.

If each agent can determine its local state from its sequence of observations, then we say the *Dec-POMDP* is *Locally Fully-observable*.

A finite horizon Goal-oriented *Dec-MDP* is a *Dec-MDP* with the following conditions (taken from [5]):

1. There exist a set of states $G \subset S$ of global goal states. At least one state of G must be reachable by some joint policy
2. The process ends at time T

3. All actions in A incur a cost, $C(a_i) < 0$. For simplicity we assume the cost is dependent only upon the action.
4. The global reward is $R(s, < a_1, \ldots, a_n >, s') = \sum_{i=1}^{n} C(a_i)$
5. If at time T the system is in a state $s \in G$ there is an additional reward $JR(s) \in \Re$ that is awarded to the system for reaching a global goal state.

A *GO-Dec-MDP* has *uniform cost* when the cost of all the actions are the same. There is also a NOP action that an agent can perform which has cost 0 and does not change the state.

2.1   Finding the Optimal Policy

The main question is, how hard is it to find a joint policy that maximizes the expected total return over the finite horizon? Bernstein et. al. in ([4]) have shown that deciding whether there exists a joint policy with at least a certain value, via an off-line algorithm, is NEXP-Complete for *Dec-POMDP* and *Dec-MDP* where $n \geq 2$.

The work in *Dec-POMDP* s has looked at finding a joint policy *offline*. This means that the model is known, and as many simulations as needed can be run. While during the search for the policy the model is known, during the execution of the policy the agents will not know the entire model. The joint policy that is to be found must take into account the constraints of the agents during execution of the policy.

Goldman et. al. study the complexity of various subclasses of the *Dec-POMDP* problem in [5]. Table 1 summarizes the results from [5]

Table 1. Complexity of *Dec-POMDP* and related models. The third column indicates where this result was obtained. The lemmas and section 3 refer to [5]. The NBCLG property will be examined in Section 2.3.

Model	Complexity	
Dec-POMDP	NEXP-C	[4]
Dec-MDP	NEXP-C	[4]
IO,IT Dec-MDP	NP-C	Lemma 4
IO, IT Dec-POMDP	NEXP	Section 3
GO-Dec-MDP	NEXP-C	Lemma 3
IO, IT, GO-Dec-MDP, 1 goal	P-C	Lemma 5
IO, IT, GO-Dec-MDP, NBCLG	P-C	Lemma 6

2.2   Mapping the Language Convergence Problem to a *Dec-POMDP*

The *Dec-POMDP* is an appropriate model to use to study the language convergence problem because the *Dec-POMDP* explicitly models decentralized control. There is a global goal - that of the entire population having the same language, but only local control - each agent independently decides what action to take. A *Dec-POMDP* explicitly models this situation, as there is a global reward based on the state of the system, yet the dynamics of the system are based on the aggregate decisions of each agent.

The first issue when mapping language convergence scenarios to *Dec-POMDP* is how to represent a language. We look at a simple type of language, represented as an association matrix. A language is considered as a mapping from meanings (objects, actions) to signals (words). There are many ways this mapping could be represented (for instance the mapping is a continuos function from the space of meanings to the space of objects in[2]) but one popular way is as an association matrix. The rows of an association matrix correspond to meanings, and the columns to words. An entry at row i and column j denotes the association between meaning i and word j. In this work we do not constrain languages to be represented as association matrices. We do, however, require that the number of different languages be finite. An association matrix can represent synonyms (multiple entries in a row) and homonyms (multiple entries in a column).

When representing a language as an association matrix, agents can change their language by modifying the association values in the matrix. In this work we assume that each agent has a finite set of actions that can modify its language.

In the language convergence case we want to reward the population when all the agents have the same language. We can do this by setting the state space to reflect the languages used by the agents. Let α be a set of n agents. Each agent can use a language from the finite set L of languages. At each point in time, every agent will be using a particular language from L. Let l_i denote the language of agent i. We set the state of the *Dec-POMDP* at time t to be the aggregate of the languages for each agent: $s_t = <l_1, l_2, \ldots, l_n>$. Thus the state set will be L^n.

An optimal joint-policy is a policy that, when the agents use it, will result in quick convergence to a high value state. We can formally specify this by setting the reward function to reward quick convergence. We constrain the reward function to be independent of the actions and the previous state, $R(s, a_1, \ldots, a_n, s') = R(s')$. Every state will have a small negative reward, except for the states where every agent has the same language. These states will have a high positive reward.

$$R(s' = \langle l_1^{t+1}, \ldots, l_n^{t+1} \rangle) = \begin{cases} 1 & if \ l_1^{t+1} = l_2^{t+1} = \ldots = l_n^{t+1} \\ -\epsilon & otherwise \end{cases}$$

Where l_i^{t+1} is the language used by agent i at time $t+1$ and ϵ is a small positive constant.

Under this reward function the policy that maximizes reward will minimize the number of states to get to a converged state from s_0.

We can also model situations in which specific languages have different rewards. As a means of communicating information, a language must be *effective* (allow agents to communicate all important meanings),*efficient* (computable, tractable) and *shared* (each agent must be able to understand each other).

We can assign to each language an objective value based on its effectiveness and efficiency. The reward function will give a higher reward to a state where all the agents have converged on a more effective and efficient language.

The set of actions that each agent can execute will depend on the type of languages that each agent can use. In this paper we will assume that each agent has the same action set, A. The transition probability function P will define the effects of executing an action.

The observations possible to an agent also have a large effect on the computational complexity of finding a solution. In this paper we will assume that each agent has the same set of observations, Ω.

2.3   Four Language Convergence Scenarios

In this section we go over four different language convergence scenarios. They differ from each other in terms of how much information each agent has about itself and the other agents in the population.

Single Goal Oriented. One of the simplest cases of a convergence problem occurs when there is only one language that has a positive reward. In this simple case, each agent can only observe its own language. In addition, let us assume that the action the agents can execute will change their language in some form, but the effect of the action does not depend upon the language of the other agents. In this situation each agent is moving in the state space trying to find the language that every agent will converge upon. This can be mapped to a uniform cost IO, IT *GO-Dec-MDP* .

An example of this situation would be where a language is represented as an association matrix where each row can only have a single entry. This means that each language does not contain any synonyms or homonyms. The set of actions would be the set of row swaps - that is we swap the meanings for two words.

Assume that s^g is the single global goal state that the agents want to converge to. Then since we are using an uniform cost *GO-Dec-MDP* , every state except s^g will have a negative reward.

Since the effect of the actions will only change the language of the agent that executed the action; and since the effect is determined only by the agents language and not on the other agents languages, the system satisfies the independent transition property.

Since the observations of a single agent depend only upon its own language the system satisfies the independent observation property.

Thus, we have a uniform cost, independent transition, independent observation, jointly-fully observable *GO-Dec-POMDP* . By Lemma 5 of [5] deciding this problem is P-Complete.

Intuitively this result makes sense, even though each agent only observes its own language. The policy for each agent can be determined independently of the policies for every other agent. Since it is known that all the agents will eventually reach the single goal state (since that is the only state that provides a positive reward), we can decompose this problem into n separate MDP's (assuming that the *Dec-POMDP* is locally fully observable).

Many Goal-Oriented. We can extend the Single Goal-Oriented case to involve multiple languages that the population can converge upon. This is more realistic since there are often multiple languages that a population can converge upon.

This situation is difficult because each agent might choose to pursue a different goal. Each agent will have to coordinate with the other agents to choose the same goal, which will lead to a very complex space of policies to search in.

Goldman and Zilberstein outline the *No Benefit to Change Local Goals* (NBCLG) property, which, if satisfied, will allow the system to be decomposed into a set of MDP. If a system satisfies the NBCLG property, it is P-Complete (Lemma 6 of [5]

The NBCLG property basically makes sure that it never is beneficial for an agent, while executing the optimal joint policy, to switch which goal state to go to. For instance, suppose that there is an optimal joint policy to one of the goal states. This can be computed for each agent by constructing a MDP for each agent to its component of the joint goal state. While executing the optimal joint policy an agent might veer from the optimal route (since the effects of actions are probabilistic, there is a chance that this might happen). If the NBCLG property is satisfied then even when an agent veers off the optimal route, it is guaranteed that the agent will not switch to another goal state.

Satisfying the NBCLG property depends upon the structure of the transition probability function. Verifying that a system satisfies this property would be quite difficult as well, since we would have to compute the value of changing goals at every intermediate step.

Teacher-Student. In many situations agents change their language via a *language game* ([6]). In a language game, a speaker and a hearer agent are drawn from the pool of agents. The agents interact with each other, exchanging words or sentences from their language. After the agents interact, either the hearer or both the speaker and hearer change their language based on the communicative success of the interaction.

What makes this situation different is that the language of an agent changes based on the language of another agent. In our previous examples, each agent modified its language independent of the languages of the other agents.

We can model this situation in a *Dec-POMDP* by having the actions correspond to the execution of a language game with a particular agent. There will be n actions, one for initiating a language game with each agent. The effects of these actions are to change the language based upon the language of the agent executing the action as well as the agent that is chosen to talk with.

In this case the independent transition property does not hold. The probability of an agents state at time $t + 1$ depends on both participants of the language game. The *Dec-MDP* still has independent observations though. Thus this situation can be modeled as an IO, *Dec-MDP* .

The complexity of an IO, *Dec-MDP* has not been studied yet. The complexity of an IO, *Dec-MDP* is bounded by the complexity of an IO,IT *Dec-MDP* (NP-Complete by Lemma 4 of [5]) and the complexity of a *Dec-MDP* (NEXP-Complete)

Teacher-Student with Population Observation. We can extend the previous Teacher-Student case by having each agent observe not just its own language, but the languages of the whole population. This situation will be mapped to a *Dec-MDP* .

Instead of having each agent observe its own language, we can allow the agents to sample the languages of the the entire population. In this case, the observations are dependent upon the state of the entire population and not just on the state of the current agent.

For instance, the observation of an agent might be of the language that is the most used. Or else, altering Ω to be the set of natural numbers the observation can be the number of agent using the same language as the observing agent. In either of these situations the observations depend upon the state of the population and not the partial view of the agent.

In this case, the language convergence problem is mapped to a *Dec-MDP* . The complexity of finding an optimal solution to a *Dec-MDP* is NEXP-Complete.

3   Complexity of Language Convergence

The four situations outlined above varied widely in terms of complexity. What makes the different situations easy or difficult to solve? The key is the level of uncertainty present in the system. There are two levels of uncertainty present, the first is the agents uncertainty of its own state, and the second is the agents uncertainty about the state of the other agents. Both of these factors affect the complexity of finding an optimal solution. Uncertainty about the agents state means that there will be an exponential number of possible policies that must be searched. Uncertainty about the state of other agents affects the size of the joint policy space that must be searched through.

In the general case the local policy of each agent will be a mapping from sequences of observations to actions. The policy must be from sequences of observations to actions because the agent is uncertain about the state that it is in. This means that there are $|A|^{\Omega^T}$ possible policies for the agent (where T is the finite horizon).

On the other hand, when the agent has knowledge of its state, the size of the policy can be substantially reduced. See [5] for more details.

While uncertainty about the local state of an agent affects the size of a policy, uncertainty about the state of other agents affects the number of policies that must be searched. If each agent knew the state of all the other agents then we could just model this as a MDP or POMDP and solve it. But since each agent does not know the state of the other agents we have to search through the combinations of policies.

In the single goal oriented case, each agent knew with certainty its current state. Lemma 1 of [5] proves that an IO,IT *Dec-MDP* is locally fully observable. This means that the size of the space of policies that need to be searched can be reduced because we don't have to consider all possible sequences of observations.

Rather, a policy for an agent will be a mapping from the local states of the agent to actions. This significantly decreases the size of the space of policies to search through.

In addition, there is no need to search through a joint policy space in the single goal oriented case. Since there is only one state with a positive reward, and each agent is striving to maximize reward, it is unnecessary to consider the policies of the other agents. It is guaranteed that at some point all the agents will reach the single goal state. Because of this assumption, finding an optimal joint policy reduces to finding n different policies, one for each agent. This is much less complex than searching for a single joint policy.

We can see that in the single goal oriented case there is no uncertainty about the local state of the agent and no uncertainty about the behavior of the other agent. Thus finding a solution is $P - Complete$.

The second situation, multiple goal oriented, is very close to the first situation except that we have added uncertainty about the state of the other agents. In the multiple goal case, the agents might converge upon different goal states, thus we cannot simplify the situation to finding n different policies.

If the *Dec-POMDP* satisfies the NBCLG property, though, it is like the single goal oriented case. Finding a policy for an NBCLG satisfying *GO-Dec-MDP* is similar to finding a policy for a single goal oriented *GO-Dec-MDP* . Since we know that once a goal is chosen no agent will veer from that goal, we are free to look at each goal state, and find the optimal policy for each agent to get to its partial view of the goal state. The goal state chosen will be the one which has the highest reward. Since we know the agents will never veer from going towards this goal state, we have found the optimal policy.

The third situation is another case where the agent does not know the state of the other agents, it is similar to the multiple goal oriented case.

The fourth situation, teacher-student with population observation, provides the most complex case. In this situation each agent does not know its own state, nor does it know the state of the other agents. Thus finding a policy is computationally expensive, since each policy will have to take into account all the possible sequences of observations, and all combinations of local policies will have to be considered.

4   Related Work

A good review of many Multi-Agent System models to the language convergence problem is given in [7]. There has been some work in studying the theoretical underpinnings of MAS models. Cucker, Smale, and Zhou [2] provide a mathematical formulation for a MAS simulation. In their work, each agent gets a set of example sentences from every other agent based on a pre-specified level of interaction between the agents. They investigate the number of examples each agent must be exposed to in order for the population of agents to converge.

[4] introduces and studies the complexity of finding optimal policies for the *Dec-MDP* and *Dec-POMDP* models. In this paper they show that deciding

these problems is NEXP-Complete. In other papers they present algorithms for constrained versions of these problems.

[5] studies various modifications of the *Dec-MDP* and *Dec-POMDP* models. These variations include goal directedness, communication, and independent transitions/observations. They showed the complexity of these problems as well as specified two algorithms for the goal directedness cases. In some cases, for instance when there is a single goal state and the transitions and observations of each agent are independent of each other, the problem becomes P-Complete.

5   Future Work

While the work here focuses on theoretical bounds for finding the optimal policy off-line, it would be very interesting to see if we can use some multi-agent reinforcement learning algorithms to learn an optimal policy.

This work shows that finding the optimal policy can be quite computationally expensive. This is because the specificity of the model is quite high - all actions every agent takes must be analyzed. On the other hand population based models like the Language Dynamical Equation ([8]) are much more tractable while giving up knowledge of the specifics of agents actions.

We are investigating approaches that incorporate the best of both worlds. The creation of a model that has the generalization and tractability of the LDE but also the fine-grained control and information that a MAS model can give us.

The crucial parameter in deciding the complexity of the language convergence problem is the amount of information that an agent has about the rest of the population. In the case of a fully observable *Dec-POMDP* , each agent can know the state of every other agent, and thus the problem can decompose into n independent MDP's. Direct communication is a possible way for agents to achieve full observability, but communication usually incurs a cost. This cost might be managed by specifying an interaction topology that limits the interaction between agents. Delgado, in ([9]), shows that a set of agents can agree on the same convention even when each agent might not interact with all the other agents. This work could provide a starting point for studying how limiting the interaction of agents could still result in language convergence.

A interesting avenue for future work would study how different interaction topologies for message passing affect the rate of convergence, and the complexity of finding optimal joint policies.

6   Conclusion

In this work we have investigated the complexity of finding an optimal policy for language convergence problems. Our main contribution is in mapping instances of language convergence problems to *Dec-POMDP* 's

Four examples of language convergence problems, and their associated *Dec-POMDP* 's were shown. In the simplest case, when there is only one language that all the agents can converge upon, deciding whether an optimal policy exists is P-Complete.

At the other extreme we have a situation where the agents are playing language games and each agent must decide who to interact with at every time step. In this case, when the agents cannot fully observe what language they are currently using, deciding upon an optimal policy is NEXP-Complete.

We have argued that the increase in complexity of finding an optimal policy is based on 2 levels of uncertainty, uncertainty over an agents local state and uncertainty over the state of the other agents in the population.

By mapping instances of the language convergence problem to instances of *Dec-POMDP* 's we have been able to study the worst case complexity of finding an optimal algorithm for the agents. This provides us with an intuition on what makes the language convergence problem complex. In future work we plan on adding communication between agents thus allowing them to gain knowledge of the languages used by other agents.

References

1. Clark, A.: Magic words: How language augments human computation. In Carruthers, P., Boucher, J., eds.: LANGUAGE AND THOUGHT: INTERDISCIPLINARY THEMES. Cambridge University Press: (1998)
2. Cucker, F., Smale, S., Zhou, D.X.: Modeling language evolution. Foundations of Computational Mathematics **4**(3) (2004) 315–343
3. Steels, L.: The origins of ontologies and communication conventions in multi-agent systems. Autonomous Agents and Multi-Agent Systems **1**(2) (1998) 169–194
4. Bernstein, D.S., Givan, R., Immerman, N., Zilberstein, S.: The complexity of decentralized control of markov decision processes. Math. Oper. Res. **27**(4) (2002) 819–840
5. Goldman, C., Zilberstein, S.: Decentralized control of cooperative systems: Categorization and complexity analysis. Journal of Artificial Intelligence Research **22** (2004) 143–174
6. Steels, L.: The evolution of communication systems by adaptive agents. In Alonso, E., D.K., Kazakov, D., eds.: Adaptive Agents and Multi-Agent Systems: Adaptation and Multi-Agent Learning. LNAI 2636, Springer Verlag, Berlin (2003) 125–140
7. Wagner, K., Reggia, J., Uriagereka, J., Wilkinson, G.: Progress in the simulation of emergent communication and language. Adaptive Behavior **11** (2003) 37–69
8. Komarova, N.L.: Replicator-mutator equation, universality property and population dynamics of learning. Journal of Theoretical Biology **230**(2) (2004) 227–239
9. Delgado, J.: Emergence of social conventions in complex networks. Artificial Intelligence **141**(1-2) (2002) 171–185

Applied Adaptive Behavior

Behavioral Analysis of Mobile Robot Trajectories Using a Point Distribution Model

Pierre Roduit[1,2], Alcherio Martinoli[1], and Jacques Jacot[2]

[1] Swarm-Intelligent System Group (SWIS),
École Polytechnique Fédérale de Lausanne, 1015 Lausanne, Switzerland
[2] Laboratoire de Production Microtechnique (LPM-IPR),
École Polytechnique Fédérale de Lausanne, 1015 Lausanne, Switzerland

Abstract. In recent years, the advent of robust tracking systems has enabled behavioral analysis of individuals based on their trajectories. An analysis method based on a Point Distribution Model (PDM) is presented here. It is an unsupervised modeling of the trajectories in order to extract behavioral features. The applicability of this method has been demonstrated on trajectories of a realistically simulated mobile robot endowed with various controllers that lead to different patterns of motion. Results show that this analysis method is able to clearly classify controllers in the PDM-transformed space, an operation extremely difficult in the original space. The analysis also provides a link between the behaviors and trajectory differences.

1 Introduction

The development of vision-based tracking systems brings about an easy way to extract trajectory data. Consequently, an ever-increasing number of domains are using it for behavior and trajectory analysis, like video surveillance [11,13,5,12,8], sports analysis [1] and ethology [4]. Behavioral analysis has also been done on human trajectories in a virtual environment [14,15] or on autonomous-robot trajectories [16,17,10]. All these applications aim to classify an individual from its trajectory, to analyze the movement differences between individuals, or to create a motion model of animals or insects.

This paper addresses the development of tools to analyze the motion of robots, or more generally people or animals, by means of their trajectories. As we will explain in further detail, robots were chosen as trajectory generators for their repeatability and behavioral controllability which natural being are lacking of. For the analysis, we use a Point Distribution Model (PDM) [2], a kind of deformable template. This model was often used to detect object shapes in an image, but it can also be used for trajectory modeling [3]. It is able to take into account spatial and temporal information, but in our experiments we focused on purely spatial analysis. We are more interested in the way the individual is moving and not by the time it needs to travel a given distance.

The paper is organized as follows. In section 2, the experimental method used will be presented, followed in the next section by a description of the Point Distribution Model. Section 4 will present the results and a discussion will close the paper.

S. Nolfi et al. (Eds.): SAB 2006, LNAI 4095, pp. 819–830, 2006.

2   Experimental Method

In order to collect hundreds of trajectories in a very short time and perform a first exploration of PDMs as tool for behavioral analysis, we have decided to work with mobile robots. They may be embedded in a physical environment in the same way as natural creatures, but are more easily programmed to produce repeatable behaviors. Moreover, they can be small and therefore the experimental setup can easily fit into a room and simple video tracking systems can extract their positions. Finally, their behavioral repertoire can be quite rich and their controllers can achieve fairly high degrees of complexity, spanning from purely reactive to more deliberative behaviors.

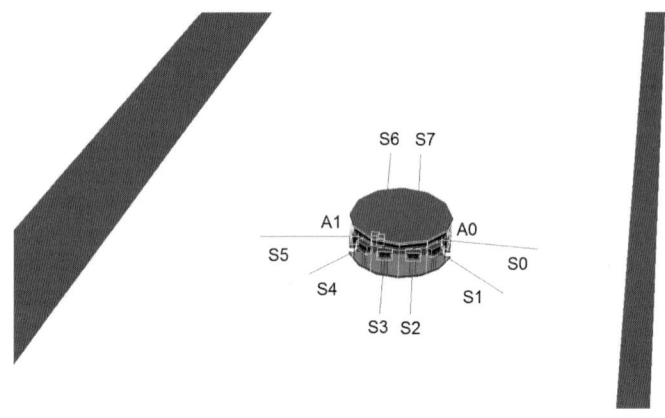

Fig. 1. Image of the simulated mobile robot, with its 8 infrared sensors (S0 to S7) and two wheels (A0 and A1). The two walls of the circuit can also be seen.

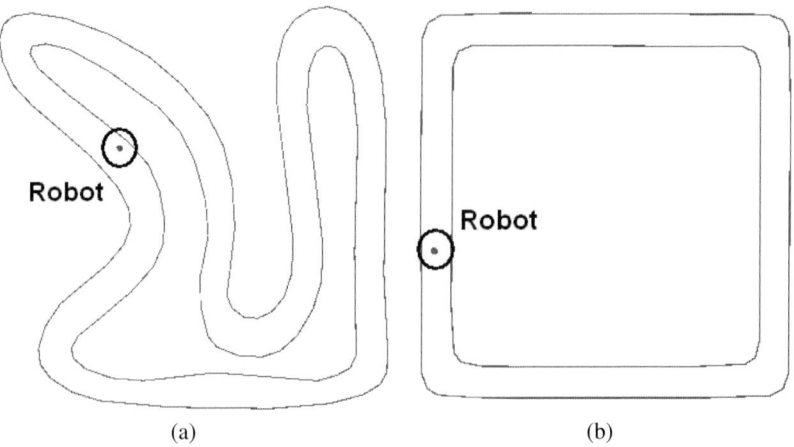

| (a) | (b) |

Fig. 2. The two circuits simulated in Webots (1 on the left, 2 on the right)

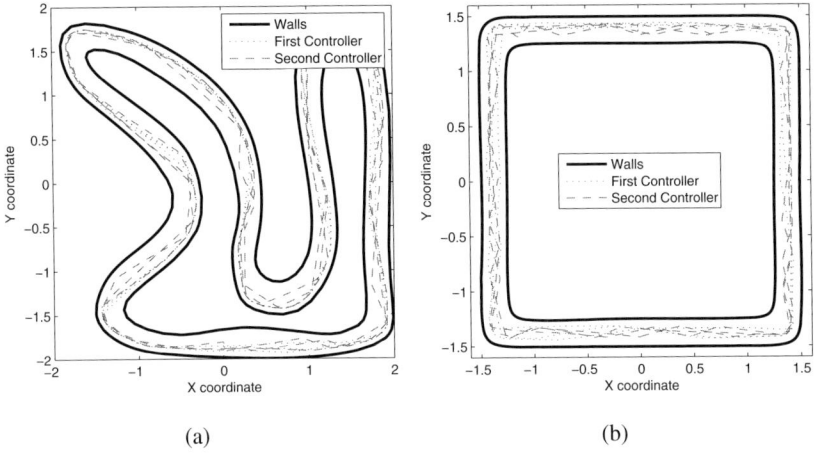

Fig. 3. 6 trajectories (3 for each controller) of the robot's movement simulated on the 2 circuits

In order to further speed up the trajectory generation, we carried out our experiments using a simulated, miniature differential-wheel robot endowed with eight proximity sensors (Fig. 2). We simulated it in *Webots* [9], a realistic simulator reproducing individual sensors and actuators with noise, nonlinearities, and dynamic effects such as slipping and friction. The resulting simulation is sufficiently faithful for the controllers to be transferred to real robots without changes and for the simulated robot behaviors to be very close to those of the real robots, as shown in several previous papers (see for instance [6]).

Figure 2 shows the shapes of the two circuits used for our simulation. They share common features such as being characterized by only one lane and a closed loop, however, the length and curvature differ between them. We use two circuits for our experiments in order to test the validity of our analysis.

We simulate the robot moving continuously within the two circuits, extracting each lap as a separate trajectory. Since a lap does not begin and end at the same point, it adds variability to the trajectories. Two different reactive controllers were implemented to drive the robot on the circuits. The two controllers move around the track at the same average speed, but using different methods of avoiding the walls. The first controller was rule-based ("if a wall is too close in front, turn away from the wall, otherwise go straight"). The second controller is essentially a Braitenberg vehicle and linearly adjusts its trajectory as a function of its proximity to a wall on the left or the right side. The robot was moving clockwise in both circuits. Both controllers continuously calculate the perception-to-action loop every 32 ms. The description of the two controllers can be found in Table 1. The two controllers were chosen for their simplicity, but the analysis has very few limitations and could easily be used with more complex controllers.

From this description, we can see that Controller 1 is characterized by essentially two discrete behaviors: go straight or turn in place. Controller 2 makes much smoother turns and its overall behavior changes as a function of the distance to the left or right

Table 1. Description of the two controllers used for the experiments

Controller 1	Controller 2
If $\quad \sum_0^2 S_i < T \Rightarrow \{^{A_0=V}_{A_1=-V}$	$S_l = \sum_0^2 S_i$
Else if $\sum_3^5 S_i < T \Rightarrow \{^{A_0=-V}_{A_1=V}$	$S_r = \sum_3^5 S_i$
Else $\quad \{^{A_0=V}_{A_1=V}$	$A_0 = V \cdot \left(1 + \frac{K \cdot (S_l - S_r)}{2 \cdot (S_l + S_r)}\right)$
	$A_1 = V \cdot \left(1 + \frac{K \cdot (S_r - S_l)}{2 \cdot (S_l + S_r)}\right)$

$S_0 \dots S_5$ are the robot sensors as shown in Figure 2
(back sensors S_6 and S_7 are not used in either of the controllers)
A_0 and A_1 are the robot actuators as shown in Figure 2
T is a constant threshold value
V is a parameter modifying the robot's overall speed
K is a parameter modifying the robot's reactivity

wall. Figure 3 shows three trajectories per controller on each of the two circuits. Slight differences can be seen between the two controllers; for example, Controller 2 makes more zigzags than Controller 1 (even if its turns are smoother, it turns more often than Controller 1). At first glance, it is not so easy for the human eye to differentiate between the raw trajectories.

2.1   Trajectory Sampling

In order to apply the *Point Distribution Model* presented in section 3, each trajectory must be sampled with the same number of points. For our previous experiments, presented in [3], the sampling of the trajectories was done with lines orthogonal to a

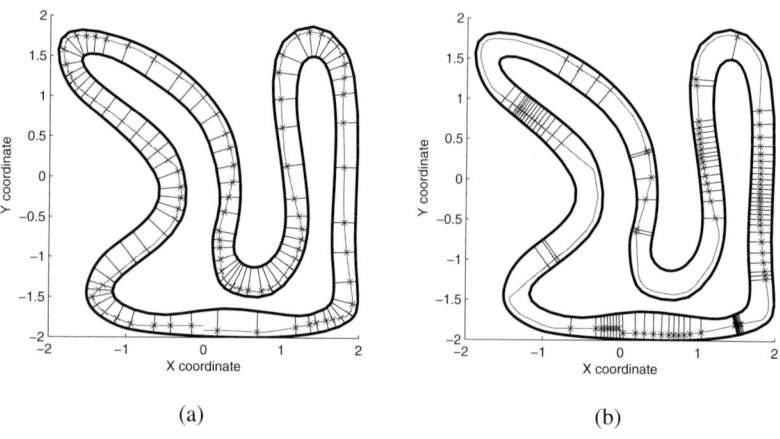

(a)   (b)

Fig. 4. Sampling of the trajectory with gates as orthogonal to the wall as possible, with more gates in the curves (SM-B, left) and more gates in the straight lines (SM-C, right)

reference trajectory that needed to be chosen. However, this solution is cumbersome and limits the number of sampling points. Therefore, we developed a new sampling methodology based on the circuit instead of a reference trajectory. The inner and outer wall of the circuit were modeled with b-splines and sampling gates were created as orthogonal as possible to the two walls. A fixed number of gates was then selected based on three different criteria: the distance between gates (Sampling Method A), the curvature (SM-B), and the linearity of the circuit (SM-C). The first criterion selects gates that are equidistant from each other (the distance is measured between the midpoints of the gates), the second places more gates in the curves (as shown in Figure 4(a)), and the last places more gates in the straight sections (Figure 4(b)). This method allows us to very easily modify the number of gates per circuit. The performance of the three placement strategies will be compared in Section 4.3.

3   Modeling of the Trajectories

We have slightly modified the representation we used in [3] to accommodate the new sampling method presented above. Each trajectory k is represented as an ordered set of N points corresponding to the intersections of the trajectory with the sampling gates. Each point can be expressed as the linear position on the ith sampling gate π_i^k. Each position π_i^k has a value between 0 (the inner wall) and 1 (the outer wall). Therefore the trajectory τ_k can be expressed as:

$$\tau_k = \left[\pi_1^k \dots \pi_N^k\right]^T. \tag{1}$$

The covariance matrix of the trajectories is

$$S = \frac{1}{K-1} \sum_{k=1}^{K} (\tau_k - \overline{\tau})(\tau_k - \overline{\tau})^T = P \cdot \Lambda \cdot P^{-1}, \tag{2}$$

where $P = [P_1 \dots P_r \dots P_R]$ is the matrix of the eigenvectors P_r, Λ the diagonal matrix containing the eigenvalues of S, K is the number of trajectories in the set, and where $R = \min(2N, K) - 1$ is the number of degrees of freedom of the set.

Each trajectory τ_k in the set can be decomposed into an average trajectory and a linear combination of deformation modes:

$$\tau_k = \overline{\tau} + P \cdot B_k \tag{3}$$

$$B_k = P^{-1}(\tau_k - \overline{\tau}). \tag{4}$$

Equations 3 and 4 correspond to the projection from the deformation space (B_k) to the trajectory space (τ_k) and the projection from the trajectory space to the deformation space, respectively.

The computation of matrix P corresponds to the Principal Component Analysis (PCA) [7] of the trajectory set. The first vector P_1 corresponds to the direction of maximal variance in the trajectory space. The second vector P_2 corresponds to the direction of maximal variance orthogonal to P_1. The other vectors are found likewise. In most cases, this construction implies that most of the deformation energy will be contained in the first few deformation modes.

The *Point Distribution Model* [2] affords the transformation from the space of the trajectories (τ_k) to the space of the modes (B_k).

3.1   Inter-cluster Distance

In this section we will describe the inter-cluster measure we used for our experiments.

Multivariate normal data tends to cluster about the mean vector, $\mu_{cluster}$, falling in an ellipsoidal cloud whose principal axes are the eigenvectors of the covariance matrix. The Mahalanobis distance, r, takes into account the covariance of the cluster, $S_{cluster}$, to calculate the distance from a point X to a cluster.

$$r = \sqrt{(X - \mu_{cluster})^T \cdot \boldsymbol{S_{cluster}}^{-1} \cdot (X - \mu_{cluster})} \tag{5}$$

If normal data is projected on a unidimensional axis, a unitary Mahalanobis distance is equivalent to a Euclidean distance of the square root of the data variance along this axis (standard deviation). Thus, the points of unitary Mahalanobis distance to a cluster forms a ellipsoid.

As a measure of distance between two clusters, we can use a combination of the two Mahalanobis distances; from the mean of one cluster to the other, and vice versa. If r_1^2 is the Mahalanobis distance from the first cluster mean to the second cluster and r_2^1 the Mahalanobis distance from the second cluster mean to the first cluster, d_{12} is a measure of the inter-cluster distance.

$$d_{12} = \frac{r_1^2 \cdot r_2^1}{r_1^2 + r_2^1} = d_{21} \tag{6}$$

A unitary inter-cluster distance is equivalent to a Euclidean distance between the two cluster means which is equal to the sum of the standard deviation of the projected multivariate cluster data on the axis connecting the two cluster means.

4   Results and Discussion

To show the performance of our method for clustering controller trajectories, we acquired 200 trajectories (100 for each controller presented in Table 1) on the two circuits (Figure 2 and 2(b)). The trajectories were then re-sampled with 100 points per trajectory, using the gate selection criterion with more gates in the curves (SM-B). Then, we model all the trajectories using the PDM presented in Section 3. Figure 5 shows the locations of the 200 trajectories in the space formed by the first two modes of the PDM for each of the two circuits; two clusters can be easily differentiated. The clear separation of the controllers shows the benefit of the PDM modeling of the trajectories. The intrinsic variance of the controllers is smaller than the distance between them. Therefore the trajectories can be clustered and hence classified.

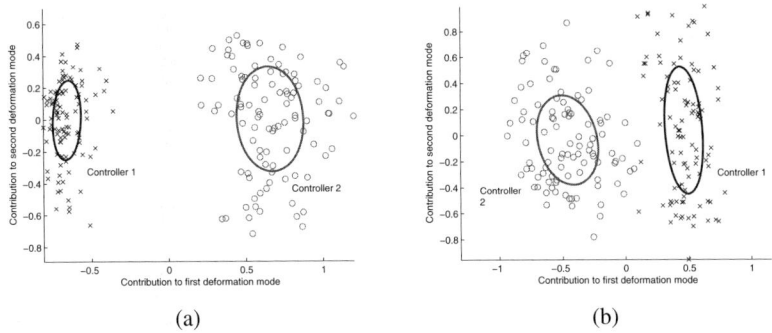

(a)   (b)

Fig. 5. Projection of all the trajectories in the space of the first two deformation modes of the PDM for the first (left) and the second (right) circuit. The ellipse of unitary Mahalanobis distance is also plotted for each controller.

4.1   Analysis Using the First Two Modes of the PDM

The separation of the first 2 clusters is narrower for the second circuit than the first. The lack of curves and the predominance of straight lines reduce the number of obstacles and therefore the number of controller reactions from which the PDM transformation can extract its data. As their straight movements are equivalent, it becomes more difficult to detect differences and therefore classify the two controllers. If we calculate the inter-cluster distance as presented in section 3.1, $d = 4.3$ for the first circuit and $d = 2.4$ for the second circuit.

4.2   Prototype Trajectories

Figure 6(a) displays the mean of each controller's trajectories. These trajectories are prototypes of each controller. Slight differences appear at certain places in the circuit; these are the places where the controllers can be differentiated.

Figure 6(b) shows the synthetic trajectories resulting from a positive or negative contribution of value 5 to the first or the second deformation mode for the first circuit. In the straight section on the right side of the circuit, we can see that the first mode is out of phase with the second mode, such that they alternate. The second mode corresponds to major trajectory variations in the straight section along the bottom of the circuit.

4.3   Variation of the Trajectory Sampling Methods and Number of Points

To evaluate the number of gates needed to achieve a good classification of the trajectories, we calculate the inter-cluster distance as a function of the number of sampling gates (from 3 to 500) for the three sampling methods described in section 2.1 and for both circuits. Figures 7(a) and 7(b) present the results of our experiments. We can see that a very small number of gates (3) is insufficient to separate the two controllers. Increasing the number of gates results in a fast gain in cluster separability, until we reached 50 gates. After this point, it has hardly any effect on the analysis, aside from requiring additional computational power for calculating the PCA. We can also see that

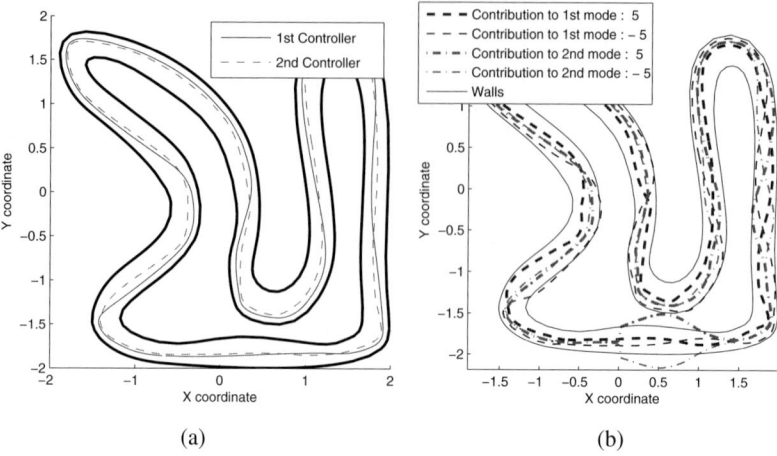

Fig. 6. On the left, prototype trajectory of the two controllers on the first circuit. On the right, synthetic trajectories resulting from a contribution of 5 to one of the first two modes (first circuit).

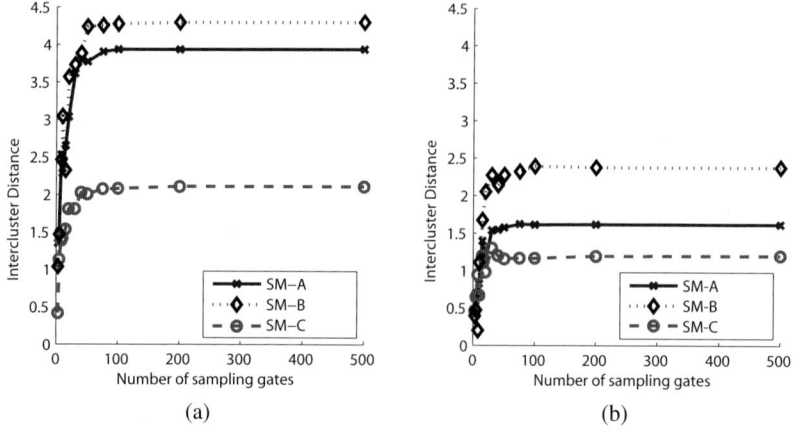

Fig. 7. Variation of the inter-cluster distance based on the first two modes as a function of the number of sampling gates, for the three methods presented in 2.1 and for two circuits (circuit 1 on the left and circuit 2 on the right)

the three gate placement strategies (equidistant, more in the curves, more in the straight sections) do not have the same influence on clustering performance. Placing gates in the straight sections is clearly the worst solution, while using gates in the curves is clearly the best solution for our experimental setup. Equidistant gates yield a solution in between the other two, but are clearly not so good as emphasizing the curves. This result can be directly traced to the structure of the controllers. They have the same behavior in the straight sections, where there is no interaction with the walls. Curves imply more interaction with the walls, and therefore elicit more behavioral differences between the two controllers. It can be foreseen that if the controllers had the same behavior in the curves, but different in the straight sections, the performance would be inverted. As another result of the controller's structure, the separability of the clusters on the second circuit is clearly worse than on the first, because there are not as many curves to separate the behavior of the two controllers.

4.4   Variation of the Robot Hardware and Software Parameters

To complete the performance evaluation of our method, we analyze the influence of possible system design choices on the resulting analysis: the controller reactivity, the overall robot speed, and the sensory range. For these experiments, we report only results obtained using Controller 2 for sake of clarity. Referring to Table 1, the reactivity corresponds to K and the overall speed to V.

Figure 8(a) shows the analysis of the variation of the sensory range from 2 to 20 centimeters. The real range of the sensors (5 cm) is shown with the small lines in Figure 2. For ranges of 10 centimeters and above, the robot will almost always be able to see both walls at the same time. Therefore, the variability of the controller will decrease significantly, as the robot will follow a path in the middle of the lane. Naturally, the greater the sensory range, the smaller the variability of the controller. The different sensory ranges are mainly separated using the first deformation mode in Figure 8(a): one variation axis (sensory range) corresponds to one dimension for classifying the different clusters.

Figure 8(b) shows the clustering of the reactivity of the controller. The factor K varies between 0.25 and 4, the reactivity increasing with K. If K is small, the robot avoids the wall with more inertia and thus oscillates much more. As as result, the dis-

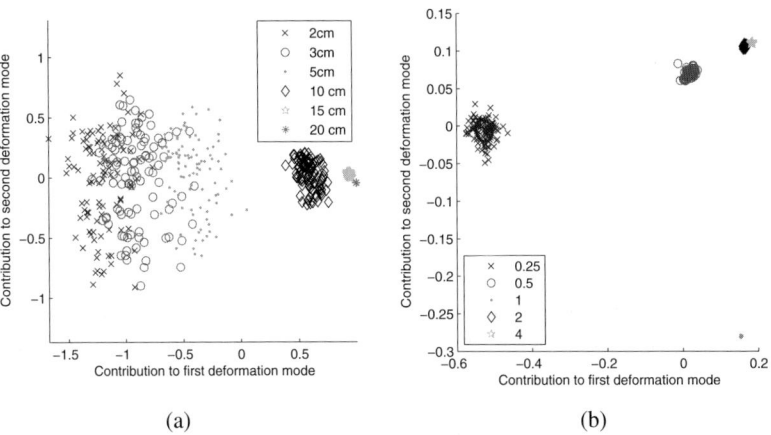

(a)   (b)

Fig. 8. On the left, analysis of the variation of the sensory range from 2 to 20 cm. On the right, analysis of the variation of the controller weights (K), for a sensory range of 20 cm.

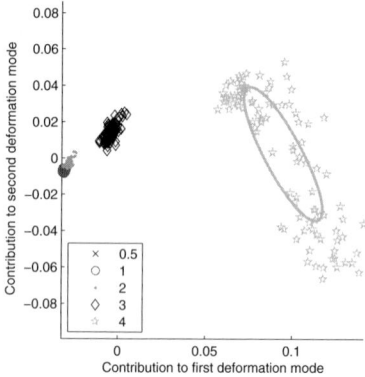

Fig. 9. Analysis of the variation of the overall robot speed (V), for sensory range of 20 cm

persion of trajectories in the mode space is much greater, as can be seen in Figure 8(b). With a sensory range of 5 centimeters and $K = 0.25$, the robot was not able to avoid colliding with the walls anymore. Therefore, this experiment was made with a sensory range of 20 centimeters ($V = 1$).

Figure 9 shows the result of the variation of the overall robot speed factor V. Similarly to the previous experiment, with the robot overall speed increased ($V = 4$), the robot was not able to avoid walls with a sensory range of 5 cm. Therefore, this experiment was performed with a sensory range of 20 cm. As with a human driver, an increase in the overall speed means that the robot pass closer to the walls before avoiding them, resulting in larger oscillations. This variability can be seen in Figure 9.

5   Conclusion and Perspectives

We have presented a method for using a PDM to analyze a robot's behavior from its trajectory on a closed circuit. Applied to trajectories of the same simulated robot driven by two different reactive controllers, it shows a complete separation of the controllers in the space of the first two deformation modes of the PDM. The controller clusters can be separated by a line, affording an easy classification of the trajectories. Quality of separation depends on the sampling method, the circuit characteristics, and the software and hardware parameters of the robot. The analysis of the prototype trajectory of each controller shows the main differences between the controllers. Variation in three other parameters, such as the robot speed, the controller reactivity, and the sensory range, implies that all of these parameters can be clustered with our method.

Even though the clustering method is not sophisticated (Principal Component Analysis is a common tool), the fact that behavioral features can be distinguished makes it very interesting. So far, only one variation axis has been analyzed at a time (circuit shape, controller description, sensory range, controller reactivity, and overall robot speed). This kind of experiment affords us to separate the trajectories using only the first two dimensions of the PDM. More complex setups might need more than two dimensions to achieve good clustering of the trajectories. Finding two combinations of hardware and software that provide different trajectories which can not be separated with our method would help us to understand its limitations. However, our goal is not to distinguish two different implementations of the same behavior, but rather to classify different behaviors.

To validate the results obtained with the trajectories of the robot simulated in Webots, we will create a similar circuit with a real robot, using a vision-based tracking system. The same analysis will be applied to real trajectories and results will be compared with the results gathered in simulation. Another challenge will be to extend our method to trajectories not bound to a closed circuit. The ultimate goal will be to model robot trajectories in an open space.

Acknowledgments

Pierre Roduit and Alcherio Martinoli are currently sponsored by two Swiss National Foundation grants (Nr. 200021-105565 and PP002-68647 respectively).

References

1. M. Bertini, A. Del Bimbo, and W. Nunziati. Highlights modeling and detection in sports videos. *Pattern Analysis and Application*, 7(4):411–421, 2005.
2. T. Cootes, C. Taylor, and D. Cooper. Active shape-models - their training and applications. *Vision and Image Understanding*, pages 38–59, 1995.
3. Yuri Lopez de Meneses, Pierre Roduit, Florian Luisier, and Jacques Jacot. Trajectory analysis for sport and video surveillance. *Electronic Letters on Computer Vision and Image Analysis*, 5(3):148–156, 2005.
4. M. Egerstedt, T. Balch, F. Dellaert, F. Delmotte, and Z. Khan. What are the ants doing ? vision-based tracking and reconstruction of control programs. pages 4193–4198. IEEE International Conference on Robotics and Automation, IEEE Computer Society, April 2005.
5. W. E. L. Grimson, C. Stauffer, R. Romano, and L. Lee. Using adaptive tracking to classify and monitor activities in a site. pages 22–29. Conference on Computer Vision and Pattern Recognition, IEEE Computer Society, June 1998.
6. A. T. Hayes, Alcherio Martinoli, and R. M. Goodman. Distributed odor source localization. In Gardner J. W. Nagle H. T. and Persaud, editors, *Special Issue in Artificial Olfaction*, volume 2, pages 260–271. IEEE Sensors Journal, 2002.
7. J. Jackson. Principal components and factor analysis: part 1. *Journal of Quality Technology*, 12:201–213, October 1980.
8. Nail Johnson and David Hogg. Representation and synthesis of behaviour using gaussian mixtures. *Image and Vision Computing*, 20:889–894, 2002.
9. Olivier Michel. Webots: Professional mobile robot simulation. *Journal of Advanced Robotics Systems*, 1(1):29–42, 2004.
10. Ulrich Nehmzov. Quantitative analysis of robot-environment interaction towards "scientific mobile robotics". *Robotics and Autonomous Systems*, 44:55–68, 2003.
11. J. Owens, A. Hunter, and E. Fletcher. Novelty detection in video surveillance using hierarchical neural networks. *Lecture Notes in Computer Science*, 2415:1249–1254, 2002.
12. Fatih Porikli. Learning object trajectory patterns by spectral clustering. volume 2, pages 1171–1174. IEEE Conference on Multimedia and Expo, IEEE, June 2004.

13. P. Remagnino, T. Tan, and K. Baker. Agent orientated annotation in model based visual surveillance. page 857. Sixth International Conference on Computer Vision, IEEE Computer Society, 1998.
14. C. Sas, G. O'Hare, and R. Reilly. A performance analysis of movement patterns. *Lecture Notes in Computer Science*, 3038:954–961, 2004.
15. C. Sas, G. O'Hare, and R. Reilly. Virtual environment trajectory analysis: a basis for navigational assistance and scene adaptivity. *Future Generation Computer Systems*, 21 (7):1157–1166, Jul 2005.
16. G. Schöner, M. Dose, and C. Engels. Dynamics of behavior : theory and applications for autonomous robot architecture. *Robotics and Autonomous Systems*, 16:213–245, 1995.
17. Thim Smithers. On quantitative performance measures of robot behaviour. *Robotics and Autonomous Systems*, 15:107–133, 1995.

Simbad: An Autonomous Robot Simulation Package for Education and Research

Louis Hugues[1] and Nicolas Bredeche[2]

[1] Ginkgo-networks, Paris, France
louis.hugues@wanadoo.fr
http://www.louis.hugues.xunuda.com/
[2] Equipe Inférence et Apprentissage, TAO / INRIA Futurs
Laboratoire de Recherche en Informatique, Bat.490,
Université de Paris-Sud, 91405 Orsay Cedex, France
bredeche@lri.fr
http://www.lri.fr/ bredeche

Abstract. Simbad is an open source Java 3d robot simulator for scientific and educational purposes. It is mainly dedicated to researchers and programmers who want a simple basis for studying Situated Artificial Intelligence, Machine Learning, and more generally AI algorithms, in the context of Autonomous Robotics and Autonomous Agents. It is is kept voluntarily readable and simple for fast implementation in the field of Research and/or Education.

Moreover, Simbad embeds two stand-alone additional packages : a Neural Network library (feed-forward NN, recurrent NN, etc.) and an Artificial Evolution Framework for Genetic Algorithm, Evolutionary Strategies and Genetic Programming. These packages are targeted towards Evolutionary Robotics.

The Simbad Package is available from http://simbad.sourceforge.net/ under the conditions of the GPL (GNU General Public Licence).

1 Introduction

This paper provides an introduction to Simbad, a new mobile robot simulator written in Java for Research and Education, and a set of tools for Evolutionary Robotics [7]. The main motivation is to provide an easy-to-use all-in-one package for Evolutionary Robotics. The Simbad package includes a mobile robot simulator with complex 3D scene modelling and simulation engine (Simbad), a Neural Network library (feed-forward and recurrent NN) as well as a complete Evolutionary Algorithm Library (Genetic Algorithm, Evolutionary Strategy, Genetic Programming with trees or graphs). All tools have been written to be efficient and easy to use by Java and/or Python programmers.

Section 2 present the state of the art in mobile robot simulation and highlights the specific characteristics of existing simulators. Section 3 describes the key motivations and characteristics of the Simbad simulator as well as its specificity compared to other simulators. This section also introduces the PicoNode and PicoEvo packages as well as implementation issues. In section 4, classic experiments are described using the Simbad packages. The last section concludes

S. Nolfi et al. (Eds.): SAB 2006, LNAI 4095, pp. 831–842, 2006.

this paper and refers to current applications of the Simbad package both for Research and Education.

2   Available Mobile Robot Simulators

Due to the spreading of the Open/GL API several tools provide some kind of 3D visualisation. However there are quite few simulators providing a complete 3D simulation and, in particular, sensing in the 3D space (vision, sonars, bumpers and lasers).

Player/Gazebo: Player and Gazebo projects have been developed at the USC (University of Southern California) [2,3]. Those two components can be used in conjunction so as to obtain a powerful multi-robot simulation in a 3D environment. The Player project provides an abstract programming framework for real robots and is widely used in the robotics community. Gazebo is a 3D simulator able to simulate several Player-based Robots. Gazebo uses ODE (Open Dynamic Engine - a rigid body dynamics simulator) to compute physicals interaction with objects. Player, Stage and Gazebo are freely available via SourceForge with both binary and source code. Player/Gazebo constitutes a powerful package but relies on several complex components and thus requires some time to learn.

Webots: [4] Is a product proposed by the Cyberbotics company from an initial basis developed by the LAMI laboratory at the EPFL (Ecole Polytechnique Federale de Lausanne). Its functionalities are very similar to those of Player/Gazebo but with the look and feel of a commercial package. As Gazebo, it uses ODE for physical simulation. The main disadvantages of Webots are its cost and the fact that source code is not available for standard users. This last point prevents experimenters to have a precise insight of the simulator behaviour.

Other tools: One can also mention the Robocup soccer simulator, EyeSim for the EyeBot Robot, SimBot and WoB [5]. Besides the 3D robot simulators, numerous tools exist which perform simulation in the 2D space. Among the very good ones are Stage simulator (which can be used in conjunction with Player instead of using Gazebo), Khepera Simulator the precursor of Webots and TeamBots, a multi robot Simulator.

3   The Simbad Package: Main Features

In this section, we describe the three main packages included in the Simbad Packages:

1. The Simbad simulator ;
2. PicoNode, the neural network library ;
3. PicoEvo, the Evolutionary Algorithm library.

3.1   The Simbad Simulator

Simbad [1] is a Java 3d multi-robots simulator developed for scientific and educational purposes. It is mainly dedicated to researchers and students who want a simple basis for studying Situated Artificial Intelligence, Machine Learning, and more generally AI algorithms, in the context of Autonomous Robotics and Autonomous Agents. Simbad enables programmers to write their own robot controllers, modify the environment and use the available sensors (or build their own). It is not intended to provide a real world simulation and is kept voluntarily readable, simple and extensible.

Simbad is written in java and requires only a standard Java development kit (version 1.4.2 or higher) and Sun Java 3d (version 1.3.1). This last one provides a well-documented 3D graph on which are constructed Simbads objects. Those two components are easily available for a wide range of platforms (Windows, Mac Os X , Linux, IRIX, AIX). The Simbad project is hosted at SourceForge and available at http://simbad.sourceforge.net/. It is *free to use and modify* under the conditions of the **GPL** (GNU General Public Licence).

The performances of the simulator are excellent and enable experimenters to use batch simulation in the context of heavy computation for learning (e.g. evolutionary algorithms). The 'Java is slow' remark is outdated and Java applications are now clearly equivalent to C++ ones [6] but do not require numerous external libraries. The simulator provides the following functionalities:

- Single or multi-robot simulation.
- Color Mono-scopic cameras.
- Contact and range sensors.
- Online or batch simulation.
- Python scripting with jython.
- Extensible user-interface.
- Simplified physical engine

Table 1 is a brief comparison of Gazebo, Webots and Simbad functionalities.

To sum it up, Simbad is very useful, and has proven to be so, for studying models in middle size projects as well as teaching in AI and Robotics classrooms while Gazebo and Webots are very good options if you are involved on a long-term project with real robots.

As for implementation, the user only provide an environment derived from the EnvironmentDescription class and a robot controller derived from Robot class. This last one has an initialisation method (*initBehavior*) and a method to be called on each simulation step (*performBehavior*). The simulator then execute the motor control orders in a similar way as a real robot controller (i.e. repetitive calls to the micro-controller). The following code shows how to create a simulator, settle the environment with a single box in it and create a robot in the environment (Figure 1 shows a comparable yet more complex environment):

```
import simbad.sim.*; import simbad.gui.*; import javax.vecmath.*;

public class MyEnv extends EnvironmentDescription {
```

Table 1. Comparaison between available simulators

	Player/Gazebo	Webots	Simbad
3D simulation	yes	yes	yes
3D vision	yes	yes	yes
multi-robots	yes	yes	yes
Physics	ODE	ODE	built-in
Source available	yes	no	yes
Freeware	yes	no	yes
Ease of use	needs expertise	quite good	easy
Platforms	win/linux/mac	win/linux/mac	win/linux/mac
Real Robots	yes	yes	limited

```
public MyEnv() {
 /* create four walls and the robot */
 Wall w1 = new Wall( new Vector3d(9, 0, 0), 19, 1, this);
 w1.rotate90(1);  add(w1);
 Wall w2 = new Wall( new Vector3d(-9, 0, 0), 19, 2, this);
 w2.rotate90(1);  add(w2);
 Wall w3 = new Wall( new Vector3d(0, 0, 9), 19, 1, this);
 add(w3);
 Wall w4 = new Wall( new Vector3d(0, 0, -9), 19, 2, this);
 add(w4);
 add(new MyRobot( new Vector3d(0, 0, 0), "my robot"));
}

public class MyRobot extends Robot {
 MyRobot (Vector3d _position, String _name )
 {    super (_position, _name);   }

 public void initBehavior()
 {  /* your init code goes here */}

 public void performBehavior()
 { /* code perfomed on each step goes here */  }
}

public static void main(String[] args)
{ Simbad frame = new Simbad( new MyEnv(), false); }
}
```

Simbad can also be used with Python, a very popular scripting language[1].

As a tutorial, a list of examples with sources is available from the menu in Simbad that gives direct access to many of Simbad features. Examples are:

- BaseDemo : demo with camera sensor, sonars and bumpers. The robot wanders and stops when it collides.
- ImagerDemo : shows how to capture the camera image , process it and display it in a dedicated window.
- AvoidersDemo : shows several robot with sonars and bumpers performing a collision avoidance behavior.

---

[1] The above example written in Python can be found on the web site.

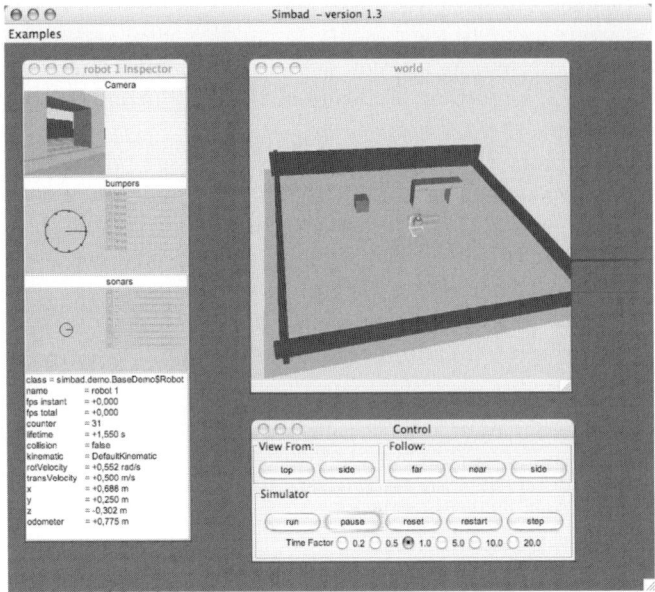

Fig. 1. Simbad interface - Simbad can be run either as a standalone program (as shown) or in batch mode with no display when simulation speed is crucial (e.g. Evolutionary Robotics setup)

- BlinkingLampDemo : the lamp on each robot blinks when the robot is approaching an obstacle.
- BumpersDemo : several robots bumping.
- DifferentialKinematicDemo: a differential drive (two wheels) kinematics demo.
- KheperaDemo : a Khepera robot demo.
- LightSearchDemo : robots search a light using light sensor.
- PickCherriesDemo : show a robot picking cherries. When touched, the cherries are removed.
- PushBallsDemo : shows a robot pushing balls.

3.2   The Neural Network Library: PicoNode

In the field of autonomous robotics, the seminal work of Rodney Brooks on Reactive Robotics and Subsomption Architecture [8] have lead the way toward a vast amount of works in the field of machine learning and robotics where the goal is to build the control function defined as : $f(sensoryinputs\{, internalstate\}) \rightarrow motorcontrol$ where the optional term *internal state* may represent some kind of memory. Popular learning approach for robot control includes Reinforcement Learning [9], Learning in Bayesian and Markov Model and Learning in Neural Network [7].

Neural Networks provide a very powerful representation framework in the scope of robot control due to their ability to handle continuous noisy data with little computation. Moreover, internal states can be represented in recurrent architectures and the important literature in this field attests for a wide range of possible controllers.

The PicoNode library provides a general graph-based representation framework along with two implementations: feed-forward and recurrent neural networks. The use of PicoNode is not limited to robot control; it has been designed so as to ease building of simple (e.g. multi-layered perceptrons) as well as less simple (e.g. N-layers recurrent nets) neural networks.

The following code illustrates how a simple feed-forward neural networks with 2 hidden units, 2 input nodes and 1 output node is built:

```
/* STEP 1 : Initializing and building a neural net */

// step 1a : create a network
FeedForwardNeuralNetwork network = new FeedForwardNeuralNetwork(
    new ActivationFunction_logisticSigmoid() );

// step 1b : create some neurons
Neuron in1 = new Neuron( network,
 new ActivationFunction_logisticSigmoid());
Neuron in2 = new Neuron( network,
 new ActivationFunction_logisticSigmoid());
Neuron hidden1 = new Neuron( network,
 new ActivationFunction_logisticSigmoid());
Neuron out1 = new Neuron( network,
 new ActivationFunction_logisticSigmoid());

// step 1c : declare I/O neurons
network.registerInputNeuron( in1 );
network.registerInputNeuron( in2 );
network.registerOutputNeuron( out1 );

// step 1d : create the topology (random weight values)
network.registerArc( new WeightedArc( in1 , hidden1 ,
                Tools.getArcWeightRandomInitValue()));
network.registerArc( new WeightedArc( in2 , hidden1 ,
                Tools.getArcWeightRandomInitValue()));
network.registerArc( new WeightedArc( hidden1 , out1,
                Tools.getArcWeightRandomInitValue()));

// step 1e : initialize the network (perform some integrity
// checks and internal encoding)
network.initNetwork();

/* STEP 2 : using the network (feed-forward signal) */

// step 2a : loading the input values (i.e. sensory inputs)
ArrayList inputValuesList = new ArrayList();
inputValuesList.add(new Double (0.5));
inputValuesList.add(new Double (0.5));

// step 2b : computing the output values (i.e. motor outputs)
network.step( inputValuesList );
System.out.println("Output value : " + out1.getOutputValue());
```

Of course, recurrent architectures may be defined as well by just adding recurrent arcs and use the available RecurrentNeuralNetwork object instead of FeedForwardNeuralNetwork.

3.3   The Evolutionary Framework: PicoEvo

PicoNode provides standard supervised learning algorithm (e.g. Back-Propagation), however such learning algorithms are usually of little use[2] in the context of sparse, noisy, delayed and asynchronous reinforcement signals that are usually part of the task of control learning in mobile robotics (e.g. a binary reward (success/failure) is provided only when the robot may reach the exit of a maze).

The Evolutionary Robotics [7] approach addresses this problem by relying on population-based stochastic optimisation algorithms, i.e. evolutionary algorithms. Such algorithms are particularly well fitted when the objective function (i.e. the task) is difficult to describe. These stochastic optimisation algorithms perform on a generational basis (i.e. optimisation at step i depends on step $i - 1$) and rely on the exploration of the space of possible solutions through a population of candidate solutions by combining selection operators (most fitted candidates are likely to survive from one generation to another) and ideally well-suited variation operators (candidates may be recombined and/or altered to diffuse supposedly good characteristics as well as to efficiently explore the search space). These algorithms have been shown to be very efficient and to achieve human-competitive results on numerous problems where standard learning algorithm are difficult to apply, which is a key advantage in robotics[3].

The second advantage of Evolutionary Algorithm consists in that such algorithms can deal with a wide range of representation formalism to be optimised, from bit vectors to tree and graphs (and thus, programs). In the scope of neural networks optimisation, it is then possible to optimise network weights (see next section) or even the whole network topology [13]. It is important to notice that even if only neural network optimisation is addressed in the scope of this paper, other representation may be optimised as well (e.g. Bayesian network, markov models, etc. – where topology learning is often an issue).

The PicoEvo library is a general Evolutionary Algorithm library that embeds several kinds of algorithms such as Genetic Algorithm, Evolutionary Strategies, tree-based and graph-based Genetic Programming. As for PicoNode, it has been conceived to ease the implementation of new operators. Then again, implementation has been done in such a way that the underlying concepts are straightforward to understand, even if the user has little knowledge in Evolutionary Algorithm (e.g. during an AI course). The typical main loop for evolving a population of vectors containing bit values is the following (the following example

---

[2] Some architectures, such as the auto-teaching network [11] and AAA [12], do nevertheless rely on both back-propagation and evolution.

[3] Note that the Simbad simulator can be launched in fast-mode when performing such experiment, i.e. no user interface, so as to compute simulation much faster than human real-time mode – indeed, Simbad has been benchmarked to run at about 15000 steps per second on a Pentium 2.8ghz with 512mo RAM under Linux (knoppix) and Windows XP for an experimental setup involving evolution of neural network weights.

concerns the classic max-one problem - the reader may refer to [10] for details
on the task and parameters):

```
/* STEP 1 - INITIALISATION */
// create an Evolution environment with a single Population

// the parameter container (may be load from file)
ParameterSet_Evolution_mulambdaES parameterSet =
  new ParameterSet_Evolution_mulambdaES();

// setup evolution parameters
parameterSet.setGenomeSize(512);
parameterSet.setMu(5);
parameterSet.setLambda(195);
parameterSet.setMutationRate(1.0/512.0);
parameterSet.setMuPlusLambda(true);
parameterSet.setGenerations(250);
parameterSet.setInitPopSize(200);

// setup evolution operators
parameterSet.setSelectionOperator(
 new SelectionOperator_MuLambda("MuLambda"));
parameterSet.setEvaluationOperator(
 new EvaluationOperator_MaxOne_StaticArray_Bit("MaxOne"));
parameterSet.setPopulationInitialisationOperator(
 new InitialisationOperator_Population_SimplePopulation("MaxOne"));
parameterSet.setIndividualInitialisationOperator(
 new InitialisationOperator_Individual_StaticArray_Bit("MaxOne"));
parameterSet.setElementInitialisationOperator(
 new InitialisationOperator_Element_StaticArray_Bit());
parameterSet.setPopulationStatisticsOperator(
 new StatisticsOperator_Population());

World myWorld = new World ("myEvolution", parameterSet);
Population_SimplePopulation maxOnePop =
  new Population_SimplePopulation("max-one population",myWorld);
myWorld.registerPopulation(maxOnePop);

maxOnePop.performInitialisation();

/* STEP 2 - RUNNING */

for ( int i=0 ; i!=myWorld.getTemplate().getGenerations() ; i++ )
    myWorld.evolveOneStep(true);
maxOnePop.displayInformation();
```

Of course, PicoEvo may also be considered as a stand-alone module for non-
robotics-related optimisation tasks (e.g. classic examples are provided such as
max-one, symbolic regression, etc.).

4   Experiments

In this section we detail two experiments that illustrates the use of the Simbad
Package. The first experiment is a straightforward use of the Simbad simulator
and relies on a classic wall-avoidance and random wandering behaviour. The
second experiment involves the PicoNode and PicoEvo packages in order to automatically build controllers for the same wander/wall-avoider task.

4.1 Experiment 1: A Simple Hand-Written Wander Behavior

In this experiment, we have implemented a simple wander behaviour. The robot we consider is equipped with Infrared and bumper sensors and two motor commands: directional and translation velocities. The code is the following:

```
public void performBehavior() {
 if (bumpers.oneHasHit()) { setTranslationalVelocity(-0.1);
  setRotationalVelocity(0.5-(0.1 * Math.random())); }
 else
  if (collisionDetected()) {
   // stop the robot
   setTranslationalVelocity(0.0);
   setRotationalVelocity(0); }
  else
   if (sonars.oneHasHit()) {
    // reads the three front quadrants
    double left = sonars.getFrontLeftQuadrantMeasurement();
    double right = sonars.getFrontRightQuadrantMeasurement();
    double front = sonars.getFrontQuadrantMeasurement();

    // if obstacle near
    if ((front  < 0.7)||(left  < 0.7)||(right  < 0.7)) {
     if (left < right)
      setRotationalVelocity(-1);
     else
      setRotationalVelocity(1);
     setTranslationalVelocity(0);
    } else {
     setRotationalVelocity(0);
     setTranslationalVelocity(0.6); }
   } else {
    setTranslationalVelocity(0.8);
    setRotationalVelocity(0); }
 }
```

As shown here, accessing sensor and motor information is very intuitive and complex behaviours may be implemented on this basis. In this setup, this code is executed 20 times per "virtual"second[4] and the simulator impact the motor commands on the robot. An example of the resulting trace is shown in figure 2.

4.2 Experiment 2: Evolving a Wander Behaviour

In this experiment, we use a Multi-Layered Perceptron with four sensory inputs (four quadrants of an Infrared sensors belt) and two motor outputs (translational and rotational velocity), i.e. same as before except that bumpers are not used. Problem properties and parameters are : 11 neurons (4 inputs, 4 hidden, 2 outputs, 1 bias) fully connected (i.e. $26+6 = 32$ weights) ; 20 individuals, (2+18)-ES, mutation rate is 0.1 ; in the same environment as previous experiment (see fig. 2). The objective for a given robot is optimal if the robot does not hit walls, maximise translation, minimise rotation and minimise wall proximity. That is :

if hit wall : $fitness_{robot_i} = 0$ else : $fitness_{robot_i} = \sum(translation_{speed} + (1 - rotational_{speed}) + (1 - max(IR_{value}))$

---

[4] As stated before, "virtual" simulation time can be accelerated more than 1000 times depending on the machine used.

Fig. 2. Trace for hand-written wander behavior

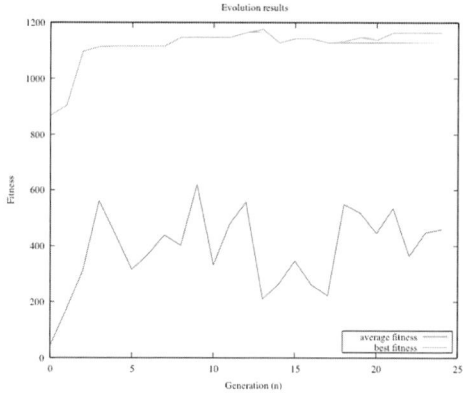

Fig. 3. Evolving a NN-based wander behaviour: Results

Results are shown in fig. 3 – Due to the task simplicity, the best individual quickly achieve near-optimal control and performance keeps on improving over time. This is of course a very basic, yet classic, experiment. This basic Evolutionary Robotics introduction task is taken from the classic reference [7] to illustrate the simultaneous use of Simbad, PicoNode and PicoEvo.

Due to the necessarily limited size of this paper, only simple experiments are described to illustrate the use of the Simbad package. However, the reader may refer to [14,15] which present Evolutionary Robotics experiments performed using elements from this package.

5   Conclusion

In this paper, we have presented the Simbad package for mobile robot simulation and evolutionary robotics. This package provides three stand-alone important components:

- Simbad : an extensible mobile robot simulator which can handle complex 3D scene and physics-based interaction;
- PicoNode : a library for graph-based controller such as (but not limited to) feed-forward and recurrent neural networks;
- PicoEvo : an Evolutionary Algorithm library which includes Genetic Algorithm, Genetic Programming (tree and graphs) and other popular approach in the field of population-based stochastic optimisation algorithms;

The Simbad package provides a powerful and easy-to-use framework written in Java for researchers and teachers in Evolutionary Robotics. To date, several works in the field of Evolutionary Robotics have already been achieved using the Simbad package [14, 15] and current works relies on this package. Moreover, the Simbad package is also used in the scope of Education in the last year of IFIPS, the engineering school at Universite Paris-Sud (France), as well as at Ecole Polytechnique (France), for a set of courses and projects on Reactive and Evolutionary Robotics on simulated and real Khepera robots.

The Simbad package is freely distributed with sources and provides an ideal solution for both researchers and teachers in the field of Autonomous and Evolutionary mobile robotics. Current on-going works on the Simbad Package includes extensions of the Simbad simulator towards easy switching capability from simulation to in situ robotics and of the PicoNode and PicoEvo libraries to other machine learning algorithms (Reinforcement Learning, topology learning in bayesian network for robot control, dynamical systems modelling, etc.).

Acknowledgment

The authors would like to thank Cedric Hartland, Justine Lebrun and Thomas Delquie for their contributions to the Simbad framework. This work was supported in part by the PASCAL Network of Excellence and by a CNRS TCAN grant (Nerobot project).

References

1. Simbad Simulator - http://simbad.sourceforge.net
2. B. Gerkey, R. T. Vaughan and A. Howard. The Player/Stage Project: Tools for Multi-Robot and Distributed Sensor Systems. Proceedings of the 11th International Conference on Advanced Robotics, 2003.
3. N. Koenig and A. Howard. Design and Use Paradigms for Gazebo, An Open-Source Multi-Robot Simulator. IEEE/RSJ International Conference on Intelligent Robots and Systems (IROS), Sendai, Japan, 2004.
4. Webots - http://www.cyberbotics.com/
5. http://www.lri.fr/~mary/WoB
6. P.Lewis and U. Neumann. Performance of Java versus C++. http://www.idiom. com/~zilla/Computer/javaCbenchmark.html
7. S. Nolfi and D. Floreano. Evolutionary Robotics. MIT Press, 2000.
8. R. Brooks. A Robust Layered Control System for a Mobile Robot. IEEE Journal of Robotics and Automation, Vol. 2, No. 1, pp.14-23 1986.

9. R. Sutton and A. Barto. Reinforcement Learning. MIT Press, 1998.
10. D. Goldberg. Genetic Algorithms in search. Addison-Wesley, 1989.
11. S. Nolfi and D. Parisi. Auto-teaching : networks that develop their own teaching input. Proceedings of the second ECAL, 1993.
12. N. Godzik, M. Schoenauer and M. Sebag. Robustness in the long run : Auto-teaching vs. Anticipation in Evolutionary Robotics. Proceedings of PPSN, 2004.
13. F. Gruau. Modular Genetic Neural Networks for six-legged Locomotion. Proceedings of Artificial Evolution (EA), 1995.
14. N. Bredeche and L. Hugues. Speeding up learning with Dynamic Environment Shaping in Evolutionary Robotics. Proceedings of the fifth International Workshop of Epigenetics Robotics (EpiRob), 2005.
15. N. Bredeche and L. Hugues. Evolutionary Robotics : incremental learning of sequential behavior. Proceedings of the fourth IEEE International Conference on Developmental Learning, 2005.

Comparing Robot Controllers
Through System Identification

Ulrich Nehmzow[1], Otar Akanyeti[1], Roberto Iglesias[2],
Theocharis Kyriacou[1], and S.A. Billings[3]

[1] Department of Computer Science, University of Essex, UK
[2] Electronics and Computer Science,
University of Santiago de Compostela, Spain
[3] Dept. of Automatic Control and Systems Engineering,
University of Sheffield, UK

Abstract. In mobile robotics, it is common to find different control programs designed to achieve a particular robot task. It is often necessary to compare the performance of such controllers. So far this is usually done qualitatively, because of a lack of quantitative behaviour analysis methods.

In this paper we present a novel approach to compare robot control codes *quantitatively*, based on system identification. Using the NARMAX system identification process, we "translate" the original behaviour into a transparent, analysable mathematical model of the original behaviour. We then use statistical methods and sensitivity analysis to compare models quantitatively.

We demonstrate our approach by comparing two different robot control programs, which were designed to drive a Magellan Pro robot through door-like openings.

1 Introduction

In mobile robotics it is common to find different control programs, developed for the same behaviour, typically developed through an empirical trial-and-error process of iterative refinement.

It is often interesting to compare and analyse such different controllers for the same behaviour. However, because of a lack of quantitative descriptors of behaviour such analyses are usually qualitative. As the number of controllers developed for identical tasks increases, the need for fair comparisons based on quantitative measures also becomes more important, and to find quantitative answers to questions like "What are the advantages and disadvantages of each controller?", "Which one is more stable and efficient?", "Will they work in completely different environments or crash even with slight modifications?" will enhance our understanding of robot-environment interaction.

Realistic mobile robot control programs tend to be so complex that their direct analysis is impossible — the "meaning" of thousands of lines of code is not clear to a human observer. In this paper we therefore propose the alternative approach of "identifying" the behaviour in question, using transparent polynomial models

S. Nolfi et al. (Eds.): SAB 2006, LNAI 4095, pp. 843–854, 2006.

obtained through NARMAX system identification [1], to ascertain the validity of the model, and to analyse the model instead of the original behaviour, both qualitatively and quantitatively.

2   Controller Identification

Motivation and Related Work. The main benefits of "system identification" (*i.e.* mathematical modelling) are

1. Efficiency: The obtained models are polynomials. They are fast in execution and occupy little space in memory.
2. Transparency: The transparent model structure simplifies mathematical analysis
3. Portability: Polynomial models are universal, in the sense that they can be quickly incorporated into any robot programming language.

Our models are obtained based on NARMAX (Nonlinear Auto-Regressive Moving Average model with eXogenous inputs) system identification [1]. This identification process has already been applied in modelling sensor-motor couplings of robot controllers [2, 4]. In [5] we demonstrate how easy it is to exchange the generated models between different robot platforms.

After the identification of the controllers it is relatively straightforward to evaluate the models and to compare them quantitatively. The work presented in [3] gives examples of the characterisation of models of robot behaviour.

The system identification process can be divided into two stages: First the robot is driven by the original robot controller which we wish to analyse. While the robot is moving, we log sensor and motor information to model the relationship between the robot's sensor perception and motor responses.

In the second stage, a nonlinear polynomial NARMAX model (see below) is estimated. This model relates input sensor values to output actuator signals and subsequently can be used to control the robot.

NARMAX Modelling. The NARMAX modelling approach is a parameter estimation methodology for identifying the important model terms and associated parameters of unknown nonlinear dynamic systems. For multiple input, single output noiseless systems this model takes the form:

$$
\begin{aligned}
y(n) = f(&u_1(n), u_1(n-1), u_1(n-2), \cdots, u_1(n-N_u), u_1(n)^2, u_1(n-1)^2, \\
&u_1(n-2)^2, \cdots, u_1(n-N_u)^2, \cdots, u_1(n)^l, u_1(n-1)^l, u_1(n-2)^l, \cdots, \\
&u_1(n-N_u)^l, u_2(n), u_2(n-1), u_2(n-2), \cdots, u_2(n-N_u), u_2(n)^2, \\
&u_2(n-1)^2, u_2(n-2)^2, \cdots, u_2(n-N_u)^2, \cdots, u_2(n)^l, u_2(n-1)^l, \\
&u_2(n-2)^l, \cdots, u_2(n-N_u)^l, \cdots, u_d(n), u_d(n-1), u_d(n-2), \cdots, \\
&u_d(n-N_u), u_d(n)^2, u_d(n-1)^2, u_d(n-2)^2, \cdots, u_d(n-N_u)^2, \cdots, \\
&u_d(n)^l, u_d(n-1)^l, u_d(n-2)^l, \cdots, u_d(n-N_u)^l, y(n-1), y(n-2), \cdots, \\
&y(n-N_y), y(n-1)^2, y(n-2)^2, \cdots, y(n-N_y)^2, \cdots, y(n-1)^l, \\
&y(n-2)^l, \cdots, y(n-N_y)^l)
\end{aligned}
$$

where $y(n)$ and $\boldsymbol{u}(n)$ are the sampled output and input signals at time n respectively, N_y and N_u are the regression orders of the output and input respectively and d is the input dimension. $f()$ is a non-linear function, this is typically taken to be a polynomial or wavelet multi-resolution expansion of the arguments. The degree l of the polynomial is the highest sum of powers in any of its terms.

The NARMAX methodology breaks the modelling problem into the following steps: i) structure detection, ii) parameter estimation, iii) model validation, iv) prediction and v) analysis.

Logged data is first split into two sets (usually of equal size). The first, the *estimation data set*, is used to determine model structure and parameters. The remaining data set, the *validation data set*, is subsequently used to validate the model.

3   Experimental Scenario: Door Traversal

The example presented in this paper demonstrates how system identification can be used to model sensor-motor couplings of two different robot controllers designed to achieve the same particular task: the episodic task of "door traversal", where each episode comprises the movement of the robot from a starting position to a final position, traversing a door-like opening *en route*.

In this paper, we compare two different controllers for this task. "Model 1" is a hard-wired laser controller which maps the laser perception of the robot onto action, using a behaviour based approach. For "model 2", the robot was driven manually by a human operator. While these two original behaviours were being executed, we logged all sensory perceptions, motor responses and the robot's precise position, using a camera logging system, every 160 ms.

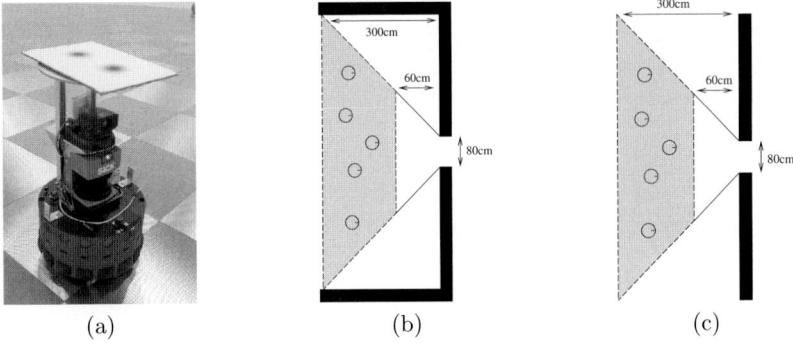

(a)   (b)   (c)

Fig. 1. The *Magellan Pro* robot used in the experiments (a). Training environments for the "Hard-Wired Laser Controller" (model 1, centre) and "Manual Control" (model 2, right). Environment (b) has extra walls on the wings of the door-like opening. The initial positions of the robot were within the shaded area indicated in each figure; door openings were twice the robot's diameter (80cm).

Each controller was then modelled through non-linear polynomial functions (NARMAX), so that the resulting two models identify the two robot controllers by relating robot perception to action. To test the validity of the two models, we then let the models themselves control the robot.

The models of the two different robot controllers were deliberately trained in different environments (figure 1 (b,c)), with different translational velocities and different sensor information, to simulate the natural differences which might exist between different implementations of the same task. All experiments were conducted with a Magellan Pro robot (figure 1 (a)) in the Robotics Arena at Essex University.

3.1   Model 1: Hardwired Laser Control

The first control program used a hardwired, behaviour-based control strategy that mapped laser perception to steering speed of the robot. The translational velocity of robot was kept constant at 0.2 m/s, the laser perception of the robot was encoded in 36 sectors by taking the median value of every 5 degree interval over a semi-circle. The model of the robot's angular velocity ω, obtained through NARMAX system identification, contains 84 terms and is given in table 1.

Table 1. Model 1: NARMAX model of the angular velocity ω as a function of laser perception for the "hard-wired laser controller". $d_1, \cdots, d_{29}$ are laser bins, coarse coded by taking the median value over each 5-degree interval.

$$\omega = 0.574 + 0.045 * d_1(t) + 0.001 * d_6(t) + 0.032 * d_7(t) - 0.017 * d_8(t)$$
$$+0.113 * d_9(t) - 0.099 * d_{10}(t) - 0.011 * d_{11}(t) - 0.006 * d_{12}(t) - 0.020 * d_{13}(t)$$
$$+0.020 * d_{14}(t) - 0.015 * d_{15}(t) - 0.016 * d_{16}(t) - 0.046 * d_{20}(t) - 0.009 * d_{21}(t)$$
$$+0.007 * d_{22}(t) - 0.035 * d_{23} - 0.017 * d_{24}(t) - 0.056 * d_{25}(t) + 0.026 * d_{26}(t)$$
$$+0.039 * d_{27}(t) - 0.003 * d_{28}(t) + 0.044 * d_{29}(t) - 0.018 * d_{30}(t) - 0.144 * d_{31}(t)$$
$$+0.001 * d_{32}(t) - 0.023 * d_1^2(t) - 0.001 * d_6^2(t) - 0.011 * d_7^2(t) + 0.003 * d_8^2(t)$$
$$-0.010 * d_{13}^2(t) - 0.001 * d_{15}^2(t) - 0.005 * d_{16}^2(t) - 0.001 * d_{17}^2(t)$$
$$-0.002 * d_{19}^2(t) + 0.002 * d_{20}^2(t) - 0.001 * d_{21}^2(t) - 0.001 * d_{22}^2(t)$$
$$+0.002 * d_{23}^2(t) + 0.002 * d_{24}^2(t) + 0.008 * d_{25}^2(t) - 0.004 * d_{27}^2(t)$$
$$-0.001 * d_{28}^2(t) + 0.019 * d_{31}^2(t) - 0.001 * d_{35}^2(t) + 0.001 * d_{36}^2(t)$$
$$-0.001 * d_1(t) * d_7(t) - 0.001 * d_1(t) * d_{18}(t) + 0.001 * d_1(t) * d_{20}(t)$$
$$+0.003 * d_1(t) * d_{22}(t) + 0.006 * d_1(t) * d_{23}(t) - 0.003 * d_1(t) * d_{25}(t)$$
$$-0.001 * d_1(t) * d_{32}(t) + 0.003 * d_1(t) * d_{33}(t) + 0.007 * d_1(t) * d_{34}(t)$$
$$-0.002 * d_1(t) * d_{36}(t) + 0.003 * d_7(t) * d_{16}(t) - 0.001 * d_8(t) * d_{10}(t)$$
$$-0.006 * d_9(t) * d_{14}(t) - 0.003 * d_9(t) * d_{17}(t) - 0.009 * d_9(t) * d_{18}(t)$$
$$-0.002 * d_9(t) * d_{28}(t) + 0.013 * d_{10}(t) * d_{13}(t) - 0.002 * d_{11}(t) * d_{16}(t)$$
$$+0.004 * d_{11}(t) * d_{18}(t) - 0.001 * d_{12}(t) * d_{15}(t) + 0.001 * d_{12}(t) * d_{16}(t)$$
$$+0.002 * d_{12}(t) * d_{18}(t) + 0.002 * d_{13}(t) * d_{15}(t) + 0.011 * d_{13}(t) * d_{16}(t)$$
$$-0.003 * d_{14}(t) * d_{16}(t) + 0.002 * d_{14}(t) * d_{23}(t) + 0.001 * d_{14}(t) * d_{36}(t)$$
$$+0.002 * d_{15}(t) * d_{21}(t) + 0.003 * d_{16}(t) * d_{19}(t) + 0.002 * d_{16}(t) * d_{25}(t)$$
$$+0.002 * d_{17}(t) * d_{28}(t) + 0.001 * d_{18}(t) * d_{31}(t) + 0.001 * d_{20}(t) * d_{21}(t)$$
$$+0.002 * d_{20}(t) * d_{28}(t) + 0.003 * d_{20}(t) * d_{31}(t) - 0.004 * d_{21}(t) * d_{26}(t)$$
$$+0.005 * d_{21}(t) * d_{30}(t) - 0.005 * d_{25}(t) * d_{29}(t)$$

Qualitative Model Evaluation. Figure 2 shows the trajectories of the robot under the control of the original hard-wired laser controller and its NARMAX model respectively. In both cases the robot was started from 36 different initial positions, and completed the task successfully in each case.

(a)   (b)

Fig. 2. Robot trajectories under control by the original "Hard-wired Laser Controller" (a) and under control by its model, "model 1" (b). Note that side walls of the environment can not be seen in the figures because they were outside the field of view of the camera.

3.2   Model 2: Manual Control

The second control strategy used was to drive the robot through the opening manually. Here, the translational velocity of robot was kept constant at 0.07 m/s. For sensor information (*i.e.* model input), the values delivered by the laser scanner were averaged in twelve sectors of 15 degrees each, to obtain a twelve dimensional vector of laser distances. These laser bins as well as the 16 sonar sensor values were inverted before they were used to obtain the model, so that large readings indicate close-by objects. The identified model of the angular velocity ω, which contained 35 terms, is given in table 2.

Qualitative Model Evaluation. Figure 3 shows the trajectories of the robot under the control of human operator and its NARMAX model respectively. In both cases the robot was started with 41 different initial positions, and passed through the door-like gap successfully each time.

4   Comparison of the Two Models

Figures 2 and 3 show that on a qualitative level both models are good representations of the original. Small differences in trajectory between "original" and "model-controlled" behaviour we attribute to natural fluctuations in robot-environment interaction.

Table 2. Model 2: NARMAX model of the angular velocity ω as a function of sensor perception for the door traversal behaviour under manual control. $s_{10}, \cdots, s_{16}$ are the inverted and normalised sonar readings ($s_i' = (1/s_i - 0.25)/19.75$), while $d_1, \cdots, d_6$ are the inverted and normalised laser bins $d_i' = (1/d_i - 0.12)/19.88$. Taken from [4].

$$\omega(t) = 0.010 - 1.633 * d_1'(t) - 2.482 * d_2'(t) + 0.171 * d_3'(t) + 0.977 * d_4'(t)$$
$$-1.033 * d_5'(t) + 1.947 * d_6'(t) + 0.331 * s_{13}'(t) - 1.257 * s_{15}'(t) + 12.639 * d_1'^2(t)$$
$$+16.474 * d_2'^2(t) + 28.175 * s_{15}'^2(t) + 80.032 * s_{16}'^2(t) + 14.403 * d_1'(t) * d_3'(t)$$
$$-209.752 * d_1'(t) * d_5'(t) - 5.583 * d_1'(t) * d_6'(t) + 178.641 * d_1'(t) * s_{11}'(t)$$
$$-126.311 * d_1'(t) * s_{16}'(t) + 1.662 * d_2'(t) * d_3'(t) + 225.522 * d_2'(t) * d_5'(t)$$
$$-173.078 * d_2'(t) * s_{11}'(t) + 25.348 * d_3'(t) * s_{12}'(t) - 24.699 * d_3'(t) * s_{15}'(t)$$
$$+100.242 * d_4'(t) * d_6'(t) - 17.954 * d_4'(t) * s_{12}'(t) - 3.886 * d_4'(t) * s_{15}'(t)$$
$$-173.255 * d_5'(t) * s_{11}'(t) + 40.926 * d_5'(t) * s_{15}'(t) - 73.090 * d_5'(t) * s_{16}'(t)$$
$$-144.247 * d_6'(t) * s_{12}'(t) - 57.092 * d_6'(t) * s_{13}'(t)) + 36.413 * d_6'(t) * s_{14}'(t)$$
$$-55.085 * s_{11}'(t) * s_{14}'(t) + 28.286 * s_{12}'(t) * s_{15}'(t) - 11.211 * s_{14}'(t) * s_{16}'(t)$$

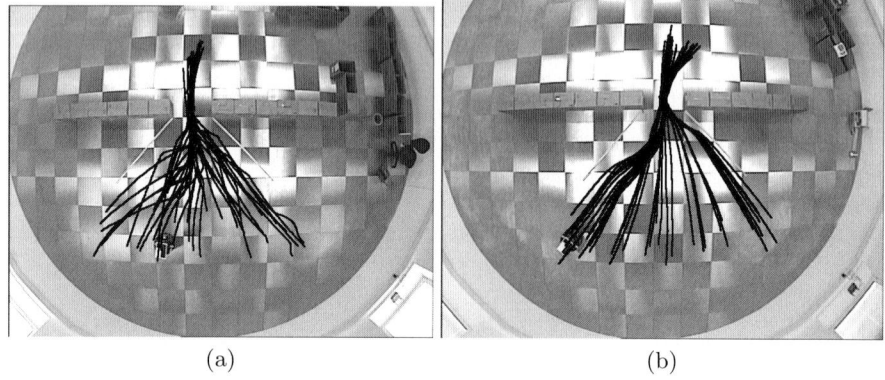

(a)   (b)

Fig. 3. Robot Trajectories under under "manual control" (a) and under control of its model, "model 2" (b). Taken from [4].

It is interesting to see that "Model 1" has significantly more terms (83) than "Model 2" (35), and therefore requires more memory and computing power resources. Immediately obvious is also that model 2 is simpler, in that it only uses information from sensors found on the right side of the robot. In the following, we were interested in investigating the differences between the models of the two door-traversal behaviours further.

Testing and Evaluation Setup. We first tested the controllers in four different environments (figure 4) to reveal differences in the behaviour of the robot when controlled by the two models. To minimise the influence of constant errors, model 1 and model 2 were selected randomly to control the robot.

In the first scenario (figure 4 (1)), the robot was driven by "Model 1" in the training environment of "Model 2". After this, the two controllers were tested in a completely different environment, where two door-like openings with equal

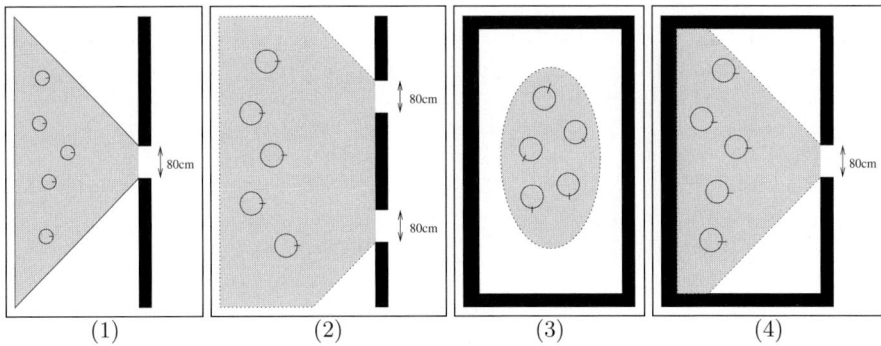

Fig. 4. Four different test environments used to compare the two models qualitatively. 1) Single-door test scenario, 2) Double-door test scenario, 3) Completely-enclosed test scenario, 4) Enclosure with a single door. The initial positions of the robot are within the shaded area indicated in each figure. All openings found in the environment had a width of two robot diameters (80cm).

width were presented at the same time (figure 4 (2)). This demonstrated how each opening interfered with the robot's behaviour. It also revealed the dominant sensors.

In the third and fourth scenarios (figure 4 (3) and (4)) we tried to measure the effect of a gap found in an environment. We assumed that in this application openings would be the dominant environmental factors in robot-environment interaction. Therefore the models were tested first in a completely closed environment, subsequently we introduced a gap to observe the resulting change in behaviour.

4.1 Qualitative Comparison

As can be seen in figure 5, when we remove the side walls from the environment, the robot under the control of "Model 1" has some problems in centreing itself while passing through the gap. We realised that side walls make the robot turn sharper while approaching to the door-like opening, which made it easier for robot to centre itself while passing through the gap.

For the double-door test scenario, it is interesting to see how two gaps interfere with each other for the behaviour of robot under the control of "Model 1". If the robot was started from the middle region between two gaps, it generally collided with the wall, without turning towards any of the gaps. If the robot pointed at one of the gaps initially, it was attracted towards it, but the other gap still effected the behaviour and the robot could not centre itself enough to pass through the gap (figure 6 (a)).

When we look at the trajectories of robot under the control of "Model 2" (figure 6 (b)), we observe that the two gaps do not interfere with each other. This is a predictable outcome, as "Model 2" uses sensors only from the right side of the robot. Therefore, if the robot detects the gap found on the right side, it

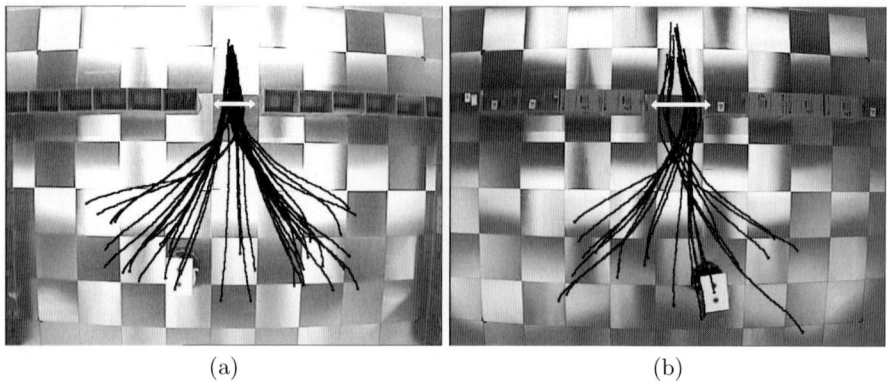

(a)   (b)

Fig. 5. Robot trajectories under the control of "Model 1" in the training environment of "Model 1" (a), and in the training environment of "Model 2" (b). The arrows in both figures indicate how much the robot deviated from the centre point of the opening while passing through it. Note that side walls of the environment in (A) can not be seen in the figure because of the limited range of camera view.

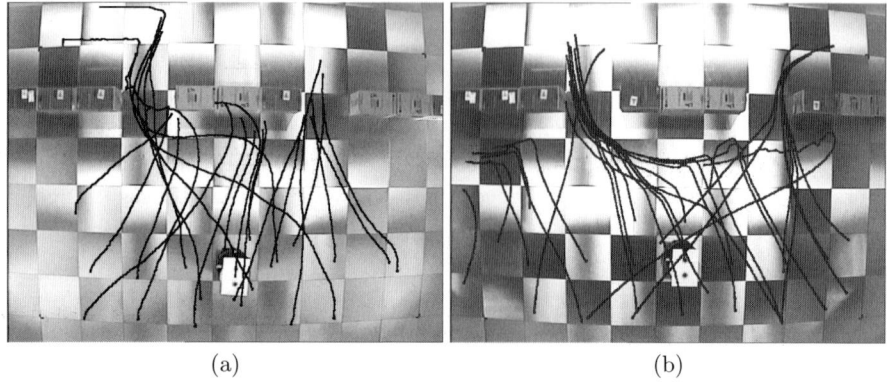

(a)   (b)

Fig. 6. Robot Trajectories in the "double door" test scenario under the control of "Model 1" (a), and controlled by "Model 2" (b).

passes through it. If it doesn't, it automatically turns to the left, finds the other gap and traverses it.

In the completely-enclosed environment, "Model 1" turned the robot constantly to the left. Figure 7 (a) shows a sample trajectory of robot under the control of "Model 1". When we introduced a gap to the environment, we observed that the gap has the dominating effect on the controller behaviour. Although the model can not pass through the gap successfully in every trial, we still observe a general tendency of robot towards the gap (figure 7 (b)).

Finally, "Model 2" showed very unstable characteristics in the last two environments. When we look at the trajectories closely we observe oscillations in

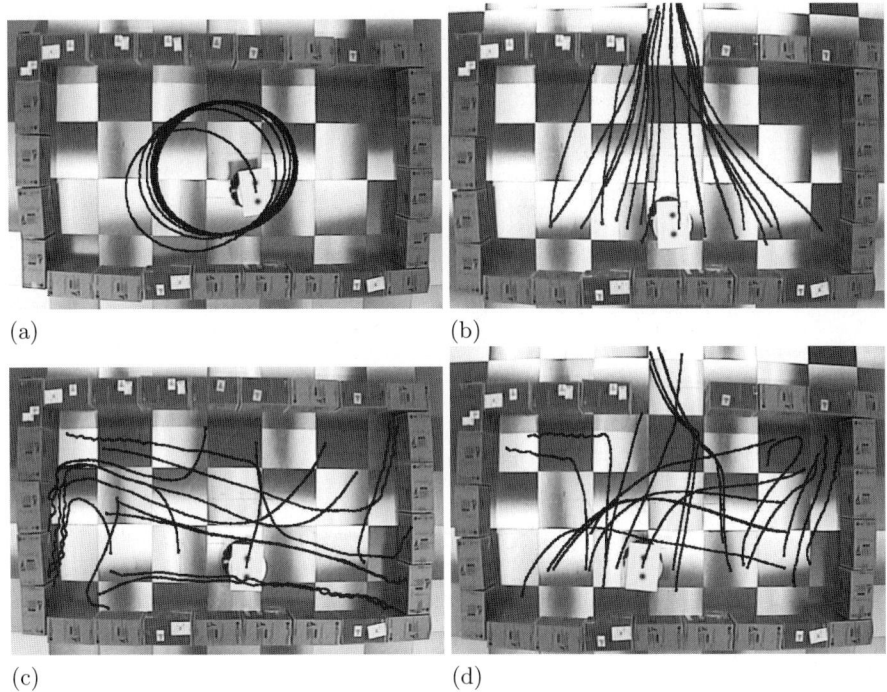

(a)   (b)

(c)   (d)

Fig. 7. Robot trajectories under control of "Model 1" in the "Completely enclosed environment" (a) and when an opening is introduced to the enclosed environment (b). Robot trajectories under control of "Model 2" in the "Completely enclosed environment" (c), and when an opening is introduced to the enclosed environment (d).

the robot's behaviour, especially when the robot is close to corners. The overall model generally looks like a right wall follower when the robot is far from the corners. As it comes closer, the variation in angular velocity of the robot increases and it bumps into the corners (figure 7 (c)). We can also see that for "Model 2", the influence of the gap on the robot behaviour is not as big as the influence on "Model 1" (figure 7 (d)).

4.2   Quantitative Comparison

When we look at the behaviours of the robot qualitatively in different test environments, we see that the responses of the two models differ. We also observe that both models do not produce a *general* door-traversal behaviour, but that they fail in environments that differ considerably from the training environment. In this section, we extended our work by comparing the two models quantitatively, based on a "hardware in the loop simulation" process.

Hardware in the Loop Simulation. In the first quantitative analysis we wanted to see if the two models produce similar outputs for the same inputs. We

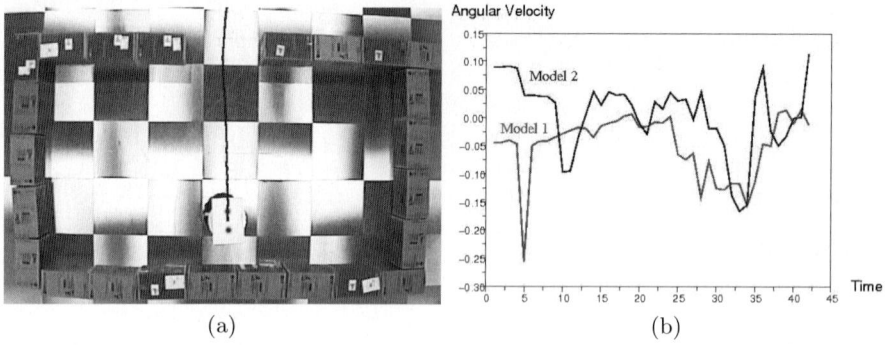

(a)   (b)

Fig. 8. Trajectory along which sensor data for "hardware in the loop" simulation was taken (a), and steering velocity graphs of both models for the given trajectory (b). There is no significant correlation between the two model's angular velocities (r_S=0.078, not sig.,p>5%).

Table 3. Spearman rank correlation coefficients between "model 1" and "model 2" responses to identical real world input data

Environment	r_S	Stat. Sig.
1	0.08	not sig. (p>0.05)
2	-0.09	sig. (p<0.05)
3	0.27	sig. (p<0.05)
4	0.02	not sig. (p>0.05)

therefore fed the two models with same sensory perception of the robot, taken from the real test environments, and then compared the corresponding outputs of the models quantitatively in order to see if there is a significant correlation between the responses of the two models.

For each of the four test environments, we therefore first drove the robot manually to traverse the door from different initial positions and logged the sensory perception of the robot at every 250ms. We then fed identical logged sensor data to the two models individually, and computed the resulting angular velocity vectors. As an example, figure 8(b) shows the steering velocity graphs which the robot would attain along the sample trajectory given in figure 8(a)).

To assess the agreement between both model responses to identical input vectors, we computed the Spearman rank correlation coefficient r_S [6] between the two model outputs for each test environment. The results are given in table 3, even in cases where there is a statistically significant correlation coefficient they are very low, indicating quantitatively that the two models generate different robot behaviour.

Stability and Efficiency Analysis. The fact that the controller is known as a transparent polynomial simplifies stability and efficiency analysis, as follows.

Efficiency. The efficiency of the controllers can be evaluated by measuring their execution time on the robot and memory space required to represent the model. Obviously, a polynomial consisting of fewer terms will execute faster and occupy less memory space. It therefore becomes immediately obvious that "Model 1", obtained through the hardwired laser-guided door traversal, is considerably less efficient (83 model terms) than "Model 2", obtained through manual training (35 model terms).

BIBO stability. A controller is considered stable if it produces bounded output for a bounded input. With today's robot programming techniques, the only way to determine if a controller is stable or not is to check the response of the controller for the entire input space and make sure that the output never goes to $\pm\infty$. Typically, robot programmers try to address this issue by writing extra safety routines in the control code which constrain the controller's output within an acceptable interval, without formally analysing for controller stability.

   With system identification it is possible to analyse the stability of the controllers without searching the entire input space. The identified polynomials presented in this paper can be classified as "zero order open loop controllers", because they don't incorporate any feedback from the output (open loop control), and don't use any lags related to past input values (no difference equations in the polynomial). For this kind of polynomials it has been shown that their responses are bounded for bounded inputs [7], which proves the stability of the controllers.

5   Summary and Conclusion

It is sometimes useful in mobile robotics to be able to describe and analyse the behaviour of a mobile robot quantitatively, for example when one is interested in establishing the differences between different robot controllers. In the experiments presented in this paper, we compared two controllers that accomplish the same task (door traversal), one implemented by a hardwired, "traditional" control program and the other through manual control.

   In order to compare these two behaviours, we "identified" both behaviours by determining transparent, analysable non-linear polynomial mappings between sensor perception and motor response, and subsequently comparing the two models with each other. Such transfer to a common descriptor of behaviour has a number of advantages.

   Once transparent mathematical models of sensor motor behaviour are available, they can be analysed quantitatively. In the experiments presented in this paper, we used a hardware-in-the-loop simulation to demonstrate that the hardwired and the manual control strategy differ from each other noticeably. We furthermore evaluated the controllers in terms of efficiency and stability, again revealing considerable differences between the two models in terms of efficiency.

Future Work. Further analysis methods, not demonstrated here, include analysing the models' spectra, detailed stability analysis, or sensitivity analysis [8, 9, 10]. Such detailed mathematical analysis of the polynomial models is part of ongoing research at the universities of Sheffield and Essex.

Acknowledgements

The authors thank Hugo Vieira Neto for his contributions to the preparation of this paper. The authors also thank the following institutions for their support: The RobotMODIC project is supported by the Engineering and Physical Sciences Research Council under grant GR/S30955/01. Roberto Iglesias is supported through the research grants PGIDIT04TIC206011PR, TIC2003-09400-C04-03 and TIN2005-03844.

References

[1] Korenberg, M., Billings, S. A., Liu, Y. P., and McIlroy, P. J., Orthogonal Parameter Estimation Algorithm for Non-Linear Stochastic Systems, *Int. J. Control* , 1998, vol 48(1):193-210.

[2] Iglesias, R., Kyriacou, T., Nehmzow, U., and Billings, S. A., Task Identification and Characterisation in Mobile Robotics, *Proc. "Towards Autonomous Robotic Systems"*, TAROS, Essex 2004.

[3] Iglesias, R., Kyriacou, T., Nehmzow, U., and Billings, S. A., Modelling and characterisation of a mobile robot's operation, *Proc. "CAEPIA 2005, 11th Conference of the Spanish association for Artificial Intelligence"*, Santiago De Compostela 2005.

[4] Iglesias, R., Nehmzow, U., Kyriacou, T., and Billings, S. A., Programming Through System Identification and Training, *Proc. European Conference on Mobile Robotics* , Ancona 2005.

[5] Kyriacou, T., Nehmzow, U., Iglesias, R., and Billings, S. A., Cross-Platform Programming Through System Identification, *Proc. "Towards Autonomous Robotic Systems"*, TAROS, London 2005.

[6] Barnard, C., Gilbert, F., and McGregor, P., *Asking questions in Biology, key Skills for Practical Assessments and Project Work* , Pearson Education Limited, 2001.

[7] P.C. Dorf and R.H. Bishop, Modern Control Systems; Addison Wesley 1995

[8] Max D. Morris, Factorial sampling plans for preliminary computational experiments, *Technometrics*, v.33 n.2, p.161-174, May 1991.

[9] A. Saltelli, K. Chan and E.M. Scott, *Sensitivity Analysis*, John Wiley 2000

[10] I.M. Sobol, Sensitivity Estimates for Nonlinear Mathematical Models, *J.Mathematical Modelling and Computational Experiment*, Vol 1, pp. 407-414, 1993.

Adaptive Fuzzy Sliding Mode Controller for the Snorkel Underwater Vehicle

Eduardo Sebastián

Centro de Astrobiología (CAB), Grupo de Robótica y Exploración Planetaria,
Ctra. Ajalvir Km. 4, Torrejón de Ardoz. Madrid, Spain
sebastianme@inta.es

Abstract. This paper address the kinematic variables control problem for the low-speed manoeuvring of a low cost and underactuated underwater vehicle. Control of underwater vehicles is not simple, mainly due to the non-linear and coupled character of plant equations, the lack of a precise model of vehicle dynamics and parameters, as well as the appearance of internal and external perturbations. The proposed methodology is an approach that makes use of a pioneering algorithm in underwater vehicles, based on the fusion of a robust or sliding mode controller and an adaptive fuzzy system, including the advantages of both systems. The main property of this methodology is that it relaxes the required knowledge of vehicle model, reducing the cost of its design.

1 Introduction

Underwater vehicles have replaced human beings, especially in dangerous or precise tasks, making the design of automatic and precise navigation and control systems necessary. The problem of underwater vehicles control is difficult because of the unknown non-linear hydrodynamics effects, and parameter uncertainties.

The problem analysed in this paper, the low-speed control of the kinematic variables of an underactuated underwater vehicle, can be defined as follow. Given an unknown underwater vehicle plant and a continuous bounded time-varying velocity and/or position references, design a controller that ensures that the plant states converge asymptotically to the kinematic references.

Most dynamically positioned marine vehicles in used today employ PD or PID controllers for each kinematic variable. Moreover, PID control cannot dynamically compensate for unmodeled vehicle hydrodynamic forces or unknown disturbances [10]. To avoid this problem only a reduce number of commercial vehicles employ model-based controllers, because they require a plant model with unknown parameters which are difficult to estimate with accuracy. From this point of view, most of the proposed control schemes take into account the uncertainty in the model by resorting to an adaptive strategy or a robust approach. Thus, a significant number of studies have employed linearized plant approximations [1], [2] and [3].

In the area of modern control, relatively few studies directly address decoupled non-linear plant model for underwater vehicles. In [4] the authors report non-linear sliding mode control for surge, sway and yaw movements. Also, adaptive versions of sliding controllers have been implemented [5] and [6], reducing effectively model

S. Nolfi et al. (Eds.): SAB 2006, LNAI 4095, pp. 855–866, 2006.

uncertainty and control activity, and maintaining robustness without sacrificing performance. In [7] a state linearization control is studied, based on the known model of the plant. A step forward is an adaptive non-linear controller [8], which presents a problem of sensitivity to noise in the measurement of the kinematic variables. In [9] a modification of the non-linear adaptive control law is presented, in which in the adaptation process the velocity and position measurements are replaced by their input references. In [10] the authors compare, using experimental trial, some of the previously reported controllers, based on a decoupled non-linear plant model of the JHUROV vehicle. In the area of intelligent control, [11] proposes a neural control, using a nonparametric and adaptive recursive algorithm control that do not require knowledge of the plant dynamics. Finally, [12] proposes a fuzzy controller with 14 rules for depth control of an UUV.

This paper studies a pioneering algorithm in underwater vehicles, which is based on the work and results developed in [13], about adaptive fuzzy sliding mode control (AFSM). Euler angles and body fix reference frame are used to describe a semi-decoupled non-linear plant model of the underactuated Snorkel vehicle, a UUV to show controller performance. The controller is based on the fusion of a sliding mode controller and an adaptive fuzzy system, and exhibits adaptive and robust features. The main advantage of the proposed theory versus previous studies is that it employs a nonparametric adaptive technique that requires a minimum knowledge of plant dynamics, being only necessary a theoretical and simple model of it. A Lyapunov-like stability analysis of the control algorithm is developed, ensuring the stability of the adaptation process and the convergence to the references.

The paper is organized as follows; section 2 introduces the dynamic equations of the Snorkel vehicle. In section 3 the AFSM controller and its theoretical demonstrations of stability are presented. Section 4 is dedicated to the experiments setup. Section 5 shows a series of real experiments results, and the performance of the controller is described and compared. Finally, section 6 summarizes the results.

2   Dynamical Model

Dynamic modelling of UUV uses a finite dimensional approximation in which plant parameters enter linearly into the non-linear differential equations of motion. In this work a Newton-Euler formulation and a non-inertial reference system have been selected. Euler angles do not present singularities in the Snorkel vehicle, due to its moderate pitch and roll motion. Equation (1) represents the most reported [9] dynamic equations of 6-DOF, represented in compact form,

$$M\dot{v} + C(v)v + D(v)v + g(\eta) = \tau \ , \tag{1}$$

where $M \in \mathfrak{R}^{6 \times 6}$ mass matrix that includes rigid body and added mass and satisfies $M = M^T > O$ and $\dot{M} = 0$; $C(v) \in \mathfrak{R}^{6 \times 6}$ matrix of Coriolis and centrifugal terms including added mass and satisfies $C(v) = -C(v)^T$; $D(v) \in \mathfrak{R}^{6 \times 6}$ matrix of friction and hydrodynamic damping terms; $g(\eta) \in \mathfrak{R}^6$ vector of gravitational and buoyancy generalized forces; $\eta \in \Re^3$ is the vector of Euler angles; $v \in \Re^6$ is the vector of vehicle velocities in its six DOF, relative to the fluid and in a body-fixed reference frame and $\tau \in \Re^6$ is the driver vector considering vehicle's thrusters position.

Actually there is no an exact model to describe the value of matrix and vectors of (1). From AFSM controller capabilities, it is adopted a simplify vehicle model [9], that consider null values for: off diagonal entries of the damping matrix $D(v)$, with only linear and/or quadratic terms, inertial products, the tethered dynamics, as well as assuming a constant added mass. Thus, the equation (1) can be simplified [10] and divided in each of the semi-decoupled single DOF dynamical equations (2).

$$\ddot{x}_i = f_i(\xi) + g_i(\xi)\tau_i , \qquad (2)$$

where $f_i(\xi) = \dfrac{1}{m_i}\left[-c_i(v) - X_{|\dot{x}_i|\dot{x}_i}|\dot{x}_i|\dot{x}_i - g_i(\eta) - d_i\right]$, $g_i(\xi) = 1/m_i$, and for each DOF i, τ_i is the control force or moment, m_i is the effective inertia, $c_i(v)$ are the Coriolis and centripetal terms, $X_{|\dot{x}_i|\dot{x}_i}$ is the quadratic hydrodynamic drag coefficient, $g_i(\eta)$ is the buoyancy and weight term, d_i is a term that represents unmodeled dynamics and perturbations, and x_i, $\dot{x}_i$ and $\ddot{x}_i$ are the velocity and acceleration of the vehicle.

3 Fuzzy Sliding Mode Control Algorithm (AFSM)

In this section the equations and a stability analysis of the resulting close loop of the AFSM controller is presented. The controller shares the control law with a pure sliding mode controller (SM). Since functions $f(\xi)$ and $g(\xi)$ of the vehicle model are partially unknown and non-linear, a set of fuzzy functions to estimate them are proposed, being the control diagram of the overall system shown in the Fig. 1.

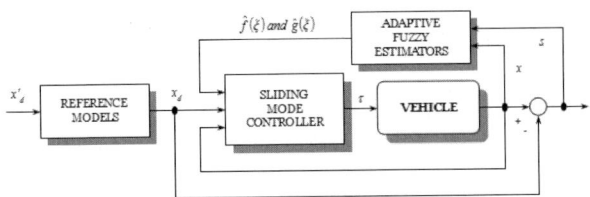

Fig. 1. Adaptive fuzzy sliding mode controller diagram

3.1 Sliding Mode Controller

The SM controller for a single DOF takes the form,

$$\tau = \hat{g}(\xi)^{-1}\left[\hat{f}(\xi) - \lambda\tilde{\dot{x}} + \ddot{x}_d - \eta_\Delta sat(s/b)\right] , \qquad (3)$$

where λ, b and η_Λ are positive definite constant, s is the sliding surface defined as $s = \dot{\tilde{x}} + \lambda\tilde{x}$, where $\tilde{x} = x - x_d$ and $\dot{\tilde{x}} = \dot{x} - \dot{x}_d$, $\hat{g}(\xi)$ and $\hat{f}(\xi)$ are the estimation of the respective functions, and τ is the control action.

The evolution of the sliding controller can be divided in two phases: The approximation phase, where $s \neq 0$, and the sliding phase when $s = 0$. The right election of the parameter η_Λ, based on the uncertainty boundaries of system functions and perturbations, allows the designer to ensure that error vectors $\tilde{x}$ and $\dot{\tilde{x}}$ change from the approximation phase to the sliding phase. Once on the surface, it is ensured that the system follows the input references, in presence of uncertainties, with a time constant of value, $\frac{1}{\lambda}$. The saturation function is defined as

$$sat(s/b) = \begin{cases} -1 & si & s/b \leq -1 \\ s/b & si & -1 < s/b \leq 1 \\ 1 & si & s/b > 1 \end{cases}$$

where b is the thickness of the boundary layer,

and its function is to avoid the chattering effect [8].

3.2 Fuzzy Adaptive System

A fuzzy system may be used like a non-linear universal approximator [15], due to its ability to introduce verbal information, and its capacity to uniformly approximate any real and continuous function with different degrees of precision. In general, good verbal information can help to establish initial conditions, and so faster adaptation will take place.

Thus, by using a Sugeno-like fuzzy system, with a singleton fuzzification strategy, product interface and media defuzzification, the functions $f(\xi)$ and $g(\xi)$ are parameterized by fuzzy logic systems as,

$$\hat{f}(\xi \mid \boldsymbol{\theta}_f) = \boldsymbol{\theta}_f^T \zeta(\xi), \quad \hat{g}(\xi \mid \boldsymbol{\theta}_g) = \boldsymbol{\theta}_g^T \zeta(\xi), \tag{4}$$

where $\zeta(\xi) = (\zeta^1(\xi), \dots, \zeta^m(\xi))^T$, with $\zeta^j(\xi) = \dfrac{\prod\limits_{i=1}^{n} \mu_{A_i^j}(\xi_i)}{\sum\limits_{j=1}^{m}\left[\prod\limits_{i=1}^{n} \mu_{A_i^j}(\xi_i)\right]}$ and $\mu_{A_i^j}(\xi_i)$ are the

membership functions of the fuzzy variable ξ_i, which are supposed to be fixed. And the elements $\boldsymbol{\theta}_f^T$ and $\boldsymbol{\theta}_g^T$, which take the form $(y^1, \dots, y^m)^T$, can be adaptively tuned till they reach the optimal values, $\boldsymbol{\theta}_f^*$ and $\boldsymbol{\theta}_g^*$.

3.2.1 Adaptation Law
The adaptive functions will be tuned by the next parameter adaptation algorithm [13],

$$\Sigma_1 : \dot{\boldsymbol{\theta}}_f = \begin{cases} r_1 s \zeta(\xi) & if \ (\left|\boldsymbol{\theta}_f\right| < M_f) \ or \ (\left|\boldsymbol{\theta}_f\right| = M_f \ and \ s\boldsymbol{\theta}_f^T \zeta(\xi) \leq 0) \\ P\{r_1 s \zeta(\xi)\} & if \ (\left|\boldsymbol{\theta}_f\right| = M_f \ and \ s\boldsymbol{\theta}_f^T \zeta(\xi) > 0) \end{cases}, \tag{5a}$$

$$\Sigma_2 : \dot{\theta}_g \big|_{\theta_{gi}>\epsilon} = \begin{cases} r_2 s\zeta(\xi)\tau & if \quad (|\theta_g|<M_g) \quad or \quad (|\theta_g|=M_g \quad and \quad s\theta_g{}^T\zeta(\xi)\tau \le 0), \\ P\{r_2 s\zeta(\xi)\tau\} & if \quad (|\theta_g|=M_g \quad and \quad s\theta_g{}^T\zeta(\xi)\tau > 0) \end{cases} \tag{5b}$$

$$\Sigma_3 : \dot{\theta}_{gj} \big|_{\theta_{gj}=\epsilon} = \begin{cases} r_2 s\zeta_j(\xi)\tau & if \quad s\zeta_j(\xi)\tau > 0 \\ 0 & if \quad s\zeta_j(\xi)\tau \le 0 \end{cases}, \tag{5c}$$

where r_1 and r_2 are positive constants which define adaptation velocity, M_f and M_g are positive constants which fix the maximum value of the second norm of θ_f and θ_g respectively, and ϵ specifies the minimum value of θ_g elements, $\zeta_j(\xi)$ is the jth element of $\zeta(\xi)$, θ_{gij} is the jth element of θ_{gi}, and projection operators are defined

as $P\{r_1 s\zeta(\xi)\} = r_1 s\zeta(\xi) - r_1 s \dfrac{\theta_f \theta_f{}^T \zeta(\xi)}{|\theta_f|^2}$, $P\{r_2 s\zeta(\xi)u\} = r_2 s\zeta(\xi)\tau - r_2 s \dfrac{\theta_g \theta_g{}^T \zeta(\xi)\tau}{|\theta_g|^2}$.

Theorem [13]. For a non-linear system (2), consider the controller (3). If the parameter adaptation algorithm (5) is applied, then the system can guarantee that: (a) the parameters are bounded, and (b) closed loop signals are bounded and tracking error converges asymptotically to zero under the assumption of a fuzzy integrable approximation error.

The proof of s, θ_f and θ_g boundedness is shown in [13]. Thus if the reference signal x_d is bounded, the system state variable x will be bounded, and that both the velocity tracking error and the time derivative of the parameters estimates converge asymptotically to zero. However, absent additional arguments, it cannot be claim either $lim_{t\to\infty}|x(t)| = 0$, $lim_{t\to\infty}|s(t)| = 0$ or that $lim_{t\to\infty}|\tilde{\theta}_f| = 0$ and $lim_{t\to\infty}|\tilde{\theta}_g| = 0$.

4   Experimental Setup

The Snorkel vehicle, Fig. 2, developed at the Centro de Astrobiología is a tethered remotely operated UUV. The vehicle reduced cost conditions instruments and methods, limiting the identification experiments. The main goal of the vehicle is to carry out a scientific and autonomous inspection task in the Tinto River. The Snorkel vehicle has a dry mass of 75Kg and its dimensions are 0.7m long, 0.5m wide and 0.5m high. Actuation is provided by four DC electric motors, two of them are placed

Table 1. Vehicle instrumentation and its parameters

Variable	Sensor	Company	Precision	Update rate
Angular velocity	ENV-05D gyroscopes	Murata	0.14°/sec	100msec
Depth	600Kpa pressure sensor	Bosh	5cm	100msec
Heave velocity	Differential presume	Bosh	1cm/sec	100msec
Roll, pitch and yaw	HMR3000	Honeywell	0.1°	50msec
Heave, surge and sway	DVL	Sontek	1mm	100msec

in a horizontal plane while the others in a vertical one. Additionally, it is equipped with a distributed electronic architecture, and a low cost sensorial system, Table 1. A complete description of the Snorkel UUV is reported in [16].

The experiments have been carried out in a small tank of 1.8m of diameter and 2m of depth. Based on these limits and trying to avoid bumping with tank walls, the close loop tests only study the controller behaviour in yaw and heave movements, however the results can be extended to the surge DOF. Additionally, while an experiment is done in one DOF, the references for the rest of the two controllable DOF are zero.

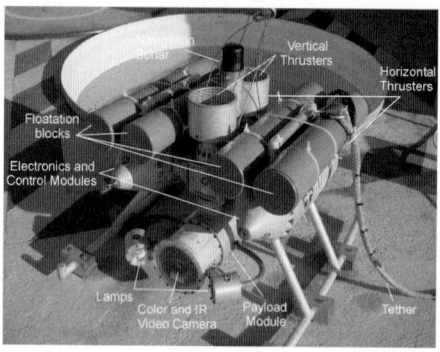

Fig. 2. Snorkel robot image and modules description

4.1 Definition of Control Parameters

Firstly, it must be pointed out that a reference model, implemented by a first order Butterworth low past filter, has been introduced to smooth references, trying to obtain a reasonable control effort.

Several parameters have to be fixed for the sliding part of the controller. The value of $\eta_{\Delta i}$ (3), must be established from the maximum value of functions $f_i(\xi)$, $g_i(\xi)$ and d_i uncertainty, and the initial value of the sliding surface [16]. Lastly, the thickness of the boundary layer, b_i, has been fixed using numerical simulations.

In order to define fuzzy estimators, firstly it is necessary to fix which of the kinematic variables are used for each estimator, based on the partial knowledge of the variables which determine vehicle dynamics. This knowledge can be obtained from the theoretical model of the vehicle, or from the a priori knowledge that an operator possesses of the system. From here, it has been decided to take the relations of Table 2 [16], where $\mu_{A_i}(k)$ are the membership functions of the k variable, and $\dot{x}_u$, $\dot{x}_v$, $\dot{x}_w$ and $\dot{x}_w$ are the surge, sway, heave and yaw velocities. The dependence of $\hat{g}_i(\xi)$ with respect to the velocity of each DOF allows us to keep the adaptation process active. Also, the number and kind of the membership functions have to be defined [16].

Other specific parameters of the fuzzy estimators are related to the adjustment function of the output consequents. The value of constants M_{fi} and M_{gi}, has been fixed following the criteria of doubling the theoretical values of $\hat{f}_i(\xi)$ and $\hat{g}_i(\xi)$ functions. Additionally, the criterion chosen to fix the values of ϵ_i is to multiply the same theoretical value of $\hat{g}_i(\xi)$ by ½ [16]. Adaptation velocities, r_1 and r_2, have been fixed by data analysis of numerical simulations, Table 2.

Table 2. AFSM control parameters for each controllable DOF

DOF	$f(\zeta)$	$g(\zeta)$	λ	μ_A	b	r_1	r_2
Surge	$\mu_{A_1}(\dot{x}_u)$, $\mu_A(\dot{x}_v)$, $\mu_{A_2}(\dot{x}_r)$	$\mu_{A_2}(\dot{x}_u)$	0.3	0.38	6	0.2	0.005
Heave	$\mu_{A_1}(\dot{x}_w)$	$\mu_{A_2}(\dot{x}_w)$	0.15	0.13	6	0.2	0.01
Yaw	$\mu_{A_1}(\dot{x}_r)$, $\mu_{A_2}(\dot{x}_u)$, $\mu_A(\dot{x}_v)$	$\mu_{A_1}(\dot{x}_r)$	0.3	0.55	1.75	10	0.005

5 Experimental Results

This section reports comparative experiments of the AFSM controller with a PD and a pure sliding mode controller (SM). The parameters of the PD controller have been obtain by identification with the AFSM control constants, eliminating the $f(\xi)$ function, while the SM controller share the control law with the AFSM controller, but in this case $f(\xi)$ and $\hat{g}(\xi)$ functions are theoretically estimated. In order to implement the controllers a digital version with an Euler integration algorithm and a sample period of 100msec has been used. The figures correspond to the period after the initial adaptation process of $\hat{f}(\xi)$ and $\hat{g}(\xi)$ functions.

5.1 Velocity Reference and Adaptation Capabilities

The first section reports a performance comparison, while a square yaw velocity reference of 10°/sec of amplitude, and 40sec of period is tracked. Real and reference positions are obtain by integrating the real and reference velocities respectively.

From an analysis of the Fig. 3 we observe that the tracking of the input reference for the AFSM controller is nearly perfect, with a reasonable control effort, in spite of the oscillatory behaviour when the velocity is closed to zero. The oscillation is caused by an error in the on line algorithm of offset adjustment of the gyroscope signals.

Fig. 4 shows analytically a comparison with the PD and SM controllers [10]. It concludes that AFSM controller presents the smaller velocity error, measure as $\tilde{x} = mean(|\dot{x}_d - \dot{x}|)$, while its control effort calculated as $\tau_{TOTAL} = mean(|\tau_d|)$, is only slightly higher than the control effort of the PD controller. This is due to the adaptation capabilities of the AFSM controller, based on its fuzzy estimators and the right adaptation law. The performance of a model-based controller, as the SM controller, depends entirely on the accuracy of the dynamic plant model used in the designing of the controller. This section also corroborates the lack of accuracy of the theoretical determined dynamical plant model for the Snorkel UUV.

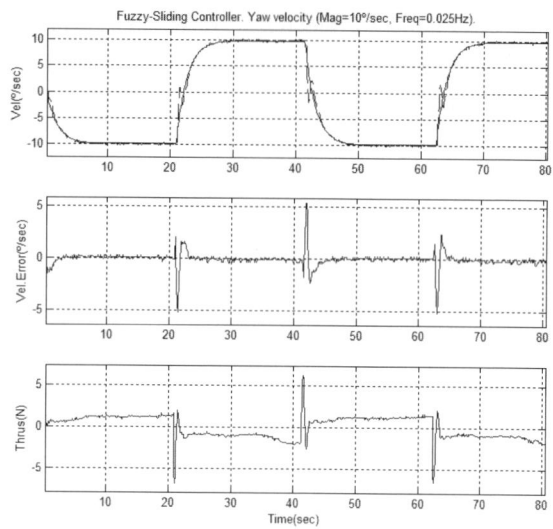

Fig. 3. Plot of AFSM controller in the yaw DOF. (Top) actual yaw velocity $\dot{x}_r$(- -) and reference $\dot{x}_{rd}$(-). (Medium) Velocity tracking error. (Bottom) Thrust of a horizontal thruster. Square yaw velocity reference of amplitude 10°/sec, and period 40sec.

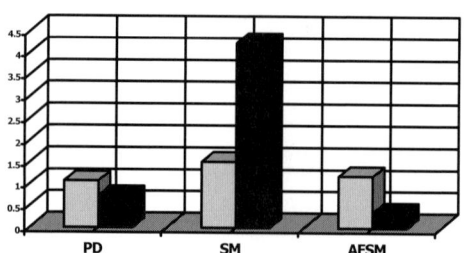

Fig. 4. Plot of velocity tracking error, $\tilde{\dot{x}}$ [°/sec] (black) and control effort, τ_{TOTAL} [N] (grey) for PD, SM and AFSM. Square yaw velocity reference of amplitude 10°/sec, and period 40sec.

5.2 Thrusters Saturation with Position Reference

The second set of tests attempts to show the influence of thrusters saturation in velocity and position tracking errors, while tracking a square position input reference for the yaw angle of 50° of amplitude, 70° of offset and 40sec of period.

It is evident that there are some deficits in the tracking of position and velocity references, for the AFSM controller, Fig. 5. Nevertheless, the final values of the yaw angle and velocity are reached, in spite of the appearance of overshoot. The deficit is due to the excessively fast and large velocity reference, which generates fast and high thrust references, witch cannot be achieved by the propulsion system [17].

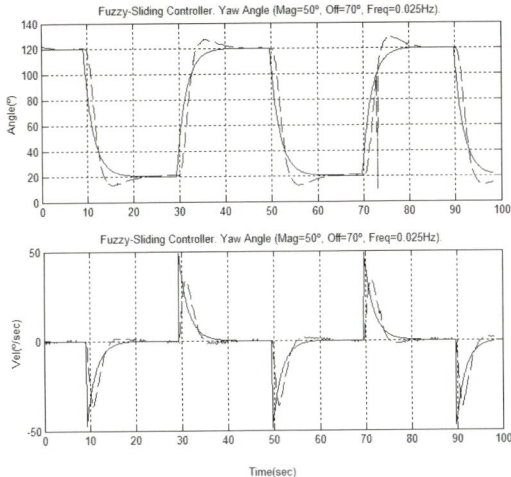

Fig. 5. Plot of AFSM controller in the yaw DOF. (Top) actual yaw angle x_r(- -) and reference x_{rd}(-). (Bottom) actual yaw velocity $\dot{x}_r$ (- -) and reference $\dot{x}_{rd}$ (-). Square yaw angle reference of amplitude 50°, offset 70°, and period 40sec.

Fig. 6 shows directly the comparison among the PD, SM and AFSM controllers, using the same reference. It can be concluded that PD controller presents the smaller velocity and angle tracking errors, this last one measure as $\tilde{x} = mean(|x_d - x|)$[10], as well as slightly smaller control effort that the AFSM controller. This advantage is due to its better performance under the saturation of the thrusters. Thrusters saturation represents an unmodeled discontinuous dynamics [10], which affects negatively on adaptive controllers, because they attempt to estimate the parameters values of that ill-structured plant model. The worst performance of the SM controller is justified due to the lack of accuracy of the theoretical model of the Snorkel vehicle.

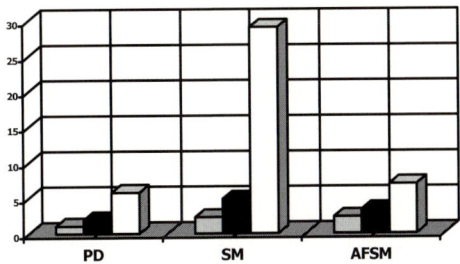

Fig. 6. Plot of angle tracking error, $\tilde{x}$ [°] (white), velocity tracking error, $\dot{\tilde{x}}$ [°/sec](black) and control effort, τ_{TOTAL} [N] (grey) for PD, SM and AFSM controllers. Square yaw angle reference of amplitude 50°, offset 70°, and period 40sec.

5.3 Adaptation Capabilities and Noise in Measurements

The last section attempts to show controllers performance, while tracking a square depth position reference of 0.3m of amplitude, 0.7 of offset and 40sec of period, in the presence of a change in the buoyancy of the vehicle.

In Fig. 7, the better performance of the AFSM control (Bottom) can be observed. This control technique is capable of reaching the permanent regime of input reference, while the other control techniques are not able to do it. Despite this, the transitory response does not track the reference due to two different reasons: the high quantification noise level associated with the measurements of depth and heave velocity, Table 1, and the high rates of depth reference that at the same time generates high values of velocity references, that can not be followed by the propulsion system [17]. A possible solution to this problem is to obtain the position references from the integration of velocity references, which could be saturated to reasonable values for the propulsion system. This better performance of the AFSM controller is based on the adaptation capability of the controller.

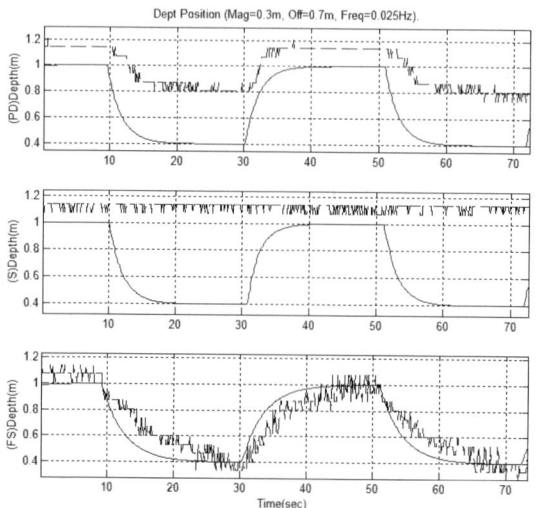

Fig. 7. Plot of actual depth x_w(- -) and reference x_{wd} (-) for (Top)PD controller, (Medium)SM controller and (Bottom) AFSM controller. Square depth reference of amplitude 0.3m, offset 0.7m, and period 40sec.

7 Conclusion

The AFSM controller, applied for the very first time in a UUV, allows us to consider the non-linearity of the system adapting to uncertainty in model plant and parameters. This feature permits the designer to work with minimum knowledge of the model, avoiding a tedious process of identification, helping to reduce the design cost. The theoretical and practical stability of the AFSM controller has been shown, assuring the convergence of the system to the input references, with a reasonable control effort.

The controller is capable of incorporating and compensating the dynamic problems and the perturbations of underwater vehicles. Also, it generates systems that are simple to implement and interpret.

The proposed controller could be defined as a combination of an adaptive and a robust system. In this way, it presents the advantages of robust control like the capability of adapting to rapid variation of the parameters, perturbations, noise from unmodeled dynamics, and theoretical insensibility to errors of the state measurements and its derivatives. Also, it presents the advantages of adaptive systems, like no requirement for prior and precise knowledge of uncertainty, reducing the required knowledge of system boundaries of uncertainty, and the capacity of improving the output performance as the system adapts. The fuzzy adaptive part of the controller permits us to relax the design conditions of the sliding part. Additionally, one of the restrictions of the fuzzy adaptive part is the low speed of the parametric adjustment. Nevertheless, this lack it is compensated by the sliding part of the controller.

The experiments of the close-loop performance of the studied controllers corroborate the theoretical predictions. Moreover, the experiments suggest that the AFSM controller is a valid method to be applied in underwater vehicles that outperform the PD and SM controllers using velocity trajectories. Thruster's saturation significantly degrades the performance of AFSM controller, while PD controller shows better performance under this circumstance. The success of a simple model based SM controller rely on the plant model parameters to be exactly correct, where as AFSM based on its adaptation capabilities is not affected by inaccuracy in theoretical plant model. Noise is another factor that affects significantly the performance of the SM controller, and less seriously of the AFSM controller.

References

1. Goheen, K., Jefferys, E.R.: On the adaptive control of remotely operated underwater vehicles, Journal Adaptive Control and signal processing, Vol. 4. (1990) 287-297.
2. Koivo, H.N.: A multivariable self-tuning controller, Automatica, 16(4), 1980, 351-366.
3. Hsu, L., Costa, R.R., Lizaralde, F., Cunha, J.P.V.S.: Dynamic positioning of remotely operated underwater vehicles, IEEE Robot Automatic Magazine, Vol 7. (2000) 21-31.
4. Yoerger, D.R., Slotine, J.J.: Adaptive sliding control of an experimental underwater vehicle. IEEE International conference on Robotics and Automation, (1991) 2746-2751.
5. Yoerger, D.R., Slotine, J.J.: Robust Trajectory Control of Underwater Vehicles. IEEE Journal of Oceanic Engineering, Vol. 10. (1985) 462-470.
6. Cristi, R., Papoulias, F.A., Healey, A.: Adaptive sliding control mode of autonomous underwater vehicles in the dive plane. IEEE Journal of Oceanic Engineering, Vol. 15. (1991) 462-470.
7. T.I. Fossen, T.I., Paulsen, M.: Adaptive Feedback linearization Applied to Steering of Ships. Proc. of the IEEE Conference on Control Applications, (1991) 1088-1093.
8. Slotine, J.J., Li, W.: Applied non-linear control, Prentice Hall, Englewood Cliffs, NJ, (1991).
9. Fossen, T.I., Fjellstad, O.E.: Robust Adaptive Control of Underwater Vehicles: A Comparative Study. Proc. of the IFAC Workshop on Control Applications in Marine Systems, (1995) 66-74.

10. Smallwood, D.A., Whitcomb, L.L.: Model-based Dynamic positioning of underwater Robotic Vehicles: Theory and Experiment. IEEE Journal of Oceanic Engineering, Vol. 29. (2004) January.
11. Yuh, J.: Learning control for Underwater Robotics Vehicles. IEEE Control System Magazine, Vol. 14. (1994) 39-46.
12. DeBitetto, P.A.: Fuzzy logic for depth control for unmanned undersea vehicles, Proc. of Symposium of Autonomous Underwater Vehicle Technology, (1994).
13. Wang, J., Get S.S., Lee, T.H.: Adaptive Fuzzy Sliding Mode Control of a Class of Non-linear Systems. Proc. of the 3rd Asian Control Conference, (2000).
14. Fossen, T.I.: Underwater vehicle dynamics, John Wiley & Sons Ltd, Baffins Lane, Chichester (1994).
15. Wang, L.X..: Adaptive Fuzzy Systems and Control, Prentice Hall, Englewood Cliff, NJ (1994).
16. Sebastián, E.: Control y navegación semi-autónoma de un robot subacuático para la inspección de entornos desconocidos, doctoral diss., Universidad de Alcalá, Departamento de Electrónica, (2005).
17. Manfredi, J.A., Sebastián, E., Gomez-Elvira, J., Martín, J., Torres J.: Snorkel: vehículo subacuático para la exploración del río Tinto. Proc. of XXII Jornadas de Automática, (2001).

Author Index

Lecture Notes in Artificial Intelligence (LNAI)

Vol. 3946: T.R. Roth-Berghofer, S. Schulz, D.B. Leake (Eds.), Modeling and Retrieval of Context. XI, 149 pages. 2006.

Vol. 3944: J. Quiñonero-Candela, I. Dagan, B. Magnini, F. d'Alché-Buc (Eds.), Machine Learning Challenges. XIII, 462 pages. 2006.

Vol. 3930: D.S. Yeung, Z.-Q. Liu, X.-Z. Wang, H. Yan (Eds.), Advances in Machine Learning and Cybernetics. XXI, 1110 pages. 2006.

Vol. 3918: W.K. Ng, M. Kitsuregawa, J. Li, K. Chang (Eds.), Advances in Knowledge Discovery and Data Mining. XXIV, 879 pages. 2006.

Vol. 3913: O. Boissier, J. Padget, V. Dignum, G. Lindemann, E. Matson, S. Ossowski, J.S. Sichman, J. Vázquez-Salceda (Eds.), Coordination, Organizations, Institutions, and Norms in Multi-Agent Systems. XII, 259 pages. 2006.

Vol. 3910: S.A. Brueckner, G.D.M. Serugendo, D. Hales, F. Zambonelli (Eds.), Engineering Self-Organising Systems. XII, 245 pages. 2006.

Vol. 3904: M. Baldoni, U. Endriss, A. Omicini, P. Torroni (Eds.), Declarative Agent Languages and Technologies III. XII, 245 pages. 2006.

Vol. 3900: F. Toni, P. Torroni (Eds.), Computational Logic in Multi-Agent Systems. XVII, 427 pages. 2006.

Vol. 3899: S. Frintrop, VOCUS: A Visual Attention System for Object Detection and Goal-Directed Search. XIV, 216 pages. 2006.

Vol. 3898: K. Tuyls, P.J. 't Hoen, K. Verbeeck, S. Sen (Eds.), Learning and Adaption in Multi-Agent Systems. X, 217 pages. 2006.

Vol. 3891: J.S. Sichman, L. Antunes (Eds.), Multi-Agent-Based Simulation VI. X, 191 pages. 2006.

Vol. 3890: S.G. Thompson, R. Ghanea-Hercock (Eds.), Defence Applications of Multi-Agent Systems. XII, 141 pages. 2006.

Vol. 3885: V. Torra, Y. Narukawa, A. Valls, J. Domingo-Ferrer (Eds.), Modeling Decisions for Artificial Intelligence. XII, 374 pages. 2006.

Vol. 3881: S. Gibet, N. Courty, J.-F. Kamp (Eds.), Gesture in Human-Computer Interaction and Simulation. XIII, 344 pages. 2006.

Vol. 3874: R. Missaoui, J. Schmidt (Eds.), Formal Concept Analysis. X, 309 pages. 2006.

Vol. 3873: L. Maicher, J. Park (Eds.), Charting the Topic Maps Research and Applications Landscape. VIII, 281 pages. 2006.

Vol. 3864: Y. Cai, J. Abascal (Eds.), Ambient Intelligence in Everyday Life. XII, 323 pages. 2006.

Vol. 3863: M. Kohlhase (Ed.), Mathematical Knowledge Management. XI, 405 pages. 2006.

Vol. 3862: R.H. Bordini, M. Dastani, J. Dix, A.E.F. Seghrouchni (Eds.), Programming Multi-Agent Systems. XIV, 267 pages. 2006.

Vol. 3849: I. Bloch, A. Petrosino, A.G.B. Tettamanzi (Eds.), Fuzzy Logic and Applications. XIV, 438 pages. 2006.

Vol. 3848: J.-F. Boulicaut, L. De Raedt, H. Mannila (Eds.), Constraint-Based Mining and Inductive Databases. X, 401 pages. 2006.

Vol. 3847: K.P. Jantke, A. Lunzer, N. Spyratos, Y. Tanaka (Eds.), Federation over the Web. X, 215 pages. 2006.

Vol. 3835: G. Sutcliffe, A. Voronkov (Eds.), Logic for Programming, Artificial Intelligence, and Reasoning. XIV, 744 pages. 2005.

Vol. 3830: D. Weyns, H. V.D. Parunak, F. Michel (Eds.), Environments for Multi-Agent Systems II. VIII, 291 pages. 2006.

Vol. 3817: M. Faundez-Zanuy, L. Janer, A. Esposito, A. Satue-Villar, J. Roure, V. Espinosa-Duro (Eds.), Nonlinear Analyses and Algorithms for Speech Processing. XII, 380 pages. 2006.

Vol. 3814: M. Maybury, O. Stock, W. Wahlster (Eds.), Intelligent Technologies for Interactive Entertainment. XV, 342 pages. 2005.

Vol. 3809: S. Zhang, R. Jarvis (Eds.), AI 2005: Advances in Artificial Intelligence. XXVII, 1344 pages. 2005.

Vol. 3808: C. Bento, A. Cardoso, G. Dias (Eds.), Progress in Artificial Intelligence. XVIII, 704 pages. 2005.

Vol. 3802: Y. Hao, J. Liu, Y.-P. Wang, Y.-m. Cheung, H. Yin, L. Jiao, J. Ma, Y.-C. Jiao (Eds.), Computational Intelligence and Security, Part II. XLII, 1166 pages. 2005.

Vol. 3801: Y. Hao, J. Liu, Y.-P. Wang, Y.-m. Cheung, H. Yin, L. Jiao, J. Ma, Y.-C. Jiao (Eds.), Computational Intelligence and Security, Part I. XLI, 1122 pages. 2005.

Vol. 3789: A. Gelbukh, Á. de Albornoz, H. Terashima-Marín (Eds.), MICAI 2005: Advances in Artificial Intelligence. XXVI, 1198 pages. 2005.

Vol. 3782: K.-D. Althoff, A. Dengel, R. Bergmann, M. Nick, T.R. Roth-Berghofer (Eds.), Professional Knowledge Management. XXIII, 739 pages. 2005.

Vol. 3763: H. Hong, D. Wang (Eds.), Automated Deduction in Geometry. X, 213 pages. 2006.

Vol. 3755: G.J. Williams, S.J. Simoff (Eds.), Data Mining. XI, 331 pages. 2006.

Vol. 3735: A. Hoffmann, H. Motoda, T. Scheffer (Eds.), Discovery Science. XVI, 400 pages. 2005.

Vol. 3734: S. Jain, H.U. Simon, E. Tomita (Eds.), Algorithmic Learning Theory. XII, 490 pages. 2005.

Vol. 3721: A.M. Jorge, L. Torgo, P.B. Brazdil, R. Camacho, J. Gama (Eds.), Knowledge Discovery in Databases: PKDD 2005. XXIII, 719 pages. 2005.

Vol. 3720: J. Gama, R. Camacho, P.B. Brazdil, A.M. Jorge, L. Torgo (Eds.), Machine Learning: ECML 2005. XXIII, 769 pages. 2005.

Vol. 3717: B. Gramlich (Ed.), Frontiers of Combining Systems. X, 321 pages. 2005.

Vol. 3702: B. Beckert (Ed.), Automated Reasoning with Analytic Tableaux and Related Methods. XIII, 343 pages. 2005.

Vol. 3698: U. Furbach (Ed.), KI 2005: Advances in Artificial Intelligence. XIII, 409 pages. 2005.

Vol. 3690: M. Pěchouček, P. Petta, L.Z. Varga (Eds.), Multi-Agent Systems and Applications IV. XVII, 667 pages. 2005.